The Protein Protocols Handbook

The Protein Protocols Handbook

Edited by

John M. Walker

University of Hertfordshire, Hatfield, UK

HUMANA PRESS  TOTOWA, NEW JERSEY

999 Riverview Drive, Suite 208
Totowa, New Jersey 07512

For additional copies, pricing for bulk purchases, and/or other information about other Humana titles, contact Humana at the above address or at any of the following numbers: Tel.: 201-256-1699; Fax: 201-256-8341; E-mail: humana@interramp.com

This publication is printed on acid-free paper. ∞
ANSI Z39.48-1984 (American Standards Institute) Permanence of Paper for Printed Library Materials.

Printed in the United States of America. 10 9 8 7 6 5 4 3 2

ISBN 0-89603-338-4 (hardcover)
ISBN 0-89603-339-2 (combbound)

Preface

In *The Protein Protocols Handbook,* I have attempted to provide a cross-section of analytical techniques commonly used for proteins and peptides, thus providing a benchtop manual and guide both for those who are new to the protein chemistry laboratory and for those more established workers who wish to use a technique for the first time.

We each, of course, have our own favorite, commonly used gel system, gel-staining method, blotting method, and so on; I'm sure you will find yours here. However, I have also described a variety of alternatives for many of these techniques; though they may not be superior to the methods you commonly use, they may nevertheless be more appropriate in a particular situation. Only by knowing the range of techniques that are available to you, and the strengths and limitations of these techniques, will you be able to choose the method that best suits your purpose.

All chapters are written in the same format as that used in the *Methods in Molecular Biology* series. Each chapter opens with a description of the basic theory behind the method being described. The Materials section lists all the chemicals, reagents, buffers, and other materials necessary for carrying out the protocol. Since the principal goal of the book is to provide experimentalists with a full account of the practical steps necessary for carrying out each protocol successfully, the Methods section contains detailed step-by-step descriptions of every protocol that should result in the successful execution of each method. The Notes section complements the Methods material by indicating how best to deal with any problem or difficulty that may arise when using a given technique, and how to go about making the widest variety of modifications or alterations to the protocol.

In general, I have avoided analytical techniques that require expensive specialist hardware. Such techniques are described in specialist volumes in the *Methods in Molecular Biology* series (vol. 60, *Protein NMR Techniques;* vol. 56, *Crystallographic Methods and Protocols*; vol. 52, *Capillary Electrophoresis Guidebook*; vol. 40, *Protein Stability and Folding*; vol. 22, *Microscopy, Optical Microscopy, and Macroscopic Techniques;* vol. 17, *Spectroscopic Methods and Analyses*). The main exception has been the introduction of some techniques that involve the use of mass spectrometry. The recent availability of benchtop machines has made this technique available to a much wider range of workers than might previously have been possible, and thus mass spectrometry is fast becoming a routine analytical method for the protein chemist. However, for those who require a more detailed and in-depth description of the exciting new applications of this technique to the analysis of proteins and peptides, extensive coverage is provided in *Methods in Molecular Biology,* vol. 61, *Protein and Peptide Analysis by Mass Spectrometry*. For those of you who require guidance on protein and peptide purification, the subject is extensively covered in *Methods in Molecular Biology,* vol. 59, *Protein Purification Protocols*.

John M. Walker

Contents

PART V: PROTEIN/PEPTIDE CHARACTERIZATION

Contributors

ALASTAIR AITKEN • *National Institute for Medical Research, Mill Hill, London, UK*
ROBERT E. AKINS • *Department of Medical Cell Biology, Nemours Research Program, AI duPont Institute, Wilmington, DE*
SALLY ANN AMERO • *Department of Biology, Washington University, St. Louis, MO*
GRAHAM S. BAILEY • *Department of Biological and Chemical Sciences, University of Essex, Colchester, UK*
SALVADOR BARTOLOMÉ • *Department de Bioquímica i Biologia Molecular, Universitat Autònoma de Barcelona, Bellaterra (Barcelona), Spain*
DAVID J. BEGLEY • *Biomedical Sciences Division, King's College, London, UK*
ANTONIO BERMÚDEZ • *Department de Bioquímica i Biologia Molecular, Universitat Autònoma de Barcelona, Bellaterra (Barcelona), Spain*
MAHESH K. BHALGAT • *Molecular Probes Inc., Eugene, OR*
SYLVIE BOURASSA • *Eastern Quebec Peptide Sequencing Facility, CHUL Research Center and Laval University, Ste-Foy, Québec, Canada*
J. MARK CARTER • *Cytogen, Princeton, NJ*
FRANCA CASAGRANDA • *CSIRO Division of Biomolecular Engineering, Victoria, Australia; Present address: European Molecular Biology Laboratory, Heidelberg, Germany*
JUNG-KAP CHOI • *College of Pharmacy, Chonnam National University, Kwangju, Korea*
ANTONELLA CIRCOLO • *Maxwell Finland Laboratory for Infectious Diseases, Boston City Hospital, Boston University School of Medicine, Boston, MA*
JOHN COLYER • *Department of Biochemistry and Molecular Biology, University of Leeds, UK*
JOAN-RAMON DABÁN • *Department de Bioquímica i Biologia Molecular, Universitat Autònoma de Barcelona, Bellaterra (Barcelona), Spain*
JAMES R. DAVIE • *Department of Biochemistry and Molecular Biology, University of Manitoba, Winnipeg, Manitoba, Canada*
MICHAEL J. DAVIES • *Department of Biochemistry and Molecular Biology, University College London, UK*
GENEVIÈVE P. DELCUVE • *Department of Biochemistry and Molecular Biology, University of Manitoba, Winnipeg, Manitoba, Canada*
SERGE DESNOYERS • *Eastern Quebec Peptide Sequencing Facility, CHUL Research Center and Laval University, Ste-Foy, Québec, Canada*
MONIQUE DIANO • *Lab de Genetique, CNRS, Marseille, France*
MICHAEL J. DUNN • *Department of Cardiothoracic Surgery, National Heart and Lung Institute, Heart Science Centre, Harefield Hospital, Middlesex, UK*
SARAH C. R. ELGIN • *Department of Biology, Washington University, St. Louis, MO*

JOSEPH FERNANDEZ • *Protein/DNA Technology Center, The Rockefeller University, New York*
MERCEDES FERRERAS • *Departimiento de Bioquímica y Biología Molecular, Universidad Complutense, Madrid, Spain*
SUSAN J. FOWLER • *Amersham International plc., Amersham, UK*
THOMAS D. FRIEDRICH • *Department of Microbiology, Immunology, and Molecular Genetics, The Albany Medical College, Albany, NY*
JUAN M. GARCÍA-SEGURA • *Departimiento de Bioquímica y Biología Molecular, Universidad Complutense, Madrid, Spain*
OLIVIER GOLAZ • *Central Clinical Chemistry Laboratory, University Hospital of Geneva, Switzerland*
MOHAMMAD T. GOODARZI • *Department of Clinical Biochemistry and Metabolic Medicine, The Medical School, University of Newcastle, UK*
MORAG A. GRASSIE • *Division of Biochemistry and Molecular Biology, Institute of Biomedical and Life Sciences, University of Glasgow, Scotland, UK*
PATRICIA GRAVEL • *Clinical Research Unit, Psychiatric Institutions of Geneva, Switzerland*
SUNITA GULATI • *Maxwell Finland Laboratory for Infectious Diseases, Boston City Hospital, Boston University School of Medicine, Boston, MA*
GRAEME R. GUY • *Institute of Molecular and Cell Biology, National University of Singapore*
ROSARIA P. HAUGLAND • *Molecular Probes Inc., Eugene, OR*
LUCY F. HENLEY • *Division of Life Sciences, King's College, University of London, UK*
HEE-YOUN HONG • *College of Pharmacy, Chonnam National University, Kwangju, Korea*
MARTIN HORST • *Department of Biochemistry, Biozentrum of the University of Basel, Switzerland*
ELIZABETH F. HOUNSELL • *Department of Biochemistry and Molecular Biology, University College London, UK*
G. BRENT IRVINE • *School of Biology and Biochemistry, Queen's University of Belfast, Ireland*
THARAPPEL C. JAMES • *Department of Biology, Washington University, St. Louis, MO*
PAUL JENÖ • *Department of Biochemistry, Biozentrum of the University of Basel, Switzerland*
RALPH C. JUDD • *Division of Biological Sciences, University of Montana, Missoula, MT*
NICHOLAS J. KRUGER • *Department of Plant Sciences, University of Oxford, UK*
JUDITH LAFFIN • *Department of Microbiology, Immunology, and Molecular Genetics, The Albany Medical College, Albany, NY*
WILLIAM J. LAROCHELLE • *Laboratory of Cellular and Molecular Biology, National Cancer Institute, National Institutes of Health, Bethesda, MD*
MICHÈLE LEARMONTH • *National Institute for Medical Research, Mill Hill, London, UK*
ANDRÉ LE BIVIC • *Lab de Genetique, CNRS, Marseille, France*
JOHN M. LEHMAN • *Department of Microbiology, Immunology, and Molecular Genetics, The Albany Medical College, Albany, NY*
FAN LIN • *Department of Anatomy and Cell Biology, School of Veterinary Medicine, Louisiana State University, Baton Rouge, LA*

PHILIP S. LOW • *Department of Chemistry, Purdue University, West Lafayette, IN*
HARRY R. MATTHEWS • *Department of Biological Chemistry, University of California, Davis, CA*
PHILIP N. MCFADDEN • *Department of Biochemistry and Biophysics, Oregon State University, Corvallis, OR*
GRAEME MILLIGAN • *Division of Biochemistry and Molecular Biology, Institute of Biomedical and Life Sciences, University of Glasgow, Scotland, UK*
SHEENAH M. MISCHE • *Protein/DNA Technology Center, The Rockefeller University, New York*
TSUGUO MIZUOCHI • *Department of Biochemistry and Molecular Biology, University College London, UK*
HOLGER J. MØLLER • *Department of Clinical Biochemistry, KH University Hospital, Nørrebrogade, Århus, Denmark*
GLENN E. MORRIS • *MRIC, NE Wales Institute, Deeside, Clwyd, UK*
ULF NEUMANN • *Medizinische Hochschule Hannover, Zentrum Biochem/ Physiologische Chem, Hannover, Germany*
NGUYEN THI MAN • *MRIC, NE Wales Institute, Deeside, Clwyd, UK*
MARK PAGE • *MEDEVA, Vaccine Research Unit, Department of Biochemistry, Imperial College of Science, Technology, and Medicine, London, UK*
ROBIN J. PHILP • *Institute of Molecular and Cell Biology, National University of Singapore*
GUY G. POIRIER • *Eastern Quebec Peptide Sequencing Facility, CHUL Research Center and Laval University, Ste-Foy, Québec, Canada*
JEFFREY W. POLLARD • *Departments of Developmental and Molecular Biology/ Obstetrics and Gynecology, Albert Einstein College of Medicine, Bronx, NY*
JORGEN H. POULSEN • *Department of Clinical Biochemistry, KH University Hospital, Nørrebrogade, Århus, Denmark*
THOMAS J. PRITCHETT • *Beckman Instruments, Fullerton, CA*
F. ANDREW RAY • *Department of Microbiology, Immunology, and Molecular Genetics, The Albany Medical College, Albany, NY*
DOUGLAS D. ROOT • *Department of Chemistry and Biochemistry, University of Texas at Austin, TX*
JOHN RUSH • *Biotechnology Resource Laboratory, Yale University, New Haven, CT*
SUZY M. SAMANDAR • *Biotechnology Resource Laboratory, Yale University, New Haven, CT*
MELISSA SAYLOR • *Biotechnology Resource Laboratory, Yale University, New Haven, CT*
BRYAN JOHN SMITH • *Celltech Therapeutics Ltd., Slough, UK*
KEVIN D. SMITH • *Department of Pharmaceutical Sciences, University of Strathclyde, Glasgow, Scotland, UK*
WAYNE R. SPRINGER • *Medical Center, Department of Veterans Affairs, San Diego, CA*
KATHRYN L. STONE • *Biotechnology Resource Laboratory, Yale University, New Haven, CT*
DONALD F. SUMMERS • *Department of Microbiology and Molecular Genetics, University of California, College of Medicine, Irvine, CA*
PATRICIA J. SWEENEY • *Division of Biosciences, University of Hertfordshire, Hatfield, UK*

Boguslaw Szewczyk • *Department of Microbiology and Molecular Genetics, College of Medicine, University of California, Irvine, CA*
Dan S. Tawfik • *Department of Chemical Immunology, Weizmann Institute of Science, Rehovot, Israel; Present address: Centre for Protein Engineering, Medical Research Council Centre, Cambridge, UK*
Robin Thorpe • *National Institute for Biological Standards and Control, Potters Bar, UK*
Christopher F. Thurston • *Division of Life Sciences, King's College, University of London, UK*
Rocky S. Tuan • *Department of Orthopedic Surgery, Thomas Jefferson University, Philadelphia, PA*
Graham A. Turner • *Department of Clinical Biochemistry and Metabolic Medicine, The Medical School, University of Newcastle, UK*
John M. Walker • *Division of Biosciences, University of Hertfordshire, Hatfield, UK*
Kuan Wang • *Department of Chemistry and Biochemistry, University of Texas at Austin, TX*
Malcom Ward • *Structural Chemistry Department, Glaxo Research and Development, Hertfordshire, UK*
Jakob H. Waterborg • *School of Biological Sciences, University of Missouri-Kansas City, Kansas City, MO*
Darin J. Weber • *Department of Biochemistry and Biophysics, Oregon State University, Corvallis, OR*
Kenneth R. Williams • *Biotechnology Resource Laboratory, Yale University, New Haven, CT*
John F. K. Wilshire • *CSIRO Division of Biomolecular Engineering, Victoria, Australia*
Julia S. Winder • *Division of Biosciences, University of Hertfordshire, Hatfield, UK*
G. Brian Wisdom • *Division of Biochemistry, School of Biology and Biochemistry, The Queen's University, Medical Biology Centre, Belfast, UK*
Gary E. Wise • *Department of Anatomy and Cell Biology, School of Veterinary Medicine, Louisiana State University, Baton Rouge, LA*
Gyurng-Soo Yoo • *College of Pharmacy, Chonnam National University, Kwangju, Korea*
Wendy W. You • *Molecular Probes Inc., Eugene, OR*
Jie Yuan • *Department of Chemistry, Purdue University, West Lafayette, IN*

PART I

QUANTITATION OF PROTEINS

1

Protein Determination by UV Absorption

Alastair Aitken and Michèle Learmonth

1. Introduction

1.1. Near UV Absorbance (280 nm)

Quantitation of the amount of protein in a solution is possible in a simple spectrometer. Absorption of radiation in the near UV by proteins depends on the Tyr and Trp content (and to a very small extent on the amount of Phe and disulfide bonds). Therefore the A_{280} varies greatly between different proteins (for a 1 mg/mL solution, from 0 up to 4 for some tyrosine-rich wool proteins, although most values are in the range 0.5–1.5 *[1]*). The advantages of this method are that it is simple, and the sample is recoverable. The method has some disadvantages, including interference from other chromophores, and the specific absorption value for a given protein must be determined. The extinction of nucleic acid in the 280-nm region may be as much as 10 times that of protein at their same wavelength, and hence, a few percent of nucleic acid can greatly influence the absorption.

1.2. Far UV Absorbance

The peptide bond absorbs strongly in the far UV with a maximum at about 190 nm. This very strong absorption of proteins at these wavelengths has been used in protein determination. Because of the difficulties caused by absorption by oxygen and the low output of conventional spectrophotometers at this wavelength, measurements are more conveniently made at 205 nm, where the absorbance is about half that at 190 nm. Most proteins have extinction coefficients at 205 nm for a 1 mg/mL solution of 30–35 and between 20 and 24 at 210 nm *(2)*.

Various side chains, including those of Trp, Phe, Tyr, His, Cys, Met, and Arg (in that descending order), make contributions to the A_{205} *(3)*.

The advantages of this method include simplicity and sensitivity. As in the method outlined in Section 3.1. the sample is recoverable and in addition there is little variation in response between different proteins, permitting near-absolute determination of protein. Disadvantages of this method include the necessity for accurate calibration of the spectrophotometer in the far UV. Many buffers and other components, such as heme or pyridoxal groups, absorb strongly in this region.

From: The Protein Protocols Handbook
Edited by: J. M. Walker Humana Press Inc., Totowa, NJ

2. Materials

1. 0.1M K_2SO_4 (pH 7.0).
2. 5 mM potassium phosphate buffer, pH 7.0.
3. Nonionic detergent (0.01% Brij 35)
4. Guanidinium-HCl.
5. 0.2-µm Millipore (Watford, UK) filter.
6. UV-visible spectrometer: The hydrogen lamp should be selected for maximum intensity at the particular wavelength.
7. Cuvets, quartz, for <215 nm.

3. Methods

3.1. Estimation of Protein by Near UV Absorbance (280 nm)

1. A reliable spectrophotometer is necessary. The protein solution must be diluted in the buffer to a concentration that is well within the accurate range of the instrument (*see* Notes 1 and 2).
2. The protein solution to be measured can be in a wide range of buffers, so it is usually no problem to find one that is appropriate for the protein which may already be in a particular buffer required for a purification step or assay for enzyme activity, for example (*see* Notes 3 and 4).
3. Measure the absorbance of the protein solution at 280 nm, using quartz cuvets or cuvets that are known to be transparent to this wavelength, filled with a volume of solution sufficient to cover the aperture through which the light beam passes.
4. The value obtained will depend on the path length of the cuvet. If not 1 cm, it must be adjusted by the appropriate factor. The Beer-Lambert law states that:

$$A \text{ (absorbance)} = \varepsilon\, c\, l \quad (1)$$

where ε = extinction coefficient, c = concentration in mol/L and l = optical path length in cm. Therefore, if ε is known, measurement of A gives the concentration directly, ε is normally quoted for a 1-cm path length.

5. The actual value of UV absorbance for a given protein must be determined by some absolute method, e.g., calculated from the amino acid composition, which can be determined by amino acid analysis *(4)*. The UV absorbance for a protein is then calculated according to the following formula:

$$A_{280} \text{ (1 mg/mL)} = (5690n_w + 1280n_y + 120n_c)/M \quad (2)$$

where n_w, n_y, and n_c are the numbers of Trp, Tyr, and Cys residues in the polypeptide of mass M and 5690, 1280 and 120 are the respective extinction coefficients for these residues (*see* Note 5).

3.2. Estimation of Protein by Far UV Absorbance

1. The protein solution is diluted with a sodium chloride solution (0.9% w/v) until the extinction at 215 nm is <1.5 (*see* Notes 1 and 6).
2. Alternatively, dilute the sample in another non-UV-absorbing buffer such as 0.1M K_2SO_4, containing 5 mM potassium phosphate buffer adjusted to pH 7.0 (*see* Note 6).
3. Measure the absorbances at the appropriate wavelengths (either A_{280} and A_{205}, or A_{225} and A_{215}, depending on the formula to be applied), using a spectrometer fitted with a hydrogen lamp that is accurate at these wavelengths, using quartz cuvets filled with a volume of solution sufficient to cover the aperture through which the light beam passes (details in Section 3.1.).

4. The A_{205} for a 1 mg/mL solution of protein ($A_{205}^{1\ mg/mL}$) can be calculated within ±2%, according to the empirical formula proposed by Scopes *(2)* (*see* Notes 7–10):

$$A_{205}^{1\ mg/mL} = 27 + 120\ (A_{280}/A_{205}) \quad (3)$$

5. Alternatively, measurements may be made at longer wavelengths *(5)*:

$$\text{Protein concentration } (\mu g/mL) = 144\ (A_{215} - A_{225}) \quad (4)$$

The extinction at 225 nm is subtracted from that at 215 nm; the difference multiplied by 144 gives the protein concentration in the sample in μg/mL. With a particular protein under specific conditions accurate measurements of concentration to within 5 μg/L are possible.

4. Notes

1. It is best to measure absorbances in the range 0.05–1.0 (between 10 and 90% of the incident radiation). At around 0.3 absorbance (50% absorption), the accuracy is greatest.
2. Bovine serum albumin is frequently used as a protein standard; 1 mg/mL has an A_{280} of 0.66.
3. If the solution is turbid, the apparent A_{280} will be increased by light scattering. Filtration (through a 0.2-μm Millipore filter) or clarification of the solution by centrifugation can be carried out. For turbid solutions, a convenient approximate correction can be applied by subtracting the A_{310} (proteins do not normally absorb at this wavelength unless they contain particular chromophores) from the A_{280}.
4. At low concentrations, protein can be lost from solution by adsorption on the cuvet; the high ionic strength helps to prevent this. Inclusion of a nonionic detergent (0.01% Brij 35) in the buffer may also help to prevent these losses.
5. The presence of nonprotein chromophores (e.g., heme, pyridoxal) can increase A_{280}. If nucleic acids are present (which absorb strongly at 260 nm), the following formula can be applied. This gives an accurate estimate of the protein content by removing the contribution to absorbance by nucleotides at 280 nm, by measuring the A_{260} which is largely owing to the latter *(6)*.

$$\text{Protein (mg/mL)} = 1.55\ A_{280} - 0.76\ A_{260} \quad (5)$$

Other formulae (using similar principles of absorbance differences) employed to determine protein in the possible presence of nucleic acids are the following *(7,8)*:

$$\text{Protein (mg/mL)} = (A_{235} - A_{280})/2.51 \quad (6)$$

$$\text{Protein (mg/mL)} = 0.183\ A_{230} - 0.075.8\ A_{260} \quad (7)$$

6. Protein solutions obey Beer-Lambert's Law at 215 nm provided the extinction is <2.0.
7. Strictly speaking, this value applies to the protein in 6*M* guanidinium-HCl, but the value in buffer is generally within 10% of this value, and the relative absorbances in guanidinium-HCl and buffer can be easily determined by parallel dilutions from a stock solution.
8. Sodium chloride, ammonium sulfate, borate, phosphate, and Tris do not interfere, whereas 0.1*M* acetate, succinate, citrate, phthalate, and barbiturate show high absorption at 215 nm.
9. The absorption of proteins in the range 215–225 nm is practically independent of pH between pH values 4–8.
10. The specific extinction coefficient of a number of proteins and peptides at 205 nm and 210 nm *(3)* has been determined. The average extinction coefficient for a 1 mg/mL solution of 40 serum proteins at 210 nm is 20.5 ± 0.14. At this wavelength, a protein concentration of 2 μg/mL gives $A = 0.04$ *(5)*.

References

1. Kirschenbaum, D. M. (1975) Molar absorptivity and A1%/1 cm values for proteins at selected wavelengths of the ultraviolet and visible regions. *Anal. Biochem.* **68,** 465–484.
2. Scopes, R. K. (1974) Measurement of protein by spectrometry at 205 nm. *Anal. Biochem.* **59,** 277–282.
3. Goldfarb, A. R., Saidel, L. J., and Mosovich, E. (1951) The ultraviolet absorption spectra of proteins. *J. Biol. Chem.* **193,** 397–404.
4. Gill, S. C. and von Hippel, P. H. (1989) Calculation of protein extinction coefficients from amino acid sequence data. *Anal. Biochem.* **182,** 319–326.
5. Waddell, W. J. (1956) A simple UV spectrophotometric method for the determination of protein. *J. Lab. Clin. Med.* **48,** 311–314.
6. Layne, E. (1957) Spectrophotornetric and turbidimetric methods for measuring proteins. *Methods Enzymol.* **3,** 447–454.
7. Whitaker, J. R. and Granum, P. E. (1980) An absolute method for protein determination based on difference in absorbance at 235 and 280 nm. *Anal. Biochem.* **109,** 156–159.
8. Kalb, V. F. and Bernlohr, R. W. (1977). A new spectrophotometric assay for protein in cell extracts. *Anal. Biochem.* **82,** 362–371.

2

The Lowry Method for Protein Quantitation

Jakob H. Waterborg and Harry R. Matthews

1. Introduction

The most accurate method of determining protein concentration is probably acid hydrolysis followed by amino acid analysis. Most other methods are sensitive to the amino acid composition of the protein, and absolute concentrations cannot be obtained. The procedure of Lowry et al. *(1)* is no exception, but its sensitivity is moderately constant from protein to protein, and it has been so widely used that Lowry protein estimations are a completely acceptable alternative to a rigorous absolute determination in almost all circumstances where protein mixtures or crude extracts are involved.

The method is based on both the Biuret reaction, where the peptide bonds of proteins react with copper under alkaline conditions producing Cu^{+}, which reacts with the Folin reagent, and the Folin–Ciocalteau reaction, which is poorly understood but in essence phosphomolybdotungstate is reduced to heteropolymolybdenum blue by the copper-catalyzed oxidation of aromatic amino acids. The reactions result in a strong blue color, which depends partly on the tyrosine and tryptophan content. The method is sensitive down to about 0.01 mg of protein/mL, and is best used on solutions with concentrations in the range 0.01–1.0 mg/mL of protein.

2. Materials

1. Complex-forming reagent: Prepare immediately before use by mixing the following three stock solutions A, B, and C in the proportion 100:1:1 (v:v:v), respectively.
   Solution A: 2% (w/v) Na_2CO_3 in distilled water.
   Solution B: 1% (w/v) $CuSO_4 \cdot 5H_2O$ in distilled water.
   Solution C: 2% (w/v) sodium potassium tartrate in distilled water.
2. 2*N* NaOH.
3. Folin reagent (commercially available): Use at 1*N* concentration.
4. Standards: Use a stock solution of standard protein (e.g., bovine serum albumin fraction V) containing 4 mg/mL protein in distilled water stored frozen at –20°C. Prepare standards by diluting the stock solution with distilled water as follows:

Stock solution, µL	0	1.25	2.50	6.25	12.5	25.0	62.5	125	250
Water, µL	500	499	498	494	488	475	438	375	250
Protein conc., µg/mL	0	10	20	50	100	200	500	1000	2000

From: The Protein Protocols Handbook
Edited by: J. M. Walker Humana Press Inc., Totowa, NJ

3. Method

1. To 0.1 mL of sample or standard (*see* Notes 1–3), add 0.1 mL of 2*N* NaOH. Hydrolyze at 100°C for 10 min in a heating block or boiling water bath.
2. Cool the hydrolyzate to room temperature and add 1 mL of freshly mixed complex-forming reagent. Let the solution stand at room temperature for 10 min (*see* Notes 4 and 5).
3. Add 0.1 mL of Folin reagent, using a vortex mixer, and let the mixture stand at room temperature for 30–60 min (do not exceed 60 min) (*see* Note 6).
4. Read the absorbance at 750 nm if the protein concentration was below 500 µg/mL or at 550 nm if the protein concentration was between 100 and 2000 µg/mL.
5. Plot a standard curve of absorbance as a function of initial protein concentration and use it to determine the unknown protein concentrations (*see* Notes 7–12).

4. Notes

1. If the sample is available as a precipitate, then dissolve the precipitate in 2*N* NaOH and hydrolyze as in step 1. Carry 0.2-mL aliquots of the hydrolyzate forward to step 2.
2. Whole cells or other complex samples may need pretreatment, as described for the Burton assay for DNA *(2)*. For example, the PCA/ethanol precipitate from extraction I may be used directly for the Lowry assay, or the pellets remaining after the PCA hydrolysis step (step 3 of the Burton assay) may be used for Lowry. In this latter case, both DNA and protein concentration may be obtained from the same sample.
3. Peterson *(3)* has described a precipitation step that allows the separation of the protein sample from interfering substances and also consequently concentrates the protein sample, allowing the determination of proteins in dilute solution. Peterson's precipitation step is as follows:
   a. Add 0.1 mL of 0.15% deoxycholate to 1.0 mL of protein sample.
   b. Vortex, and stand at room temperature for 10 min.
   c. Add 0.1 mL of 72% TCA, vortex, and centrifuge at 1000–3000*g* for 30 min.
   d. Decant the supernatant and treat the pellet as described in Note 1.
4. The reaction is very pH-dependent, and it is therefore important to maintain the pH between 10 and 10.5. Take care, therefore, when analyzing samples that are in strong buffer outside this range.
5. The incubation period is not critical and can vary from 10 min to several hours without affecting the final absorbance.
6. The vortex step is critical for obtaining reproducible results. The Folin reagent is only reactive for a short time under these alkaline conditions, being unstable in alkali, and great care should therefore be taken to ensure thorough mixing.
7. The assay is not linear at higher concentrations. Ensure, therefore, that you are analyzing your sample on the linear portion of the calibration curve.
8. A set of standards is needed with each group of assays, preferably in duplicate. Duplicate or triplicate unknowns are recommended.
9. One disadvantage of the Lowry method is the fact that a range of substances interferes with this assay, including buffers, drugs, nucleic acids, and sugars. The effect of some of these agents is shown in Table 1 in Chapter 3. In many cases, the effects of these agents can be minimized by diluting them out, assuming that the protein concentration is sufficiently high to still be detected after dilution. When interfering compounds are involved, it is, of course, important to run an appropriate blank. Interference caused by detergents, sucrose, and EDTA can be eliminated by the addition of SDS *(4)*.
10. Modifications to this basic assay have been reported that increase the sensitivity of the reaction. If the Folin reagent is added in two portions, vortexing between each addition, a

20% increase in sensitivity is achieved *(5)*. The addition of dithiothreitol 3 min after the addition of the Folin reagent increases the sensitivity by 50% *(6)*.

11. The amount of color produced in this assay by any given protein (or mixture of proteins) is dependent on the amino acid composition of the protein(s) (*see* Introduction). Therefore, two different proteins, each for example at concentrations of 1 mg/mL, can give different color yields in this assay. It must be appreciated, therefore, that using BSA (or any other protein for that matter) as a standard only gives an approximate measure of the protein concentration. The only time when this method gives an absolute value for protein concentration is when the protein being analyzed is also used to construct the standard curve. The most accurate way to determine the concentration of any protein solution is amino acid analysis (*see* Chapters 77 and 78).
12. A means of speeding up this assay using a microwave oven is described in Chapter 5.

References

1. Lowry, O. H., Rosebrough, N. J., Farr, A. L., and Randall, R. J. (1951) Protein measurement with the Folin phenol reagent. *J. Biol. Chem.* **193,** 265–275.
2. Waterborg, J. H. and Matthews, H. R. (1984) The Burton Assay for DNA, in *Methods in Molecular Biology, vol. 2: Nucleic Acids* (Walker, J. M., ed.), Humana, Totowa, NJ, pp. 1–3.
3. Peterson, G. L. (1983) Determination of total protein. *Methods Enzymol.* **91,** 95–121.
4. Markwell, M. A. K., Haas, S. M., Tolbert, N. E., and Bieber, L. L. (1981) Protein determination in membrane and lipoprotein samples. *Methods Enzymol.* **72,** 296–303.
5. Hess, H. H., Lees, M. B., and Derr, J. E. (1978) A linear Lowry-Folin assay for both water-soluble and sodium dodecyl sulfate-solubilized proteins. *Anal. Biochem.* **85,** 295–300.
6. Larson, E., Howlett, B., and Jagendorf, A. (1986) Artificial reductant enhancement of the Lowry method for protein determination. *Anal. Biochem.* **155,** 243–248.

3

The Bicinchoninic Acid (BCA) Assay for Protein Quantitation

John M. Walker

1. Introduction

The bicinchoninic acid (BCA) assay, first described by Smith et al. *(1)* is similar to the Lowry assay, since it also depends on the conversion of Cu^{2+} to Cu^{+} under alkaline conditions (*see* Chapter 2). The Cu^{+} is then detected by reaction with BCA. The two assays are of similar sensitivity, but since BCA is stable under alkali conditions, this assay has the advantage that it can be carried out as a one-step process compared to the two steps needed in the Lowry assay. The reaction results in the development of an intense purple color with an absorbance maximum at 562 nm. Since the production of Cu^{+} in this assay is a function of protein concentration and incubation time, the protein content of unknown samples may be determined spectrophotometrically by comparison with known protein standards. A further advantage of the BCA assay is that it is generally more tolerant to the presence of compounds that interfere with the Lowry assay. In particular it is not affected by a range of detergents and denaturing agents such as urea and guanidinium chloride, although it is more sensitive to the presence of reducing sugars. Both a standard assay (0.1–1.0 mg protein/mL) and a microassay (0.5–10 μg protein/mL) are described.

2. Materials

2.1. Standard Assay

1. Reagent A: sodium bicinchoninate (0.1 g), $Na_2CO_3 \cdot H_2O$ (2.0 g), sodium tartrate (dihydrate) (0.16 g), NaOH (0.4 g), $NaHCO_3$ (0.95 g), made up to 100 mL. If necessary, adjust the pH to 11.25 with $NaHCO_3$ or NaOH (*see* Note 1).
2. Reagent B: $CuSO_4 \cdot 5H_2O$ (0.4 g) in 10 mL of water (*see* Note 1).
3. Standard working reagent (SWR): Mix 100 vol of regent A with 2 vol of reagent B. The solution is apple green in color and is stable at room temperature for 1 wk.

2.2. Microassay

1. Reagent A: $Na_2CO_3 \cdot H_2O$ (0.8 g), NaOH (1.6 g), sodium tartrate (dihydrate) (1.6 g), made up to 100 mL with water, and adjusted to pH 11.25 with 10*M* NaOH.
2. Reagent B: BCA (4.0 g) in 100 mL of water.
3. Reagent C: $CuSO_4 \cdot 5H_2O$ (0.4 g) in 10 mL of water.

From: The Protein Protocols Handbook
Edited by: *J. M. Walker Humana Press Inc., Totowa, NJ*

4. Standard working reagent (SWR): Mix 1 vol of reagent C with 25 vol of reagent B, then add 26 vol of reagent A.

3. Methods

3.1. Standard Assay

1. To a 100-µL aqueous sample containing 10–100 µg protein, add 2 mL of SWR. Incubate at 60°C for 30 min (*see* Note 2).
2. Cool the sample to room temperature, then measure the absorbance at 562 nm (*see* Note 3).
3. A calibration curve can be constructed using dilutions of a stock 1 mg/mL solution of bovine serum albumin (BSA) (*see* Note 4).

3.2. Microassay

1. To 1.0 mL of aqueous protein solution containing 0.5–1.0 µg of protein/mL, add 1 mL of SWR.
2. Incubate at 60°C for 1 h.
3. Cool, and read the absorbance at 562 nm.

4. Notes

1. Reagents A and B are stable indefinitely at room temperature. They may be purchased ready prepared from Pierce, Rockford, IL.
2. The sensitivity of the assay can be increased by incubating the samples longer. Alternatively, if the color is becoming too dark, heating can be stopped earlier. Take care to treat standard samples similarly.
3. Following the heating step, the color developed is stable for at least 1 h.
4. Note, that like the Lowry assay, response to the BCA assay is dependent on the amino acid composition of the protein, and therefore an absolute concentration of protein cannot be determined. The BSA standard curve can only therefore be used to compare the relative protein concentration of similar protein solutions.
5. Some reagents interfere with the BCA assay, but nothing like as many as with the Lowry assay (*see* Table 1). The presence of lipids gives excessively high absorbances with this assay *(2)*. Variations produced by buffers with sulfhydryl agents and detergents have been described *(3)*.
6. Since the method relies on the use of Cu^{2+}, the presence of chelating agents such as EDTA will of course severely interfere with the method. However, it may be possible to overcome such problems by diluting the sample as long as the protein concentration remains sufficiently high to be measurable. Similarly, dilution may be a way of coping with *any* agent that interferes with the assay (*see* Table 1). In each case it is of course necesary to run an appropriate control sample to allow for any residual color development. A modification of the assay has been described that overcomes lipid interference when measuring lipoprotein protein content *(4)*.
7. A modification of the BCA assay, utilizing a microwave oven, has been described that allows protein determination in a matter of seconds *(see* Chapter 5).
8. A method has been described for eliminating interfering compounds such as thiols and reducing sugars in this assay. Proteins are bound to nylon membranes and exhaustively washed to remove interfering compounds; then the BCA assay is carried out on the membrane-bound protein *(5)*.

Table 1
Effect of Selected Potential Interfering Compounds[a]

Sample (50 µg BSA) in the following	BCA assay (µg BSA found) Water blank corrected	BCA assay (µg BSA found) Interference blank corrected	Lowry assay (µg BSA found) Water blank corrected	Lowry assay (µg BSA found) Interference blank corrected
50 µg BSA in water (reference)	50.00	—	50.00	—
0.1*N* HCl	50.70	50.80	44.20	43.80
0.1*N* NaOH	49.00	49.40	50.60	50.60
0.2% Sodium azide	51.10	50.90	49.20	49.00
0.02% Sodium azide	51.10	51.00	49.50	49.60
1.0*M* Sodium chloride	51.30	51.10	50.20	50.10
100 m*M* EDTA (4 Na)	No color		138.50	5.10
50 m*M* EDTA (4 Na)	28.00	29.40	96.70	6.80
10 m*M* EDTA (4 Na)	48.80	49.10	33.60	12.70
50 m*M* EDTA (4 Na), pH 11.25	31.50	32.80	72.30	5.00
4.0*M* Guanidine HCl	48.30	46.90	Precipitated	
3.0*M* Urea	51.30	50.10	53.20	45.00
1.0%Triton X-100	50.20	49.80	Precipitated	
1.0% SDS (lauryl)	49.20	48.90	Precipitated	
1.0% Brij 35	51.00	50.90	Precipitated	
1.0% Lubrol	50.70	50.70	Precipitated	
1.0% Chaps	49.90	49.50	Precipitated	
1.0% Chapso	51.80	51.00	Precipitated	
1.0% Octyl glucoside	50.90	50.80	Precipitated	
40.0% Sucrose	55.40	48.70	4.90	28.90
10.0% Sucrose	52.50	50.50	42.90	41.10
1.0% Sucrose	51.30	51.20	48.40	48.10
100 m*M* Glucose	245.00	57.10	68.10	61.70
50 mM Glucose	144.00	47.70	62.70	58.40
10 m*M* Glucose	70.00	49.10	52.60	51.20
0.2*M* Sorbitol	42.90	37.80	63.70	31.00
0.2*M* Sorbitol, pH 11.25	40.70	36.20	68.60	26.60
1.0*M* Glycine	No color		7.30	7.70
1.0*M* Glycine, pH 11	50.70	48.90	32.50	27.90
0.5*M* Tris	36.20	32.90	10.20	8.80
0.25*M* Tris	46.60	44.00	27.90	28.10
0.1*M* Tris	50.80	49.60	38.90	38.90
0.25*M* Tris, pH 11.25	52.00	50.30	40.80	40.80
20.0% Ammonium sulfate	5.60	1.20	Precipitated	
10.0% Ammonium sulfate	16.00	12.00	Precipitated	
3.0% Ammonium sulfate	44.90	42.00	21.20	21.40
10.0% Ammonium sulfate, pH 11	48.10	45.20	32.60	32.80
2.0*M* Sodium acetate, pH 5.5	35.50	34.50	5.40	3.30
0.2*M* Sodium acetate, pH 5.5	50.80	50.40	47.50	47.60
1.0*M* Sodium phosphate	37.10	36.20	7.30	5.30
0.1*M* Sodium phosphate	50.80	50.40	46.60	46.60
0.1*M* Cesium bicarbonate	49.50	49.70	Precipitated	

[a]Reproduced from ref. *1* with permission from Academic Press Inc.

References

1. Smith, P. K., Krohn, R. I., Hermanson, G. T., Mallia, A. K., Gartner, F. H., Provenzano, M. D., Fujimoto, E. K., Goeke, N. M., Olson, B. J., and Klenk, D. C. (1985) Measurement of protein using bicinchoninic acid. *Anal. Biochem.* **150,** 76–85.
2. Kessler, R. J. and Fanestil, D. D. (1986) Interference by lipids in the determination of protein using bicinchoninic acid. *Anal. Biochem.* **159,** 138–142.
3. Hill, H. D. and Straka, J. G. (1988) Protein determination using bicinchoninic acid in the presence of sulfhydryl reagents. *Anal. Biochem.* **170,** 203–208.
4. Morton, R. E. and Evans, T. A. (1992) Modification of the BCA protein assay to eliminate lipid interference in determining lipoprotein protein content. *Anal. Biochem.* **204,** 332–334.
5. Gates, R. E. (1991) Elimination of interfering substances in the presence of detergent in the bicinchoninic acid protein assay. *Anal. Biochem.* **196,** 290–295.

4

The Bradford Method for Protein Quantitation

Nicholas J. Kruger

1. Introduction

A rapid and accurate method for the estimation of protein concentration is essential in many fields of protein study. An assay originally described by Bradford *(1)* has become the preferred method for quantifying protein in many laboratories. This technique is simpler, faster, and more sensitive than the Lowry method. Moreover, when compared with the Lowry method, it is subject to less interference by common reagents and nonprotein components of biological samples (*see* Note 1).

The Bradford assay relies on the binding of the dye Coomassie blue G250 to protein. Detailed studies indicate that the free dye can exist in four different ionic forms for which the pK_a values are 1.15, 1.82, and 12.4 *(2)*. The more cationic red and green forms of the dye, which predominate in the acidic assay reagent solution, have absorbance maxima at 470 and 650 nm, respectively. In contrast, the more anionic blue form of the dye, which binds to protein, has an absorbance maximum at 590 nm. Thus, the quantity of protein can be estimated by determining the amount of dye in the blue ionic form. This is usually achieved by measuring the absorbance of the solution at 595 nm (*see* Note 2).

The dye appears to bind most readily to arginyl and lysyl residues of proteins (but does not bind to the free amino acids) *(3,4)*. This specificity can lead to variation in the response of the assay to different proteins, which is the main drawback of the method. The original Bradford assay shows large variation in response between different proteins *(5–7)*. Several modifications to the method have been developed to overcome this problem (*see* Note 3). However, these changes generally result in a less robust assay that is often more susceptible to interference by other chemicals. Consequently, the original method devised by Bradford remains the most convenient and widely used formulation. Two types of assay are described here: the standard assay, which is suitable for measuring between 10 and 100 µg protein, and the microassay, for detecting between 1 and 10 µg protein.

2. Materials

1. Reagent: The assay reagent is made by dissolving 100 mg of Coomassie blue G250 in 50 mL of 95% ethanol. The solution is then mixed with 100 mL of 85% phosphoric acid and made up to 1 L with distilled water (*see* Note 4).

From: The Protein Protocols Handbook
Edited by: J. M. Walker Humana Press Inc., Totowa, NJ

The reagent should be filtered through Whatman No. 1 filter paper and then stored in an amber bottle at room temperature. It is stable for several weeks. However, during this time, dye may precipitate from solution, so the stored reagent should be filtered before use.

2. Protein standard (*see* Note 5): Bovine γ-globulin at a concentration of 1 mg/mL (100 μg/mL for the microassay) in distilled water is used as a stock solution. This should be stored frozen at –20°C. Since the moisture content of solid protein may vary during storage, the precise concentration of protein in the standard solution should be determined from its absorbance at 280 nm. The absorbance of a 1 mg/mL solution of γ-globulin, in a 1-cm light path, is 1.35. The corresponding values for two alternative protein standards, bovine serum albumin and ovalbumin, are 0.66 and 0.75, respectively.
3. Plastic and glassware used in the assay should be absolutely clean and detergent free. Quartz (silica) spectrophotometer cuvets should not be used, as the dye binds to this material. Traces of dye bound to glassware or plastic can be removed by rinsing with methanol or detergent solution.

3. Methods

3.1. Standard Assay Method

1. Pipet between 10 and 100 μg of protein in 100 μL total volume into a test tube. If the approximate sample concentration is unknown, assay a range of dilutions (1, 1/10, 1/100, 1/1000). Prepare duplicates of each sample.
2. For the calibration curve, pipet duplicate volumes of 10, 20, 40, 60, 80, and 100 μL of 1 mg/mL γ-globulin standard solution into test tubes, and make each up to 100 μL with distilled water. Pipet 100 μL of distilled water into an additional tube to provide the reagent blank.
3. Add 5 mL of protein reagent to each tube, and mix well by inversion or gentle vortexing. Avoid foaming, which will lead to poor reproducibility.
4. Measure the A_{595} of the samples and standards against the reagent blank between 2 min and 1 h after mixing (*see* Note 6). The 100-μg standard should give an A_{595} value of about 0.4. The standard curve is not linear, and the precise absorbance varies depending on the age of the assay reagent. Consequently, it is essential to construct a calibration curve for each set of assays (*see* Note 7).

3.2. Microassay Method

This form of the assay is more sensitive to protein. Consequently, it is useful when the amount of the unknown protein is limited (*see also* Note 8).

1. Pipet duplicate samples containing between 1 and 10 μg in a total volume of 100 μL into 1.5-mL polyethylene microfuge tubes. If the approximate sample concentration is unknown, assay a range of dilutions (1, 1/10, 1/100, 1/1000).
2. For the calibration curve, pipet duplicate volumes of 10, 20, 40, 60, 80, and 100 μL of 100 μg/mL γ-globulin standard solution into microfuge tubes, and adjust the volume to 100 μL with water. Pipet 100 μL of distilled water into a tube for the reagent blank.
3. Add 1 mL of protein reagent to each tube, and mix gently, but thoroughly. Measure the absorbance of each sample between 2 and 60 min after addition of the protein reagent. The A_{595} value of a sample containing 10 μg γ-globulin is 0.45. Figure 1 shows the response of three common protein standards using the microassay method.

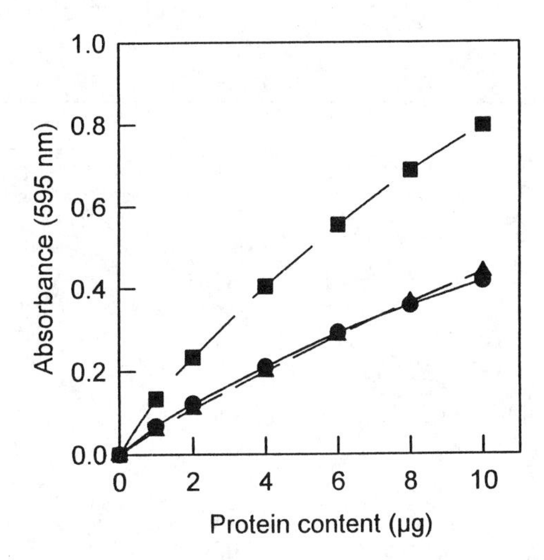

Fig. 1. Variation in the response of proteins in the Bradford assay. The extent of protein–dye complex formation was determined for bovine serum albumin (■), γ-globulin (●), and ovalbumin (▲) using the microassay. Each value is the mean of four determinations. These data allow comparisons to be made between estimates of protein content obtained using these protein standards.

4. Notes

1. The Bradford assay is relatively free from interference by most commonly used biochemical reagents. However, a few chemicals may significantly alter the absorbance of the reagent blank or modify the response of proteins to the dye (Table 1). The materials that are most likely to cause problems in biological extracts are detergents and ampholyte *(3,8)*. These should be removed from the sample solution, for example, by gel filtration or dialysis. Alternatively, they should be included in the reagent blank and calibration standards at the same concentration as that found in the sample. The presence of base in the assay increases absorbance by shifting the equilibrium of the free dye toward the anionic form. This may present problems when measuring protein content in concentrated basic buffers *(3)*. Guanidine hydrochloride and sodium ascorbate compete with dye for protein, leading to underestimation of the protein content *(3)*.
2. Binding of protein to Coomassie blue G250 shifts the absorbance maximum of the blue ionic form of the dye from 590 nm to 620 nm *(2)*. Although it might, therefore, appear more sensible to measure the absorbance at the higher wavelength, this is not advisable. At the usual pH of the assay, an appreciable proportion of the dye is in the green form (λ max = 650 nm), which interferes with absorbance measurement of the dye–protein complex at 620 nm. The absorbance is normally measured at 595 nm, as this was previously believed to be the absorbance maximum for the blue form of the dye *(3)*.
3. The assay technique described here is subject to variation in sensitivity between individual proteins (*see* Table 2). Several modifications have been suggested that reduce this variability *(5–7,9)*. Generally, these rely on increasing either the dye content or the pH of the solution. In one variation, adjusting the pH by adding NaOH to the reagent improves the sensitivity of the assay and greatly reduces the variation observed with different proteins *(7)*. (This is presumably due to an increase in the proportion of free dye in the

Table 1
Effects of Common Reagents on the Bradford Assay

	Absorbance at 600 nm	
Compound	Blank	5 μg Immunoglobulin
Control	0.005	0.264
0.02% SDS	0.003	0.250
0.1% SDS	0.042[a]	0.059[a]
0.1% Triton	0.000	0.278
0.5% Triton	0.051[a]	0.311[a]
1*M* 2-Mercaptoethanol	0.006	0.273
1*M* Sucrose	0.008	0.261
4*M* Urea	0.008	0.261
4*M* NaCl	–0.015	0.207[a]
Glycerol	0.014	0.238[a]
0.1*M* HEPES, pH 7.0	0.003	0.268
0.1*M* Tris, pH 7.5	–0.008	0.261
0.1*M* Citrate, pH 5.0	0.015	0.249
10 m*M* EDTA	0.007	0.235[a]
1*M* $(NH_4)_2SO_4$	0.002	0.269

Data were obtained by mixing 5 μL of sample with 5 μL of the specified compound before adding 200 μL of dye-reagent.

[a]Measurements that differ from the control by more than 0.02 absorbance unit for blank values or more than 10% for the samples containing protein. Data taken from ref. *7*.

Table 2
Comparison of the Response of Different Proteins in the Bradford Assay

	Relative absorbance	
Protein	Assay 1	Assay 2
Myelin basic protein	139	—
Histone	130	175
Cytochrome c	128	142
Bovine serum albumin	100	100
Insulin	89	—
Transferrin	82	—
Lysozyme	73	—
α-Chymotrypsinogen	55	—
Soybean trypsin inhibitor	52	23
Ovalbumin	49	23
γ-Globulin	48	55
β-Lactoglobulin A	20	—
Trypsin	18	15
Aprotinin	13	—
Gelatin	—	5
Gramicidin S	5	—

For each protein, the response is expressed relative to that of the same concentration of bovine serum albumin. The data for Assays 1 and 2 are recalculated from refs. *5* and *7*, respectively.

blue form, the ionic species that reacts with protein.) However, the optimum pH is critically dependent on the source and concentration of the dye (*see* Note 4). Moreover, the modified assay is far more sensitive to interference from detergents in the sample.

Particular care should be taken when measuring the protein content of membrane fractions. The conventional assay consistently underestimates the amount of protein in membrane-rich samples. Pretreatment of the samples with membrane-disrupting agents, such as NaOH or detergents may reduce this problem, but the results should be treated with caution *(10)*.

4. The amount of soluble dye in Coomassie blue G250 varies considerably between sources, and suppliers' figures for dye purity are not a reliable estimate of the Coomassie blue G250 content *(11)*. Generally, Serva blue G is believed to have the greatest dye content and should be used in the modified assays discussed in Note 3. However, the quality of the dye is not critical for routine protein determination using the method described in this chapter. The data presented in Fig. 1 were obtained using Coomassie brilliant blue G (C.I. 42655; product code B-0770, Sigma Chemical Co., St. Louis, MO).
5. Whenever possible, the protein used to construct the calibration curve should be the same as that being determined. Often this is impractical, and the dye response of a sample is quantified relative to that of a "generic" protein. Bovine serum albumin is commonly used as the protein standard because it is inexpensive and readily available in a pure form. The major argument for using this protein is that it allows the results to be compared directly with those of the many previous studies that have used bovine serum albumin as a standard. However, it suffers from the disadvantage of exhibiting an unusually large dye response in the Bradford assay, and thus, may underestimate the protein content of a sample. Increasingly, bovine γ-globulin is being advanced as a more suitable general standard, since the dye binding capacity of this protein is closer to the mean of those proteins that have been compared (Table 2). Because of the variation in response between different proteins, it is essential to specify the protein standard used when reporting measurements of protein amounts using the Bradford assay.
6. Generally, it is preferable to use a single new disposable polystyrene semi-microcuvet that is discarded after a series of absorbance measurements. Rinse the cuvet with reagent before use, zero the spectrophotometer on the reagent blank, and then do not remove the cuvet from the machine. Replace the sample in the cuvet gently using a disposable polyethylene pipet.
7. The standard curve is nonlinear at high protein levels, because the amount of free dye becomes depleted. If this presents problems, the linearity of the assay can be improved by plotting the ratio of absorbances at 595 and 465 nm, which corrects for depletion of the free dye *(12)*.
8. For routine measurement of the protein content of many samples, the microassay may be adapted for use with a microplate reader *(7,13)*. The total volume of the modified assay is limited to 210 μL by reducing the volume of each component. Ensure effective mixing of the assay components by pipeting up to 10 μL of the protein sample into each well before adding 200 μL of the dye reagent.

References

1. Bradford, M. M. (1976) A rapid and sensitive method for the quantitation of microgram quantities of protein utilizing the principle of protein-dye binding. *Anal. Biochem.* **72,** 248–254.
2. Chial, H. J., Thompson, H. B., and Splittgerber, A. G. (1993) A spectral study of the charge forms of Coomassie blue G. *Anal. Biochem.* **209,** 258–266.

3. Compton, S. J. and Jones, C. G (1985) Mechanism of dye response and interference in the Bradford protein assay. *Anal. Biochem.* **151,** 369–374.
4. Congdon, R. W., Muth, G. W., and Splittgerber, A. G. (1993) The binding interaction of Coomassie blue with proteins. *Anal. Biochem.* **213,** 407–413.
5. Friendenauer, S. and Berlet, H. H. (1989) Sensitivity and variability of the Bradford protein assay in the presence of detergents. *Anal. Biochem.* **178,** 263–268.
6. Reade, S. M. and Northcote, D. H. (1981) Minimization of variation in the response to different proteins of the Coomassie blue G dye-binding assay for protein. *Anal. Biochem.* **116,** 53–64.
7. Stoscheck, C. M. (1990) Increased uniformity in the response of the Coomassie blue protein assay to different proteins. *Anal. Biochem.* **184,** 111–116.
8. Spector, T. (1978) Refinement of the Coomassie blue method of protein quantitation. A simple and linear spectrophotometric assay for <0.5 to 50 μg of protein. *Anal. Biochem.* **86,** 142–146.
9. Peterson, G. L. (1983) Coomassie blue dye binding protein quantitation method, in *Methods in Enzymology,* vol. 91 (Hirs, C. H. W. and Timasheff, S. N., eds.), Academic, New York, pp. 95–119.
10. Kirazov, L. P., Venkov, L. G., and Kirazov, E. P. (1993) Comparison of the Lowry and the Bradford protein assays as applied for protein estimation of membrane-containing fractions. *Anal. Biochem.* **208,** 44–48.
11. Wilson, C. M. (1979) Studies and critique of Amido Black 10B, Coomassie blue R and Fast green FCF as stains for proteins after polyacrylamide gel electrophoresis. *Anal. Biochem.* **96,** 263–278.
12. Sedmak, J. J. and Grossberg, S. E. (1977) A rapid, sensitive and versatile assay for protein using Coomassie brilliant blue G250. *Anal. Biochem.* **79,** 544–552.
13. Redinbaugh, M. G. and Campbell, W. H. (1985) Adaptation of the dye-binding protein assay to microtiter plates. *Anal. Biochem.* **147,** 144–147.

5

Ultrafast Protein Determinations Using Microwave Enhancement

Robert E. Akins and Rocky S. Tuan

1. Introduction

In this chapter, we describe modifications of existing protein assays that take advantage of microwave irradiation to reduce assay incubation times from the standard 15–60 min down to just seconds *(1)*. Adaptations based on two standard protein assays will be described:

1. The classic method of Lowry et al. (ref. *2* and *see* Chapter 2), which involves intensification of the biuret reaction through the addition of Folin-Ciocalteau phenol reagent; and
2. The recently developed method of Smith et al. *(3)*, which involves intensification of the biuret assay through the addition of bicinchoninic acid (BCA) (*see* Chapter 3).

Performing incubations in a 2.45–GHz microwave field (i.e., in a microwave oven) for 10–20 s results in rapid, reliable, and reproducible protein determinations. We have provided here background information concerning protein assays in general and the microwave enhanced techniques in particular. The use of microwave exposure as an aid in the preparation of chemical and biological samples is well established; interested readers are directed to Kok and Boon *(4)* for excellent discussions concerning the application of microwave technologies in biological research.

1.1. Microwave Assay

Household microwave ovens expose materials to nonionizing electromagnetic radiation at a frequency of about 2.45 GHz (i.e., 2.45 billion oscillations/s). Such exposures have been applied to a myriad of scientific purposes including tissue fixation, histological staining, immunostaining, PCR, and many others. The practical and theoretical aspects of many microwave techniques are summarized by Kok and Boon *(4)*, and the reader is directed to this reference for excellent discussions concerning general procedures and theoretical background.

Microwave ovens are conceptually simple and remarkably safe devices (*see* Note 1). Typically, a magnetron generator produces microwaves that are directed toward the sample chamber by a wave guide. The beam is generally homogenized by a "mode-stirrer," consisting of a reflecting fan with angled blades that scatter the beam as it

From: The Protein Protocols Handbook
Edited by: J. M. Walker Humana Press Inc., Totowa, NJ

passes. The side walls of the chamber are made of a microwave reflective material, and the microwaves are thereby contained within the defined volume of the oven (*see* Note 2). Specimens irradiated in a microwave oven absorb a portion of the microwave energy depending on specific interactions between the constituent molecules of the sample and the oscillating field.

As microwaves pass through specimens, the molecules in that specimen are exposed to a continuously changing electromagnetic field. This field is often represented as a sine wave with amplitude related to the intensity of the field at a particular point over time and wavelength related to the period of oscillation. Ionically polarized molecules (or dipoles) will align with the imposed electromagnetic field and will tend to rotate as the sequential peaks and troughs of the oscillating "wave" pass. Higher frequencies would, therefore, tend to cause faster molecular rotations. At a point, a given molecule will no longer be able to reorient quickly enough to align with the rapidly changing field, and it will cease spinning. There exists, then, a distinct relationship between microwave frequency and the "molecular-size" of the dipoles that it will affect. This relationship is important, and at the 2.45–GHz frequency used in conventional microwave ovens, only small molecules may be expected to rotate; specifically, water molecules rotate easily in microwave ovens but proteins do not (*see* Note 3).

For most microwave oven functions, a portion of the rotational energy of the water molecules in a sample dissipates as heat. Since these molecular rotations occur throughout exposed samples, microwave ovens provide extremely efficient heating, and the effects of microwaves are generally attributed to changes in local temperature. At some as yet undetermined level, however, microwave exposure causes an acceleration in the rate of reaction product formation in the protein assays discussed here. This dramatic acceleration is independent of the change in temperature, and our observations have suggested that microwave-based heating is not the principal means of reaction acceleration.

1.2. Microwave Enhanced Protein Determinations

Modifications of the procedures of Lowry (Chapter 2) and Smith (Chapter 3) to include microwave irradiation result in the generation of linear standard curves. Figure 1 illustrates typical standard curves for examples of each assay using bovine serum albumin (BSA). Both standard curves are linear across a practical range of protein concentrations.

One interesting difference between the two microwave enhanced assays concerns the relationship between irradiation time and assay sensitivity. In the *DC Protein Assay*, a modification of Lowry's *(2)* procedure supplied by Bio-Rad Laboratories, Inc. (Hercules, CA), illustrated in Fig. 1, a colorimetric end point was reached after 10 s of microwave irradiation; no further color development occurred in the samples. This end point was identical to that achieved in a 15-min room temperature control assay.

The *BCA* Protein Assay*, a version of Smith's assay *(3)* supplied by Pierce Chemical Co. (Rockford, IL), afforded some flexibility in assay sensitivity because the formation of detectable reaction product was a function of the duration of microwave exposure. Figure 2 shows the rate of reaction product formation for three different concentrations of BSA as a function of microwave exposure in a BCA assay. Absorbance values increased for each BSA concentration as a second order function of irradiation time. A linear standard curve could be generated from BSA dilutions that were

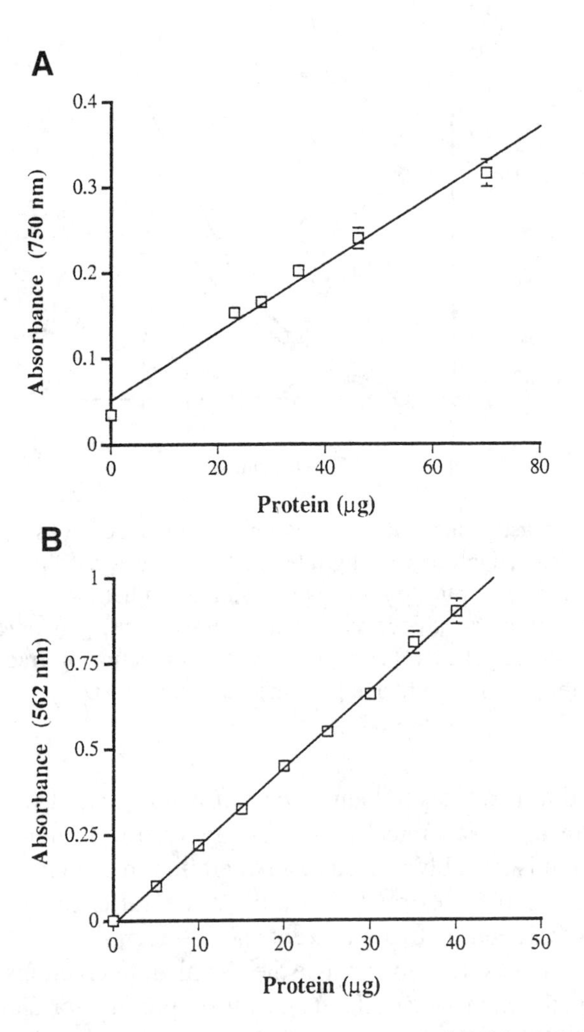

Fig. 1. Typical standard curves for microwave Lowry and BCA assays. Standard curves were generated with microwave protocols using BSA (Sigma) dissolved in water. BSA samples in 100-μL vol were prepared in triplicate for each assay and were combined with reagent as described in Notes 1 and 2. Tubes and a water load (total vol 100 mL, *see* Note 5) were placed in the center of a microwave oven. Samples for the Lowry assay were irradiated for 10 s; samples for the BCA assay were irradiated for 20 s. Results in **A** show a linear standard curve generated using a microwave Lowry assay. Results in **B** show a linear standard curve generated using a microwave BCA assay. Values presented are means ± SD.

irradiated for any specific time; longer irradiation times yielded more steeply sloped standard curves. In practice, then, the duration of microwave exposure can be selected to correspond to a desired sensitivity range with longer times being more suitable for lower protein concentrations. In contrast to the *DC Protein Assay*, the BCA microwave procedure described here is more sensitive than a standard, room temperature assay, and we have used it for most applications (*see* Note 4).

Figure 3 illustrates that the dramatic effects of microwave exposure cannot be mimicked by external heating. These results are surprising since microwave effects are

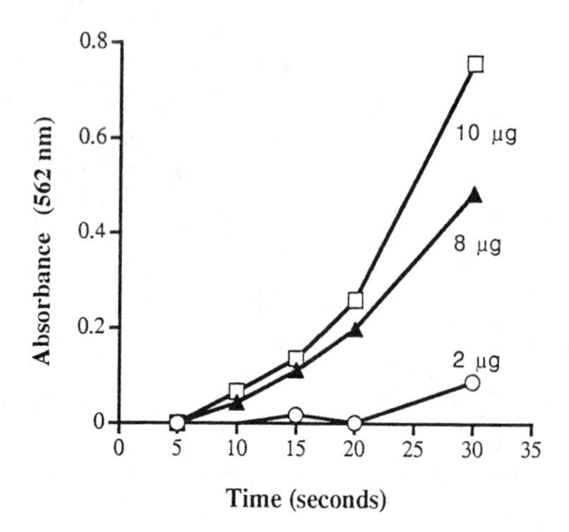

Fig. 2. Effect of Increasing microwave irradiation time on the BCA assay. Three amounts of BSA (Sigma) were prepared in water: 10 µg/tube (□), 8 µg/tube (▲), 2 µg/tube (○). Each time point was determined from triplicate samples in a single irradiation trial with the water load replaced between determinations. Each assay time resulted in the generation of a linear standard curve; the slope of each standard curve increased as a function of irradiation time. Values were normalized to the 5-s time point and presented as means ± SD.

generally attributed to increases in temperature. It is not clear at what level(s) microwaves interact with the biochemical processes involved in protein estimation; however, the acceleration is possibly related to an alteration in solvent/solute interactions. As the solvent water molecules rotate, specific structural changes may occur in the system such that interactions between solvent and solute molecules (or among the solvent molecules themselves) tend to enhance the chemical interactions between the protein and the assay components to accelerate the rate of product formation. For example, water molecules rotating in a microwave field may no longer be available to form hydrogen bonds within the solvent/solute structure. Clearly, the nature and mechanism of nonthermal microwave effects need to be studied further.

2. Materials

2.1. Lowry Assay (see *Chapter 2*)

Lowry reagents are available from commercial sources. Assay reagents were routinely purchased from Bio-Rad Laboratories in the form of a detergent-compatible Lowry kit (*DC Protein Assay*). Assay reactions are typically carried out in polystyrene Rohren tubes (Sarstedt, Inc., Newtown, NC). Tubes were placed in a plastic test tube rack at the center of a suitable microwave oven *(see the following)* along with a beaker containing approx 100 mL of H_2O (*see* Note 5).

2.2. BCA Assay (see *Chapter 3*)

BCA protein reagent is available from commercial sources. Assay reagents are routinely purchased from Pierce Chemical Co. in the form of a *BCA* Protein Assay* kit. As with the Lowry assay, reactions were typically carried out in polystyrene Rohren

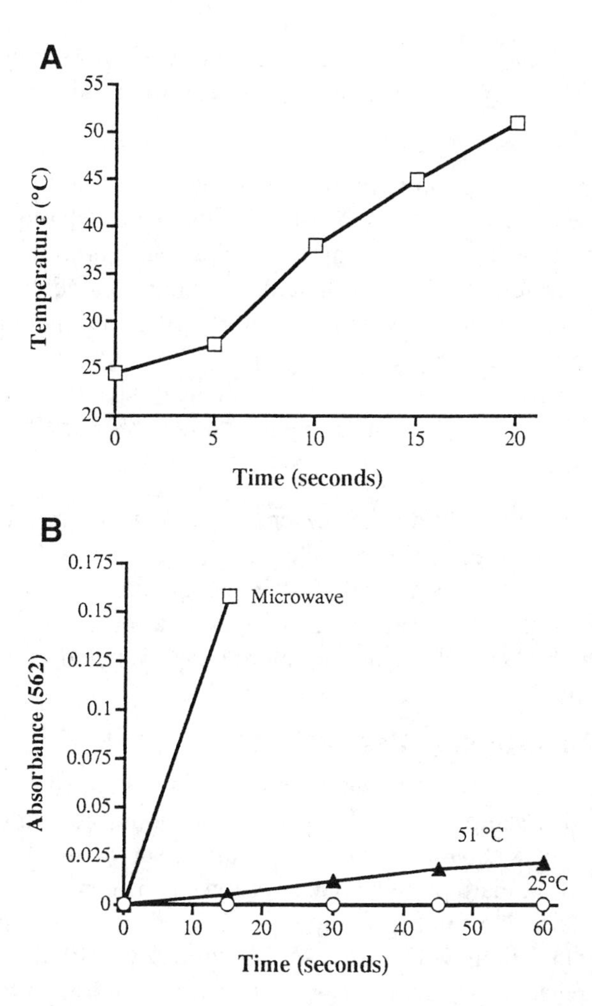

Fig. 3. Comparison of microwave irradiation with incubation at elevated temperature. **(A)** The change in temperature of BCA assay samples containing bovine serum albumin (BSA, Sigma). Temperatures reached 51°C during a typical 20-s irradiation. Since the assays are carried out in an open system, temperatures near 100°C would cause sample boil-over and should be avoided. **(B)** Development of reaction product under three different conditions: microwave irradiation (□), incubation at 51°C (▲), and incubation at 25°C (○). Incubation at 51°C, the maximal temperature reached during a 20-s irradiation, did not mirror microwave irradiation.

tubes (Sarstedt). Tubes were placed in a plastic test tube rack at the center of a suitable microwave oven *(see* Section 3.2.) along with a beaker containing approx 100 mL of H_2O (*see* Note 5).

3. Methods

3.1. Sample Preparation

Sample preparation should be carried out as specified by the manufacturers. Generally, samples are solubilized in a noninterfering buffer (*see* Note 3) so that the final protein concentration falls within the desired range (*see* Note 4). Samples should be either filtered or centrifuged to remove any debris prior to protein determination.

3.2. Selection of Microwave Oven

Microwave ovens differ substantially in their suitability for these assays. Desirable attributes include fine control of irradiation time, a chamber size large enough to easily accommodate the desired number of samples, and a configuration that results in a homogenous field of irradiation so that all samples within the central volume of the oven receive a uniform microwave dose (*see* Note 4). Samples should be placed in a nonmetallic test tube rack in the center of the oven. A volume of room temperature water is included in the oven chamber as well so that the total amount of fluid (samples + additional water) is constant from one assay to another (*see* Note 5).

3.3. Sample Irradiation

Once the samples are placed into the center of the microwave chamber, close the door and irradiate the samples. Using the *DC Protein Assay* , a 10-s irradiation was optimal as a replacement for the standard 15–min incubation. Using the *BCA* Protein Assay*, a 20-s irradiation has proved adequate in most situations. We have found it most convenient to use the highest setting on the microwave oven and to control exposure using an accurate timer.

3.4. Reading and Interpreting Assay Results

After irradiation, the absorbance of each sample and standard should be determined spectrophotometrically. Samples from Lowry assays should be read at 750 nm and samples from BCA assays should be read at 562 nm. A standard curve of absorbance values as a function of standard protein concentration can be generated easily and used to determine the protein levels in the unknown samples. It is recommended that standard curves be generated along with each assay to avoid any difficulties that may arise from differences in reagents or alterations in total microwave exposure.

3.5. Summary

In summary, the microwave BCA protein assay protocol is as follows:

1. Combine samples and BSA standards with BCA assay reagent in polystyrene tubes;
2. Place samples into an all plastic test tube rack in the center of a microwave oven along with a beaker containing a volume of room temperature water sufficient to make the total volume of liquid in the chamber 100 mL;
3. Irradiate samples for 20 s on the highest microwave setting; and
4. Measure A_{562} for each sample and determine protein concentrations based on a BSA calibration curve.

The microwave Lowry assay protocol is virtually identical:

1. Combine samples and Lowry assay reagents in polystyrene tubes;
2. Place samples into an all plastic test tube rack in the center of a microwave oven along with a beaker containing a volume of room temperature water sufficient to make the total volume of liquid in the chamber 100 mL;
3. Irradiate samples for 10 s on the highest microwave setting; and
4. Measure A_{750} for each sample and determine protein concentrations based on a BSA calibration curve.

Microwave protein assays are suitable for all situations where standard assays are presently used. The ability to determine accurate protein concentrations in so little time should greatly facilitate routine assays and improve efficiency when protocols require protein determination at multiple intermediate steps. Similar microwave techniques have also been applied as time-saving and efficiency-enhancing procedures by several authors (*see* ref. *4*). We have used microwave assays to generate chromatograms during protein purification and for general protein determinations (e.g., before electrophoretic analysis). The assays consistently yield reliable results that are comparable to those obtained by standard protocols. The modifications we present here are very easily adapted to most commercially available microwave ovens and, in the case of the BCA assay, can be adjusted to cover a wide range of protein concentrations. Since the duration of the assays is so short, it is possible to try several irradiation times and water loads to select the specific conditions required by the particular microwave and samples to be used. Microwave enhanced protein estimations should prove to be extremely useful in laboratories currently doing standard Smith *(3)* or Lowry *(2)* based protein determinations.

4. Notes

1. Although microwave radiation is nonionizing, precautions should be taken to avoid direct irradiation of parts of the body. Microwaves can penetrate the skin and cause significant tissue damage in relatively short periods of time. Most contemporary microwave ovens are remarkably safe and leakage is unlikely; however, periodic assessment of microwave containment within the microwave chamber should be carried out. Perhaps more dangerous than radiation effects are potential problems caused by the rapid heating of irradiated samples. Care should be taken when removing samples from the microwave chamber to avoid getting burned, sealed containers should not be irradiated as they may explode, and metal objects should be excluded from the chamber to avoid sparking. Users should consult their equipment manuals and institute safety offices prior to using microwave ovens. A detailed discussion of microwave hazards is included in ref. *4*.
2. Microwave ovens use a nominal frequency of 2.45 GHz. The energy put into the microwave chamber is actually a range of frequencies around 2.45 GHz. As the waves in the oven reflect off of the metal chamber walls, "hot" and "cold" spots may be set up by the constructive and destructive interference of the waves. The positions of these "hot" and "cold" spots is a function of the physical design of the oven chamber and the electrical properties of the materials contained in the oven. The unevenness of microwave fields can be minimized by mode stirring *(see the preceeding)*, appropriate oven configuration, or by rotating the specimen in the chamber during irradiation. If the design of a particular microwave oven does not provide a relatively uniform irradiation volume, it may not be useful for the assays outlined here.

   The suitability of a particular microwave oven may be tested easily using a number of tubes with known concentrations of protein. Uneven irradiation patterns will be detected by significant differences in the color development for a given protein concentration as a function of position within the oven chamber. We have had success with several microwave ovens including a 0.8 cu ft, 600W, General Electric oven, model JEM18F001 and a 1.3 cu ft, 650W, Whirlpool oven, model RJM7450.
3. The frequency used in household microwave ovens is actually below the resonance frequency for water. Above the 2.45 GHz used in household microwave ovens, water molecules are capable of rotating faster and of absorbing substantially more microwave

energy. Too much absorption, however, is undesirable. It is possible that the outer layers of a sample may absorb energy so efficiently that the interior portions receive substantially less energy. The resulting uneven exposure may have adverse effects.

It is important to note that molecules other than water may absorb microwave energy. If a compound added to a microwave enhanced assay absorbs strongly near 2.45 GHz, uneven sample exposures may result because of the presence of the compound. In addition, compounds that are degraded or converted during a microwave procedure, or compounds that have altered interactions with other assay components during irradiation, may substantially affect assay results. Although it may be possible to predict which materials would interfere with a given assay by considering the relevant chemical and electrical characteristics of the constituent compounds, potential interference is most easily assessed empirically by directly determining the effects of a given additive on the accuracy and sensitivity of the standard microwave assay.

4. Assay sensitivity may be improved by increasing irradiation time. The time required may be determined quickly by using BSA test solutions in the range of protein concentrations expected until desirable A_{562} values were obtained. By increasing microwave exposure time, it is possible to substantially increase assay sensitivity while keeping irradiation times below 60 s. The ease with which sensitivity may be adjusted within extremely short time frames places the microwave BCA assay among the quickest, most flexible assays available for protein determinations.
5. The addition of a volume of water to the microwave chamber, such that the total volume contained in the microwave chamber is constant from one assay to the next, allows irradiation conditions to be controlled easily from assay to assay. The additional water acts as a load on the oven and absorbs some of the microwave energy. Since the total amount of water remains constant from one assay to the next, the amount of energy absorbed also remains constant. The time of irradiation, therefore, becomes independent of the number of samples included in the assay.

Acknowledgments

This work was supported in part by funds from NASA-SBRA and AHA 9406244S (REA), as well as NIH HD 15822 and NIH HD 29937 (RST).

References

1. Akins, R. E. and Tuan, R. S. (1992) Measurement of protein in 20 seconds using a microwave BCA assay. *Biotechniques* **12,** 496–499.
2. Lowry, O. H., Rosebrough, N. J., Farr, A. L., and Randall, R. J. (1951) Protein measurement with the Folin phenol reagent. *J. Biol. Chem.* **193,** 265–275.
3. Smith, P. K., Krohn, R. I., Hermanson, G. T., Mallia, A. K., Gartner, F. H., Provenzano, M. D., Fujimoto, E. K., Goeke, N. M., Olson, B. J., and Klenk, D. C. (1985) Measurement of protein using bicinchoninic acid. *Analyt. Biochem.* **150,** 76–85.
4. Kok, L. P. and Boon, M. E. (1992) *Microwave Cookbook for Microscopists.* Coulomb, Leyden, Leiden, Netherlands.

6

Quantitation of Tryptophan in Proteins

Alastair Aitken and Michèle Learmonth

1. Introduction

1.1. Hydrolysis Followed by Amino Acid Analysis

Accurate measurement of the amount of tryptophan in a sample is problematic, since it is completely destroyed under normal conditions employed for the complete hydrolysis of proteins. Strong acid is ordinarily the method of choice, and constant boiling hydrochloric acid, 6*M*, is most frequently used. The reaction is usually carried out in evacuated sealed tubes or under N_2 at 110°C for 18–96 h. Under these conditions, peptide bonds are quantitatively hydrolyzed (although relatively long periods are required for the complete hydrolysis of bonds to valine, leucine, and isoleucine). As well as complete destruction of tryptophan, small losses of serine and threonine occur, which are corrected for. The advantages of amino acid analysis include the measurement of absolute amounts of protein, provided that the sample is not contaminated by other proteins. However, it may be a disadvantage if an automated amino acid analyzer is not readily available. Acid hydrolysis in the presence of 6*N* HCl, containing 0.5–6% (v/v) thioglycolic, acid at 110°C for 24–72 h, *in vacuo* will result in greatly improved tryptophan yields *(1)*, although most commonly, hydrolysis in the presence of the acids described in Section 3.1. may result in almost quantitative recovery of tryptophan.

Alkaline hydrolysis followed by amino acid analysis is also used for the estimation of tryptophan. The complete hydrolysis of proteins is achieved with 2–4*M* sodium hydroxide at 100°C for 4–8 h. This is of limited application for routine analysis, because cysteine, serine, threonine, and arginine are destroyed in the process, and partial destruction by deamination of other amino acids occurs.

The complete enzymatic hydrolysis of proteins (where tryptophan would be quantitatively recovered) is difficult, because most enzymes attack only specific peptide bonds rapidly. Often a combination of enzymes is employed (such as "Pronase") and extended time periods are required (*see* Chapter 76). A further complication of this method is possible contamination resulting from autodigestion of the enzymes.

1.2. Measurement of Tryptophan Content by UV

The absorption of protein solutions in the UV is the result of tryptophan and tyrosine (and to a very minor, and negligible, extent phenylalanine and cysteine). The absorp-

From: The Protein Protocols Handbook
Edited by: J. M. Walker Humana Press Inc., Totowa, NJ

tion maximum will depend on the pH of the solution, and spectrophotometric measurements are usually made in alkaline solutions. Absorption curves for tryptophan and tyrosine show that at the points of intersection, 257 and 294 nm, the extinction values are proportional to the total tryptophan + tyrosine content. Measurements are normally made at 294.4 nm, since this is close to the maximum in the tyrosine curve (where $\Delta\varepsilon/\Delta\lambda$, the change in extinction with wavelength, is minimal), and in conjunction with the extinction at 280 nm (where $\Delta\varepsilon/\Delta\lambda$ is minimal for tryptophan), the concentrations of each of the two amino acids may be calculated. This is the method of Goodwin and Morton *(2)*.

2. Materials

1. 3*M* *p*-toluenesulfonic acid.
2. 0.2% tryptamine 3-[2-Aminoethyl] indole (Pierce, Chester, UK).
3. 3*M* mercaptoethanesulfonic acid (Pierce).
4. 1*M* NaOH.

3. Methods

3.1. Quantitation of Tryptophan by Acid Hydrolysis

1. To the protein dried in a Pyrex glass tube (1.2 × 6 cm or similar, in which a constriction has been made by heating in an oxygen/gas flame) is added 1 mL of 3*M* p-toluenesulphonic acid, containing 0.2% tryptamine (0.2% 3-[2-aminoethyl] indole) *(3)*.
2. The solution is sealed under vacuum and heated in an oven for 24–72 h at 110°C, *in vacuo*.
3. Alternatively, the acid used may be 3*M* mercaptoethanesulfonic acid, The sample is hydrolyzed for a similar time and temperature *(4)*.
4. The tube is allowed to cool and cracked open with a heated glass rod held against a horizontal scratch made in the side of the tube.
5. The acid is taken to near neutrality by carefully adding 2 mL of 1*M* NaOH. An aliquot of the solution (which is still acid) is mixed with the amino acid analyzer loading buffer.
6. Following this hydrolysis, quantitative analysis is carried out for each of the amino acids on a suitable automated instrument.

3.2. Alkaline Hydrolysis

1. To the protein dried in a Pyrex glass tube (as above, Section 3.1. step 1) 0.5 mL of 3*M* sodium hydroxide is added.
2. The solution is sealed under vacuum and heated in an oven for 4–8 h at 100°C, *in vacuo*.
3. After cooling and cracking open, the alkali is neutralized carefully with an equivalent amount of 1*M* HCl. An aliquot of the solution is mixed with the amino acid analyzer loading buffer and analyzed (as above, Section 3.1. step 6).

3.3. Measurement of Tryptophan Content by UV

1. The protein is made 0.1*M* in NaOH.
2. Measure the absorbance at 294.4 and 280 nm in cuvets (transparent to this wavelength, i.e., quartz) in a spectrometer.
3. The amount of tryptophan (w) is estimated from the relative absorbances at these wavelengths by the method of Goodwin and Morton *(2)* shown in Eq. (1), where x = total mol/L, w = tryptophan mol/L, and $(x - w)$ = tyrosine mol/L. ε_y = Molar extinction of tyrosine in 0.1*M* alkali at 280 nm = 1576. ε_w = Molar extinction of tryptophan in 0.1*M* alkali at 280 nm = 5225.

Also, x is measured from $E_{294.4}$ (the molar extinction at this wavelength). This is 2375 for both Tyr and Trp (since their absorption curves intersect at this wavelength). An accurate reading of absorbance at one other wavelength is then sufficient to determine the relative amounts of these amino acids.

$$E_{280} = w\,\varepsilon_w + (x - w)\varepsilon_y \quad (1)$$

Therefore:

$$w = (E_{280} - x\,\varepsilon_y)/(\varepsilon_w - \varepsilon_y) \quad (2)$$

4. An alternative method of obtaining the ratios of Tyr and Trp is to use the formulae derived by Beaven and Holiday *(5)*.

$$M_{Tyr} = (0.592\,K_{294} - 0.263\,K_{280}) \times 10^{-3} \quad (3)$$
$$M_{Trp} = (0.263\,K_{280} - 0.170\,K_{294}) \times 10^{-3} \quad (4)$$

where M_{Tyr} and M_{Trp} are the moles of tyrosine and tryptophan in 1 g of protein, and K_{294} and K_{280} are the extinction coefficients of the protein in 0.1*M* alkali at 294 and 280 nm.

Extinction values can be substituted for the K values to give the molar ratio of tyrosine to tryptophan according to the formula:-

$$M_{Tyr}/M_{Trp} = (0.592\,E_{294} - 0.263\,E_{280}/0.263\,E_{280} - 0.170\,E_{294}) \quad (5)$$

4. Notes

1. The extinction of nucleic acid in the 280-nm region may be as much as 10 times that of protein at the same wavelength, and hence a few percent of nucleic acid can greatly influence the absorption.
2. In this analysis, the tyrosine estimate may be high and that of tryptophan low. If amino acid analysis indicates absence of tyrosine, tryptophan is more accurately determined at its maximum, 280.5 nm.
3. Absorption by most proteins in 0.1*M* NaOH solution decreases at longer wavelengths into the region 330–450 nm, where tyrosine and tryptophan do not absorb. Suitable blanks for 294 and 280 nm are therefore obtained by measuring extinctions at 320 and 360 nm and extrapolating back to 294 and 280 nm.
4. In proteins, in a peptide bond, the maximum of the free amino acids is shifted by 1–3 nm to a longer wavelength, and pure peptides containing tyrosine and tryptophan residues are better standards than the free amino acids. A source of error may be owing to turbidity in the solution, and if a protein shows a tendency to denature, it is advisable to treat with a low amount of proteolytic enzyme to obtain a clear solution.

References

1. Matsubara, H. and Sasaki, R. M. (1969) High recovery of tryptophan from acid hydrolysates of proteins. *Biochem. Biophys. Res. Commun.* **35,** 175–181.
2. Goodwin, T. W. and Morton, R. A. (1946) The spectrophotometric determination of tyrosine and tryptophan in proteins. *Biochem. J.* **40,** 628–632.
3. Liu, T.-Y. and Chang, Y. H. (1971) Hydrolysis of proteins with *p*-toluenesulphonic acid. *J. Biol. Chem.* **246,** 2842–2848.
4. Penke, B., Ferenczi, R., and Kovacs, K. (1974) A new acid hydrolyisis method for determining tryptophan in peptides and proteins. *Anal. Biochem.* **60,** 45–50.
5. Beaven, G. H. and Holiday, E. R. (1952) Utraviolet absorption spectra of proteins and amino acids. *Adv. Protein Chem.* **7,** 319.

7

Protein Quantitation Using Flow Cytometry

F. Andrew Ray, Thomas D. Friedrich, Judith Laffin, and John M. Lehman

1. Introduction

It is often desirable to know how much of a particular protein is produced by a particular cell. For example, is this protein expressed at low levels throughout a population of cells, or is this protein expressed at very high levels in some cells and not in others? Does the expression of the protein change in a particular cell-cycle compartment? Technical protocols that require cell lysis, such as immunoblotting *(1)*, allow the mean expression level of a given protein to be compared between populations of cells. It is even possible to use the immunoblotting protocol to determine the amount of a given protein expressed by a population of cells, if the protein has been purified for use as a standard. However, cell lysis obliterates the potential for single-cell quantitation of proteins.

Individual cells stained with fluorescent molecules, focused into a thin stream by hydrodynamic pressure and passed through an intersecting laser beam, can be detected by properly aligned photomultiplier tubes. This procedure, known as flow cytometry, allows the fluorescence generated by the laser excitation of fluorescent molecules to be collected at rates of hundreds to thousands of events/second (e.g., *see* ref. *2)*. By targeting protein molecules with fluorescently labeled antibodies, flow cytometry can be used to determine relative quantities of specific proteins in individual cells of a population *(3–7)*. Counterstaining with a molecule that binds specifically to DNA and fluoresces at a different wavelength allows the simultaneous quantitation of protein and determination of cell-cycle position.

A technique to detect and quantitate a viral protein expressed in mammalian cells is presented here. Cells expressing the protein of interest are grown, collected, and fixed. The cells are then reacted with an MAb to the protein of interest. Subsequently, the cells are reacted with a second antibody that is covalently modified with fluoresceine isothiocyanate (FITC). The cellular DNA is then stained with propidium iodide (PI). Filtered samples of cells flow through a focused 488-nm Argon ion laser beam. FITC and PI are both excited to fluoresce at 488 nm. Appropriate filters allow emitted fluorescence from FITC and PI to be resolved and collected on separate photomultipliers simultaneously. FITC fluorescence is proportional to the amount of the specific protein, and PI fluorescence is proportional to DNA content. In karyotypically stable cells, the position of a given cell in the cell cycle is resolved by its DNA content, and there-

From: The Protein Protocols Handbook
Edited by: J. M. Walker Humana Press Inc., Totowa, NJ

fore the amount of the protein can be correlated to cell-cycle position (Fig. 1). Using flow cytometry, the relative amount of protein/cell can be quantitated for thousands of cells in a few minutes.

2. Materials

2.1. Cell Strain and Antibodies

1. For the purpose of this chapter, cells that express a viral protein were used. Human fibroblasts were transfected with a plasmid encoding the SV40 T-antigen and a second plasmid encoding the *neo* gene *(8)*. G418^r clones of fibroblasts expressing large T-antigen were selected and expanded. The T-antigen protein has 708 amino acids, with several MAbs available to various epitopes along the molecule. PAb 101 is specific to the carboxy terminal end of the SV40 T-antigen and is a mouse MAb available from ATCC *(9)*.
2. The secondary antibody is an FITC-conjugated goat antimouse g-globulin fraction (high fluorescent; Antibodies Inc., Davis, CA).

2.2. Solutions for Fixation and Staining

All aqueous solutions and reagents should be sterilized by autoclaving, or sterile filtration and sterility should be maintained.

1. Phosphate-buffered saline (PBS): Dissolve 8.0 g NaCl, 0.2 g KCl, 1.2 g Na_2HPO_4, and 0.2 g KH_2PO_4 in distilled deionized H_2O, and adjust the pH to 7.4. Adjust the volume to 1 L.
2. Trypsin/EDTA (Gibco BRL).
3. Methanol –20°C.
4. Wash solution (WS): Heat inactivate 100 mL normal goat serum (*see* Note 1) for 60 min at 56°C and combine with 900 mL PBS. Add 20 µL Triton X-100, and 1.0 g sodium azide. Be extremely cautious when handling sodium azide.
5. PI: Add 1.0 mg PI (Calbiochem-Behring, La Jolla, CA) to 100 mL PBS. Add 20 µL Triton X-100 and 0.1 g Na azide, and dissolve. Store at 4°C. Protect this solution from light.
6. RNase: Add 100 mg RNase A (Sigma, St. Louis, MO) to 100 mL PBS. Add 100 µL Triton X-100. Place in a boiling water bath for 1 h.

3. Methods

3.1. Fixation

1. Aspirate culture media thoroughly.
2. Add 5 mL trypsin/EDTA/75-cm^2 flask, and bathe cells for approx 30 s (*see* Note 2). Aspirate trypsin/EDTA solution.
3. After the cells have detached (*see* Note 3), resuspend them in 5 mL culture media containing serum, and centrifuge at 100*g* for 10 min at 4°C. Prior to centrifugation, set aside a small aliquot of cells for cell number determination.
4. Aspirate all of the media, and resuspend the pellet in 100 µL PBS. Then add 900 µL of –20°C methanol (*see* Note 4).
5. Store the fixed cells at –20°C (*see* Note 5).

3.2. Titration of the Primary Antibody

Titrate the primary antibody (*see* Note 6) by using known positive (high-level expressing) and negative (low-level expressing) cell lines or strains. The goal is to maximize the difference between positive and negative populations of cells while

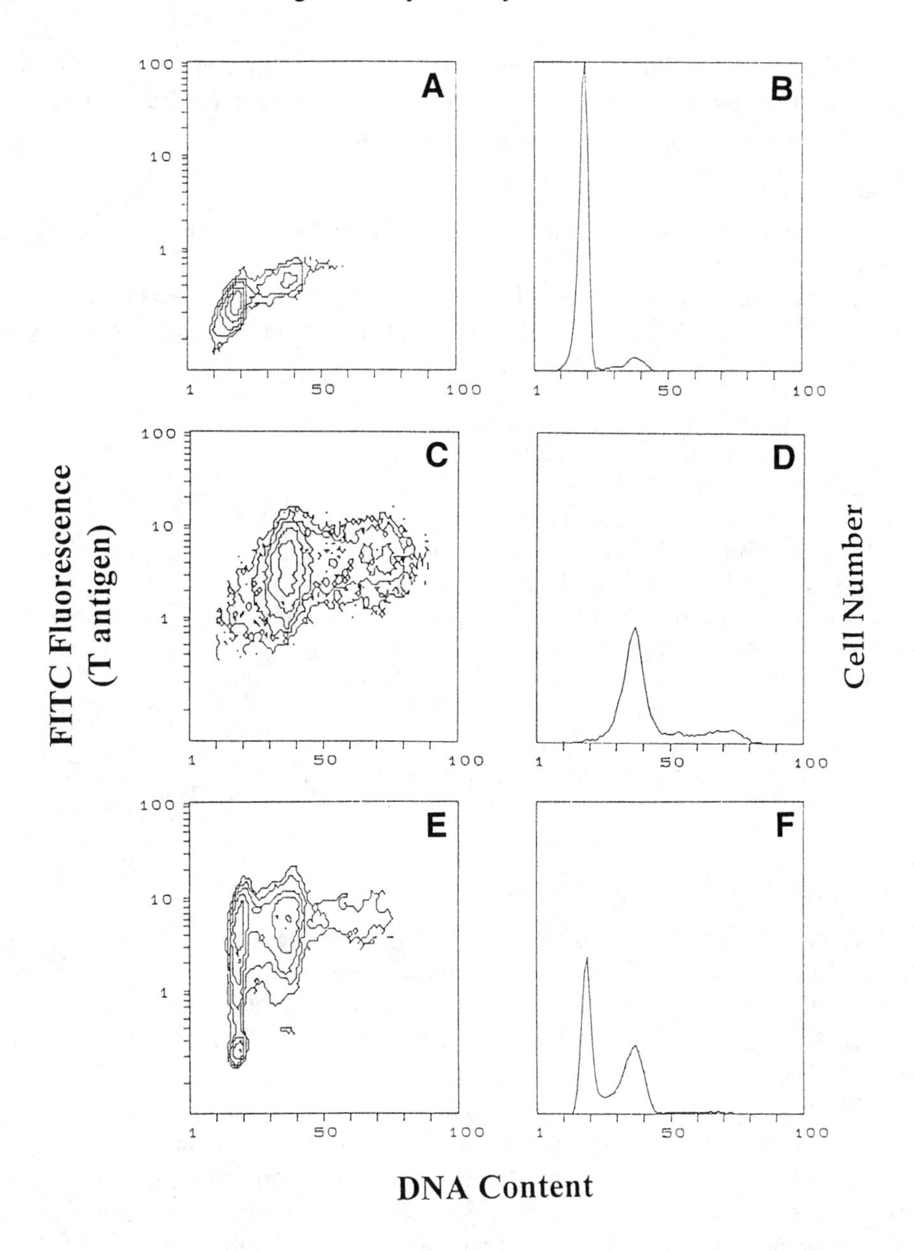

Fig. 1. Flow cytometric comparison of expressed SV40 T-antigen. **(A,C,E)** Bivariate distribution of cells with FITC-fluorescence (T-antigen) on the *Y*-axis and PI fluorescence (DNA content) on the *X*-axis. The number of events (cell number) is portrayed in the third dimension as a contour map. **(B,D,F)** Univariate distribution of number of events vs DNA content. (A,B) T-antigen-negative parental human fibroblasts. (C,D) T-antigen-positive clonal population of cells derived from the same fibroblasts by transfection. Note virtually all of the cells have above background FITC fluorescence and are thus T-antigen-positive. (E,F) Another T-antigen-positive clonal population of cells derived from the same fibroblasts by transfection. In these cells, there is a significant fraction of cells with background FITC fluorescence. Note that the cells that have a tetraploid DNA content also express high levels of T-antigen, suggesting a cause–effect relationship. This correlation would not be observable using immunoblotting.

assuring the primary antibody is present in excess. Select three dilutions of primary antibody and three dilutions of secondary antibody, and determine the maximum difference with the least nonspecific (background) staining.

3.3. Staining

1. Transfer 1.5×10^6 cells to a 1.5-mL microcentrifuge tube, and centrifuge at 4000*g* in a microcentrifuge for 15 s (*see* Note 7).
2. Aspirate fixative, and resuspend the cells in 1.0 mL PBS at 4°C. Repeat the centrifugation step.
3. Aspirate PBS, and add 0.5 mL primary antibody at the appropriate dilution. Mix by gently vortexing.
4. Incubate at 4°C overnight (*see* Note 8).
5. Repeat centrifugation, and aspirate solution.
6. Add 0.5 mL WS to the cell pellet, and mix by gently vortexing.
7. Repeat steps 5 and 6.
8. Repeat centrifugation, and aspirate solution.
9. Add 0.5 mL of the FITC-labeled secondary antibody at the appropriate dilution to each sample, and mix by gently vortexing.
10. Incubate the samples at 37°C for 2 h with gentle mixing, and minimize exposure to light.
11. Repeat the washing steps 5–7.
12. Repeat centrifugation, and aspirate solution.
13. Add 0.5 mL RNase, and mix by gently vortexing.
14. Incubate at 37°C for 30 min.
15. Add 0.5 mL PI, bringing the solution volume to 1 mL. Mix by gently vortexing.
16. Filter each sample through a 53-μm mesh nylon grid (Nitex HC3-53, Tetko Inc., Elmsford, NY) (*see* Note 9).

3.4. Flow Cytometry and Analysis

Many commercially available flow cytometer systems are capable of these measurements (*see* Note 10). Using 20 mW of power tuned to 488 nm, minimize the coefficient of variation (CV) for the green photomultiplier (*see* Note 11) to $<1.0\%$ by aligning the instrument with 2.0-μm fluorescent microspheres (cat. no. 09847, Polysciences, Inc., Warrington, PA). Likewise, minimize the CV for red fluorescence (*see* Note 12) to $<6.0\%$ with PI-stained lymphocytes (*see* Note 13).

Set data acquisition to trigger on a photomultiplier collecting unfiltered 90° light scatter. Set gates to collect data representing single-cells, and to eliminate data representing subcellular debris and cell aggregations. Compare red fluorescence to light scatter, and set gate 1 to eliminate subcellular debris (low red fluorescence) and clumps (excessive light scatter). Then compare the peak height and area of red fluorescence and eliminate doublets (off axis owing to biphasic peak height). Collect and display red fluorescence vs green fluorescence (*see* Note 14). Collect data from at least 10,000 cells. Include appropriate control samples (*see* Note 15), save data, and analyze. Representative data are shown in Fig. 1.

4. Notes

1. Goat serum decreases nonspecific binding of antibodies.
2. Each sample requires an initial cell number of 1.5×10^6 because cells will be lost during fixation and staining. Final cell number is approx 1×10^6. Minimize tube size, and use polypropylene tubes to minimize cell loss.

3. Carefully monitor detachment to avoid clumping.
4. Cell concentration can be adjusted at the time of fixing by adding the appropriate amounts of PBS (10%) and then methanol (90%). The method of fixation can dramatically alter the results depending on the protein of interest.
5. Fixed cells have been stored at –20°C for >1 yr with negligible loss of T-antigen.
6. Prior to testing titrations in flow, titrate antibody on cells grown and fixed to glass coverslips to determine the range of titration. To assure primary antibody is present in excess, two samples are treated with the same primary antibody-containing solution successively and compared by flow. Other than primary antibody treatment, the two samples are treated identically. The fluorescence signal from the second population of cells should not be significantly decreased when compared to the first.
7. A swinging bucket microcentrifuge is preferable because fewer cells are lost.
8. Adjust time and temperature for maximum signal-to-noise ratio. Gentle rocking is recommended.
9. The samples should be refiltered if reanalyzed on the flow cytometer as clumps will form. Samples have been reanalyzed up to 1 wk later, but resolution is generally decreased.
10. We use the Cytofluorograph II Model H.H. and the 2151 data-analysis system (Ortho Diagnostics, Inc., Westwood, MA) with an air-cooled argon laser (model 532, Omnichrome, Chino, CA).
11. Use a 535-nm band pass filter for this photomultiplier.
12. Use a 640-nm long pass filter for this photomultiplier.
13. For example, wash an aliquot of 1×10^6 mouse spleen cells (previously fixed in methanol and stored at –20°C) with PBS, and resuspend in 0.5 mL RNase and 0.5 mL PI. Incubate 30 min at 37°C, filter, and analyze as described.
14. DNA content recorded on the X-axis is generally displayed in a linear scale. T-antigen, expressed over several orders of magnitude, is displayed in a logarithmic scale. The specificity of the antibodies, amount of antigen, and quantum yield of the fluorochrome determine the data-collection mode.
15. Include as controls, cells that are known to express high and low levels of the protein of interest. Omit FITC from cells that express low levels to determine autofluorescence levels.

References

1. Towbin, H. T., Staehelin, T., and Gordon, J. (1979) Electrophoretic transfer of proteins from polyacrylamide gels to nitrocellulose sheets: procedure and some applications. *Proc. Natl. Acad. Sci. USA* **76,** 4350–4354.
2. Shapiro, H. M. (1995) *Practical Flow Cytometry.* Wiley-Liss, New York.
3. Jacobberger, J. W., Fogleman, D., and Lehman, J. M. (1986) Analysis of intracellular antigens by flow cytometry. *Cytometry* **7,** 356–364.
4. Gong, J., Li, X., Traganos, F., and Darzynkiewicz, Z. (1994) Expression of G_1 and G_2 cyclins measured in individual cells by multiparameter flow cytometry: a new tool in the analysis of the cell cycle. *Cell Prolif.* **27,** 357–371.
5. Laffin, J. and Lehman, J. M. (1994) Detection of intracellular virus and viral products. *Methods Cell Biol.* **41,** 543–557.
6. Larsen, J. K., Christensen, I. J., Christiansen, J., and Mortensen, B. T. (1991) Washless double staining of unfixed nuclei for flow cytometric analysis of DNA and a nuclear antigen (Ki-67 or bromodeoxyuridine). *Cytometry* **12,** 429–437.

7. Teague, K. and El-Naggar, A. (1994) Comparative flow cytometric analysis of proliferating cell nuclear antigen (PCNA) antibodies in human solid neoplasms. *Cytometry* **15,** 21–27.
8. Ray, F. A., Peabody, D. S., Cooper, J. L., Cram, L. S., and Kraemer, P. M. (1990) SV40 T-antigen *alone* drives karyotype instability which precedes neoplastic transformation of human diploid fibroblasts. *J. Cell. Biochem.* **42,** 13–31.
9. Gurney, E. G., Harrison, R. O., and Fenno, J. (1980) Monoclonal antibodies against simian virus 40 T-antigens: evidence for distinct subclasses of large T-antigen and for similarities among nonviral T-antigens. *J. Virol.* **34,** 752–763.

8

Copper Iodide Staining of Proteins on Solid Phases

Douglas D. Root and Kuan Wang

1. Introduction

Copper iodide staining and silver enhancement are recent implementations of copper-based protein assays, and are designed to quantitate proteins adsorbed to solid surfaces, such as nitrocellulose, nylon, polyvinylenedifluoride, and polystyrene *(1–5)*. The binding of cupric ions to the backbone of proteins under alkaline conditions and their consequent reduction to the cuprous state are the basis of popular assays of protein in solution including the biuret, Lowry, and bicinchoninic acid methods (*see* Chapters 2 and 3). In the case of copper iodide staining, it is thought that the protein binds copper iodide under highly alkaline conditions. Sensitivity, speed, reversibility, low cost, and lack of known interfering substances (including nucleic acid) are among the virtues of this protein assay *(4–6)*. One interesting use of copper iodide staining is the quantification of protein adsorbed to microtiter plates *(5)*. This information is particularly useful for quantitative ELISA and protein binding when radioactive measurements are inconvenient or undesirable (e.g., chemical modification affects the properties of the protein of interest). The precision of the determination of protein adsorbed to the microtiter plate by copper iodide staining is typically about 10–15%. The high sensitivity of copper iodide staining (about 50 pg/mm^2) may be increased an additional 10-fold by a silver-enhancement procedure, allowing the detection of protein down to about 5 pg/mm^2 *(6)*. Protein concentrations may be estimated from copper iodide staining from very dilute protein solutions or when only small amounts of a precious protein are available.

2. Materials

2.1. Copper Iodide Staining

1. Prepare the copper iodide staining reagent by mixing $CuSO_4{\cdot}5H_2O$ (12 g), KI (20 g), and potassium sodium tartrate (36 g) with 80 mL of distilled water in a glass beaker (*see* Note 1). As the slurry is vigorously stirred, solid NaOH (10 g) is slowly added. The suspension becomes warmer and changes color from brown to green to dark blue. After the NaOH is completely dissolved, the beaker is allowed to cool at room temperature for 30 min without stirring to allow the brownish-red precipitate to settle. Then 70 mL of solution are aspirated from the top to leave approx 50 mL of reagent with precipitate. The reagent is stable and may be stored in a sealed bottle at room temperature for at least 1 mo or at 4°C for at least 1 yr (*see* Note 2).

From: The Protein Protocols Handbook
Edited by: J. M. Walker Humana Press Inc., Totowa, NJ

2. Prepare the copper iodide stain-remover solution with 0.19 g $Na_4EDTA \cdot H_2O$, 0.28 g $NaH_2PO_4 \cdot H_2O$, and 2.14 g $Na_2HPO_4 \cdot 7H_2O$ in 100 mL deionized water.
3. Nitrocellulose (e.g., BA85, Schleicher & Schuell, Keene, NH), polyvinylenedifluoride (e.g., PVDF, Millipore Corporation, Bedford, MA), or nylon blotting paper (e.g., Zeta probe, Bio-Rad Laboratories, Hercules, CA; *see* Note 3).

2.2. Silver-Enhanced Copper Staining (SECS)

1. Prepare the silver-enhancing reagent just prior to use by dissolving 0.1 g $AgNO_3$, 0.1 g NH_4NO_3, and 7 μL of 5% (v/v in water) β-mercaptoethanol in 100 mL distilled water. After the other components are dissolved and immediately before use, add 2.5 g Na_2CO_3.
2. Prepare the SECS stain remover by dissolving 33.2 g KI in 100 mL distilled water (final concentration, 2*M* KI).

2.3. Copper Iodide Microtiter Plate Assay

1. Polystyrene 96-well microtiter plates (e.g., Nunc Immuno-Plate, Denmark; Titertek 76-381-04, McLean, VA; or Immulon 1 and 2, Dynatech Laboratories, Chantilly, VA).
2. Nitrocellulose membranes (e.g., BA85, Schleicher & Schuell, Keene, NH).
3. Household 3-in-one lubricating oil (Boyle-Midway, Inc., New York).
4. A standard single-hole puncher (6-mm diameter).
5. A densitometer is required, such as a flatbed scanner or video camera and framegrabber with image analysis software (*see* Note 4).

3. Methods

3.1. Copper Iodide Staining

1. Stir the copper iodide staining reagent vigorously at room temperature immediately prior to use. The reagent should be a fine slurry.
2. Rock the copper iodide staining reagent over the dried protein blot (or Western blot) for at least 2 min (but not more than 5 min; *see* Note 5) as the reddish-brown bands appear on the blot.
3. Gently dip the stained blot up and down in three beakers of deionized water and then allow the blot to dry (*see* Note 6).
4. The stained blot then may be quantified by densitometry (*see* Note 7) and photographed for documentation.
5. The staining pattern is stable at room temperature for at least 1 yr.
6. If greater sensitivity is desired, the blot may be used directly in the silver-enhanced copper staining (SECS) procedure (*see* Section 3.2.).
7. If subsequent immunostaining on the same blot is required, destain for 15 min with gentle agitation in the copper iodide stain remover prior to immunostaining.

3.2. SECS

1. Rock a nitrocellulose protein blot stained with copper iodide (*see* Section 3.1.) for 5 min in freshly prepared silver-enhancing reagent until the bands become dark black.
2. Dip the blot in a 1 L beaker of deionized water, and store in the dark to dry (to prevent background development).
3. The blot then may be quantified by densitometry and photographed for documentation.

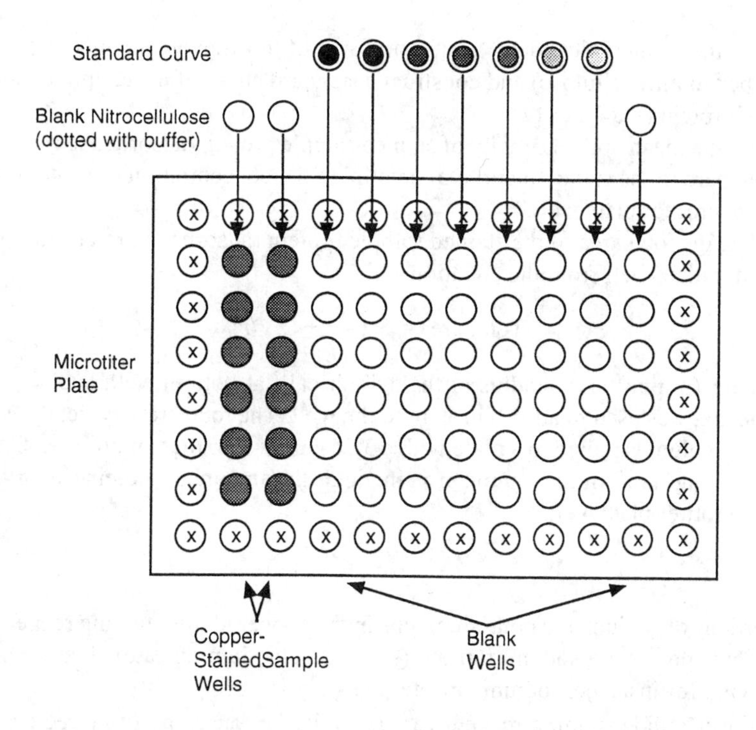

Fig. 1. Schematic diagram of the assembly of a microtiter plate for the copper iodide staining assay. Avoid wells marked with "X" because of possible distortions from edge effects.

4. If subsequent immunostaining on the same blot is required, destain for 30 min with gentle agitation in the SECS stain remover prior to immunostaining (*see* Note 8).

3.3. Copper Iodide Microtiter Plate Assay

1. Adsorb the protein of interest to duplicate microtiter plates, and note the volume (V) that was used to adsorb the protein in each well (*see* Note 9). One of the microtiter plates is for copper iodide staining, and the other is for quantitative ELISA or binding experiments. The microtiter plate for copper iodide staining contains protein adsorbed to only a few of the wells (e.g., 4–16 wells; *see* Fig. 1 and Note 10).
2. Create a standard curve by dot blotting approx 5–100 ng of the protein of interest/5 μL drop onto nitrocellulose paper, and allow to air-dry. Blanks are dotted with equal volumes of buffers.
3. Stain both the nitrocellulose paper and the microtiter plate with adsorbed protein by the copper iodide staining procedure (*see* Section 3.1.).
4. Use the hole puncher to excise stained dots from the nitrocellulose membrane, and place them stained side down into blank wells on the microtiter plate (*see* Note 11).
5. Use the hole puncher to excise blank nitrocellulose circles, and place them in microtiter plate wells containing copper iodide-stained protein and also in some blank wells to determine the background density (*see* Fig. 1).
6. For transmittance densitometry, first add 5 μL/well of household three-in-one oil to make the nitrocellulose translucent, thus reducing the background. For reflectance densitometry, the bottom of the microtiter plate may be scanned directly (the stained sides of the nitrocellulose must all be face down in the wells).

7. Measure the mean optical density and total area of the known amounts of stained protein (adsorbed to nitrocellulose) and construct a standard curve of mean optical density vs ng/ mm^2 of protein (*see* Note 12).
8. Measure the mean optical density of stained sample protein (adsorbed to microtiter plate), and compare to the standard curve to determine the concentration of protein (in ng/mm^2) on the microtiter plate well.
9. Calculate the total area of the stained sample protein (adsorbed to microtiter plate) on the microtiter plate well from the equation:

$$\text{Total area} = 3.14r^2 + (2V/r)$$

in which r (in mm) is the radius of the cylindrical flat-bottom well, and V (in μL) is the volume that was used to adsorb protein to the well. The total area (typically 94.7 mm^2 for a 100-μL volume applied to a plate with a 3.25-mm well radius) multiplied by the surface density (in ng/mm^2) of the sample protein yields the amount of protein (in ng) adsorbed to each microtiter plate well.

4. Notes

1. Proportions of sodium and potassium ions in the copper iodide staining reagent are important. Thus, potassium sodium tartrate (sodium potassium tartrate) should not be substituted with, for instance, sodium tartrate.
2. The copper iodide staining reagent can generally be reused two to three times, but will eventually become less sensitive.
3. Copper iodide staining reagent stains proteins adsorbed on most solid-phase adsorbents, including nitrocellulose, nylon, polyvinylenedifluoride, and polystyrene. However, only nitrocellulose and polystyrene have been tested so far for quantitation.
4. Microtiter plate readers are to be avoided for quantitative measurements, because they are usually not sensitive enough (copper iodide staining yields OD ≤ 0.1) and do not sample a large enough area of the stained surface to detect any nonuniformity in the staining density.
5. Exceeding a staining time of 5 min can both damage nitrocellulose membranes and lead to solubilization of adsorbed protein. A staining time of 2 min is optimal.
6. Washing of microtiter plates should be handled gently by dipping in beakers of deionized water. Vigorous washing procedures often lead to nonuniform protein distribution and consequent uneven staining of microtiter plates.
7. An example of a low-cost densitometer is a desktop flatbed scanner, color Apple Macintosh computer, and NIH Image software (public domain, by Wayne Rasband; further details are available by personal communication).
8. SECS may be removed by concentrations of KI that are <2*M,* but will require longer incubations (e.g., 90 min for 0.5*M* KI).
9. The wells on the edge of microtiter plates should be avoided for quantitative measurements because they tend to yield less accurate numbers.
10. Nitrocellulose quantitatively binds most proteins that are dot-blotted onto it, and retains them well throughout copper iodide staining.
11. If there is a problem with static repulsion, the household three-in-one oil (*see* Section 3.3., steps 5 and 6) may be applied to the microtiter plate well to release the static charge before placing nitrocellulose circles in the wells (provided that transmission densitometry will be used).
12. Quantitative measurements of copper iodide staining should be done at least in triplicate because of the relatively high (10–15%) standard deviation of the results.

References

1. Jenzano, J. W., Hogan, S. L., Noyes, C.M., Featherstone, G. L., and Lundblad, R. L. (1986) Comparison of five techniques for the determination of protein content in mixed human saliva. *Anal. Biochem.* **159,** 370–376.
2. Lowry, O. H., Rosebrough, N. J., Farr, A. L., and Randall, R. J. (1951) Protein measurement with the Folin phenol reagent. *J. Biol. Chem.* **193,** 265–275.
3. Smith, P. K., Krohn, R. I., Hermanson, G. T., Mania, A. K., Gartner, F. H., Provenzano, M. D., Fujimoto, E. K., Goeke, N. M., Olson, B. J., and Klenk, D. C. (1985) Measurement of protein using bicinchoninic acid. *Anal. Biochem.* **150,** 76–85.
4. Root, D. D. and Reisler, E. (1989) Copper iodide staining of protein blots on nitrocellulose membranes. *Anal. Biochem.* **181,** 250–253.
5. Root, D. D. and Reisler, E. (1990) Copper iodide staining and determination of proteins adsorbed to microtiter plates. *Anal. Biochem.* **186,** 69–73.
6. Root, D. D. and Wang, K. (1993) Silver-enhanced copper staining of protein blots. *Anal. Biochem.* **209,** 15–19.

9

Kinetic Silver Staining of Proteins Adsorbed to Microtiter Plates

Douglas D. Root and Kuan Wang

1. Introduction

The quantification of proteins adsorbed on microtiter plates is useful for quantitative ELISA and protein interaction assays. One nonradioactive procedure, copper iodide staining, has been described *(1* and *see* Chapter 8). A kinetic silver staining method for measuring the amount of adsorbed protein in a microtiter plate has been developed. This assay has a sensitivity similar to copper iodide staining (5–150 ng/well), but much higher precision (<5%; *2).* The kinetic silver staining assay uses the silver staining reagent developed originally by Gottlieb and Chavko for nucleic acids in agarose gels *(3).* Quantification is based on the time required for staining to reach a fixed optical density. This time-based assay has very little protein-to-protein variation, but is sensitive to interfering substances *(2).*

2. Materials

1. Polystyrene 96-well microtiter plates (e.g., Titertek 76–381–04, McLean, VA).
2. Kinetic silver staining reagent:
   a. Reagent A: 0.2% (w/v) $AgNO_3$, 0.2% (w/v) NH_4NO_3, 1% (w/v) tungstosilicic acid, and 0.3% (v/v) formaldehyde (from a 37% stock solution in water) in distilled water. Store in the dark at room temperature.
   b. Reagent B: 5% (w/v) Na_2CO_3 in distilled water.
3. A microtiter plate reader is required (e.g., EIA Autoreader, model EL310, Biotek Instruments, Burlington, VT).

3. Method

1. Adsorb the protein of interest to duplicate microtiter plates (*see* Note 1). Wash the microtiter plate profusely with distilled water after the adsorption (*see* Note 2). One of the microtiter plates is used for the kinetic silver staining assay and the other is for quantitative ELISA or binding experiments. The microtiter plate for the kinetic silver staining assay contains protein adsorbed to only a few of the wells (e.g., 4–16 wells).
2. Prepare a known concentration of the standard protein in distilled water (e.g., by dialysis) for use in a standard curve (*see* Notes 2 and 3).
3. Apply varying concentrations (e.g., 50–1000 ng/mL) of the standard protein to the blank wells on the microtiter plate at the same volume (50–200 µL) that was used to adsorb

From: The Protein Protocols Handbook
Edited by: J. M. Walker Humana Press Inc., Totowa, NJ

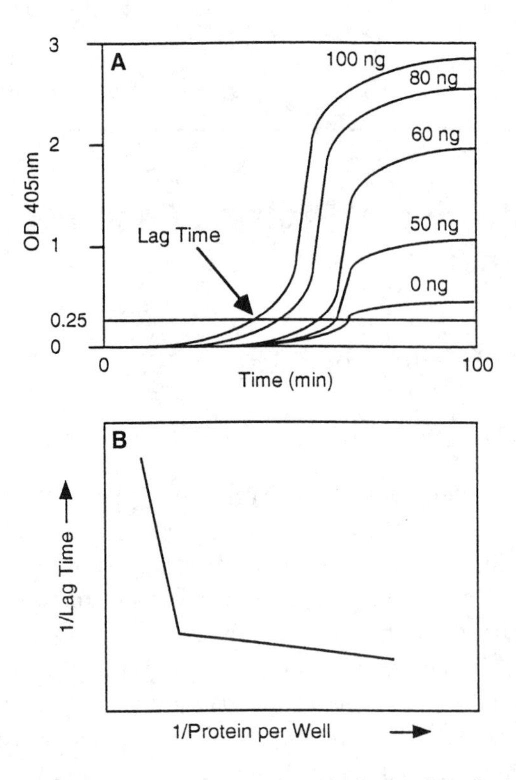

Fig. 1. Representative plots of standard curve data. Raw data for varying standard protein mass per well **(A).** Double reciprocal plot of standard curve **(B).**

protein in step 1. Cover the microtiter plate with a tissue to avoid dust from settling in the wells, and allow the protein to air-dry for several days (*see* Note 4). Do not wash the plate after this step!

4. Mix equal volumes of reagents A and B immediately before use. A fixed volume (e.g., 100 μL) of the mixture is added quickly to the wells on the microtiter plate that were adsorbed with protein in steps 1 and 3, and to blank control wells. The time for the addition of reagent to each well should be noted as time zero. All wells are filled in <10 min from the time of mixing reagents A and B; otherwise, excessive silver development in the reagent may cause a high background.
5. Read optical densities of each well with a microtiter plate reader at 405 nm (*see* Note 5). The time elapsed from time zero for each well to reach 0.25 OD (lag time) is noted as the lag time.
6. The lag time is plotted against the mass/well of adsorbed protein in the standard curve (step 3), which typically yields an inverse sigmoidal shape curve. Comparison of the lag times of the sample wells to the standard curve allows the determination of the total mass of protein in the sample well. The standard curve may be linearized to a sharply biphasic shape curve by plotting 1/lag time vs 1/protein/well (*see* Fig. 1).

4. Notes

1. The wells on the edge of microtiter plates should be avoided for quantitative measurements, because they tend to yield less accurate numbers.

Table 1
Compounds Known to Interfere with Kinetic Silver Staining

Compounds	Concentration in the staining reagent
TCA	>0.01 m*M*
Glucose	>1 m*M*
β-Mercaptoethanol	>0.1 m*M*
DTT	>0.01 m*M*
KOH	>0.01 m*M*
EDTA	>1 m*M*
Phosphate	>0.01 m*M*
SDS	>1 m*M*
Triton X-100	>0.0005%
Ammonium sulfate	>1 m*M*
Urea	>1 m*M*
Imidazole	>0.01 m*M*
Tris	>1 m*M*
NaCl	>0.01 m*M*
KCl	>0.01 m*M*
Guanidine HCl	>0.01 m*M*

2. Washing of microtiter plates is essential, since residual buffer reagents may interfere with silver staining (*see* Table 1). The washing is performed by gently by dipping in beakers of deionized water. Vigorous washing was avoided for fear of losing adsorbed protein.
3. Kinetic silver staining shows little protein-to-protein variation (<30% over six tested proteins). It is possible to estimate the total amount of adsorbed protein or mixture of proteins using a standard (e.g., bovine albumin) that can be easily dissolved in water and measured for concentration.
4. Kinetic silver staining detects proteins bound to polystyrene and apparently not protein in solution. Thus, the quantitative adsorption of the standard protein in distilled water (to avoid interference, *see* Note 2) to polystyrene by drying is necessary. As the binding capacity of polystyrene is exceeded (typically about 100–200 ng/well, but depends on both the microtiter plate and the protein of interest), kinetic silver staining does not detect further increases in protein mass per well.
5. Kinetic silver staining is based on measurements of light scattering. Thus, other wavelengths may be used, but the corresponding OD reading will be different and may require optimization.

References

1. Root, D. D. and Reisier, E. (1990) Copper iodide staining and determination of proteins adsorbed to microtiter plates. *Anal. Biochem.* **186,** 69–73.
2. Root, D. D. and Wang, K. (1993) Kinetic silverstaining and calibration of proteins adsorbed to microtiter plates. *Anal. Biochem.* **209,** 354–359.
3. Gottlieb, M. and Chavko, M. (1987) Silver staining of native and denatured eucaryotic DNA in agarose gels. *Anal. Biochem.* **165,** 33–37.

Part II

Electrophoresis of Proteins and Peptides and Detection in Gels

10

Nondenaturing Polyacrylamide Gel Electrophoresis of Proteins

John M. Walker

1. Introduction

SDS-PAGE (Chapter 11) is probably the most commonly used gel electrophoretic system for analyzing proteins. However, it should be stressed that this method separates denatured protein. Sometimes one needs to analyze native, nondenatured proteins, particularly if wanting to identify a protein in the gel by its biological activity (for example, enzyme activity, receptor binding, antibody binding, and so on). On such occasions it is necessary to use a nondenaturing system such as described in this chapter. For example, when purifying an enzyme, a single major band on a gel would suggest a pure enzyme. However this band could still be a contaminant; the enzyme could be present as a weaker (even nonstaining) band on the same gel. Only by showing that the major band had enzyme activity would you be convinced that this band corresponded to your enzyme. The method described here is based on the gel system first described by Davis *(1)*. To enhance resolution a stacking gel can be included (*see* Chapter 11 for the theory behind the stacking gel system).

2. Materials

1. Stock acrylamide solution: 30 g acrylamide, 0.8 g *bis*-acrylamide. Make up to 100 mL in distilled water and filter. Stable at 4°C for months (*see* Note 1). **Care: Acrylamide Monomer Is a Neurotoxin.** Take care in handling acrylamide (wear gloves) and avoid breathing in acrylamide dust when weighing out.
2. Separating gel buffer: 1.5*M* Tris-HCl, pH 8.8.
3. Stacking gel buffer: 0.5*M* Tris-HCl, pH 6.8.
4. 10% Ammonium persulfate in water.
5. *N,N,N',N'*-tetramethylethylenediamine (TEMED).
6. Sample buffer (5X). Mix the following:
   a. 15.5 mL of 1*M* Tris-HCl pH 6.8;
   b. 2.5 mL of a 1% solution of bromophenol blue;
   c. 7 mL of water; and
   d. 25 mL of glycerol.

   Solid samples can be dissolved directly in 1X sample buffer. Samples already in solution should be diluted accordingly with 5X sample buffer to give a solution that is 1X sample buffer. Do not use protein solutions that are in a strong buffer that is not near to pH

From: The Protein Protocols Handbook
Edited by: J. M. Walker Humana Press Inc., Totowa, NJ

6.8 as it is important that the sample is at the correct pH. For these samples it will be necessary to dialyze against 1X sample buffer.

7. Electrophoresis buffer: Dissolve 3.0 g of Tris base and 14.4 g of glycine in water and adjust the volume to 1 L. The final pH should be 8.3.
8. Protein stain: 0.25 g Coomassie brilliant blue R250 (or PAGE blue 83), 125 mL methanol, 25 mL glacial acetic acid, and 100 mL water. Dissolve the dye in the methanol component first, then add the acid and water. Dye solubility is a problem if a different order is used. Filter the solution if you are concerned about dye solubility. For best results do not reuse the stain.
9. Destaining solution: 100 mL methanol, 100 mL glacial acetic acid, and 800 mL water.
10. A microsyringe for loading samples.

3. Method

1. Set up the gel cassette.
2. To prepare the separating gel (*see* Note 2) mix the following in a Buchner flask: 7.5 mL stock acrylamide solution, 7.5 mL separating gel buffer, 14.85 mL water, and 150 µL 10% ammonium persulfate.

   "Degas" this solution under vacuum for about 30 s. This degassing step is necessary to remove dissolved air from the solution, since oxygen can inhibit the polymerization step. Also, if the solution has not been degassed to some extent, bubbles can form in the gel during polymerization, which will ruin the gel. Bubble formation is more of a problem in the higher percentage gels where more heat is liberated during polymerization.
3. Add 15 µL of TEMED and gently swirl the flask to ensure even mixing. The addition of TEMED will initiate the polymerization reaction, and although it will take about 20 min for the gel to set, this time can vary depending on room temperature, so it is advisable to work fairly quickly at this stage.
4. Using a Pasteur (or larger) pipet, transfer the separating gel mixture to the gel cassette by running the solution carefully down one edge between the glass plates. Continue adding this solution until it reaches a position 1 cm from the bottom of the sample loading comb.
5. To ensure that the gel sets with a smooth surface, *very carefully* run distilled water down one edge into the cassette using a Pasteur pipet. Because of the great difference in density between the water and the gel solution, the water will spread across the surface of the gel without serious mixing. Continue adding water until a layer about 2 mm exists on top of the gel solution.
6. The gel can now be left to set. When set, a very clear refractive index change can be seen between the polymerized gel and overlaying water.
7. While the separating gel is setting, prepare the following stacking gel solution. Mix the following quantities in a Buchner flask: 1.5 mL stock acrylamide solution, 3.0 mL stacking gel buffer, 7.4 mL water, and 100 µL 10% ammonium persulfate. Degas this solution as before.
8. When the separating gel has set, pour off the overlaying water. Add 15 µL of TEMED to the stacking gel solution and use some (~2 mL) of this solution to wash the surface of the polymerized gel. Discard this wash, then add the stacking gel solution to the gel cassette until the solution reaches the cutaway edge of the gel plate. Place the well-forming comb into this solution and leave to set. This will take about 30 min. Refractive index changes around the comb indicate that the gel has set. It is useful at this stage to mark the positions of the bottoms of the wells on the glass plates with a marker pen.
9. Carefully remove the comb from the stacking gel, remove any spacer from the bottom of the gel cassette, and assemble the cassette in the electrophoresis tank. Fill the top reservoir with electrophoresis buffer ensuring that the buffer fully fills the sample loading wells,

and look for any leaks from the top tank. If there are no leaks, fill the bottom tank with electrophoresis buffer, then tilt the apparatus to dispel any bubbles caught under the gel.

10. Samples can now be loaded onto the gel. Place the syringe needle through the buffer and locate it just above the bottom of the well. Slowly deliver the sample (~5–20 μL) into the well. The dense sample solvent ensures that the sample settles to the bottom of the loading well. Continue in this way to fill all the wells with unknowns or standards, and record the samples loaded.
11. The power pack is now connected to the apparatus and a current of 20–25 mA passed through the gel (constant current) (*see* Note 3). Ensure that the electrodes are arranged so that the proteins are running to the anode (*see* Note 4). In the first few minutes the samples will be seen to concentrate as a sharp band as it moves through the stacking gel. (It is actually the bromophenol blue that one is observing, not the protein but, of course, the protein is stacking in the same way.) Continue electrophoresis until the bromophenol blue reaches the bottom of the gel. This will usually take about 3 h. Electrophoresis can now be stopped and the gel removed from the cassette. Remove the stacking gel and immerse the separating gel in stain solution, or proceed to step 13 if you wish to detect enzyme activity (*see* Notes 5 and 6).
12. Staining should be carried out, with shaking, for a minimum of 2 h and preferably overnight. When the stain is replaced with destain, stronger bands will be immediately apparent and weaker bands will appear as the gel destains. Destaining can be speeded up by using a foam bung, such as those used in microbiological flasks. Place the bung in the destain and squeeze it a few times to expel air bubbles and ensure the bung is fully wetted. The bung rapidly absorbs dye, thus speeding up the destaining process.
13. If proteins are to be detected by their biological activity, duplicate samples should be run. One set of samples should be stained for protein and the other set for activity. Most commonly one would be looking for enzyme activity in the gel. This is achieved by washing the gel in an appropriate enzyme substrate solution that results in a colored product appearing in the gel at the site of the enzyme activity (*see* Note 7).

4. Notes

1. The stock acrylamide used here is the same as used for SDS gels (*see* Chapter 11) and may already be availabe in your laboratory.
2. The system described here is for a 7.5% acrylamide gel, which was originally described for the separation of serum proteins *(1)*. Since separation in this system depends on both the native charge on the protein *and* separation according to size owing to frictional drag as the proteins move through the gel, it is not possible to predict the electrophoretic behavior of a given protein the way that one can on an SDS gel, where separation is based on size alone. A 7.5% gel is a good starting point for unknown proteins. Proteins of mol wt >100,000 should be separated in 3–5% gels. Gels in the range 5–10% will separate proteins in the range 20,000–150,000, and 10–15% gels will separate proteins in the range 10,000–80,000. The separation of smaller polypeptides is described in Chapter 16. To alter the acrylamide concentration, adjust the volume of stock acrylamide solution in Section 3., step 2 accordingly, and increase/decrease the water component to allow for the change in volume. For example, to make a 5% gel change the stock acrylamide to 5 mL and increase the water to 17.35 mL. The final volume is still 30 mL, so 5 mL of the 30% stock acrylamide solution has been diluted in 30 mL to give a 5% acrylamide solution.
3. Because one is separating native proteins, it is important that the gel does not heat up too much, since this could denature the protein in the gel. It is advisable therefore to run the gel in the cold room, or to circulate the buffer through a cooling coil in ice. (Many gel apparatus are designed such that the electrode buffer cools the gel plates.) If heating is

thought to be a problem it is also worthwhile to try running the gel at a lower current for a longer time.

4. This separating gel system is run at pH 8.8. At this pH most proteins will have a negative charge and will run to the anode. However, it must be noted that any basic proteins will migrate in the opposite direction and will be lost from the gel. Basic proteins are best analyzed under acid conditions, as described in Chapters 14 and 15.
5. It is important to note that concentration in the stacking gel *may* cause aggregation and precipitation of proteins. Also, the pH of the stacking gel (pH 6.8) may affect the activity of the protein of interest. If this is thought to be a problem (e.g., the protein cannot be detected on the gel), prepare the gel without a stacking gel. Resolution of proteins will not be quite so good, but will be sufficient for most uses.
6. If the buffer system described here is unsuitable (e.g., the protein of interest does not electrophorese into the gel because it has the incorrect charge, or precipitates in the buffer, or the buffer is incompatible with your detection system) then one can try different buffer systems (without a stacking gel). A comprehensive list of alternative buffer systems has been published *(2)*.
7. The most convenient substrates for detecting enzymes in gels are small molecules that freely diffuse into the gel and are converted by the enzyme to a colored or fluorescent product within the gel. However, for many enzymes such convenient substrates do not exist, and it is necessary to design a linked assay where one includes an enzyme together with the substrate such that the products of the enzymatic reaction of interest is converted to a detectable product by the enzyme included with the substrate. Such linked assays may require the use of up to two or three enzymes and substrates to produce a detectable product. In these cases the product is usually formed on the surface of the gel because the coupling enzymes cannot easily diffuse into the gel. In this case the zymogram technique is used where the substrate mix is added to a cooled (but not solidified) solution of agarose (1%) in the appropriate buffer. This is quickly poured over the solid gel where it quickly sets on the gel. The product of the enzyme assay is therefore formed at the gel–gel interface and does not get washed away. A number of review articles have been published which described methods for detecting enzymes in gels *(3–7)*. A very useful list also appears as an appendix in ref. *8*.

References

1. Davis, B. J. (1964) Disc electrophoresis II—method and application to human serum proteins. *Ann. NY Acad. Sci.* **121,** 404–427.
2. Andrews, A. T. (1986) *Electrophoresis: Theory, Techniques, and Biochem-ical and Clinical Applications.* Clarendon, Oxford, UK.
3. Shaw, C. R. and Prasad, R. (1970) Gel electrophoresis of enzymes—a compilation of recipes. *Biochem. Genet.* **4,** 297–320.
4. Shaw, C. R. and Koen, A. L. (1968) Starch gel zone electrophoresis of enzymes, in *Chromatographic and Electrophoretic Techniques,* vol. 2 (Smith, I., ed.), Heinemann, London, pp. 332–359.
5. Harris, H. and Hopkinson, D. A. (eds.) (1976) *Handbook of Enzyme Electrophoresis in Human Genetics.* North-Holland, Amsterdam.
6. Gabriel, O. (1971) Locating enymes on gels, in *Methods in Enzymology,* vol. 22 (Colowick, S. P. and Kaplan, N. O., eds.), Academic, New York, p. 578.
7. Gabriel, O. and Gersten, D. M. (1992) Staining for enzymatic activity after gel electrophoresis. I. *Analyt. Biochem.* **203,** 1–21.
8. Hames, B. D. and Rickwood, D. (1990) *Gel Electrophoresis of Proteins,* 2nd ed., IRL, Oxford and Washington.

11

SDS Polyacrylamide Gel Electrophoresis of Proteins

John M. Walker

1. Introduction

SDS-PAGE is the most widely used method for qualitatively analyzing protein mixtures. It is particularly useful for monitoring protein purification, and because the method is based on the separation of proteins according to size, the method can also be used to determine the relative molecular mass of proteins (*see* Note 14).

1.1. Formation of Polyacrylamide Gels

Crosslinked polyacrylamide gels are formed from the polymerization of acrylamide monomer in the presence of smaller amounts of *N,N'*-methylene-bis-acrylamide (normally referred to as "bis-acrylamide") (Fig. 1). Note that bis-acrylamide is essentially two acrylamide molecules linked by a methylene group and is used as a crosslinking agent. Acrylamide monomer is polymerized in a head-to-tail fashion into long chains, and occasionally a bis-acrylamide molecule is built into the growing chain, thus introducing a second site for chain extension. Proceeding in this way, a crosslinked matrix of fairly well-defined structure is formed (Fig. 1). The polymerization of acrylamide is an example of free-radical catalysis, and is initiated by the addition of ammonium persulfate and the base *N,N,N',N'*-tetramethylenediamine (TEMED). TEMED catalyzes the decomposition of the persulfate ion to give a free radical (i.e., a molecule with an unpaired electron):

$$S_2O_8^{2-} + e^- \rightarrow SO_4^{2-} + SO_4^{-\bullet} \quad (1)$$

If this free radical is represented as $R^\bullet$ (where the dot represents an unpaired electron) and M as an acrylamide monomer molecule, then the polymerization can be represented as follows:

$$\begin{aligned} R^\bullet + M &\rightarrow RM^\bullet \\ RM^\bullet + M &\rightarrow RMM^\bullet \\ RMM^\bullet + M &\rightarrow RMMM^\bullet, \text{and so forth} \end{aligned} \quad (2)$$

In this way, long chains of acrylamide are built up, being crosslinked by the introduction of the occasional bis-acrylamide molecule into the growing chain. Oxygen "mops up" free radicals, and therefore the gel mixture is normally degassed (the solutions are briefly placed under vacuum to remove loosely dissolved oxygen) prior to addition of the catalyst.

From: The Protein Protocols Handbook
Edited by: J. M. Walker Humana Press Inc., Totowa, NJ

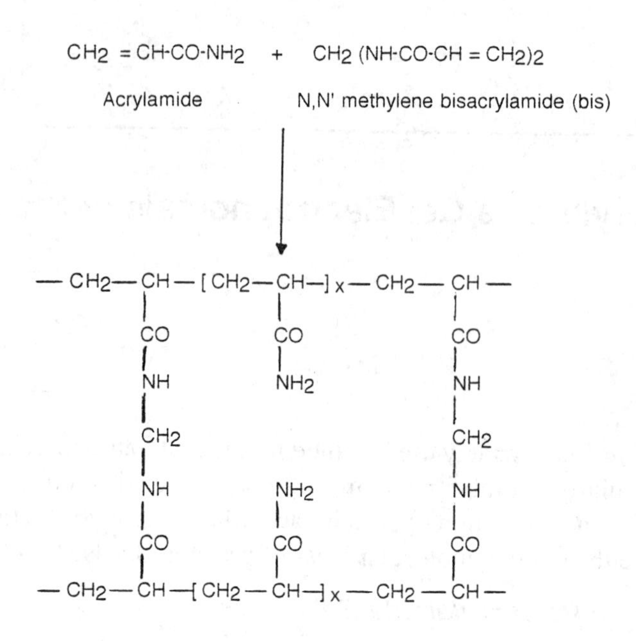

Fig. 1. Polymerization of acrylamide.

1.2. The Use of Stacking Gels

For both SDS and buffer gels samples may be applied directly to the top of the gel in which protein separation is to occur (the separating gel). However, in these cases, the sharpness of the protein bands produced in the gel is limited by the size (volume) of the sample applied to the gel. Basically the separated bands will be as broad (or broader, owing to diffusion) as the sample band applied to the gel. For some work, this may be acceptable, but most workers require better resolution than this. This can be achieved by polymerizing a short stacking gel on top of the separating gel. The purpose of this stacking gel is to concentrate the protein sample into a sharp band before it enters the main separating gel, thus giving sharper protein bands in the separating gel. This modification allows relatively large sample volumes to be applied to the gel without any loss of resolution. The stacking gel has a very large pore size (4% acrylamide) which allows the proteins to move freely and concentrate, or stack under the effect of the electric field. Sample concentration is produced by isotachophoresis of the sample in the stacking gel. The band-sharpening effect (isotachophoresis) relies on the fact that the negatively charged glycinate ions (in the reservoir buffer) have a lower electrophoretic mobility than the protein–SDS complexes. which in turn, have lower mobility than the Cl^- ions if they are in a region of higher field strength. Field strength is inversely proportional to conductivity, which is proportional to concentration. The result is that the three species of interest adjust their concentrations so that $[Cl^-] >$ [protein-SDS] > [glycinate]. There are only a small quantity of protein–SDS complexes, so they concentrate in a very tight band between the glycinate and Cl^- ion boundaries. Once the glycinate reaches the separating gel, it becomes more fully ionized in the higher pH environment and its mobility increases. (The pH of the stacking gel is 6.8 and that of the separating gel is 8.8.) Thus, the interface between glycinate and the

Cl^- ions leaves behind the protein–SDS complexes, which are left to electrophorese at their own rates. A more detailed description of the theory of isotachophoresis and electrophoresis generally is given in ref. *(1)*.

1.3. SDS-PAGE

Samples to be run on SDS-PAGE are first boiled for 5 min in sample buffer containing β-mercaptoethanol and SDS. The mercaptoethanol reduces any disulfide bridges present that are holding together the protein tertiary structure. SDS (CH_3-$[CH_2]_{10}$ - $CH_2OSO_3^-Na^+$) is an anionic detergent and binds strongly to, and denatures, the protein. Each protein in the mixture is therefore fully denatured by this treatment and opens up into a rod-shaped structure with a series of negatively charged SDS molecules along the polypeptide chain. On average, one SDS molecule binds for every two amino acid residues. The original native charge on the molecule is therefore completely swamped by the SDS molecules. The sample buffer also contains an ionizable tracking dye usually bromophenol blue that allows the electrophoretic run to be monitored, and sucrose or glycerol which gives the sample solution density, thus allowing the sample to settle easily through the electrophoresis buffer to the bottom when injected into the loading well. When the main separating gel (normally about 10 cm long) has been poured between the glass plates and allowed to set, a shorter (approx 2 cm) stacking gel is poured on top of the separating gel, and it is into this gel that the wells are formed and the proteins loaded. Once all samples are loaded, a current is passed through the gel. Once the protein samples have passed through the stacking gel and have entered the separating gel, the negatively charged protein–SDS complexes continue to move toward the anode, and because they have the same charge per unit length they travel into the separating gel under the applied electric field with the same mobility. However, as they pass through the separating gel the proteins separate, owing to the molecular sieving properties of the gel. Quite simply, the smaller the protein, the more easily it can pass through the pores of the gel, whereas large proteins are successively retarded by frictional resistance owing to the sieving effect of the gel. Being a small molecule, the bromophenol blue dye is totally unretarded and therefore indicates the electrophoresis front. When the dye reaches the bottom of the gel the current is turned off and the gel is removed from between the glass plates, shaken in an appropriate stain solution (usually Coomassie brilliant blue) for a few hours, and then washed in destain solution overnight. The destain solution removes unbound background dye from the gel, leaving stained proteins visible as blue bands on a clear background. A typical gel would take 1–1.5 h to prepare and set, 3 h to run at 30 mA, and have a staining time of 2–3 h with an overnight destain. Vertical slab gels are invariably run since this allows up to 20 different samples to be loaded onto a single gel.

2. Materials

1. Stock acrylamide solution: 30% acrylamide, 0.8% bis-acrylamide. Filter through Whatman No. 1 filter paper and store at 4°C (*see* Note 1).
2. Buffers:
   a. 1.875*M* Tris-HCl, pH 8.8.
   b. 0.6*M* Tris-HCl, pH 6.8.
3. 10% Ammonium persulfate. Make fresh.

4. 10% SDS (*see* Note 2).
5. TEMED.
6. Electrophoresis buffer: Tris (12 g), glycine (57.6 g), and SDS (2.0 g). Make up to 2 L with water. No pH adjustment is necessary.
7. Sample buffer (*see* Notes 3 and 4):

0.6*M* Tris-HCl, pH 6.8	5.0 mL
SDS	0.5 g
Sucrose	5.0 g
β-Mercaptoethanol	0.25 mL
Bromophenol blue, 0.5% stock	5.0 mL

Make up to 50 mL with distilled water.
8. Protein stain: 0.1% Coomassie brilliant blue R250 in 50% methanol, 10% glacial acetic acid. Dissolve the dye in the methanol and water component first, and then add the acetic acid. Filter the final solution through Whatman No. 1 filter paper if necessary.
9. Destain: 10% methanol, 7% glacial acetic acid.
10. Microsyringe for loading samples. Micropipet tips that are drawn out to give a fine tip are also commercially available.

3. Method

The system of buffers used in the gel system described below is that of Laemmli *(2)*.

1. Samples to be run are first denatured in sample buffer by heating to 95–100°C for 5 min (*see* Note 3).
2. Clean the internal surfaces of the gel plates with detergent or methylated spirits, dry, then join the gel plates together to form the cassette, and clamp it in a vertical position. The exact manner of forming the cassette will depend on the type of design being used.
3. Mix the following in a 250-mL Buchner flask (*see* Note 5):

	For 15% gels	For 10% gels
1.875*M* Tris-HCl, pH 8.8	8.0 mL	8.0 mL
Water	11.4 mL	18.1 mL
Stock acrylamide	20.0 mL	13.3 mL
10% SDS	0.4 mL	0.4 mL
Ammonium persulfate (10%)	0.2 mL	0.2 mL

4. "Degas" this solution under vacuum for about 30 s. Some frothing will be observed, and one should not worry if some of the froth is lost down the vacuum tube: you are only losing a very small amount of liquid (*see* Note 6).
5. Add 14 μL of TEMED, and gently swirl the flask to ensure even mixing. The addition of TEMED will initiate the polymerization reaction and although it will take about 15 min for the gel to set, this time can vary depending on room temperature, so it is advisable to work fairly quickly at this stage.
6. Using a Pasteur (or larger) pipet transfer this separating gel mixture to the gel cassette by running the solution carefully down one edge between the glass plates. Continue adding this solution until it reaches a position 1 cm from the bottom of the comb that will form the loading wells. Once this is completed, you will find excess gel solution remaining in your flask. Dispose of this in an appropriate waste container **not** down the sink.
7. To ensure that the gel sets with a smooth surface **very carefully** run distilled water down one edge into the cassette using a Pasteur pipet. Because of the great difference in density between the water and the gel solution the water will spread across the surface of the gel

without serious mixing. Continue adding water until a layer of about 2 mm exists on top of the gel solution (*see* Notes 7 and 8).

8. The gel can now be left to set. As the gel sets, heat is evolved and can be detected by carefully touching the gel plates. When set, a very clear refractive index change can be seen between the polymerized gel and overlaying water.
9. While the separating gel is setting prepare the following stacking gel (4°C) solution. Mix the following in a 100-mL Buchner flask (*see* Notes 8 and 9):

0.6*M* Tris-HCl, pH 6.8	1.0 mL
Stock acrylamide	1.35 mL
Water	7.5 mL
10% SDS	0.1 mL
Ammonium persulfate (10%)	0.05 mL

   Degas this solution as before.

10. When the separating gel has set, pour off the overlaying water. Add 14 μL of TEMED to the stacking gel solution and use some (~2 mL) of this solution to wash the surface of the polymerized gel. Discard this wash, and then add the stacking gel solution to the gel cassette until the solution reaches the cutaway edge of the gel plate. Place the well-forming comb into this solution, and leave to set. This will take about 20 min. Refractive index changes around the comb indicate that the gel has set. It is useful at this stage to mark the positions of the bottoms of the wells on the glass plates with a marker pen to facilitate loading of the samples (*see also* Note 9).
11. Carefully remove the comb from the stacking gel, and then rinse out any nonpolymerized acrylamide solution from the wells using electrophoresis buffer. Remove any spacer from the bottom of the gel cassette, and assemble the cassette in the electrophoresis tank. Fill the top reservoir with electrophoresis buffer, and look for any leaks from the top tank. If there are no leaks fill the bottom tank with electrophoresis buffer, and then tilt the apparatus to dispel any bubbles caught under the gel.
12. Samples can now be loaded onto the gel. Place the syringe needle through the buffer and locate it just above the bottom of the well. Slowly deliver the sample into the well. Five- to 10-μL samples are appropriate for most gels. The dense sample buffer ensures that the sample settles to the bottom of the loading well (*see* Note 10). Continue in this way to fill all the wells with unknowns or standards, and record the samples loaded.
13. Connect the power pack to the apparatus, and pass a current of 30 mA through the gel (constant current). Ensure your electrodes have correct polarity: all proteins will travel to the anode (+). In the first few minutes, the samples will be seen to concentrate as a sharp band as it moves through the stacking gel. (It is actually the bromophenol blue that one is observing not the protein, but of course the protein is stacking in the same way.) Continue electrophoresis until the bromophenol blue reaches the bottom of the gel. This will take 2.5–3.0 h (*see* Note 11).
14. Dismantle the gel apparatus, pry open the gel plates, remove the gel, discard the stacking gel, and place the separating gel in stain solution.
15. Staining should be carried out with shaking, for a minimum of 2 h. When the stain is replaced with destain, stronger bands will be immediately apparent, and weaker bands will appear as the gel destains (*see* Notes 12 and 13).

4. Notes

1. Acrylamide is a potential neurotoxin and should be treated with great care. Its effects are cumulative, and therefore, regular users are at greatest risk. In particular, take care when weighing out acrylamide. Do this in a fume hood, and wear an appropriate face mask.

2. SDS come out of solution at low temperature, and this can even occur in a relatively cold laboratory. If this happens, simply warm up the bottle in a water bath. Store at room temperature.
3. Solid samples can be dissolved directly in sample buffer. Pure proteins or simple mixtures should be dissolved at 1–0.5 mg/mL. For more complex samples suitable concentrations must be determined by trial and error. For samples already in solution dilute them with an equal volume of double-strength sample buffer. Do not use protein solutions that are in a strong buffer, that is, not near pH 6.5, since it is important that the sample be at the correct pH. For these samples, it will be necessary to dialyze them first. Should the sample solvent turn from blue to yellow, this is a clear indication that your sample is acidic.
4. The β-mercaptoethanol is essential for disrupting disulfide bridges in proteins. However, exposure to oxygen in the air means that the reducing power of β-mercaptoethanol in the sample buffer decreases with time. Every couple of weeks, therefore, mercaptoethanol should be added to the stock solution or the solution remade. Similarly protein samples that have been prepared in sample buffer and stored frozen should, before being rerun at a later date, have further mercaptoethanol added.
5. Typically, the separating gel used by most workers is a 15% polyacrylamide gel. This give a gel of a certain pore size in which proteins of relative molecular mass (M_r) 10,000 move through the gel relatively unhindered, whereas proteins of 100,000 can only just enter the pores of this gel. Gels of 15% polyacrylamide are therefore useful for separating proteins in the range of 100,000–10,000. However, a protein of 150,000 for example, would be unable to enter a 15% gel. In this case, a larger-pored gel (e.g., a 10% or even 7.5% gel) would be used so that the protein could now enter the gel, and be stained and identified. It is obvious, therefore, that the choice of gel to be used depends on the size of the protein being studied. If proteins covering a wide range of mol-wt values need to be separated, then the use of a gradient gel is more appropriate (*see* Chapter 12).
6. Degassing helps prevent oxygen in the solution from "mopping up" free radicals and inhibiting polymerization although this problem could be overcome by the alternative approach of increasing the concentration of catalyst. However, the polymerization process is an exothermic one. For 15% gels, the heat liberated can result in the formation of small air bubbles in the gel (this is not usually a problem for gels of 10% or less where much less heat is liberated). It is advisable to carry out degassing as a matter of routine.
7. An alternative approach is to add a water-immiscible organic solvent, such as isobutanol, to the top of the gel. Less caution is obviously needed when adding this, although if using this approach, this step should be carried out in a fume cupboard, not in the open laboratory.
8. To save time some workers prefer to add the stacking gel solution directly and carefully to the top of the separating gel, i.e., the overlaying step (step 7) is omitted, the stacking gel solution itself providing the role of the overlaying solution.
9. Some workers include a small amount of bromophenol blue in this gel mix. This give a stacking gel that has a pale blue color, thus allowing the loading wells to be easily identified.
10. Even if the sample is loaded with too much vigor, such that it mixes extensively with the buffer in the well, this is not a problem, since the stacking gel system will still concentrate the sample.
11. When analyzing a sample for the first time, it is sensible to stop the run when the dye reaches the bottom of the gel, because there may be low mol-wt proteins that are running close to the dye, and these would be lost if electrophoresis was continued after the dye had run off the end of the gel. However, often one will find that the proteins being separated are only in the top two-thirds of the gel. In this case, in future runs, the dye would be run

off the bottom of the gel, and electrophoresis carried out for a further 30 min to 1 h to allow proteins to separate across the full length of the gel thus increasing the separation of bands.

12. Normally, destain solution needs to be replaced at regular intervals since a simple equilibrium is quickly set up between the concentration of stain in the gel and destain solution, after which no further destaining takes place. To speed up this process and also save on destain solution, it is convenient to place some solid material in with the destain that will absorb the Coomassie dye as it elutes from the gel. We use a foam bung such as that used in culture flasks (ensure it is well wetted by expelling all air in the bung by squeezing it many times in the destain solution), although many other materials can be used (e.g., polystyrene packaging foam).
13. It is generally accepted that a very faint protein band detected by Coomassie brilliant blue, is equivalent to about 0.1 μg (100 ng) of protein. Such sensitivity is suitable for many people's work. However if no protein bands are observed or greater staining is required, then silver staining (Chapter 35) can be further carried out on the gel.
14. Because the principle of this technique is the separation of proteins based on size differences, by running calibration proteins of known molecular weight on the same gel run as your unknown protein, the molecular weight of the unknown protein can be determined. For most proteins a plot of $\log_{10}$ molecular mass vs relative mobility provides a straight line graph, although one must be aware that for any given gel concentration this relationship is only linear over a limited range of molecular masses. As an approximate guide, using the system described here, the linear relationship is true over the following ranges: 15% acrylamide, 10,000–50,000; 10% acrylamide 15,000–70,000; 5% acrylamide 60,000–200,000. It should be stressed that this relationship only holds true for proteins that bind SDS in a constant weight ratio. This is true of many proteins but some proteins for example, highly basic proteins, may run differently than would be expected on the basis of their known molecular weight. In the case of the histones, which are highly basic proteins, they migrate more slowly than expected, presumably because of a reduced overall negative charge on the protein owing to their high proportion of positively-charged amino acids. Glycoproteins also tend to run anomalously presumably because the SDS only binds to the polypeptide part of the molecule.

    To determine the molecular weight of an unknown protein the relative mobilities (R_f) of the standard proteins are determined and a graph of log molecular weight vs R_f plotted.

$$R_f = \text{(distance migrated by protein/distance migrated by dye)} \quad (3)$$

    Mixtures of standard mol-wt markers for use on SDS gels are available from a range of suppliers. The R_f of the unknown protein is then determined and the logMW (and hence molecular weight) determined from the graph. A more detailed description of protein mol-wt determination on SDS gels is described in refs. *1* and *3*.

References

1. Deyl, Z. (1979) *Electrophoresis: A Survey of Techniques and Applications. Part A Techniques.* Elsevier, Amsterdam.
2. Laemmli, U. K. (1970) Cleavage of structural proteins during the assembly of the head of bacteriophage T4. *Nature* **227,** 680–685.
3. Hames, B. D. and Rickwood, D. (eds.) (1990) *Gel Electrophoresis of Proteins—A Practical Approach.* IRL, Oxford University Press, Oxford.

12

Gradient SDS Polyacrylamide Gel Electrophoresis of Proteins

John M. Walker

1. Introduction

The preparation of fixed-concentration polyacrylamide gels has been described in Chapters 10 and 11. However, the use of polyacrylamide gels that have a gradient of increasing acrylamide concentration (and hence decreasing pore size) can sometimes have advantages over fixed-concentration acrylamide gels. During electrophoresis in gradient gels, proteins migrate until the decreasing pore size impedes further progress. Once the "pore limit" is reached, the protein banding pattern does not change appreciably with time, although migration does not cease completely *(1)*. There are two main advantages of gradient gels over linear gels.

First, a much greater range of protein M_r values can be separated than on a fixed-percentage gel. In a complex mixture, very low-mol-wt proteins travel freely through the gel to begin with, and start to resolve when they reach the smaller pore size toward the lower part of the gel. Much larger proteins, on the other hand, can still enter the gel but start to separate immediately owing to the sieving effect of the gel. The second advantage of gradient gels is that proteins with very similar M_r values may be resolved, which otherwise cannot resolve in fixed percentage gels. As each protein moves through the gel, the pore size become smaller until the protein reaches its pore size limit. The pore size in the gel is now too small to allow passage of the protein, and the protein sample stacks up at this point as a sharp band. A similar-sized protein, but with slightly lower M_r, will be able to travel a little further through the gel before reaching its pore size limit, at which point it will form a sharp band. These two proteins, of slightly different M_r values, therefore separate as two, close, sharp bands.

The usual limits of gradient gels are 3–30% acrylamide in linear or concave gradients. The choice of range will of course depend on the size of proteins being fractionated. The system described here is for a 5–20% linear gradient using SDS polyacrylamide gel electrophoresis. The theory of SDS polyacrylamide gel electrophoresis has been decribed in Chapter 11.

2. Materials

1. Stock acrylamide solution: 30% acrylamide, 0.8% bis-acrylamide. Dissolve 75 g of acrylamide and 2.0 g of *N,N'*-methylene *bis*-acrylamide in about 150 mL of water. Filter and make the volume to 250 mL. Store at 4°C. The solution is stable for months.

From: The Protein Protocols Handbook
Edited by: J. M. Walker Humana Press Inc., Totowa, NJ

2. Buffers:
   a. 1.875*M* Tris-HCl, pH 8.8.
   b. 0.6*M* Tris-HCl, pH 6.8.
   Store at 4°C.
3. Ammonium persulfate solution (10% [w/v]). Make fresh as required.
4. SDS solution (10% [w/v]). Stable at room temperature. In cold conditions, the SDS can come out of solution, but may be redissolved by warming.
5. *N,N,N',N'*-Tetramethylene diamine (TEMED).
6. Gradient forming apparatus (*see* Fig. 1). Reservoirs with dimensions of 2.5 cm id and 5.0 cm height are suitable. The two reservoirs of the gradient former should be linked by flexible tubing to allow them to be moved independently. This is necessary since although equal volumes are placed in each reservoir, the solutions differ in their densities and the relative positions of A and B have to be adjusted to balance the two solutions when the connecting clamp is opened (*see* Note 3).

3. Method

1. Prepare the following solutions:

	Solution A, mL	Solution B, mL
1.875*M* Tris-HCl, pH 8.8	3.0	3.0
Water	9.3	0.6
Stock acrylamide, 30%	2.5	10.0
10% SDS	0.15	0.15
Ammonium persulfate (10%)	0.05	0.05
Sucrose	—	2.2 g (equivalent to 1.2 mL volume)

2. Degas each solution under vacuum for about 30 s and then, when you are ready to form the gradient, add TEMED (12 µL) to each solution.
3. Once the TEMED is added and mixed in, pour solutions A and B into the appropriate reservoirs (*see* Fig. 1.)
4. *With the stirrer stirring,* fractionally open the connection between A and B and adjust the relative heights of A and B such that there is no flow of liquid between the two reservoirs (easily seen because of the difference in densities). Do not worry if there is some mixing between reservoirs—this is inevitable.
5. When the levels are balanced, completely open the connection between A and B, turn the pump on, and fill the gel apparatus by running the gel solution down one edge of the gel slab. Surprisingly, very little mixing within the gradient occurs using this method. A pump speed of about 5 mL/min is suitable. If a pump is not available, the gradient may be run into the gel under gravity.
6. When the level of the gel reaches about 3 cm from the top of the gel slab, connect the pump to distilled water, reduce pump speed, and overlay the gel with 2–3 mm of water.
7. The gradient gel is now left to set for 30 min. Remember to rinse out the gradient former before the remaining gel solution sets in it.
8. When the separating gel has set, prepare a stacking gel by mixing the following:
   a. 1.0 mL 0.6*M* Tris-HCl, pH 6.8;
   b. 1.35 mL Stock acrylamide;
   c. 7.5 mL Water;
   d. 0.1 mL 10% SDS;
   e. 0.05 mL Ammonium persulfate (10%).

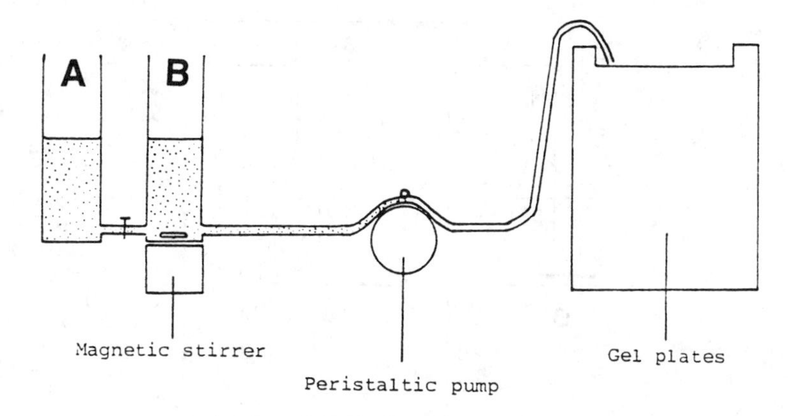

Fig. 1. Gradient forming apparatus.

9. Degas this mixture under vacuum for 30 s and then add TEMED (12 μL).
10. Pour off the water overlayering the gel and wash the gel surface with about 2 mL of stacking gel solution and then discard this solution.
11. The gel slab is now filled to the top of the plates with stacking gel solution and the well-forming comb placed in position (*see* Chapter 11).
12. When the stacking gel has set (~15 min), carefully remove the comb. The gel is now ready for running. The conditions of running and sample preparation are exactly as described for SDS gel electrophoresis in Chapter 11.

4. Notes

1. The total volume of liquid in reservoirs A and B should be chosen such that it approximates to the volume available between the gel plates. However, allowance must be made for some liquid remaining in the reservoirs and tubing.
2. As well as a gradient in acrylamide concentration, a density gradient of sucrose (glycerol could also be used) is included to minimize mixing by convectional disturbances caused by heat evolved during polymerization. Some workers avoid this problem by also including a gradient of ammonium persulfate to ensure that polymerization occurs first at the top of the gel, progressing to the bottom. However, we have not found this to be necessary in our laboratory.
3. The production of a linear gradient has been described in this chapter. However, the same gradient mixed can be used to produce a concave (exponential) gradient. This concave gradient provides a very shallow gradient in the top half of the gel such that the percentage of acrylamide only varies from about 5–7% over the first half of the gel. The gradient then increases much more rapidly from 7–20% over the next half of the gel. The shallow part of the gradient allows high-mol-wt proteins of similar size to sufficiently resolve while at the same time still allowing lower mol-wt proteins to separate lower down the gradient. To produce a concave gradient, place 7.5 mL of solution B in reservoir B, then tightly stopper this reservoir with a rubber bung. Equalize the pressure in the chamber by briefly inserting a syringe needle through the bung. Now place 22.5 mL of solution A in reservoir A, open the connector between the two chambers, and commence pouring the gel. The volume of reservoir B will be seen to remain constant as liquid for reservoir A is drawn into this reservoir and diluted.

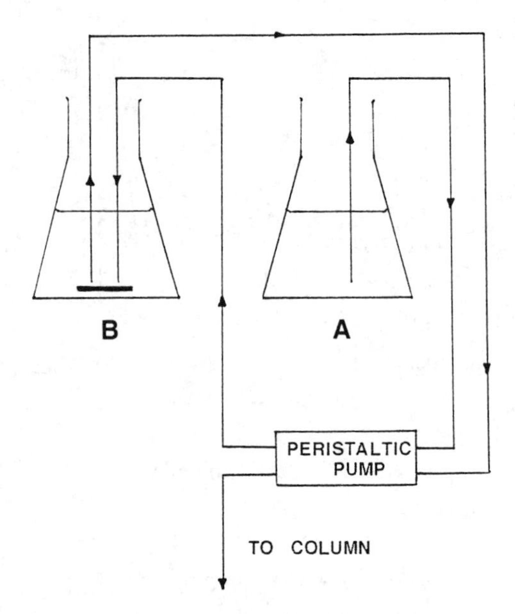

Fig. 2. Diagrammatic representation of a method for producing a gradient using a two-channel peristaltic pump. Reservoir B has the high percentage acrylamide concentration, reservoir A the lower.

4. We have described the production of a linear gradient using a purpose built gradient mixer. However, it is not necessary to purchase this since the simple arrangement, shown in Fig. 2 using just flasks or beakers, a stirrer, and a dual channel peristaltic pump, can just as easily be used.

Reference

1. Margolis, J. and Kenrick, K. G. (1967) *Nature (London),* **214,** 1334.

13

Cetyltrimethylammonium Bromide Discontinuous Gel Electrophoresis of Proteins

M_r*-Based Separation of Proteins with Retained Native Activity*

Robert E. Akins and Rocky S. Tuan

1. Introduction

This chapter describes a novel method of electrophoresis that allows the fine separation of proteins to be carried out with the retention of native activity. The system combines discontinuous gel electrophoresis in an arginine/*N*-Tris (hydroxymethyl) methylglycine) (Tricine) buffer with sample solubilization in cetyltrimethylammonium bromide (CTAB). Because the components that distinguish this system are **C**TAB, **a**rginine, and **T**ricine and because CTAB is a **cat**ionic detergent, we refer to this method as CAT gel electrophoresis *(1,2)*. Proteins separated on CAT gels appear as discrete bands, and their mobility is a logarithmic function of M_r across a broad range of molecular weights. After CAT electrophoresis, many proteins retain high enough levels of native activity to be detected, and gel bands may be detected by both M_r and protein-specific activities. In this chapter, we provide a description of the procedures for preparing and running CAT gels. We also provide some technical background information on the basic principles of CAT gel operation and some points to keep in mind when considering the CAT system.

1.1. Technical Background

The electrophoretic method of Laemmli *(3)* is among the most common of laboratory procedures. It is based on observations made by Shapiro et al. *(4)* and Weber and Osborn *(5)*, which showed that sodium dodecyl sulfate (SDS) could be used for the separation of many proteins based on molecular size. In Laemmli's method, SDS solubilization was combined with a discontinuous gel system using a glycine/Tris buffer, as detailed by Ornstein *(6)* and Davis *(7)* (*see* Chapter 11). Typically, SDS-discontinuous gel electrophoresis results in the dissociation of protein complexes into denatured subunits and separation of these subunits into discrete bands. Since the mobility of proteins on SDS gels is related to molecular size, many researchers have come to rely on SDS gels for the convenient assignment of protein subunit M_r.

Unfortunately, it is difficult to assess the biological activity of proteins treated with SDS: proteins prepared for SDS gel electrophoresis are dissociated from native complexes and are significantly denatured. Several proteins have been shown to renature to

From: The Protein Protocols Handbook
Edited by: J. M. Walker Humana Press Inc., Totowa, NJ

an active form after removal of SDS *(8,9)*; however, this method is inconvenient and potentially unreliable. A preferred method for determining native protein activity after electrophoresis involves the use of nonionic detergents like Triton X 100 (Tx-100) *(10)*; however, proteins do not separate based on molecular size. The assignment of M_r in the nonionic Tx-100 system requires the determination of mobilities at several different gel concentrations and "Ferguson analysis" *(11–13)*. The CAT gel system combines the most useful aspects of the SDS and Tx-100 systems by allowing the separation of proteins based on M_r with the retention of native activity.

Previous studies have described the use of CTAB and the related detergent tetradecyltrimethylammonium bromide (TTAB), in electrophoretic procedures for the determination of M_r *(14–18)*. In addition, as early as 1965, it was noted that certain proteins retained significant levels of enzymatic activity after solubilization in CTAB *(19)*. A more recent report further demonstrated that some proteins even retained enzymatic activity after electrophoretic separation in CTAB *(14)*. Based on the observed characteristics of CTAB and CTAB-based gel systems, we developed the CAT gel system.

In contrast to previous CTAB-based gel methods, the CAT system is discontinuous and allows proteins to be "stacked" prior to separation (*see* refs. *6* and *7*). CAT gel electrophoresis is a generally useful method for the separation of proteins with the retention of native activity. It is also an excellent alternative to SDS-based systems for the assignment of protein M_r (*see* Note 1).

1.2. Basic Principles of CAT Gel Operation

The CAT gel system is comprised of two gel matrices and several buffer components in sequence. A diagram of the CAT gel system is shown in Fig. 1. In an applied electric field, the positive charge of the CTAB–protein complexes causes them to migrate toward the negatively charged cathode at the bottom of the system. The arginine component of the tank buffer also migrates toward the cathode; however, arginine is a *zwitterion*, and its net charge is a function of pH. The arginine is positively charged at the pH values used in the tank buffer, but the pH values of the stacking gel and sample buffer are closer to the pI of arginine, and the arginine will have a correspondingly lower net positive charge as it migrates from the tank buffer into these areas. Therefore, the interface zone between the upper tank buffer and the stacking gel/sample buffer contains a region of high electric field strength where the sodium ions in the stacking gel/sample buffer (Tricine-NaOH) move ahead of the reduced mobility arginine ions (Tricine-arginine). In order to carry the electric current, the CTAB-coated proteins migrate more quickly in this interface zone than in the sodium-containing zone just below. As the interface advances, the proteins "stack," because the trailing edge of the applied sample catches up with the leading edge. When the cathodically migrating interface zone reaches the separating gel, the arginine once again becomes highly charged owing to a drop in the pH relative to the stacking gel. Because of the sieving action of the matrix, the compressed bands of stacked proteins differentially migrate through the separating gel based on size.

Two features of CTAB-based gels set them apart from standard SDS-based electrophoretic methods. First, proteins separated in CTAB gels migrate as a function of log M_r across a much broader range of molecular weights than do proteins separated in

TANK BUFFER	--------ANODE--------- 25 mM Tricine pH 8.2 0.1% CTAB 14 mM Arginine Free Base
SAMPLE BUFFER	10 mM Tricine-NaOH pH 8.8 1% CTAB 10% Glycerol
STACKING GEL	0.7% Agarose 125 mM Tricine-NaOH pH 9.96 0.1% CTAB
SEPARATING GEL	6% Polyacrylamide 375 mM Tricine-NaOH pH 7.96
TANK BUFFER	25 mM Tricine pH 8.2 0.1% CTAB 14 mM Arginine Free Base -------CATHODE-------

Fig. 1. Diagram of a CAT gel. CAT gels begin at the top with the anode immersed in tank buffer and end at the bottom with the cathode immersed in additional tank buffer. The tank buffer solution contains CTAB, arginine, and Tricine. Between the tank buffers are the stacking gel and the separating gel. The gels are made up of acrylamide polymers in a Tricine-NaOH-buffered solution. Prior to electrophoresis, protein samples are solubilized in a sample buffer that contains CTAB, to solubilize the protein sample, Tricine-NaOH, to maintain pH, and carry current and glycerol, to increase specific gravity. Proteins solubilized in sample buffer are typically layered under the upper tank buffer and directly onto the stacking gel. *See* Note 3 for a listing of some physical characteristics of the CAT gel components.

SDS gels. As shown in Fig. 2, a plot of relative migration distance, as a function of known log M_r of standard proteins, results in a straight line. Because of the consistent relationship between M_r and distance migrated, the relative molecular weights of unknown proteins can be determined. CAT gels may be especially useful for the assignment of M_r to small proteins or for the comparison of proteins with very different molecular weights. Second, the retention of significant levels of native activity in CAT gels allows electrophoretic profiles to be assessed *in situ* for native activities without additional steps to ensure protein renaturation (*see* Note 2). Taken together, these two characteristics of CAT gels make them an attractive alternative to standard electrophoretic systems.

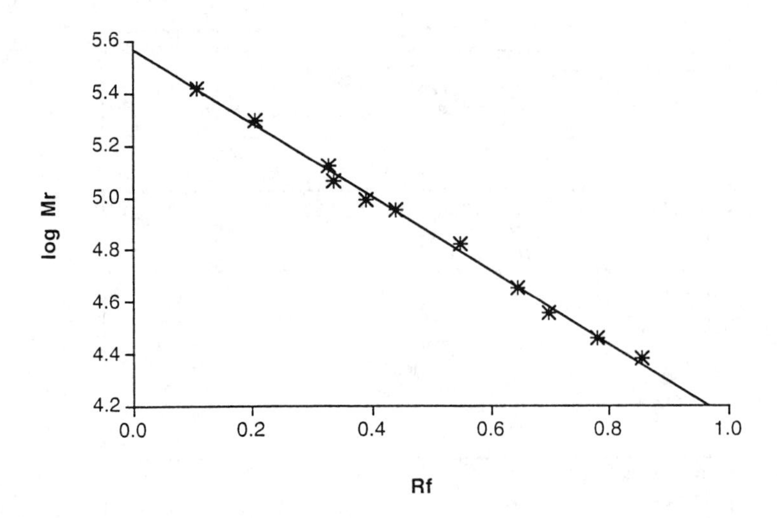

Fig. 2. Mobility of proteins in a CAT gel as a function of M_r. A mixture of proteins fractionated in a CAT gel with a 6% T acrylamide separator and a 0.7% agarose stacker was visualized by CBB R-250 staining. Relative mobilities (R_f) were calculated as distance migrated divided by total distance to the salt/dye front and were plotted against the known M_r values for each protein band. The plot is linear across the entire range ($R^2 > 0.99$). Protein bands included trypsinogen (24 kDa), carbonic anhydrase (29 kDa), glyceraldehyde-3-phospate dehydrogenase (36 kDa), ovalbumin (45 kDa monomer and 90 kDa dimer), bovine serum albumin (66 kDa monomer and 132, 198, and 264 kDa multimers), phosphorylase-B (97.4 kDa), and β-galactosidase (116 kDa). *See* Note 2 concerning the comparison of R_f values from different gels.

2. Materials

1. CAT tank buffer: One liter of 5X tank buffer may be prepared using CTAB, Tricine, and arginine free base. First, prepare 80 mL of a 1*M* arginine free base solution by dissolving 13.94 g in distilled water. Next, dissolve 22.40 g of Tricine in 900 mL of distilled water; add 5 g of CTAB, and stir until completely dissolved. Using the 1*M* arginine solution, titrate the Tricine/CTAB solution until it reaches pH 8.2. Approximately 75 mL of 1*M* arginine solution will be required/L of CAT tank buffer. Since Tricine solutions change pH with changes in temperature, the tank buffer should be prepared at the expected temperature of use (typically 10–15°C). Finally, add distilled water to 1000 mL. Store the CAT tank buffer at room temperature. Prior to use, prepare 1X tank buffer by diluting 200 mL of the 5X stock to 1000 mL using distilled water of the appropriate temperature (usually 10–15°C); filter the 1X tank buffer through #1 Whatman filter paper to remove any particulate material. The 1X CAT tank buffer may be stored cold, but it should not be reused (*see* Note 5). Note that CTAB is corrosive, and care should be taken when handling CTAB powder or CTAB solutions: avoid inhalation or skin contact as advised by the supplier.
2. CAT stacking gel buffer: Prepare a 500 m*M* Tricine-NaOH by dissolving 22.4 g of Tricine in 200 mL of distilled water. Add NaOH until the pH of the solution reaches 10.0. Bring the solution to a total volume of 250 mL using distilled water. As with all Tricine solutions, the pH of CAT stacking gel buffer should be determined at the expected temperature of use. The CAT stacking gel buffer should be stored at room temperature to avoid any precipitation that may occur during long-term cold storage.

3. CAT separating gel buffer: Prepare a 1.5*M* Tricine-NaOH solution by dissolving 134.4 g of Tricine in 400 mL of distilled water. Add NaOH until the pH of the solution reaches 8.0. Bring the solution to a total volume of 500 mL using distilled water. As with all Tricine solutions, the pH of CAT separating gel buffer should be determined at the expected temperature of use. The CAT separating gel buffer should be stored at room temperature to avoid any precipitation that may occur during long-term cold storage.
4. CAT sample buffer: Dilute 0.67 mL of CAT separating gel buffer to approx 80 mL with distilled water; to this add 10 mL of glycerol and 1 g of CTAB. Mix the solution until all the components are dissolved, and adjust the pH to 8.8 using NaOH. Bring the solution to a final volume of 100 mL using distilled water. In some cases, it may be helpful to add a low-mol-wt cationic dye that will be visible during electrophoresis: 10 μL of a saturated aqueous solution of crystal violet may be added/mL of sample buffer. Note that CTAB is corrosive, and care should be taken when handling CTAB powder or CTAB solutions: avoid inhalation or skin contact as advised by the supplier. Store CAT sample buffer at room temperature to avoid precipitation of the components.
5. Acrylamide stock solution: A 40% acrylamide stock solution may be prepared by combining 38.93 g of ultrapure acrylamide with 1.07 g of bis-acrylamide in a total of 100 mL of distilled water. The final solution is 40%T (w/v) and 2.67%C (w/w). The "%T" and "%C" values indicate that the total amount of acrylamide in solution is 40 g/100 mL and that the amount of bis-acrylamide included is 2.67% of the total acrylamide by weight. The acrylamide stock solution should be stored in the refrigerator. Unpolymerized acrylamide is very toxic, and great care should be taken when handling acrylamide powders and solutions: Follow all precautions indicated by the supplier, including the wearing of gloves and a particle mask during preparation of acrylamide solutions.
6. Agarose stock solution: A ready-to-use agarose stacking gel solution may be prepared by combining 25 mL of CAT stacking gel buffer, 0.1 g CTAB, and 0.7 g of electrophoresis-grade agarose distilled to a final volume of 100 mL. Mix the components well, and, if necessary, adjust the pH to 10.0. Heat the solution in a microwave oven to melt the agarose, and swirl the solution to mix thoroughly. Divide the agarose stock solution into 10 aliquots, and store at 4°C until ready to use.
7. 10% Ammonium persulfate (AP): Dissolve 0.1 g of ammonium persulfate in 1 mL of distilled water. Make just prior to use.
8. Water saturated isobutanol: Combine equal volumes of isobutanol and distilled water. Mix well, and allow the two phases to separate: the water-saturated isobutanol will be the upper layer. Store at room temperature in a clear container so that the interface is visible.
9. CAT gel fixative: Combine 40 mL of distilled water, 10 mL of acetic acid, and 50 mL of methanol; mix well. Store CAT gel fixative in a tightly sealed container at room temperature.
10. Coomassie brilliant blue stain (CBB): Combine 40 mL of distilled water with 10 mL of acetic acid and 50 mL of methanol. Add 0.25 g of CBB R-250, and dissolve with stirring (usually overnight). Filter the solution through #1 Whatman paper to remove any particulate material. Store at room temperature in a tightly sealed container.
11. CBB Destain: Combine 437.5 mL of distilled water, 37.5 mL of acetic acid, and 25 mL of methanol. Mix well, and stored in a closed container at room temperature.
12. Electrophoresis apparatus: A suitable electrophoresis apparatus and power supply are required to run CAT gels. It is desirable to set aside combs, spacers, gel plates, and buffer tanks to use specifically with CAT gels; however, if the same apparatus is to be used alternately for CAT gels and SDS gels, it is necessary to clean it thoroughly between each use. Often, the first CAT gel run in an apparatus dedicated to SDS gels will have a smeared appearance with indistinct bands. This smearing is the result of residual SDS, and subse-

quent CAT electrophoretic runs will resolve protein bands distinctly. This smearing may be somewhat avoided by soaking the gel apparatus and gel plates in CAT tank buffer prior to a final rinse in distilled water at the final step in the cleaning process.

The selection of an electrophoresis apparatus to be used for CAT gels should be based on a consideration of the electrical configuration of the system. Because molecular bromine (Br_2) will form at the anode, the anode should be located away from the top of the gel (*see* Note 5). In addition, it is important to realize that CAT gels are "upside-down" relative to SDS gels: proteins migrate to opposite electrodes in the two systems. Some electrophoresis apparatus are intentionally designed for use with SDS, and the anode (usually the electrode with the red lead) may be fixed at the bottom of the gel, whereas the cathode (usually the electrode with the black lead) is fixed at the top of the gel. If such an apparatus is used, the red lead wire should be plugged into the black outlet on the power supply, and the black lead should be plugged into the red outlet on the power supply. Crossing the wires in this fashion ensures that the CTAB-coated proteins in the CAT system will run into the gel and not into the tank buffer.

3. Method

The methods for the preparation and running of CAT gels are similar to other familiar electrophoretic techniques. In this section, we will describe the basic methods for preparing samples, casting gels, loading and running gels, visualizing protein bands, and transferring proteins to nitrocellulose (or other) membranes. We will emphasize the differences between CAT gels and other systems. To provide the best results, the recommendations of the manufacturer should be followed concerning the assembly of the apparatus and the casting of discontinuous gels.

3.1. Preparing Samples

1. Protein samples should be prepared at room temperature immediately prior to loading the gel. Typically, tissue fragments, cells, or protein pellets are resuspended in 1.5-mL microfuge tubes using CAT sample buffer (*see* Note 6). CAT sample buffer may also be used to solubilize cultured cells or minced tissues directly. In each case, the samples should be spun in a microfuge for 0.5 min at 16,000*g* to pellet any debris or insoluble material prior to loading the gel. Good results have been obtained when the final concentration of protein in CAT sample buffer is between 1 and 5 mg/mL; however, the preferred concentration of protein will vary depending on the sample and the particular protein of interest. A series of protein dilutions should be done to determine the optimal solubilization conditions for a particular application.

3.2. Casting CAT Separating Gels

1. Assemble the gel plates and spacers in the gel casting stand as described by the manufacturer.
2. Prepare a separating gel solution by combining the 40%T acrylamide, CAT Separating gel buffer, and distilled water in the ratios indicated in Table 1. Mix the solution by swirling with the introduction of as little air as possible (oxygen inhibits the reactions necessary to accomplish acrylamide polymerization, *see* Note 7).
3. Degas the solution by applying a moderate vacuum for 5–10 min: the vacuum generated by an aspirator is generally sufficient.
4. Add 10% AP and TEMED to the solution as indicated in Table 1, and swirl the solution gently to mix. Note that insufficient mixing will result in the formation of a nonhomogeneous gel, but that vigorous mixing will introduce oxygen into the mixture.

Table 1
Preparation of Acrylamide Solutions for CAT Gels

Regent	4%T, mL	6%T, mL	8%T, mL	10%T, mL
40%T Acrylamide	1.00	1.50	2.00	2.50
Tricine buffer	2.50	2.50	2.50	2.50
Distilled water	6.39	5.89	5.39	4.89
Degas solution				
10% AP	0.10	0.10	0.10	0.10
TEMED	0.01	0.01	0.01	0.01

Volumes indicated are in milliliters required to prepare 10 mL of the desired solution. Solutions should be degassed prior to the addition of the crosslinking agents, AP and TEMED.

5. Carefully pour the gel mixture into the gel plates to the desired volume; remember to leave room for the stacking gel and comb.
6. Finally, layer a small amount of water-saturated isobutanol onto the top of the gel. The isobutanol layer reduces the penetration of atmospheric oxygen into the surface of the gel and causes the formation of an even gel surface. Allow polymerization of the separating gel to proceed for at least 60 min to assure complete crosslinking; then pour off the isobutanol, and rinse the surface of the separating gel with distilled water.

3.3. Casting CAT Stacking Gels

Two different types of gel stackers are routinely used with CAT gels. For gel histochemical analyses, or where subsequent protein activity assays will be performed, stacking gels made from agarose have provided the best results.

1. Slowly melt a tube of agarose stock solution in a microwave oven; avoid vigorous heating of the solution, since boiling will cause foaming to occur and may result in air pockets in the finished gel.
2. Insert the gel comb into the apparatus, and cast the stacking gel directly onto the surface of the acrylamide separating gel. Allow the agarose to cool thoroughly before removing the comb (*see* Note 8).
3. As an alternative to agarose stacking gels, low%T acrylamide stackers may also be used. To prepare an acrylamide stacking gel, combine the 40%T acrylamide stock, CAT separating gel buffer (0.5*M* Tricine-NaOH, pH 10.0), and distilled water in the ratios indicated in Table 1. Typically, a 4%T stacking gel is used. Degas the solution by applying a moderate vacuum for 5–10 min. Next, add 10% AP and TEMED to the solution as indicated in Table 1, and swirl the solution gently to mix. Insert the gel comb and cast the stacking gel directly onto the surface of the acrylamide separating gel. Do not use water-saturated isobutanol with stacking gels! It will accumulate between the comb and the gel, and cause poorly defined wells to form. Allow the stacking gel to polymerize completely before removing the comb.

3.4. Loading and Running CAT Gels

1. After the stacking and separating gels are completely polymerized, add 1X CAT tank buffer to the gel apparatus so that the gel wells are filled with buffer prior to adding the samples.

2. Next, using a Hamilton syringe (or other appropriate loading device), carefully layer the samples into the wells. Add the samples slowly and smoothly to avoid mixing them with the tank buffer, and fill any unused wells with CAT sample buffer. The amount of sample to load on a given gel depends on several factors: the size of the well, the concentration of protein in the sample, staining or detection method, and so forth. It is generally useful to run several dilutions of each sample to ensure optimal loading. Check that the electrophoresis apparatus has been assembled to the manufacturer's specifications, and then attach the electrodes to a power supply. Remember that in CAT gels, proteins run toward the negative electrode, which is generally indicated by a black-colored receptacle on power supplies.
3. Turn the current on, and apply 100 V to the gel. For a single minigel (approx 80 mm across, 90 mm long, and 0.8 mm thick), 100 V will result in an initial current of approx 25 mA. Excessive current flow through the gel should be avoided, since it will cause heating.
4. When the front of the migrating system reaches the separating gel, turn the power supply up to 150 V until the front approaches the bottom of the gel. The total time to run a CAT gel should be around 45–60 min for minigels or 4–6 h for full-size gels.

3.5. Visualization of Proteins

1. As with any electrophoretic method, proteins run in CAT gels may be visualized by a variety of staining techniques. A simple method to stain for total protein may be carried out by first soaking the gel for 15 min in CAT gel fixative, followed by soaking the gel into CBB stain until it is thoroughly infiltrated. Infiltration can take as little as 5 min for thin (0.8 mm), low-percentage (6%T) gels or as long as 1 h for thick (1.5 mm), high-percentage (12%T) gels. When the gel has a uniform deep blue appearance, it should be transferred to CBB destain.
2. Destain the gel until protein bands are clearly visible (*see* Note 9). It is necessary to observe the gel periodically during the destaining procedure, since the destain will eventually remove dye from the protein bands as well as the background. Optimally, CBB staining by this method will detect about 0.1–0.5 µg of protein/protein track; when necessary, gels containing low amounts of total protein may be silver-stained (*see* Note 10). Note that the CBB stain may be retained and stored in a closed container for reuse.
3. In addition to total protein staining by the CBB method, enzyme activities may be detected by a variety of histochemical methods. The individual protocol for protein or enzyme detection will, of course, vary depending on the selected assay (*see* Note 11). Generally, when CAT gels are to be stained for enzyme activity, they should be rinsed in the specific reaction buffer prior to the addition of substrate or detection reagent. The CAT gel system provides an extremely flexible method for the analysis of protein mixtures by a variety of direct and indirect gel staining methods.

4. Notes

1. SDS vs CTAB: CTAB and SDS are very different detergents. CTAB is a cationic detergent, and proteins solubilized in CTAB are positively charged; SDS is an anionic detergent, and proteins solubilized in SDS are negatively charged. In terms of electrophoretic migration, proteins in CTAB gels migrate toward the cathode (black electrode), and proteins in SDS gels run toward the anode (red electrode). SDS is not compatible with the CAT gel system, and samples previously prepared for SDS-PAGE are not suitable for subsequent CAT gel electrophoresis. Also, the buffer components of the typical SDS-PAGE system, Tris and glycine, are not compatible with the CAT system: Tricine and arginine should be used with CAT gels.

Although the detergents are different, protein banding patterns seen in CAT gels are generally similar to those seen when using SDS-PAGE. R_f values of proteins fractionated by CAT electrophoresis are consistently lower than R_f values determined on the same%T SDS-PAGE, i.e., a particular protein will run nearer the top of a CAT gel than it does in a similar%T SDS gel. As a rule of thumb, a CAT gel with a 4%T stacker and a 6%T separator results in electrophoretograms similar to an SDS gel with a 4%T stacker and 8%T separator. Differences between the protein banding patterns seen in CAT gels and SDS gels are usually attributable to subunit associations: multisubunit or self-associating proteins are dissociated to a higher degree in SDS than in CTAB, and multimeric forms are more commonly seen in CAT gels than in SDS gels.

2. Detergent solubilization and protein activity: Many proteins separated on CAT gels may be subsequently identified based on native activity, and under the conditions presented here, CTAB may be considered a nondenaturing detergent. Denaturants generally alter the native conformation of proteins to such an extent that activity is abolished; such is the case when using high levels of SDS. Sample preparation for SDS-based gels typically results in a binding of 1.4 g of SDS/1 g of denatured protein across many types of proteins *(20)*, and it is this consistent ratio that allows proteins to be electrophoretically separated by log M_r. Interestingly, at lower concentrations of SDS, another stable protein binding state also exists (0.4 g/1 g of protein) which reportedly does not cause massive protein denaturation *(20)*. In fact, Tyagi et al. *(21)* have shown that low amounts of SDS (0.02%) combined with pore-limit electrophoresis could be used for the simultaneous determination of M_r and native activity. The existence of detergent–protein complexes, which exhibit consistent binding ratios without protein denaturation, represents an exciting prospect for the development of new electrophoretic techniques: protein M_r and activity may be identified by any of a variety of methods selected for applicability to a specific system.
3. Comparing CAT gels: The comparison of different CAT gel electrophoretograms depends on using the same separating gel, stacking gel, and sample buffer for each determination. An increase in the acrylamide%T in the separating gel will cause bands to shift toward the top of the gel, and high%T gels are more suitable for the separation of low M_r proteins. In addition, the use of acrylamide or high-percentage agarose stackers will lead to the determination of R_f values that are internally consistent, but uniformly lower than those determined in an identical gel with a low-percentage agarose stacker. This effect is likely owing to some separation of proteins in the stacking gel, but, nonetheless, to compare R_f values among CAT gels, the stacker of each gel should be the same. Similarly, any changes in sample preparation (for example, heating the sample before loading to dissociate protein subunits) or the sample buffer used (for example, the addition of salt or urea to increase sample solubilization) often precludes direct comparisons to standard CAT gel electrophoretograms.
4. Characteristics of system components: The CAT gel system is designed around the detergent CTAB. The other system components were selected based on the cationic charge of CTAB and the desire to operate the gel near neutral pH. Some of the important physical characteristics of system components are summarized here; the values reported are from information supplied by manufacturers and *Data for Biochemical Research (22)*.

In solution, CTAB exists in both monomer and micelle forms. CTAB has a monomer mol wt of 365 Dalton. At room temperature and in low-ionic-strength solutions (<0.05*M* Na^+), CTAB has a critical micelle concentration (CMC) of about 0.04% (≈1 m*M*). The CMC may be defined as the concentration of detergent monomer that may be achieved in solution before micellization occurs; it depends on temperature, ionic strength, pH, and the presence of other solutes. The actual CMC of CTAB in the solutions used in the CAT gel system is not known. In low-ionic-strength solutions (< $0.1M$ Na^+), the mol wt of CTAB micelles is approx 62 kDa with about 170 monomers/micelle. The solubility of CTAB is also a function of ionic conditions and solution temperature, and CTAB will precipitate from CAT sample buffer at temperatures below 10–15°C.

Arginine is an amino acid used in CAT tank buffer to carry current toward the cathode. Arginine is a zwitterion. It has a mol wt of 174.2 Dalton and contains three pH-sensitive charge groups (pK_a =1.8, 9.0, 12.5) with an isoelectric point (pI) near pH 10.8. Arginine was selected as the zwitterionic stacking agent because of its basic pI: at near-neutral pH levels, arginine will be positively charged and will migrate toward the cathode when an electric current is applied. Arginine free base is used in CAT tank buffer, and the proper pH of the CAT tank buffer is arrived at by mixing an acidic solution of Tricine with a basic solution of arginine.

Tricine functions to maintain the desired pH levels throughout the system. Tricine has a mol wt of 179.2 Dalton, and a pK_a of 8.15 (there is also a second $pK_a \approx 3$). During electrophoresis, Tricine also functions as a counterion to carry current toward the anode during electrophoresis. Tricine-buffered solutions will tend to change pH as temperature changes (a drop in temperature from 25 to 4°C will result in a shift in pH of about 0.5 U). Also, at the pH used in CAT stacking gels and CAT sample buffer, Tricine has a relatively low buffering capacity; therefore, the pH of the stacker and sample buffer should be confirmed, especially when any additions or alterations are made to the standard recipes.

5. Bromine drip: CTAB is an ionic compound comprised of both cetyltrimethylammonium cations and bromide anions. During electrophoresis, bromide anions migrate toward the anode at the top of the CAT system. Since electrons are removed at the anode, molecular bromine (Br_2) is formed at the anode. Under standard conditions, Br_2 is a dense, highly reactive liquid. While running CAT gels, a small amount of Br_2 will drip from the anode, and, if the anode is located directly above the top of the gel, Br_2 may drip onto the samples. The "bromine drip" problem may be avoided if the apparatus used in CAT gel electrophoresis is configured so that the anode is away from (or even below) the top of the gel. Substitution of the bromide anion during CTAB preparation (perhaps with chloride to make CTACl) would be useful; however, the authors know of no high-quality commercial source. It should also be noted that the accumulation of reactive Br_2 during electrophoresis may preclude the reuse of tank buffer.
6. CAT sample buffer: The CTAB component of CAT sample buffer precipitates at low temperature (below 10–15°C). Samples should be prepared at room temperature immediately prior to use. Protease inhibitors, for example, phenylmethylsulfonyl fluoride (PMSF), may be added to the sample buffer to inhibit endogenous protease activity. The potential effects of any sample buffer additives (including PMSF) on the enzyme of interest should be assessed in solution before using the additive. Also, to avoid contaminating samples with "finger proteins" from the experimenter's hands, gloves should be worn when handling samples or sample buffer.
7. Acrylamide CAT gels: The polymerization of acrylamide generally involves the production of acrylamide free radicals by the combined action of ammonium persulfate and TEMED. Oxygen inhibits acrylamide polymerization by acting as a trap for the ammonium persulfate and TEMED intermediate free radicals that are necessary to accomplish chemical crosslinking. It is important to degas acrylamide solutions prior to the addition of crosslinking agents. CTAB itself interferes with gel polymerization as well and should not be included with acrylamide solutions; in the case of agarose stacking gels, however, CTAB may be included. Also, we have noticed a slight increase in the time required for acrylamide polymerization in the presence of Tricine buffer, but Tricine does not appar-

ently affect gel performance. Although gelation will occur in about 10–20 min, polymerization should be allowed to go to completion (1–2 h) before running the gel. Finally, to optimize gel performance and reproducibility, both degassing and polymerization should be done at room temperature: cooling a polymerizing gel does not accelerate gel formation, and warming a polymerizing gel may cause brittle matrices to form.

The presence of residual crosslinking agents in the acrylamide matrix may result in the inhibition of some protein activities. An alternative method of gel formation that avoids the use of reactive chemical crosslinkers involves photoactivation of riboflavin to initiate acrylamide polymerization (*see* ref. *23*). We have had success using acrylamide separating gels polymerized with ammonium persulfate and TEMED in combination with agarose stacking gels; however, in some instances, riboflavin polymerized gels may be useful.

8. Agarose CAT gels: When using agarose stackers, remove the comb carefully to avoid creating a partial vacuum in the wells. To avoid pulling the agarose stacker away from the underlying acrylamide separating gel, wiggle the comb so that air or liquid fills the sample wells as the comb is lifted out.
9. Smeared gels: During electrophoresis, proteins run as mixed micelles combined with CTAB and other solution components. The presence of other detergents or high levels of lipid in the sample solution may result in a heterogenous population of protein-containing micelles. In such cases, protein bands may appear indistinct or smeared. Samples that have been previously solubilized in another detergent or that contain high levels of lipid may not be suitable for subsequent CAT gel analysis. Also, precipitation may occur when CTAB is mixed with polyanions (e.g., SDS micelles or nucleic acids). Samples containing high levels of nucleic acid or SDS may not be suitable for CAT gel analysis owing to precipitation in the sample buffer. Furthermore, if the gel apparatus to be used for CAT gels is also routinely used for SDS-based gels, the apparatus should be thoroughly cleaned to remove traces of SDS prior to CAT gel analysis, especially from the gel plates. Interaction between SDS and CTAB during electrophoresis may cause the formation of heterogenous mixed micelles or the precipitation of the components, and will invariably result in smeared gels. Although it is not always possible, it is preferable to have a separate apparatus dedicated to CAT gel use in order to avoid this problem.
10. Staining gels and transfer membranes: In addition to the basic CBB staining of protein bands, CAT gels may be stained by a variety of other techniques. Silver staining *(24)* allows the detection of even trace amounts of proteins (1–10 ng), including any fingerprints or smudges that would be otherwise undetected. In silver staining it is, therefore, essential to wear gloves throughout the procedure, even when cleaning the glassware prior to assembly of the apparatus. When the CAT gel is done, stain it with CBB to visualize proteins. (It is a good idea to take a photograph of the gel at this point, but be careful when handling it to avoid smudging the surface.) Next, place the gel into a 10% glutaraldehyde solution and soak in a fume hood for 30 min with gentle shaking. Rinse the gel well with distilled water for at least 2 h (preferably overnight) changing the water frequently; the gel will swell and become soft and somewhat difficult to handle while it is in the water. Combine 1.4 mL NH_4OH (concentrated solution ≈14.8*M*) and 21.0 mL 0.36% NaOH (made fresh); add approx 4.0 mL of 19.4% $AgNO_3$ (made fresh) with constant swirling. A brown precipitate will form as the $AgNO_3$ is added, but it will quickly disappear; if it does not, a small amount of additional NH_4OH (just enough to dissolve the precipitate) may be added. Soak the gel in the ammoniacal/silver solution for 10 min with shaking. Pour off the ammoniacal/silver solution and precipitate the silver with HCl. Transfer the gel to a fresh dish of distilled water and rinse for 5 min with two or three changes of water. Decant the water and add a freshly prepared solution of 0.005% citric acid and 0.019% formaldehyde

(commercial preparations of formaldehyde are 37% solutions); bands will become visible at this point. Stop the silver-staining reaction by quickly rinsing the gel in water followed by soaking in a solution of 10% acetic acid and 20% methanol. Silver-staining takes a little practice to do well.

Transfer blots may be stained by a variety of conventional techniques. Two rapid and simple procedures are as follows. (1) Rinse the membrane briefly in phosphate-buffered saline containing 0.1% Tween 20 (PBS/Tw); soak the membrane in 0.1% solution of India ink in PBS/Tw until bands are detected (10 to 15 min.); rinse the blot in PBS/Tw. (2) Rinse the membrane briefly in phosphate-buffered saline (PBS); soak the blot in 0.2% Ponceau-S (3-hydroxy-4-(2-sulfo-4(sulfo-phenylazo)phenylazo)-2,7-naphthalene disulfonic acid) in 3% trichloroacetic acid and 3% sulfosalicylic acid for 10 min.; rinse the blot in PBS until bands appear. Ponceau S can be stored as a 10X stock solution and diluted with distilled water just prior to use. Staining with Ponceau-S is reversible, so blots should be marked at the position of mol-wt markers before continuing. When higher sensitivity is required to visualize bands, transfer membranes may be stained with ISS Gold Blot (Integrated Separation Systems, Natick, MA) (*see* Chapter 44).

11. Enzyme activities and protein banding patterns: The histochemical detection of proteins in CAT gels is generally a straightforward process of soaking the gel in reaction buffer, so that the necessary compounds penetrate the gel, followed by the addition of substrate; however, not all enzymes retain detectable levels of activity when solubilized in CTAB or when they are run on CAT gels. It is a good idea to check the relative activity of the protein of interest in a CTAB-containing solution vs standard reaction buffer prior to investing time and effort into running a CAT gel. Of the proteins that do retain detectable the levels of activity, there are substantial differences in the level of retained activity after solubilization in CTAB. There is also the possibility that the protein of interest will run anomalously in CAT gels. Anomalous migration may occur owing to differential CTAB binding, conformational differences in the protein/CTAB mixed micelle, or the presence of previously unrecognized subunits or associated proteins. Unexpected enzyme histochemistry patterns in samples with the expected CBB protein staining pattern may reflect the presence of cofactors or other protein/protein interactions that are necessary for activity and should be interpreted accordingly.

    It should also be pointed out that the intensity of a histochemical or binding assay does not necessarily reflect the actual level of enzyme in the sample. The rate of histochemical reaction within the gel matrix is not necessarily linear with respect to the amount of enzyme present: measured product also depends on the amount of substrate present and the ratio of product to substrate over time as well as the response of the detection apparatus. Reactions should be performed with an excess of substrate at the outset and should preferably be calibrated based on the varying level of a standard activity. Also, as with all detergents, CTAB may differentially solubilize proteins in a given sample, depending on the overall solution conditions. The detergent/protein solution is a complex equilibrium that may be shifted, depending on the level of other materials (proteins, lipids, salts, and so on) in the solution. When seeking to compare protein profiles on CAT gels, it is advisable to prepare samples from the same initial buffer and to solubilize at a uniform CTAB-to-protein ratio.

5. Conclusion

Gel electrophoresis of proteins is a powerful and flexible technique. There are many excellent references for general information concerning the theory behind electrophoresis. One good source of information is Hames and Rickwood *(25)*. The CAT gel electrophoresis system presented here efficiently stacks and separates a wide range of proteins as a function of M_r and preserves native enzymatic activity to such a degree that it allows the identification of protein bands based on native activity. The nature of the interaction between CTAB and native proteins allows the formation of complexes in which the amount of CTAB present is related to the size of the protein moiety. Based on characterizations of detergent/protein interactions that indicate consistent levels of detergent binding without massive denaturation *(10,26–28)*, the retention of activity in certain detergent/protein complexes is expected, and CTAB is likely to represent a class of ionic detergents that allow the electrophoretic separation of native proteins by M_r. Since the level of retained activity after solubilization in CTAB varies depending on the protein of interest, and a given protein will retain varying amounts of measurable activity depending on the detergent used (*see* ref. *1*, for example), a battery of detergents may be tried prior to selecting the desired electrophoresis system. In general, cationic detergents (e.g., the quartenary ammoniums like CTAB, TTAB, and so forth) may be used in the arginine/Tricine buffer system described above; anionic detergents (e.g., alkyl sulfates and sulfonates) may be substituted for SDS in the familiar glycine/Tris buffer system *(29)*.

The CAT gel system and its related cationic and anionic gel systems provide useful adjuncts to existing biochemical techniques. These systems allow the electrophoretic separation of proteins based on log M_r with the retention of native activity. The ability to detect native protein activities, binding characteristics, or associations and to assign accurate M_r values in a single procedure greatly enhances the ability of researchers to analyze proteins and protein mixtures.

Acknowledgment

This work has been supported in part by funds from Nemours Research Programs (to R. E. A.), NASA (SBRA93-15, to R. E. A.), the Delaware Affiliate of the American Heart Association (9406244S, to R. E. A.), and the NIH (HD29937, ES07005, and DE11327 to R. S. T.).

References

1. Akins, R. E., Levin, P., and Tuan, R. S. (1992) Cetyltrimethylammonium bromide discontinuous gel electrophoresis: M_r-based separation of proteins with retention of enzymatic activity. *Anal. Biochem.* **202,** 172–178.
2. Akins, R. E. and Tuan, R. S. (1994) Separation of proteins using cetyltrimethylammonium bromide discontinuous gel electrophoresis. *Mol. Biotech.* **1,** 211–228.
3. Laemmli, U. K. (1970) Cleavage of structural proteins during the assembly of the head of bacteriophage T4. *Nature* **227,** 680–685.
4. Shapiro, A. L., Vinuela, E., and Maizel, J. V. (1967) Molecular weight estimation of polypeptide chains by electrophoresis in SDS-polyacrylamide gels. *Biochem. Biophys. Res. Commun.* **28,** 815–820.
5. Weber, K. and Osborn, M. (1969) The reliability of molecular weight determination by dodecyl sulfate-polyacrylamide electrophoresis. *J. Biol. Chem.* **244,** 4406–4412.
6. Ornstein, L. (1964) Disc electrophoresis I: Background and theory. *Ann. NY Acad. Sci.* **121,** 321–349.

7. Davis, B. J. (1964) Disc electrophoresis II: Method and application to human serum proteins. *Ann. NY Acad. Sci.* **121,** 404–427.
8. Manrow, R. E. and Dottin, R. P. (1980) Renaturation and localization of enzymes in polyacrylamide gels: Studies with UDP-glucose pyrophosphorylase of Dictyostelium. *Proc. Natl. Acad. Sci. USA* **77,** 730–734.
9. Scheele, G. A. (1982) Two-dimensional electrophoresis in basic and clinical research, as exemplified by studies on the exocrine pancreas. *Clin. Chem.* **28,** 1056–1061.
10. Hearing, V. J., Klingler, W. G., Ekel, T. M., and Montague, P. M. (1976) Molecular weight estimation of Triton X-100 solubilized proteins by polyacrylamide gel electrophoresis. *Anal. Biochem.* **126,** 154–164.
11. Ferguson, K. (1964) Starch gel electrophoresis—Application to the classification of pituitary proteins and polypeptides. *Metabolism* **13,** 985–1002.
12. Hedrick, J. L. and Smith A. J. (1968) Size and charge isomer separation and estimation of molecular weights of proteins by disc gel electrophoresis. *Arch. Biochem. Biophys.* **126,** 154–164.
13. Tuan, R. S. and Knowles, K. (1984) Calcium activated ATPase in the chick embryonic chorioaliantoic membrane: Identification and topographic relationship with the calcium-biding protein. *J. Biol. Chem.* **259,** 2754–2763.
14. Akin, D., Shapira, R., and Kinkade, J. M. (1985) The determination of molecular weights of biologically active proteins by cetyltrimethylammonium bromide-polyacrylamide gel electrophoresis. *Anal. Biochem.* **145,** 170–176.
15. Eley, M. H., Burns, P. C., Kannapell, C. C., and Campbell, P. S. (1979) Cetyltrimethylammonium bromide polyacrylamide gel electrophoresis: Estimation of protein subunit molecular weights using cationic detergents. *Anal. Biochem.* **92,** 411–419.
16. Marjanen, L. A. and Ryrie, I. J. (1974) Molecular weight determinations of hydrophilic proteins by cationic detergent electrophoresis: Application to membrane proteins. *Biochem. Biophys. Acta* **37,** 442–450.
17. Panyim, S., Thitiponganich, R., and Supatimusro, D. (1977) A simplified gel electrophoretic system and its validity for molecular weight determinations of protein-cetyltrimethylammonium complexes. *Anal. Biochem.* **81,** 320–327.
18. Schick, M. (1975) Influence of cationic detergent on electrophoresis in polyacrylamide gel. *Anal. Biochem.* **63,** 345–349.
19. Spencer, M. and Poole, F. (1965) On the origin of crystallizable RNA from yeast. *J. Mol. Biol.* **11,** 314–326.
20. Reynolds, J. A. and Tanford, C. (1970) The gross conformation of protein-sodium dodedcyl sulfate complexes. *J. Biol. Chem.* **245,** 5161–5165.
21. Tyagi, R. K., Babu, B. R., and Datta, K. (1993) Simultaneous determination of native and subunit molecular weights of proteins by pore limit electrophoresis and restricted use of sodium dodecyl sulfate. *Electrophoresis* **14,** 826–828.
22. Dawson, R. M. C., Elliott, D. C., Elliott, W. H., and Jones, K. M. (1986) *Data for Biochemical Research.* Clarendon Press, Oxford.
23. Bio-Rad (1987) "Bio-Rad Technical Bulletin #1156: Acrylamide Polymerization—A Practical Approach." Bio-Rad Laboratories, Richmond, CA.
24. Oakley, B. R., Kirsch, D. R., and Morris, N. R. (1980) A simplified ultrasensitive silver stain for detecting proteins in polyacrylamide gels. *Anal. Biochem.* **105,** 361–363.
25. Hames, B. D. and Rickwood, D. (1990) *Gel Electrophoresis of Proteins: A Practical Approach.* IRL, London.
26. Tanford, C., Nozaki, Y., Reynolds, J. A., and Makino, S. (1974) Molecular characterization of proteins in detergent solutions. *Biochemistry* **13,** 2369–2376.

27. Reynolds, J. A., Herbert, S., Polet, H., and Steinhardt, J. (1967) The binding of divers detergent anions to bovine serum albumin. *Biochemistry* **6,** 937–947.
28. Ray, A., Reynolds, J. A., Polet, H., and Steinhardt, J. (1966) Binding of large organic anions and neutral molecules by native bovine serum albumin. *Biochemistry* **5,** 2606–2616.
29. Akins, R. E. and Tuan, R. S. (1992) Electrophoretic techniques for the M_r-based separation of proteins with retention of native activity. *Mol. Biol. Cell* **3,** 185a.

14

Acetic Acid-Urea Polyacrylamide Gel Electrophoresis of Basic Proteins

Jakob H. Waterborg

1. Introduction

Panyim and Chalkley described in 1969 a continuous acetic acid-urea (AU) gel system that could separate very similar basic proteins based on differences in size and effective charge *(1)*. For instance, unmodified histone H4 can be separated from its monoacetylated or monophosphorylated forms *(2)*. At the acidic pH (3.0) of this gel system, basic proteins with a high isoelectric point will clearly have a net positive charge that will be the major determinant of electrophoretic mobility. If one of these positive charges is removed, e.g., by in vivo acetylation of one of the positively charged ε-amino lysine side chain residues in the small histone H4 protein (102 residues), a significant decrease in effective gel mobility is observed. Similarly, addition of a phosphate moiety decreases the net positive charge of the protein during gel electrophoresis by one. Separation of histone H4 forms that differ by a single positive charge are baseline separated in a 30-cm gel with peak-to-peak separation of 5 mm or more. The separation by size is typically sufficient to separate histone H4 from other core histones that in most organisms are slightly larger, and to separate the much larger histone H1 forms from all core histones and often from each other. A size difference of 10% generally suffices for separation of two homogeneous protein forms. However, this may not be the case if postsynthetic modifications by phosphorylation or acetylation introduce charge-based protein heterogeneities. Thus, separation between similarly sized and charged proteins, e.g., the partially acetylated H2A, H2B, and H3 histones of most organisms, can typically only be achieved by inclusion of a nonionic detergent like Triton X100 (*see* Chapter 15).

In 1980, Bonner and coworkers introduced a discontinuous acetic acid-urea-Triton (AUT) variation that avoids the necessity for exhaustive pre-electrophoresis *(3)*. Omission of Triton from this method creates the high capacity and high resolution AU gel electrophoresis protocol described below. Figure 1 shows an example of the possibilities and limitations of the AU gel system (Fig. 1B) when a commercial preparation of calf thymus histones (Worthington, Freehold, NJ), fractionated by reversed-phase HPLC (Fig. 1A), is analyzed. It is clear that, in the unfractionated sample, histone H1 polypeptides could readily be resolved from the core histones that would largely

From: The Protein Protocols Handbook
Edited by: J. M. Walker Humana Press Inc., Totowa, NJ

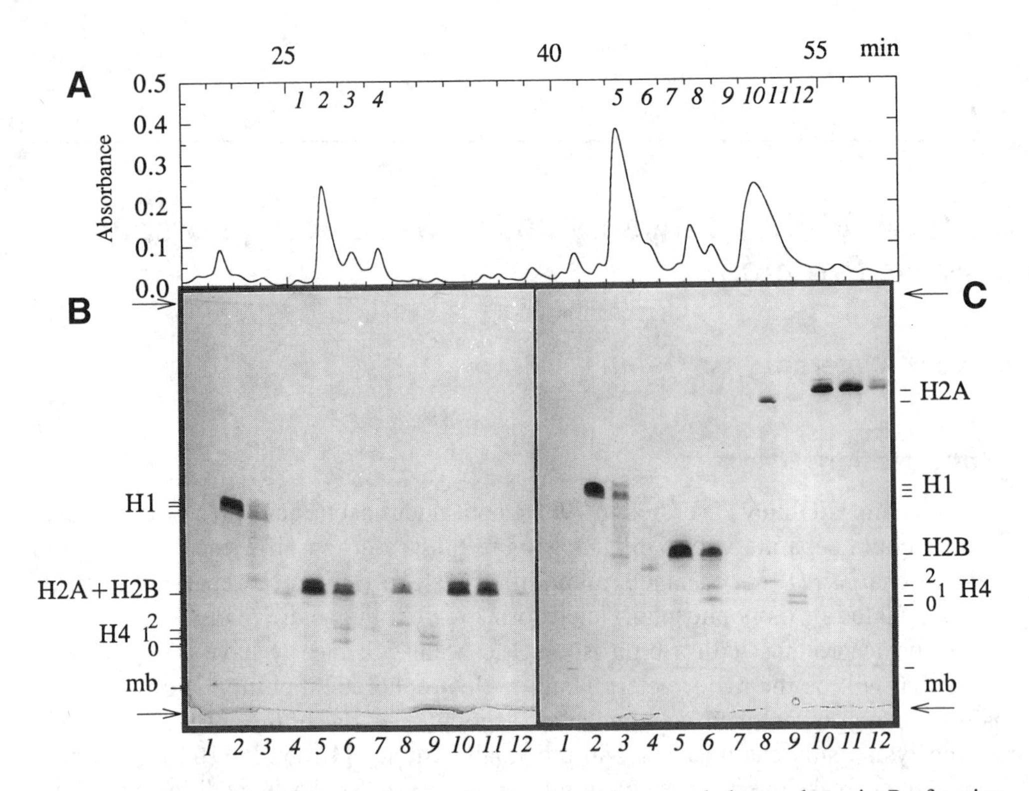

Fig. 1. Analysis of calf thymus histones by acetic acid-urea gel electrophoresis. Prefractionation of a commercial preparation of total histones (Worthington) was used to demonstrate separation of otherwise basic polypeptides in acetic acid-urea gel electrophoresis. This histone preparation was more than 5 yr old, stored at 4°C, but undesiccated and under normal air. Oxidation of histone H2B at methionines is visible. **(A)** Fractionation of 0.75 mg of histones on a 4.5 mm × 25 cm Zorbax ProteinPlus reversed-phase HPLC column, developed at 1 mL/min and 0.1% trifluoroacetic acid by a 90-min gradient from 30 to 60% acetonitrile in water, and monitored by absorbance at 214 nm. The elution profile is that obtained between 20 and 60 min of chromatography, and excludes the histone H3 variant forms, which elute between 68 and 77 min. Fractions of 1.5 mL were collected, lyophilized, and solubilized in sample buffer, as described. The fractions used in gel analysis are indicated by the lane numbers in **B** and **C**. **(B)** Electrophoresis in a 15-cm-long AU gel of histone fractions 1–12 of A. The top and bottom edges of the gel are marked by arrows. The electrophoretic gel front is marked by a residual trace of methylene blue (mb). Ten% of each fraction was loaded in a separate gel lane. Electrophoresis was for 6 h at constant power with a starting condition of 135 V and 20 mA. The gel was stained overnight in Coomassie, destained in 4 h with two aliquots of polyurethane foam, and photographed. The band resolution in this gel is typical for AU gels where band sharpness is typically less than in Laemmli-type SDS gel electrophoresis. The heterogeneous histone H1 band (lane 2) contains several histone H1 variant polypeptides with, as fastest component, histone H1°. The major band in lane 5 is histone H2B, primarily in nonacetylated form, but with a significant amount of monoacetylated derivative. These forms elute together, closely followed by two methionine-oxidation artifacts of histone H2B, the two fastest bands in lane 6. Distinct forms of H2A are seen in lanes 8 and 10–12, and non-, mono-, and di-acetylated (0–2) histone H4 in lane 9. **(C)** Electrophoresis, under identical conditions and with the same samples, in parallel to the AU gel of B, except that 8 m*M* Triton X-100 had been added to the separating gel.

coelectrophorese, as shown for H2A and H2B forms (Fig. 1B). Only a change of the gel system to one with Triton would allow electrophoretic separation, even in the short (15-cm) gels shown (Fig. 1C). The visibly acetylated pattern of histone H4 would be separated from the other core histones in either gel system, but coelectrophoresis of histone H4 (lane 9) with histone H2B methionine-oxidation artifacts (lane 6) would occur in both gel systems.

2. Materials

1. Vertical gel apparatus for short (15-cm) or long (30-cm) slab gels: A gel electrophoresis apparatus that allows gel polymerization between the glass plates with spacers without necessarily being assembled in the apparatus is preferable. This facilitates the even and complete photopolymerization of the acrylamide gel. In this type of apparatus, the glass-gel sandwich is typically clamped to the lower buffer reservoir, which acts as a stand, after which the upper buffer reservoir is clamped to the top of the gel assembly.
   Details of the procedure are described for a fairly standard and flexible gel apparatus that uses two rectangular glass plates (4 mm thick standard plate glass with sanded edges), 21 cm wide and 32.5 and 35.5 cm long, respectively. The Plexiglass bottom buffer reservoir with platinum electrode is 22.5 cm wide with three sides 5 cm high and one of the long sides 12.5 cm high. The glass plates are clamped to this side. The upper buffer reservoir with platinum electrode and a similar buffer capacity is 18 cm wide with one long side enlarged to measure 21 cm wide by 10 cm high. It contains a cutout of 18 cm wide and 3.5 cm high that allows access of the upper reservoir buffer to the top of the gel. The 21-cm-wide Plexiglass plate is masked with 5-mm-thick closed-cell neoprene tape (weather strip) and provides a clamping ridge for attachment to the top of the glass-gel sandwich.
2. Spacers and combs are cut from 1 mm Teflon sheeting. High-efficiency fluorography may benefit from 0.5-mm spacers. Teflon up to 3 mm thick is less easy to cut, but yields very high capacity gels (*see* Note 1).
   Two side spacers (1.5 × 35 cm) and one bottom spacer (0.5 × 24 cm) are required. Added to the top of the side spacers is 3 cm adhesive, closed-cell neoprene tape (weather strip, 14 mm wide and 5-mm thick). This is not required if a more expensive glass plate with "rabbit ears" is used instead of the rectangular shorter plate.
   Combs have teeth 5–10 mm wide and 25–50 mm long, separated by gaps of at least 2.5 mm. For the detailed protocol described, a 15-cm-wide comb with 20 teeth of 5 × 30 mm is used.
3. Vaseline pure petroleum jelly.
4. Acrylamide stock solution: 60% (w/v) acrylamide, highest quality available, in water (*see* Note 2). The acrylamide is dissolved by stirring. Application of heat should be avoided, if possible, to prevent generating acrylic acid. The solution can be kept at least for 3 mo on the laboratory shelf at room temperature. Storage at 4°C can exceed 2 yr without detectable effects.
5. *N,N'*-methylenebis(acrylamide) stock solution: 2.5% (w/v) in water (*see* Note 2).
6. Glacial acetic acid (HAc): 17.5*M*.
7. Concentrated ammonium hydroxide: NH_4OH, 28–30%, approx 15*M*.
8. *N,N,N',N'*-tetramethylenediamine (TEMED), stored at 4°C.
9. Riboflavin-5'-phosphate (R5P) solution: 0.006% (w/v) in water. This solution is stable for more than 6 mo if kept dark and stored at 4°C.
10. Urea, ultrapure quality.

11. Sidearm suction flasks with stoppers, magnetic stirrer, stirrer bar, and water-aspirator vacuum; measuring cylinders with silicon-rubber stoppers; pipets, and pipeting bulbs or mechanical pipeting aids; plastic syringes, 1 and 5 mL, with 20-gage needles.
12. Fluorescent light box with diffuser for even light output and with the possibility to stand vertically. Light intensity should equal or exceed 5 klx at a distance of 5–10 cm. A high-quality X-ray viewing light box with three 40 W bulbs typically will meet this specification.
13. Aluminum foil.
14. Electrophoresis power supply with constant voltage mode at 300–500 V with up to 50 mA current, preferably with a constant power mode option.
15. Urea stock solution: 8*M* urea in water. An aliquot of 40 mL with 1 g of mixed-bed resin (Bio-Rad [Hercules, CA] AG 501-X8) can be used repeatedly over a period of months if refrozen and stored between use at –20°C (*see* Note 3).
16. Phenolphthalein indicator solution: 1% (w/v) in 95% ethanol, stored indefinitely at room temperature in a closed tube.
17. Dithiothreitol (DTT, Cleland's reagent) is stored at 4°C and is weighed freshly for each use.
18. Methylene blue running front indicator dye solution: 2% (w/v) in sample buffer (*see* Section 3., steps 19–21).
19. Reference histones: total calf thymus histones (Worthington), stored dry at 4°C indefinitely or at –20°C in 100-µL aliquots of 5 mg/mL in sample buffer for up to 1 year (*see* Section 3, steps 19–21).
20. Electrophoresis buffer: 1*M* acetic acid, 0.1*M* glycine (*see* Note 4). This solution can be made in bulk and stored indefinitely at room temperature.
21. Destaining solution: 20% (v/v) methanol, 7% (v/v) acetic acid in water.
22. Staining solution: dissolve freshly 0.5 g Coomassie brilliant blue R250 in 500 mL destaining solution for overnight gel staining (*see* Note 5). For rapid staining within the hour, the dye concentration should be increased to 1% (w/v). If the dye incompletely dissolves, the solution should be filtered through Whatman no. 1 paper to prevent staining artifacts.
23. Glass tray for gel staining and destaining.
24. Rotary or alternating table-top shaker.
25. Destaining aids: polyurethane foam for Coomassie or Bio-Rad ion-exchange resin AG1-X8 (20–50 mesh) for Amido black-stained gels.

3. Method

1. Assemble a sandwich of two clean glass plates with two side spacers and a bottom spacer, lightly greased with Vaseline to obtain a good seal, clamped along all sides with 2–in. binder clamps. The triangular shape of these clamps facilitates the vertical, free-standing position of the gel assembly a few centimeters in front of the vertical light box.
2. Separating gel solution: Pipet into a 100-mL measuring cylinder 17.5 mL acrylamide stock solution, 2.8 mL bis-acrylamide stock solution, 4.2 mL glacial acetic acid, and 0.23 mL concentrated ammonium hydroxide (*see* Notes 2 and 6).
3. Add 33.6 g urea, and add distilled water to a total vol of 65 mL.
4. Stopper the measuring cylinder, and place on a rotary mixer until all urea has dissolved. Add water to 65 mL, if necessary.
5. Transfer this solution to a 200-mL sidearm flask with magnetic stir bar on a magnetic stirrer. While stirring vigorously, stopper the flask and apply water-aspirator vacuum. Initially, a cloud of small bubbles of dissolved gas arises, which clears after just a few seconds. Terminate vacuum immediately to prevent excessive loss of ammonia.
6. Add 0.35 mL TEMED and 4.67 mL R5P (*see* Note 7), mix, and pipet immediately between the glass plates to a marking line 5 cm below the top of the shorter plate (*see* Notes 2 and 8).

7. Carefully apply 1 mL distilled water from a 1-mL syringe with needle along one of the glass plates to the top of the separating gel solution to obtain a flat separation surface.
8. Switch the light box on, and place a reflective layer of aluminum foil behind the gel to increase light intensity and homogeneity (*see* Note 1). Gel polymerization becomes detectable within 2 min and is complete in 15–30 min.
9. Switch the light box off, completely drain the water from between the plates, and insert the comb 2.5 cm between the glass plates. The tops of the teeth should always remain above the top of the short glass plate.
10. Stacking gel solution, made in parallel to steps 2–4: into a 25-mL measuring cylinder, pipet 1.34 mL acrylamide stock solution, 1.28 mL bis-acrylamide stock solution, 1.14 mL glacial acetic acid, and 0.07 mL concentrated ammonium hydroxide (*see* Notes 2 and 6).
11. Add 9.6 g urea, and add distilled water to a total vol of 18.6 mL.
12. Stopper the measuring cylinder, and place on a rotary mixer until all urea has dissolved. Add water to 18.6 mL, if necessary.
13. Once the separating gel has polymerized, transfer the stacking gel solution to a 50-mL sidearm flask with stir bar and degas as under step 5.
14. Add 0.1 mL TEMED and 1.3 mL R5P, mix, and pipet between the plates between the comb teeth. Displace air bubbles.
15. Switch the light box on, and allow complete gel polymerization in 30–60 min.
16. Prepare sample buffer freshly when the separation gel is polymerizing (step 8). The preferred protein sample is a salt-free lyophilizate (*see* Note 9). Determine the approximate volume of sample buffer required, depending on the number of samples.
17. Weigh DTT into a sample buffer preparation tube for a final concentration of 1*M*, i.e., 7.7 mg/mL.
18. Per 7.7 mg DTT add 0.9 mL 8*M* urea stock solution, 0.05 mL phenolphthalein, and 0.05 mL NH_4OH to the tube to obtain the intensely pink sample buffer.
19. Add 0.05 mL sample buffer/sample tube with lyophilized protein to be analyzed in one gel lane (*see* Note 10). To assure full reduction of all proteins by DTT, the pH must be above 8.0. If the pink phenolphtalein color disappears owing to residual acid in the sample, a few microliters of concentrated ammonium hydroxide should be added to reach an alkaline pH.
20. Limit the time for sample solubilization and reduction to 5 min at room temperature to minimize the possibility of protein modification at alkaline pH by reactive urea side reactions, e.g., by modification of cysteine residues by cyanate.
21. Acidify the sample by adding 1/20 vol of glacial acetic acid.
22. To each sample add 2 µL methylene blue running front dye (*see* Note 11).
23. Prepare appropriate reference protein samples: to 2 and 6 µL reference histone solution with 10 and 30 µg total calf thymus histones, add 40 µL sample buffer (step 18), 2.5 µL glacial acetic acid, and 2 µL methylene blue.
24. When stacking gel polymerization is complete, remove the comb. Rinse the wells momentarily with water, and drain to remove residual unpolymerized gel solution. The high urea concentration of such residual liquid interferes with the tight application of samples.
25. If a bottom spacer is used to seal the bottom of the gel assembly, remove this as well as any residual Vaseline on the lower surface of the gel.
26. Clamp the gel assembly into the electrophoresis apparatus, and fill the lower buffer reservoir with electrophoresis buffer.
27. Use a 5-mL syringe with a bent syringe needle to displace any air bubbles from the bottom of the gel.
28. Samples can be applied to individual sample wells by any micropipeter with plastic disposable tip or by Hamilton microsyringe (rinsed with water between samples) (*see*

Note 12). Pipet each sample solution against the long glass plate, and let it run to the bottom of the well. For the combination of comb and gel dimensions listed, 50 µL sample will reach a height of 1 cm (*see* Note 8).

29. Apply reference samples in the outer lanes, which frequently show a slight loss of resolution owing to edge effects.
30. Gently overlayer the samples with electrophoresis buffer, dispensed from a 5-mL syringe fitted with a 21-gage needle until all wells are full.
31. Fill the upper buffer reservoir with electrophoresis buffer.
32. Attach the electrical leads between power supply and electrophoresis system: the + lead to the upper and the – lead to the lower reservoir. Note that this is opposite to the SDS gel electrophoresis configuration. Remember, basic proteins like histones are positively charged and will move toward the cathode (negative electrode).
33. Long (30-cm) gels require 15–20 h of electrophoresis at 300 V in constant voltage mode. They are most easily run overnight. For maximum resolution and stacking capacity, the initial current through a 1-mm-thick and 18-cm-wide gel should not exceed 25 mA. Gel electrophoresis is completed in the shortest amount of time in constant power mode with limits of 300 V, 25 mA, and 5 W. The current will drop toward completion of electrophoresis to 6 mA at 300 V.

    Short (15-cm) gels are run at 250 V in constant voltage mode with a similar maximal current, or in constant power mode starting at 25 mA. In the latter example, electrophoresis starts at 25 mA and 135 V, and is complete in 5.5 h at 13 mA and 290 V (*see* Note 13).
34. Electrophoresis is complete just before the methylene blue dye exits the gel. Obviously, electrophoresis may be terminated if lesser band resolution is acceptable or may be prolonged to enhance separation of basic proteins with low gel mobilities, e.g., histone H1 variants or phosphorylated forms of histone H1.
35. Open the glass-gel sandwich, and place the separating gel into staining solution, which is gently agitated continuously overnight on a shaker (*see* Note 14).
36. Decant the staining solution. The gel can be given a very short rinse in water to remove all residual staining solution.
37. Place the gel in ample destaining solution. Diffusion of unbound Coomassie dye from the gel is facilitated by the addition of polyurethane foam as an absorbent for free Coomassie dye. To avoid overdestaining and potential loss of protein from the gel (*see* Note 14), destaining aids in limited amounts are added to only the first and second destaining solutions. Final destaining is done in the absence of any destaining aids.
38. Record the protein pattern of the gel on film and possibly by densitometry. Subsequently, the gel may be discarded, dried, eluted (*see* Chapter 32), blotted (*see* Chapter 40), or prepared for autoradiography or fluorography (*see* Chapter 36).

4. Notes

1. Because of absorbance of the light that initiates gel polymerization, gels thicker than 1.5 mm tend to polymerize better near the light source and produce protein bands that are not perpendicular to the gel surface. For very thick gels, two high-intensity light boxes, placed at either side of the gel assembly, may be required for optimal gel polymerization and resolution.
2. All (bis)acrylamide solutions are potent neurotoxins and should be dispensed by mechanical pipeting devices.
3. Storage of urea solutions at –20°C minimizes creation of ionic contaminants like cyanate. The mixed-bed resin assures that any ions formed are removed. Care should be taken to exclude resin beads from solution taken, e.g., by filtration through Whatman no. 1 paper.

4. The stacking ions between which the positively charged proteins and peptides are compressed within the stacking gel during the initial phase of gel electrophoresis are NH_4^+ within the gel compartment and glycine$^+$ in the electrophoresis buffer. Chloride ions interfere with the discontinuous stacking system. This requires that protein samples (preferably) be free of chloride salts, and that glycine base rather than glycine salt be used in the electrophoresis buffer.
5. Amido black is an alternate staining dye that stains less intensely and destains much slower than Coomassie, but is the better stain for peptides shorter than 30–50 residues.
6. The separating and stacking gels contain 15 vs 4% acrylamide and 0.1 vs 0.16% *N,N'*-methylene(bis-acrylamide), respectively, in 1*M* acetic acid, 0.5% TEMED, 50 m*M* NH_4OH, 8*M* urea, and 0.0004% riboflavin-5'-phosphate.

   We have observed that 8*M* urea produces the highest resolution of histones in these gels when Triton X-100 is present (*see* Chapter 15). Equal or superior resolution of basic proteins has been reported for AU gels when the urea concentration is reduced to 5*M*.
7. Acrylamide is photo-polymerized with riboflavin or riboflavin-5'-phosphate as initiator, because the ions generated by ammonium persulfate-initiated gel polymerization, as used for SDS polyacrylamide gels, interfere with stacking (*see* Note 4).
8. The height of stacking gel below the comb determines the volume of samples that can be applied and fully stacked before destacking at the surface of the separating gel occurs. In our experience, 2.5-cm stacking gel height suffices for samples that almost completely fill equally long sample wells. In general, the single blue line of completely stacked proteins and methylene blue dye should be established 1 cm above the separating gel surface. Thus, 1–1.5 cm stacking gel height will suffice for small-volume samples.
9. Salt-free samples are routinely prepared by exhaustive dialysis against 2.5% (v/v) acetic acid in 3500 mol-wt cutoff dialysis membranes, followed by freezing at –70°C and lyophilization in conical polypropylene tubes (1.5, 15, or 50 mL, filled up to half of nominal capacity) with caps punctured by 21-gage needle stabs. This method gives essentially quantitative recovery of histones, even if very dilute.

   Alternatively, basic proteins can be precipitated with trichloroacetic acid or acetone, acidified by hydrochloric acid to 0.02*N*, provided that excess salt and acidity are removed by multiple washes with acetone.

   Solutions of basic proteins can be used directly, provided that the solution is free of salts that interfere with gel electrophoresis (*see* Note 11) and that the concentration of protein is high enough to compensate for the 1.8-fold dilution that occurs during sample preparation. Add 480 mg urea, 0.05 mL phenolphthalein, 0.05 mL concentrated ammonium hydroxide and 0.05 mL 1*M* DTT (freshly prepared)/mL sample. If not pink, add more ammonium hydroxide. Leave for 5 min at room temperature. Add 0.05 mL glacial acetic acid. Measure an aliquot for one gel lane and continue at Section 3., step 22.
10. The amount of protein that can be analyzed in one gel lane depends highly on the complexity of the protein composition. As a guideline, 5–50 μg of total calf thymus histones with five major proteins (modified to varying extent) represent the range between very lightly to heavily Coomassie-stained individual protein bands in 1-mm-thick, 30-cm-long gels using 5-mm-wide comb teeth.
11. Methylene blue is a single blue dye that remains in the gel discontinuity stack of 15% acrylamide separating gels (*see* Note 4). Methyl green is an alternate dye marker that contains methylene blue together with yellow and green dye components that remain together in discontinuous mode but that, in continuous gel electrophoresis, show progressively slower gel mobilities (*see* Note 15).
12. As an alternative to Section 3., steps 28–31, electrophoresis buffer is added to the upper buffer reservoir, and all sample wells are filled. Samples are layered under the buffer when dispensed by a Hamilton microsyringe near the bottom of each well.

13. Long gels used in overnight electrophoresis are made on the day that electrophoresis is started. The time for preparation of short gels may prevent electrophoresis on the same day. The nature of the stacking system of the gel (*see* Note 4) allows one to prepare a gel on day 1, to store it at room temperature overnight, and to initiate electrophoresis on the morning of d 2. Storage should not be refrigerated to prevent precipitation of urea. Storage should not be under electrophoresis buffer, as glycine would start to diffuse into the gel and destroy the stacking capability of the system. We routinely store short gels overnight once polymerization is complete (*see* Section 3., step 15) and before the comb is removed. Saran Wrap is used to prevent exposed gel surfaces from drying.
14. Be warned that small and strongly basic proteins like histones are not fixed effectively inside 15% acrylamide gels in methanol-acetic acid without Coomassie. Comparison of identical gels, one fixed and stained as described and the other placed first in destain solution alone for several h, followed by regular staining by Coomassie, reveals that 90% or more of core histones are lost from the gel. We speculate that the Coomassie dye helps to retain histones within the gel matrix. This is consistent with the observation that gradual loss of Coomassie intensity of histone bands is observed on exhaustive removal of soluble dye.
15. Gel pre-electrophoresis until the methylene blue dye, and thus all ammonium ions (*see* Note 4), have exited the separating gel converts this gel system into a continuous one. This option can be used to separate small proteins and peptides that do not destack at the boundary with the separating gel. Although this option is available, one should consider alternatives, such as increasing the acrylamide concentration of the separating gel. West and coworkers have developed a system with 40–50% polyacrylamide gels that is similar to the one described here and that has been optimized for the separation of small basic peptides *(4)*.

References

1. Panyim, S. and Chalkley, R. (1969) High resolution acrylamide gel electrophoresis of histones. *Arch. Biochem. Biophys.* **130,** 337–346.
2. Ruiz-Carrillo, A., Wangh, L. J., and Allfrey, V. G. (1975) Processing of newly synthesized histone molecules. Nascent histone H4 chains are reversibly phosphorylated and acetylated. *Science* **190,** 117–128.
3. Bonner, W. M., West, M. H. P., and Stedman, J. D. (1980) Two-dimensional gel analysis of histones in acid extracts of nuclei, cells, and tissues. *Eur. J. Biochem.* **109,** 17–23.
4. West, M. H. P., Wu, R. S., and Bonner, W. M. (1984) Polyacrylamide gel electrophoresis of small peptides. *Electrophoresis* **5,** 133–138.

15

Acid-Urea-Triton Polyacrylamide Gels for Histones

Jakob H. Waterborg

1. Introduction

Acid-urea polyacrylamide gels are capable of separating basic histone proteins provided they differ sufficiently in size and/or effective charge (*see* Chapter 14). Separation between similarly sized and charged H2A, H2B, and H3 forms of most organisms can typically not be achieved. Zweidler discovered that core histones but not linker histones or any other known protein (*see* Note 1) bind the nonionic detergent Triton *(1)*. This limits the usability of detergent addition to the separation of core histones only, unless they must be separated from other basic proteins in an acid-urea gel electrophoresis environment. The binding of Triton to a core histone increases the effective mass of the protein within the gel without affecting its charge, and thus reduces its mobility during electrophoresis. Separation between most or all core histone proteins of diverse species can virtually always be obtained by adjusting concentrations of Triton and of urea, which appears to act as a counter-acting, dissociating agent *(2)*. Experimentally, an optimal balance can be determined by gradient gel electrophoresis with a gradient of urea *(3)* or Triton *(4)*. The Triton gradient protocol in the discontinuous gel system, developed by Bonner and coworkers *(5)*, is described in Section 3. It has a distinct advantage over the urea gradient protocol. Generally, it can identify a core histone protein band as belonging to histone H4, H2B, H3, or H2A. In this order, apparent affinities for Triton X-100 increase sharply *(4,6,7)*. An example of such a separation of a crude mixture of histones with nonhistone proteins from a tobacco callus culture is shown in Fig. 1A. In addition, a detailed working protocol for a long acid-urea-Triton (AUT) gel at 9 m*M* Triton and 8*M* urea is provided. It describes the protocol used extensively in my laboratory for the analysis of core histones, especially of histone H3, dicots *(6)*, monocots *(7)*, and the green alga *Chlamydomonas (8)*. Figure 1B shows an example of the differentially acetylated histone H3 variant proteins of tobacco, purified by reversed-phase HPLC *(6)*. The protocol description directly parallels the acid-urea gel protocol described in Chapter 14, which also provides details for the use of different gel dimensions.

2. Materials

1. Vertical gel apparatus for long (30 cm) slab gels. A gel electrophoresis apparatus that allows gel polymerization between the glass plates with spacers without being assembled

From: The Protein Protocols Handbook
Edited by: J. M. Walker Humana Press Inc., Totowa, NJ

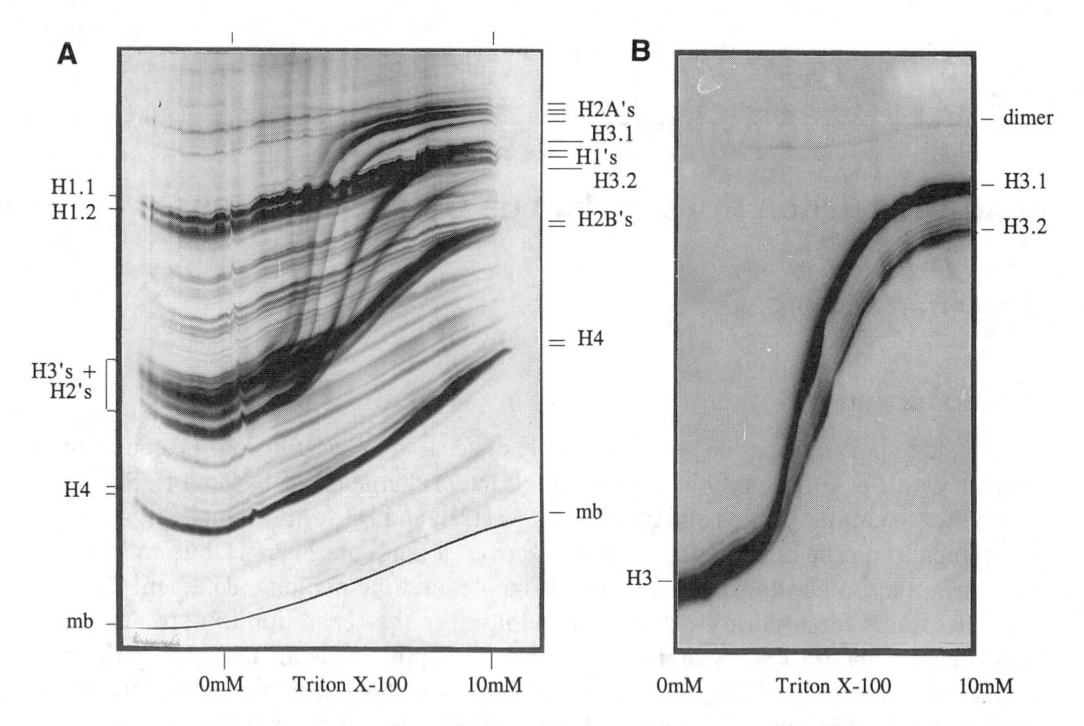

Fig. 1. Acid-urea gradient gel electrophoresis of tobacco histones. **(A)** A crude preparation of basic proteins, extracted from callus cultures of tobacco *(6)*, was electrophoresed on the gradient AUT gel and stained with Coomassie, as described. Marks at top and bottom indicate the joints between the gel compartments, from left to right, 0 m*M* Triton, 0–10 m*M* Triton gradient, and 10 m*M* Triton. The buffer front is marked by the methylene blue dye (mb). The identified histone bands are marked: two variants for histone H1, histone H4 with detectable monoacetylation, two histone H2B variants, two histone H3 variants (H3.1 and the more highly acetylated histone H3.2), and, at least four histone H2A variant forms. Note that the Triton affinity of the core histones increases in the typical order of H4, H2B, H3, and H2A. **(B)** Tobacco histone H3, purified by reversed-phase HPLC as a mixture of low acetylated histone H3.1 and highly acetylated histone H3.2, on a gradient AUT gel. The 0 and 10 m*M* Triton concentrations coincide with the edge of the figure. The presence of a small amount of histone H3 dimer is indicated.

in the apparatus is required. This allows the even and complete photo-polymerization of the acrylamide gel compartments in both orientations used. In this type of apparatus, the glass-gel sandwich is typically clamped to the lower buffer reservoir, which acts as a stand, after which the upper buffer reservoir is clamped to the top of the gel assembly.

Details of the procedure are described for a fairly standard and flexible gel apparatus that uses two rectangular glass plates (4-mm-thick standard plate glass with sanded edges), 21 cm wide and 32.5 and 35.5 cm long, respectively. The plexiglass bottom buffer reservoir with platinum electrode is 22.5 cm wide with three sides 5 cm high and one of the long sides 12.5 cm high. To this side the glass plates are clamped. The upper buffer reservoir with platinum electrode and with a similar buffer capacity is 18 cm wide with one long side enlarged to measure 21 cm wide by 10 cm high. It contains a cut-out of 18 cm wide and 3.5 cm high, which allows access of the upper reservoir buffer to the top of the gel. The 21-cm-wide plexiglass plate is masked with 5-mm-thick closed-cell neoprene tape (weather strip), and provides a clamping ridge for attachment to the top of the glass-gel sandwich.

2. Spacers and combs are cut from 1 mm Teflon sheeting. Required are two side spacers (1.5 × 35 cm), one bottom spacer (0.5 × 24 cm), and, for the gradient gel, one temporary spacer (1.5 × 21 cm). Added to the top of the side spacers is 3-cm adhesive, closed-cell neoprene tape (weather strip, 14 mm wide and 5 mm thick). A 15-cm-wide comb with 20 teeth of 5 × 30 mm, cut from a rectangle of 15 × 5 cm Teflon, is used with the teeth pointing down for the regular gel and with the teeth pointing up as a block comb for the gradient gel.
3. Vaseline pure petroleum jelly.
4. Acrylamide stock solution: 60% (w/v) acrylamide, highest quality available, in water (*see* Note 2). The acrylamide is dissolved by stirring. Application of heat should be avoided, if possible, to prevent generating acrylic acid. The solution can be kept for at least 3 mo on the laboratory shelf at room temperature. Storage at 4°C can exceed 2 yr without detectable effects.
5. N,N'-methylenebis(acrylamide) stock solution: 2.5% (w/v) in water (*see* Note 2).
6. Glacial acetic acid (HAc):17.5*M*.
7. Concentrated ammonium hydroxide: NH_4OH, 28–30%, approx 15*M*.
8. Triton X-100 stock solution: 25% (w/v) in water (0.4*M*) is used, as it is much easier to dispense accurately than 100%.
9. N,N,N',N'-tetramethylenediamine (TEMED), stored at 4°C.
10. Riboflavin-5'-phosphate solutions: 0.006% (R5P) and 0.06% (R5P-hi) (w/v) in water. The lower concentration solution is used for the regular gel and for all stacking gels. It is stable for more than 6 mo if kept dark and stored at 4°C. Riboflavin-5'-phosphate readily dissolves at the higher concentration, which is used for the gradient gel formulation, but it cannot be stored for more than 1 d at room temperature in the dark, as precipitates form readily upon storage at 4°C.
11. Glycerol.
12. Urea, ultrapure quality.
13. Side-arm suction flasks with stoppers, magnetic stirrer, stirrer bar, and water-aspirator vacuum. Measuring cylinders with silicon-rubber stoppers. Pipets and pipetting bulbs or mechanical pipetting aids. Plastic syringes, 1 and 5 mL, with 20-gage needles.
14. Fluorescent light box with diffuser for even light output and with the possibility to stand vertically. Light intensity should equal or exceed 5 kLux at a distance of 5–10 cm. A high-quality X-ray viewing light box with three 40-W bulbs typically will meet this specification.
15. Aluminum foil.
16. Gradient mixer to prepare a linear concentration gradient with a volume of 30 mL. We use with success the 50 mL Jule gradient maker (Research Products International Corp., Mt. Prospect, IL) with two 25-mL reservoirs.
17. Electrophoresis power supply with constant Voltage mode at 300–500 V with up to 50 mA current, preferably with a constant power mode option.
18. Urea stock solution: 8*M* urea in water. An aliquot of 40 mL with 1 g of mixed-bed resin (Bio-Rad [Hercules, CA] AG 501-X8) can be used repeatedly over a period of months if refrozen and stored between use at –20°C (*see* Note 3).
19. Phenolphthalein indicator solution: 1% (w/v) in 95% ethanol, stored indefinitely at room temperature in a closed tube.

20. Dithiothreitol (DTT; Cleland's reagent) is stored at 4°C and is weighed freshly for each use.
21. Methylene blue running front indicator dye solution: 2% (w/v) in sample buffer (*see* Section 3., steps 10–12).
22. Reference histones: total calf thymus histones (Worthington, Freehold, NJ), stored dry at 4°C indefinitely or at –20°C in 100-µL aliquots of 5 mg/mL in sample buffer for up to 1 yr (*see* Section 3., steps 10–12).
23. Electrophoresis buffer: 1*M* acetic acid, 0.1*M* glycine (*see* Note 4). This solution can be made in bulk and stored indefinitely at room temperature.
24. Destaining solution: 20% (v/v) methanol, 7% (v/v) acetic acid in water.
25. Staining solution: dissolve freshly 0.5 g Coomassie brilliant blue R-250 in 500 mL destaining solution. If the dye incompletely dissolves, the solution should be filtered through Whatman #1 paper to prevent staining artifacts.
26. Glass tray for gel staining and destaining.
27. Rotary or alternating table top shaker.
28. Destairnng aid: polyurethane foam.

3. Method

1. For the AUT regular gel, follow steps a. through i. and continue at step 3.
   a. Assemble a sandwich of two clean glass plates with 2 side spacers and a bottom spacer, lightly greased with Vaseline to obtain a good seal, clamped along all sides with 2-in. binder clamps. The triangular shape of these clamps facilitates the vertical, free-standing position of the gel assembly a few centimeters in front of the vertical light box (Fig. 2A).
   b. Separating gel solution: pipet into a 100-mL measuring cylinder 17.5 mL acrylamide stock solution, 2.8 mL bisacrylamide stock solution, 4.2 mL glacial acetic acid, and 0.23 mL concentrated ammonium hydroxide solutions (*see* Note 5).
   c. Add 33.6 g urea and add distilled water to a total volume of 63.5 mL.
   d. Stopper the measuring cylinder and place on a rotary mixer until all urea has dissolved. Add water to 63.5 mL, if necessary.
   e. Transfer this solution to a 200-mL sidearm flask with a magnetic stir bar on a magnetic stirrer. While stirring vigorously, stopper the flask and apply water-aspirator vacuum. Initially, a cloud of small bubbles of dissolved gas arises, which clears after just a few seconds. Terminate vacuum immediately to prevent excessive loss of ammonia.
   f. Add 1.575 mL Triton, 0.35 mL TEMED and 4.67 mL R5P (*see* Note 6), mix and pipet immediately between the glass plates to a marking line 5 cm below the top of the shorter plate (*see* Note 7).
   g. Carefully apply 1 mL distilled water (from a 1-mL syringe with needle along one of the glass plates) to the top of the separating gel solution to obtain a flat separation surface.
   h. Switch the light box on and place a reflective layer of aluminum foil behind the gel to increase light intensity and homogeneity. Gel polymerization becomes detectable within 2 min and is complete in 15–30 min.
   i. Switch the light box off, completely drain the water from between the plates, and insert the comb 2.5 cm between the glass plates. The tops of the teeth should always remain above the top of the short glass plate.
2. For the Triton gradient gel, follow steps a–q and continue at step 3.
   a. Assemble a sandwich of two clean glass plates with one side spacer, a bottom spacer, and a temporary spacer, clamped with large binder clamps (Fig. 2B). The spacers are

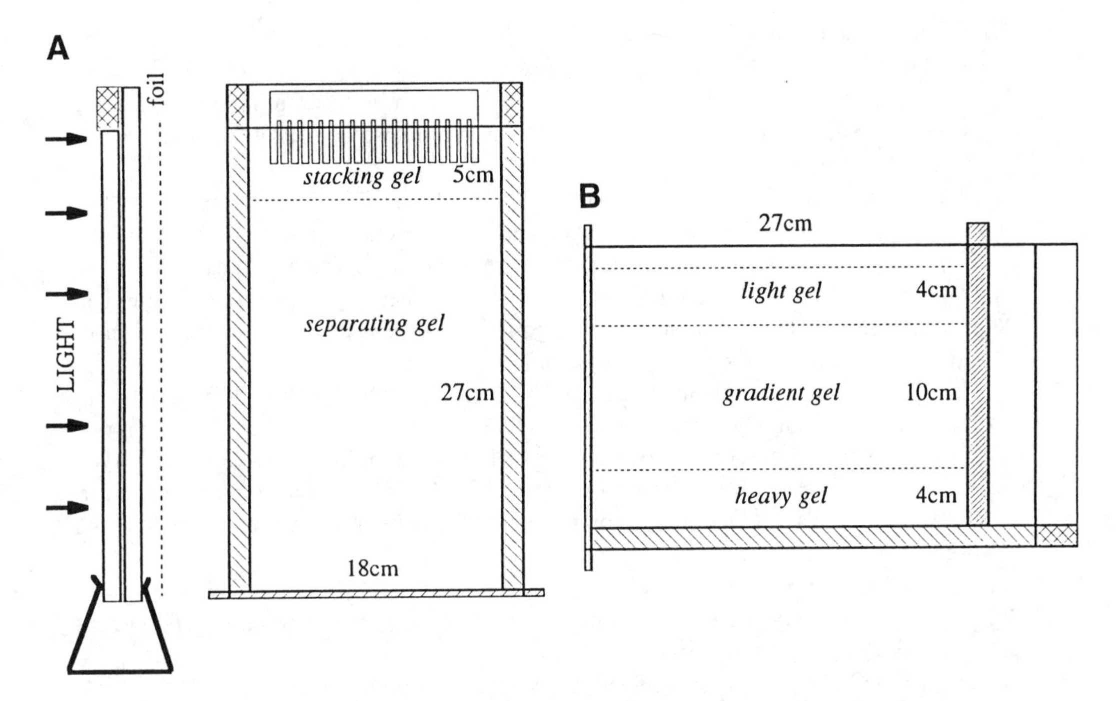

Fig. 2. (A) Side and front view of a long AUT gel assembly for 20 samples. The Teflon spacers are marked by diagonal lines and the neoprene blocks by crosshatching. One clamp is shown in the side view to demonstrate the independent vertical stand of this assembly and to represent the clamps all around the assembly at all spacer locations. **(B)** Front view of the AUT gradient assembly as used when the separation gel partitions are formed and polymerized.

lightly greased with Vaseline to obtain a good seal, but the amount of grease on the temporary spacer should be as low as possible. No grease should be present on the side of this spacer facing the buffer compartment to assure a flawless destacking surface.

Place magic marker guidance lines at 4, 14, and 18 cm above the long side spacer (Fig. 2B). Place the assembly horizontally a few centimeters in front of the vertical light box and add a reflective layer of aluminum behind the gel.

b. Heavy gel separating gel solution: pipet into a 50-mL graduated, capped polypropylene tube, 4.0 mL glycerol, 10 mL acrylamide stock solution, 1.6 mL bisacrylamide stock solution, 2.4 mL glacial acetic acid, and 0.13 mL concentrated ammonium hydroxide. Add 19.2 g urea. Add distilled water to a total volume of 38.5 mL (*see* Note 5).
c. Light gel separating gel solution: pipet into a 50 mL graduated, capped polypropylene tube, 10 mL acrylamide stock solution, 1.6 mL bisacrylamide stock solution, 2.4 mL glacial acetic acid, and 0.13 mL concentrated ammonium hydroxide. Add 19.2 g urea. Add distilled water to a total volume of 39.5 mL.
d. Place the closed tubes on a rotary mixer until all urea has dissolved. Add water to the required final volume, if necessary.
e. Transfer each solution to a 100-mL sidearm flask with magnetic stir bar on a magnetic stirrer. While stirring vigorously, stopper the flask and apply water-aspirator vacuum. Initially, a cloud of small bubbles of dissolved gas arises, which clears after just a few seconds. Terminate vacuum immediately to prevent excessive loss of ammonia.

f. Return the gel solutions after degassing to the capped tubes.
g. Add to the heavy gel solution 1.0 mL Triton.
h. Wrap each tube into aluminum foil to protect the gel solution from light (*see* Note 6).
i. Prepare freshly concentrated (0.06%) riboflavin-5'-phosphate (R5P-hi) solution (*see* Section 2., part 10).
j. Add to each gel solution 0.20 mL TEMED and 0.27 mL R5P-hi, and mix.
k. Under conditions of darkness (or very reduced light levels), pipet approx 12 mL heavy gel solution between the plates, up to the 4 cm heavy gel surface mark (Fig. 2B).
l. Turn the light box on for 2 min only. This allows the heavy gel to initiate polymerization and gelling, but retains sufficient unpolymerized acrylamide to fuse this gel partition completely into the gel layered on top.
m. In darkness, set up the gradient maker with 15 mL heavy gel and 15 mL light gel solutions and create slowly, over a period of at least several minutes, a linear gradient that flows slowly and carefully between the glass plates. Empirically, we have observed that a 10% increase over the calculated volume of the 27 × 10 × 0.1 cm gradient partition will create a gradient that is complete near the marker line between gradient and light partition (Fig. 2B).
n. Turn the light box on for 2 min only.
o. In darkness, slowly add light gel solution on top of the partially polymerized gradient gel until the upper marker line.
p. Switch the light box on for complete gel polymerization in 15–30 min.
q. Remove the temporary spacer from the assembly. Insert a slightly greased spacer along the polymerized light gel. Reclamp the assembly. Insert the 20-well comb in the middle and upside down as a block comb for 2.5 cm below the edge of the shorter glass plate. Reposition the assembly vertically between light box and aluminum foil (Fig. 2A).

3. Stacking gel solution, made in parallel to steps 1b–d or steps 2b–d: pipet into a 25-mL measuring cylinder 1.34 mL acrylamide stock solution, 1.28 mL bisacrylamide stock solution, 1.14 mL glacial acetic acid, and 0.07 mL concentrated ammonium hydroxide (*see* Note 5).
4. Add 9.6 g urea and add distilled water to a total volume of 18.6 mL.
5. Stopper the measuring cylinder and place on a rotary mixer until all urea has dissolved. Add water to 18.6 mL, if necessary.
6. Once the separating gel has polymerized, transfer the stacking gel solution to a 50-mL sidearm flask with stir bar and degas as in step 1e.
7. Add 0.1 mL TEMED and 1.3 mL R5P, mix, and pipet between the plates between the comb teeth.
8. Displace air bubbles, especially in the gradient assembly where residual Vaseline may interfere with gel solution flow. In this case, all (mini)bubbles should be carefully removed from the separation gel surface that will act as the destacking boundary.
9. Switch the light box on and allow complete gel polymerization in 30–60 min.
10. Sample buffer is freshly prepared when the separation gel is polymerizing. The preferred protein sample is a salt-free lyophilizate (*see* Note 8). For regular gels, determine the approximate volume of sample buffer required, depending on the number of samples. For a gradient gel, 0.5–1.5 mL sample buffer appears optimal.
11. Weigh DTT into a sample buffer preparation tube for a final concentration of 1*M*, i.e., 7.7 mg/mL.
12. Per 7.7 mg DTT, add 0.9 mL 8*M* urea stock solution, 0.05 mL phenolphthalein, and 0.05 mL concentrated ammonium hydroxide to the tube to obtain the intensely pink sample buffer.

13. Add 0.05 mL sample buffer per sample tube with lyophilized protein to be analyzed in one gel lane (*see* Note 9). Add 1 mL sample buffer to sample for one gradient gel.

    To assure full reduction of all proteins by DTT, the pH must be above 8.0. If the pink phenolphthalein color disappears owing to residual acid in the sample, a few microliters of concentrated ammonium hydroxide should be added to reach an alkaline pH.
14. Limit the time for sample solubilization and reduction to 5 min at room temperature to minimize the possibility of protein modification at alkaline pH by reactive urea side reactions, e.g., by modification of cysteines by cyanate.
15. Acidify the sample by addition of 1/20 vol of glacial acetic acid.
16. Add methylene blue running front dye: 2 µL per gel lane, or 50 µL for a gradient gel.
17. For a regular AUT gel, prepare reference histone samples: to 2 and 6 µL reference histone solution with 10 and 30 µg total calf thymus histones, one adds 40 µL sample buffer (step 12), 2.5 µL glacial acetic acid, and 2 µL methylene blue.
18. When stacking gel polymerization is complete, remove the comb. Rinse each well momentarily with water and drain to remove residual unpolymerized gel solution. The high urea concentration of such residual liquid interferes with the tight application of samples.
19. Remove the bottom spacer from the gel assembly together with any residual Vaseline on the lower surface of the gel.
20. Clamp the gel assembly into the electrophoresis apparatus and fill the lower buffer reservoir with electrophoresis buffer.
21. Use a 5-mL syringe with a bent syringe needle to displace any air bubbles from the bottom of the gel.
22. For regular gel, follow steps a–c and continue at step 24.
    a. Individual samples can be applied to sample wells by any micropipetter with plastic disposable tip or by Hamilton microsyringe (rinsed with water between samples). Pipet each sample solution against the long glass plate and let it run to the bottom of the well (*see* Note 10). For the combination of comb and gel dimensions listed, 50 µL sample will reach a height of 1 cm.
    b. Apply reference samples in the outer lanes, which frequently show a slight loss of resolution owing to edge effects.
    c. Gently overlayer the samples with electrophoresis buffer, dispensed from a 5-mL syringe fitted with a 21-gage needle until all wells are full.
23. For gradient gel, follow steps a–c and continue at step 24.
    a. Fill the block well with electrophoresis buffer.
    b. Use a level to confirm that the bottom of the preparative well is exactly horizontal.
    c. Distribute the total sample evenly across the width of the well using a 250-µL Hamilton microsyringe. Limited mixing of sample and electrophoresis buffer will facilitate even loading and is easily dealt with by the strong stacking capability of the gel system (*see* Note 11).
24. Fill the upper buffer reservoir with electrophoresis buffer.
25. Attach the electrical leads between power supply and electrophoresis system: the + lead to the upper and the – lead to the lower reservoir. Note that this is opposite to the SDS gel electrophoresis configuration. Remember, basic proteins like histones are positively charged and will move toward the cathode (– electrode).
26. Long (30 cm) gels require 15–20 h of electrophoresis at 300 V in constant Voltage mode. They are most easily run overnight. For maximum resolution and stacking capacity, the initial current through a 1-mm-thick and 18-cm-wide gel should not exceed 25 mA. Gel electrophoresis is completed in the shortest amount of time in constant power mode with limits of 300 V, 25 mA, and 5 W. The current will drop toward completion of electrophoresis to 6 mA at 300 V.

Gradient gel electrophoresis typically takes a few more hours. At 300 V constant Voltage, the electrophoretic current is reduced, in particular at the heavy side of the gel owing to the reduced water concentration caused by the inclusion of glycerol. The ion front with methylene blue dye reflects this in a distinct curvature (Fig. IA).

27. Electrophoresis is complete just before the methylene blue dye exits the gel. Obviously, electrophoresis may be terminated earlier if lesser band resolution is acceptable, or may be prolonged to enhance separation of histones with low gel mobilities, e.g., histone H3 variants.
28. Open the glass-gel sandwich and place the separating gel into staining solution, which is gently agitated continuously overnight on a shaker (*see* Note 12).
29. Decant the staining solution. The gel can be given a very short rinse in water to remove all residual staining solution.
30. Place the gel in ample destaining solution. Diffusion of unbound Coomassie dye from the gel is facilitated by the addition of polyurethane foam as an absorbent for free Coomassie dye. To avoid over-destaining and potential loss of protein from the gel, polyurethane foam in limited amounts is added to only the first and second destaining solutions. Final destaining is done in the absence of any foam.
31. Record the protein pattern of the gel on film (Fig. 1) and, possibly, by densitometry. Subsequently the gel may be discarded, dried, or prepared for autoradiography or fluorography (*see* Chapter 36).

4. Notes

1. It is unknown why only core histones show affinity for detergents such as Triton X-100 in an acetic acid-urea environment. To date, no other protein has been shown to behave in a similar manner. I have interpreted this observation based on the known and unique structure of core histones, as recently determined by X-ray crystallography *(9,10)*. Core histones contain a unique protein folding motif, the histone clasp structure. This helix-turn-helix-turn-helix fold spans 65 residues and is the basis for the binary matches between histones H3 and H4 and between H2A and H2B in the nucleosome octamer *(9,11)*. In our studies of histone H3, variant proteins from HeLa, *Physarum (12)*, *Chlamydomonas (8)*, and several plant species *(6,7)*, we have observed characteristic differences in Triton affinity, depending on sequence differences at one or two residues whose side chains are mapped to the inside of the histone fold. In this light, we interpret the characteristic affinity of a core histone protein for Triton as a measure of binding of Triton within the histone clasp (Waterborg, unpublished results).
2. All (bis)acrylamide solutions are potent neurotoxins and should be dispensed by mechanical pipetting devices.
3. Storage of urea solutions at –20°C minimizes creation of ionic contaminants like cyanate. The mixed-bed resin assures that any ions formed are removed. Care should be taken to exclude resin beads from solution taken, e.g., by filtration through Whatman #1 paper.
4. The stacking ions between which the positively charged proteins and peptides are compressed within the stacking gel during the initial phase of gel electrophoresis are NH_4^+ within the gel compartment and glycine$^+$ in the electrophoresis buffer. Chloride ions interfere with the discontinuous stacking system. This requires that protein samples should preferably be free of chloride salts, and that glycine base rather than glycine salt should be used in the electrophoresis buffer.
5. The separating and stacking gels contain 15 vs 4% acrylamide and 0.1 vs 0.16% N,N'-methylene(bisacrylamide), respectively, in 1*M* acetic acid, 0.5% TEMED, 50 m*M* NH_4OH,

8*M* urea, and 0.0004% riboflavin-5'-phosphate. The concentration of Triton X-100 in the separating gel of the regular system is 9 m*M*, optimal for the separation desired in our research for histone H3 variant forms of plants and algae.

In the gradient system, glycerol and Triton concentrations change in parallel: 10% (v/v) glycerol and 10 m*M* Triton in the heavy gel (Fig. 2B) and a gradient between 10 and 0% (v/v) glycerol and between 10 and 0 m*M* in the gradient from heavy to light (Fig. 2B).

6. Acrylamide is photo-polymerized with riboflavin or riboflavin-5'-phosphate as initiator because the ions generated by ammoniumpersulfate-initiated gel polymerization, as used for SDS polyacrylamide gels, interfere with stacking (*see* Note 4).
7. The height of stacking gel below the comb determines the volume of samples that can be applied and fully stacked before destacking at the surface of the separating gel occurs. In our experience, a 2.5-cm stacking gel height suffices for samples that almost completely fill equally long sample wells. In general, the single blue line of completely stacked proteins and methylene blue dye should be established 1 cm above the separating gel surface. Thus, a 1–1.5-cm stacking gel height will suffice for small volume samples.
8. Salt-free samples are routinely prepared by exhaustive dialysis against 2.5% (v/v) acetic acid in 3500 mol wt cut-off dialysis membranes, followed by freezing at –70°C and lyophilization in conical polypropylene tubes (1.5, 15, or 50 mL, filled up to half of nominal capacity) with caps punctured by 21-gage needle stabs. This method gives essentially quantitative recovery of histones, even if very dilute. (For alternative methods, *see* Chapter 1.)
9. The amount of protein that can be analyzed in one gel lane depends highly on the complexity of the protein composition. As a guideline, 5–50 µg of total calf thymus histones with five major proteins (modified to varying extent) represent the range between very lightly to heavily Coomassie stained individual protein bands in 1-mm-thick, 30-cm-long gels using 5-mm-wide comb teeth.

   The optimal amount of protein for a gradient gel also depends on the number of protein species that must be analyzed. In general, a gradient using the block comb used, equivalent to the width of 30 gel lanes, should be loaded with 30 times the sample for one lane.
10. As an alternative to Section 3., step 22a–c, electrophoresis buffer is added to the upper buffer reservoir and all sample wells are filled. Samples are layered under the buffer when dispensed by a Hamilton microsyringe near the bottom of each well.
11. The preparative well of a gradient gel should not be loaded with sample prior to the addition of electrophoresis buffer. Uneven distribution of sample cannot be avoided when buffer is added.
12. For unknown reasons, the polyacrylamide gel below the buffer front tends to stick tightly to the glass in an almost crystalline fashion. This may cause tearing of a gradient gel at the heavy gel side below the buffer front. Since attachment to one of the gel plates is typically much stronger, one can release the gel without problems by immersing the glass plate, with gel attached, in the staining solution.

References

1. Zweidler, A. (1978) Resolution of histones by polyacrylamide gel electrophoresis in presence of nonionic detergents. *Methods Cell Biol.* **17,** 223–233.
2. Urban, M. K., Franklin, S. G., and Zweidler, A. (1979) Isolation and characterization of the histone variants in chicken erythrocytes. *Biochemistry* **18,** 3952–3960.
3. Schwager, S. L. U., Brandt, W. F., and Von Holt, C. (1983) The isolation of isohistones by preparative gel electrophoresis from embryos of the sea urchin *Parechinus angulosus. Biochim. Biophys. Acta* **741,** 315–321.

4. Waterborg, J. H., Harrington, R. E., and Winicov, I. (1987) Histone variants and acetylated species from the alfalfa plant *Medicago sativa. Arch. Biochem. Biophys.* **256,** 167–178.
5. Bonner, W. M., West, M. H. P., and Stedman, J. D. (1980) Two-dimensional gel analysis of histones in acid extracts of nuclei, cells, and tissues. *Eur. J. Biochem.* **109,** 17–23.
6. Waterborg, J. H. (1992) Existence of two histone H3 variants in dicotyledonous plants and correlation between their acetylation and plant genome size. *Plant Mol. Biol.* **18,** 181–187.
7. Waterborg, J. H. (1991) Multiplicity of histone H3 variants in wheat, barley, rice and maize. *Plant Physiol.* **96,** 453–458.
8. Waterborg, J. H., Robertson, A. J., Tatar, D. L., Borza, C. M., and Davie, J. R. (1995) Histones of *Chlamydomonas reinhardtii:* Synthesis, acetylation and methylation. *Plant Physiol.* **109,** 393–407.
9. Arents, G., Burlingame, R. W., Wang, B. C., Love, W. E., and Moudrianakis, E. N. (1991) The nucleosomal core histone octamer at 3.1 Å resolution. A tripartite protein assembly and a left-handed superhelix. *Proc. Natl. Acad. Sci. USA* **88,** 10,138–10,148.
10. Ramakrishnan, V. (1994) Histone structure. *Curr. Opinion Struct. Biol.* **4,** 44–50.
11. Arents, G. and Moudrianakis, E. N. (1993) Topography of the histone octamer surface: repeating structural motifs utilized in the docking of nucleosomal DNA. *Proc. Natl. Acad. Sci. USA* **90,** 10,489–10,493.
12. Waterborg, J. H. and Matthews, H. R. (1983) Patterns of histone acetylation in the cell cycle of *Physarum polycephalum. Biochemistry* **22,** 1489–1496.

16

SDS-Polyacrylamide Gel Electrophoresis of Peptides

Ralph C. Judd

1. Introduction

Sodium dodecyl sulfate-polyacrylamide gel electrophoresis (SDS-PAGE) has proven to be among the most useful tools yet developed in the area of molecular biology. The discontinuous buffer system, first described by Laemmli *(1),* has made it possible to separate, visualize, and compare readily the component parts of complex mixtures of molecules (e.g., tissues, cells). SDS-PAGE separation of proteins and peptides makes it possible to quantify the amount of a particular protein/peptide in a sample, obtain fairly reliable molecular mass information, and, by combining SDS-PAGE with immunoelectroblotting, evaluate the antigenicity of proteins and peptides. SDS-PAGE is both a powerful separation system and a reliable preparative purification technique *(2*; and *see* Chapter 11*)*.

Parameters influencing the resolution of proteins or peptides separated by SDS-PAGE include the ratio of acrylamide to crosslinker (bis-acrylamide), the percentage of acrylamide/crosslinker used to form the stacking and separation gels, the pH of (and the components in) the stacking and separation buffers, and the method of sample preparation. Systems employing glycine in the running buffers (e.g., Laemmli *[1],* Dreyfuss et al. *[3]*) can resolve proteins ranging in molecular mass from over 200,000 Daltons (200 kDa) down to about 3 kDa. Separation of proteins and peptides below 3 kDa necessitates slightly different procedures to obtain reliable molecular masses and to prevent band broadening. Further, the increased use of SDS-PAGE to purify proteins and peptides for N-terminal sequence analysis demands that glycine, which interferes significantly with automated sequence technology, be replaced with noninterfering buffer components.

This chapter describes a modification of the tricine gel system of Schagger and von Jagow *(4)* by which peptides as small as 500 Daltons can be separated. This makes it possible to use SDS-PAGE peptide mapping (*see* Chapter 74), epitope mapping *(5),* and protein and peptide separation for N-terminal sequence analyses *(6)* when extremely small peptide fragments are to be studied. Since all forms of SDS-PAGE are denaturing, they are unsuitable for separation of proteins or peptides to be used in functional analyses (e.g., enzymes, receptors).

From: The Protein Protocols Handbook
Edited by: J. M. Walker Humana Press Inc., Totowa, NJ

2. Materials

2.1. Equipment

1. SDS-PAGE gel apparatus.
2. Power pack.
3. Blotting apparatus.

2.2. Reagents

1. Separating/spacer gel acrylamide (1X crosslinker): 48 g acrylamide, 1.5 g *N,N'*-methylene-bis-acrylamide. Bring to 100 mL, and then filter through qualitative paper to remove cloudiness (*see* Note 1).
2. Separating gel acrylamide (2X crosslinker): 48 g acrylamide, 3 g *N,N'*-methylenebis-acrylamide. Bring to 100 mL, and then filter through qualitative paper to remove cloudiness (*see* Note 2).
3. Stacking gel acrylamide: 30 g acrylamide, 0.8 g *N,N'*-methylene-bis-acrylamide. Bring to 100 mL, and then filter through qualitative paper to remove cloudiness.
4. Separating/spacer gel buffer: 3*M* Trizma base, 0.3% sodium dodecyl sulfate (*see* Note 3). Bring to pH 8.9 with HCl.
5. Stacking gel buffer: 1*M* Tris-HCl, pH 6.8.
6. Cathode (top) running buffer (10X stock): 1*M* Trizma base, 1*M* tricine, 1% SDS (*see* Note 3). Dilute 1:10 immediately before use. Do not adjust pH; it will be about 8.25.
7. Anode (bottom) buffer (10X stock): 2*M* Trizma base. Bring to pH 8.9 with HCl. Dilute 1:10 immediately before use.
8. 0.2*M* tetrasodium EDTA.
9. 10% ammonium persulfate.
10. TEMED.
11. Glycerol.
12. Fixer/destainer: 25% isopropanol, 7% glacial acetic acid in dH_2O (v/v/v).
13. 1% Coomassie brilliant blue (CBB) in fixer/destainer (w/v).
14. Sample solubilization buffer: 2 mL 10% SDS (w/v) in dH_2O, 1.0 mL glycerol, 0.625 mL 1*M* Tris-HCl, pH 6.8, 6 mL dH_2O, bromphenol blue to color.
15. Dithiothreitol.
16. 2% Agarose.
17. Molecular-mass markers, e.g., low-mol-wt kit (Bio-Rad, Hercules, CA), or equivalent, and peptide molecular-mass markers (Pharmacia Inc., Piscataway, NJ), or equivalent.
18. PVDF (nylon) membranes.
19. Methanol.
20. Blotting transfer buffer: 20 m*M* phosphate buffer, pH 8.0: 94.7 mL 0.2*M* Na_2HPO_4 stock, 5.3 mL 0.2*M* NaH_2PO_4 stock in 900 mL H_2O.
21. Filter paper for blotting (Whatman No. 1), or equivalent.
22. Distilled water (dH_2O).

2.3. Gel Recipes

2.3.1. Separating Gel Recipe

Add reagents in order given (*see* Note 4): 6.7 mL water, 10 mL separating/spacer gel buffer, 10 mL separating/spacer gel acrylamide (1X or 2X crosslinker), 3.2 mL glycerol, 10 μL TEMED, 100 μL 10% ammonium persulfate.

2.3.2. Spacer Gel Recipe

Add reagents in order given (*see* Note 4): 6.9 mL water, 5.0 mL separating/spacer gel buffer, 3.0 mL separating/spacer gel acrylamide (1X crosslinker only), 5 µL TEMED, 50 µL 10% ammonium persulfate.

2.3.3. Stacking Gel Recipe

Add reagents in order given (*see* Note 4): 10.3 mL water, 1.9 mL stacking gel buffer, 2.5 mL stacking gel acrylamide, 150 µL EDTA, 7.5 µL TEMED, 150 µL 10% ammonium persulfate.

3. Methods

3.1. Sample Solubilization

1. Boil samples in sample solubilization buffer for 10–30 min. Solubilize sample at 1 mg/mL and run 1–2 µL/lane (1–2 µg/lane) (*see* Note 5). For sequence analysis, as much sample as is practical should be separated.

3.2. Gel Preparation/Electrophoresis

1. Assemble the gel apparatus (*see* Note 6). Make two marks on the front plate to identify top of separating gel and top of spacer gel (*see* Note 7). Assuming a well depth of 12 mm, the top of the separating gel should be 3.5 cm down from the top of the back plate, and the spacer gel should be 2 cm down from the top of the back plate, leaving a stacking gel of 8 mm (*see* Note 8).
2. Combine the reagents to make the separating gel, mix gently, and pipet the solution between the plates to lowest mark on the plate. Overlay the gel solution with 2 mL of dH_2O by gently running the dH_2O down the center of the inside of the front plate. Allow the gel to polymerize for about 20 min. When polymerized, the water–gel interface will be obvious.
3. Pour off the water, and dry between the plates with filter paper. Do not touch the surface of the separating gel with the paper. Combine the reagents to make the spacer gel, mix gently, and pipet the solution between the plates to second mark on the plate. Overlay the solution with 2 mL of dH_2O by gently running the dH_2O down the center of the inside of the front plate. Allow the gel to polymerize for about 20 min. When polymerized, the water–gel interface will be obvious.
4. Pour off the water, and dry between the plates with filter paper. Do not touch the surface of spacer gel with the paper. Combine the reagents to make the stacking gel and mix gently. Place the well-forming comb between the plates, leaving one end slightly higher than the other. Slowly add the stacking gel solution at the raised end (this allows air bubbles to be pushed up and out from under the comb teeth). When the solution reaches the top of the back plate, gently push the comb all the way down. Check to be sure that no air pockets are trapped beneath the comb. Allow the gel to polymerize for about 20 min.
5. When the stacking gel has polymerized, carefully remove the comb. Straighten any wells that might be crooked with a straightened metal paper clip. Remove the acrylamide at each edge to the depth of the wells. This helps prevent "smiling" of the samples at the edge of the gel. Seal the edges of the gel with 2% agarose.
6. Add freshly diluted cathode running buffer to the top chamber of the gel apparatus until it is 5–10 mm above the top of the gel. Squirt running buffer into each well with a Pasteur pipet to flush out any unpolymerized acrylamide. Check the lower chamber to ensure that no cathode running buffer is leaking from the top chamber, and then fill the bottom cham-

ber with anode buffer. Remove any air bubbles from the under edge of the gel with a bent-tip Pasteur pipet. The gel is now ready for sample loading.

7. After loading the samples and the molecular-mass markers, connect leads from the power pack to the gel apparatus (the negative lead goes on the top, and the positive lead goes on the bottom). Gels can be run on constant current, constant voltage, or constant power settings. When using the constant current setting, run the gel at 50 mA. The voltage will be between 50 and 100 V at the beginning, and will slowly increase during the run. For a constant voltage setting, begin the electrophoresis at 50 mA. As the run progresses, the amperage will decrease, so adjust the amperage to 50 mA several times during the run or the electrophoresis will be very slow. If running on constant power, set between 5 and 7 W. Voltage and current will vary to maintain the wattage setting. Each system varies, so empirical information should be used to modify the electrophoresis conditions so that electrophoresis is completed in about 4 h (*see* Note 9).
8. When the dye front reaches the bottom of the gel, turn off the power, disassemble the gel apparatus, and place the gel in 200–300 mL of fixer/destainer. Gently shake for 16 h (*see* Note 10). Pour off spent fixer/destainer, and add CBB. Gently shake for 30 min. Destain the gel in several changes of fixer/destainer until the background is almost clear. Then place the gel in dH_2O, and gently mix until the background is completely clear. The peptide bands will become a deep purple-blue. The gel can now be photographed or dried. To store the gel wet, soak the gel in 7% glacial acetic acid for 1 h, and seal in a plastic bag.

Figure 1 demonstrates the molecular mass range of separation of a 1X crosslinker tricine gel. Whole-cell (WC) lysates and 1X and 2X purified (*see* Chapter 73) 44 kDa proteins of *Neisseria gonorrhoeae*, Bio-Rad low-mol-wt markers (mw), and Pharmacia peptide markers (pep mw) were separated and stained with CBB. The top of the gel in this figure is at the spacer gel–separating gel interface. Proteins larger than about 100 kDa remained trapped at the spacer gel–separating gel interface, resulting in the bulging of the outside lanes. Smaller proteins all migrated into the gel, but many remained tightly bunched at the top of the separating gel. The effective separation range is below 40 kDa. Comparison of this figure with Fig. 1 in Chapter 74, which shows gonococcal whole cells and the two mol-wt marker preparations separated in a standard 15% Laemmli gel *(1)*, demonstrates the tremendous resolving power of low-mol-wt components by this tricine gel system.

3.3. Blotting of Peptides

Separated peptides can be electroblotted to PVDF membranes for sequencing or immunological analyses (*see* Note 11).

1. Before the electrophoresis is complete, prepare enough of the 20 m*M* sodium phosphate transfer buffer, pH 8.0 (*see* Note 12) to fill the blotting chamber (usually 2–4 L). Degas about 1 L of transfer buffer for at least 15 min before use. Cut two sheets of filter paper to fit blotting apparatus, and cut a piece of PVDF membrane a little larger than the gel. Place the PVDF membrane in 10 mL of methanol until it is wet (this takes only a few seconds), and then place the membrane in 100 mL of degassed transfer buffer.
2. Following electrophoresis, remove the gel from the gel apparatus, and place it on blotting filter paper that is submersed in the degassed transfer buffer. Immediately overlay the exposed side of the gel with the wetted PVDF membrane, being sure to remove all air pockets between the gel and the membrane. Overlay the PVDF membrane with another piece of blotting filter paper, and place the gel "sandwich" into the blotting chamber using the appropriate spacers and holders.

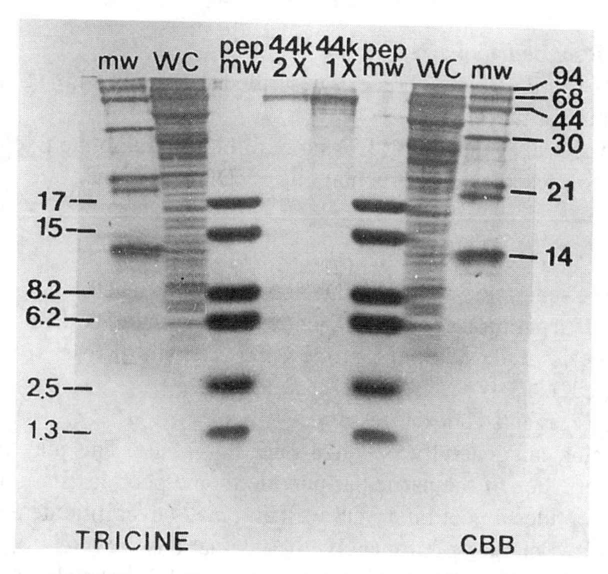

Fig. 1. WC lysates and 1X and 2X purified 44 kDa protein of *Neisseria gonorrhoeae* (*see* Chapter 73, Section 3.1.), Bio-Rad mw (1 µg of each protein), and Pharmacia pep mw (to which the 1.3-kDa protein kinase C substrate peptide [Sigma, St. Louis, MO] was added) (3 µg of each peptide), separated in a 1X crosslinker tricine gel, fixed, and stained with CBB. Molecular masses are given in thousands of daltons.

3. Connect the power pack electrodes to the blotting chamber (the positive electrode goes on the side of the gel having the PVDF membrane). Electrophorese for 16 h at 25 V, 0.8 A. Each system varies, so settings may be somewhat different than those described here.
4. Following electroblotting, disconnect the power, disassemble the blotting chamber, and remove the PVDF membrane from the gel (*see* Note 13). The PVDF membrane can be processed for immunological analyses or placed in CBB in fixer/destainer to stain the transferred peptides. Remove excess stain by shaking the membrane in several changes of fixer/destainer until background is white. Peptide bands can be excised, rinsed in dH_2O, dried, and subjected to N-terminal sequencing.

3.4. Modifications for Peptide Sequencing

Peptides to be used in N-terminal sequence analyses must be protected from oxidation, which can block the N-terminus. Several simple precautions can help prevent this common problem.

1. Prepare the separation and spacer gel the day before electrophoresing the peptides. After pouring, overlay the spacer gel with several milliliters of dH_2O, and allow the gel to stand overnight at room temperature (*see* Note 14).
2. On the next day, pour off the water, and dry between the plates with filter paper. Do not touch the gel with the filter paper. Prepare the stacking gel, but use half the amount of 10% ammonium persulfate (*see* Note 15). Pipet the stacking gel solution between the plates as described above, and allow the stacking gel to polymerize for at least 1 h. Add running buffers as described above, adding 1–2 mg of dithiothreitol to both the upper and lower chambers to scavenge any oxidizers from the buffers and gel.
3. Pre-electrophorese the gel for 15 min. then turn off the power, and load the samples and molecular-mass markers.

4. Run the gel as described above.
5. Blot the peptides to a PVDF membrane as described above, but add 1–2 mg of dithiothreitol to the blot transfer buffer.
6. Fix and stain as above, again adding 1–2 mg of dithiothreitol to the fixer/destainer, CBB, and dH_2O used to rinse the peptide-containing PVDF membrane.

4. Notes

1. Working range of separation about 40 kDa down to about 1 kDa.
2. Working range of separation about 20 kDa down to less than 500 Dalton.
3. Use electrophoresis grade SDS. If peptide bands remain diffuse, try SDS from BDH (Poole, Dorset, UK).
4. Degassing of gel reagents is **not** necessary.
5. Coomassie staining can generally visualize a band of 0.5 μg. This may vary considerably based on the properties of the particular peptide (some peptides stain poorly with Coomassie). Some peptides do not bind SDS well and may never migrate exactly right when compared to mol-wt markers. Fortunately, these situations are rare.
6. Protocols are designed for a standard 13 cm × 11 cm × 1.5 mm slab gel. Dimensions and reagent volumes can be proportionally adjusted to accommodate other gel dimensions.
7. Permanent marks with a diamond pencil can be made on the back of the back plate if the plate is dedicated to this gel system.
8. The depth of the spacer gel can be varied from 1 to 2 cm. Trial and error is the only way to determine the appropriate dimension for each system.
9. It is wise to feel the front plate several times during the electrophoresis to check for overheating. The plate will become pleasantly warm as the run progresses. If it becomes too warm, the plates might break, so turn down the power!
10. Standard-sized gels can be fixed in as little as 4 h with shaking.
11. It is best to blot peptides to PVDF membranes rather than nitrocellulose membranes, since small peptides tend to pass through nitrocellulose without binding. Moreover, peptides immobilized on PVDF membranes can be directly sequenced in automated instrumentation equipped with a "blot cartridge" *(6)*.
12. The pH of the transfer buffer can be varied from 5.7 to 8.0 if transfer is inefficient at pH 8.0 *(7)*.
13. Wear disposable gloves when handling membranes.
14. Do not refrigerate the gel. It will contract and pull away from the plates, resulting in leaks and poor resolution.
15. Do not pour the stacking gel the day before electrophoresis. It will shrink, allowing the samples to leak from the wells.

Acknowledgments

The author thanks Joan Strange for her assistance in developing this system, Pam Gannon for her assistance, and the Public Health Service, NIH, NIAID (grants RO1 AI21236) and UM Research Grant Program for their continued support.

References

1. Laemmli, U. K. (1970) Cleavage of structural proteins during the assembly of the head of bacteriophage T4. *Nature* **227,** 680–695.
2. Judd, R. C. (1988) Purification of outer membrane proteins of the Gram negative bacterium *Neisseria gonorrhoeae. Anal. Biochem.* **173,** 307–316.

3. Dreyfuss, G., Adam, S. A., and Choi, Y. D. (1984) Physical change in cytoplasmic messenger ribonucleoproteins in cells treated with inhibitors of mRNA transcription. *Mol. Cell. Biol.* **4,** 415–423.
4. Schagger, H. and von Jagow, G. (1987) Tricine-sodium dodecyl sulfate-polyacrylamide gel electrophoresis for the separation of proteins in the range from 1 to 100 kDa. *Anal. Biochem.* **166,** 368–397.
5. Judd, R. C. (1986) Evidence for N-terminal exposure of the PIA subclass of protein I of *Neisseria gonorrhoeae. Infect. Immunol.* **54,** 408–414.
6. Moos, M., Jr. and Nguyen, N. Y. (1988) Reproducible high-yield sequencing of proteins electrophoretically separated and transferred to an inert support. *J. Biol. Chem.* **263,** 6005–6008.
7. Stoll, V. S. and Blanchard, J. S. (1990) Buffers: principles and practice, in *Methods in Enzymology,* vol. 182, *A Guide to Protein Purification* (Deutscher, M. P., ed.), Academic, San Diego, CA, pp. 24–38.

17

Isoelectric Focusing of Proteins in Ultra-Thin Polyacrylamide Gels

John M. Walker

1. Introduction

Isoelectric focusing (IEF) is an electrophoretic method for the separation of proteins, according to their isoelectric points (pI), in a stabilized pH gradient. The method involves casting a layer of support media (usually a polyacrylamide gel but agarose can also be used) containing a mixture of carrier ampholytes (low-mol-wt synthetic polyamino-polycarboxylic acids). When using a polyacrylamide gel, a low percentage gel (~4%) is used since this has a large pore size, which thus allows proteins to move freely under the applied electrical field without hindrance. When an electric field is applied across such a gel, the carrier ampholytes arrange themselves in order of increasing pI from the anode to the cathode. Each carrier ampholyte maintains a local pH corresponding to its pI and thus a uniform pH gradient is created across the gel. If a protein sample is applied to the surface of the gel, where it will diffuse into the gel, it will also migrate under the influence of the electric field until it reaches the region of the gradient where the pH corresponds to its isoelectric point. At this pH, the protein will have no net charge and will therefore become stationary at this point. Should the protein diffuse slightly toward the anode from this point, it will gain a weak positive charge and migrate back towards the cathode, to its position of zero charge. Similarly diffusion toward the cathode results in a weak negative charge that will direct the protein back to the same position. The protein is therefore trapped or "focused" at the pH value where it has zero charge. Proteins are therefore separated according to their charge, and not size as with SDS gel electrophoresis. In practice the protein samples are loaded onto the gel before the pH gradient is formed. When a voltage difference is applied, protein migration and pH gradient formation occur simultaneously.

Traditionally, 1–2 mm thick isoelectric focusing gels have been used by research workers, but the relatively high cost of ampholytes makes this a fairly expensive procedure if a number of gels are to be run. However, the introduction of thin-layer isoelectric focusing (where gels of only 0.15 mm thickness are prepared, using a layer of electrical insulation tape as the "spacer" between the gel plate) has considerably reduced the cost of preparing IEF gels, and such gels are therefore described in this chapter.

From: The Protein Protocols Handbook
Edited by: *J. M. Walker* *Humana Press Inc., Totowa, NJ*

Additional advantages of the ultra-thin gels over the thicker traditional gels are the need for less material for analysis, and much quicker staining and destaining times. Also, a permanent record can be obtained by leaving the gel to dry in the air, i.e., there is no need for complex gel-drying facilities. The tremendous resolution obtained with IEF can be further enhanced by combinations with SDS gel electrophoresis in the form of 2D gel electrophoresis. Various 2D gel systems are described in Chapters 20–22.

2. Materials

1. Stock acrylamide solution: acrylamide (3.88 g), bis-acrylamide (0.12 g), sucrose (10.0 g). Dissolve these components in 80 mL of water. This solution may be prepared some days before being required and stored at 4°C (*see* Note 1).
2. Riboflavin solution: This should be made fresh, as required. Stir 10 mg riboflavin in 100 mL water for 20 min. Stand to allow undissolved material to settle out (or briefly centrifuge) (*see* Note 2).
3. Ampholytes: pH range 3.5–9.5 (*see* Note 3).
4. Electrode wicks: 22 × 0.6 cm strips of Whatman No. 17 filter paper.
5. Sample loading strips: 0.5 cm square pieces of Whatman No. 1 filter paper, or similar.
6. Anolyte: 1.0*M* H_3PO_4.
7. Catholyte: 1.0*M* NaOH.
8. Fixing solution: Mix 150 mL of methanol and 350 mL of distilled water. Add 17.5 g sulfosalicylic acid and 57.5 g trichloroacetic acid.
9. Protein stain: 0.1% Coomassie brilliant blue R250 in 50% methanol, 10% glacial acetic acid. N.B. Dissolve the stain in the methanol component first.
10. Glass plates: 22 × 12 cm. These should preferably be of 1 mm glass (to facilitate cooling), but 2 mm glass will suffice.
11. PVC electric insulation tape: The thickness of this tape should be about 0.15 mm and can be checked with a micrometer. The tape we use is actually 0.135 mm.
12. A bright light source.

3. Methods

1. Thoroughly clean the surfaces of two glass plates, first with detergent and then methylated spirit. It is *essential* for this method that the glass plates are clean.
2. To prepare the gel mold, stick strips of insulation tape, 0.5 cm wide, along the four edges of one glass plate. Do not overlap the tape at any stage. Small gaps at the join are acceptable, but do *not* overlap the tape at corners as this will effectively double the thickness of the spacer at this point.
3. To prepare the gel solution, mix the following: 9.0 mL acrylamide, 0.4 mL ampholyte solution, and 60.0 µL riboflavin solution.

   N.B. Since acrylamide monomer is believed to be neurotoxic, steps 4–6 must be carried out wearing protective gloves.
4. Place the glass mold in a spillage tray and transfer ALL the gel solution with a Pasteur pipet along one of the short edges of the glass mold. The gel solution will be seen to spread slowly toward the middle of the plate.
5. Take the second glass plate and place one of its short edges on the taped edge of the mold, adjacent to the gel solution. Gradually lower the top plate and allow the solution to spread across the mold. Take care not to trap any air bubbles. If this happens, carefully raise the top plate to remove the bubble and lower it again.

6. When the two plates are together, press the edges firmly together (NOT the middle) and discard the excess acrylamide solution spilled in the tray. Place clips around the edges of the plate and thoroughly clean the plate to remove excess acrylamide solution using a wet tissue.
7 Place the gel mold on a light box (*see* Note 4) and leave for at least 3 h to allow polymerization (*see* Note 5). Gel molds may be stacked at least three deep on the light box during polymerization. Polymerized gels may then be stored at 4°C for at least 2 mo, or used immediately. If plates are to be used immediately they should be placed at 4°C for ~15 min, since this makes the separation of the plates easier.
8. Place the gel mold on the bench and remove the top glass plate by inserting a scalpel blade between the two plates and *slowly* twisting to remove the top plate. (N.B. Protect eyes at this stage.) The gel will normally be stuck to the side that contains the insulation tape. Do *not* remove the tape. Adhesion of the gel to the tape helps fix the gel to the plate and prevents the gel coming off in the staining/destaining steps (*see* Note 6).
9. Carefully clean the underneath of the gel plate and place it on the cooling plate of the electrophoresis tank. Cooling water at 10°C should be passing through the cooling plate.
10. Down the full length of one of the longer sides of the gel lay electrode wicks, uniformly saturated with either 1.0*M* phosphoric acid (anode) or 1.0*M* NaOH (cathode) (*see* Note 7).
11. Samples are loaded by laying filter paper squares (Whatman No. 1, 0.5 × 0.5 cm), wetted with the protein sample, onto the gel surface. Leave 0.5 cm gaps between each sample. The filter papers are prewetted with 5–7 μL of sample and applied across the width of the gel (*see* Notes 8–10).
12. When all the samples are loaded, place a platinum electrode on each wick. Some commercial apparatus employ a small perspex plate along which the platinum electrodes are stretched and held taut. Good contact between the electrode and wick is maintained by applying a weight to the perspex plate.
13. Apply a potential difference of 500 V across the plate. This should give current of about 4–6 mA. After 10 min increase the current to 1000 V and then to 1500 V after a further 10 min. A current of ~4–6 mA should be flowing in the gel at this stage, but this will slowly decrease with time as the gel focuses.
14. When the gel has been running for about 1 h, turn off the power and carefully remove the sample papers with a pair of tweezers. Most of the protein samples will have electrophoresed off the papers and be in the gel by now (*see* Note 11). Continue with a voltage of 1500 V for a further 1 h. During the period of electrophoresis, colored samples (myoglobin, cytochrome c, hemoglobin in any blood samples, and so on) will be seen to move through the gel and focus as sharp bands (*see* Note 12).
15. At the end of 2 h of electrophoresis, a current of about 0.5 mA will be detected (*see* Note 13). Remove the gel plate from the apparatus and *gently* wash it in fixing solution (200 mL) for 20 min (overvigorous washing can cause the gel to come away from the glass plate).

    Some precipitated protein bands should be observable in the gel at this stage (*see* Note 14). Pour off the fixing solution and wash the gel in destaining solution (100 mL) for 2 min and discard this solution (*see* Note 15). Add protein stain and gently agitate the gel for about 10 min. Pour off and discard the protein stain and wash the gel in destaining solution. Stained protein bands in the gel may be visualized on a light box and a permanent record may be obtained by simply leaving the gel to dry out on the glass plate overnight (*see* Note 16).
16. Should you wish to stain your gel for enzyme activity rather than staining for total protein, immediately following electrophoresis gently agitate the gel in an appropriate substrate solution (*see* Note 7, Chapter 10).

4. Notes

1. The sucrose is present to improve the mechanical stability of the gel. It also greatly reduces pH gradient drift.
2. The procedure described here uses photopolymerization to form the polyacrylamide gel. In the presence of light, riboflavin breaks down to give a free radical which initiates polymerization (Details of acrylamide polymerization are given in the introduction to Chapter 11). Ammonium persulphate /TEMED can be used to polymerize gels for IEF but there is always a danger that artefactual results can be produced by persulfate oxidation of proteins or ampholytes. Also, high levels of persulfate in the gel can cause distortion of protein bands (*see* Note 10).
3. This broad pH range is generally used because it allows one to look at the totality of proteins in a sample (but note that very basic proteins will run off the gel). However, ampholytes are available in a number of different pH ranges (e.g., 4–6, 5–7, 4–8, and so on) and can be used to expand the separation of proteins in a particular pH range. This is necessary when trying to resolve proteins with very similar pI values. Using the narrower ranges it is possible to separate proteins that differ in their pI values by as little as 0.01 of a pH unit.
4. Ensure that your light box does not generate much heat as this can dry out the gel quite easily by evaporation through any small gaps at the joints of the electrical insulation tape. If your light box is a warm one, stand it on its side and stand the gels adjacent to the box.
5. It is not at all obvious when a gel has set. However, if there are any small bubbles on the gel (these can occur particularly around the edges of the tape) polymerization can be observed by holding the gel up to the light and observing a "halo" around the bubble. This is caused by a region of unpolymerized acrylamide around the bubble that has been prevented from polymerizing by oxygen in the bubble. It is often convenient to introduce a small bubble at the end of the gel to help observe polymerization.
6. If the gel stays on the sheet of glass that does not contain the tape, discard the gel. Although usable for electrofocusing you will invariably find that the gel comes off the glass and rolls up into an unmanageable "scroll" during staining/destaining. To ensure that the gel adheres to the glass plate that has the tape on it, it is often useful to siliconize (e.g., with trimethyl silane) the upper glass plate before pouring the gel.
7. The strips must be fully wetted but must not leave a puddle of liquid when laid on the gel. (Note that in some apparatus designs application of the electrode applies pressure to the wicks, which can expel liquid.)
8. The filter paper must be fully wetted but should have no surplus liquid on the surface. When loaded on the gel this can lead to puddles of liquid on the gel surface which distorts electrophoresis in this region. The most appropriate volume depends on the absorbancy of the filter paper being used but about 5 µL is normally appropriate. For pure proteins load approx 0.5–1.0 µg of protein. The loading for complex mixtures will have to be done by trial and error.
9. Theoretically, samples can be loaded anywhere between the anode or cathode. However, if one knows approximately where the bands will focus it is best not to load the samples at this point since this can cause some distortion of the band. Similarly, protein stability is a consideration. For example, if a particular protein is easily denatured at acid pH values, then cathodal application would be appropriate.
10. The most common problem with IEF is distortions in the pH gradient. This produces wavy bands in the focused pattern. Causes are various, including poor electrical contact between electrode and electrode strips, variations in slab thickness, uneven wetting of electrode strips, and insufficient cooling leading to hot spots. However, the most common

cause is excessive salt in the sample. If necessary, therefore, samples should be desalted by gel filtration or dialysis before running the gel.

11. Although not absolutely essential, removal of sample strips at this stage is encouraged since bands that focus in the region of these strips can be distorted if strips are not removed. Take care not to make a hole in the gel when removing the strips. Use blunt tweezers (forceps) rather than pointed ones. When originally loading the samples it can be advantageous to leave one corner of the filter strip slightly raised from the surface to facilitate later removal with tweezers.
12. It is indeed good idea to include two or three blood samples in any run to act as markers and to confirm that electrophoresis is proceeding satisfactorily. Samples should be prepared by diluting a drop of blood approx 1:100 with distilled water to effect lysis of the erythrocytes. This solution should be pale cherry in color. During electrophoresis, the red hemoglobin will be seen to electrophorese off the filter paper into the gel and ultimately focus in the central region of the gel (pH 3.5–10 range). If samples are loaded from each end of the gel, when they have both focused in the sample place in the middle of the gel one can be fairly certain that isoelectrofocusing is occurring and indeed that the run is probably complete.
13. Theoretically, when the gel is fully focused, there should be no charged species to carry a current in the gel. In practice there is always a slow drift of buffer in the gel (electroendomosis) resulting in a small (~0.5 mA) current even when gels are fully focused. Blood samples (*see* Note 9) loaded as markers can provide additional confirmation that focusing is completed.
14. It is not possible to stain the IEF gel with protein stain immediately following electrophoresis since the ampholytes will stain giving a uniformly blue gel. The fixing step allows the separated proteins to be precipitated in the gel, while washing out the still soluble ampholytes.
15. This brief wash is important. If stain is added to the gel still wet with fixing solution, a certain amount of protein stain will precipitate out. This brief washing step prevents this.
16. If you wish to determine the isoelectric point of a protein in a sample, then the easiest way is to run a mixture of proteins of known pI in an adjacent track (such mixtures are commercially available). Some commercially available kits comprise totally colored compounds that also allows one to monitor the focusing as it occurs. However, it is just as easy to prepare ones own mixture from individual purified proteins. When stained, plot a graph of protein pI vs distance from an electrode to give a calibration graph. The distance moved by the unknown protein is also measured and its pI read from the graph. Alternatively, a blank track can be left adjacent to the sample. This is cut out prior to staining the gel and cut into 1 mm slices. Each slice is then homogenized in 1 mL of water and the pH of the resultant solution measured with a micro electrode. In this way a pH vs distance calibration graph is again produced.

18

Isoelectric Focusing Under Denaturing Conditions

Christopher F. Thurston and Lucy F. Henley

1. Introduction

The introduction of methods for the electrophoretic analysis of proteins under denaturing conditions *(1,2)* is something of a landmark in the development of methodologies for the analysis of proteins. Not only was analysis giving good correlation to molecular weight possible with small samples of complex mixtures of proteins, but in addition, analysis of normally insoluble proteins became possible in a relatively simple manner. These methods were based on solubilizing protein in sodium dodecyl sulfate (SDS) such that the intrinsic charge differences between proteins were overwhelmed by the relatively strong charge of the dodecyl sulfate ion, which resulted in very good correlation (for most, but not all, proteins) between electrophoretic mobility and molecular weight (*see* Chapter 11).

It is precisely the property of proteins masked by dodecyl sulfate that is exploited in isoelectric focusing (IEF) analysis, and consequently if denatured protein is to be analyzed by IEF, an alternative method of solubilization is required. We describe here the IEF analysis of protein samples solubilized by the use of 8*M* urea, a powerful denaturant that is uncharged, based on the method of O'Farrell *(3)*.

Complete denaturation of many proteins involves, in addition, the reduction of inter- and/or intramolecular disulfide bonds that is accomplished by inclusion of a ithiol reagent such as 2-mercaptoethanol or dithiothreitol. This is simple to do for electrophoretic analysis because the dissociated/unfolded polypeptides are stabilized in this state by SDS, but here again the requirements of IEF are different The redox state of a protein is affected by thiol reagents and so consequently is their behavior during IEF analysis. We therefore show an example of such effects and give conditions for maintaining protein in the fully reduced state during the focusing process (*see* Figs. 1 and 2).

2. Materials

1. Silane A174 (γ-methacryloxypropyltrimethoxysilane).
2. Glacial acetic acid, analytical grade.
3. Urea. Use the highest purity available. "Ultrapure" from Schwartz-Mann is suitable. Although urea solutions can be deionized with a mixed-bed ion-exchange resin, the procedure requires solid urea at least for sample preparation, so it must be recrystallized if purification is attempted.

From: The Protein Protocols Handbook
Edited by: J. M. Walker Humana Press Inc., Totowa, NJ

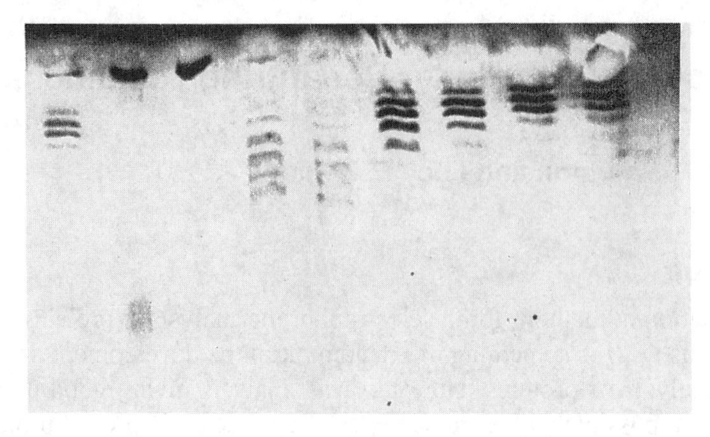

Fig. 1. Isoelectric focusing gel stained with Coomassie blue. All tracks were loaded with 10 μg pure isocitrate lyase. The concentration of dithiothreitol in the samples is shown above each track. Asterisks indicate tracks loaded with immunoprecipitated isocitrate lyase. The anti-isocitrate lyase immunoglobulins present in these samples did not focus as distinct bands and hence, cannot be seen in the photograph. The heavily stained bands near the top of the photograph in tracks containing no dithiothreitol are artifactual; in the absence of reducing agent, much of the protein commonly precipitates in the position of the applicator strip or along one end of the strip. The region of the gel shown is from about pH 6 (top) to about pH 4 (bottom) (from ref. *5*, with permission).

4. Nonidet NP40: 10% (v/v) solution in distilled water.
5. 30% Acrylamide: 28.38% (w/v) acrylamide and 1.62% (w/v) *N,N'*-methylene bisacrylamide, dissolved in distilled water, filtered through a 0.22-μm pore-size membrane filter and stored in the dark at 4°C. This reagent can also be contaminated with charged impurities (acrylic acid), which can be removed by recrystallization or treatment with ion-exchange resin, but there are now highly purified grades available from most suppliers.
6. Ampholytes: Use premixed preparations to give the desired pH range, such as Pharmacia "Pharmalyte."
7. *N,N,N',N'*-Tetramethylethylenediamine (TEMED).
8. Ammonium persulfate. Use analytical grade and store the solid at 4°C. The solid can decompose if heated, without change in appearance, so do not rely on a bottle of uncertain provenance.
9. 10 m*M* Phosphoric acid.
10. 20 m*M* NaOH.
11. Fixing solution (TCA/SSA): 0.7*M* trichloroacetic acid, 0.16*M* sulfosalicylic acid.
12. Destain solution: glacial acetic acid:ethanol:water; 1:2.5:6.5 by volume.
13. Staining solution: 2.5 g Coomassie blue R250, 100 mL glacial acetic acid, 500 mL ethanol, 400 mL distilled water. Filter through Whatman No. 1 paper on the day of use. (Dissolve Coomassie in ethanol component first.)
14. 1*M* Sodium salicylate or a commercial water-miscible fluorographic solution such as "Amplify" from Amersham-Searle or EN^{3}HANCE™ from New England Nuclear (Dupont).
15. Flatbed isoelectric focusing apparatus with provision for cooling the gel-carrying plate.

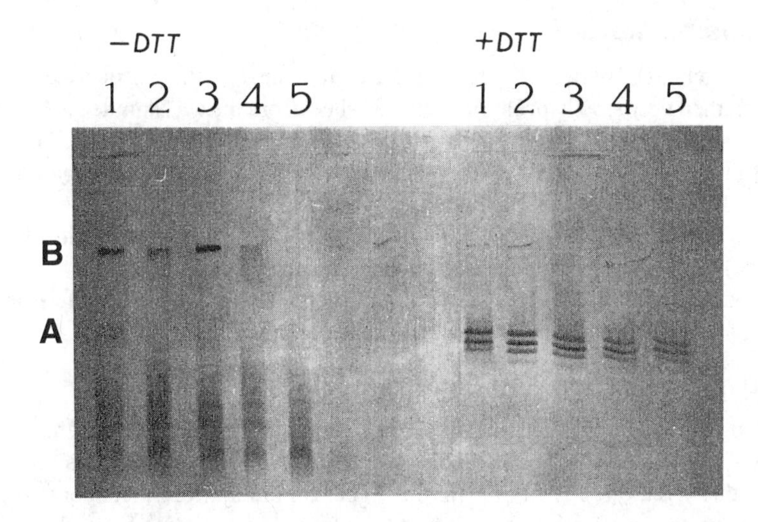

Fig. 2. Analysis of immunoprecipitated isocitrate lyase by isoelectric focusing under denaturing conditions. Samples taken from a culture labeled by growth on [^{35}S]-sulfate at 0, 2, 4, 7, and 11 h after transfer to "deadapting" conditions. With increasing time of sampling, the amount of radioactivity precipitable with anti-isocitrate lyase decreases from 11,000 cpm at 0 h (track 1) to 5000 cpm at 11 h (track 5). Fully reduced isocitrate lyase focuses as two main bands and one or two minor bands, average p*I*, 5.6, as seen when the samples were focused in the presence of 2*M* dithiothreitol (+DTT). In the absence of dithiothreitol (-DTT), only the initial sample contained detectable, fully reduced isocitrate lyase (marked A), and in successive samples the enzyme becomes progressively more oxidized focusing at progressively more acidic p*I* values. Bands marked B are precipitated protein at the position of the applicator strips. The pH range shown in the fluorograph is approx from pH 8 (top) to pH 3.5 (bottom) (from ref. *5,* with permission).

3. Method

3.1. Casting the Gel

1. Adjust the pH of 1 L of distilled water to 3.5 with glacial acetic acid and mix in 4 mL of Silane A174 by stirring for 15 min. Soak a glass gel-plate in the solution for 1 h, rinse with distilled water, and dry in air (*see* Note 2).
2. Use a few drops of ethanol to adhere a clean plastic sheet to a second glass plate (*see* Note 2).
3. Mount the silanized plate and the plate-backed plastic sheet separated by a 1-mm casting gasket, with the open corner of the gasket uppermost.
4. Prepare 20 mL of gel-forming solution in a sidearm flask as follows: 9.6 g urea, 4 mL distilled water, 4 mL 10% nonidet, 2.66 mL acrylamide solution, and 0.5 mL ampholyte mixture are gently swirled to dissolve the urea (*see* Note 1). Place an oversize rubber bung over the top of the flask and apply a vacuum from a water pump via the sidearm for 10 min with occasional further swirling to thoroughly degas the mixture. Allow air to re-enter the flask slowly (*see* Note 3).
5. Add 22 µL of TEMED and 16 µL of freshly prepared 10% (w/v) ammonium persulfate, swirl briefly to mix, take up into a disposable syringe, and deliver immediately into the gel-casting box from step 3 above. Completely fill the system and cover the opening in the casting gasket with parafilm. The mixture should polymerize within about 2 h, but may conveniently be left overnight at this stage (*see* Note 4).

3.2 Prefocusing the Gel

1. When the gel has polymerized, unclamp the plates, separate the plastic sheet from its backing plate, then very slowly peel off the plastic sheet from the acrylamide gel layer. Lift away the casting gasket and remove any liquid from the edge of the gel with clean tissue.
2. Mount the gel-plate on the cooling plate of the isoelectric focusing apparatus. Circulate water at 10°C through the system.
3. Apply electrode wicks, soaked in 10 m*M* phosphoric acid (anode) and 20 m*M* NaOH (cathode) to the edges of the gel and connect the electrodes.
4. Using a dc power source, set to give constant voltage, apply 200 V for 15 min. 300 V for 30 min, and 400 V for 30 min, to establish the pH gradient within the gel.

3.3. Sample Preparation and Electrophoresis

1. To samples in solution, add solid urea to give a concentration of 8*M* then mix with an equal vol of 8*M* urea, 2% (w/v) Nonidet NP4D, 2% (w/v) ampholytes. Protein precipitates may be dissolved directly in this mixture. The sample mixture must contain a relatively high concentration of protein since the maximum volume that can be loaded is 35 µL. The sample mixture may in addition contain 2*M* dithiothreitol (*see* Note 5).
2. Disconnect the power supply from the prefocused gel and place paper sample applicators near the cathode end of the gel (*see* Note 6). Deliver the sample (preferably no more than 10 µL) onto each applicator and focus for 30 min at 500 V. Disconnect the power supply and remove the paper sample applicators from the gel with flat-ended forceps. Focus for a further 90 min at 800 V.
3. When focusing is finished, remove the electrode wicks and cut a track of 5-mm square sections from the position of the anode to the position of the cathode. Soak each section overnight in 0.25 mL distilled water and measure the pH with a microelectrode without removing the slice of gel.
4. Fix for 1 h in TCA/SSA (this step also removes ampholytes that would otherwise stain with Coomassie blue), and wash for 30 min in destain solution. Stain with Coomassie blue with constant agitation for 2–16 h. Destain with repeated changes of destain solution until the background of the gel is cleared of stain.
5. For detection of radioactive bands on the gel, it is necessary to detach the gel from the silanized glass plate. By the time the gel is fully destained, it has usually started to detach at the edges. Hold the plate upside down over a tray containing destain solution, and separate the gel from the plate with a sharp scalpel. Soak the gel in 1*M* sodium salicylate or a commercial fluor for 2 h, and dry down onto Whatman 3-mm paper and expose to X-ray film.

4. Notes

1. The quantities given are for a gel 230 × 115 mm. Plates of this size, plastic sheets, and casting gaskets are obtained from Pharmacia. The Pharmacia manual *Isoelectric Focusing—Principles and Methods* is an extremely useful reference work for both the background and practicalities of isoelectric focusing.
2. The thickness of the gel (1 mm) is optimal for focusing, but makes handling difficult. The polymerized gel is covalently linked to the glass plate by the silane in order to prevent creepage and/or buckling during focusing and is polymerized against a plastic sheet rather than a second glass plate, since it is easier to detach from the polymerized gel after casting.
3. For a usable focusing pH range of 4–9, use 0.4 mL pH 3–10 ampholytes and 0.1 mL of pH 4–6 ampholytes. Note that the electrode solutions described are for this pH range and are not applicable to all ranges.

4. The polymerization of acrylamide mixtures is inhibited by oxygen and impurities. Shaking up the solution to mix it must be avoided, particularly after degassing. If the dispensing syringe is large enough to hold all of the mixture, the process of filling the syringe is probably sufficient to mix the final additions. If the gel does not set, try degassing longer or try a different bottle of ammonium persulfate, and if still unsuccessful, double the amount of TEMED included. If this does not work, the reagents are not pure enough. If the gel sets unevenly, the glass plate was not clean.
5. The small sample volume allowed makes precipitated samples desirable. Chloroform/methanol precipitation *(4)* is a good way of doing this, and has the additional advantage of removing lipid. The description given here is for small volumes, such as might be processed for electrophoretic analysis, but the volumes may be scaled up in proportion 10-fold, using 15 mL Corex tubes. All steps are perforrned at room temperature. Take 150 µL of protein solution in a 1.5-mL microfuge tube, add 600 L methanol, mix by three or four inversions of the capped tube, and centrifuge for 10 s. Add 150 µL chloroform, blend briefly on a vortex mixer, and centrifuge for 10 s. Add 450 µL of distilled water and blend thoroughly on a vortex mixer. Centrifuge for 10 s to separate the two liquid phases that formed on addition of the water. Carefully remove the upper phase and discard. Do not attempt to remove it all, but leave a residual layer 1–3 mm deep, because most of the protein is at the interface. Add 450 µL methanol, vortex mix briefly, and centrifuge for 4 min. After centrifugation, a white protein pellet should be visible. Aspirate off and discard (as chlorinated solvent) the bulk of the supernatant. Remove remaining supernatant and dry the pellet in a vacuum desiccator or under a stream of dry N_2. Pellets dissolve readily and without heating in sample buffers used for electrophoresis or IEF and may be stored indefinitely at –20°C.

   If the sample contains proteins that are susceptible to oxidation, clear resolution may only be obtained if the samples are made 2*M* in dithiothreitol (*see* Figs. 1 and 2). There is a need for this very high concentration to prevent migration of protein(s) and thiol reagent in different directions during focusing from depleting the protein(s) entirely of reducing agent.
6. The best position for sample loading strips has to be determined by trial and error. The samples we have run all contain proteins that precipitate at acidic pH, so it has been preferable to place the applicators in the pH 8 region of the (prefocused) gel.

References

1. Weber, K. and Osborne, M. (1969) The reliability of molecular weight determinations by dodecyl sulfate-polyacrylamide gel electrophoresis. *J. Biol. Chem.* **244,** 4406–4412.
2. Laemmli, U. K. (1970) Cleavage of structural proteins during the assembly of the head of bacteriophage T4. *Nature* **227,** 680–685.
3. O'Farrell, P. H. (1975) High resolution two-dimensional electrophoresis of proteins. *J. Biol. Chem.* **250,** 4007–4021.
4. Wessel, D. and Flugge, U. I. (1984) A method for the quantitative recovery of protein in dilute solution in the presence of detergents and lipids. *Anal. Biochem.* **138,** 141–143.
5. Henley, L. F. and Thurston, C. F. (1986) The disappearance of isocitrate lyase from the green alga *Clorella fusca* studied by immunoprecipitation. *Arch. Microbiol.* **145,** 266–271.

19

Radioisotopic Labeling of Proteins for Polyacrylamide Gel Electrophoresis

Jeffrey W. Pollard

1. Introduction

The metabolic labeling of proteins with radiolabeled amino acids, either to determine the rate of synthesis or to identify proteins following separation by gel electrophoresis, at first sight appears trivial. However, there are a number of important methodological considerations to be aware of in these procedures. Principal among these is the problem of intracellular amino acid pools and the effects of changing these on the metabolism of the cell. Fortunately, amino acid intracellular pools equilibrate rapidly with the external media, maximally within 5 min for valine and usually within 1 min for amino acids, such as leucine. Thus, providing a large extracellular pool of nonradioactive amino acids is maintained, the radiolabeled essential amino acid is introduced at low concentrations, and labeling continues for at least 10 times the pool equilibration time, accurate protein synthetic rates can be determined. This experimental design, however, can rarely be achieved if labeling of proteins to high specific activity is required, for example, for detection of all but a few of the most abundant cellular proteins following separation by two-dimensional gel electrophoresis. Usually, in these cases, the external amino acid, often methionine and/or cysteine, is removed and substituted with a low concentration of very high specific activity cognate amino acid. This manipulation, however, will inhibit the initiation of ribosomes onto mRNA *(1)*, which, in turn, results in the selective translation of high-efficiency mRNAs *(2)* and, therefore, the detection of a biased profile of proteins. Furthermore, the reduction in the rate of ribosome initiation will result in a reduced level of radioactive incorporation into proteins. Considerable care, therefore, should be taken in extrapolating the results obtained to physiological circumstances. Thus, as in all experiments, careful thought should be given to the experimental end point. For example, it is often better to reduce the media concentration of the amino acid used to label to a one-tenth or one-hundredth concentration (media amino acids are always available in excess) and label for several hours or to equilibrium, rather than to use short pulses given in the absence of the external amino acid. These conditions, however, would not suffice if the experimenter wishes to analyze proteins synthesized rap-

From: The Protein Protocols Handbook
Edited by: J. M. Walker Humana Press Inc., Totowa, NJ

idly in response to a changed external stimuli, such as the addition of a growth factor. Bearing these caveats in mind, this chapter will describe simple and relatively robust methods for metabolically radiolabeling proteins in preparation for separation by gel electrophoresis followed by detection by autoradiography.

The isotope chosen for labeling the proteins must, of course, also correspond to the question to be investigated, but those in general use are the β-emitters: ^{3}H, ^{14}C, ^{32}P, and ^{35}S. The maximum energies and half-lives are: 0.019 MeV (12.28 yr), 0.156 MeV (5730 yr), 1.706 MeV (14.3 d), and 0.167 MeV (87.4 d), respectively. Tritium is a popular isotope for biological use, potentially available at high specific activity, but its low energy precludes direct detection by autoradiography on X-ray film and, thus, gels need to be either treated for fluorography or quantified by gel slicing and scintillation counting. Consequently, the higher energy [^{14}C]- and [^{35}S]-labeled amino acids are usually used as protein labels since they can be directly detected by X-ray film autoradiography or on a phosphorimager. The low specific activity of [^{14}C]-labeled amino acids, however, often limits their usefulness, with the result that [^{35}S]-methionine or [^{35}S]-cysteine labeling, either alone or in combination, is more commonly used. Nevertheless, it is worth noting that Bravo and Celis *(3)* were able to detect 28% more proteins by labeling with a mixture of 16 [^{14}C]-amino acids than could be detected with [^{35}S]-methionine and that such labeling with [^{14}C]-amino acids allows the long-term storage of data on gels. Unfortunately, the cost in isotopes demanded by [^{14}C]-labeling is often prohibitive.

The techniques described in this chapter may easily be modified to encompass other uses, such as labeling with the isotopes ^{59}Fe, ^{67}Cu, ^{125}I, or ^{131}I for specific studies of particular proteins, or for labeling with sugar precursors, e.g., [^{3}H]-glucosamine, [^{3}H]-mannose, or [^{3}H]-fucose to detect glycoproteins, or with [^{32}P]-orthophosphate for detection of phosphoproteins.

2. Materials

1. Tissue-culture medium: The author uses a medium rich in amino acids, α-minimal essential medium, but lacking in methionine. Store at 4°C. Any defined tissue-culture medium lacking any single or combination of amino acids may be used (Note 1). Prior to use, supplement with 10% (or whatever is appropriate) dialyzed fetal calf serum (DFCS) and warm to 37°C.
2. DFCS: serum is serially dialyzed against two changes of 500-fold excess phosphate-buffered saline (PBS) in order to remove amino acids. Store in aliquots at –20°C. Do not store this dialyzed serum at 4°C, because endogenous proteolytic activity will result in a relatively high concentration of amino acids.
3. PBS (Ca^{2+} and Mg^{2+}-free): 0.14*M* NaCl, 2.7 m*M* KCl, 1.5 m*M* KH_2PO_4, 8.1 m*M* Na_2HPO_4.
4. 0.1% (w/v) Trypsin in PBS citrate (PBS containing 20 m*M* monosodium citrate).
5. Lysis buffer: 9.5*M* urea, 2% v/v Nonidet, 2% v/v ampholines (1.6%, pH range 5–7.0; 0.4%, pH range 3.5–10.0, 5% v/v β-mercaptoethanol. Stored frozen at –20°C in 0.5-mL aliquots *(4)*.
6. Isotopes: [^{35}S]-Methionine, [^{35}P]-orthophosphate and [^{3}H]-amino acids are used as supplied. [^{14}C]-amino acids are lyophilized and resuspended in a medium lacking amino acids at 500 μCi/mL.

3. Methods

3.1. Cells Growing in Monolayers

1. Plate the cells directly into 96-well dishes at about 2000 cells/well in 0.25 mL of growth medium. Leave them to attach for at least 5 h, but preferably overnight (*see* Note 2).
2. Remove the media, wash the cells with 0.5 mL of medium lacking methionine, and add 0.1 mL medium supplemented with 10% v/v DFCS, lacking methionine, but containing 100 μCi [^{35}S]-methionine and 1 mg/L unlabeled methionine to each well (*see* Note 3).
3. Label the cells for up to 20 h at 37°C in a humidified incubator (wrap the microtiter plates in cling film to reduce evaporation), and ensure that the plates are not tilted.
4. Following incubation, remove the medium (taking care to dispose of it correctly, since it contains a large amount of radioactivity), wash the cells twice with ice-cold PBS (*see* Note 4) to remove serum proteins, and if a total cell extract for two-dimensional gel electrophoresis is required, lyse the cells with 20 μL lysis buffer (*see* Note 5). The samples may be stored in small vials or in the microtiter wells at –70°C, or loaded directly onto isoelectric focusing gels *(4)*.
5. Alternatively, cells may be trypsinized and processed according to the analytical technique required. To retrieve the cells, add 200 μL of ice cold 0.1% trypsin, and incubate on ice until cells begin to detach (approx 15–20 min for most fibroblasts). Neutralize the trypsin with 20 μL of calf serum, and pipet gently to detach the cells completely. Retrieve the cells by certifugation in a microfuge and Wash them twice with ice-cold PBS before processing or dissolving the cell pellet in lysis buffer.
6. Cells may also be adequately labeled for up to about 4 h in a medium completely lacking unlabeled methionine (*see* Note 6), but under these circumstances, equilibrium labeling may not be achieved.
7. It is difficult to maintain tight physiological control of cells growing in microwells, and the cell number obtained from such cultures is often too small for adequate fractionation. Thus, when either an increased cell yield or better control of cell physiology, for example, adding a mitogenic growth factor or serum to stationary cultures, is required, then cells growing in larger flasks (25–125 cm^2) may be labeled in medium containing 10% DFCS and methionine at 1 mg/L and 100–200 μCi of [^{35}S]-methionine (2 mL of medium to a 25-cm^2 flask or prorated according to flask size). In this case, pre-equilibrate flasks with a 95% air/5% CO_2, mix, and incubate with occasional rocking to prevent desiccation of the cells. After labeling, wash the cells twice with PBS and collect cells either by scraping with a rubber policeman, or by trypsinization using 2 mL trypsin/25-cm^2 flask at 4°C, and process as before (*see* Notes 7 and 8).

3.2. Cells Growing in Suspension

1. Collect cells growing in suspension normally in the logarithmic stage (*see* Note 2) by centrifugation (300*g* for 3.5 min), wash them in methionine-free medium, and regain the cells by centrifugation.
2. Resuspend the cells at about 25% of their saturation density in 5 mL medium containing 10% DFCS lacking methionine, but supplemented with 400 μCi mL [^{35}S]-methionine in a 15-mL Falcon round-bottom plastic snap-cap tubes previously gassed with a 5% CO_2/95% air mixture (*see* Note 9). Agitate the cell suspension with a small magnetic flea. This agitation can be achieved either by using a magnetically stirred water bath or a magnetic stirrer set up in an incubator. Alternatively cells can be agitated by rolling in a temperature

regulated roller apparatus. But remember that temperature equilibration is achieved most rapidly in a water bath.

3. Label cells for 1–4 h at 37°C, collect the cells by centrifugation, and wash twice with PBS before processing as before (*see* Note 8).

3.3. Tissues

1. Excise 100–200 mg of the tissue to be labeled, blot it briefly onto filter paper, and chop it into small pieces (approx 2 mm diameter). These may be washed briefly in PBS to reduce the amount of blood contamination. Alternatively, if possible, tissues can big perfused with saline prior to isolation.
2. Place these tissue pieces into 1 mL of tissue-culture medium containing 10% DFCS, lacking methlionine, but supplemented with 200 μCi/mL [^{35}S]-methionine in glass scintillation vials gassed with a 5% CO_2/95% air mixture.
3. Place the vials in a shaking water bath at 37°C, and label for 1 h; thereafter, process the sample as appropriate.

4. Notes

1. Never label cells in a basic salt solution, such as PBS. This treatment will dramatically alter the metabolic state of the cell and disrupt internal pool concentrations such that they exhibit rapidly altering specific activities. Ideally, labeling should be performed in preconditioned medium in order to effect minimal metabolic disturbance, but this is rarely achievable for the labeling of proteins for gel electrophoretic separation. Furthermore, because protein synthesis is so temperature-sensitive (Note 3), always make sure medium is prewarmed to the appropriate temperature, usually 37°C, used for labeling.
2. Try as best as possible not to shift cells, especially normal cells, from stationary to exponential phase of growth or vice versa. This treatment results in significant shifts of protein synthesis rates and the profiles of proteins synthesized. Do not subject cells to too high a centrifugal force since this also perturbs cellular metabolism.
3. The handling of cells to be labeled with [^{3}H]- or [^{14}C]-amino acids or [^{32}P]-orthophosphate is identical to that described for [^{35}S]-methionine. The [^{3}H]- or [^{14}C] amino acids are exposed to cells in 0.1 mL of medium at 500 μCi/mL in medium lacking the appropriate amino acids or to [^{32}P] orthophosphate at 2 mCi/mL in medium lacking phosphate. Similarly, substrates such as [^{3}H]-glucosamine or [^{3}H]-mannose, can be used to radiolabel glycoproteins *(6)*.
4. Since protein synthesis involves many enzymatic reactions, flooding with ice-cold PBS immediately inhibits protein synthesis to very low levels (<2%) and allows very precise control over timing of incubations.
5. To determine the number of counts incorporated, removes a small aliquot (1–5 μL) from the final sample, add it to 0.5 mL of water containing 10 μg of bovine serum albumin, and precipitate it with 0.5 mL of ice-cold 20% (w/v) trichloroacetic acid (TCA). Allow the precipitate to develop for 10 min and then collect it onto Whatman glass fiber disks under vacuum followed by four washings of 5 mL of 5% (w/v) TCA. Wash the disks finally with ethanol, dry them, and count them in a compatible scintillation fluid. Process ^{3}H-labeled samples by overnight digestion with a commercial tissue solubilizer (or 1*M* KOH) before counting. Determine ^{32}P by Cerenkov counting.
6. There are many variations on a theme for labeling proteins. Labeling can be very short to detect rapidly synthesized proteins, or to equilibrium to determine the steady-state level of the protein. Similarly, pulse-chase experiments can be performed by removing the label-

ing medium and replacing it with the same medium lacking the radioactive amino acid and containing a 500-fold excess of the nonradioactive labeling amino acid.

7. Accurate protein synthetic rates over short periods may be determined concurrently in parallel flasks by measuring the rate of incorporation of a mixture of three [^{3}H]- or [^{14}C]-labeled essential amino acids (used at 1 or 0.2 μCi/mL, respectively) into acid-insoluble counts per cell. In this case, following labeling, aspirate the radioactive medium, wash the cells with ice-cold PBS, and trypsinize the cells with 2 mL of 0.1% trypsin (for a 25-cm^2 flask) on ice. After cells have begun to detach (10–15 min), neutralize the trypsin with 200 μL of calf serum, pipet the cells up and down five times, and transfer them to a centrifuge tube. Wash the flasks with a further 2 mL of trypsin, combine the supernatants, and regain the cells by centrifugation at 300*g* for 3.5 min. Remove the supernatant, dry the tube walls with a cotton bud, and resuspend the cells in 2 mL of PBS containing 0.2% FCS. Pipet up and down to disperse the cells, remove 0.2 mL for determination of cell number on a Coulter counter (0.2 to 7.8 mL PBS + 0.2% calf serum to give a 1/40 dilution in 15-mL snap-cap round-bottom Falcon tubes) or hemocytometer. Precipitate the remaining cell suspension with an equal volume of 20% (w/v) TCA and process for scintillation counting as described in Note 5. Alternatively, prior to precipitation, an aliquot can be removed to determine the protein concentration.
8. As discussed in the Section 1., there are considerable pitfalls in measuring the true rate of protein synthesis using isotopic methods. Accurate protein synthesis measurements may be achieved only when very small amounts of radioactive precursors are added to reduced volumes of the same conditioned growth medium removed from the growing culture, which contains large amounts of unlabeled precursors (*see* ref. *7* for a full discussion and relevant references). These conditions for maintaining protein synthetic rates at steady-state levels are clearly not met when fresh medium, often lacking methionine, is used to label proteins for gel electrophoresis! Considerable care should therefore be exercised in standardizing growth and labeling conditions for any experiments involving comparison of different samples and in relating the interpretation of the data to normal physiological conditions. Often a compromise has to be effected between high levels of incorporation and the physiological constraints of maintaining precursor pool sizes.
9. For determination of protein synthetic rates parallel tubes can be incubated with a mixture of [^{3}H]- or [^{14}C]-labeled amino acids at 1 or 0.2 μCi/mL, respectively in 5 mL of media. At appropriate times, remove 1 mL, and add it to 9 mL of ice-cold growth medium lacking serum. Regain the cells by centrifugation. Thereafter, process the cells as described in Note 7.
10. The only other problems that may be encountered, providing care is taken over sterility, pH and temperature regulation, is the toxicity of isotopes. This is rarely a problem for a day's labeling, but could potentially be so if proteins are labeled for longer. Fibroblasts will survive in 0.5 μCi ^{32}P/mL in medium containing 0.2 m*M* phosphate and in mutant isolation, 1.5 dpm of [^{3}H]-amino acid incorporation/cell is considered lethal, but only after a period in the cold for accumulation of radioactive damage. Thus, toxicity will not be a problem if the above labeling procedures are followed.
11. The procedures given above will result in [^{35}S]-methionine-labeled proteins with specific activities of around 10^5 cpm/μg of protein, but obviously, the optimal conditions, and resultant specific activity of proteins will depend on the cell type, growth conditions and specific activity of the isotope. It is also worth noting that, following two-dimensional gel electrophoresis, a 1-mm^2 spot containing 3 dpm of [^{35}S]-methionine may be readily detected (>0.01 OD above background) after 1 wk, and the phosphorimager allows even greater sensitivity.

Acknowledgments

This chapter was prepared while the author's research was supported by grants from NIH DK/CA48960, HD/AI 30280, and the Albert Einstein Cancer Center grant P30-CA1330. J. W. P. is a Monique-Weill Caulier Scholar.

References

1. Pollard, J. W., Galpine, A. R., and Clemens, M. J. (1989) A novel role for aminoacyl-tRNA synthetases in the regulation of polypeptide chain initiation. *Eur. J. Biochem.* **182,** 1–9.
2. Lodish, H. F. (1974) Model for the regulation of mRNA translation applied to haemoglobin. *Nature* **251,** 385–388.
3. Bravo, R. and Celis, J. E. (1982) Up-dated catalogue of HeLa cell proteins. Percentages and characteristics of the major cell polypeptides labelled with a mixture of 16 [^{14}C]-labelled amino acids. *Clin. Chem.* **28,** 766–781.
4. Parker, J., Pollard, J. W., Friesen, J. D., and Stanners, C. P. (1978) Stuttering: High-level mistranslation in animal and bacterial cells. *Proc. Natl. Acad. Sci. USA* **75,** 1091–1095.
5. Pollard, J. W. (1990) Basic cell culture, in *Methods in Molecular Biology, vol. 5: Animal Cell Culture* (Pollard, J. W. and Walker, J. M., eds.), Humana, Clifton, NJ.
6. Bradshaw, J. P., Hatton, J., and White, P. A. (1985) The hormonal control of protein *N*-glycosylation in the developing mammary gland and its effect upon transferrin synthesis and secretion. *Biochim. Biophys. Acta* **847,** 344–351.
7. Stanners, C. P., Adams, M. E., Harkins, J. L., and Pollard, J. W. (1979) Transformed cells have lost control of ribosome number through the growth cycle. *J. Cell. Physiol.* **100,** 127–138.

20

Two-Dimensional PAGE Using Carrier Ampholyte pH Gradients in the First Dimension

Patricia Gravel and Olivier Golaz

1. Introduction

Two-dimensional polyacrylamide gel electrophoresis (2D PAGE) is one of the most powerful tools for the separation of complex protein mixtures. This method separates individual proteins and polypeptide chains according to their isoelectric point and molecular weight. This chapter describes the protocol for 2D PAGE using carrier ampholyte pH gradient gel for the isoelectric focusing (IEF) separation (pH gradient 3.5–10.0) and a polyacrylamide gradient gel for the second dimension *(1,2)*.

2. Materials

2.1. Preparation of Samples

1. Lysis solution A: 10% (w/v) Sodium dodecyl sulfate (SDS) and 2.32% (w/v) 1,4-dithioerythritol (DTE) in distilled water (dH_2O).
2. Lysis solution B: 5.4 g Urea (9.0*M*) (must be solubilized in water at warm temperature around 35°C), 0.1 g DTE (65 m*M*), 0.4 g cholamidopropyldimethylhydroxypropane sulfonate (CHAPS) (65 m*M*), and 0.5 mL of ampholytes, pH range 3.5–10.0 (5% v/v), made to 10 mL with dH_2O.

These solutions can be aliquoted and stored at –20°C for many months.

2.2. IEF

1. IEF is performed with the Tube Cell Model 175 (Bio-Rad, Hercules, CA) (*see* Fig. 1) and with glass capillary tubes (1.0–1.4 mm internal diameter and 210 mm long). Ampholytes, pH 4.0–8.0, and Ampholytes, pH 3.5–10.0 are from BDH (Poole, UK).
2. Stock solution of acrylamide: 30% (w/v) acrylamide, 0.8% (w/v) PDA (piperazine diacrylamide) in dH_2O. This solution should be stored in the dark at 4°C for 1–2 mo.
3. Cathodic buffer: 20 m*M* NaOH in dH_2O. This solution should be made fresh.
4. Anodic buffer: 6 m*M* H_3PO_4. This solution should be made fresh.
5. Ammonium persulfate (APS) stock solution: 10% (w/v) APS. This solution should be stored at 4°C, protected from light, and made fresh every 2–3 wk.
6. SDS stock solution: 10% (w/v) SDS. This solution can be stored at room temperature.
7. Bromophenol blue stock solution: 0.05% (w/v) bromophenol blue. This solution can be stored at room temperature.

From: The Protein Protocols Handbook
Edited by: *J. M. Walker Humana Press Inc., Totowa, NJ*

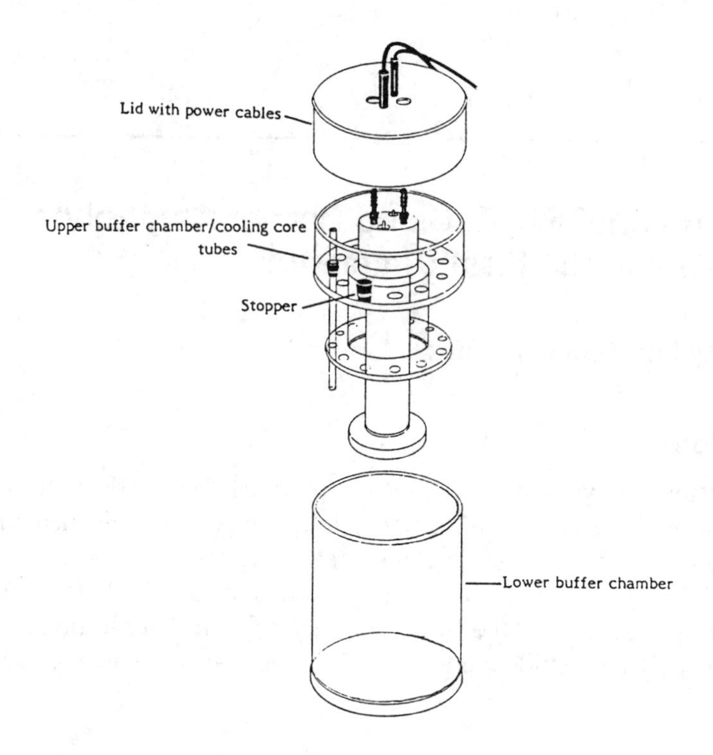

Fig. 1. Tube Cell Model 175 (Bio-Rad) for the IEF with carrier ampholyte pH gradient capillary gels.

8. Capillary gel equilibration buffer:
   20 mL 0.5*M* Tris-HCl, pH 6.8.
   40 mL 10% (w/v) SDS stock solution.
   8 mL 0.05% (w/v) bromophenol blue stock solution.
   72 mL dH_2O.

This solution can be stored at room temperature for 2–3 mo.

2.3. SDS Polyacrylamide Gel Electrophoresis (SDS-PAGE)

The Protean II chamber (Bio-Rad) is employed by us for SDS-PAGE. The gels (160 × 200 × 1.5 mm) are cast in the Protean II casting chamber (Bio-Rad). The gradient former is model 395 (Bio-Rad).

1. Running buffer: 50 m*M* Tris base, 384 m*M* glycine, 0.1% (w/v) SDS in dH_2O. For 1 L: 6 g Tris base, 28.8 g glycine, and 1 g SDS. Do not adjust pH. Fresh solution is made up for the upper tank. The lower tank running buffer can be retained for more than 6 mo with the addition of 0.02% (w/v) sodium azide.
2. Sodium thiosulfate stock solution: 5% (w/v) anhydrous sodium thiosulfate in dH_2O. This solution should be stored at 4°C. (*See* Note 1.)
3. Silver staining: Proteins in the 2D gel are stained with silver. We used the ammoniacal silver nitrate method described by Oakley et al. *(3)* and modified by Hochstrasser and Merril *(4)* and Rabilloud *(5)* (*see also* Chapter 35).

3. Method

3.1. Preparation of Protein Samples

1. Pellets of cells or tissue should be resuspended in 100 µL of 10 m*M* Tris-HCl, pH 7.4, and sonicated on ice for 30 s. Add 1 vol of lysis solution B and mix. The mixed solution should not be warmed up above room temperature. Samples can be loaded directly onto the gels or stored at –80°C until needed. (*See* Note 2 for plasma sample preparation and Note 3 for optimal sample loading.)

3.2. IEF

1. Draw a line at 16 cm on clean and dry glass capillary tubes (capillary should be cleaned with sulfochromic acid to eliminate all deposits). The remaining 0.5 cm of the capillary tubes is used to load the sample. Place each tube in a small glass test tube (tube of 5 mL) that will be filled with the IEF gel solution. Connect the tops of each glass capillary tube to flexible plastic tubes joined together with a 1-mL plastic syringe.
2. At least 12 capillary gels can be cast with 11.5 mL of IEF gel solution (800 µL of solution are needed per capillary).
   a. Prepare 1 mL of CHAPS 30% (w/v) and NP-40 (nonidet P-40) 10% (w/v) (0.3 g CHAPS and 0.1 g NP40), and degas for 5 min.
   b. Separately, prepare a second solution: 10 g urea (7 mL of water are added to dissolve the urea at warm temperature, around 35°C), 2.5 mL of acrylamide stock solution. 0.6 mL of ampholytes, pH 4.0–8.0, 0.4 mL of ampholytes, pH 3.5–10.0, and 20 µL of TEMED.
   c. Mix the first solution (CHAPS, NP-40) with the second one. Degas the mixture, and add 40 µL of APS 10% (w/v) stock solution. Pipet 1 mL of this IEF gel solution into the glass test tubes (along the side walls in order to prevent the formation of air bubbles in the solution). Fill up the capillary tubes by slowly pulling the syringe (up to the height of 16 cm).
   d. After 2 h of polymerization at room temperature, pull the capillary tubes out of the glass test tubes. Clean and gently rub the bottom of each capillary gel with Parafilm.
3. Fill the lower chamber with the anodic solution. Wet the external faces of the capillary tubes with water, and insert them in the IEF chamber. Load the samples on the top of the capillary (cathodic side). Generally, 30–40 µL of the final diluted sample are loaded using a 25-µL Hamilton syringe (*see* Notes 3 and 4).
4. Lay down the cathodic buffer solution at the top of the sample in the capillary tube, and then fill the upper chamber. Connect the upper chamber to the cathode and the lower chamber to the anode.
5. Electrical conditions for IEF are 200 V for 2 h, 500 V for 5 h, and 1000 V overnight (16 h) at room temperature (*see* Note 5).
6. After IEF, remove the capillary tubes from the tank and force the gels out from the glass tube with a 1-mL syringe that is connected to a pipet tip and filled with water (*see* Note 6). Put the extruded capillary gels on the higher glass plate of the polyacrylamide gel (*see* Section 3.3.). Residual water around the capillary gel should be soaked up with a filter paper. Put 140 µL of IEF equilibration buffer down on the capillary gel. Immediately push the capillary gels between the glass plates of the polyacrylamide gel using a small spatula. Place the cathodic side (basic end) of the capillary gel at the right side of the polyacrylamide gel. Care should be taken to avoid the entrapment of air bubbles between the capillary and the polyacrylamide gels. It is not necessary to seal the capillary gels with agarose solution, but the contact between the capillary gel and the 9–16% polyacrylamide gradient gel should be very tight.

3.3. SDS-PAGE

1. In order to separate the majority of proteins, a 9–16% (w/v) polyacrylamide gradient gel is used for the second dimension. The gel is made as follows with 60 mL of solution for a 9–16% (w/v) gel: 30 mL of 9% (w/v) acrylamide solution and 30 mL of 16% (w/v) acrylamide solution. Sixty milliliters are necessary to cast one gel of 1.5 × 200 × 160 mm (*see* Note 7).

   Light solution:

   9 mL of acrylamide stock solution (30% v/v).
   7.5 mL 1.5*M* Tris-HCl, pH 8.8 (25% v/v).
   150 µL 5% sodium thiosulfate solution (0.5% v/v) (*see* Note 2).
   15 µL TEMED (0.05% v/v).
   13.2 mL water.
   This solution is degassed, and then 150 µL of 10% APS solution (0.5% v/v) are added.

   Heavy solution:

   16 mL of acrylamide stock solution (53% v/v).
   7.5 mL 1.5*M* Tris-HCl, pH 8.8 (25% v/v).
   150 µL of 5% sodium thiosulfate solution (0.5% v/v).
   15 µL TEMED (0.05% v/v).
   6.2 mL water.
   This solution is degassed, and then 150 µL of 10% APS solution (0.5% v/v) are added.

   The gradient gel is formed using the Model 395 gradient Former (Bio-Rad) and a peristaltic pump. Immediately after the casting, the gels are gently overlayered with a water-saturated 2-butanol solution using a 1-mL syringe. This procedure avoids the contact of the gel with air and allows one to obtain a regular surface of the gel. Caution should be taken to avoid mixing the 2-butanol with the gel solution. The gels are stored overnight at room temperature to assure complete polymerization.
2. Prior to the second-dimension separation, extensively wash the top of the gels with deionized water to remove any remaining 2-butanol. Remove the excess of water by suction with a syringe.
3. After the transfer of the capillary gel on top of the gradient polyacrylamide gel (*see* Section 3.2., step 6), fill the upper and the lower reservoirs with the running buffer, and apply a constant current of 40 mA/gel. The separation usually requires 5 h. The temperature in the lower tank buffer is maintained at 10°C during the run.
4. At the end of the run, turn off the power, rinse the gel briefly in distilled water for a few seconds, and process for gel staining. A typical gel pattern of plasma proteins stained with the ammoniacal silver nitrate method is shown in Fig. 2 *(4,5)*.

4. Notes

1. The addition of thiosulfate to the gel delays the appearance of background staining with the ammoniacal silver nitrate method *(4,5)*.
2. Human plasma proteins have been efficiently separated with the following sample preparation: 5 µL of plasma (containing approx 400 µg of proteins) are solubilized in 10 µL of lysis solution A (SDS-DTE) and heated in boiling water for 5 min. After cooling for 2 min at room temperature, 485 µL of lysis solution B are added. For silver-stained gels, 30–40 µL of the final diluted sample are loaded on the top (cathodic side) of the capillary gel.
3. The best separation of complex protein mixtures is performed when <100 µg of proteins are loaded onto the capillary gel. Overloading may cause streaking and inadequate resolution of spots. The use of a highly sensitive staining method (silver staining) allows one to

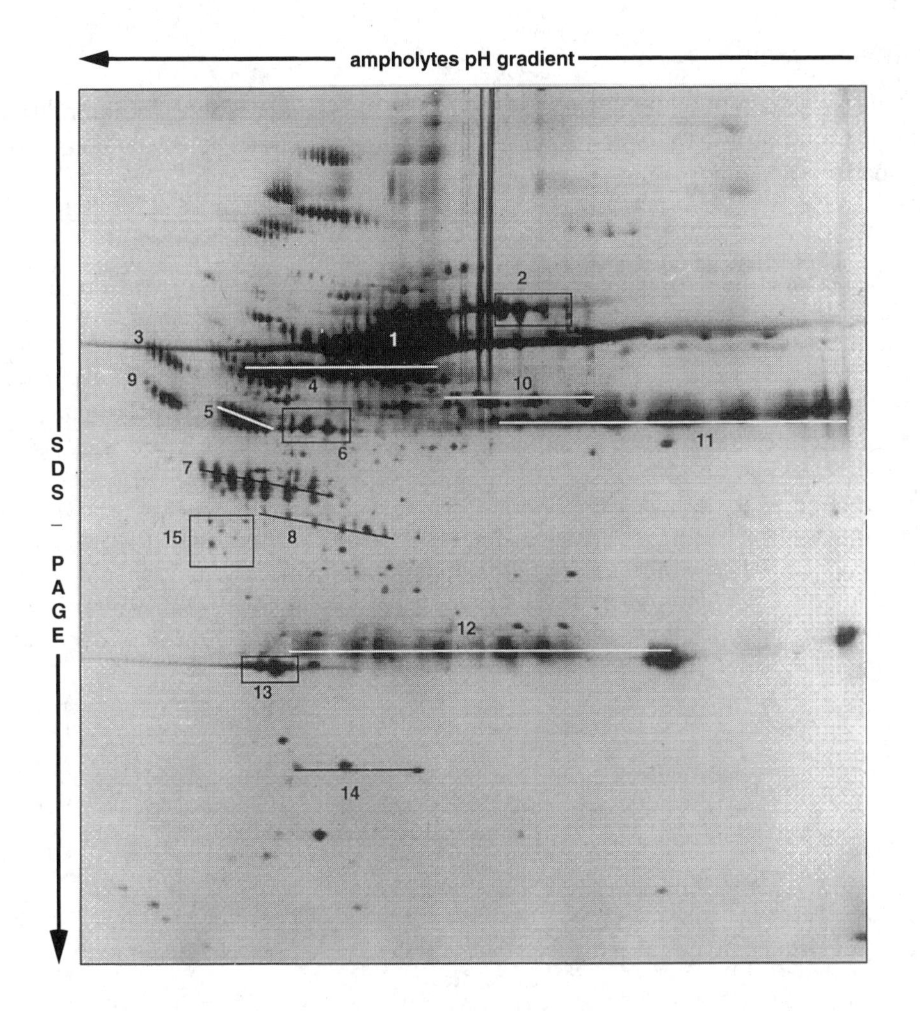

Fig. 2. Silver-stained plasma proteins separated by 2D PAGE using carrier ampholyte pH gradient in the first dimension. (1) albumin, (2) transferrin, (3) α1-antichymotrypsin, (4) IgA α-chain, (5) α1-antitrypsin, (6) fibrinogen γ-chain, (7) haptoglobin β-chain, (8) haptoglobin cleaved β-chain, (9) α2-HS-glycoprotein, (10) fibrinogen β-chain, (11) IgG γ-chain, (12) Ig light chain, (13) apolipoprotein A-1, (14) haptoglobin α2-chain, (15) apolipoprotein J.

apply a low amount of proteins onto the IEF gel. For silver-stained gel, concentrations up to 25–40 µg/gel are enough.

4. During the sample loading, it is very important to avoid the formation of air bubbles between the gel and the sample. After the loading, the Hamilton syringe should be withdrawn slowly along the side wall.
5. If more than 20 kVh are applied during the IEF, more cathodic drift occurs and the protein pattern is not stationary.
6. Low pressure on the 1-mL syringe should be exerted to extrude the capillary gel. If too much pressure is applied, small lumps will form on the gel.
7. Precise determination of the volume of solution necessary to cast the gels should first be done by measuring the volume of distilled water required to fill one resolving gel or the casting chamber.

Acknowledgments

The authors are grateful to Marianne Gex-Fabry and Claude Walzer for their discussions and revision of the chapter. P. G. received support from The Swiss Foundation for Research on Alcohol (FSRA).

References

1. Hochstrasser, D. F., Harrington, M. G., Hochstrasser, A. C., Miller, M. J., and Merril, C. R. (1988) Methods for increasing the resolution of two-dimensional protein electrophoresis. *Anal. Biochem.* **173,** 424–435.
2. Golaz, O. G., Walzer, C., Hochstrasser, D., Bjellqvist, B., Turler, H., and Balant, L. (1992) Red blood cell protein map: a comparison between carrier-ampholyte pH gradient and immobilized pH gradient, and identification of four red blood cell enzymes. *Appl. Theor. Electrophoresis* **3,** 77–82.
3. Oakley, B. R., Kirsch, D. R., and Morris, N. R. (1990) A simplified ultrasensitive silver stain for detecting proteins in polyacrylamide gels. *Anal. Biochem.* **105,** 361–363.
4. Hochstrasser, D. F. and Merril, C. R. (1988) Catalysts for polyacrylamide gel polymerization and detection of proteins by silver staining. *Appl. Theor. Electrophoresis* **1,** 35–40.
5. Rabilloud, T. (1992) A comparison between low background silver diammine and silver nitrate protein stains. *Electrophoresis* **13,** 429–439.

21

Two-Dimensional PAGE Using Immobilized pH Gradients

Graeme R. Guy and Robin J. Philp

1. Introduction

Since its introduction some 20 years ago by O'Farrell *(1)*, two-dimensional polyacrylamide gel electrophoresis (2D PAGE) has remained an extremely powerful method for the separation of complex protein mixtures, such as whole-cell lysates. In the first dimension, proteins are separated by charge using isoelectric focusing (IEF) and, in the second, by size employing sodium dodecyl sulfate (SDS)-PAGE. The combination of these two techniques results in a highly resolving separation that can be used for many applications, such as gene regulation studies, comparison of protein phosphorylation, and protein purification. At higher loadings, identification of individual proteins can be carried out using such methods as microsequencing or mass spectrometry.

The IEF step is normally carried out using carrier ampholyte (CA)-based IEF (*see* Chapter 20). In this system, CA are mixed with the acrylamide gel and a suitable buffer and cast in glass capillaries. The pH gradient is formed by prefocusing the gel in the presence of an electric field, thus causing the ampholytes to align into the required orientation. Samples are then loaded and "focused" during the separation. CA-IEF, however, has been noted to suffer from several drawbacks, namely, discontinuities in the pH gradient, which causes gaps in the separation, and an overall drift in the gradient causing loss of resolution in the basic region of the gel. Normally, to separate very basic proteins, a nonequilibrium pH gradient electrophoresis (NEPHGE) gel is run, allowing a reversal of the polarity used in standard CA-IEF and providing extended resolution of proteins in this area. This is carried out in addition to the standard separation. In addition to this, some difficulties in achieving higher protein loadings, necessary for the identification of proteins, have been found, notably resulting in excessive streaking of protein spots.

Immobilized pH gradient-IEF (IPG-IEF), introduced by Gorg *(2)*, is a more recent addition to the 2D PAGE technology. In this system, the IEF gels are cast using ampholytes, which form an integral part of the gel during polymerization and therefore circumvent some of the problems associated with CA-IEF. This may be particularly useful when there is a requirement to separate basic proteins, thus replacing the need to run NEPHGE gels.

From: The Protein Protocols Handbook
Edited by: *J. M. Walker Humana Press Inc., Totowa, NJ*

The gels are formed on a plastic backing sheet that is cut into strips, allowing a batch of identical strips to be prepared, dehydrated, and stored frozen until required. Unlike the CA-IEF gels, which are run vertically, IPG strips are applied to a flat-bed apparatus allowing an even temperature distribution along the whole length of the gel. Since the pH gradient is formed during the casting, prefocusing is unnecessary. It has been found that higher total protein loadings are achievable with IPG gels, and this has been particularly significant in the identification of low abundant proteins present in cell extracts *(3)*. This could possibly be attributed to the longer focusing times that can be used with these gels.

The use of precast IPG gels, as with the CA gels, can be "interfaced" to most SDS-PAGE systems for the separation in the second-dimension. This includes the flat-bed electrophoresis system or the more common vertical system.

Several manufacturers offer complete systems allowing reproducible 2D gels to be run with the minimum of experience. Both CA- and IPG-based IEF provide an excellent means of separating proteins in the first dimension of 2D PAGE. The nature of the proteins being separated, rather, dictate the type of separation to use. Commercially available IPG-IEF strips are prepared by rehydrating in the presence of a suitable buffer, after which they are installed in alignment guides. Samples are loaded by using cups attached to each gel, and up to 80 µL can be applied/strip.

The following description of the use of IPG-IEF is based on the Pharmacia Immobiline system, being used routinely in this laboratory.

2. Materials

1. IPG-IEF strips can be obtained from Pharmacia Biotech, Uppsala, Sweden (Immobiline Dry Strip, pH 4.0–7.0 or 3.0–10.0, 180 mm long, 3 mm wide, and 0.5 mm thickness).
2. Reswelling solution, make fresh (50 mL required): 8*M* urea, 0.5% v/v Triton X-100 (Scintran, BDH, UK), 10 m*M* DTT (Calbiochem, USA), 0.5% v/v CA, pH 3.0–1.0 (Pharmalytes, Sigma, St. Louis, MO), and Milli-Q water (Millipore, Bedford, MA) to 50 mL.
3. Sample buffer (make 48 mL, store at room temperature): 8*M* urea, 0.5% v/v Triton X-100, 0.01% w/v bromophenol blue, and Milli-Q water to 48 mL. Just prior to use, for each 1 mL of sample buffer used, add 40 µL of β-mercaptoethaol and 40 µL of CA, pH 3.0–10.0 (Pharmalytes, Sigma) (*see* Note 5).
4. The electrophoresis system used in this work (Multiphor II Electrophoresis System, including the Dry Strip Kit and Reswelling cassette) is available from Pharmacia Biotech, (Uppsala, Sweden). The methods and protocols associated with this type of separation have been developed by the manufacturer.
5. Paraffin oil (GPR, BDH, UK).
6. Equilibation buffer (store stock at room temperature): 50 m*M*, Tris-HCl, pH 6.8, 6*M* urea, 33% v/v glycerol (Ultra Pure, BRL), 1% w/v SDS (Biochemical, BDH), 0.01% w/v bromophenol blue, and Milli-Q water to 1000 mL. Prior to use, make required volume of equilibration buffer, 20 m*M* with DTT.
7. Equilibration tubes: Using 10-mL Falcon plastic serological pipets, snap off the tapered tip and the narrow plugged section resulting in a parallel sided tube. These can be sealed with Parafilm at one end. The IPG strip and equilibration buffer can be put in through the other end, and finally sealed with Parafilm for equilibration to be carried out.
8. Agarose (for gel overlay): 0.5% w/v agarose (Gibco-BRL), 25 m*M*, Tris-HCl, pH 6.8, 192 m*M* glycine (Sigma), and 0.1% w/v SDS in Milli-Q water.

3. Method

1. Remove plastic cover strip from the surface of the gel and place, with the plastic backing downward, onto one-half of the glass reswelling cassette. Place other half of cassette over the strips, and secure with the clips. The reswelling solution is then poured between the glass plates until the strips are covered. This is left at room temperature for a minimum of 6 h.
2. Align gel strips in the Dry Strip kit holder, and place buffer wicks moistened with water at each end. The electrodes are placed across the ends of the strips on the buffer wicks. The positive electrode is placed at the anodic end of the gel. The temperature of the water bath is set to 15°C.
3. Sample application cups are attached to the holder and pushed gently, but firmly, onto the surface of the gel.
4. Light paraffin oil is poured over the gel strips and allowed to fill up to the level of the sample cups.
5. Cell extract pellets are dissolved at the required concentration in the sample buffer and up to 80 µL added to each sample cup. The stock solution may crystalize out because of the high urea concentration. If this does occur, warm the bottle in a 37°C water bath for 30 min with occasional swirling. Do not heat above this temperature.
6. Carefully overlay the sample with paraffin oil using a pipet.
7. Connect leads to the power supply, and run under the following conditions:

Voltage	mA	W	Time, h	Vh
500	1.0	5	3	1500
2000	1.0	5	3	6000
3500	1.0	5	15	52,500

   This gives a total of 60,000 Vh.
8. Remove gels from apparatus and allow oil to drain briefly. Place strips in the equilibration tubes or other suitable container. Add about 8 mL of the equilibration buffer and seal. Agitate gently for 10–15 min.
9. Remove gel strips, and gently blot the edges on some wet 3MM filter paper. Position gel strips between the plates of the second-dimension gel with the edge in contact all the way along the width of the gel to minimize air bubbles between the gel strip and the SDS-PAGE gel.
10. Using a pipet, carefully overlay the gel with the agarose, and allow to set for several min. Add running buffer, and run second-dimension gel until the dye front is within 5 mm of the bottom.

4. Notes

1. Up to 12 strips can be equilibrated and run in the first dimension. The limiting factor is the number of second-dimension gels. It is advised to cover the cassette with Saran Wrap to prevent evaporation, especially if it is being left overnight.
2. After removing the strips from rehydration, carefully blot excess solution from the surface of the gel with water-soaked 3MM paper.
3. It is advised to check carefully for leaks after the sample cups have been fitted. As the paraffin oil is added, any leaks will show up, as the cups will slowly fill from below. If this happens, readjust sample cup and use a pipet to remove the oil. Watch carefully to see that the leak has been stopped.
4. Do not add paraffin oil above the top level of the sample cups.

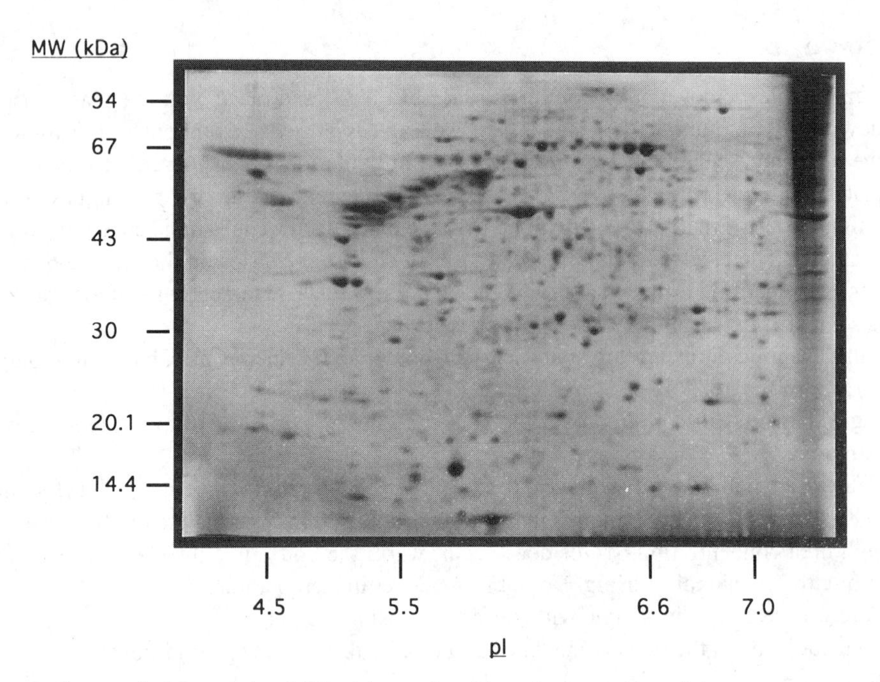

Fig. 1. Coomasie blue-stained 2D polyacrylamide gel separation of 1 mg total protein extracted from MRC-5 cells (Human lung fibroblast). Cell lysis was done using an SDS-based buffer followed by acetone precipitation of the protein. A precast IPG gel, pH 4.0–7.0, was used in the first dimension, and the sample was loaded at the cathode. The second-dimension was a 12.5% SDS-PAGE gel of dimensions 20 cm × 16 cm × 0.75 mm and run using a conventional vertical gel apparatus. Selected spots were identified by microsequencing following in-gel LysC digestion and HPLC separation of subsequent peptides.

5. Sample solubility is one of the keys to successful separations in 2D PAGE. Care should be taken to ensure no insoluble material is added to the sample cup. Briefly spin the sample in a bench-top centrifuge to pellet any insoluble material. SDS can be tolerated in this system, but it may be necessary to load the sample at the cathode if it is present.

   Cell lysis using SDS-based buffers can be used if a step, such as acetone precipitation, is carried out to remove as much of the SDS as possible. Again, loading at the cathode may be necessary.
6. Overlaying the sample is important to prevent the gel from drying after the sample has run into the gel as well as preventing oxidation of the sample.
7. The running conditions given are a guideline. Extended run times may be necessary for some separations, but this is a suitable starting point. The length of run is rather determined by the power supply. If a higher voltage can be reached, then the time can be shortened. The lower voltage at the beginning allows the sample to run into the gel before the voltage is increased to allow full separation. The bromophenol blue will be seen to migrate along the strip (if loaded at the anode) quite soon into the run. As it reaches the cathode, it will be seen to turn yellow owing to the acidic environment.
8. If the second-dimension gels are not being run at this point, then the strips can be frozen at –20°C.
9. When the stack of the second-dimension gel is cast, fill to within 5 mm from the top of the plates, and overlay with water or water-saturated butanol. This provides a flat interface for

the IPG strip to lie on. The IPG strip can then be slipped between the plates and allowed to make good contact with the full width of the gel. Ensure that the IPG strip fits between the two spacers of the second-dimension gel. If it is a bit tight, then cut off some of the excess plastic backing at the end of each strip. Note which end is the anode. Overlaying with agarose is not always necessary, but to prevent the gel strip from lifting as the running buffer is added, it is a good idea to include it.

10. There is usually some slight traces of insoluble material left at the origin of the IPG strip after running the first dimension. This is often resolubilized during the second-dimension and runs as a streak down one side of the gel. This does help sometimes in obtaining the correct orientation of the gel for assessing the results. An example of this can be seen in Fig. 1, where the sample was loaded at the cathode.

References

1. O'Farrell, P. H. (1975) High resolution two-dimensional electrophoresis of proteins. *J. Biol. Chem.* **250**, 4007–4021.
2. Gorg, A. (1992) Two-dimensional electrophoresis. *Nature* **349**, 545,546.
3. Hanash, S. M., Strahler, J. R., Neel, J. V., Hailat, N., Melhem, R., Keim, D., Zhu, X. X., Wagner, D., Gage, D. A., and Watson, J. T. (1991) *Proc. Natl. Acad. Sci. USA* **88**, 5709–5713.

22

Two-Dimensional PAGE Using Flat-Bed IEF in the First Dimension

Robin J. Philp

1. Introduction

Two-dimensional polyacrylamide gel electrophoresis (2D PAGE) of proteins involves the use of two independent separation techniques resulting in the ability to resolve very complex mixtures. Normally, isoelectric focusing (IEF) and sodium dodecyl sulfate- polyacrylamide gel electrophoresis (SDS-PAGE) are used in the first and second-dimensions, respectively *(1)*. Traditionally, the first dimension, IEF, has been run by casting the gels in glass capillaries that are subsequently loaded into a tank vertically, with an upper and lower reservoir. Samples can be applied to the space above the gel, and separation occurs by applying an electric field across the two ends of the gel. After the required separation time, the gels are removed by extruding them from the capillaries using a water-filled syringe to displace them directly into a suitable equilibration buffer.

An alternative to the vertical IEF method is the use of horizontal or flat-bed electrophoresis. The apparatus for flat-bed electrophoresis consists of a hollow glass or ceramic "bed," which can be cooled by circulating water or other suitable coolant through internal ducting. Gels are cast as a slab on a plastic backing sheet and are then cut into individual strips, which are applied to the flat horizontal surface. Samples are applied using small reservoirs or cups, which are held onto the surface of the gel and can be positioned at any point along the length of the gel, allowing loading to be done at either the anode or cathode. Unlike the capillary IEF gels, flat-bed IEF gels can be transferred directly to a suitable equilibration buffer after running and prior to loading onto the second-dimension gel. The plastic backing allows the transfer to the second-dimension to be carried out with minimal stretching or distortion.

Both carrier ampholyte (CA) and immobilized pH gradient (IPG) *(2)* IEF gels can be used on a flat-bed apparatus, although IPG gels have shown much popularity for the IEF step of 2D PAGE in recent years. The choice of which to use is ultimately dependent on the type of analysis being carried out, as well as specific factors relating to the sample, such as high protein loading or the ability to cast a narrow pH gradient (*see* other chapters in this volume). The use and treatment of the two types of gels are very similar with one or two exceptions: CA gels require prefocusing to allow the formation

From: The Protein Protocols Handbook
Edited by: J. M. Walker Humana Press Inc., Totowa, NJ

of the pH gradient, and with prolonged focusing times, CA-IEF gels are subject to a loss of resolution in the basic region owing to cathodic drift. IPG gels can be purchased in a precast, dehydrated form, which is stored frozen. This has the advantage of not only the convenience factor, but also that batches of identical gels can be stored allowing run-to-run reproducibility.

Flat-bed IEF gels are compatible with both flat-bed or vertical electrophoresis when running the second-dimension SDS-PAGE separation.

2. Materials

1. Precast, dehydrated IPG gels for flat-bed IEF can be obtained from Pharmacia Biotech, Uppsala, Sweden. These are available as single strips in two lengths or as a multiwell slab gel.
2. Apparatus to run flat-bed IEF can be purchased from Pharmacia Biotech. Methods and protocols for this type of separation have been developed by the manufacturer.
3. Reswelling solution, mae fresh (50 mL required): 8*M* urea (Ultra Pure BRL, Gaithersburg, MD), 0.5% v/v Triton X-100 (Scintran, BDH, UK), 10 m*M* DTT (Calbiochem, La Jolla, CA), 0.5% v/v CA, pH 3.0–10.0 (Pharmalytes, Sigma, St. Louis, MO), and Milli-Q water (Millipore, Bedford, MA) to 50 mL.
4. Sample buffer (make 48 mL, store at room temperature): 8*M* urea, 0.5% v/v Triton X-100, 0.01% w/v bromophenol blue and Milli-Q water to 48 mL. Just prior to use, for each 1 mL of sample buffer used, add 40 μL of β-mercaptoethanol and 40 μL of CA, pH 3.0–10.0 (Pharmalytes, Sigma) (*see* Note 5).
5. Equilibration buffer (store stock at room temperature): 50m*M* Tris-HCl, pH 6.8, 6*M* urea, 33% v/v glycerol (Ultra Pure, BRL), 1% w/v SDS (Biochemical, BDH), 0.01% w/v bromophenol blue, and Milli-Q water to 1000 mL. Prior to use make required volume of equilibration buffer 20 m*M* with DTT.
6. Equilibration tubes: Using 10-mL Falcon plastic serological pipets, snap off the tapered tip and the narrow plugged section, resulting in a parallel-sided tube. These can be sealed with Parafilm at one end, the IPG strip and equilibration buffer can be put in through the other end, and finally sealed with Parafilm for equilibration to be carried out.
7. Agarose (for gel overlay): 0.5% w/v agarose (Gibco-BRL), 25 m*M* Tris-HCl, pH 6.8, 192 m*M* glycine (Sigma), and 0.1% w/v SDS in Milli-Q water.

3. Method

1. Remove plastic cover strip from the surface of the gel, and place, with the plastic backing downward, onto one-half of the glass reswelling cassette. Place other half of cassette over the strips, and secure with the clips. The reswelling solution is then poured between the glass plates until the strips are covered. This is left at room temperature for a minimum of 6 h.
2. Align gel strips in the Dry Strip kit holder, and place buffer wicks moistened with water at each end. The electrodes are placed across the ends of the strips on the buffer wicks. The positive electrode is placed at the anodic end of the gel. The temperature of the water bath is set to 15°C.
3. Sample application cups are attached to the holder and pushed gently, but firmly, onto the surface of the gel.
4. Light paraffin oil is poured over the gel strips and allowed to fill up to the level of the sample cups.
5. Cell extract pellets are dissolved at the required concentration in the sample buffer, and up to 80 μL added to each sample cup. The stock solution may crystallize out because of the

high urea concentration. If this does occur, warm the bottle in a 37°C water bath for 30 min with occasional swirling. Do not heat above this temperature.

6. Carefully overlay the sample with paraffin oil using a pipet.
7. Connect leads to the power supply, and run under the following conditions:

Voltage	mA	W	Time, h	Vh
500	1.0	5	3	1500
2000	1.0	5	3	6000
3500	1.0	5	15	52,500

This gives a total of 60,000 Vh.

8. Remove gels from apparatus and allow oil to drain briefly. Place strips in the equilibration tubes or other suitable container. Add about 8 mL of the equilibration buffer and seal. Agitate gently for 10–15 min.
9. Remove gel strips, and gently blot the edges on some wet 3MM filter paper. Position gel strips between the plates of the second-dimension gel with the edge in contact all the way along the width of the gel to minimize air bubbles between the gel strip and the SDS-PAGE gel.
10. Using a pipet, carefully overlay the gel with the agarose, and allow to set for several min. Add running buffer, and run second-dimension gel until the dye front is within 5 mm of the bottom.

4. Notes

1. Up to 12 strips can be equilibrated and run in the first dimension. The limiting factor is the number of second-dimension gels. It is advised to cover the cassette with Saran Wrap to prevent evaporation, especially if it is being left overnight.
2. After removing the strips from rehydration, carefully blot excess solution from the surface of the gel with water-soaked 3MM paper.
3. It is advised to check carefully for leaks after the sample cups have been fitted. As the paraffin oil is added, any leaks will show up, as the cups will slowly fill from below. If this happens, readjust sample cup and use a pipet to remove the oil. Watch carefully to see that the leak has been stopped.
4. Do not add paraffin oil above the top level of the sample cups.
5. Sample solubility is one of the keys to successful separations in 2D PAGE. Care should be taken to ensure no insoluble material is added to the sample cup. Briefly spin the sample in a bench-top centrifuge to pellet any insoluble material. SDS can be tolerated in this system, although less so with CA-IEF, but it may be necessary to load the sample at the cathode if it is present.

   Cell lysis using SDS-based buffers can be used, although it will be necessary to include a protein precipitation step (e.g., with acetone) to remove as much SDS as possible. Again, loading at the cathode may be necessary.
6. Overlaying the sample is important to prevent the gel drying after the sample has run into the gel as well as preventing oxidation of the sample.
7. The running conditions given are a guideline. Extended run times may be necessary for some separations, but this is a suitable starting point. The length of run is rather determined by the power supply. If a higher voltage can be reached, then the time can be shortened. The lower voltage at the beginning allows the sample to run into the gel before the voltage is increased to allow full separation. The bromophenol blue will be seen to migrate along the strip (if loaded at the anode) quite soon into the run. As it reaches the cathode, it will be seen to turn yellow owing to the acidic environment.

8. If the second-dimension gels are not being run at this point, then the strips can be frozen at –20°C. However, some evidence shows a reduced number of protein spots are seen after freezing.
9. When the stack of the second-dimension gel is cast, fill to within 5 mm from the top of the plates, and overlay with water or water-saturated butanol. This provides a flat interface for the gel strip to lie on. The gel strip can then be slipped between the plates and allowed to make good contact with the full width of the gel. Ensure that the gel strip fits between the two spacers of the second-dimension gel. If it is a bit tight, then cut off some of the excess plastic backing at the end of each strip. Note which end is the anode. Overlaying with agarose is not always necessary, but to prevent the gel strip from lifting as the running buffer is added, it is a good idea to include it.
10. There is usually some slight traces of insoluble material left at the origin of the gel strip after running the first dimension. This is often resolubilized during the second-dimension and runs as a streak down one side of the gel. This does help sometimes in obtaining the correct orientation of the gel for assessing the results.

References

1. O'Farrell, P. H. (1975) High resolution two-dimensional electrophoresis of proteins. *J. Biol. Chem.* **250**, 4007–4021.
2. Gorg, A. (1992) Two-dimensional electrophoresis. *Nature* **349**, 545,546.

23

Free Zone Capillary Electrophoresis

David J. Begley

1. Introduction

Electrophoresis within silica capillaries is a relatively new technique, with the first publications appearing in the 1970s followed by a rapid expansion and establishment of the technique in the 1980s. Commercial machines for the application of capillary electrophoresis appeared in the late 1980s and now a number of machines are manufactured.

All of the commercially available machines function in essentially the same way, although there is of course individual variation. This account of capillary electrophoresis is tailored not to any particular machine, but is written in a manner that describes the basic theory and method in a way that can be easily varied to suit a particular application. This chapter contains a detailed description of a capillary electrophoresis apparatus and some theory of the operation, together with protocols and notes for performing separations.

High-performance capillary electrophoresis (HPCE) can be used to separate a wide variety of solutes, both charged and uncharged, and is particularly suited to the separation of small peptides and proteins. Separations can be performed under a variety of conditions with free zone capillary electrophoresis (FZCE), where the solutes being separated are in simple solution in buffer. A variant of the technique is capillary isoelectric focusing (CIF), where the ampholyte and sample are premixed and loaded into the capillary before application of a voltage. A pH gradient is formed, and separation of samples according to their isoelectric point (pI) takes place within the capillary. This chapter will be essentially confined to FZCE separations, since these have found the greatest application in the biological sciences *(1–6)*.

Capillary electrophoresis is normally carried out in fused silica capillaries with internal diameters between 10 and 100 μm. Capillary length may be varied with a longer capillary producing a greater spatial separation between two solutes of similar mobility. A popular capillary for a variety of separations would be one of 25-μm internal diameter and 20 cm in length. A potential difference is normally applied across the two ends of the capillary with a power source. Electrophoresis is usually conducted under constant voltage conditions with the current finding its own level. Typical operating parameters under differing conditions are shown in Table 1. Separation under

From: The Protein Protocols Handbook
Edited by: J. M. Walker Humana Press Inc., Totowa, NJ

Table 1
Operating Parameters for Capillary Electrophoresis Under Differing Conditions[a]

Buffer	pH	Voltage, kV	Current, μA	Resistance, Meg Ω	Power, watts
Phosphate 100 m*M*	2.5	8.0	12.6	634.9	0.101
Phosphate 100 m*M*	4.4	8.0	25.25	316.8	0.202

[a]Note that the resistance approximately halves with the change in pH, and the current and the power dissipation (heat production) double.

normal conditions produces significant amounts of heat, but this is not usually a problem that affects the quality of a separation or leads to denaturation of protein, since the large surface area of a capillary in relation to its volume efficiently dissipates the heat produced. Power sources for use in capillary electrophoresis normally supply voltages up to 30 kV or more.

1.1. Apparatus and Theory of Operation

A typical layout for a capillary electrophoresis apparatus is shown in Fig. 1. The apparatus consists of the silica capillary, which is usually coiled to make it more manageable. The ends of the capillary project into two buffer reservoirs that contain silver electrodes. A facility for changing the polarity of these two electrodes usually exists, although for the majority of applications, the left-hand electrode on the figure will be the anode and the right-hand the cathode.

When a potential difference is applied across the ends of the capillary and because the walls of the capillary have a standing charge (the ζ potential), an electroendo-osmotic flow of water is produced from anode to cathode. This is shown diagrammatically in Fig. 2. Thus, migration of a positively charged solute from anode to cathode along the capillary is partially produced by the voltage gradient applied and partly the result of electroendo-osmotic flow. This electroendo-osmotic flow makes it possible to achieve the separation of uncharged solutes, such as steroids, in a silica capillary. With uncharged solutes, the electroendo-osmotic flow of water provides the propulsive force, and solutes of similar molecular weight are separated by a combination of differences in their mass and also partly by a chromatographic interaction with the capillary wall. For a charged species, the apparent rate of migration (U_{app}) is the product of the electrophoretic mobility (U), plus the rate at which the solute is propelled by the electroendo-osmotic flow (U_{eo}), and is related to the mass/charge ratio of the solute (Fig. 2).

Electroendo-osmosis can be greatly reduced by coating the inner surface of the capillary with a thin layer of polyacrylamide. This layer of polyacrylamide abolishes the ζ potential and reduces the electroendo-osmotic flow to virtually zero. Under these conditions U_{app} and U are virtually identical, and the mobility of solutes in the electrical gradient is a product of the mass charge ratio. Highly charged solutes with a small mass exhibit the greatest mobility. The separation of charged solutes is improved in coated capillaries, because when there is fluid flow in the capillary, the solute front is ellipti-

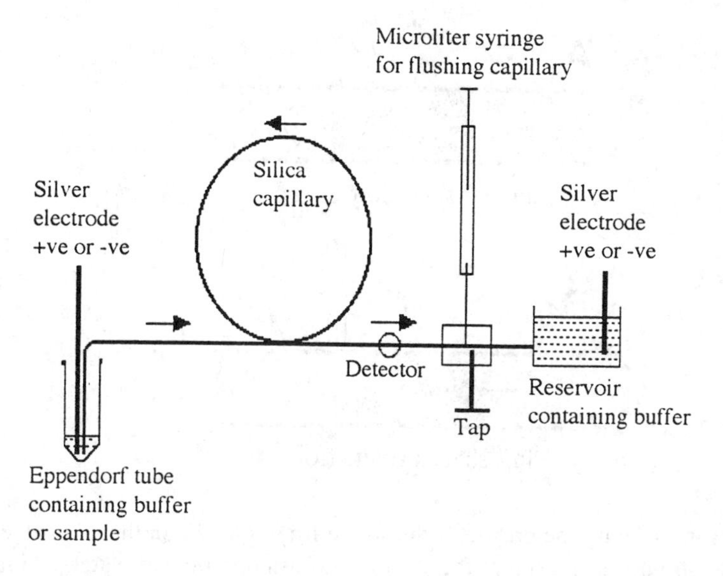

Fig. 1. A typical layout for a high performance capillary electrophoresis apparatus, (HPCE). For electrophoretic loading, an Eppendorf tube containing the sample is offered up to the left-hand electrode, and the capillary and a loading voltage applied for a set time. With suitable Eppendorf tubes, the sample contained in loading buffer may be as little as 10 μL. The tube containing the sample is then substituted with one containing the running buffer and the run conducted. After the run, the tap can be closed and the capillary back-flushed with running buffer between runs. At pH 2.5 with positively charged analyses, the left-hand electrode will normally be the anode and the right-hand electrode the cathode.

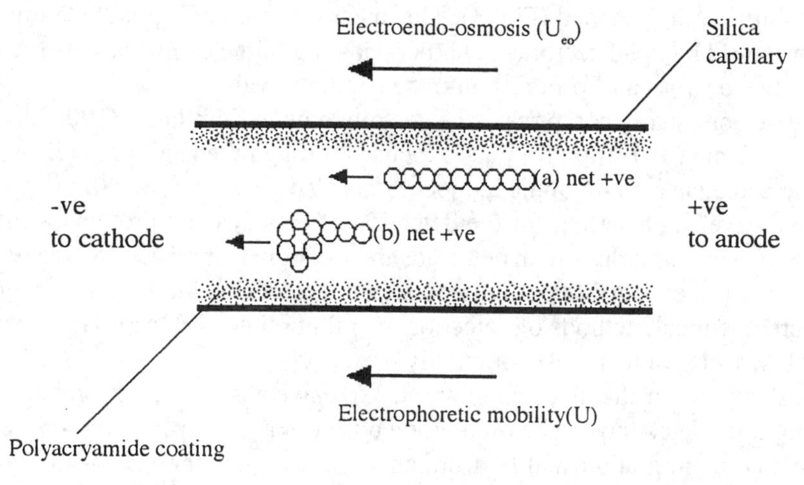

Fig. 2. Migration of solute in the silica capillary. The ζ potential on the wall of the capillary produces an electroendo-osmotic flow (U_{eo}) of water from the anode to cathode. Coating the capillary internally with polyacrylamide reduces this standing charge and reduces the magnitude of U_{eo}. The electrophoretic mobility of a charged solute (U) is dependent on the mass/charge ratio and the applied potential difference. Thus the actual rate of migration (U_{app}) is equal to $U + U_{eo}$. With coated capillaries, the degree of electroendo-osmosis (U_{eo}) is minimized For two solutes of equal charge and mass, such as (a) and (b). The configuration of the molecule can also influence the rate of migration (U) with (b) exhibiting a greater molecular sieving and/or chromatographic effect than (a). (*See also* Fig. 4.)

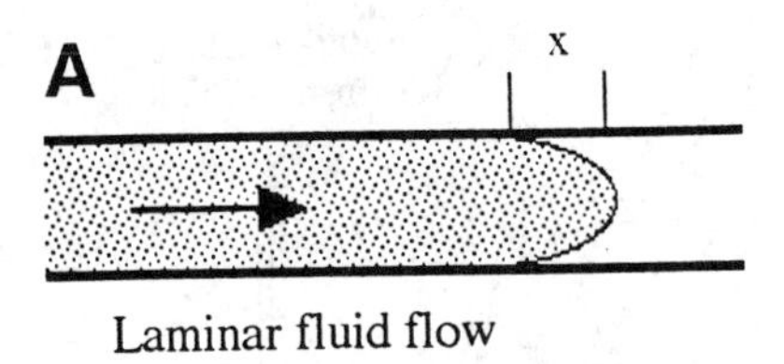

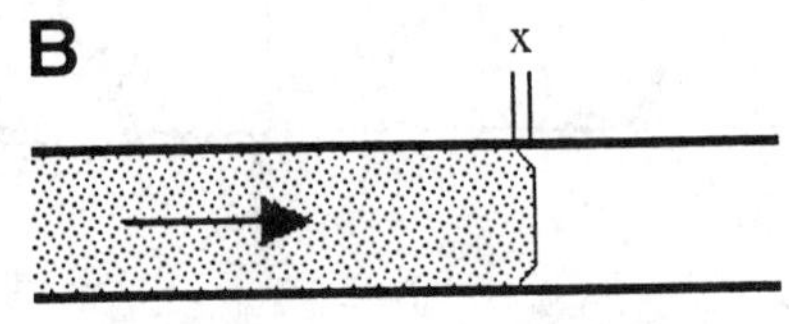

Fig. 3. Factors influencing the profile of the solute front. **(A)** When there is solvent movement through the capillary as a result of the electroendo-osmotic flow of water, a laminar flow profile is set up where the solvent near the wall of the capillary moves more slowly as a result of solvent drag. **(B)** In coated capillaries where the solutes are migrating through the solvent down an electrical gradient, this effect is minimized and the solute front is plug-shaped, and thus produces a sharp well-defined zone (x).

cal in shape as a result of laminar flow in the capillary, whereas when the propulsive force is purely the electrical gradient the solute is migrating through the solvent and the solute front is "plug" shaped (Fig. 3). This means that solute spread is minimized and peaks produced by the detector are sharper. Coating of the capillaries also reduces any nonspecific adsorption of proteins onto the capillary wall.

The detector usually comprises a beam from a deuterium lamp (190–380 nm; UV) or tungsten lamp (380–800 nm) that is focused through the capillary. The wavelength can be selected via a holographic monochromator to give a bandwidth of approx 5 nm. A typical wavelength setting for the detection of peptides and proteins would be 200 nm. The transmitted light is then detected, and the signal amplified to give typical full-scale deflections of 1.0–0.005 AUFS on a chart recorder. Fluorescence, conductivity, and electrochemical methods of detection are all possible with capillary electrophoresis, but UV absorption is most commonly used.

Almost any buffer that does not have an excessive absorption of light at the chosen wavelength for detection can be used. For a buffer of a given pH, if the solute has a net positive charge, migration will be from anode to cathode, the rate of migration for a given molecular mass being proportional to the charge. Thus, if the buffer pH is close to the isoelectric point (pI) of the solute in question, the rate of migration through the capillary will be slower. This effect can be used to advantage to separate similar molecules with similar isoelectric points, for example, isoforms of a protein or enzyme. A buffer with a pH close to the pI of a solute, in this manner, accentuates the relative differences in mobility of solutes with a similar pI with respect to their elution time. The time taken for the solute to move through the capillary to the detector is called the elution time (T_e), and depends on the mass/charge ratio under the conditions of separa-

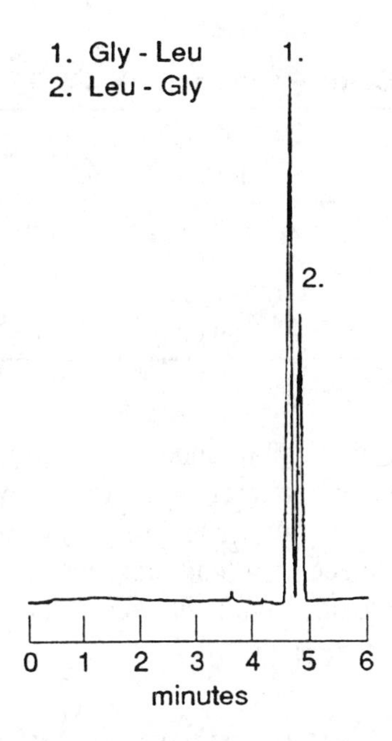

Fig. 4. Separation of two dipeptides glycyl-L-leucine and leucyl-glygine. These two peptides can be separated with free zone capillary electrophoresis. Their slightly different configurations allow separation to be achieved by a combination of molecular sieving and a chromatographic interaction with the walls of the capillary. Sample: Mixture of glycyl-L-leucine and leucyl-glycine, 25 μg/mL. Buffer: Phosphate/HCl, pH 2.5. Load: 10 m*M* buffer, 4 kV, 4 s, 2.2–2.6 μA. Run: 100 m*M* buffer, 8 kV, 13.8 μA. Detection: 200 nm.

tion and the length of the capillary and the voltage gradient, (i.e., the driving force through the capillary). Figure 4 illustrates the powerful resolution that can be obtained with FZCE. Typical elution times for some proteins in relation to their physical characteristics are shown in Table 2.

The sample can be introduced into the capillary by:

1. Displacement loading (sometimes called vacuum or pressure injection);
2. By electrokinetic loading (sometimes called electroendo-osmotic loading); and
3. Electrophoretic loading.

With displacement loading, the sample is injected directly into the capillary via a suitable manifold or drawn into the capillary by applying a negative pressure at the far end. Both of these methods suffer from problems with reproducibility. Electrokinetic loading is achieved with uncoated capillaries, and the sample is moved into the capillary by applying a potential difference that induces an electrophoretic movement of the solutes and an electroendo-osmotic flow. It is a combination of these two effects that moves sample into the capillary. Only with coated capillaries is pure electrophoretic loading achieved, where the electrophoretic mobility alone moves the solute into the capillary. It is important to appreciate that the amount of solute that moves into the capillary in a given time and, with a given potential difference, is proportional to

Table 2
Elution Time, T_e, for Some Proteins at pH 4.4

Protein	T_e	Mol wt, Daltons	pI	Mean radius, nm
IgG	8.4	150,000	6.4–8.9	5.34
Transthyretin (prealbumin)	10.2	61,000	4.7	3.25
Albumin	15	69,000	4.8–4.9	3.58

the electrophoretic mobility of that solute under the loading conditions. Thus, with a mixture of solutes of differing electrophoretic mobilities more of the more mobile solute is loaded relative to the solutes of lesser mobility. With electrokinetic loading, this effect is reduced and, with displacement loading, does not occur. The protocols that follow assume electrophoretic loading with coated capillaries.

2. Materials

1. Buffer: For the separation of peptides and proteins, a typical buffer would be 100 m*M* phosphate/HCl buffer of pH 2.5. At this pH, all proteins and peptides will have a net positive charge and will migrate from anode to cathode. The buffer should contain 0.1% hydroxypropylmethylcellulose (HPMC). This has the effect of protecting the polyacrylamide coating of the capillary and also increases the viscosity of the buffer, which slightly retards the movement of solute through the capillary and accentuates separation in relation to mass and shape. To make this buffer, first dissolve sufficient HPMC in HPLC-grade water in a volumetric flask. This is a slow process. Do not warm the solution excessively, since this degrades the HPMC. Then add sufficient sodium di-hydrogen phosphate for a 100 m*M* solution, adjust the pH to 2.5 with dilute HCl, and bring to volume. This buffer will store in the refrigerator at 4°C for a week or so, but check the pH and look out for the growth of bugs!
2. Sample: Dissolve the protein or peptide or a mixture of proteins or peptides to be separated in a sample of 10 m*M* phosphate buffer, pH 2.5 (i.e., one-tenth of the concentration of the separating buffer). A sample concentration of 50 μg/mL is a good starting point. This is assuming that the sample that you wish to separate is available in crystalline form in a bottle. Biological samples present a little more difficulty and often need special attention (*see also* Note 1).

3. Methods

3.1. Loading the Sample

To load the sample electrophoretically, an Eppendorf tube containing the sample, which may be as little as 10 μL in a suitably sized tube, is offered up to the cathode and capillary tip. A potential difference, for example, 8 kV, is applied for 8 s, and the sample is loaded. In a mixture, the amount of each solute loaded into the capillary is proportional to the electrophoretic mobility, the loading time, and the voltage applied, and is thus proportionally different for each solute. This is shown graphically for bovine serum albumin (BSA) in Fig. 5. Solutes will have a greater mobility in a buffer of lower ionic concentration than the running buffer. Hence, the loading buffer should be 10

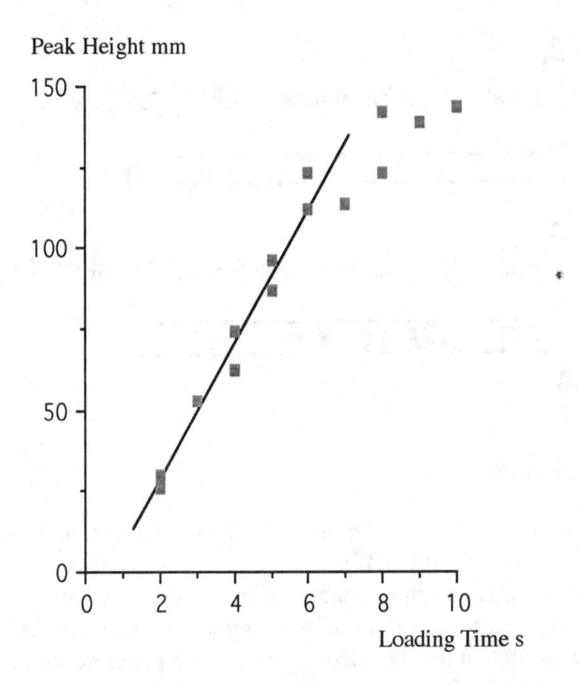

Fig. 5. Effect of time on electrophoretic loading and peak height for bovine serum albumin. Sample: Bovine serum albumin, 50 µg/mL (Sigma, A7888) Buffer: Phosphate/HCl, pH 2.5. Load: 10 m*M* buffer, 8.0 kV, 11.8–12.6 µA. Run: 100 m*M* buffer, 7.0 kV, 8.8–11.2 µA. Detection wavelength: 200 nm. Note that the amount of BSA loaded declines with longer loading times.

m*M,* and the running buffer 100 m*M*. Thus, a greater difference between the ionic concentration of the running and loading buffer will produce a greater loading of the sample. With electrophoretic loading, this phenomenon also has the useful effect of also causing the solutes to slow down once they enter the more concentrated buffer at the start of the capillary. This results in a "stacking" of the solutes at the beginning of the capillary (*see* Fig. 6) and means that all of the solutes are starting their migration essentially from the start of the capillary. The end result is a better resolution of the individual peaks in a sample, since the zone concentrations may be increased by a factor of ten. This does not occur with electrokinetic or displacement loading, where a substantial length of the capillary will contain the sample, prior to the run. For a particular sample, the experimenter must vary the loading current and loading time until an optimal loading and separation of the solutes is achieved. For electrophoretic loading a good starting point for a peptide or protein is 8 kv for 8 s and to then vary first loading time and then voltage until loading is optimized. A voltage of 8 kV with a 25 µm × 20 cm capillary with pH 2.5 phosphate buffer will give a loading current of approx 11–14 µA. The current should be repeatable from sample to sample, when loaded under the same conditions.

3.2. Running the Separation

The run can be initially carried out at the same potential difference as the loading voltage. The run should be carried out with a constant voltage set on the power supply

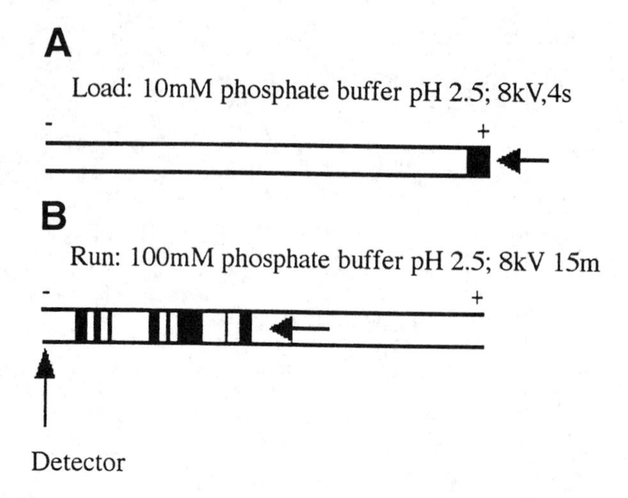

Fig. 6. Loading and running the separation. **(A)** When the sample is loaded from 10 m*M* buffer into a capillary containing 100 m*M* buffer, the effect is to stack the sample at the start of the capillary. **(B)** When the separating voltage is applied, the positively charged solutes in the mixture migrate toward the cathode at a rate predominantly determined by their mass-charge ratio to form discrete zones that are detected as absorbance peaks at the detector.

and, at 8 kV, will give approximately the same current as the loading conditions (*see* Table 1). Elution times for different solutes of course vary widely according to mass and charge. However, few peaks yielding useful information will elute after 25 min or so, as a result of zone spreading and distortion of the peaks. The voltage can be increased to speed the migration of slow peaks, or reduced, to slow down fast-moving solutes and improve the resolution between peaks with a similar T_e. Small peptides with mol wt of 1 kDa or less usually elute within 15 min, whereas a complex biological sample may still be producing peaks after 20 min or so. Careful experimentation with the separating voltage must be conducted to optimize the separation. In reality, this means a careful balance between the loading conditions and the separating conditions to produce the best separation of the solutes of interest. In a complex mixture, as with the technique of high-performance liquid chromatography (HPLC), these conditions are often a compromise, and it proves impossible to separate all of the contents of a complex mixture with equal resolution.

3.3. Quantifying the Results

The best method for determining the quantity of material represented by a peak on the chart recorder is to measure the peak height. Peak area with HPCE techniques is not reliable as a measure of the quantity of solute, since a rapidly migrating solute will pass the detector quickly and produce a narrow peak, and a slowly moving solute will produce a wider peak. Thus, peak area is related to speed of migration as well as the quantity of material present. This effect is obviously most pronounced with coated capillaries, where electroendo-osmosis is reduced to a minimum. A graph relating peak height to the quantity of albumin loaded is shown in Fig. 7. The apparatus must be calibrated with known quantities of the solute being determined to give the relationship between peak height and quantity for that solute.

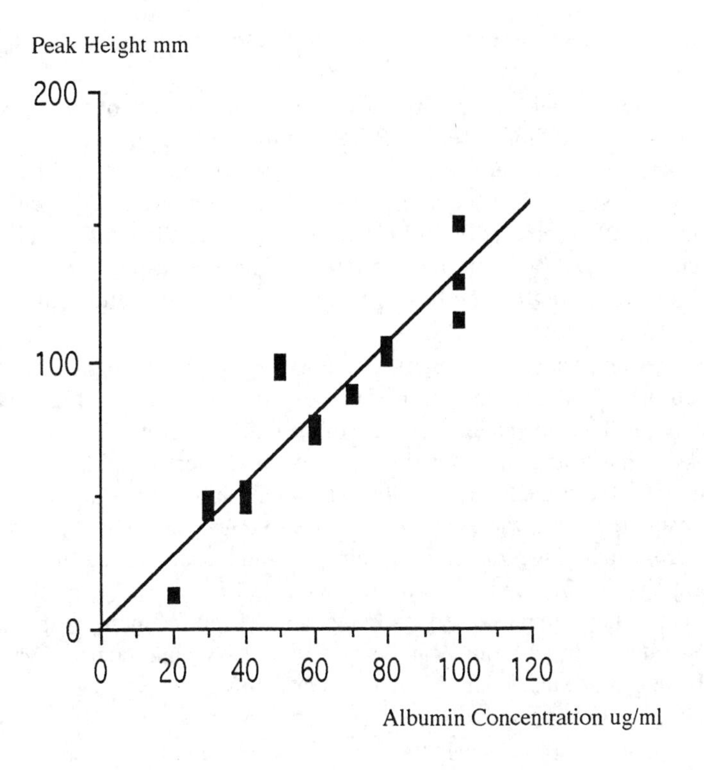

Fig. 7. Relationship between concentration of bovine serum albumin in sample and peak height. Sample: Bovine serum albumin, 20–100 μg/mL (Sigma, A7888). Buffer: Phosphate/HCl, pH 2.5. Load: 10 m*M* buffer, 8.0 kV, 8 s, 11.4–12.6 μA. Run: 100 m*M* buffer, 7 kV, 10.6–12.0 μA. Detection: 200 nm $T_e = 4.31 \pm 0.03$ min ($n = 16$).

4. Notes

1. Sample preparation: Biological samples are often complex mixtures of solutes, and a careful pretreatment of the sample is usually required prior to any analysis *(7)*.
   a. Desalting. Samples of biological fluids will contain the ionic constituents of extracellular fluid at a total concentration of 290 mosM. This will be more concentrated than the running buffer, and thus for the reasons stated earlier, electrokinetic and electrophoretic loading under these conditions will be poor. When samples containing a high concentration of protein and peptide are available in reasonable volume, for example saliva, plasma, and seminal fluid, simply diluting the sample with HPLC-grade water will suffice to lower the salt concentration to a level where loading becomes efficient. If the sample has a low concentration of peptide or protein, for example, cerebrospinal fluid or aqueous humor, or when only a few microliters of sample are available, dilution will be ineffective and a different strategy must be employed.

   It is possible to desalt the sample by passing it through a desalting resin, such as AG11-A8 (Bio-Rad), or a similar ion-retarding or ion-exchange resin. If the sample volume is just a few microliters it can be passed through a resin by centrifugation. A small amount of resin may be loaded into a disposable filter holder with luer fittings containing a membrane suitable for aqueous solvents, such as an ACRO LC3A. The sample can then be loaded onto the resin, and the whole assembly put into an Eppendorf tube and centrifuged. The sample can then be retrieved from the Eppendorf tube with virtually 100% recovery.

Some luer filter assemblies are fitted with a membrane that contains a reversed-phase sorbent similar to those used in HPLC columns (SM-2 membrane, Bio-Rad). A sample of peptides in aqueous solution can be passed through the filter, and the peptides will bind strongly to the hydrophobic membrane. Remaining salts can be washed from the membrane with HPLC-grade water, and the peptides eluted with a water/acetonitrile or water/methanol mixture. The eluted peptides can then be recovered by evaporating the eluting solvent (freeze-drying) and redissolving in loading buffer.

b. Removing some components of a complex mixture: With some biological samples, the number of absorbing components in the UV range is so great that individual peaks often cannot be resolved adequately, because of the high background absorbance masking the peaks of interest. The removal of most of the material that is not of interest from the sample is the strategy to apply here. The same approach as is used in Note 1a above can be applied using various absorbents layered into the ACRO-type filter. For example, hydroxyapatite, which binds proteins *(8)*, can be put into the filter assembly. This will selectively remove proteins from a complex mixture. If the proteins are the solute of interest, they can be subsequently recovered from the hydroxyapatite. Obviously, a large number of variations can be used with this technique using various immobilized ligands on beads, such as antibodies or lectins and a wide variety of liquid chromatography column packings.

c. Extraction of peptides from tissue: Peptides in tissue samples can be extracted by homogenizing the tissue in 1*N* HCl and centrifuging. The supernatant containing the peptides can be further purified by passing the sample through a SEP-PAK (Waters Associates) sample preparation cartridge *(9)*. The peptides can then be eluted with a water/acetonitrile mixture and freeze-dried before redissolving in loading buffer.

2. Care of the capillary: The silica capillaries are remarkably robust, and it is in fact quite difficult to damage them. A capillary should perform several hundred separations before it needs replacing. The greatest danger to the capillary is blockage either as the result of particulate material in the sample being drawn into the capillary or because solute has crystallized in the capillary lumen. To guard against the former, all samples should be centrifuged before any attempt is made to load them, and if very fine suspended material is present, the sample should be passed through a 0.45-μ filter. Crystallization of a solute in the capillary is most likely to occur when the solute passes from buffer of one composition to another, i.e., from the loading buffer into the running buffer. This effect should be borne in mind when dealing with solutes of limited solubility. If the capillary blocks, some attempts to unblock it may be successful. Back-flush the capillary with buffer under pressure. This will not damage the capillary and may shift the blockage. Some authorities have been known to use an HPLC pump to do this producing pressures in excess of 3000 psi to shift the blockage! Remember that the blockage is most likely to be at the end of the capillary where the run is started, so it is important to back-flush. Blockage as the result of crystallization can sometimes clear with time, so do not be too efficient in throwing blocked capillaries away. With time, the crystals can slowly dissolve and the blockage disappear. This process can sometimes be helped by gently heating the capillary in a laboratory oven. Remember you have nothing to lose but your capillary, and a blocked capillary is useless.

Capillary performance and life can be enhanced by regular flushing with a 0.5% SDS (sodium dodecylsulfate/laurylsulfate) solution after use. Capillaries should be stored between use in the refrigerator at 4°C and filled with distilled water. These treatments will greatly minimize the chance of a capillary blocking while being stored.

It is often a good idea to do a blank run for about 20 min each day when the apparatus is first set up. The voltage and current can then be checked for stability. This procedure also seems to "run in" the capillary for the day's work.

3. Optimizing the separation:
   a. Running conditions: Suggestions as to the initial loading and running conditions to select have been given above. However, careful experimentation with the settings on the power source can often improve a separation enormously. The best practice is to optimize the loading first by altering the voltage and the time. Try increasing and decreasing the voltage initially, and then altering the time. In general, with most solutes, shorter loading times give the best results. Several runs should be performed at various potential differences until the best separation for the solute(s) of interest are obtained.
   b. Detection wavelength: As the solutes are migrating in an electrical field, if the current is switched off, they will stop moving. Thus, when a solute of interest is in the UV light path, the power can be switched off and the solute is immobilized. The wavelength setting on the monochromator can then be adjusted until the absorbance is optimized. In this way, an absorbance spectrum for an unknown solute separated from a mixture can be obtained. Remember that the silica capillary will also have an absorbance spectrum, so the best approach is to obtain this spectrum first with the capillary filled with buffer and to then subtract this spectrum from that of the capillary containing the solute of interest.
4. Special techniques: A particular technique, which may be useful if a sample contains a mixture of both charged and noncharged solutes to be separated, is that of micellar electrokinetic capillary Chromatography (MECC) *(10–12)*, where separation is achieved by a combination of both electrophoretic migration and micellar partitioning. Micelles are formed by adding SDS 0.05*M* (approx 1.5%) to the buffer. The SDS will solubilize hydrophobic species and form ion pairs with those of opposite charge. The separation is usually performed at pH 7.0 and with uncoated capillaries so as to induce electroendo-osmotic flow. Components of the sample have differing degrees of interaction with the micelles formed by the detergent action of the SDS. The micelles formed by SDS are anionic in nature and will therefore migrate in the opposite direction to the induced electroendo-osmotic flow, in a countercurrent manner, thus setting up the partitioning effect. The migration of the various solutes in a sample therefore ranges among the rates of electroendo-osmotic flow, electrophoretic migration, and the opposing rate of micelle migration. MECC has the advantage of allowing both charged and noncharged solutes to be separated in a single procedure, and has a far better resolution than an uncoated capillary alone. However, because the electroendo-osmotic flow and micelle movement are in opposite directions, not all of the components of a mixture may be drawn past the detector. To optimize the separation, it may be necessary to modify the composition of the buffer to increase or decrease the rate of electroendo-osmosis. Other buffer additives *(13)* are worth experimenting with, for example, detergents such as Triton X-100 and disaggregating agents, such as urea.

Acknowledgments

The author would like to thank Bio-Rad Laboratories UK Ltd., for the grant of an HPE 100 capillary electrophoresis apparatus and for the donation of research materials to the laboratory.

Further Reading

Altria, K. D., ed. (1996) *Methods in Molecular Biology, vol. 52: Capillary Electrophoresis Guidebook: Principles, Operation, and Applications.* Humana, Totowa, NJ.

Horvath, C. and Nikelly, J. G., eds. (1990) Analytical biotechnology: capillary electrophoresis and chromatography. *Am. Chem. Soc. Symp. Series* **434,** American Chemical Society, Washington, D.C.

References

1. Zhu, M., Hansen, D. L., Burd, S., and Gannon, F. (1989) Factors affecting free zone electrophoresis and isoelectric focusing in capillary electrophoresis. *J. Chromatography* **480,** 311–319.
2. Hjerten, S., Elenberg, K., Kilar, F., Liao, J-L., Chen, A. J. C., Siebert, C. J., and Zhu, M. (1987) Carrier-free zone electrophoresis, displacement electrophoresis and isoelectric focusing in a high-performance electrophoresis apparatus. *J. Chromatography* **403,** 47–61.
3. Mazzeo, J. R. and Krull, S. (1991) Capillary isoelectric focusing of proteins in un-coated fused-silica capillaries using polymeric additives. *Anal. Chem.* **63,** 2852–2857.
4. Krull S. and Mazzeo, J. R. (1992) Capillary electrophoresis: the promise and the practice. *Nature* **357,** 92–94.
5. Karger, B. L., Cohen, A. S., and Guttman, A. (1989) High-performance capillary electrophoresis in the biological sciences. *J. Chromatography* **492,** 585–614.
6. Higashami, T., Fuchigami, T, Imasak, T., and Ishibashi, N. (1992) Determination of amino acids by capillary zone electrophoresis based on semiconductor laser fluorescence detection. *Anal. Chem.* **64,** 711–714.
7. McDowall, R. D. (1989) Sample preparation for biomedical analysis. *J. Chromatography* **492,** 3–58.
8. Gorbunoff, M. J. and Timasheff, S. N. (1984) The interaction of proteins with hydroxyapatite: III Mechanism. *Anal. Biochem.* **136,** 440–445.
9. Advis, J. P., Hernandez, L., and Guzman, N. A. (1989) Analysis of brain neuropeptides by capillary electrophoresis: determination of luteinizing hormone-releasing hormone from ovine hypothalamus. *Peptide Research* **2,** 389–394.
10. Terabe, S. and Isemura, T. (1990) Ion-exchange electrokinetic chromatography with polymer ions for the separation of isomeric ions having identical electrophoretic mobilities. *Anal. Chem.* **62,** 650–652.
11. Ghowsi, K., Foley, J. P., and Gale, R. J. (1990) Micellar electrokinetic capillary chromatography theory based on electrochemical parameters: optimisation for three models of operation. *Anal. Chem.* **62,** 2714–2721.
12. Foley, J. P. (1990) Optimization of micellar electrophoretic chromatography. *Anal. Chem.* **62,** 1302–1308.
13. Burton, D., Sepaniak, M., and Maskarinec, M. (1987) Evaluation of the use of various surfactants in micellar electrokinetic capillary chromatography. *J. Chromatogr. Sci.* **24,** 347–351.

24

Capillary Isoelectric Focusing with Electro-Osmotic Flow Mobilization

Thomas J. Pritchett

1. Introduction

In isoelectric focusing (IEF), proteins migrate, under the influence of an electrical field, through a pH gradient created by carrier ampholytes (CA). Separation is based on the characteristic isoelectric point (pI) of each protein, which is the pH at which the protein has zero net charge, and thus zero mobility in the electrical field.

Traditionally, IEF has been performed using slab gel techniques, with detection by means of silver or Coomassie blue staining. Although satisfactory separations can be obtained with these methods, slab gels have several limitations, including long analysis times, difficulty of automation, and lack of quantitative capability.

IEF performed by capillary electrophoresis (cIEF) with electro-osmotic flow (EOF) mobilization is a recently developed technique that has already shown great potential as a means of analyzing and characterizing rDNA proteins *(1,2),* monoclonal antibodies (MAb) *(3,4),* hemoglobins *(5),* and other proteins of interest to academic researchers and the biotechnology industry. It offers advantages, including rapid analysis (often <10 min), ease of automation, direct detection by UV absorbance, quantitative capability, and direct transfer of data to a computerized data station. In addition, satisfactory separations can often be achieved using standard conditions with little or no methods development.

Capillary IEF techniques can be divided into two types, two-step and one-step, depending on whether or not EOF (the bulk flow of ions that is present anytime the capillary has a charged inner surface) has been eliminated. In the two-step methods, focusing takes place in a capillary that has been modified so that EOF is essentially absent. The focused zones must then be mobilized past the detector or the detector must be moved (not currently a commercially available option). Mobilization is accomplished by either chemical or hydrodynamic means. If chemical mobilization is being used, the anodic buffer is replaced with the cathodic buffer (or vice versa), or salt is added to either buffer *(6).* Hydrodynamic mobilization uses either pressure or vacuum for mobilization of the focused proteins *(7–9).*

This chapter presents the procedure for a one-step cIEF method, first reported by Mazzeo and Krull *(10),* in which reduced (but still present) EOF is used to mobilize the focused proteins. This method offers several advantages over the two-step methods.

From: The Protein Protocols Handbook
Edited by: J. M. Walker Humana Press Inc., Totowa, NJ

The one-step method is simpler than the two-step, and thus easier to learn and to automate. In performing the one-step method, there is no need for the analyst to decide when to stop focusing and start mobilization, as is the case with the two-step methods.

1.1. Overview of the Method

Briefly, the method is performed as follows. (*See* Fig. 1 for a sample electropherogram.)

1. The CE is set up with a coated capillary to operate in reversed-polarity mode (cathode [–] at the inlet). The catholyte, 20 m*M* NaOH, is placed at the capillary inlet, and 10 m*M* phosphoric acid, the anolyte, is placed at the outlet.
2. The capillary is rinsed with 10 m*M* phosphoric acid, after which the entire capillary is filled with the standard and sample protein(s) plus an ampholyte solution containing N,N,N',N'-tetramethylethylenediamine (TEMED) as a gradient extender and hydroxypropylmethylcellulose (HPMC) to provide viscosity and further reduce the EOF. The TEMED concentration is adjusted so that when an electrical field is applied, the ampholyte pH gradient forms in the 6–10 cm between the detector and the capillary outlet.
3. On application of the electrical field (usually 100–500 V/cm), focusing and mobilization back toward the inlet (and past the detector) occur simultaneously, thus avoiding the need for a second mobilization step. The current will quickly (within 30 s) spike to its highest value (usually 10–30 µA) and then decline, at first rapidly and then more slowly to a value of 4–6 µA. Focusing is probably complete when the slope of the current decrease curve levels off.
4. Detection is by absorbance at 280 nm. Because the polycarboxylic acid-containing ampholytes have significant absorbance in the lower UV region, detection at wavelengths below 254 nm is usually not practical. Detection of proteins and peptides at low concentrations (<10 µg/mL) is often possible (as long, of course, as they contain sufficient aromatic amino acids) because 50- to 200-fold concentration of the protein occurs during focusing.

2. Materials

1. Capillary Electrophoresis Instrument (commercial or lab-made) with UV detector and 280-nm filter. Detector-to-outlet distance should be 6–10 cm (for instruments with outlet-to-detector distances of <6 cm, *see* Note 5).
2. Capillary with inner surface coated to reduce EOF (e.g., Beckman [Fullerton, CA] eCAP Neutral Capillary, Supelco [Bellefonte, PA] H-150 capillary, or J&W [Folsom, CA] DB-1 capillary). Capillary id can be either 50 or 75 µm. Capillary length should be 27–47 cm. Longer capillaries may be used, but the amount of TEMED may need to be increased to prevent basic proteins from focusing on the wrong side of the detector.
3. 3.0–10.0 Ampholytes (e.g., Pharmalyte or Servalyte).
4. TEMED (Bio-Rad, Hercules, CA).
5. HPMC, 4000 cps (Sigma, St. Louis, MO).
6. Purified water.
7. Phosphoric acid.
8. 1.0*N* sodium hydroxide solution.
9. 0.45-µm Filters.
10. pI reference standards spanning a pH range of about 5.0–10.0 (e.g., horse heart cytochrome C, pI 9.6; lens culinaris [lentil] lectin, pI 8.8, 8.6, 8.2; horse heart myoglobin, pI 7.2, 6.8; human erythrocyte carbonic anhydrase I, pI 6.6; bovine erythrocyte carbonic anhydrase II, pI 5.9; bovine milk β-lactoglobulin A, pI 5.1). Available from Sigma.

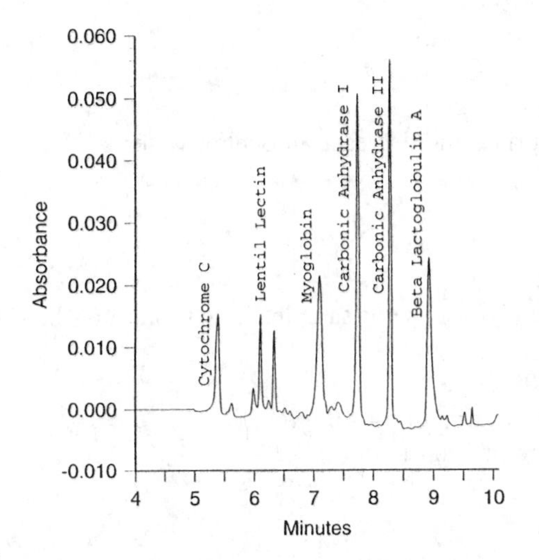

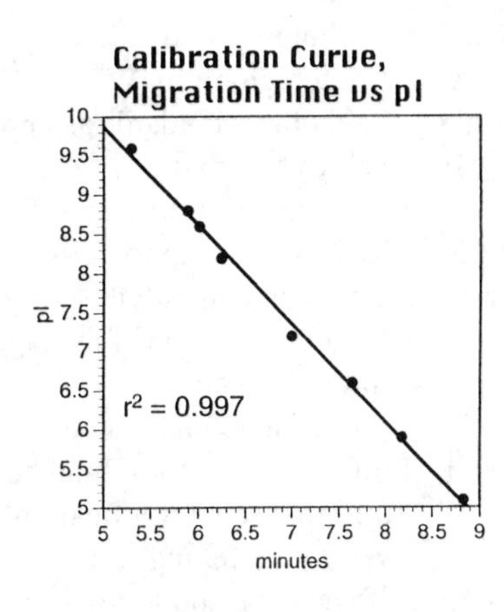

Fig. 1. C8 Capillary IEF analysis of pI standards. Conditions were as follows: sample concentration, 50 μg/mL each protein; instrument, Beckman P/ACE Series 5000; capillary, Supelco H-150, 50 μm id × 47 cm total length (40 cm inlet to detector, 7 cm outlet to detector); E, 125 V/cm; run temperature, 23°C. Isoelectric points of the standards are given in the text in Section 2., item 10.

3. Method

1. Prepare stock and working solutions.
   a. 1% (w/v) HPMC: add 1 g of HPMC (4000 cps) to approx 90 mL purified water; stir vigorously (2–4 h, or overnight) to dissolve; add purified water for a final volume of 100 mL.
   b. pI Standards: add sufficient purified water to each standard for a concentration of 4.0 mg/mL; divide each standard into separate 10-μL aliquots, label, date, and store at –80°C.
   c. Anode buffer (anolyte), 10 m*M* phosphoric acid.
   d. Cathode buffer (catholyte), 20 m*M* sodium hydroxide.
   e. 2X cIEF ampholyte solution (2 mL):
      i. To a 15-mL polypropylene centrifuge tube, add:
         1. Purified water 0.970 mL
         2. 1% HPMC 0.800 mL
         3. 3–10 Ampholytes 0.200 mL
         4. TEMED 0.030 mL
      ii. Mix well by vortexing.
      iii. Filter through 0.45 μm Acrodisk, or other 0.45-μm filter.
      iv. Store at 2–8°C for up to 2 wk.
2. Prepare reference standards in ampholyte solution (200 μL final volume; final concentration of pI standards, 50 μg/mL).
   a. To a 0.5 mL microcentrifuge tube, add:
      i. 2X cIEF ampholyte solution 100.0 μL
      ii. Purified water 85.0 μL
      iii. Cytochrome C 2.5 μL
      iv. Lentil Lectin 2.5 μL
      v. Myoglobin 2.5 μL

vi. Carbonic anhydrase I 2.5 µL
vii. Carbonic anhydrase II 2.5 µL
viii. β-Lactoglobulin A 2.5 µL
ix. Add other standards as appropriate and decrease amount of water.

b. Mix well by vortexing.
c. Centrifuge at >10,000*g* for 1–2 min.
d. Store at 2–8°C for up to 48 h.

3. Prepare sample in ampholyte solution:
   a. To a 0.5-mL microcentrifuge tube, add (to conserve sample, these values may be scaled down proportionally):
      i. 2X cIEF ampholyte solution, 100 µL.
      ii. Protein sample for a final concentration of 50–100 µg/mL.
      iii. Purified water for a final volume of 200 µL.
   b. Mix well by vortexing.
   c. Centrifuge at >10,000*g* for 1–2 min.
   d. Store at 2–8°C for up to 48 h.
4. Analyze standards and samples in separate cIEF runs using the following conditions (the purpose of these separate runs is to identify standards that will not comigrate with the sample):

Instrument polarity	Cathode (–) at the capillary inlet: anode (+) at the outlet
Cathode solution	20 m*M* sodium hydroxide
Anode solution	10 m*M* phosphoric acid
Temperature	23°C
Prerun capillary conditioning	1–2 min (at 20 psi) pressure rinse with 10 m*M* phosphoric acid. Adjust time for instruments with lower or higher pressure or for capillaries longer than 47 cm
Injection	1–2 min (at 20 psi) pressure rinse with sample in ampholyte solution
Applied voltage	10 kV (may need to be adjusted; *see* Section 4.)
Detection	UV absorbance at 280 nm
Postrun rinse	5 min pressure rinse (20 psi) with purified water

5. Determine pI of sample:
   a. Based on the results of the standards and sample run individually, mix appropriate standards, sample, ampholytes, and analyze by cIEF as described in Steps 1–4 above.
   b. Plot migration time vs pI for the standard proteins. Interpolate the pI of the sample from the plot.

4. Notes

1. Quickstart: When analyzing a protein for the first time, the following steps are suggested to arrive quickly at optimal conditions.
   a. Set the sample concentration to about 100 µg/mL. This will result in a final concentration of 50 µg/mL (after sample is mixed 1:1 with the 2X ampholyte solution). This should give an acceptable signal (if a 1 mg/mL solution of the protein gives a signal of at least 0.5–1 AU_{280} in a spectrophotometer with a 1 cm path length). This fairly low concentration also minimizes the chance of protein precipitation.

   b. Determine the voltage that gives an initial current spike of about 20 µA for basic and neutral proteins, and 30–40 µA for acidic proteins. Time is saved if this is done manually through the front panel or direct control. The proper voltage will depend on the ionic concentration of the sample, which may be as high as 200 m*M* (before mixing with ampholyte solution).
   c. Fine tune the voltage/current spike until an acceptable separation is obtained. If problems are encountered, *see* Note 4.
2. Optimization: If further fine tuning is required, try some or all of the following, depending on your needs:
   a. Protein concentration may be varied from <10 µg/mL to over 200 µg/mL depending on the physical characteristics of the protein (e.g., solubility and absorbance at 280 nm).
   b. Additives may be included. Additives that have been used to advantage include nonionic surfactants at concentrations (be sure to stay below the critical micellar concentration) of 0.1–2% (e.g., Brij-35, reduced Triton X-100, Nonidet P-40), zwitterionic surfactants (e.g., CHAPS, sulfobetanes), and organic modifiers (e.g., glycerol, or 10–40 % v/v ethylene glycol). Urea (6–8*M*) can be used to maintain protein solubility, but at the cost of protein denaturation. When using urea, care must be taken (e.g., addition of ion scavengers to stock solutions) to avoid pH dependent carbamylation of α and ε amino groups, which will alter the protein's pI.
   c. The ampholyte brand and/or pH range may be changed. Several companies (e.g., Serva [Hauppauge, NY], Bio-Rad, Pharmacia [Piscataway, NJ]) make ampholytes, and one of these brands may have advantages for a specific protein. In addition, narrow-range ampholytes will improve resolution, at the expense of pH range. Many analysts are making custom blends of wide and narrow-range ampholytes to suit their needs.
   d. The run temperature may be varied. Higher resolution may be obtained by lowering the temperature (e.g., to 20°C). Greater solubility may be obtained by raising the temperature to 25–30°C.
   e. The viscosity of the solution may be changed by varying the amount of HPMC, the centapoise (cps) value of the HPMC used, or by using another viscosity modifier (e.g., plain methylcellulose).
   f. The ionic strength of the solution can be varied. Better linearity of the standards is sometimes achieved if NaCl is added to a final concentration of 10–20 m*M*.
   g. Try increasing or decreasing the TEMED concentration.
3. If you are separating acidic proteins (especially with pI below 5), you will probably need to do some or all of the following. For proteins with pI below 4, you will probably need to use pressure/voltage mobilization *(7–9)*.
   a. Use a higher concentration of phosphoric acid for the anode solution to reduce anodic drift. Concentrations of up to 100 m*M* have been used.
   b. Use higher field strengths so that an initial current spike of 30–40 µA is obtained. This may cause loss of resolution for basic and neutral proteins. If basic, neutral, and acid proteins must be separated in the same run, experiment with the voltage ramp technique below. If this does not work, consider switching to pressure/voltage mobilization *(7–9)*.
   c. Try voltage ramping. After about 3–5 min, ramp the voltage (e.g., to 25 kV over 10 min). This will increase the EOF for mobilization of the acidic proteins against the anodic drift. Experiment with the timing and value of the ramp until the desired results are obtained.
4. Troubleshooting.
   a. No peaks observed: make sure polarity is correctly set; make sure that the initial current spike is sufficiently high (usually >8–10 µA); check capillary window, detector lamp; check for clogged capillary.

b. Very low current (initial current spike of <8 μA): check for clogged capillary; increase the electrical field.
c. Spikes in UV trace of electropherogram and/or capillaries frequently clogging: this is the result of protein precipitation, an often-encountered problem in cIEF; decrease protein concentration; decrease the electrical field; use additives (*see* Note 2); try increasing the run temperature.
d. Nonlinear plot of migration time vs pI: replace cathode and anode solutions; make up fresh ampholyte solution; try a different ampholyte range; incorporate additives to prevent protein–protein interactions; try different standards; try different capillaries (*see also* Optimization, 2F, above).
e. Poor resolution: decrease electrical field; use narrow-range ampholytes instead of wide range, try additives as in Note 2.
f. Broad peaks for acidic proteins: a main cause of this is anodic drift; *see* Note 3 *above* for suggestions.

5. For instruments with very short (<6 cm) outlet-to-detector distances, this method may need to be run so that focusing occurs in the portion of the capillary between the inlet and the detector (as in ref. *5*). In this case, the polarity should be set so that the anode (+) is at the inlet and the cathode (–) is at the outlet. The anolyte (10 m*M* phosphoric acid) should be placed at the inlet and the catholyte (20 m*M* NaOH) at the outlet. The amount of TEMED should be reduced to 5–10 μL/200 mL. All other procedures are identical.

References

1. Yowell, G., Fazio, S., and Vivilecchia, R. (1993) The analysis of recombinant granulocyte macrophage colony stimulating factor (GM-CSF) by capillary isoelectric focusing. *Beckman Application Bulletin* A-1744.
2. Yowell, G. G., Fazio, S. D., and Vivilecchia, R. V. (1993) Analysis of recombinant granulocyte macrophage colony stimulating factor dosage form by capillary electrophoresis, capillary isoelectric focusing and high performance liquid chromatography. *J. Chromatog.* **652,** 215–224.
3. Pritchett, T. (1994) Qualitative and quantitative analysis of monoclonal antibodies by one-step capillary isoelectric focusing. Beckman Application Information Publication No. A-1769.
4. Pritchett., T. (1994) Quantitative capillary electrophoresis of monoclonal antibodies using the neutral capillary and the SDS 14–200 method development kits. Beckman Application Information Publication No. A-1772-A.
5. Molteni, S., Frischknecht, H., and Thorman, W. (1994) Application of dynamic capillary isoelectric focusing to the analysis of human hemoglobin variants. *Electrophoresis* **15,** 22–30.
6. Hjertén, S. (1992) Isoelectric focusing in capillaries, in *Capillary Electrophoresis, Theory and Practice* (Grossman, P. D. and Colburn, J. C., eds.), Academic, San Diego, CA, pp. 191–214.
7. Nolan, J. (1993) Capillary isoelectric focusing using pressure/voltage mobilization. *Beckman Application Bulletin* A-1750.
8. Hempe, J. M. (1994) Hemoglobin analysis by capillary isoelectric focusing. Beckman Application Information Publication No. A-1771-A.
9. Hempe, J. M. and Craver, R. D. (1994) Quantification of hemoglobin variants by capillary isoelectric focusing. *J. Clin. Chem.* **40,** 2288–2295.
10. Mazzeo, J. R. and Krull, I. S. (1991) Capillary isoelectric focusing of proteins in uncoated fused-silica capillaries using polymeric additives. *Anal. Chem.* **63,** 2852–2857.

25

Quantification of Radiolabeled Proteins in Polyacrylamide Gels

Wayne R. Springer

1. Introduction

Autoradiography is often used to detect and quantify radiolabeled proteins present after separation by polyacrylamide gel electrophoresis (PAGE) (*see* Chapter 36). The method, however, requires relatively high levels of radioactivity when weak β-emitters, such as tritium, are to be detected. In addition, lengthy exposures requiring the use of fluorescent enhancers are often required. Recent developments in detection of proteins using silver staining *(1,2)* have added to the problem because of the fact that tritium emissions are quenched by the silver *(2)*. Since for many metabolic labeling studies, tritium labeled precursors are often the only ones available, it seemed useful to develop a method that would overcome these drawbacks. The method the author developed involves the use of a cleavable crosslinking agent in the polyacrylamide gels that allows the solubilization of the protein for quantification by scintillation counting. Although developed for tritium *(3)*, the method works well with any covalently bound label, as demonstrated here with ^{35}S. Resolution is as good as or better than autoradiography (*3* and Fig. 1), turnaround time can be greatly reduced, and quantification is more easily accomplished.

2. Materials

Reagents should be of high quality, particularly the sodium dodecyl sulfate (SDS) to obtain the best resolution, glycerol to eliminate extraneous bands, and ammonium persulfate to obtain proper polymerization. Many manufacturers supply reagents designed for use in polyacrylamide gels. These should be purchased whenever practical. Unless otherwize noted, solutions may be stored indefinitely at room temperature.

2.1. SDS-Polyacrylamide Gels

1. Acrylamide-DATD: Acrylamide (45 g) and *N,N'*-dialyltartardiamide (DATD 4.5 g) are dissolved in water to a final volume of 100 mL. Water should be added slowly and time allowed for the crystals to dissolve.
2. Acrylamide-bis: Acrylamide (45 g) and *N,N'*-methylenebis(acrylamide) (bis-acrylamide, 1.8 g), are dissolved as in item 1 in water to a final volume of 100 mL.
3. Tris I: 0.285*M* Tris-HCl, pH 6.8.

From: The Protein Protocols Handbook
Edited by: J. M. Walker Humana Press Inc., Totowa, NJ

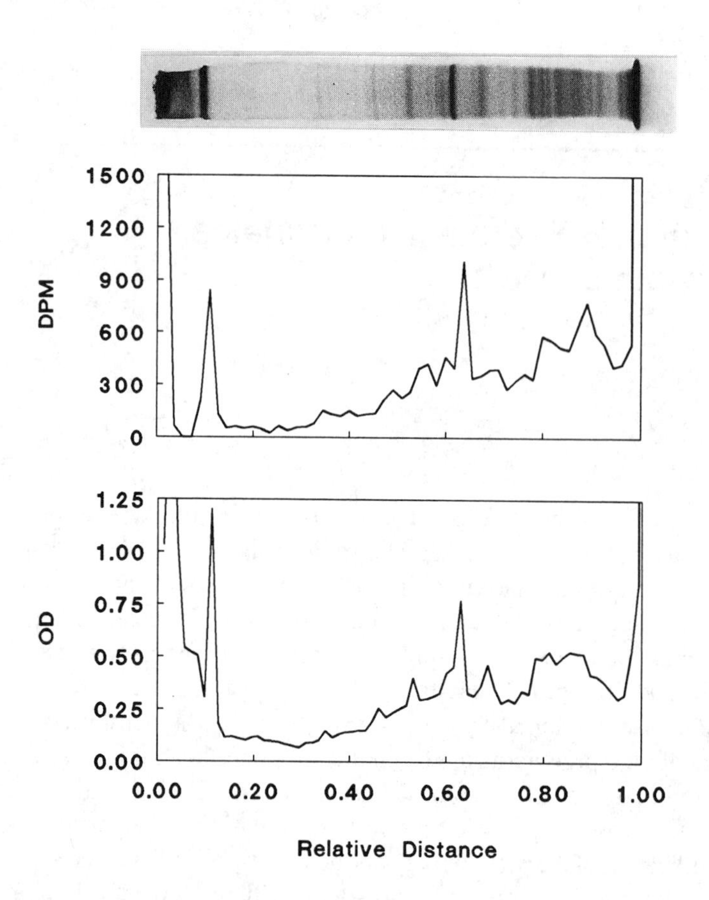

Fig. 1. Comparison of an autoradiogram, densitometry scan, and radioactivity in gel slices from a mixture of labeled proteins separated by PAGE. Identical aliquots of ^{35}S-methionine-labeled proteins from the membranes of the cellular slime mold, *D. purpureum,* were separated using the methods described in Section 3.1. One lane was fixed, soaked in Flouro-Hance (Research Products International Corp., Mount Prospect, IL), dried, and autoradiographed (photograph). Another lane was cut from the gel, sliced into 1-mm pieces, solubilized, and counted using a Tracor Mark III liquid scintillation counter with automatic quench correction (TM Analytic, Elk Grove Village, IL) as described in Section 3.3. The resultant disintigrations per minute (DPM) were plotted vs the relative distance from the top of the running gel (top figure). The autoradiogram (photograph) was scanned relative to the top of the running gel using white light on a Transidyne RFT densitometer (bottom figure).

4. Tris II: 1.5*M* Tris-HCl, pH 8.8, 0.4% SDS.
5. Tris III: 0.5*M* Tris-HCl pH 6.8, 0.4% SDS.
6. Ammonium persulfate (APS) solution: A small amount of ammonium persulfate is weighed out and water is added to make it 100 mg/mL. This should be made fresh the day of the preparation of the gel. The ammonium persulfate crystals should be stored in the refrigerator and warmed before opening.
7. *N,N,N',N'*-tetramethylethylenediamine (TEMED) is used neat as supplied by the manufacturer.
8. 10X Running buffer: Tris base (30.2 g), glycine (144.1 g), and SDS (10.00 g) are made up to 1 L with distilled water.

9. Sample buffer: The following are mixed together and stored at 4°C: 1.5 mL 20% (w/v) SDS, 1.5 mL glycerol, 0.75 mL 2-mercaptoethanol, 0.15 mL 0.2% (w/v) Bromophenol blue, and 1.1 mL Tris I.

2.2. Silver Staining of Gels

All solutions must be made with good-quality water.

1. Solution A: methanol:acetic acid:water (50:10:40).
2. Solution B: methanol:acetic acid:water (5:7:88).
3. Solution C: 10% (v/v) glutaraldehyde.
4. 10X Silver nitrate: 1% (w/v) silver nitrate stored in brown glass bottle.
5. Developer: Just before use, 25 μL of 37% formaldehyde are added to freshly made 3% (w/v) sodium carbonate.
6. Stop bath: 2.3*M* Citric acid (48.3 g/100 mL).

2.3. Solubilization and Counting of Gel Slices

1. Solubilizer: 2% sodium metaperiodate.
2. Scintillation fluid: Ecolume (+) (ICN Biomedical, Inc., Irvine, CA) was used to develop the method. *See* Section 4. for other scintillants.

3. Methods

The general requirements to construct, prepare, and run SDS-PAGE gels as described by Laemmli *(4)* are given in Chapter 11. Presented here are the recipes and solutions developed in the author's laboratory for this particular technique. For most of the methods described a preferred preparation of a standard gel can be substituted, except for the requirement of the replacement of DATD for bis at a ratio of 1:10 DATD: acrylamide in the original formulation.

3.1. SDS-Polyacrylamide Gels

Sufficient medium for one 8 × 10 × 0.75 to 1.0-cm gel can be made by combining stock solutions and various amounts of acrylamide-DATD and water to achieve the required percentage of acrylamide (*see* Note 1) by using the quantities listed in Table 1. The stacking gel is that described by Laemmli *(4)* and is formed by combining 0.55 mL of acrylamide-bis with 1.25 mL of Tris III, 3.2 mL of water, 0.015 mL of APS, and 0.005 mL of TEMED.

1. For both the running and the stacking gel, degas the solutions by applying a vacuum for approx 30 s before adding the TEMED.
2. After the addition of the TEMED, pour the gels, insert the comb in the case of the stacking gel, and overlay quickly with 0.1% SDS to provide good polymerization.
3. Prepare protein samples by dissolving two parts of the protein sample in one part of sample buffer and heating to 100°C for 2 min.
4. Run gels at 200 V constant voltage for 45 min to 1 h or until the dye front reaches the bottom of the plate.

3.2. Silver Staining

The method used is that of Morrissey *(1)*.

1. Remove gels from the plates, and immerse in solution A for 15 min.
2. Transfer to solution B for 15 min and then solution C for 15 min, all while gently shaking.

Table 1
Recipe for Various Percentages of SDS-PAGE Gels

	Percentage of acrylamide		
	7.5	10	12.5
Stock solution			
Acrylamide-DATD, mL	0.96	1.33	1.66
Tris II, mL	1.50	1.50	1.50
Water, mL	3.50	3.13	2.80
APS, mL	0.03	0.03	0.03
TEMED, mL	0.003	0.003	0.003

3. At this point, the gel can be rinsed in glass-distilled water for 2 h to overnight (*see* Note 2).
4. After the water rinse, add fresh water and enough crystalline dithiotreitol (DTT) to make the solution 5 μg/mL.
5. After 15 min, remove the DTT, and add 0.1% silver nitrate made fresh from the 1% stock solution. Shake for 15 min.
6. Rapidly rinse the gel with a small amount of water followed by two 5–10 mL rinses with developer followed by the remainder of the developer.
7. Watch the gel carefully, and add stop bath to the gel and developer when the desired darkness of the bands is reached.
8. Store the stopped gel in water until the next step.

3.3. Slicing and Counting of Gel Slices

1. Remove individual lanes from the gel for slicing by cutting with a knife or spatula.
2. Cut each lane into uniform slices, or cut identified bands in the gel (*see* Note 3).
3. Place each slice into a glass scintillation vial, and add 0.5 mL of 2% sodium metaperiodate solution.
4. Shake the vials for 30 min to dissolve the gel.
5. Add a 10-mL aliquot of scintillation fluid to the vial, cool the vial, and count in a refrigerated scintillation counter (*see* Notes 4–6).

4. Notes

4.1. SDS-Polyacrylamide Gels

1. Acrylamide-DATD gels behave quite similarly to acrylamide-bis gels, except for the fact that for a given percentage of acrylamide, the relative mobility of all proteins are reduced in the DATD gel. In other words, a DATD gel runs like a higher percentage acrylamide-bis gel.

4.2. Silver Staining

2. The water rinse can be reduced to 1 h, if the water is changed at 10–15 min intervals. The author found in practice that the amount of DTT was not particularly critical, and routinely added the tip of a microspatula of crystals to approx 30 mL of water.

4.3. Slicing and Counting of Gel Slices

3. The method seems relatively insensitive to gel volume, as measured by changes in efficiency *(3)* over the range of 5–100 mm^3 for tritium and an even larger range for ^{14}C or ^{35}S.

Larger pieces of gel can be used if one is comparing relative amounts of label, such as in a pulse chase or other timed incorporation, but longer times and more metaperiodate may be required to dissolve the gel. Recovery of label from the gel is in the 80–90% range for proteins from 10–100 kDa *(3)*. In most cases, the label will remain with the protein, but in the case of periodate-sensitive carbohydrates on glycoproteins, it may be released from the protein. This, however, does not prevent the quantification. It just does not allow the solubilized labeled protein to be recovered for other manipulations.

4. It is necessary to cool the vials in the counting chamber before counting in order to eliminate occasional chemiluminescence. The cause of this phenomenon was not explored, but one should determine whether this occurs when using other scintillants or counters.
5. Figure 1 shows the results of a typical experiment using ^{35}S-methionine to label proteins metabolically from the cellular slime mold, *Dictyostelium purpureum,* and quantify them using the method described or by autoradiography. As can be seen, the resolution of the method is comparable to that of the autoradiogram (*see* ref. *3* for similar results using tritium). The time to process the lane by the method described here was approx 8 h, whereas the results of the autoradiogram took more than 3 d to obtain. Examination of the stained gel suggests that, if one were interested in a particular protein, it would be fairly easy to isolate the slice of gel containing that protein for quantification. This makes this method extremely useful for comparing incorporation into a single protein over time as in pulse/chase experiments *(3)*.
6. The author has found that as little as 400 dpm of tritium associated with a protein could be detected *(3),* which makes this method particularly useful for scarce proteins or small samples.

References

1. Morrissey, J. H. (1981) Silver stain for proteins in polyacrylamide gels: a modified procedure with enhanced uniform sensitivity. *Anal. Biochem.* **117,** 307–310.
2. Van Keuren, M. L. Goldman, D., and Merril, C. R. (1981) Detection of radioactively labeled proteins is quenched by silver staining methods: quenching is minimal for ^{14}C and partially reversible for ^{3}H with a photochemical stain. *Anal. Biochem.* **116,** 248–255.
3. Springer, W. R. (1991) A method for quantifying radioactivity associated with protein in silver-stained polyacrylamide gels. *Anal. Biochem.* **195,** 172–176.
4. Laemmli, U. K. (1970) Cleavage of structural proteins during assembly of the head of bacteriophage T4. *Nature* **227,** 680–685.

26

Quantification of Proteins by Staining in Polyacrylamide Gels

Bryan John Smith

1. Introduction

Quantification of proteins is a common challenge. There are various methods described for estimation of the amount of protein present in a sample as described in Chapters 1–9. These quantify total protein present, but do not quantify one protein in a mixture of several. For this, the mixture must be resolved. Liquid chromatography achieves this, and the various proteins may be quantified by their absorbance at 220 or 280 nm. Microgram to submicrogram amounts of protein can be analyzed in this way, using microbore HPLC. Problems may occur if particular species chromatograph poorly (such as hydrophobic polypeptides on reverse-phase chromatography). An alternative is polyacrylamide gel electrophoresis (PAGE) as the mixture-resolving step, followed by protein staining and densitometry. Exceptions are small peptides that are not successfully resolved and stained in gels. Microgram to submicrogram amounts of protein (of over a few kilodaltons in size) may be quantified in this way. Although not every protein stain is best suited to this purpose, the method described herein is suitable for quantification of microgram-to-submicrogram amounts of proteins on sodium dodecyl sulfate (SDS) polyacrylamide gels *(1,2)*.

2. Materials

1. A suitable scanning densitometer, e.g., a spectrophotometer with scanning capabilities, or the Molecular Dynamics 300A with Image Quant software. This apparatus scans transparent objects, such as wet gels, gels dried between transparent films, or photographic film. It then digitizes the images, which may then be analyzed. Either transparent or opaque objects may likewize be converted to digitized images in an apparatus, such as the Flogen IS-100, which utilizes a digitizing video camera to generate an image that can be processed for quantification.
2. Protein stain (*see* Note 1): 0.2% (w/v) Procion blue MX-2G-125 (Kemtex Services Ltd., Manchester, UK). Dissolve 0.4 g dye in 100 mL methanol, and then add 20 mL glacial acetic acid and 80 mL distilled water. Make fresh each time. Although this dye has no ingredient currently known to be hazardous, it is reactive and has been associated with respiratory sensitization. Avoid inhalation, contact with skin or eyes, and ingestion. Use masks and other protective clothing.

From: The Protein Protocols Handbook
Edited by: J. M. Walker Humana Press Inc., Totowa, NJ

3. Destaining solution: Mix together 100 mL methanol, 100 mL glacial acetic water, and 800 mL distilled water.

3. Method

1. At the end of electrophoresis, immerse the gel in Procion blue MX-2G-125 stain, and gently agitate until the dye has fully penetrated the gel. This time varies with the gel type (e.g., 1.5 h for a 0.5-mm thick SDS polyacrylamide [15%T] gel slab), but cannot really be overdone.
2. At the end of the staining period, decolorize the background by immersion in destain, with agitation and a change of destain whenever the destain becomes deeply colored. This passive destaining may take 48 h even for a 0.5-mm thick gel. Thicker gels will take longer.
3. Measure the extent of dark blue dye bound by each band by scanning densitometry. Compare the dye bound by a sample with that by standard proteins run and stained in parallel with the sample—ideally on the same gel.

4. Notes

1. Procion blue MX-2G-125 dye possesses a dichlorotriazine group, which reacts with hydroxyl and amino groups on proteins. Other dyes in the Procion series have a similar group, combined with a variety of chromophoric groups that provide a range of colors. The dyes have wide use in the dying of textiles and other materials, and have found use in biochemistry as reagents for making dye-affinity matrices for use in preparation of proteins. Procion navy MXRB was previously used for quantitative staining of proteins *(1,2)*, but is no longer produced. Procion blue MX-2G-125 is dark blue in color, as was Procion navy MXRB, and it is used in the same way. Other Procion dyes may be used, e.g., Procion blue MX-4GD. Alternative names for Procion blue MX-2G-125 are Reactive blue 109 and Ostazin blue SLG. Procion blue MX-4GD is also known as Reactive blue 168.
2. It is a requirement of this method that dye is bound stoichiometrically to polypeptide over a usefully wide range of sample size. Procion blue MX-2G-125 fulfills this requirement—linear-scale plots of binding to proteins are linear up to at least 1 μg/mm^2 of gel of 1-mm thickness.

   The lower limit of detection is on the order of 100 ng/mm^2. This limit is somewhat variable, since different proteins take up dye to different extents, as was previously observed for Procion navy MXRB *(2)*.

   As an example, duplicate samples of horse myoglobin (Sigma, St. Louis, MO), code no. M-1882) were used to construct a standard curve after staining by Procion blue MX-2G-125 and densitometry as a Molecular Dynamics 300A with ImageQuant software. The best-fit straight line had the following equation:

$$y = 4.204x + 0.0799 \tag{1}$$

   where y = peak area and x = protein, μg/mm^2 gel of 1-mm thickness. The correlation coefficient is $r = 0.998$.

   Staining of rabbit albumin was approximately one-third less, however, with a standard curve conforming to the equation:

$$y = 2.7783x - 0.0553 \quad \text{with } r = 0.999 \tag{2}$$

   This presumably reflects the different chemical natures of the two proteins.

The blue dye Procion blue MX-4GD also reacts with proteins stoichiometrically, but only to about one-third the extent of Procion blue MX-2G125. This may be seen by comparing the above standard curve equations with the following for Procion blue MX-4GD staining of horse myoglobin:

$$y = 1.5195x + 0.0564 \quad \text{with } r = 0.994 \qquad (3)$$

and for rabbit albumin:

$$y = 0.8345x + 0.0168 \quad \text{with } r = 0.993 \qquad (4)$$

3. The Procion blue MX-2G-125 method is straightforward, but has the drawback of poorer sensitivity than comes with the use of the commonly used noncolloidal Coomassie dyes, such as PAGE blue 83. However, as discussed by Neuhoff et al. *(3),* staining by noncolloidal Coomassie dye is difficult to control—it may not fully penetrate and stain dense bands of protein, it may demonstrate a metachromatic effect (whereby the protein–dye complex may show any of a range of colors from blue through purple to red), and it may be variably or even completely decolorized by excessive destaining procedures (since the dye does not bind covalently). As a consequence, stoichiometric binding of Coomassie dye to protein is commonly not achieved.
4. A number of alternative staining methods have been discussed in the literature. Notably, Neuhoff et al. *(3)* made a thorough study of various factors affecting protein staining by Coomassie brilliant blue G250 (Color Index no. 42655; 0.1% [w/v] 2% [w/v] phosphoric acid, 6% [w/v] ammonium sulfate. This colloidal stain is in the order of 100-fold more sensitive than is the Procion navy MXRB stain described here. The colloidal stain does not stain the background in a gel and so does not require destaining. In our own lab we have used a Coomassie brilliant blue G250-perchloric acid staining procedure for quantification of protein on gels. The method derives from Reisner et al. *(4)*. The stain is brilliant blue G, 0.04% w/v in perchloric acid, 3.5% w/v, and can be obtained ready to use from Sigma (Poole, UK). After electrophoresis the gel is submerged in the stain with gentle shaking for 90 min. The gel is then washed in several changes of water until the background is completely clear (taking up to 20 h). Heavily loaded samples show up during staining, but during destaining the blue staining of the proteins becomes accentuated. Bands of just a few tens of ng are visible on a 1-mm-thick gel (i.e., the lower limit of detection is less than 10 ng/mm^2). Different proteins bind to the dye to different extents, however: horse myoglobin is stained twice as heavily as is bovine serum albumin. The staining may be quantitative, or nearly so, up to large loadings of 20 µg or more per band. Horse myoglobin and bovine serum albumin have proven not to give completely quantitative staining by this method, but standard curves from these proteins are only gently curved: for myoglobin,

$$y = -0.36313x^2 + 5.4271x + 0.13531 \quad \text{with } r = 0.999 \qquad (5)$$

and for albumin,

$$y = -0.13808x^2 + 3.5252x + 0.28827 \quad \text{with } r = 0.999 \qquad (6)$$

where y = peak area and x = loading (note: not quoted per mm^2 since band size and shape vary greatly over the wide range of loadings possible).

The staining may be close enough to being quantitative to make it a useful method. Performance is markedly variable. For example, duplicate loadings of samples on separate gels, electrophoresed and stained in parallel, have differed in staining achieved by a factor of 1.5, for reasons unknown. This makes it essential to run standards and samples on the same gel.

Syrovy and Hodny *(5)* reviewed this and other methods, including covalent modification of proteins by dye (such as Remazol brilliant blue R) prior to electrophoresis, and elution of dye (such as Coomassie brilliant blue R or Fast green) from stained protein bands, so as to allow estimation by spectrophotometry (*see* Chapter 27 for further details of this approach). Recently, O'Keefe *(6)* described a method whereby cysteine residues are labeled with the thiol-specific reagent monobromobimane. On transillumination (at $\lambda = 302$ nm), labeled bands fluoresce. Photograph negatives of the gel may be scanned for densitometry. The method requires quantitative reaction of cysteines (totally lacking in some polypeptides), and a nominal detection limit of 10 pmol of cysteine is claimed. Another recently described method is a multistep method that generates negatively stained bands *(7*; and *see* Chapter 29*)*. The first step generates a white background of precipitated zinc salt, against which clear bands of protein (where precipitation is inhibited) may be viewed by dark-field illumination. The background may subsequently be colored by incubation with tolidine. Sensitivity is good. The method was described for 12 and 15%T gels, but the degree of zinc salt precipitation and subsequent toning is dependent on %T, and at closer to 20%T, the toning procedure may completely destain the gel. The gel may be scanned after the first step (giving a white background), the negative staining being approx quantitative (for horse myoglobin) from 100 ng/mm^2 or less to 2 µg/mm^2 or more. This method is fully described in Chapter 29.

5. Quantification of protein spots on 2D gels is an important area. Autoradiography is commonly used for this purpose, but silver staining can attain similar sensitivity (on the order of 0.05 ng/mm^2). There are many silver-staining methods described, but as indeed with organic dyes, the details of the mechanisms involved are obscure. For various reasons, proteins vary widely in their stainability. Some do not stain at all, others may show a metachromatic effect. Use of silver staining for quantification is complicated by other factors, too, such as the difficulty in obtaining staining throughout the gel, not just at the surface. This subject is reviewed at greater length in ref. *4,* but in summary, it may be said that although silver-stain methods have great sensitivity, they are extremely problematical for quantification purposes.
6. Best quantification is achieved after having achieved good electrophoresis. Adapt the electrophoresis as necessary to achieve good resolution, lack of any band smearing or tailing, and lack of retention of sample at the top of the gel by aggregation. For stains where penetration of dense bands may be a problem, avoid dense bands by reducing the size of the loaded sample and/or electrophoresing the band further down the length of the gel (to disperse the band further) and/or use lower %T gel. Thin gels, of 1-mm thickness or less, allow easier penetration of dye throughout their thickness. Gradient gels have a gradient of pore size that may cause variation of band density and background staining. Use nongradient gels if this is a problem.
7. For absolute quantification of a band on a gel, accurate pipeting is required, as is a set of standard protein solutions which cover the concentration range expected of the experimental sample. Run and stain these standard solutions at the same time and if possible on the same gel as the experimental sample in order to reduce possible variations in band resolution or staining, background destaining, and so on. Ideally, the standard protein should be the same as that to be quantified, but if, as is commonly the case, this is impossible, then another protein may be used (while recognizing that this protein may bind a different amount of dye from that by the experimental protein, so that the final estimate obtained may be in error). The standard protein should have similar electrophoretic mobility to the proteins of interest, so that any effect, such as dye penetration, owing to pore size, is similar. Make the standard protein solution by dissolving a relatively large and

accurately weighed amount of dry protein in water or buffer, and dilute this solution as required. If possible, check the concentration of this standard solution by alternative means (e.g., amino acid analysis). Standardize treatment of samples in preparation for electrophoresis—treating of sample solutions prior to SDS-PAGE may cause sample concentration by evaporation of water, for instance. To minimize this problem, heat in small (0.5-mL) capped Eppendorf tubes, cool, and briefly centrifuge to take condensed water back down to the sample in the bottom of the tube.

8. When analyzing the results of scanning, construct a curve of absorbance vs protein concentration from standard samples, and compare the experimental sample(s) with this. Construct a standard curve for each experiment.
9. If comparing the abundance of one protein species with others in the same sample, standards are not required, provided that no species is so abundant that dye binding becomes saturated. Be aware that such relative estimates are approximate, since different proteins bind dye to different extents.
10. It is sometimes observed in electrophoresis that band shape and width are irregular—heavily loading a gel can generate a broad band, which may interfere with the running of neighboring bands, for instance. It is necessary to include all of such a band for most accurate results.
11. Avoid damage to the gel; a crack can show artificially as an absorbing band (or peak) on the scan. Gels of low %T are difficult to handle without damage, but they may be made tougher (and smaller) by equilibration in aqueous ethanol, say, 40% (v/v) ethanol in water for 1 h or so. Too much ethanol may cause the gel to become opaque, but if this occurs, merely rehydrate the gel in a lower % (v/v) ethanol solution. A gradient gel may assume a slightly trapezoid shape on shrinkage in ethanol solutions—this makes scanning tracks down the length of the gel more difficult. When scanning, eliminate dust, trapped air bubbles, and liquid droplets, all of which contribute to noisy baseline.
12. Methods have been described for quantification of submicrogram amounts of proteins that have been transferred from gels to polyvinylidene difluoride or similar membrane *(e.g., 8)*. Note, however, that transfer from gel to membrane need not (indeed, usually does not) proceed with 100% yield, so that results do not necessarily accurately reflect the content of the original sample.
13. Methods have been described for quantification of protein in sample solvent prior to electrophoresis and can provide an estimate of total amount of protein in the sample *(9,10)*.

References

1. Harris, M. R. and Smith, B. J. (1983) A qualitative and quantitative study of subfractions of the histone H1° in various mammalian tissues. *Biochem J.* **211,** 763–766.
2. Smith, B. J., Toogood, C. I. A., and Johns, E. W. (1980) Quantitative staining of submicrogram amounts of histone and high-mobility group proteins on sodium dodecylsulphate-polyacrylamide gels. *J. Chromatog.* **200,** 200–205.
3. Neuhoff, V., Stamm, R., Pardowitz, I., Arold, N., Ehhardt, W., and Taube, D. (1990) Essential problems in quantification of proteins following colloidal staining with Coomassie Brilliant Blue dyes in polyacrylamide gels, and their solution. *Electrophoresis* **11,** 101–117.
4. Reisner, A. H., Nemes, P., and Bucholtz, C. (1975) The use of Coomassie brilliant blue G250 perchloric acid solution for staining in electrophoresis and isoelectric focusing on polyacrylamide gels. *Anal. Biochem.* **64,** 509–516.
5. Syrovy, I. and Hodny, Z. (1991) Staining and quantification of proteins separated by PAGE. *J. Chromatog.* **569,** 175–196.

6. O'Keefe, D. O. (1994) Quantitative electrophoretic analysis of proteins labeled with monobromobimane. *Anal. Biochem.* **222,** 86–94.
7. Ferreras, M., Gavilanes, J. G., and Garcia-Segura, J. M. (1993) A permanent Zn^{2+} reverse staining method for the detection and quantification of proteins in polyacrylamide gels. *Anal. Biochem.* **213,** 206–212.
8. Patton, W. F., Lam, L., Su, Q., Lui, M., Erdjument-Bromage, H., and Tempst, P. (1994) Metal chelates as reversible stains for detection of electroblotted proteins: application to protein microsequencing and immune-blotting. *Anal. Biochem.* **220,** 324–335.
9. Dráber, P. (1991) Quantification of proteins in sample buffer for sodium dodecyl sulfate-polyacrylamide gel electrophoresis using colloidal silver. *Electrophoresis* **12,** 453–456.
10. Henkel, A. W. and Bieger, S. C. (1994) Quantification of proteins dissolved in an electrophoresis sample buffer. *Anal. Biochem.* **223,** 329–331.

27

Quantitation of Proteins Separated by Electrophoresis Using Coomassie Brilliant Blue

Ulf Neumann

1 Introduction

Different electrophoretic techniques are widely used for analyzing mixtures of proteins. Most important today are sodium dodecyl sulfate (SDS) gel electrophoresis, isoelectric focusing, and two-dimensional techniques. It is often desired to determine the quantity of protein in single or multiple bands or spots. Probably the most common method for this purpose is densitometric scanning *(1,2)*. Other approaches are based on dye elution with a wide variety of solvents, such as pyridine *(3)*, methanol *(4)*, NaOH *(5)*, or alcoholic solutions of detergents *(6)*, followed by photometric determination of dye concentration. Other techniques rely on solubilization of gels prepared with special crosslinkers *(7,8)*. Most of these methods overcome some of the problems of densitometry, but show some limitations and drawbacks, such as the use of toxic solvents, limited stability of the dye in the solvent or limited range of protein concentration, no background correction, or limitation to special electrophoretic techniques.

This chapter describes a method for the quantitation of electrophoretically separated proteins stained with Coomassie brilliant blue (CBB) G250. The method allows highly sensitive and reproducible quantitation of proteins, suitable for SDS electrophoresis, isoelectric focusing, and two-dimensional electrophoresis. A method for background correction is included. The method was originally described in ref. *9*.

Staining is performed with purified CBB G250 in colloidal solution after a fixation step in trichloroacetic acid (TCA) solution. This is a modified method of ref. *10*. With the use of purified dye and the additional fixation step, it is possible to detect as little as 20 ng, depending on the nature of the protein.

For quantitation, single or multiple bands or spots are cut out of the gel, and the bound dye is eluted from the gel with a 3% (w/v) solution of SDS. The amount of eluted dye is determined by measurement of absorbance. Background correction is achieved by determining the bound dye per gram of gel. Quantitation is linear at least in the range from 50 ng to 10 µg and highly reproducible even under nonoptimized conditions, as shown for purified proteins and complex mixtures in ref. *9*. The method described in this chapter (staining and quantitation) has been tested for standard SDS electrophoresis, isoelectric focusing, and two-dimensional electrophoresis. It can easily be adapted to a variety of other electrophoretic techniques.

From: The Protein Protocols Handbook
Edited by: J. M. Walker Humana Press Inc., Totowa, NJ

2. Materials

All chemicals should be of analytical grade. Solutions should be made up with distilled water. All solutions are stable at room temperature. The staining solution should be stored in the dark.

1. Fixing Solution: 20% TCA.
2. Staining solution: 800 mg of purified dye are dissolved in 400 mL H_2O then 400 mL of 1*M* H_2SO_4 is added, and the solution stirred for 3 h. After filtration, 90 mL of 10*M* KOH and 120 mL of 100% (w/v) TCA are added to the filtrate in this order (*see* Note 1).
3. Eluent: 3% SDS.
4. Other materials needed are electrophoresis equipment and reagents suitable for the desired technique, photometer capable of measuring at 610 nm, cuvets (standard and micro), analytical balance, incubating device at 37°C, and reaction tubes with caps.

3. Method

1. Dye purification: CBB G250 is purified according to the method of Neuhoff et al. *(11)*: 4 g of dye are dissolved in 250 mL of 7.5% acetic acid and warmed up to 70°C. The dye is precipitated by addition of 80 g of solid ammonium sulfate. The supernatant is discarded, and the precipitated dye is dissolved in 100 mL of hot methanol. Two hundred milliliters of aceton are added to remove remaining traces of ammonium sulfate. After 12 h at 4°C, the solution is filtered and the solvent evaporated. The typical yield is 80–85% of purified dye. Only purified dye should be used for the staining solution (*see* Note 2).
2. Electrophoresis: Electrophoresis should be carried out as usual. To achieve good data for background correction, some of the lanes of the gel should be left free (*see* Note 3).
3. Protein fixation: Directly after electrophoresis, the gel is placed in an appropriate volume of the fixing solution and agitated for at least 1 h (*see* Note 4).
4. Staining: After fixation, the gel is transferred to an appropriate volume of the staining solution. Gels should be stained for at least 12 h, even though bands can be detected after about half an hour. The staining time should be kept constant (*see* Notes 5 and 6).
5. Destaining: Background staining is removed by washing the gels in water until a clear background is achieved.
6. Gel slicing: After destaining, the gel is thoroughly drained, and the bands or spots of interest are cut out of the gel by use of a scalpel. The gel pieces are weighed on an analytical balance and transferred to a reaction tube. Some pieces of gel that contain no protein (free lanes) are handled in the same way (*see* Note 7).
7. Dye elution and measurement: To each reaction tube an appropriate volume of the eluent is added. The tubes are sealed and incubated for 24 h at 37°C. After incubation, the optical density of the solutions is determined at 610 nm against the solvent (*see* Notes 8 and 9).
8. Calculation of background correction. Even though background staining is very low with this staining technique, it is necessary to make corrections for background staining. Since, after 12 h of staining, background is very even throughout a gel of constant concentration, the background staining per gram of gel is determined. The specific background staining is:

$$A_B = (A_{610}/m_b\ [g]) \cdot (V_{elut}\ [mL]/1\ [mL]) \qquad (1)$$

where A_B is the specific background staining, A_{610} the measured absorbance for the gel piece containing no protein, V_{elut} the elution volume, and m_b the mass of the gel piece containing no protein.

The background corrected value is:

$$A_{corr} = A_{610} \cdot (V_{elut}\,[mL]/1\,[mL]) - m_p\,[g] \cdot A_B \qquad (2)$$

where A_{corr} is the background corrected absorbance, A_{610} the measured absorbance for the gel piece containing protein, and m_p the mass of the gel piece containing protein (*see* Note 10).

9. Protein quantitation: To correlate the obtained values for background-corrected absorbance with the amount of protein, a calibration curve must be prepared. Since different proteins show different binding capacities for CBB G250, the protein under investigation or a very similar one should be used as a standard for the calibration curve. For the calibration curve, different amounts of the standard are subjected to electrophoresis, fixed, stained, cut, and eluted as described. The calibration curve is constructed by plotting background-corrected absorbance values against the amount of protein (*see* Notes 11 and 12).

4. Notes

1. The staining solution is a dark green colloidal solution. For nonquantitative work, it can be reused as long as the pH stays <1.0. For quantitative work, the solution should be discarded after use. The solution is stable for at least several months when stored in the dark at room temperature.
2. Purification of CBB G250 is essential to achieve high staining sensitivity. Best results are reached with SERVA Blue G. which is known to be the purest source of CBB G250 *(12)*. However, even this product contains impurities that can be removed completely by the purification step. After purification, the dye shows no impurities when tested by thin-layer chromatography. When the dye is precipitated with ammonium sulfate, the impurities remain dissolved in the supernatant, which is a bright green solution. The acetone step in the procedure may be omitted. Traces of ammonium sulfate cause no harm in the staining solution. Gloves should be worn when handling CBB G250, since the dye binds strongly to skin.
3. Electrophoresis can be carried out as usual. It is useful to leave some of the lanes of the gel free. These lanes are used to obtain data for background correction. Do not use the lanes at the edge of the gel for this purpose, since background staining is stronger at the edges of the gel.
4. The fixation step is essential for highly sensitive staining. The gel should stay in the fixing solution for at least 1 h, but may be left in the solution for a longer time if necessary. The gel should be gently shaken during fixation. The fixing solution should not be reused, since buffer substances, carrier ampholytes, and other unwanted substances accumulate and may disturb the fixing process.
5. During staining, the gel should be gently shaken. Major bands can be detected after about 30 min. Staining time should be at least 12 h and should be kept constant since staining with CBB G250 in colloidal acidic solutions does not reach a true equilibrium. After more than 12 h, the increase in staining intensity is extremely low. In this staining solution, bands appear green against a low background and turn to blue during destaining. For destaining, the gel is washed in water until a clear background is achieved. Normally gels can be destained within 20 min; higher concentrated gels (gradient gels for example) need prolonged destaining times.
6. Staining with CBB G250 as described is very sensitive. The procedure is a modification of the method of Blakesley and Boezi *(10)*. The fixation step was added and purified dye was used. These modifications led to higher sensitivity and lower background staining. Depending on the nature of the protein, as little as 20 ng of protein could be detected visually after destaining, compared to 1 μg as given by those authors *(10)*.

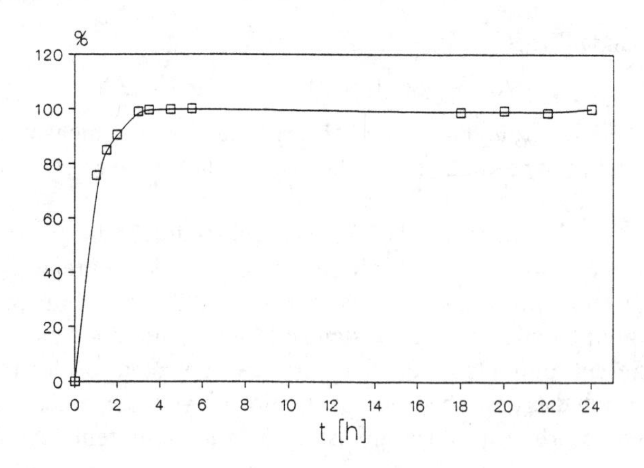

Fig. 1. Kinetics of dye elution. Absorbance after 24 h was set to 100%.

7. Bands or spots for quantitation can be cut out of the gel with a scalpel. For regular formed slices, automatic gel slicers are available. For better handling, thin focusing gels backed on a support can be dried without cracking and then cut with higher accuracy.
8. The eluent is a 3% (w/v) solution of SDS in water. This solution elutes the dye quantitatively. The eluent does not elute any protein from the gel. In this solution, absorbance of CBB G250 is a linear function of dye concentration over a wide range. The concentration of SDS is not critical. Concentrations up to 5% SDS were tested and no significant differences for all concentrations above 1.5% (w/v) SDS were found, neither in elution properties nor in dye absorbance behavior. CBB G250 shows solvatochromic behavior in aqueous solutions of SDS. With increasing concentrations of SDS, the absorbance maximum undergoes a batochromic shift from 580 nm in water to 610 nm in solutions of SDS with concentrations >1.5%. Absorbance increases strongly with this batochromic shift.
9. For elution, an appropriate volume of eluent is added to the weighed gel slice. The tube is sealed and incubated at 37°C for 24 h. For small gel slices (single bands or spots), 500 μL of eluent are adequate. Whole lanes can be eluted with 5 mL of eluent. Since the dye is highly soluble in the eluent, the volume should be chosen as small as possible to achieve higher absorbance. The volume is limited by cuvet size and photometer requirements. Elution is fast at 37°C. Figure 1 shows the elution kinetics. After 3 h 99% of the dye is eluted from a 10% gel. At lower temperatures or higher gel concentrations, elution kinetics are smaller, but under no condition did complete elution take more than 24 h. With these results, elution time can be varied depending on used gel concentrations and temperature. Following incubation, the absorbance of the solutions is measured against the solvent at 610 nm.
10. Even though background staining is very low when CBB G250 is used in colloidal solution, it is necessary to make corrections for background staining to achieve highly reproducible results. Since, after 12 h of staining, background is very even throughout a gel of constant concentration, the background staining per gram of gel can be determined. For this purpose, a piece of gel that contains no protein is cut out of the gel, drained, and weighed. After dye elution, the background staining per gram of gel is calculated. Gel pieces containing protein are treated the same way, and the appropriate amount of background staining is subtracted. This method allows background correction without care about the size of the gel piece used to determine the background. If gels with uneven background (e.g., gradient gels) are to be investigated, it is necessary to determine background using a blank cut from a region with similar background staining. Routinely one or

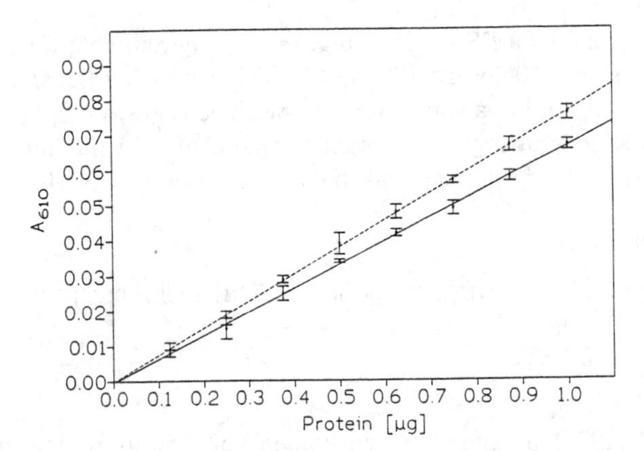

Fig. 2. Sample standard curves for two different proteins (BSA [solid line] and lysozyme [dashed line]) separated by SDS electrophoresis.

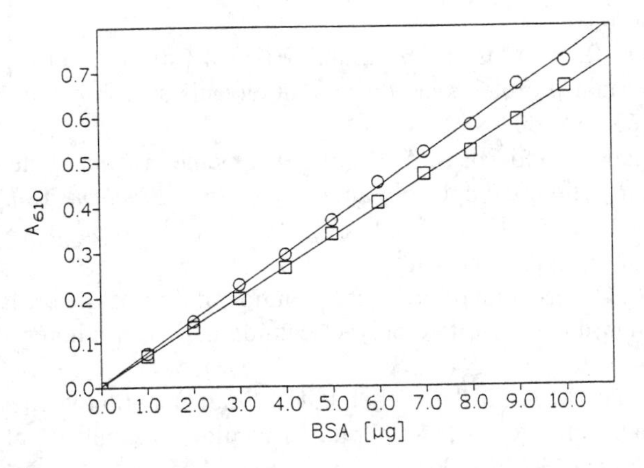

Fig. 3. Sample standard curves for BSA separated by two different electrophoretic techniques. SDS electrophoresis (□) and isoelectric focusing (○).

two lanes of the gel are left without protein. These lanes are cut into three pieces after electrophoresis, stained, and eluted. Background values are determined, and the mean of these values is used as the specific background for the whole gel. With this method, highly reproducible protein quantitation is obtained even for very small amounts of protein.

11. To correlate the obtained values for background-corrected absorbance with the amount of protein, calibration curves must be prepared. Staining intensity varies with a range of parameters. It depends on the nature of the protein, electrophoretic technique, gel concentration, and staining conditions (actual dye concentration, staining time, temperature). Calibration curves should therefore be made using either the protein under investigation or one with similiar electrophoretic and staining behavior.
12. Protein quantitation using the method described is linear at least in the range from 50 ng to 10 μg, an amount of protein larger than the typical load for analytical gels. The lower limit represents typical amounts of proteins in single spots of two-dimensional gels. Figure 2 shows standard curve samples for two different proteins in the submicrogram range. The response for BSA is 0.066 ± 0.005/μg and 0.077 ± 0.006/μg for lysozyme. Figure 3 shows

sample standard curves for BSA in the range of 1–10 μg for two different electrophoretic techniques. The slope is 0.066 ± 0.005/μg for SDS electrophoresis and 0.073 ± 0.009/μg for isoelectric focusing. For a single protein, the slope is always higher for focusing gels compared to SDS gels owing to the larger pore size of focusing gels, leading to better staining. For SDS gels, the slope seems to increase for proteins with high mobility *(9)*.

Acknowledgment

The author thanks Nicole Brauer for her skillful technical assistance during the development of this method.

References

1 Fishbein, W. N. (1972) Quantitative Densitometry of 1-50 μg protein in acrylamide gel slabs with Coomassie Blue. *Anal. Biochem.* **46,** 388–401.
2. Neuhoff, V., Arold, N., Taube, D., and Ehrhardt, W. (1988) Improved staining of proteins in polyacrylamide gels including isoelectric focusing gels with clear background at nanogram sensitivity using Coomassie Brillant Blue G-250 and R-250. *Electrophoresis* **9,** 255–262.
3. Fenner, C., Traut, R. R., Mason, D. T., and Wikman-Coffelt, J. (1975) Quantitation of Coomassie Blue stained proteins in polyacrylamide gels based on analysis of eluted dye. *Anal. Biochem.* **63,** 595–602.
4. Wong, P., Barbeau, A., and Roses, A. D. (1985) A method to quantitate Coomassie Blue-stained proteins in cylindrical polyacrylamide gels. *Anal. Biochem.* **150,** 288–293.
5. Martini, O. H. W., Kruppa, J., and Temkin, R. (1980) in *Electrophoresis 79* (Radola, B. J., ed.), de Gruyter, Berlin, pp. 241–248.
6. Ball, E. H. (1986) Quantitation of proteins by elution of Coomassie brillant blue R stained bands after sodium dodecyl sulfate-polyacrylamide gel electrophoresis. *Anal. Biochem.* **155,** 23–27.
7. Young, R. B., Orcutt, M., and Blauwiekel, P. B. (1980) Quantitative measurement of protein mass and radioactivity in *N,N'*-diallyltartardiamide crosslinked polyacrylamide slab gels. *Anal. Biochem.* **108,** 202–206.
8. Neumann, U., Khalaf, H., and Rimpler, M. (1992) Quantitation of proteins separated in *N,N'*-1,2-dihydroxyethylenebisacrylamide-crosslinked polyacrylamide gels. *Anal. Biochem.* **206,** 1–5.
9. Neumann, U., Khalaf, H., and Rimpler, M. (1994) Quantitation of electrophoretically separated proteins in the submicrogram range by dye elution. *Electrophoresis* **15,** 916–921.
10. Blakesley, R. W. and Boezi, J. A. (1977) A new staining technique for proteins in polyacrylamide gels using Coomassie brillant blue G 250. *Anal. Biochem.* **82,** 580–582.
11. Neuhoff, V., Stamm, R., and Eibl, H. (1985) Clear background and highly sensitive protein staining with Coomassie Blue dyes in polyacrylamide gels: a systematic analysis. *Electrophoresis* **6,** 427–448.
12. Wilson, C. M. (1979) Studies and critique of Amido Black 10B, Coomassie Blue R, and Fast Green FCF as stains for proteins after polyacrylamide gel electrophoresis. *Anal. Biochem.* **96,** 263–278.

28

Rapid Staining of Proteins in Polyacrylamide Gels with Nile Red

Joan-Ramon Daban, Salvador Bartolomé, and Antonio Bermúdez

1. Introduction

Sodium dodecyl sulfate-polyacrylamide gel electrophoresis (SDS-PAGE) is one of the most powerful methods for protein analysis *(1,2)*. Unfortunately, the typical procedures for the detection of protein bands after SDS-PAGE, using the visible dye Coomassie blue and silver staining, have several time-consuming steps and require the fixation of proteins in the gel. This chapter describes a rapid and very simple method for protein staining in SDS gels developed in our laboratory *(3,4)*. The method is based on the fluorescent properties of the hydrophobic dye Nile red (9-diethylamino-5H-benzo[α]phenoxazine-5-one; *see* Fig. 1), and allows the detection of <0.1 μg of unfixed protein per band about 6 min after the electrophoretic separation. Furthermore, it has been shown elsewhere *(5)* that, in contrast to the current staining methods, Nile red staining does not preclude the direct electroblotting of protein bands, and does not interfere with further sequencing and immunodetection analysis.

Nile red was considered a fluorescent lipid probe, because this dye shows a high fluorescence and intense blue shifts in the presence of neutral lipids and lipoproteins *(6)*. We have shown that Nile red can also interact with SDS micelles and proteins complexed with SDS *(3)*. Nile red is nearly insoluble in water, but is soluble and shows a high increase in the fluorescence intensity in nonpolar solvents and in the presence of pure SDS micelles and SDS–protein complexes. In the absence of SDS, Nile red can interact with some proteins in solution, but the observed fluorescence is extremely dependent on the hydrophobic characteristics of the proteins investigated *(6,7)*. In contrast, Nile red has similar fluorescence properties in solutions containing different kinds of proteins associated with SDS, suggesting that this detergent induces the formation of structures having equivalent hydrophobic properties independent of the different initial structures of native proteins *(3)*. In agreement with this, X-ray scattering and cryoelectron microscopy results have shown that proteins having different properties adopt a uniform necklace-like structure when complexed with SDS *(8)*. In this structure the polypeptide chain is most likely situated at the interface between the hydrocarbon core and the sulfate groups of the SDS micelles dispersed along the unfolded protein molecule.

From: The Protein Protocols Handbook
Edited by: J. M. Walker Humana Press Inc., Totowa, NJ

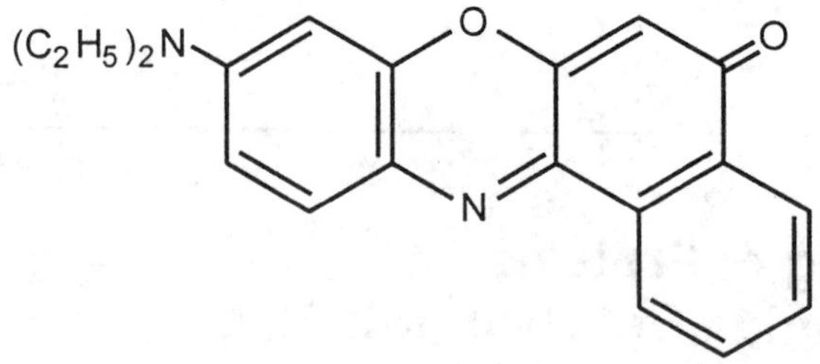

Fig. 1. Structure of the noncovalent hydrophobic dye Nile red.

The enhancement of Nile red fluorescence observed with different SDS–protein complexes occurs at SDS concentration lower than the critical micelle concentration of this detergent in the typical Tris-glycine buffer used in SDS-PAGE *(3)*. Thus, for Nile red staining of SDS-polyacrylamide gels *(4)*, electrophoresis is performed in the presence of 0.05% SDS instead of the typical SDS concentration (0.1%) used in current SDS-PAGE protocols. This concentration of SDS is high enough to maintain the stability of the SDS–protein complexes in the bands, but is lower than SDS critical micelle concentration and consequently precludes the formation of pure detergent micelles in the gel *(4)*. The staining of these modified gels with Nile red produces very high fluorescence intensity in the SDS–protein bands and low background fluorescence (*see* Fig. 2). Furthermore, under these conditions (*see* details in Section 3.), most of the proteins separated in SDS gels show similar values of the fluorescence intensity per unit mass.

2. Materials

All solutions should be prepared using electrophoresis-grade reagents and deionized water and stored at room temperature (exceptions are indicated). Wear gloves to handle all reagents and solutions, and do not pipet by mouth. Collect and dispose all waste according to good laboratory practice and waste-disposal regulations.

1. Nile red: Concentrated stock (0.4 mg/mL) in dimethyl sulfoxide (DMSO). This solution is stable for at least 3 mo when stored at room temperature in a glass bottle wrapped in aluminum foil to prevent damage by light. Handle this solution with care. DMSO is flammable, and in addition, this solvent may facilitate the passage of water-insoluble and potentially hazardous chemicals, such Nile red, through the skin. Nile red can be obtained from Sigma (St. Louis, MO) or Eastman Kodak (Rochester, NY); an electrophoresis-grade formulation of Nile red (SYBR Red Protein Gel Stain) is available commercially from Molecular Probes (Eugene, OR).
2. The acrylamide stock solution and the resolving and stacking gel buffers are prepared as described in Chapter 11.
3. 2X Sample buffer: 4% (w/v) SDS, 20% (v/v) glycerol, 10% (v/v) 2-mercaptoethanol, 0.125*M* Tris-HCl, pH 6.8; bromophenol blue (0.05% [w/v]) can be added as tracking dye.
4. 10X Electrophoresis buffer: 0.5% (w/v) SDS, 0.25*M* Tris, 1.92*M* glycine, pH ~8.3 (do not adjust the pH of this solution).
5. Plastic boxes for gel staining. Use opaque polypropylene containers (e.g., 21 × 20 × 7.5 and 14 × 9.5 × 7 cm for large [20 × 16 × 0.15 cm] and small [8 × 7.3 × 0.075 or 0.15 cm]

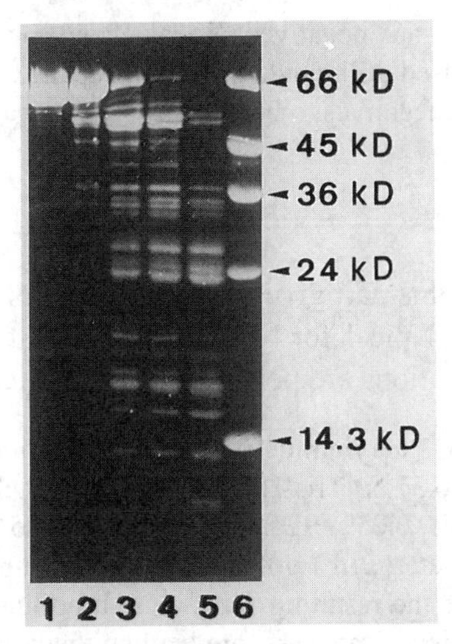

Fig. 2. Example of Nile red staining of different proteins and peptides in 0.05% SDS-15% polyacrylamide gels. The protein mol-wt markers (lane 6, from top to bottom: BSA, ovalbumin, glyceraldehyde-3-phosphate dehydrogenase, trypsinogen, and lysozyme), and BSA were digested with increasing amounts of trypsin (lanes 1–5), then stained for 5 min with a solution of Nile red in water prepared by quick dilution of a stock solution of this dye in DMSO (*see* Section 3.).

gels, respectively) with a close-fitting lid to allow intense agitation without spilling the staining solution.

6. Orbital shaker for the agitation of the plastic boxes during gel staining.
7. Transilluminator equipped with mid-range ultraviolet (UV) bulbs (~300 nm) to excite Nile red *(3,4)*: A transilluminator with a cooling fan (e.g., Foto UV 300 [Fotodyne Inc., Harland, WI] or similar) is very convenient to prevent thermal damage of the gel when long exposures are necessary (*see* Section 3., step 11). The glass filter of the transilluminator should be replaced when it loses transmission efficiency owing to the solarization produced by long periods of UV irradiation. UV light is dangerous to skin and particularly to eyes. UV-blocking goggles and a full face shield and protective gloves and clothing should be worn when the stained gel is examined using the transilluminator. For the photography, we place the transilluminator with the gel and the photographic camera inside a homemade cabinet with opaque (UV-blocking) curtains to prevent operator exposure to UV light.
8. Photographic camera (e.g., Polaroid [Cambridge, MA] MP-4 camera or other Polaroid systems). The camera and transilluminator should be placed in a darkroom.
9. Optical filters: Use the Wratten (Eastman Kodak) filters numbers 9 (yellow) and 16 (orange) to eliminate the UV and visible light from the transilluminator. Place the two filters together in the filter holder of the camera so that filter 16 is on top of filter 9 (i.e., filter 16 should be facing toward the camera lens). Store the filters in the dark, and protect them from heat, intense light sources, and humidity.
10. Photographic films: The following Polaroid instant films can be used for the photography of the red bands (maximum emission at ~640 nm *[3]*) seen after staining: 667 (3000 ASA, panchromatic, black-and-white positive film), 665 (80 ASA, panchromatic, black-and-white positive/negative film), and 669 (80 ASA, color, positive film). Store the films at 4°C.
11. Negative-clearing solution: 18% (w/v) sodium sulfite.

12. Wetting agent for Polaroid 665 negatives (Kodak Photo-Flo diluted at 1:600).
13. Densitometer: We have used a Shimadzu (Tokyo, Japan) CS-9000 densitometer for the scanning of photographic negatives; other densitometers available commercially can also be used.

3. Method

The method described in this part gives all the details for staining SDS-polyacrylamide gels with Nile red. *See* Note 1 for Nile red staining of gels without SDS. Unless otherwise indicated, all operations are performed at room temperature.

1. Typically we prepare 15% acrylamide-0.4% bisacrylamide separating gel containing 0.05% SDS (*see* Note 2), 0.375*M* Tris-HCl, pH 8.8. The stacking gel contains 6% acrylamide–0.16% bisacrylamide, 0.05% SDS (*see* Note 2), 0.125*M* Tris-HCl, pH 6.8.
2. Dissolve the proteins in water, add 1 volume of the 2X sample buffer (Section 2., step 3; *see* Note 3), and incubate the resulting samples in a boiling water bath for 3 min. The samples are kept at room temperature before loading them into the gel (*see* Note 4).
3. Place the gel sandwiched between the glass plates in the electrophoresis apparatus, fill the electrode reservoirs with 1X electrophoresis buffer (*see* Note 2), and rinse the wells of the gel with this buffer.
4. Load the protein samples (about 20 and 10 μL in large and small gels, respectively), carry out electrophoresis, and, at the end of the run, remove the gel sandwich from the electrophoresis apparatus and place it on a flat surface.
5. Wearing gloves, remove the upper glass plate (use a spatula), excise the stacking gel and the bottom part of the separating gel (*see* Note 5), and transfer the gel to a plastic box.
6. Place 10 mL of the concentrated Nile red (0.4 mg/mL) staining solution in DMSO in the bottom of a dry glass Erlenmeyer flask, and add quickly 490 mL of deionized water (*see* Note 6). These volumes are required for staining large gels; use an Erlenmeyer flask with a capacity of 1 L to facilitate the rapid mixing of the concentrated Nile red solution and water (*see* step 7). For small gels, mix 2 mL of the concentrated Nile red solution in DMSO and 98 mL of water, in an Erlenmeyer flask with a capacity of 0.5 L.
7. Immediately after the addition of water, agitate the resulting solution vigorously (in a circular way) for 3 s (*see* Note 6).
8. Pour the resulting staining solution very quickly into the plastic box containing the gel, put the lid on, and agitate very vigorously (*see* Note 6) using an orbital shaker (at about 300 rpm) for 5 min (*see* Note 7).
9. Discard the staining solution, and rinse the gel with deionized water (4 times; about 10 s each time) to remove completely the excess Nile red precipitated during the staining of the gel (*see* Note 8).
10. Wearing gloves, remove the gel from the plastic box, and place it on the UV transilluminator. Turn off the room lights, turn on the transilluminator, and examine the protein bands, which fluoresce light red under UV light (*see* Note 9). Turn off the transilluminator immediately after the visualization of the bands (*see* Note 10).
11. Focus the camera with the help of lateral illumination with a white lamp, place the optical filters indicated in Section 2., step 9 in front of the camera lens, and, in the dark, turn on the transilluminator and photograph the gel. Typical exposure times (at aperture f/4.5) are: 5 s for Polaroid 667 and 4.5 min for Polaroid 665 and 669 (*see* Note 10). Finally, turn off the transilluminator.
12. Develop the different Polaroid films for the time indicated by the manufacturer (*see* Note 11). Spread the Polaroid print coater on the surface of the 665 positive immediately after

development. (The positive prints of the 667 and 669 films do not require coating.) The 665 negatives can be stored temporarily in water but, before definitive storage, immerse the negatives in 18% sodium sulfite, and agitate gently for about 1 min, wash in running water for 5 min, dip in a solution of wetting agent (about 5 s), and finally dry in air (*see* Note 12).

13. Scan the photographic negatives to determine the intensity of the Nile red-stained protein bands if a quantitative analysis is required (*see* Note 13).

4. Notes

1. Isoelectric focusing gels do not contain SDS and should be treated with this detergent after the electrophoretic run in order to generate the hydrophobic SDS–protein complexes specifically stained by Nile red *(5)*. In general, for systems without SDS, we recommend an extensive gel washing (20 min in the case of 0.75-mm-thick isoelectric focusing [IEF] gels) with 0.05% SDS, 0.025*M* Tris, 0.192*M* glycine, pH 8.3, after electrophoresis. The gel equilibrated in this buffer can be stained with Nile red following steps 6–9 of Section 3.
2. In order to reduce the background fluorescence after the staining with Nile red, it is necessary to preclude the formation of pure SDS micelles in the gel (*see* Section 1.). Thus, use 0.05% SDS to prepare both the separating and stacking gels and the electrophoresis buffer. This concentration is lower than the critical micelle concentration of this detergent (~0.1% *[3]*), but is high enough to allow the formation of the normal SDS–protein complexes that are specifically stained by Nile red *(4)*.
3. Use 2% SDS in the sample buffer in order to be sure that all protein samples are completely saturated with SDS. Lower concentrations of SDS in the sample buffer can produce only a partial saturation of proteins (in particular in highly concentrated samples), and, consequently, the electrophoretic bands could have anomalous electrophoretic mobilities. The excess SDS (uncomplexed by proteins) present in the sample buffer migrates faster than the proteins and forms a broad band at the bottom of the gel (*see* Note 5 and ref. *4*).
4. Storage of protein samples prepared as indicated in Section 3., step 2 at 4°C (or at lower temperatures) causes the precipitation of the SDS present in the solution. These samples should be incubated in a boiling water bath to redissolve SDS before using them for electrophoresis.
5. The bottom part of the gel (i.e., from about 0.5 cm above the bromophenol blue band to the end) should be excised before the staining of the gel. Otherwise, after the addition of Nile red, the lower part of the gel produces a broad band with intense fluorescence. This band is presumably caused by the association of Nile red with the excess SDS used in the sample buffer (*see* Note 3). In the case of long runs, bromophenol blue and the excess SDS band diffuse into the buffer of the lower reservoir, and it is not necessary to excise the gel bottom.
6. Nile red is very stable when dissolved in DMSO (*see* Section 2., item 1), but this dye precipitates in aqueous solutions. Since the precipitation of Nile red in water is a rapid process and this dye is only active for the staining of SDS–protein bands before it is completely precipitated *(4)*, to obtain satisfactory results, special care should be taken in the following points:
   a. Use a dry flask (to avoid any initial precipitation of the dye owing to the presence of water).
   b. Work very rapidly in steps 6–9 of Section 3.
   c. Perform all the intense agitations indicated in these steps (in order to favor as much as possible the dispersion of the dye).
7. The staining time is the same for large and small gels.

8. After the water rinsing indicated in step 9 of Section 3., the staining process is completely finished and the gel can be photographed immediately. It is not necessary, however, to examine and photograph the gel just after staining. Nile red-stained bands are stable, and the gel can be kept in the plastic box immersed in water for 1–2 h before photography.
9. Faint bands that are not visible by direct observation of the transilluminated gel can be clearly seen in the photographic image. Very faint bands (containing as little as 20 ng of protein) can be detected by long exposure. To obtain this high sensitivity, it is necessary to have sharp bands; broad bands reduce considerably the sensitivity.
10. Nile red is sensitive to intense UV irradiation *(3)*, but has a photochemical stability high enough to allow gel staining without being necessary to introduce complex precautions in the protocol (*see* Section 3., steps 6–9). For long-term storage, solutions containing this dye are kept in the dark (*see* Section 2., step 1). Transillumination of the gel for more than a few minutes produces a significant loss of fluorescence intensity. Thus, transillumination time must be reduced as much as possible both during visualization and photography. The short exposure times required for Polaroid film 667 allow one to make several photographs with different exposure times if necessary. With films requiring longer exposures (Polaroid 665 and 669), only the first photograph from each gel shows the maximum intensity in the fluorescent bands.
11. The development time of Polaroid films is dependent on the film temperature. Store the films at 4°C (*see* Section 2., step 10), but allow them to equilibrate at room temperature before use.
12. The relatively large area (7.3 × 9.5 cm) of the Polaroid 665 negative is very convenient for further densitometric measurements (*see* Note 13). Furthermore, this negative can be used for making prints with an adequate level of contrast (*see*, for instance, the photograph presented in Fig. 2).
13. In quantitative analyses, care has to be taken to ensure that the film has a linear response for the amounts of protein under study. Use different amounts (in the same range as the analyzed protein) of an internal standard to obtain an exposure time producing a linear film response. Furthermore, the internal standard is necessary to normalize the results obtained with different gels, under different electrophoretic conditions, and with different exposure and development times. Nile red can be considered a general stain for proteins separated in SDS gels *(4)*. However, proteins with prosthetic groups, such as catalase, and proteins having anomalous SDS binding properties, such as histone H5, show atypical values of fluorescence intensity after staining with Nile red *(4)*.

Acknowledgment

This work was supported in part by grant PB92-0602 from the Dirección General de Investigación Científica y Técnica. A. Bermúdez was supported by a predoctoral fellowship from the Generalitat de Catalunya.

References

1. Andrews, A. T. (1986) *Electrophoresis. Theory, Techniques, and Biochemical and Clinical Applications.* Oxford University Press, Oxford, UK.
2. Zewert, T. E. and Harrington, M. G. (1993) Protein electrophoresis. *Curr. Opin. Biotechnol.* **4,** 3–8.
3. Daban, J.-R., Samsó, M., and Bartolomé, S. (1991) Use of Nile red as a fluorescent probe for the study of the hydrophobic properties of protein-sodium dodecyl sulfate complexes in solution. *Anal. Biochem.* **199,** 162–168.

4. Daban, J.-R., Bartolomé, S., and Samsó, M. (1991) Use of the hydrophobic probe Nile red for the fluorescent staining of protein bands in sodium dodecyl sulfate-polyacrylamide gels. *Anal. Biochem.* **199,** 169–174.
5. Bermúdez, A., Daban, J.-R., Garcia, J. R., and Mendez, E. (1994) Direct blotting, sequencing and immunodetection of proteins after five-minute staining of SDS and SDS-treated IEF gels with Nile red. *BioTechniques* **16,** 621–624.
6. Greenspan, P. and Fowler, S. D. (1985) Spectrofluorometric studies of the lipid probe Nile red. *J. Lipid Res.* **26,** 781–789.
7. Sackett, D. L. and Wolff, J. (1987) Nile red as a polarity-sensitive fluorescent probe of hydrophobic protein surfaces. *Anal. Biochem.* **167,** 228–234.
8. Samsó, M., Daban, J.-R., Hansen, S., and Jones, G. R. (1995) Evidence for sodium dodecyl sulfate/protein complexes adopting a necklace structure. *Eur. J. Biochem.* **232,** 818–824.

29

Zn^{2+}-Reverse Staining of Proteins in Polyacrylamide Gels

Juan M. García-Segura and Mercedes Ferreras

1. Introduction

Unlike the traditional methods for staining of polyacrylamide gels (e.g., Coomassie blue, silver stain) whose purpose is to dye the protein bands *(1,2),* the purpose of reverse staining (RS) methods is to leave protein bands unstained and dye the rest of the gel instead. RS methods are usually based on the precipitation of some metal salt all along the gel, except in the positions where proteins are. The contrast between protein bands and background, and the sensitivity of a particular RS method, will depend on both the intensity of the selective precipitation and the color of the precipitate.

In general, the sensitivity of RS methods is higher than that of Coomassie blue *(3,4),* approaching in many instances that of silver staining. Taking into account that such a valuable sensitivity can be obtained just a few minutes after electrophoresis completion (what immersion of the gel in two or three solutions can take), RS methods are very suitable for routine work. In addition, RS methods do not use fixative solvents, thus making it easy for elution of "unstained proteins" from the gels.

However, objections can be raised to the reproducibility of RS methods, and especially to the loss of the electrophoretic pattern when the negatively stained gels are dried (e.g., for storage or spectroscopic scanning).

The RS method described in this chapter is based on the precipitation of Zn^{2+}-Imidazole, which has been reported to be a good way of circumventing problems with reproducibility, as well as increasing the final sensitivity *(5);* moreover, the RS method described herein embodies the modifications needed to make it suitable for all forms of protein polyacrylamide gel electrophoresis (PAGE) (i.e., isoelectric focusing [IEF], nondenaturing, and sodium dodecyl sulfate [SDS]) *(6).* The authors have studied this RS method in order to identify the origin of some artifacts that still remained, and have attempted to make the negative pattern resistant to drying *(7).* As a result, a protocol has been developed that can be subdivided into two basic procedures: RS in the strict sense, by which a white and homogeneous imidazole-zinc precipitate is formed along the gel, except in the protein bands; a toning reaction (TR), which turns this white precipitate into a deep blue, leaving the protein bands transparent and colorless.

The first procedure is recommended for routine work or when further analyses of proteins eluted from the gel are required, whereas the complete protocol (RS plus TR) is mandatory if gels should be dried, because only after toning does the electrophoretic

From: The Protein Protocols Handbook
Edited by: J. M. Walker Humana Press Inc., Totowa, NJ

pattern become permanent. In any case, the same high sensitivity is reached. Thus, 10 ng of protein/band can easily be visualized, and even 5 ng can become distinguishable with some practice.

After drying, the reverse-stained gels can conveniently be scanned, showing then a linear relationship between the area of a protein band and the respective amount of protein loaded. Moreover, in the range of 10–100 ng of protein/band, such a relationship seems to be the same irrespective of the protein chemical composition.

2. Materials

All solutions are prepared from reagent-grade chemicals and can be stored at room temperature.

2.1. Reverse Staining

1. RS-I: 1% (w/v) sodium carbonate.
2. RS-II: 0.2*M* imidazole containing 0.1% SDS (w/v).
3. RS-III: 0.2*M* zinc sulfate.

2.2. Toning Reaction

1. TR-I: 0.2% (w/v) Potassium ferricyanide.
2. TR-II: 0.2% (w/v) *o*-Tolidine (Sigma, St. Louis, MO; practical grade) in 40% (v/v) methanol. This solution should be prepared by first dissolving *o*-tolidine in methanol, and then adding the required volume of water to reach the specified percentages of *o*-tolidine and methanol. For instance, to prepare 1 L of this solution, dissolve 2 g of *o*-tolidine in 400 mL of methanol and, once dissolved, add water up to 1 L. It has been shown that *o*-tolidine is a carcinogenic agent in rats *(8)*. Therefore, care must be exercised when handling. The wearing of gloves throughout the procedure, including the preparation of stock solutions, is strongly recommended.
3. TR-III: 0.36*N* sulfuric acid.

3. Method

The following reverse staining protocol is optimized for polyacrylamide gels that have been prepared and run as described.

3.1. PAGE

Running gels containing 12.5–15.0% acrylamide and stacking gels containing 4.0% acrylamide are electrophoresed in the well-known buffer system of Laemmli (*9; see* Chapter 11) by using a MiniPROTEAN II (Bio-Rad, Hercules, CA), electrophoresis cell provided with 0.75-mm thick spacers and comb (for different gel specifications, *see* Note 18). Electrophoreses are performed at 25 mA/gel just until the tracking dye (0.1% bromophenol blue) leaves the gel.

3.2. Staining Protocol

All the following steps should be performed under gentle orbital agitation (50–60 rpm). One hundred milliliters of the respective solutions are employed at each step, provided that the container allows this volume to cover the gel completely.

3.2.1. RS

1. Immediately after electrophoresis, soak the gel for 5 min in RS-I solution (*see* Note 1).
2. Soak the gel in RS-II solution for 15 min (*see* Notes 2 and 3).

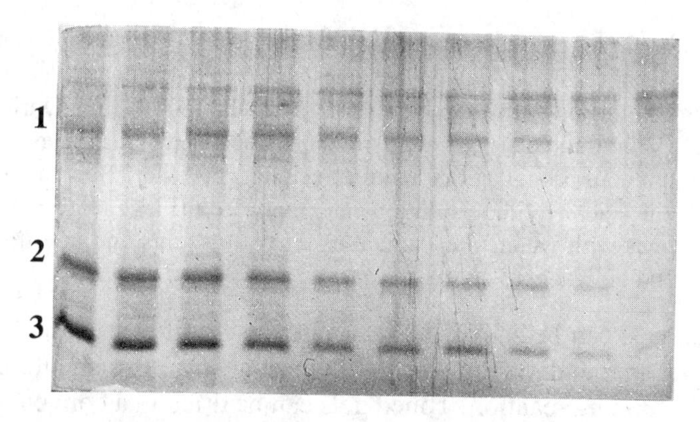

Fig. 1. RS of a gel containing 15% acrylamide (SDS-PAGE). Different amounts of an evenly formulated mixture of commercial proteins (1, ovalbumin from hen egg white; 2, soybean trypsin inhibitor; and 3, lysozyme from hen egg white) were applied in each well. From left to right, the amount of each protein was 1 μg, 750 ng, 500 ng, 250 ng, 100 ng, 75 ng, 50 ng, 25 ng, 10 ng, and 5 ng. Some artificial bands can be seen in the region 50–70 kDa, their staining contrast being inverse to the decrease of protein loading. These artifacts, sometimes also seen in silver-stained gels *(1)*, are owing to contaminants that are reduced by the 2-mercaptoethanol of the sample buffer. Hence, the lesser the amount of loaded protein, the higher the excess of reducing agent and, thus, the intensity of the artifact. The gel was photographed by putting it against a black surface.

3. Wash the gel with water for 10 s, trying to eliminate the foam from the detergent.
4. Soak the gel in RS-III solution (*see* Note 4). In a few seconds, a white background starts to develop while the protein bands remain transparent (*see* Note 5). **Caution!:** This step must not be extended longer than 40 s (*see* Note 6). Therefore, the Zn^{2+} solution should rapidly be poured off at 30–35 s.
5. Eliminate the excess of Zn^{2+} by washing the gel with abundant water: 2 × 10 s plus 2 × 5 min (*see* Note 7).

At this stage, the reverse-stained gel can be stored in water if the TR is to be postponed, or when toning is not going to be performed, for instance, when the gel is intended for recovery of the reverse-stained proteins (*see* Note 8). The next TR can be done at will, when one positively decides not to elute proteins from reverse-stained gels. Figure 1 illustrates the RS that can be obtained when the procedure just described is performed.

3.2.2. TR

1. Prepare the solution required for the next step. Mix equal volumes of TR-I and TR-II solutions (*see* Note 9). This can be done directly in the container.
2. Soak the gel in the above solution for 4 min. During this step, the white background of the gel will become brown-yellowish (*see* Note 10), whereas the protein bands will turn from colorless to yellow (*see* Note 11). Also at this step, a lightweight precipitate appears at the solution surface, but this does not affect the final results, since the orbital agitation should avoid its deposition in the gel surface (*see* Note 12).

   During this incubation time, the solution required for step 4 should be prepared by mixing equal volumes of TR-II and TR-III solutions (*see* Note 13).
3. Rinse the gel with water for 10 s, trying to eliminate any trace of the precipitate (*see* Note 12).

4. Soak the gel in the solution just prepared (TR-II + TR-III) (*see* Note 14). During the first 2 min, the brown background of the gel becomes blue (*see* Note 15), while the yellow protein bands become colorless again (*see* Note 16), giving the final toning. During this time, the solution turns yellow and develops at its surface the white precipitate mentioned in Note 13. Under the continuous orbital agitation, this precipitate does not affect the final result.
5. After 5 min, the acid *o*-tolidine solution can be poured off (*see* Note 17). Then wash the gel several times with water to eliminate excess of reagents and any trace of the white precipitate that could adhere on the surface.

At this stage, gels can be stored in water with no significant losses in their toning. However, it is recommended to proceed with drying, since this should be the purpose of performing the toning reaction. Toned gels can be dried in a conventional way, and then scanned for quantification purposes (*see* Note 19). Once again, it should be stressed that toned gels are not valid for protein elution (*see* Note 20). Figure 2 shows a reverse-stained and toned gel as resulting after application of the whole protocol (Sections 3.2.1. and 3.2.2.) herein described.

4. Notes

1. The main purpose of this step is to make the gel alkaline enough for a good imidazole-zinc precipitation (*see* Note 5). At the same time, residual components of the running buffer are washed out, thereby improving the uniformity of the imidazole-zinc precipitate. Concerning the first point, although the pK_a of imidazole is 6.95 *(10),* solutions of some commercial imidazole preparations give a pH in the 9.5–10.0 range owing to trace impurities. In such cases, it is not necessary to incubate gels previously with sodium carbonate. Nevertheless, this first incubation for 5 min should be kept, in order to wash out excess running buffer in the gel, especially glycine, which interferes with the formation of the imidazole-zinc precipitate, thereby affecting the uniformity of the background. For such a purpose, water can be used instead of RS-I solution.
2. The presence of SDS in RS-II solution improves the RS, and makes this method valid even for gels run without SDS (e.g., IEF gels) *(6)*. In fact, a role for SDS in the mechanism of RS has recently been suggested *(11)*. This is the longest step in the RS procedure, and in order to shorten it, one is always tempted to reduce this incubation time. Although early reports on imidazole-zinc RS *(6)* claimed shorter incubations with the imidazole solution, in the author's experience, this most probably results in poorer staining (i.e., bands positively stained, slight or uneven background). On the contrary, extending this second step will only make sense when gels thicker than 0.75 mm are used (*see* Note 18). Otherwise, the effect of a longer incubation with imidazole-SDS will be an undesirable diffusion of proteins through the polyacrylamide matrix and out of it. In the 15 min of this step, the levels of imidazole reached in the gels are enough to give a "good precipitation" in the next step.
3. The incubation with the imidazole-SDS solution has the inconvenience of foam formation resulting from the detergent. This can be reduced by using a lower rate of orbital agitation (30–40 rpm).
4. It is very advisable to perform this step in a glass container and over a black surface. In this way, the precipitation can easily be monitored as it progresses.
5. Zn^{2+} forms a white insoluble complex with imidazole. This complex redissolves at acid pH (*see* Note 6). Of course, this precipitation will also occur between the imidazole present in the gel and the Zn^{2+} of the RS-III solution. In such a situation, the precipitate will be the result of the encounter between two fronts of diffusion, that of imidazole going out of the

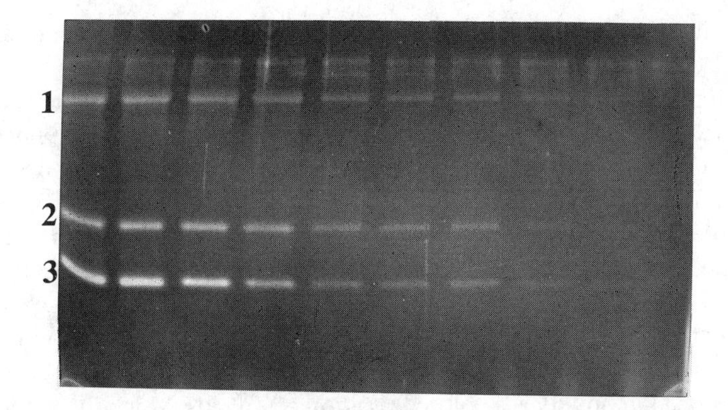

Fig. 2. RS following toning of a gel containing 15% acrylamide (SDS-PAGE). The proteins loaded and their amounts are the same as those specified in Fig. 1 (*see* legend). The gel, once toned, was dried between cellophane sheets and then photographed by placing it on a light box.

gel and that of Zn^{2+} going into it. Once the precipitate is formed, the respective diffusions of imidazole and zinc are blocked. Therefore, the white background of this RS method is just a precipitation layer formed at the gel surface, which justifies its rapid development. The exact mechanism by which protein bands remain transparent is unknown. It has been suggested that this behavior cannot be attributed to the protein itself, but to the increase in the local SDS concentration that it induces *(11)*. Although this explanation has been supported by some experimental evidences, when SDS is completely absent (gels run without SDS and RS-II solution only containing imidazole), some proteins become negatively stained while some others stain positively, indicating that SDS does not have a crucial effect. Hence, there are several factors that should additionally be taken into account:

a. Buffer properties of proteins (or SDS–protein complexes) can prevent the local pH from reaching the proper value for the precipitation to occur.
b. Proteins (or SDS–protein complexes) can chelate the zinc ion, preventing its precipitation.
c. Protein molecules in the bands can interfere with the diffusion of reactants. This latter possibility cannot be underestimated, since the rapidity of the RS (few seconds) supports the importance of the diffusion processes.

6. The white precipitate in the gel surface can be redissolved by an excess of Zn^{2+}. This fact is owing to both the acid-soluble nature of the imidazole-zinc precipitate and the acid character of Zn^{2+}. Consequently, in order to obtain a heavy precipitate in the gel surface, it seems reasonable to use a diluted $ZnSO_4$ solution instead of the strong 0.2*M* solution described. Certainly, this will yield heavier backgrounds, but at the expense of lengthening the time of gel incubation with this diluted solution. Then, imidazole will have time enough to diffuse from gels in an appreciable amount, allowing unspecific deposition of the precipitate in the protein bands (fog), in short, reducing sensitivity in terms of discrimination between background and protein bands. After detailed studies *(7)*, the author concluded that the conditions described for this step are the best compromise to obtain backgrounds that are as dense as possible, while protein bands remain transparent, that is, without developing a fog that covers all the gel surface. Of course, this step is the optimized one for our experimental conditions. More or less concentrated acrylamide gels would probably require shorter or longer incubation times, respectively (*see* Note 18). This aspect should be investigated for each particular case, but it should be emphasized that too long Zn^{2+} incubation times can have adverse effects in the final staining.

7. This step is intended for efficiently washing the excess of Zn^{2+} in order to assess that the RS is stable.
8. It has been reported that the negative electrophoretic pattern is stable in water for up to 2 yr *(12)*. Moreover, from the point of view of recovery of proteins from unstained bands, this remains quantitatively the same for 16 h after electrophoresis and RS, and for some proteins their recovery was still 66% even 2 yr later *(12)*. Zn^{2+} seems to be involved in this apparent protein immobilization by coordinating different SDS–protein micelles. In fact, a prerequisite for eluting negatively stained proteins is to chelate the metal cation *(6,12)*.
9. The solution required for the second step of the TR is not stable, and slowly develops (*see* Note 10) a dark and lightweight precipitate over its surface. Hence, it must be prepared immediately before use from the stock solutions mentioned.
10. The toning procedure involves a colored reaction of Zn^{2+} by which the white background is turned into deep blue. It is based on the redox reaction between ferricyanide and *o*-tolidine *(13):* Ferricyanide anion oxidizes *o*-tolidine up to a blue product while it is reduced to ferrocyanide (this explains why the mixture of TR-I and TR-II solutions is not stable). In the absence of Zn^{2+}, this is a very slow and incomplete reaction, but if Zn^{2+} is present, the formed ferrocyanide precipitates as $Zn_2[Fe(CN)_6]$, thus increasing the oxidation potential of ferricyanide and therefore the rate of the redox reaction. As this redox/precipitation reaction occurs, the ferricyanide/*o*-tolidine system is being adsorbed in the precipitate. This adsorption helps to intensify the final tone and is responsible for its permanent character as well. Therefore, reaction conditions should be established for allowing a rapid redox reaction that results in an intense adsorption. A factor to take into account is the requirement of acid, which acts as a catalyst of the redox reaction. Although Zn^{2+} remains precipitated in the background as a complex with imidazole, the levels of free cation are low, and, additionally, the pH is neutral owing to the buffer properties of imidazole. Thus, the redox reaction progresses very slowly, just accounting for some darkening of the background owing to the limited amounts of free Zn^{2+} released. The triggering of the redox/precipitai;ion reaction will be done in step 4 (*see* Note 14).
11. The yellow color of the protein bands is the result of ferricyanide impregnation. Zn^{2+} is bound to proteins so strongly than it is not available, even for a slow redox reaction.
12. This precipitate is the same described in Note 9, but is much more evident owing to the presence of trace amounts of Zn^{2+} released from the gel. As already mentioned, this precipitate is so lightweight that no special actions other than continuous orbital agitation are required for avoiding its deposition on the gel. Effective rinsing in step 3 is enough to get rid of it.
13. In the last step of the TR, an acid *o*-tolidine solution is employed. This solution is also unstable and tends to develop a white and lightweight precipitate. Therefore, it must be prepared just before use by mixing equal volumes of the stock *o*-tolidine solution (TR-II) and the H_2SO_4 solution (TR-III). Although the main component of this solution is H_2SO_4, which is needed for a rapid acidification of the gel (*see* Note 14), *o*-tolidine should also be present in order to obtain a deep toning. In its absence, a poor toning results, which is owing to the low rate of the redox reaction and thus a low adsorption on the zinc ferrocyanide precipitate. Attempts to increase the color density of the background additionally by also including ferricyanide in this last incubating solution are unsuccessful. If this is done, an intense precipitation outside of the gels results with a poor toning effect.
14. In order to monitor the toning process, the container, preferably of glass, should be transilluminated.
15. There are two factors that contribute to the triggering of the colored reaction in step 4:
    a. H_2SO_4 dissolves the imidazole-zinc precipitate in the background, thus locally releasing the free cation needed for the global reaction to occur, and

b. H_2SO_4 rapidly cancels the buffer properties of imidazole, allowing the acid pH needed for the redox reaction.

It should be stressed that adsorption of oxidized *o*-tolidine in the precipitate only occurs during the redox/precipitation reaction, the extent of this adsorption being responsible for the final tone. The faster the redox/precipitation reaction, the deeper the final color. Thus, final color can fluctuate among brown, green, or deep blue, depending on reactant concentrations as well as the presence of acid. Additionally, it should be taken into account that toned gel color once formed cannot be changed further. Therefore, it is neccesary to trigger the colored reaction only when diffusion of the redox pair (ferricyanide/*o*-tolidine) in the gel provides the proper concentrations for a rapid redox reaction. This is the purpose of the first step of the TR in which the neutral pH of imidazole helps to prevent any irreversible poor toning. Once the gel has the proper concentrations of reagents, the pH is rapidly decreased in step 4 by H_2SO_4, so triggering the toning. Of course, the amount of H_2SO_4 needed is very dependent on the imidazole levels in the gel. Modification of the experimental conditions herein described will require corresponding adjustment of the amount of H_2SO_4 in the last incubating solution.

16. As stated in Note 11, the yellow color of the protein bands is the result of ferricyanide impregnation, which in this step is eliminated by the adsorption of this reactant together with *o*-tolidine in the precipitate of the background. Should any residual impregnation remain after this step, giving a yellowish tone to the protein bands, it will be eliminated by the subsequent washings with water. Complexation of Zn^{2+} by proteins should be the reason that explains why toned protein bands do not result.
17. The TR reaches an end point. Therefore, there is no need to control the time of this step carefully.
18. Some considerations about specified timings: If the employed acrylamide percentages and/or gel thicknesses differ from those described under Section 3.1., some timings of the staining protocol require some modifications. As a general rule, for gels thicker than 0.75 mm, the steps should be lengthened. In the same way, the higher the acrylamide percentage, the longer the incubation times, and vice versa. The exact timings for each particular case will require some trivial investigation, but it should be stressed that the precipitation responsible for the RS has a very superficial character, and thus, has the toning of this precipitate. As a consequence, one should not look for major modifications to the timings herein reported when a different gel thickness is employed. In case such modifications seem to be necessary, steps 1 and 2 of the RS (Section 3.2.1.), and perhaps step 1 of the TR (Section 3.2.2.), should first be considered, since they are intended for gel equilibration in the respective solutions, and certainly this equilibration will be influenced by gel specifications to some extent. Also washings (steps 3 and 5 in Section 3.2.1., and steps 3 and 5 in Section 3.2.2.) would benefit of some lengthening when gels thicker than 0.75 mm are employed. Concerning zinc incubation (Section 3.2.1., step 4), which is responsible for the negative pattern development, its timing is almost undependent on gel thickness, but on the contrary, acrylamide percentages can markedly influence the final staining, making it neccesary to adapt the timing of this step to each particular case. Finally, no special actions have to be taken concerning the TR itself (step 4 of Section 3.2.2.), irrespective of gel concentration or thickness. This is an advantage of the end point that the TR reaches. Thus, this step is continued until obtaining a background tone whose contrast with unstained bands does not improve by further incubation. With dependence on the gel specifications, this final tone can slightly differ from the deep blue obtained for 12.5–15.0% acrylamide, 0.75-mm thick gels (Fig. 2).
19. Gels should be scanned in transmittance mode, because only in this mode a linear relationship between peak areas and protein amount is found. This is the behavior expected

for a staining method in which the cromophore is associated not to the protein, but to the background. In such a situation, the higher the amount of protein, the higher its transparency. As mentioned in Note 15, the color of the background can change depending on the TR conditions, but if the experimental conditions herein described are followed, a deep blue tone is obtained (Fig. 2). The absorbance spectrum of this toned background shows a maximum at 580 nm *(7)*. Thus, recording the absorbance spectrum of toned gels in those portions lacking proteins will provide good control over the reproducibility of this staining method. In addition, 580 nm should be the wavelength employed for the transmittance scanning in order to provide the densitometric measurement with the maximum sensitivity. When this is done, a linear relationship can be found between area of transmittance peak and the respective amount of protein. This linear relationship is followed in the protein range 10–100 ng and is the same for different proteins, suggesting a nonspecific protein-negative stain. Above this range, each protein shows a different behavior, and the linearity is lost. This fact has been interpreted as a consequence of differences in diffusion properties of proteins inside the gel matrix *(7)*.

20. The TR drastically decreases the efficiency of protein elution, even in the presence of chelating agents (*see* Note 8). It is unknown how *o*-tolidine and/or ferricyanide could affect the elution behavior, whether modifying proteins or not, but certainly this topic is of minor importance, since it makes no sense to tone a reverse-stained gel that is intended for protein elution. This only would waste time and money and without providing any advantages.

References

1. Garfin, D. E. (1990) One-dimensional gel electrophoresis, in *Guide to Protein Purification* (Deutscher, M. P., ed.), Academic, San Diego, CA, pp. 425–441.
2. Merril, C. R. (1990) Gel-staining techniques, in *Guide to Protein Purification* (Deutscher, M. P., ed.), Academic, San Diego, CA, pp. 477–488.
3. Lee, C., Levin, A., and Branton, D. (1987) Copper staining: a five-minute protein stain for sodium dodecyl sulfate-polyacrylamide gels. *Anal. Biochem.* **166,** 308–312.
4. Dzandu, J. K., Johnson, J. F., and Wise, G. E. (1988) Sodium dodecyl sulfate-gel electrophoresis staining of polypeptides using heavy metal salts. *Anal. Biochem.* **174,** 157–167.
5. Fernández-Patrón, C., Castellanos-Serra, L., and Rodríguez, P. (1992) Reverse staining of sodium dodecyl sulfate polyacrylamide gels by imidazole-zinc salts: sensitive detection of unmodified proteins. *Biotechniques* **12,** 564–573.
6. Ortiz, M. L., Calero, M., Fernández-Patrón, C., Castellanos, L., and Méndez, E. (1992) Imidazole-SDS-Zn reverse staining of proteins in gels containing or not SDS and microsequence of individual unmodified electroblotted proteins. *FEBS Lett.* **296,** 300–304.
7. Ferreras, M., Gavilanes, J. G., and García-Segura, J. M. (1993) A permanent Zn^{2+} reverse staining method for the detection and quantification of proteins in polyacrylamide gels. *Anal. Biochem.* **213,** 206–212.
8. Pliss, G. B. and Zabezhinskii, M. A. (1970) Carcinogenic properties of orthotoluidine (3,3'-dimethylbenzidine). *J. Natl. Cancer Inst.* **45,** 283–289.
9. Laemmli, U. K. (1970) Cleavage of structural proteins during the assembly of the head of bacteriophage T4. *Nature* **227,** 680–685.
10. Stoll, V. S. and Blanchard, J. S. (1990) Buffers: principles and practice, in *Guide to Protein Purification* (Deutscher, M. P., ed.), Academic, San Diego, CA, pp. 24–38.
11. Fernández-Patrón, C., Hardy, E., Sosa, A., Seoane, J., and Castellanos, L. (1995) Double staining of Coomassie blue-stained polyacrylamide gels by imidazole sodium dodecyl sulfate-zinc reverse staining: sensitive detection of Coomassie blue-undetected proteins. *Anal. Biochem.* **224,** 263–269.

12. Fernández-Patrón, C., Calero, M., Rodríguez Collazo, P., García, J.R., Madrazo, J., Musacchio, A., Soriano, F., Estrada, R., Frank, R., Castellanos-Serra, L. R., and Méndez, E. (1995) Protein reverse staining: high-efficiency microanalysis of unmodified proteins detected on electrophoresis gels. *Anal. Biochem.* **224,** 203–211.
13. Burriel, F., Arribas, S., Lucena, F., and Hernández, J. (1985) *Química Analítica Cualitativa,* Paraninfo, Madrid, p. 695.

30

Protein Staining with Calconcarboxylic Acid in Polyacrylamide Gels

Jung-Kap Choi, Hee-Youn Hong, and Gyurng-Soo Yoo

1. Introduction

Sodium dodecyl sulfate-polyacrylamide gel electrophoresis (SDS-PAGE) has become a highly reliable separation technique for protein characterization. Broad application of electrophoresis techniques has required the development of detection methods that can be used to visualize the proteins separated on polyacrylamide gels. A few of these methods are Coomassie brilliant blue (CBB) staining (*1* and *see* Chapter 11), silver staining (*2* and *see* Chapter 35), fluorescent staining *(3–5)*, specific enzyme visualization (*6*), and radioactive detection (*7* and *see* Chapter 25 and 36).

CBB staining is the most commonly used method owing to its proven reliability, simplicity, and economy *(8,9)*, but it lacks sensitivity compared with the silver-staining method. In addition, the staining/destaining process is time consuming *(10)*. Silver staining is the most sensitive nonradioactive protein detection method currently available, and can detect as little as 10 fg of protein *(11,12)*. However, it has several drawbacks, such as high-background staining, multiple steps, high cost of reagent, and toxicity of formaldehyde *(10)*. A sensitive method superior to CBB staining includes the use of fluorescent dyes, such as dansyl chloride *(3)*, fluorescamine *(4)*, and *o*-phthalaldehyde *(5)*. However, the fluorescent intensity diminishes with time, and UV radiation must be used to visualize stains. Moreover, for quantitative work, expensive equipment is required.

In this chapter a protein staining method using calconcarboxylic acid [1-(2-hydroxy-4-sulfo-l-naphthylazo)-2-hydroxy-3-naphthoic acid, NN] is described *(13)*. This method can be performed by both simultaneous and postelectrophoretic staining techniques. Simultaneous staining using 0.01% of NN in upper reservoir buffer eliminates the poststaining step, and thus enables detection of the proteins more rapidly and simply. In poststaining, proteins can be stained by a 30-min incubation of a gel in 40% methanol/7% acetic acid solution of 0.05% NN and destaining in 40% methanol/7% acetic acid for 40 min with agitation. NN staining can detect as little as 10 ng of bovine serum albumin (BSA) by poststaining and 25 ng by simultaneous staining, compared to 100 ng detectable by CBB poststaining. These techniques produce protein-staining patterns identical to the ones obtained by the conventional CBB staining and also work well in

From: The Protein Protocols Handbook
Edited by: J. M. Walker Humana Press Inc., Totowa, NJ

Table 1
Linearity of CBB and NN Staining for Four Purified Proteins

	Slope[b]		y-intercept[b]		Correlation coefficient[b]	
Protein[a]	A	B	A	B	A	B
BSA	14.1	10.1	13.2	6.4	0.986	0.994
OVA	8.7	7.1	2.7	1.4	0.993	0.998
G-3-P DHase	8.0	5.8	5.3	2.2	0.997	0.999
CA	15.4	13.5	12.5	8.8	0.986	0.997

[a]Proteins were separated on 12.5% polyacrylamide gel, and densities and band area were determined with computerized densitometer. Some of the data are illustrated in Fig. 1. The range of amount of proteins was 0.25–12.5 μg. The number of points measured was six (0.25, 0.5, 1.0, 2.5, 5.0, and 12.5 μg).

[b]Slopes, *y*-intercepts, and correlation coefficients were determined by linear regression analysis. A, CBB staining; B, NN staining.

nondenaturing PAGE, like in SDS-PAGE. The bands stained with NN present purple color. In addition, NN staining gives better linearity than CBB staining, although the slopes (band intensity/amount of protein) of it are somewhat lower (Table 1, Fig. 1). It suggests that NN staining is more useful than CBB staining for quantitative work of proteins.

2. Materials

2.1. Equipment

1. Slab-gel apparatus.
2. Power supply (capacity 600 V, 200 mA).
3. Boiling water bath.
4. Gel dryer and vacuum pump.
5. Plastic container.
6. Rocking shaker.

2.2. Solutions

1. Destaining solution (1.0 L): Mix 530 mL distilled water with 400 mL methanol and 70 mL glacial acetic acid.
2. Simultaneous staining solution (1.0% [w/v] NN): Dissolve 1.0 g NN (pure NN* without K_2SO_4) in 100 mL reservoir buffer. Stir until fully dissolved at 50–60°C (stable for months at 4°C) (*see* Note 7).
3. Poststaining solution (0.05% [w/v] NN): Dissolve 0.05 g NN (pure NN*) in 100 mL destaining solution. Stir until dissolved thoroughly (store at room temperature).

3. Methods

3.1. Simultaneous Staining

The method is based on the procedure of Borejdo and Flynn *(14)*, and is described for staining proteins in a 7.5% SDS-polyacrylamide gel.

*NN diluted with 100- to 200-fold K_2SO_4 has been used as an indicator for the determination of calcium in the presence of magnesium with EDTA (*see* Note 7).

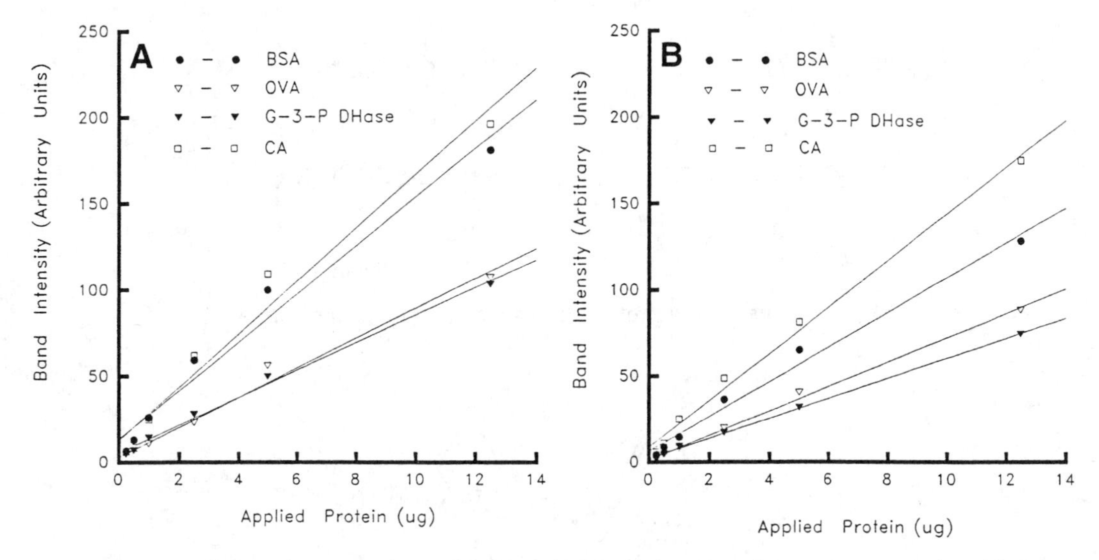

Fig. 1. Densitometric comparison of CBB and NN staining. Proteins were separated on 12.5% gels. **(A)** poststaining with 0.1% CBB; **(B)** simultaneous staining with 0.015% NN. Densitometric scanning was performed at 585 nm (CBB) and 580 nm (NN). The curves are fitted by the method of least squares. Each point represents the mean of three determinations. BSA, bovine serum albumin; OVA, ovalbumin; G-3-P DHase, glyceraldehyde-3-phosphate dehydrogenase; CA, carbonic anhydrase.

1. Electrophorese the samples for 10 min to allow protein penetration into the upper gel phase.
2. Turn off the power, and then add 1% NN dissolved in reservoir buffer to the upper reservoir buffer to give a final concentration of 0.01–0.015% NN.
3. Stir the reservoir buffer sufficiently to ensure homogeneity.
4. Resume electrophoresis.
5. Immediately after electrophoresis, remove the stained gel from the apparatus.
6. Destain in 40% methanol/7% acetic acid for 30 min. To destain completely, change destaining solution several times and agitate (*see* Notes 3–5,8).

3.2. Postelectrophoretic Staining

1. Agitate the freshly run gel in 0.05% NN dissolved in destaining solution for 30 min.
2. Pour off the staining solution and rinse the gel with changes of destaining solution (two to three times). Staining solution can be reused several times.
3. Destain in 40% methanol/7% acetic acid for 40 min with agitation (*see* Notes 3–5,8).

4. Notes

1. The protocol for poststaining is the same as that for CBB staining, except for staining/destaining times and dye used.
2. The simultaneous staining method allows one to control the intensity of stained bands reproducibly by adjusting the concentration of the dye in the upper reservoir. More than 0.02% NN in the upper reservoir buffer does not increase sensitivity and requires greater destaining times.
3. Gel staining/destaining with NN is pH dependent. Intense staining occurs at pH 1.6–4.4, and weak staining with blue-purple color is observed at alkaline pH. In excessively strong

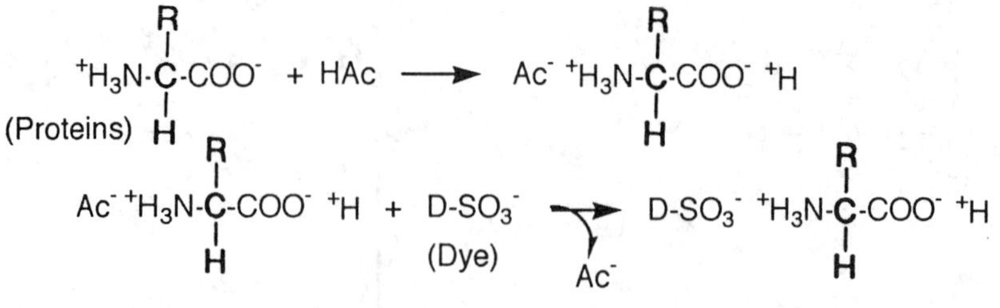

$$^{+}H_3N\text{-}\underset{H}{\overset{R}{C}}\text{-}COO^- + HAc \longrightarrow Ac^-\ ^{+}H_3N\text{-}\underset{H}{\overset{R}{C}}\text{-}COO^-\ ^{+}H$$

(Proteins)

$$Ac^-\ ^{+}H_3N\text{-}\underset{H}{\overset{R}{C}}\text{-}COO^-\ ^{+}H + D\text{-}SO_3^- \longrightarrow D\text{-}SO_3^-\ ^{+}H_3N\text{-}\underset{H}{\overset{R}{C}}\text{-}COO^-\ ^{+}H + Ac^-$$

(Dye) (Proteins-Dye)

Fig. 2. Mechanism of protein–dye interaction in acidic solution.

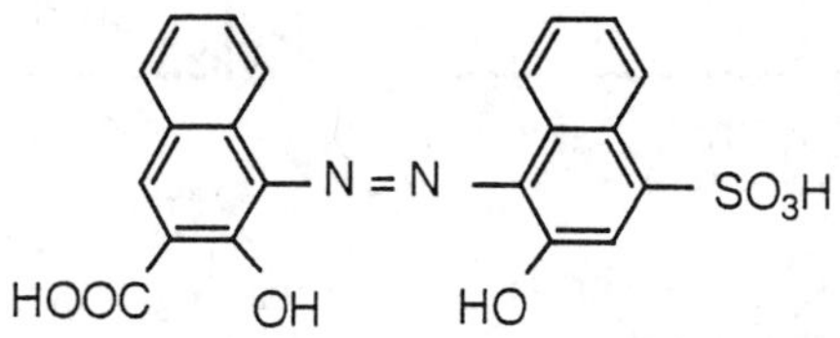

N = N
SO_3H
HOOC
OH
HO

Fig. 3. Structure of NN.

acidic solution (<pH 1.0), however, the staining effect is markedly decreased, because the solubility of the dye is decreased and, thus replacement between dye anions and acetate ions may be suppressed (*see* Fig. 2). In destaining solution, the stained band diffuses, and its intensity is decreased significantly at pH higher than 4.4. Destaining in this pH range is rather slow compared with that in strongly acidic conditions.

4. The rate of destaining speeds up with increasing methanol content; however, at high methanol content (>55%), gels are opaqued and shrunken. Addtionally, increasing the temperature of destaining solution is a great help in removing background (at 60–70°C, in 5 min), although sensitivity is a little reduced.
5. Destaining can be completed in 30 min in 7.5% polyacrylamide gels, but destaining time should be increased for 10 and 12.5% gels (50–60 min).
6. NN has several functional groups, such as hydroxyl, diazoic, carboxyl, and sulfonate groups (*see* Fig. 3). At acidic pH, NN probably forms electrostatic bonds with protonated amino groups, which are stabilized by hydrogen bonds and Van der Waals forces, as does CB *(1)*.
7. For maximal staining effect, the dye solution should be freshly prepared. The preparation of staining solution requires stirring and warming at 50–60°C since NN is poorly soluble.
8. Bands stained with NN are indefinitely stable when gels are stored in a refrigerator wrapped up in polyethylene films or dried on Whatmann No. 1 filter paper.
9. Throughout the staining/destaining processes, it is necessary to agitate the gel container using a shaker.

References

1. Fazekas de St. Groth, S., Webster, R. G., and Datyner, A. (1963) Two new staining procedures for quantitative estimation of proteins on electrophoretic strips. *Biochim. Biophys. Acta* **71,** 377–391.

2. Merril, C. R., Goldman, D., Sedman, S., and Ebert, M. (1980) Ultrasensitive stain for proteins in polyacrylamide gels shows regional variation in cerebrospinal fluid proteins. *Science* **211,** 1437,1438.
3. Schetters, H. and McLeod, B. (1979) Simultaneous isolation of major viral proteins in one step. *Anal. Biochem.* **98,** 329–334.
4. Jakowski, G. and Liew, C. C. (1980) Fluorescamine staining of nonhistone chromatin proteins as revealed by two-dimensional polyacrylamide gel electrophoresis. *Anal. Biochem.* **102,** 321–325.
5. Weiderkamm, E., Wallach, D. F. H., and Fluckiger, R. (1973) A new sensitive, rapid fluorescence technique for the determination of proteins in gel electrophoresis and in solution. *Anal. Biochem.* **54,** 102–114.
6. Gabriel, O. (1971) Locating enzymes on gels in *Methods in Enzymology,* vol. 22 (Colowick, S. P. and Kaplan, N. O., eds.), Academic, New York, pp. 363–367.
7. O'Farrell, P. H. (1975) High resolution two-dimensional electrophoresis of proteins. *J. Biol. Chem.* **250,** 4007–4021.
8. Diezel, W., Kopperschlager, G., and Hofmann, E. (1972) An improved procedure for protein staining in polyacrylamide gels with a new type of Coomassie Brilliant Blue. *Anal. Biochem.* **48,** 617–620.
9. Zehr, B. D., Savin, T. J., and Hall, R. E. (1989) A one-step low background Coomassie staining procedure for polyacrylamide gels. *Anal. Biochem.* **182,** 157–159.
10. Hames, B. D. and Rickwood, D. (1990) Analysis of gels following electrophoresis, in *Gel Electrophoresis of Proteins: A Practical Approach,* IRL, Oxford, UK, pp. 52–81.
11. Switzer, R. C., Merril, C. R., and Shifrin, S. (1979) A highly sensitive silver stain for detecting proteins and peptides in polyacrylamide gels. *Anal. Biochem.* **98,** 231–237.
12. Ohsawa, K. and Ebata, N. (1983) Silver stain for detecting 10-femtogram quantities of protein after polyacrylamide gel electrophoresis. *Anal. Biochem.* **135,** 409–415.
13. Hong, H. Y., Yoo, G. S., and Choi, J. K. (1993) Detection of proteins on polyacrylamide gels using calconcarboxylic acid. *Anal. Biochem.* **214,** 96–99.
14. Borejdo, J. and Flynn, C. (1984) Electrophoresis in the presence of Coomassie Brilliant Blue R-250 stains polyacrylamide gels during protein fractionation. *Anal. Biochem.* **140,** 84–86.

31

Detection of Proteins and Sialoglycoproteins in Polyacrylamide Gels Using Eosin Y Stain

Fan Lin and Gary E. Wise

1. Introduction

A rapid, sensitive, and reliable staining technique is essential in detection of proteins in polyacrylamide gels. Coomassie brilliant blue R-250 (CBB) is the stain that meets these criteria except for sensitivity; i.e., CBB staining requires relatively large amounts of proteins. It has been reported that the sensitivity for CBB stain in polyacrylamide gels is 0.1–0.5 μg/protein band *(1)*. This problem of relatively low staining sensitivity is often circumvented by employing silver staining techniques *(2–5)*. However, it is difficult to transfer silver stained proteins to transfer membranes unless they either are negatively stained by silver *(6)* or the positively silver-stained proteins are treated with 2X SDS sample buffer prior to transfer *(7)*. In addition, sialoglycoproteins cannot be detected by CBB and thus have to be visualized by other stains, such as the periodic acid-Schiff (PAS) reagent *(8)*, silver stains *(9)*, or silver/Coomassie blue R-250 double staining technique *(10)*.

To circumvent the deficiencies of the above staining techniques, we have developed an eosin Y staining technique *(11)*. This staining method allows one to detect proteins more rapidly than most CBB and silver staining methods. It detects a variety of proteins in amounts as little as 10 ng in polyacrylamide gels, including membrane sialoglycoproteins, and has the added advantage of the antigenicity of the stained proteins being retained. The precise mechanism by which eosin Y stains both proteins and sialoglycoproteins is not fully understood.

However, in the staining protocol described here, the fixing and developing solution of 10% acetic acid/40% methanol (pH 2.5) is strongly acidic. Under such conditions, eosin Y might be converted into its precursor form of the dihydrofluoran. Thus, a protein might be stained by means of both hydrophobic interaction between aromatic rings of eosin Y and hydrophobic sites of the protein and by hydrogen bonding between hydroxyl groups of eosin Y and the backbone of a protein. Here we describe this detailed staining protocol and technical advice in order to enable others to obtain optimal staining results.

From: The Protein Protocols Handbook
Edited by: J. M. Walker Humana Press Inc., Totowa, NJ

2. Materials

1. Eosin Y staining solution: A stock solution of 10% eosin Y (w/v) is prepared. This solution is stable at room temperature for at least 6 mo. Each 100 mL of staining solution contains 10 mL of 10% eosin Y solution, 40 mL of 100% methanol, 49.5 mL distilled deionized water, and 0.4–0.5 mL of full strength glacial acetic acid. The staining solution is made and filtered prior to use.
2. Gel fixation solution: 10% glacial acetic acid/40% methanol.
3. Gel developing solution: distilled-deionized water, 10% glacial acetic acid/40% methanol.
4. A black plastic board and a transilluminated fluorescent white light box.

3. Methods

3.1. Staining of Various Protein in SDS-PAGE

1. Immediately following electrophoresis, the SDS-polyacrylamide gel containing given proteins is fixed in 5 gel volumes of 10% glacial acetic acid/40% methanol for 10 min at room temperature with shaking and then rinsed with distilled water twice.
2. The gel is immersed with 200 mL (5–6 gel volumes) of the 1% eosin Y staining solution for 15 min at room temperature with shaking.
3. The gel is transferred to a clean glass container, quickly rinsed with distilled water and then washed with distilled water for 3 min (*see* Note 2).
4. The stained bands of the gel are developed by placing the gel in 10% acetic acid/40 methanol for about 15 s (*see* Note 3).
5. The development is stopped by immersing the gel in distilled water.
6. The gel can be kept in distilled water for at least 1 mo without fading.
7. The stained gel can be viewed either by using transilluminated fluorescent white light or by placing the gel on a black plastic board with top light illumination.

3.2. Staining of Membrane Sialoglycoproteins in SDS-PAGE

1. Immediately following electrophoresis, the SDS-polyacrylamide gel is placed in 200 mL (5–6 gel volumes) eosin Y staining solution for 45 min at room temperature with shaking.
2. The gel is quickly rinsed with distilled water and then washed with distilled water for 3 changes of 5 min each. Protein bands are visualized at the end of washing step (*see* Note 4).
3. The gel is then developed in 10% acetic acid/40% methanol for about 2 min.
4. The development of staining is terminated by washing the gel in distilled water twice.
5. The gel can be kept in distilled water for at least 1 mo without fading.
6. The stained gel can be viewed either by using transilluminated fluorescent white light or by placing the gel on a black plastic board with table top light illumination.

4. Notes

1. The 1% eosin Y staining solution needs to be made fresh and filtered prior to use. Eosin Y should be soluble in both water and methanol. Acetic acid must be added last. The final concentration of acetic acid in this staining solution is critical for staining background and sensitivity. The acetic acid should be added to solution with stirring. After adding acetic acid, the staining solution should appear to be cloudy but should not precipitate. If any precipitation occurs, it indicates that the acetic acid concentration is too high.
2. In step 3 of staining various proteins, some orange precipitation may cover a SDS-gel surface. One may gently clean up the precipitation by wiping off the gel surface with a latex glove or by using Kimwipe tissue, which will reduce the background and improve the staining sensitivity.

3. An appropriate development time will ensure yellow–orange staining bands. Prolonging development in 10% acetic acid/40% methanol often results in yellow–orange bands becoming brown bands which, in turn, will decrease the staining sensitivity.
4. It should be noted that at the end of the washing step, the proteins should appear to be yellow-orange and the sialoglycoproteins appear to be light yellow. If the protein bands are still not visualized at this point, one may prolong the washing step for another 5–10 min.
5. The eosin Y stained gel can be stored in distilled water for at least 1 mo without fading. It is not recommended to dry and store the stained gels because the intensity and resolution of protein bands are greatly decreased.
6. The eosin Y stained proteins in SDS-gels can be transferred to immobilon-P membrane without additional treatment. The antigenicity of a given protein is usually not affected by the eosin Y stain *(11)*.

References

1. Harlow, E. and Lane, D. (1988) *Antibodies,* Cold Spring Harbor Laboratory, Cold Spring Harbor, NY, pp. 649–653.
2. Switzer, R. C., Merril, C. R., and Shifrin, S. (1979) A highly sensitive silver stain for detecting proteins and peptides in polyacrylamide gels. *Anal. Biochem.* **98,** 231–237.
3. Merril, C. R., Switzer, R. C., and Van Keuren, N. L. (1979) Trace polypeptides in cellular extracts and human body fluids detected by two-dimensional electrophoresis and a highly sensitive silver stain. *Proc. Natl. Acad. Sci. USA* **76,** 4335–4339.
4. Oakley, B. R., Kirsch, D. R., and Morris, N. R. (1980) A simplified ultrasensitive silver stain for detecting proteins in polyacrylamide gels. *Anal. Biochem.* **105,** 361–363.
5. Wray, W., Boulikas, T., Wray, V. P., and Hancook, R. (1981) Silver staining of proteins in polyacrylamide gels. *Anal. Biochem.* **118,** 197–203.
6. Nalty, T. J. and Yeoman, L. C. (1988) Transfer of proteins from acrylamide gels to nitrocellulose paper after silver detection. *Immunol. Meth.* **107,** 143–149.
7. Wise, G. E. and Lin, F. (1991) Transfer of silver-stained proteins from polyacrylamide gels to polyvinylidene difluoride membranes. *J. Biochem. Biophys. Meth.* **22,** 223–231.
8. Glossmann, H. and Neville, D. M. (1971) Glycoproteins of cell surfaces. *J. Biol. Chem.* **246,** 6339–6346.
9. Mueller, T. J., Dow, A. W., and Morrison, M. (1976) Heterogeneity of the sialoglycoproteins of the normal human erythrocyte membrane. *Biochem. Biophys. Res. Commun.* **72,** 94–99
10. Dzandu, J. K., Deh, M. E., Barratt, D. L., and Wise, G. E. (1984) Detection of erythrocyte membrane proteins, sialoglycoproteins, and lipids in the same polyacrylamide gel using a double-staining technique. *Proc. Natl. Acad. Sci. USA* **81,** 1733–1737.
11. Lin, F., Fan, W., and Wise, G. E. (1991) Eosin Y staining of proteins in polyacrylamide gels. *Anal. Biochem.* **196,** 279–283.

32

Electroelution of Proteins from Polyacrylamide Gels

Paul Jenö and Martin Horst

1. Introduction

One- or two-dimensional polyacrylamide gel electrophoresis (PAGE) is one of the most versatile methods for protein separation. The resolving power of a two-dimensional O'Farrell gel *(1* and *see* Chapter 20–22) cannot be reached with today's high-performance liquid chromatographic techniques unless specific affinity interactions are being exploited for protein purification. In combination with electroblotting, analytical sodium dodecyl sulfate (SDS)-PAGE is often used to prepare proteins for N-terminal sequencing *(2)*. Recent developments have made it possible to fragment proteins electroblotted onto membranes enzymatically and to obtain sequence information from proteins that are not amenable to N-terminal Edman degradation *(3)*. For the purpose of obtaining protein sequence information, electroblotting has almost completely replaced electroelution of proteins from the polyacrylamide matrix. The reason for this is that electroeluted proteins are contaminated with Coomassie blue, buffer salts, and large amounts of SDS, making it necessary to desalt samples before subsequent Edman degradation. So why electroelution?

In our attempts to isolate components of the import machinery of yeast mitochondria, some proteins turned out to be extremely rare, and it soon became evident that conventional purification techniques were impractical to obtain these proteins in sufficient quantity and purity for protein sequencing. Furthermore, some proteins of the import machinery were extremely sensitive to proteases, and they could only be obtained intact by first denaturing mitochondria with trichloroacetic acid and then solubilizing them in boiling sample buffer for SDS-PAGE *(4)*. This in turn made the separation of several milligrams of denatured proteins necessary, which was best achieved by preparative SDS gel electrophoresis followed by electroelution of individual proteins. In this chapter, we describe an electroelution procedure that has worked well with proteins in the mol-wt range of 20–100 kDa isolated from mitochondrial membranes.

As mentioned, electroeluates contain large amounts of salts and SDS, which interfere with enzymatic fragmentation procedures or N-terminal sequencing. A number of methods are available to remove SDS from proteins, including precipitation of the detergent by organic solvents *(5,6)*, and solvent extraction with *(7)* or without ion-pairing reagents *(8)*. However, they all suffer from the disadvantage that once the detergent is removed, many proteins become virtually insoluble in buffers lacking SDS.

From: The Protein Protocols Handbook
Edited by: J. M. Walker Humana Press Inc., Totowa, NJ

Alternatively, chromatographic methods can be used to remove SDS from proteins. Simpson et al. *(9)* described desalting of electroeluates based on the finding that certain reverse-phase matrices retain proteins at high organic modifier concentrations, whereas small mol-wt compounds are not. This allows bulk separation of salts, SDS, and Coomassie blue-staining components from protein material. Based on a different stationary phase, a similar approach was recently developed *(10)* to separate protein from contaminants originating from the electroelution process. In this method, electroeluted proteins are applied to a poly(2-hydroxyethyl-aspartamide)-coated silica, which provides a polar medium for binding of proteins when equilibrated at high organic solvent concentrations *(11)*. Bound proteins are then eluted with a decreasing gradient of organic modifier, allowing recovery of protein that is free of SDS and buffer salts. Although the methodology is similar to the one described by Simpson et al. *(9)*, proteins tend to adsorb to the poly-hydroxyethyl-aspartamide matrix at lower *n*-propanol concentrations than to reverse-phase matrices, therefore minimizing the danger of irreversible protein precipitation on the stationary phase.

2. Materials

1. Electrophoresis apparatus: Preparative electrophoresis is carried out on 16 × 10-cm separating and 16 × 2-cm stacking gels of 1.5-mm thickness with the buffer system described by Laemmli *(12)*. For sample application, a preparative sample comb of 2-cm depth and 14-cm width is used. Up to 2.5 mg total protein is applied onto one preparative slab gel. Samples are electrophoresed at 15V for 14 h. Chemicals used for electrophoresis (acrylamide, *N,N'*-methylene-bis-acrylamide, ammonium persulfate, *N,N,N',N'*-tetramethylenediamine, SDS, and Coomassie blue) are electrophoresis grade and are purchased from Bio-Rad (Hercules, CA). Methanol and acetic acid used for staining are pro analysi (p.a.) grade from Merck. Unless stated, all other chemicals used are of the highest grade available.
2. Staining solution: 0.125% (w/v) Coomassie brilliant blue R250, 50% (v/v) methanol, 10% (v/v) acetic acid. Filter the staining solution over 0.2-µm Nalgene filters before use.
3. Destaining solution: 50% (v/v) methanol, 10% acetic acid.
4. Electroelution apparatus: BIOTRAP from Schleicher and Schuell (Keene, NH) ("Elutrap" trademark in the US and Canada).
5. Membranes: BT1 and BT2 membranes for electroelution (Schleicher & Schuell).
6. Electroelution buffer: 25 m*M* Tris, 192 m*M* glycine, 0.1% SDS. This buffer is prepared with NANOpure water (obtained from a Barnstead NANOpure water purification system) and electrophoresis-grade SDS (Bio-Rad).
7. Electrodialysis buffer: 15 m*M* NH_4HCO_3, 0.025% SDS. Prepare the buffer with NANOpure water and electrophoresis-grade SDS (Bio-Rad).
8. High-pressure liquid chromatography (HPLC) equipment: We are using a Merck (Piscataway, NJ) Hitachi L6200 low-pressure gradient system connected to a Merck Hitachi L4200 UV detector for operating columns of 4.6-mm diameter. Narrow-bore columns (2.1 mm id) are operated on a Pharmacia (Uppsala, Sweden) Smart micro-FPLC system.
9. Poly-hydroxyethyl-aspartamide (PHA) columns: 5-µm particle size, 20–nm pore size, 4.6 × 200 mm, or 2.1 × 200 mm (PolyLC Inc., Columbia, MD).
10. Solvents for hydrophilic interaction chromatography: solvent A: 100% *n*-propanol (HPLC-grade, Merck), 50 m*M* formic acid (Merck, analytical grade); solvent B: 50 m*M* formic acid in water (NANOpure).

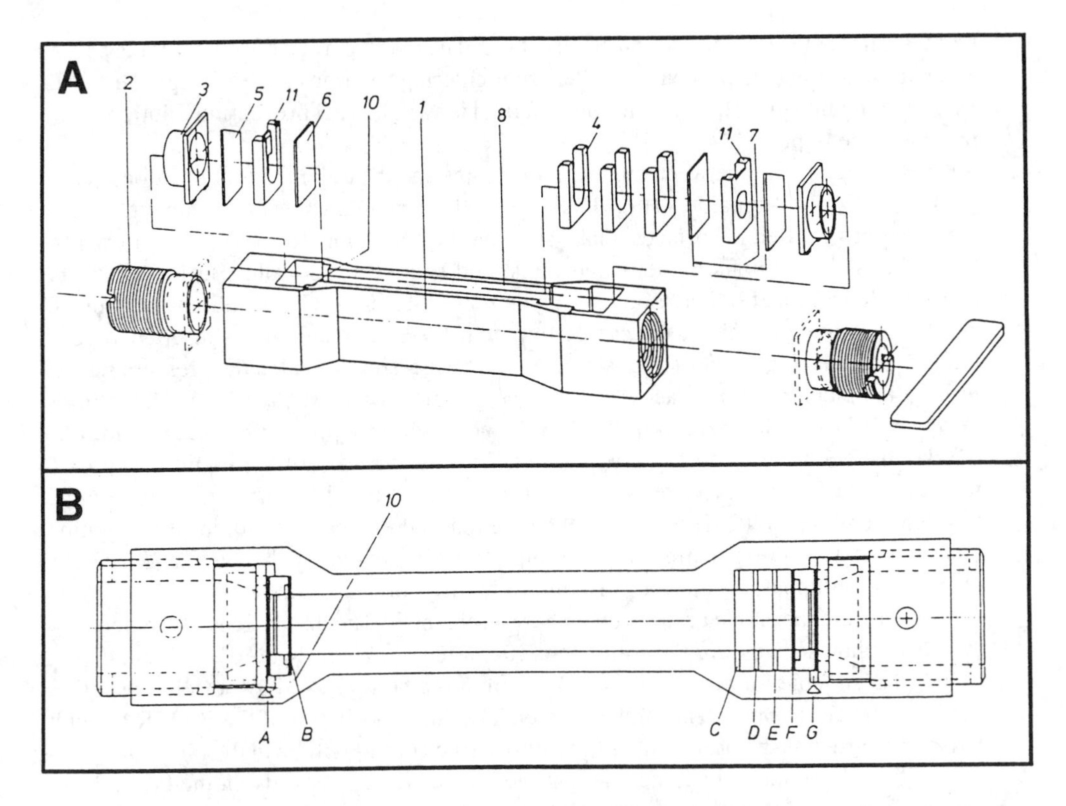

Fig. 1. Side **(A)** and top **(B)** view of the electroelution device. The apparatus is assembled by inserting a BT1 membrane at points A and G. The trap that collects the protein during the elution process is formed between points F and G. The chamber that holds the gel pieces is formed by inserting a BT2 membrane at point C. Smaller elution chambers can be made by inserting the BT2 membrane at either point D or point E. The trap inserts and membranes are fixed by clamping plates, which press the trap inserts against the cell body. 1, Cell body; 2, pressure screw; 3, clamping plate; 4, trap inserts; 5, membrane BT1; 6, membrane BT2; 7, trap chamber; 8, elution chamber; 10, mark for correct orientation of membrane BT1; 11, trap insert for membrane BT1 (modified with permission from Schleicher and Schuell).

3. Method

1. After electrophoresis, stain the gel for 15 min with Coomassie blue. In order to prevent irreversible fixation of the protein, destaining is followed on a light box and as soon as the protein of interest becomes visible, the band is cut out with a razor blade. The gel piece is washed once with 10 mL of 1*M* Tris-HCl, pH 8.0, for 5 min, followed by three 10-mL washes with water for 5 min each. Cut the gel piece with a razor blade into small cubes. Equilibrate them in 10 mL of electroelution buffer for 10 min with occasional shaking. In the meantime, assemble the electroelution apparatus.
2. The electroelution device is a block of polycarbonate (160 × 30 × 30 mm) that has an open channel along its axis (Fig. 1). An elution chamber that holds the PAGE pieces is formed with trap inserts between points C and F of the body. During the elution process, the protein is trapped into a chamber formed between points F and G (Fig. 1). The device works with two types of membranes having different ion permeabilities: The BT1 membrane retains all charged macromolecules larger than 5 kDa, whereas buffer ions can freely

permeate under the influence of an electric field. The macroporous BT2 membrane acts as a barrier that prevents particular matter from entering the trap. It also keeps the buffer from flowing into the trap when the electric field is switched off preventing dilution of the protein in the trap.

3. Slide BT1 membranes between the clamping plates and the trap inserts at positions A and G of the BIOTRAP apparatus (Fig. 1). Since the BT1 membrane is an asymmetric membrane with two different surfaces, make sure that they are mounted in the proper orientation. The BT1 membrane is delivered moist and should not dry out. Buffer should be added within 5 min after insertion of the membranes. Insert a BT2 membrane at points C and F. Smaller elution chambers can be formed if the BT2 membrane is inserted at positions D or E, making the elution of single gel pieces possible. Tighten the pressure screws to hold the membranes in place. Transfer the gel pieces with a spatula into the elution chamber formed between the two BT2 membranes inserted at positions C and F. Carefully overlay the gel pieces with electroelution buffer until the level of the liquid is approx 5 mm above the gel pieces. After some minutes, the trap is filled with buffer by seeping through membrane BT2. However, make sure that there is enough liquid in the elution chamber so that the gel pieces are completely immersed in the buffer. Place the electroelution apparatus into a horizontal electrophoresis chamber with the + mark directed toward the anode of the electrophoresis chamber. The dimensions of the horizontal electrophoresis tank are the following: 30-cm length, 20-cm width, and 7-cm depth. The T-shaped table for agarose gels is 3 cm from the bottom. Add 3 L of electroelution buffer to the electrophoresis chamber, which is enough to fill half of the BIOTRAP with buffer. Electroelute the protein for 18 h at 100 V (the current will be in the range of 70–90 mA). The volume into which the eluted protein is recovered depends on the buffer level inside the BIOTRAP and ranges from 200–800 μL.
4. Replace the electroelution buffer with 3 L electrodialysis buffer and electrodialyze the sample for 6 h at 40 V against 15 m*M* NH_4HCO_3, 0.025% SDS.
5. Remove the eluted protein from the trap. Be careful not to perforate the BT1 membrane with the pipet tip! Rinse the trap twice with 100 μL fresh electrodialysis buffer. Combine the dialysate and the washes. The solution is dried in a Savant speed vac and stored at –20°C.
6. For desalting of the eluted protein, equilibrate the PHA-column with solvent A (65% *n*-propanol, 50 m*M* formic acid). Electroeluates containing more than 5 μg of protein are desalted on 4.6 mm id columns, which are operated at 0.5 mL/min. Less than 5 μg protein are chromatographed on 2.1 mm id columns at 75 μL/min. The effluent is monitored at 280 nm.
7. Dissolve the dried protein in a small volume of water (50–100 μL). The dried SDS efficiently solubilizes the protein. *n*-propanol is added to 65% final concentration. The sample is then applied in 50-μL aliquots onto the PHA column. After each injection, a number of UV-absorbing peaks, caused by Coomassie blue components, elute from the column. It is important that these components are completely washed out before the next aliquot is injected. With this procedure, the protein is efficiently concentrated on the column inlet. After the entire sample has been applied, the gradient is initiated, which is developed in 10 min from 65% *n*-propanol/50 m*M* formic acid to 50 m*M* formic acid. The protein elutes at the end of the gradient and is now devoid of any salt or SDS (Fig. 2).
8. The desalted protein is now ready for further protein structural characterization. It can be directly subjected to automated Edman degradation. For enzymatic fragmentation, residual *n*-propanol has to be removed in the speed vac prior to adding the protease.

4. Notes

1. In order to locate the protein of interest in the gel, a staining method has to be chosen so that maximal sensitivity with minimal fixation is obtained. A number of methods exist

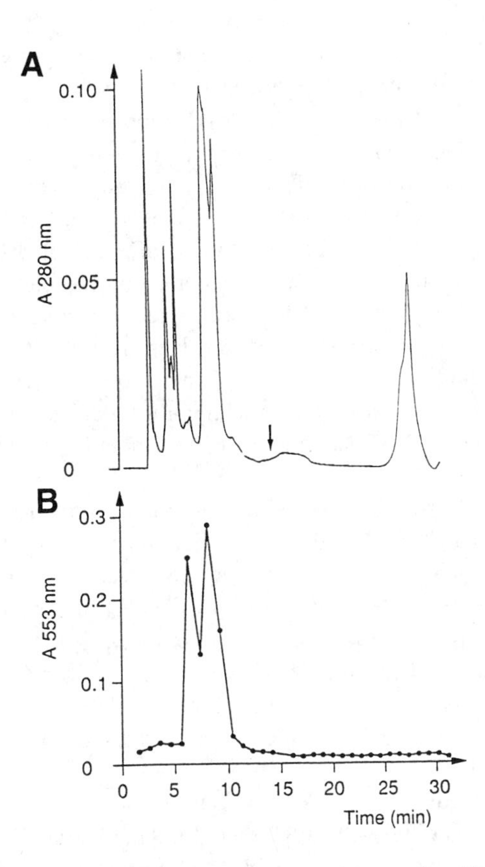

Fig. 2. Removal of SDS from an electroeluate. **(A)** 20 µg of a 45-kDa mitochondrial outer membrane protein in 50 µL was injected onto a PHA column (4.6 × 200 mm) that had been equilibrated in 70% *n*-propanol-50 m*M* formic acid. After the baseline had stabilized, the gradient was initiated (marked with an arrow). Bound protein was eluted with a 10-min linear gradient from 70% *n*-propanol-50 m*M* formic acid to 50 m*M* formic acid at a flowrate of 0.5 mL/min. The protein elutes between 27 and 30 min. **(B)** Fractions of 500 µL were collected and tested with Fuchsin red for the presence of SDS (modified with permission from ref. *10*).

to visualize proteins in the SDS-polyacrylamide matrix, such as formation of insoluble protein–SDS complexes with potassium *(13),* or precipitation of nonprotein-bound SDS by 4*M* sodium acetate *(14)*. We found staining of complex protein patterns with these methods difficult, since they tend to produce diffuse staining. Staining the gels for 15 min with Coomassie blue is sufficient to visualize also faint bands in a PAGE without fixing them irreversibly. Destaining is then done on a light box, so that the band of interest can be sliced out of the gel as soon as it becomes visible.

2. The electroelution apparatus routinely used in our laboratory is that originally described by Jacobs and Clad *(15)* and is commercially available from Schleicher and Schuell. We found this particular type of apparatus easy to handle and very reliable during routine use. The volume of the elution chamber can be adjusted depending on the volume of gel pieces used. The volume can be increased or decreased by varying the position of the BT2 membranes between positions C and F (*see* Fig. 1). By forming the smallest possible elution chamber, one can process Coomassie blue-stained bands from a single one-dimensional

analytical PAGE. With the larger elution chamber, up to five preparative gels can be processed at a time. However, other suitably constructed devices will give identical results.

3. So far we have eluted proteins in the 20–100-kDa mol-wt range from preparative SDS-polyacrylamide gels where approx 2–50 µg of protein were present in a single band. After elution and dialysis, proteins are typically recovered in volumes of between 300 and 800 µL. It is obvious that elution of proteins <10 µg becomes difficult owing to the large volume of the trap. Even though the elution buffer contains SDS, which efficiently solubilizes the protein, nonspecific adsorption to the BT1 membrane or microleaks in the trap may drastically reduce protein recovery. We therefore prefer to run several preparative gels in parallel and pool multiple protein slices until at least 10 µg of total protein are accumulated, rather than attempting to elute a single band of only a few micrograms of protein. We are currently performing experiments with elution devices of smaller dimensions in order to be able to recover Coomassie blue-stained spots from a single two-dimensional O'Farrell gel.
4. From a protein chemical point of view, electroelution of proteins into ammonium hydrogen carbonate, which can be removed by lyophilization, would be preferable. As stated by Jacobs and Clad *(15)*, high-mol-wt proteins require at least 8 h to be eluted quantitatively into the trap. Owing to its low buffering capacity, the pH of the buffer drops after 4 h, and, therefore, does not allow long elution times without frequent buffer changes. Large proteins in the mol-wt range of around 100 kDa elute very slowly from the polyacrylamide matrix and, therefore, require much longer elution times. The pH of the Tris/glycine buffer remains constant for at least 18 h and, therefore, allows much higher elution efficiencies for high-mol-wt proteins.
5. When using 0.1% SDS in the electroelution buffer, the micelles formed in front of the BT1 membrane lead to massive accumulation of SDS in the trap. Because of this, removal of the detergent is required for subsequent protein chemical work. This can be easily achieved by hydrophilic interaction chromatography with simultaneous desalting of the protein into a volatile buffer system *(10,11)*. Alternatively, the procedure suggested by Simpson et al. can be used to separate SDS from protein *(9,16)*.

Protein adsorption to a PHA-stationary phase requires careful control of the solvent composition. In order to test column performance, we use a test mixture of proteins consisting of cytochrome c, ovalbumin, and bovine serum albumin. Use of the test mixture allows us to find the minimum percentage of *n*-propanol at which the proteins adsorb to the stationary phase and at which low-mol-wt components, such as SDS, are not adsorbed. This usually occurs at 60–65% *n*-propanol concentration. Direct application of electro-eluates onto the column is not possible, since the high content of Tris and glycine in the electroeluate leads to buffer salt precipitation at *n*-propanol concentrations >50%. Exchanging the buffer salts against ammonium hydrogen carbonate makes lyophilization of the dialysate possible. The plug, which is mainly formed by SDS, can be redissolved in a minimal volume of water. No salt precipitation is observed when adding *n*-propanol to 65% final concentration.

The solvent used to elute bound proteins contains 50 m*M* formic acid. Because of the high absorption of formic acid, monitoring of the effluent can only be achieved at wavelengths >230 nm. Most proteins contain tyrosine and tryptophan residues, which can be detected at 280 nm. In these rare cases where no aromatic amino acids are present, UV-transparent solvent systems, such as *n*-propanol containing 0.05% trifluoroacetic acid (TFA) may be used. However, we found that TFA significantly reduces the lifetime of PHA columns, so exposure to TFA should be kept to an absolute minimum. Efficient removal of small-mol-wt components, such as SDS and Coomassie blue, from the protein requires that the injection volume be kept as small as possible. When using a 4.6 mm id column, 50-μL injections were found to be optimal; for 2.1-mm id columns, the injection volume is reduced to 20 μL. Larger volumes are best applied onto the column with multiple injections. It is important to note that when an electroeluate is applied with repeated injections onto a PHA column, the column should be allowed to re-equilibrate between individual injections. Otherwise loss of protein in the breakthrough of the column can occur, and the protein peak at the end of the gradient tends to be contaminated with small-mol-wt components from the electroelution procedure.

Electroeluted proteins chromatographed by hydrophilic interaction often display unsymmetrical peak shapes. This either indicates the presence of several different proteins in the eluate that are partially resolved by the stationary phase or that the protein became modified during the electrophoretic separation process. Since the main purpose of this type of chromatography is to free the eluted protein from SDS and Coomassie blue-staining components, we tend to run extremely steep gradients in order to elute even mixtures of proteins into a single peak. This allows efficient concentration of proteins into a completely volatile buffer. Further purification into single components can be achieved either by chromatographic techniques or by two-dimensional electrophoresis *(4)*.

References

1. O'Farrell, P. H. (1975) High resolution two-dimensional electrophoresis of proteins. *J. Biol. Chem.* **250,** 4007–4021.
2. Matsudaira, P. (1987) Sequence from picomole quantities of proteins electroblotted onto Polyvinylidene Difluoride membranes. *J. Biol. Chem.* **262,** 10,035–10,038.
3. Aebersold, R. H., Leavitt, J., Saavedra, R. A., Hood, L. E., and Kent, S. B. (1987) Internal amino acid sequence analysis of proteins separated by one- or two-dimensional gel electrophoresis after *in situ* protease digestion on nitrocellulose. *Proc. Natl. Acad. Sci. USA* **84,** 6970–6974.
4. Horst, M., Jenö, P., Kronidou, N. G., Bolliger, L., Oppliger, W., Scherer, P., Manning-Krieg, U., Jascur, T., and Schatz, G. (1993) Protein import into yeast mitochondria: the inner membrane import site ISP45 is the *MPI1* gene product. *EMBO J.* **12,** 3035–3041.
5. Wessel, D. and Flügge, U. I. (1984) A method for the quantitative recovery of protein in dilute solution in the presence of detergent and lipids. *Anal. Biochem.* **138,** 141–143.
6. Stearne, P. A., van Driel, I. R., Grego, B., Simpson, R. J., and Goding, J. W. (1985) The murine plasma cell antigen PC-1: purification and partial amino acid sequence. *J. Immunol.* **134,** 443–448.
7. Konigsberg, W. H. and Henderson, L. (1983) Removal of sodium dodecyl sulfate from proteins by ion-pair extraction. *Methods Enzymol.* **91,** 254–259.
8. Bosserhoff, A., Wallach, J., and Frank, R. (1989) Micropreparative separation of peptides derived from sodium dodecyl sulfate-solubilized proteins. *J. Chromatogr.* **437,** 71–77.
9. Simpson, R. J., Moritz, R. L., Nice, E. E., and Grego, B. (1987) A high-performance liquid chromatography procedure for recovering subnanomole amounts of protein from SDS-gel electroeluates for gas-phase sequence analysis. *Eur. J. Biochem.* **165,** 21–29.
10. Jenö, P., Scherer, P., Manning-Krieg, U., and Horst, M. (1993) Desalting electroeluted proteins with hydrophilic chromatography. *Anal. Biochem.* **215,** 292–298.
11. Alpert, A. J. (1990) Hydrophilic-interaction chromatography for the separation of peptides, nucleic acids and other polar compounds. *J. Chromatogr.* **499,** 177–196.
12. Laemmli, M. K. (1970) Cleavage of structural proteins during the assembly of the head of bacteriophage T4. *Nature* **227,** 680–685.

13. Hager, D. A. and Burgess, R. (1980) Elution of proteins from sodium dodecyl sulfate-polyacrylamide gels, removal of sodium dodecyl sulfate, and renaturation of enzymatic activity: results with sigma subunit of *Escherichia coli* RNA polymerase, wheat germ DNA topoisomerase, and other enzymes. *Anal. Biochem.* **109,** 76–86.
14. Higgins, R. C. and Dahmus, M. E. (1979) Rapid visualization of protein bands in preparative SDS-polyacrylamide gels. *Anal. Biochem.* **93,** 257–260.
15. Jacobs, E. and Clad, A. (1986) Electroelution of fixed and stained membrane proteins from preparative sodium dodecyl sulfate-polyacrylamide gels into a membrane trap. *Anal. Biochem.* **154,** 583–589.
16. Simpson, R. J., Ward, L. D., Reid, G. E., Batterham, M. P., and Simpson, R. L. (1989) Peptide mapping and internal sequencing of proteins electroblotted from two-dimensional gels onto polyvinylidene difluoride membranes. A chromatographic procedure for separating proteins from detergents. *J. Chromatogr.* **476,** 345–361.

33

High-Performance Electrophoresis Chromatography

Serge Desnoyers, Sylvie Bourassa, and Guy G. Poirier

1. Introduction

Protein purification is important because it gives essential data on enzyme, e.g., its catalytic activity, its interactions with other proteins or DNA, amino acid quantitation, N-terminal microsequencing, and other information. These analyses will provide important data for designing oligodeoxynucleotide probes for gene cloning and peptide synthesis for antibody production. Purified proteins can be obtained by several methods: gel-filtration chromatography, ion-exchange chromatography, immunoaffinity chromatography, immunoprecipitation, and salt precipitation *(1)*.

Micropreparative electrophoresis is a relatively new protein purification method *(2–4)* that gives fast results (usually a few hours) and yields enough purified protein (≤200 μg) to allow microsequencing either directly from the elution chamber *(5)* or by centrifugation on polyvinylenedifluoride (PVDF) membrane *(6)*. Micropreparative electrophoresis purification is based on the principle of relative mobility of proteins in sodium dodecyl sulfate-polyacrylamide gel electrophoresis (SDS-PAGE).

Originally described by Ornstein *(7)* and subsequently modified by Laemmli *(8)*, SDS-PAGE is still the most reliable analytical system for proteins. However, it can also be used as a preparative system. This system is based on the effect of an electrical field causing charged molecules to move toward the electrode of opposite polarity. The mobility of each molecule decreases as it interacts with the surrounding matrix, in the present case, polyacrylamide gel. The gel acts as a molecular sieve. The amphoteric nature of proteins implies that they will migrate toward the electrode opposite their net charge. In SDS-PAGE, the net charge of each protein is the same because the proteins are solubilized in an ionic detergent, SDS. The anion dodecyl sulfate gives each molecule of protein a negative charge. Proteins are thus separated according to their molecular weight, and small proteins move faster than larger ones. The criterion of purity is to obtain of a single band on SDS-PAGE, which is based on the principle of protein purification by micropreparative electrophoresis. Instead of stopping the run when the front of migration reaches the bottom of the gel, electrophoresis is continued until each zone is eluted from the gel. If eluate is fractionated, each zone can be collected as a purified one.

Several devices have been described for micropreparative electrophoresis *(2,9)*, and there are now commercialized apparatuses, such as the high-performance electrophore-

From: The Protein Protocols Handbook
Edited by: *J. M. Walker Humana Press Inc., Totowa, NJ*

sis chromatography (HPEC) 230A from Applied Biosystems (Foster City, CA), ELFE from Genofit (Grand Lancy, Switzerland), Prep-Cell 491 from Bio-Rad (Hercules, CA), and 1100PG from BRL (Gathersburg, MD). This chapter describes a method originally described for a slab gel apparatus *(10),* but applied on a commercialized micropreparative electrophoresis apparatus from Applied Biosystems model 230A HPEC *(11).* The HPEC apparatus is provided with an on-line UV detector, an elution module, and a fraction collector. Proteins are chromatographed in a polyacrylamide gel matrix under the influence of an electrical field. The gel column stands between the cathodic module and the anodic module. Buffers are constantly renewed by a pressurized tubule system, eliminating heat build-up and ion depletion.

The protein sample is loaded onto the top of the gel at the cathodic end. Once current is applied, the proteins are eletrophoresed under thermostated conditions. Since each separated zone is eluted from the bottom of the gel, proteins are swept through the detector by a flow of elution buffer and collected by a fraction collector. The proteins so purified can be analyzed further.

Most of the known buffer systems existing for slab gel can be applied for HPEC purification, with slight modifications *(12).* The inconvenience with the Laemmli buffer system Tris-glycine-SDS is that the sample, once purified, needs to be handled for desalting *(6)* to allow microsequencing of the purified protein, which implies a loss of material. We describe here a micropreparative electrophoretic system buffer that is completely compatible with the microsequencing of protein permitting one-step protein purification and direct microsequencing. This buffer increases the resolution of the different proteins and is faster to recover the proteins because its use makes the blotting or the desalting step unnecessary. Virtually 100% of the proteins eluted from the HPEC is recovered.

2. Materials

It is recommended that reagents of the highest grade of purity possible or of electrophoresis grades be used.

2.1. Apparatus

Separation system HPEC 230A (Applied Biosystems):

1. Gas cylinder and regulator.
2. Gel tubes.
3. Column filters (Zytex membrane).
4. Dialysis membrane.
5. Forcep.
6. Razor blade.
7. Pasteur pipet.
8. 0.5-mL Eppendorf tube with ventilation holes.

2.2. Stock Solutions and Buffers

1. Acrylamide/bis-acrylamide concentrate solution (30%T, 2.7%C): Dissolve 29.2 g of acrylamide and 0.8 g of bis-acrylamide in 80 mL of deionized (DI) water. Once completely dissolved complete at 100 mL with deionized (DI) water. Filter using a 0.22-µm filter. Store at 4°C in a brown bottle for 1 mo (*see* Note 2).

Table 1
Gel Formulation for 10 mL of Final Mixture

Solutions	7.5%T	10%T	15%T
Acrylamide-bis, mL	2.44	3.33	4.88
10X gel buffer, mL	1	1	1
Water, mL	6.44	5.6	4
10% APS, μL	70	70	70
TEMED, μL	7	7	7

2. Gel buffer (10X): 0.6*M* 2-dimethylaminoethanol (DME) solution titrated with HCl at pH 8.3. Add 16 mL of DME (straight from the commercial bottle) to 70 mL of DI water. Adjust pH to 8.3 with concentrate HCl. Make up to 100 mL with DI water. Store at 4°C in a brown bottle for 6–8 wk.
3. 10% Ammonium persulfate (APS): Add 50 mg of APS to 0.5 mL DI water. Prepare fresh before each use.
4. Upper buffer (1X): 120 m*M* DME-HCl, 150 m*M* boric acid, 0.1% SDS (*see* Note 1). Add 24 mL of DME (straight from the commercial bottle) to 1.5 L of DI water. Dissolve 9.27 g of boric acid. Once dissolved make up to 1.98 L with DI water, filter using a 0.22-μm filter, and sparge with helium for 20 min. Add 20 mL of 10% SDS; stir gently. The pH should be close to 9.3.
5. Lower and elution buffers (1X): 150 m*M* Boric acid-DME, pH 8.3. Dissolve 13.9 g of boric acid in 2.8 L of DI water and adjust pH to 8.3 with a few drops of DME (usually around 4 mL). Complete at 3 L with DI water, filter using a 0.22-μm filter, and sparge with helium for 20 min.
6. Sample buffer (2X): 0.06*M* DME-Cl, 6% SDS, 20% glycerol, 0.4% 2-mercaptoethanol. Combine 1 mL 0.6*M* DME-Cl, 3 mL 20% SDS, 4 mL 50% glycerol, and 40 μL 2-mercaptoethanol. Mix all reagents in water to a final volume of 10 mL.

3. Methods

3.1. Tube Gel Preparation

1. Gels are poured into glass tubes, which must have been thoroughly cleaned with detergent (e.g., Alconox) overnight.
2. Rinse with DI water and dry.

3.2. Dialysis Membrane

1. This is an important component because it is placed between the elution block and the lower block of the HPEC 230A, preventing proteins from going into the lower buffer tubule. Boil the membrane in DI water for 30 min. Handle the membrane with forceps. Store at 4°C in a 50:50 water:methanol solution.

3.3. Gel Preparation (see Table 1)

1. Put a plastic cap at one end of the tube and a plastic collar at the other end. Tubes must be maintained in a vertical position.
2. Prepare the required gel volume, excluding APS and TEMED, according to the number of desired gel columns to be used (one gel column = 1 mL).
3. Deaerate the gel solution for 15 min under vacuum.
4. Add catalyst (APS and TEMED) and gently swirl the solution (*see* Table 1).

5. Fill tubes using a 9-in. Pasteur pipets.
6. Remove any air bubbles.
7. Add a layer of water-saturated butanol over the gel solution.
8. Allow to polymerize for 1 h (*see* Note 3).
9. Remove the collar, and cut the protrubing portion of gel with a razor blade while leaving 2 mm of gel protruding for storage.
10. Seal the end with a plactic cap and Parafilm™. Put in a small container (Falcon 50-mL conical tube with 1 or 2 mL of water). Gels can be conserved for 2 wk in such conditions.

3.4. Buffer Installation and Circulation

1. Use buffers described in Section 2.2.
2. Fill appropriate bottles with upper, lower, and elution buffer.
3. Adjust pressure to 1–2 psi in the tubule system to allow buffers to circulate (*see* Note 4).

3.5. Gel Installation

1. Carefully remove cap and collar from the gel tube by cutting in the plastic with a razor blade. Be careful not to touch or move the gel and thus introduce an air pocket.
2. With razor blade, cut the protruding gel at the top and bottom of the gel tube, which must be as even as possible. Cut the gel under water to avoid introduction of an air pocket.
3. Wet the bottom and top ends of the gel tube with a drop of upper and lower buffer, respectively. Place a Zytex membrane on each end (they will stay in place because of the surface tension).
4. Place the gel tube between the upper and lower electrode assembly. Be sure there is no leak of running buffers.
5. Purge the tubule system with buffer to make sure there is no air bubble in the system, especially in the elution line, which goes directly through the UV detector.
6. Elution buffer flowrate should be 15–20 μL/min for good analysis and collection of the sample protein.

3.6. Sample Preparation

The sample must be dissolved in the 2X sample buffer and in the smallest volume possible (10 μL), even if the apparatus allows a sample volume of 90 and 200 μL *(13)* (*see* Note 5).

1. If the sample is solid, dissolve the proteins in 5–10 μL of DI water and mix an equal volume of 2X sample buffer. Heat at 100°C for 2 min. Place on ice 5 min, and spin at 13,000*g* for 2 min (*see* Note 6).
2. For sample of <25 μL, mix an equal volume of 2X sample buffer. Heat at 100°C for 2 min. Place on ice for 5 min and spin at 13,000*g* for 2 min.
3. For a sample of >25 μL, concentrate the sample before solubilization as a liquid (*see* Note 7).

3.7. Prerunning

Each new gel use for micropreparative electrophoresis must be prerun to remove UV-absorbing contaminant. This is done by applying a current of 1 mA for 1–2 h. As depicted in Fig. 1, it is possible to see the peaks of absorption of these contaminants in both of the systems, i.e., Tris-glycine-SDS and DME-borate. The shift in the baseline level is caused by glycinate ions for the Tris-glycine-SDS system (Fig. 1A), but in DME-borate system, there is no shift because of the presence of DME in lower and elution buffers (Fig. 1B; *see* Note 8).

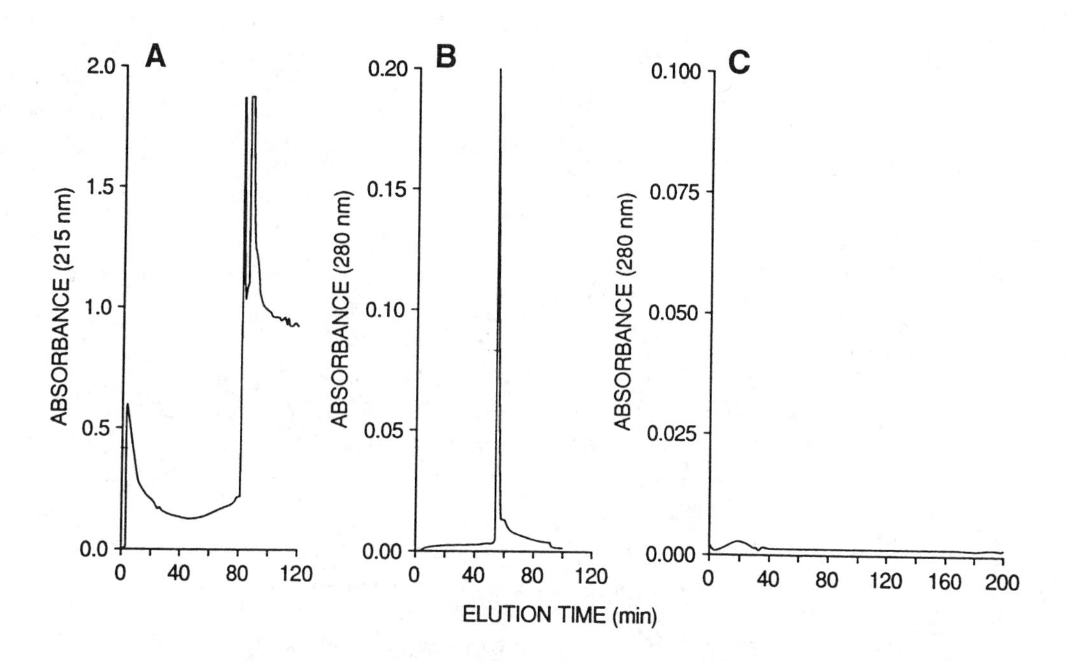

Fig. 1. Prerunning in Tris-glycine-SDS compared with DME buffers. The Tris-glycine-SDS buffer shows a typical shift in the optical density when the glycine ions replace the chloride ions **(A)**. There is no shift in the optical density with the DME buffer system, but removal of UV-absorbing matter is still visible **(B)**. The baseline in such a system is very stable **(C)**.

3.8. Electrophoresis

Electrophoresis is conducted at 1 mA after loading the sample at the top of gel column. Eluted fractions are recovered by fraction collectors in 0.5-mL conical tubes with ventilation holes (Fig. 2). Sample proteins can be further analyzed for microsequencing (Fig. 3; *see* Note 9).

4. Notes

1. The quality of SDS is important because an impure C_{12} SDS source can cause an electrophoretic pattern to change *(13)*. Bio-Rad has been proven to be of the highest quality.
2. Acrylamide is a neurotoxic agent. Wear protective gloves and mask when handling the powder and the liquid. To dispose of old liquid acrylamide solution, add an excess of catalyst (APS and TEMED), and let polymerize. Discard the solid product. Handle bisacrylamide with the same precaution as acrylamide, even though there are no data on its toxicity.
3. Polymerization of polyacrylamide is an exothermic process, and control of heat production might be important in a high percentage gel because of the possibility of bubble formation in the gel. Polymerization should then be conducted at 4°C.
4. Gas flows from the gas tank to pressurize the buffer bottles. Upper buffer flows across the cathodic electrode and out to the waste bottle. Lower buffer flows across the anodic electrode and out to waste. Elution buffer flows between the anodic end of the gel and the lower buffer.
5. Samples must be loaded in small volumes because there is no stacking gel in the HPEC system. Glycine can be eliminated and a single buffer (DME) used in place of Tris/

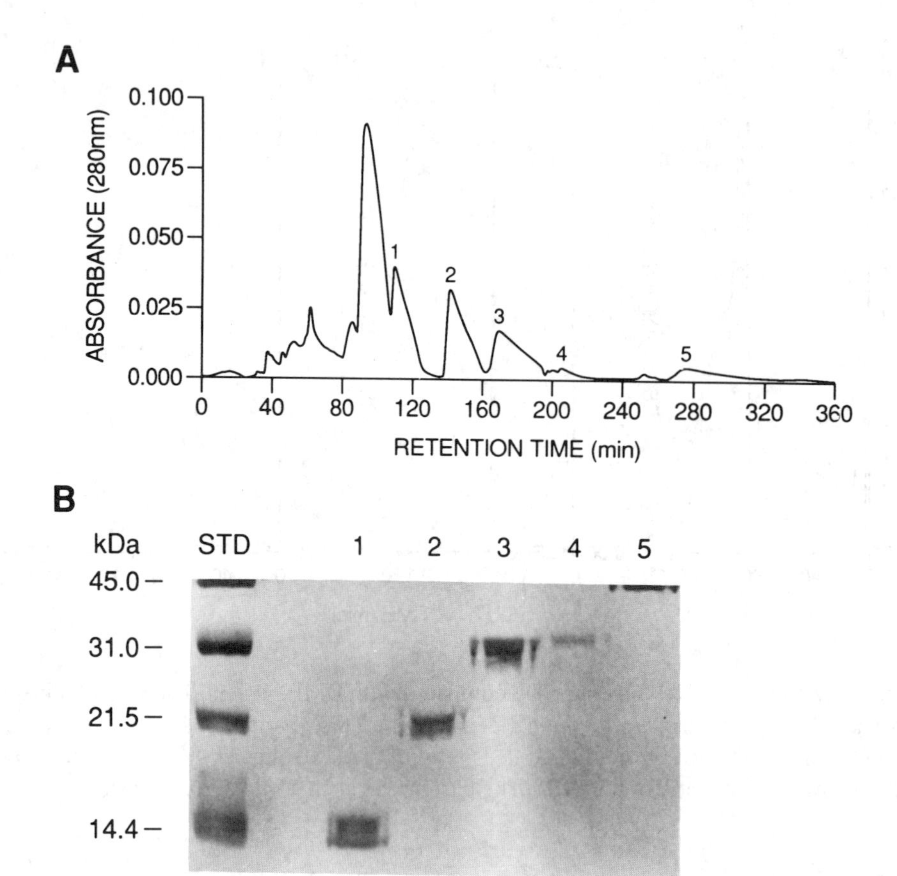

Fig. 2. Separation of four different proteins using a 2.5 × 50-mm polyacrylamide column (15%T) and DME buffer. Twenty micrograms of protein were loaded onto the column, and electrophoresis was performed for 360 min at 1 mA **(A)**. An aliquot of each peak was analyzed on a mini-SDS-PAGE and silver-stained **(B)**. 1, lysozyme; 2, soybean trypsin inhibitor; 3 and 4, carbonic anhydrase; 5, ovalbumin.

glycine. DME contains no primary amino groups and is therefore compatible with microsequencing, unlike Tris.

6. Centrifugation is important to remove particulate insoluble material from the sample. If particulate material is loaded on the gel, it may dissolve during the run and cause a "ghost" peak to appear.
7. Protein samples can be concentrated by ethanol or trichloroacetic acid precipitation or lyophilization. Precipitated proteins are usually difficult to resolubilize and require longer heat treatment. If a sample cannot be concentrated sufficiently, it can be dialyzed against 1X sample buffer without 2-mercaptoethanol, which has to be added just prior to heating. Another possibility is to achieve composition of 1X sample buffer within the sample using concentrated gel buffer, SDS, 2-mercaptoethanol, and glycerol.

   A minimum of protein must be loaded on the gel column to obtain maximum recovery. Recovery of more than 95% is achieved with 50 µg or more of protein. Loading of less protein could yield a recovery of as low as 25% after an HPEC run *(14)*.

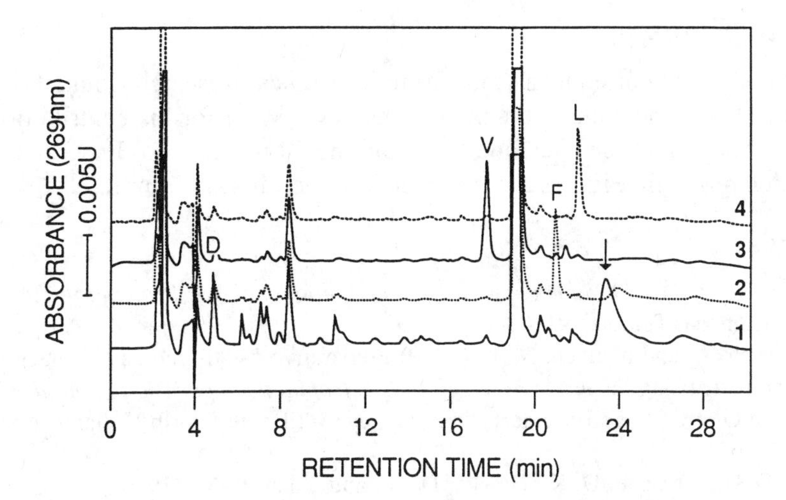

Fig. 3. Microsequencing of HPEC-DME buffer-eluted soybean trypsin inhibitor. The protein, after elution from the HPEC, was directly applied to a polybrene membrane and sequenced. The cycles are numbered in the figure. The first cycle shows a contaminant peak at 23 min (arrow), which is out of the range of separation of the amino acids. Microsequencing was performed on an ABI Model 473A.

8. Prerunning the gel column in the HPEC apparatus is necessary to establish a continuous system of ions in the gel and is a useful diagnostic tool. In the standard Tris-glycine-SDS system, the prerun showed (Fig. 1A) two peaks eluted between 80 and 100 min of the run. These two peaks are gel contaminants and must be removed before the run. As also shown, the baseline value increased compared with the initial value at the beginning of the run. This increase in baseline value was caused by glycine ions replacing chloride ions in the gel. Thus, the prerun is necessary to establish a continuous system of ions. Glycine and SDS are not recommended for Model 230A. Glycine generated high background absorbance at low wavelength and interferes with microsequencing. The SDS caused foaming to occur and disturbed the baseline. The DME buffer system is better used at 280 nm for detection, because it has a strong absorption at 215 nm, and baseline stability is thus affected.
9. To demonstrate the compatibility of the elution buffer with the microsequencing, soybean trypsin inhibitor collected after micropreparative electrophoresis was directly applied to a polybrene membrane and sequenced. The first four amino acid cycles are shown in Fig. 3: Asp (D), Phe (F), Val (V), and Leu (L). There is a single additional peak in the first cycle (arrowed), which presumably represented a contaminant. This peak did not interfere with the separation and identification of any authentic amino acids in the cycle. Since DME contains no primary amine that would be derivatized, there is less probability that these contaminant peaks come from DME. With a repetitive yield of 95.7% and an initial yield of 95.6 pmol, which is 10% of what was applied on HPEC and almost 100% of what was recovered from HPEC, this yield compared very well with sequencing results from soluble proteins. Similar results were obtained by sequencing lysozyme separated under the same conditions (data not shown). Although sample recovery on the HPEC is better in the Tris system (25% data not shown), the separation is far better in DME-Borate (Fig. 2 and ref. *14*) because the soybean trypsin inhibitor is completely separated from lysozyme, which is not the case in the Tris system.

Acknowledgments

This work was supported by a grant from the Medical Research Council of Canada (grant no. 12344). The authors thank Michael H. P. West for his contribution to the method described in this chapter, and Van Luu The for making the HPEC available for their research program. Figures are reprinted with permission from ref. *5*.

References

1. Doonan, S., ed. (1996) *Methods in Molecular Biology, Vol. 59: Protein Purification Protocols,* Humana, Totowa, NJ.
2. Chrambach, A. and Nguyen, N. Y. (1979) Preparative electrophoresis, isotachophoresis and electrofocusing on acrylamide gel, in *Electrokinetic Separation Methods* (Righetti, P. J., van Oss, C. J., and Vanderhoff, J. W., eds.), Elsevier/North Holland, Amsterdam, pp. 337–367.
3. Sheer, D. G., Yamane, D. K., Hawke, D. H., and Yuan, P.-M. (1990) The use of micropreparative electrophoresis of protein/peptide isolations for primary structure determination. *Biotechniques* **9,** 486–495.
4. Sheer, D. (1990) Sample centrifugation onto membranes for sequencing. *Anal. Biochem.* **187,** 76–83.
5. Desnoyers, S., Bourassa, S., West, M. H. P., and Poirier, G. G. (1994) One-step protein purification using micropreparative electrophoresis fully compatible with protein microsequencing. *Anal. Biochem.* **221,** 418–420.
6. Sheer, D. G. (1989) The use of PVDF membranes in sample recovery following an HPEC™ isolation bioseparation. *User Bulletin No. 4,* Applied Biosystems Inc., Foster City, CA.
7. Ornstein, L. (1964) Disc electrophoresis-I: background and theory. *Ann. NY Acad. Sci.* **121,** 321–349.
8. Laemmli, U. K. (1970) Cleavage of structural protein during the assembly of the head of bacteriophage T4. *Nature* **227,** 680–685.
9. Baumann, M. and Lauraeus, M. (1993) Purification of membrane proteins using a micropreparative gel electrophoresis apparatus: purification of subunits of the integral membrane protein *Bacillus subtilis* aa3-type quinol oxidase for low level amino acid sequence analysis. *Anal. Biochem.* **214,** 142–148.
10. Wu, R. S., Stedman, J. D., West, M. H. P., Pantazis, P., and Bonner, W. M. (1982) Discontinuous agarose electrophoretic system for the recovery of stained proteins from polyacrylamide gels. *Anal. Biochem.* **124,** 264–271.
11. Sheer, D. and Kochersperger, M. (1990) Separation and characterization of proteins in the range of 1 to 200 kDa with HPEC, in *Current Research in Protein Chemistry* (Villafranca, J. ed.), Academic, New York, pp. 245–262.
12. Sheer, D. G. (1990) The Tris-glycine-SDS system for HPEC™ applications. *User Bulletin No. 1*, Applied Biosystems Inc., Foster City, CA.
13. Margulies, M. M. and Tiffany, H. L. (1984) Importance of sodium dodecyl sulfate source to electrophoretic separations of thylakoid polypeptide. *Anal. Biochem.* **136,** 309–313.
14. (1991) Installation and use of the expanded volume (200 µL) sample block on the model 230A. *User Bulletin No. 9*, Applied Biosystems Inc., Foster City, CA.

34

Drying Polyacrylamide Gels

Bryan John Smith

1. Introduction

Gels can be fragile and difficult to handle and store without suffering damage. This can be overcome by drying, whereby they become bonded to a stabilizing medium. Furthermore, it may be necessary to dry a gel in order to obtain the most efficient detection of radioactive samples on it by autoradiography, and fluorography. In this chapter, two methods are described—the first being drying onto absorbent paper (this being suitable for storage, reflection densitometry, and autoradiography), and the second being drying between cellulose sheets (this being suitable for storage, transmission densitometry, and demonstration by overhead projection). Both methods are suitable for slab gels—tube-shaped gels are not suited to drying down.

2. Materials

2.1. Drying Onto Absorbent Paper

1. A high vacuum rotary pump: Protect the pump and its oil from acid and water by inclusion of a cold trap in the vacuum line.
2. A heat source (such as a hot air fan, infrared lamp, electric hot plate, or a steam/hot water bath).
3. A gel dryer, available from various commercial sources (such as Biometra, Maidstone, UK) or made in the laboratory, along the lines shown diagramatically in Fig. 1. Essentially, a gel dryer has the gel placed on top of absorbent paper, which in turn is supported by a firm sheet of porous polyethylene and/or a metal grill. As shown in Fig. 1, the gel is covered with plastic cling film domestic food wrapping, such as Saran Wrap. This construction is put under vacuum beneath a sheet of silicon rubber and is heated to help drive off moisture. A suitable polyethylene sheet, absorbent paper, cellophane, and other items may be obtained commercially. Whatman 3MM (Maidstone, UK) paper, or thicker, is suitable as absorbent paper.

2.2. Drying Between Transparent Sheets

1. Two rigid frames: The aperture in each should be larger than the gel(s) to be dried. The two frames are held face to face by bulldog clips placed around the edge of the frame. This equipment is available commercially (e.g., from Promega, Southampton, UK).
2. Sheets of water-permeable, transparent material, such as cellophane dialysis tubing (cut lengthwise to a single sheet) or sheets sold for this purpose (e.g., by Promega). The sheets should be larger than the frames (mentioned in step 1).

From: The Protein Protocols Handbook
Edited by: J. M. Walker Humana Press Inc., Totowa, NJ

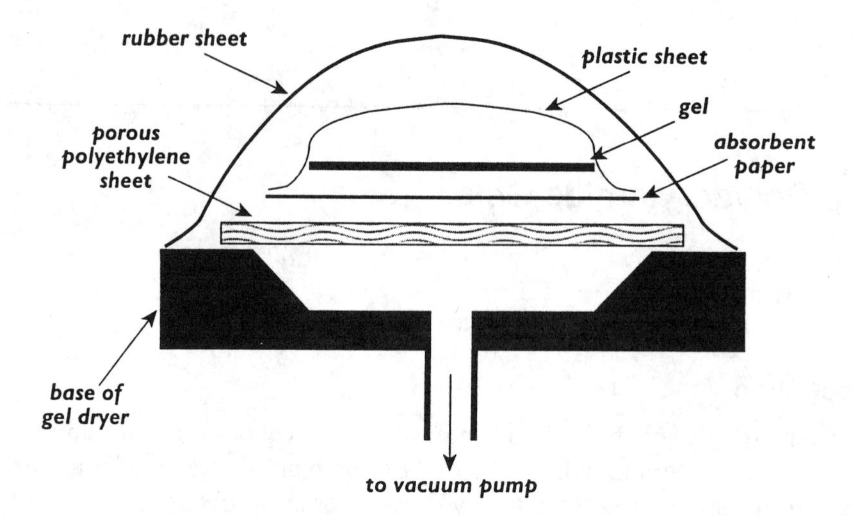

Fig. 1. The basic elements of a gel-drying apparatus. Heat may be applied from a heating element in the base of the dryer or from an infrared lamp placed above the apparatus.

3. Gel equilibration solutions:
   a. 30 mL 30% (v/v) Methanol, and 3 mL 3% (v/v) glycerol. Make to 100 mL with distilled water.
   b. 3 mL 3% (v/v) Glycerol. Make to 100 mL with distilled water.

3. Methods

3.1. Drying Onto Absorbent Paper

1. The gel will already have been run and, if carrying out fluorography, suitably treated and then washed thoroughly with distilled water (*see* Chapter 36).
2. Place the gel on a piece of absorbent paper that is slightly larger than the gel itself. Do not trap air bubbles between them. Place them, gel uppermost, on top of the porous polyethylene sheet and cover the whole with some of the nonporous cling film. Place this construction on the gel dryer base and cover over with the sheet of silicon rubber, as shown in Fig. 1.
3. Apply the vacuum, which should draw all layers tightly together. Check that there is no air leak. After about 10–15 min, apply heat (~60°C) evenly over the face of the gel.
4. Continue until the gel is dry, at which point the silicon rubber sheet over the gel should assume a completely flat appearance. The time taken to dry a gel is dependent on various factors (*see* Notes 2, 5, 6, 8, and 9), but a gel of 0.5–1 mm thickness, equilibrated with water, will take 1–2 h to dry.
5. When the gel is completely dry and bound onto the absorbent paper, it may be removed and the cling film over it peeled off and discarded. The gel may be stored in a notebook or used for autoradiography.

3.2. Drying Between Transparent Sheets

1. After electrophoresis, staining, and destaining, equilibrate the gel as follows: soak the gel with gentle agitation in equilibration solution (a) for 30 min at room temperature. Then soak the gel in solution (b) for 30 min at room temperature.

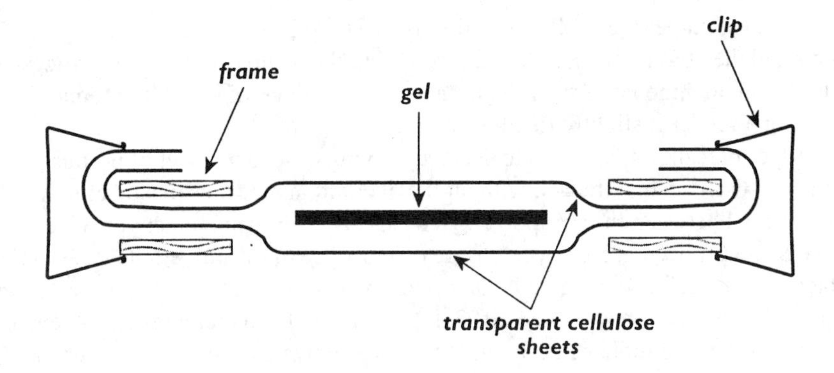

Fig. 2. Arrangement of a gel between cellulose sheets in preparation for air-drying.

2. Wet a sheet of transparent film with water. Place the gel on this sheet. Place a second wet sheet over the gel to sandwich the gel between the two sheets without trapping any air bubbles.
3. Place this "sandwich" between the two frames, fold the free edges of the sheets over the edges of the frames, and clip with bulldog clips. The gel should then be sandwiched between two taut sheets of transparent film, as indicated in Fig. 2.
4. Place the frame so that it lies horizontally, with both sides exposed to air, and leave to dry at room temperature. Drying takes overnight or longer (dependent on gel type and local atmospheric conditions).
5. After drying, remove the sheets from the drying frame and cut to size. The transparent sheets are firmly bound to the gel, which may be stored in a notebook.

4. Notes

1. The plastic cling film layered over the gel is used for two reasons. First, it prevents sticking of the gel to the rubber sheet covering it. Second, and more importantly, it reduces the likelihood of contamination of the dryer by radioactive substances from the gel. After drying, the plastic film may be discarded as radioactive waste.
2. If a suitable high vacuum pump is not available for use with the gel dryer, a good water pump may suffice instead, but in this case, the gel will take longer to dry.
3. The main problem with this method is that gels may crack up and spoil the end result. This may occur as a chronic process during the drying. Gels of higher %T acrylamide, gradient gels, and gels of greater thickness are more prone to suffer this fate. However, to alleviate this problem, the gel may be treated before drying with one of several solutions:
   a. Methanol (70% [v/v]) in water *(1):* Equilibrate the gel by soaking in this solution (0.5–1 h for thin, 1-mm, gels or longer for thick gels) and then proceed with drying. This treatment will cause the gel to shrink. High %T gels may dehydrate and go opaque. This process may be reversed with water, but if it is too extreme the gel may crack. If this is a danger, use a weaker methanol solution (e.g., 40% [v/v]). This treatment speeds up the drying process somewhat, since the methanol is driven off fairly quickly.
   b. Glycerol (1% [v/v]) and acetic acid (10% [v/v]) in water *(2):* Equilibrate the gel and proceed with drying.
   c. DMSO (2% [v/v]) and acetic acid (10% [v/v]) in water *(2):* Equilibrate the gel and proceed with drying.

Of these three, the DMSO solution is most likely to prevent cracking of difficult gels and the 70% methanol the least likely, but the former will take the longest time to dry down, and the latter the least time. Traces of glycerol or DMSO remaining in the dried gel render it slightly flexible and more resilient.

4. As a further precaution, and also to speed up drying, a second sheet of porous polyethylene may be used, so that the gel, without the overlaid, nonporous plastic cling film sheet, is sandwiched between the two polyethylene sheets. This arrangement provides a greater surface of gel for drying. However, ensure that the face of the polyethylene sheet that is in contact with the gel is very smooth. Otherwize, the gel will dry into it (as well as into the absorbent paper), and they will be difficult to separate. If this remains a problem, employ a sheet of porous cellophane between the gel and polyethylene. The cellophane may be removed after drying.
5. Another precaution is to use a lower temperature during the drying process, so that gradients of temperature and hydration through the system are less extreme. Thus, probably the best approach to drying a difficult gel, such as a 3-mm thick 10 or 15%T acrylamide gel, would be to use the DMSO (2%) soaking of the gel before drying, and two polyethylene sheets and less heat (e.g., 40°C) during the drying. Under these circumstances, a thick gel may take overnight to dry down.
6. Gel cracking may also occur as an acute phenomenon when the vacuum is released from a gel that is not completely dry. To ensure that this does not happen, determine (by trial) the time required to dry down a gel in your own gel-drying equipment. This time increases with increased %T of acrylamide, thickness, and surface area of the gel, and with decreased temperature and vacuum during the drying.
7. A problem may arise if the gel is prepared for fluorography using DMSO as solvent. The gel must be thoroughly washed in water (or other solution; *see* Note 3) to remove the DMSO. This is because DMSO has a boiling point of 189–193°C and is difficult to remove under the conditions of drying. If remaining in the gel in significant amounts, the gel will remain sticky, and photographic film that touches it may become fogged.
8. Agarose gels may be dried in the manner described. Even 3-mm agarose gels dry quickly (about 0.5 h) and without cracking when using only one polyethylene sheet and room temperature for drying. Composite weak acrylamide-agarose gels are likewise readily dried down. When dealing with gels containing agarose, beware the use of DMSO, which dissolves agarose.
9. If using EN^3HANCE (New England Nuclear, Boston, MA) in the gel, follow the manufacturer's recommendation, and do not employ temperatures above 70°C for drying, as EN^3HANCE is volatile at higher temperatures (boiling point is 117°C).
10. If the gel has been stained suitably for densitometry (*see* Chapter 26), sample bands may be quantified after recording by video camera and digitization.
11. For the method of drying between transparent sheets, the glycerol in the equilibration solution permeates the gel, renders it slightly flexible on drying, and protects it from physical damage. Cracking of the gel may still occur during the drying process itself (as mentioned in Note 3) but this gentler method has less risk of this happening.
12. The gel dried between transparent sheets is suitable for transmission densitometry and for use as a transparency for overhead projection.
13. Radioactively labeled gels may be dried between transparent sheets for autoradiography, but for weak emitters, the gel must be in direct contact with the photographic film. To allow this, one of the transparent sheets is replaced by a sheet of Saran Wrap. After drying, the Saran Wrap may be removed.
14. Either polyacrylamide or agarose gels may be dried between transparent sheets.

References

1. Joshi, S. and Haenni, A. L. (1980) Fluorographic detection of nucleic acids labeled with weak β-emitters in gels containing high acrylamide concentrations. *FEBS Lett.* **118,** 43–46.
2. Bio-Rad Inc. (1981) Model 1125B high capacity gel slab dryer for protein gels and DNA sequencing. *Bio-Rad Bulletin 1079,* Bio-Rad Inc., Hercules, CA.

35

Detection of Proteins in Polyacrylamide Gels by Silver Staining

Michael J. Dunn

1. Introduction

The versatility and resolving capacity of polyacrylamide gel electrophoresis (PAGE) has resulted in this group of methods becoming the most popular for the analysis of patterns of gene expression in a wide variety of complex systems. These techniques are often used to characterize protein purity and to monitor the various steps in a protein purification process. Moreover, with the advent of sensitive methods for chemical characterization, gel electrophoresis has become one of the most important methods of protein purification.

Coomassie brilliant blue R250 (CBB R250) has been used for many years as a general protein stain following gel electrophoresis. However, the trend toward the use of thinner gels and the need to detect small amounts of protein within single bands or spots resolved by one- or two-dimensional (1D or 2D) electrophoresis have necessitated the development of more sensitive detection methods.

The ability of silver to develop images was discovered in the mid-17th century, and this property was exploited in the development of photography, followed by its use in histological procedures. Silver staining for the detection of proteins following gel electrophoresis was first reported in 1979 by Switzer et al. *(1)*, resulting in a major increase in the sensitivity of protein detection. More than 100 publications have subsequently appeared describing variations in silver-staining methodology *(2,3)*. This group of procedures is generally accepted to be between 20 and 200 times more sensitive than methods using CBB R250, being able to detect about 0.1 ng protein/band or spot.

All silver-staining procedures depend on the reduction of ionic silver to its metallic form, but the precize mechanism involved in the staining of proteins has not been fully established. It has been proposed that silver cations complex with protein amino groups, particularly the ε-amino group of lysine *(4)*, and with sulfur residues of cysteine and methionine *(5)*. However, Gersten and his colleagues showed that "stainability" cannot be attributed entirely to specific amino acids and suggested that some element of protein structure, higher than amino acid composition, is responsible for differential silver staining *(6)*.

Silver-staining procedures can be grouped into two types of methods depending on the chemical state of the silver ion when used for impregnating the gel. The first group

From: The Protein Protocols Handbook
Edited by: J. M. Walker Humana Press Inc., Totowa, NJ

are alkaline methods based on the use of an ammoniacal silver or diamine solution, prepared by adding silver nitrate to a sodium–ammonium hydroxide mixture. Copper can be included in these diamine procedures to give increased sensitivity, possibly by a mechanism similar to that of the Biuret reaction. The silver ions complexed to proteins within the gel are subsequently developed by reduction to metallic silver with formaldehyde in an acidified environment, usually using citric acid. In the second group of methods, silver nitrate in a weakly acidic (approx pH 6.0) solution is used for gel impregnation. Development is subsequently achieved by the selective reduction of ionic silver to metallic silver by formaldehyde made alkaline with either sodium carbonate or NaOH. Any free silver nitrate must be washed out of the gel prior to development as precipitation of silver oxide will result in high background staining.

Silver stains are normally monochromatic, resulting in a dark brown image. However, if the development time is extended, dense protein zones become saturated, and color effects can be produced. Some staining methods have been designed to enhance these color effects, which were claimed to be related to the nature of the polypeptides detected *(7)*. However, it has now been established that the colors produced depend on:

1. The size of the silver particles;
2. The distribution of silver particles within the gel; and
3. The refractive index of the gel *(8)*.

Rabilloud recently compared several staining methods based on both the silver diamine and silver nitrate types of procedures *(9)*. The most rapid procedures were found to be generally less sensitive than the more time-consuming methods. Methods using glutaraldehyde pretreatment of the gel and silver diamine complex as the silvering agent were found to be the most sensitive. The method described here is based on that of Hochstrasser et al. *(10,11)*, together with modifications and technical advice that will enable an experimenter to optimize results. An example of a sodium dodecyl sulfate (SDS)-PAGE separation of the total proteins of human endothelial cells stained by this procedure is shown in Fig. 1.

2. Materials

1. All solutions should be freshly prepared, and overnight storage is not recommended. Solutions must be prepared using clean glassware and deionized, distilled water.
2. Gel fixation solution: trichloroacetic acid (TCA) solution, 20% (w/v).
3. Sensitization solution: 10% (w/v) glutaraldehyde solution.
4. Silver diamine solution: 21 mL of 0.36% (w/v) NaOH are added to 1.4 mL of 35% (w/v) ammonia and then 4 mL of 20% (w/v) silver nitrate are added dropwise with stirring. When the mixture fails to clear with the formation of a brown precipitate, further addition of a minimum amount of ammonia results in the dissolution of the precipitate. The solution is made up to 100 mL with water. The silver diamine solution is unstable and should be used within 5 min.
5. Developing solution: 2.5 mL of 1% (w/v) citric acid, 0.26 mL of 36% (w/v) formaldehyde made up to 500 mL with water.
6. Stopping solution: 40% (v/v) ethanol, 10% (v/v) acetic acid in water.
7. Farmer's reducer: 0.3% (w/v) potassium ferricyanide, 0.6% (w/v) sodium thiosulfate, 0.1% (w/v) sodium carbonate.

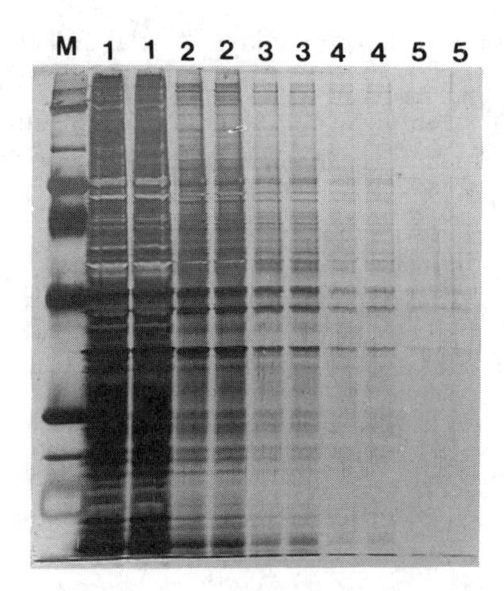

Fig. 1. Separation of total proteins of human endothelial cells by 10%T SDS-PAGE and visualized by silver staining. Molecular-weight markers were applied to lane M. The amount of total cellular proteins applied to each lane was: lane 1, 20 μg; lane 2, 10 μg; lane 3, 5 μg; lane 4, 2 μg; lane 5, 0.5 μg.

3. Methods

Note: All incubations are carried out at room temperature with gentle agitation.

3.1. Fixation

1. After electrophoresis, fix the gel immediately (*see* Note 1) in 200 mL of TCA (*see* Notes 2 and 3) for a minimum of 1 h at room temperature. High-percentage polyacrylamide and thick gels require an increased period for fixation, and overnight soaking is recommended.
2. Place the gel in 200 mL of 40% (v/v) ethanol and 10% (v/v) acetic acid in water and soak for 2 × 30 min (*see* Note 4).
3. Wash the gel in excess water for 2 × 20 min, facilitating the rehydration of the gel and the removal of methanol. An indication of rehydration is the loss of the hydrophobic nature of the gel.

3.2. Sensitization

1. Soak the gel in a 10% (w/v) glutaraldehyde solution for 30 min at room temperature (*see* Note 5).
2. Wash the gel in water for 3 × 20 min to remove excess glutaraldehyde.

3.3. Staining

1. Soak the gel in the silver diamine solution for 30 min. For thick gels (1.5 mm), it is necessary to use increased volumes so that the gels are totally immersed. Caution should be exercised in disposal of the ammoniacal silver reagent, since it decomposes on standing and may become explosive. The ammoniacal silver reagent should be treated with dilute hydrochloric acid (1*N*) prior to disposal.
2. Wash the gel (3 × 5 min) in water.

3.4. Development

1. Place the gel in developing solution. Proteins are visualized as dark brown zones within 10 min (*see* Note 6), after which the background will gradually increase (*see* Note 7). It is

important to note that the reaction displays inertia, and that staining will continue for 2–4 min after removal of the gel from the developing solution. Staining times in excess of 20 min usually result in an unacceptable high background (*see* Note 8).
2. Terminate staining by immersing the gel in stopping solution.
3. Wash the stained gel in water prior to storage or drying.

3.5. Destaining

Partial destaining of gels using Farmer's reducing reagent *(5)* is recommended for the controlled removal of background staining that obscures proper interpretation of the protein pattern.

1. Wash the stained gel in water for 5 min to remove the stop solution.
2. Place the gel in Farmer's reducer for a time dependent on the intensity of the background.
3. Terminate destaining by returning the gel to the stop solution.

4. Notes

1. Gloves should be worn at all stages when handling gels, since silver staining will detect keratin proteins from the skin.
2. Volumes of the solutions used at all stages should be sufficient such that the gel is totally immersed. If the volume of solution is insufficient for total immersion, staining will be uneven, and the gel surface can dry out.
3. A mixture of alcohol, acetic acid, and water (9:9:2) is recommended for gel fixation in many published protocols, but TCA is a better general protein fixative and its use is compatible with silver staining provided that the gel is washed well after fixation to remove the acid.
4. In addition to removing TCA, the washing step also effectively removes reagents, such as Tris, glycine, and detergents (especially SDS), that can bind silver and result in increased background staining.
5. Treatment of the gel with reducing agents, such as glutaraldehyde, prior to silver impregnation results in an increase in staining sensitivity by increasing the speed of silver reduction on the proteins.
6. If image development is allowed to proceed for too long, dense protein zones will become saturated and negative staining will occur, leading to serious problems if quantitative analysis is attempted. In addition, certain proteins stain to give yellow or red zones regardless of protein concentration, and this effect has been linked to the posttranslational modification of the proteins.
7. An inherent problem with the staining of gradient SDS-PAGE gels is uneven staining along the concentration gradient. The less concentrated polyacrylamide region develops background staining prior to the more concentrated region. A partial solution to this problem is to increase the time of staining in silver diamine (*see* Section 3.3., step 1).
8. Various chemicals used in 1D and 2D PAGE procedures can inhibit staining, whereas others impair resolution or produce artifacts. Acetic acid will inhibit staining and should be completely removed prior to the addition of silver diamine solution (Section 3.3., step 1). Glycerol, used to stabilize SDS gradient gels during casting, and urea, used as a denaturing agent in isoelectric focusing (IEF), are removed by water washes. Agarose, often used to embed rod IEF gels onto SDS-PAGE gels in 2D PAGE procedures, contains peptides that are detected by silver staining as diffuse bands and give a strong background. Tris, glycine, and detergents (especially SDS) present in electrophoresis buffers can complex with silver and must be washed out with water prior to staining. The use of

2-mercaptoethanol as a disulfide bond-reducing agent should be avoided, since it leads to the appearance of two artifactual bands at 50 and 67 kDa on the gel *(12)*.

9. Radioactively labeled proteins can be detected by silver staining prior to autoradiography or fluorography for the majority of the commonly used isotopes (^{14}C, ^{35}S, ^{32}P, ^{125}I). In the case of ^{3}H, however, silver deposition will absorb most of the emitted radiation.

References

1. Switzer, R. C., Merril, C. R., and Shifrin, S. (1979) A highly sensitive stain for detecting proteins and peptides in polyacrylamide gels. *Anal. Biochem.* **98,** 231–237.
2. Merril, C. R. (1987) Detection of proteins separated by electrophoresis, in *Advances in Electrophoresis*, vol. 1 (Chrambach, A., Dunn, M. J., and Radola, B. J., eds.), VCH Verlagsgesellschaft, Weinheim, Germany, pp. 111–139.
3. Rabilloud, T. (1990) Mechanisms of protein silver staining in polyacrylamide gels: a 10-year synthesis. *Electrophoresis* **11,** 785–794.
4. Dion, A. S. and Pomenti, A. A. (1983) Ammoniacal silver staining of proteins: Mechanism of glutaraldehyde enhancement. *Anal. Biochem.* **129,** 490–496.
5. Heukeshoven, J. and Dernick, R. (1985) Simplified method for silver staining of proteins in polyacrylamide gels and the mechanism of silver staining. *Electrophoresis* **6,** 103–112.
6. Gersten, D. M., Rodriguez, L. V., George, D. G., Johnston, D. A., and Zapolski, E. J. (1991) On the relationship of amino acid composition to silver staining of protein in electrophoresis gels: II. Peptide sequence analysis. *Electrophoresis* **12,** 409–414.
7. Sammons, D. W., Adams, L. D., and Nishizawa, E. E. (1982) Ultrasensitive silver-based color staining, of polypeptides in polyacrylamide gels. *Electrophoresis* **2,** 135–141.
8. Merril, C. R., Harasewych, M. G., and Harrington, M. G. (1986) Protein staining and detection methods, in *Gel Electrophoresis of Proteins* (Dunn, M. J., ed.), Wright, Bristol, UK, pp. 323–362.
9. Rabilloud, T. (1992) A comparison between low background silver diamine and silver nitrate protein stains. *Electrophoresis* **13,** 429–439.
10. Hochstrasser, D. F., Patchornik, A., and Merril, C. R. (1988) Development of polyacrylamide gels that improve the separation of proteins and their detection by silver staining. *Anal. Biochem.* **173,** 412–423.
11. Hochstrasser, D. F. and Merril C. R. (1988) "Catalysts" for polyacrylamide gel electrophoresis polymerization and detection of proteins by silver staining. *Appl. Theor. Electrophoresis* **1,** 35–40.
12. Guevarra, J., Johnston, D. A., Ramagli, L. S., Martin, B. A., Capitello, S., and Rodriguez, L. V. (1982) Quantitative aspects of silver deposition in proteins resolved in complex polyacrylamide gels. *Electrophoresis* **3,** 197–205.

36

Autoradiography and Fluorography of Acrylamide Gels

Antonella Circolo and Sunita Gulati

1. Introduction

Autoradiography detects the distribution of radioactivity on gels or filters by producing permanent images on photographic film. It is frequently used in a variety of experimental techniques ranging from Southern and Northern blot analysis *(1),* to visualization of radioactive proteins separated in a sodium dodecyl sulfate (SDS)-Polyacrylamide gels *(2),* to detection of nuclear factors bound to a labeled DNA probe in gel shift analysis *(3),* and to localization of DNA bands in sequencing gels *(4).*

Autoradiographic images are formed when particles emitted by radioactive isotopes encounter the emulsion of an X-ray film and cause emission of electrons from silver halide crystals that, in turn, react with positively charged silver ions, resulting in the precipitation of silver atoms and the formation of an image *(5).*

^{35}S and ^{32}P isotopes are the most commonly used isotopes for autoradiography. ^{35}S is a β-emitter of relatively low energy (0.167 MeV). Thus ^{35}S particles penetrate a film to a depth of 0.22 mm, generally sufficient to interact with the emulsion in the film, as long as care is taken to assure that the film and the source of radioactivity are in direct contact, and that no barriers are posed between the film and the gel. In addition, gels must be completely dry before autoradiography. ^{32}P is a β-emitter with an energy of 1.71 MeV. Therefore, its particles penetrate water or other materials to a depth of 6 mm, passing completely through a film. In this case, gels or filters do not need to be dry, since water will not block particles of this energy and may be covered with a clear plastic wrap before autoradiography. The efficiency of ^{32}P-emitted β-particles is enhanced when an intensifying screen is placed behind the X-ray film, because radioactive particles that pass through the film cause the screen to emit photons that sensitize the film emulsion. The use of intensifying screens results in a fivefold increased enhancement of the autoradiographic image when the exposure is performed at low temperature (–70°C). In general, calcium tungstate screens are the most suitable because they emit blue light to which X-ray films are very sensitive *(6).*

Radiation of sufficiently high energy (e.g., ^{32}P and ^{35}S) can be detected by simple autoradiography, but low energy emissions may not penetrate the coating of the film, and the most sensitive fluorography procedure is used in these cases. In fluorography, the use of fluorescent chemicals increases about 10-fold the sensitivity of detection of

From: The Protein Protocols Handbook
Edited by: J. M. Walker Humana Press Inc., Totowa, NJ

weak β-emitters (^{35}S and ^{14}C), and permits detection of radioactivity from ^{3}H, virtually undetectable with simple autoradiography. Gels are impregnated with scintillant or fluors, which come in direct contact with the low-energy particles emitted by ^{35}S, ^{14}C, and ^{3}H. In response to radiation, fluors emit photons that react with the silver halide crystals in the emulsion of the film. Because the wavelength of the emitted photons depends on the fluorescent chemical and not on the radioactive emission of the isotope, the same type of film can be used to detect radioactivity from isotopes of varying energies *(7)*.

In this chapter, we describe the most common techniques of autoradiography and fluorography, and a double silver-staining method for acrylamide gels that approaches the sensitivity of autoradiography and that may be used for those experiments in which radiolabeling of proteins is not easily obtained (*see* Note 8). Autoradiography of wet and dry gels is described in Section 3.1., and protocols for fluorography and methods for quantification of radioactive proteins in polyacrylamide gels are given in Section 3.2. Double silver staining is described in Section 3.3.

2. Materials

2.1. Autoradiography

1. X-ray film (Kodak XR [Rochester, NY], Fuji RX [Pittsburgh, PA], Amersham [Arlington Heights, IL], or equivalent).
2. Plastic wrap (e.g., Saran Wrap).
3. 3MM Whatman paper.
4. SDS-PAGE fixing solution: 46% (v/v) methanol, 46% (v/v) water, 8% (v/v) acetic acid glacial. Store in an air-tight container. It is stable for months at room temperature.

2.2. Fluorography

1. X-ray blue-sensitive film (Kodak XAR-5, Amersham, or equivalent).
2. Plastic wrap.
3. 3MM Whatman paper.
4. SDS-PAGE fixing solution (as in Section 2.1.).
5. Coomassie blue staining solution: 0.125% Coomassie brilliant blue R250 in SDS-PAGE fixing solution. Stir overnight to dissolve, filter through a Whatman paper, and store at room temperature protected from light. It is stable for several weeks. For longer storage, dissolve the Coomassie blue in 46% water and 46% methanol only, and add 8% acetic acid glacial just before use.
6. Commercially available autoradiography enhancers (acidic acid-based: En^{3}Hance from Dupont [Mount Prospect, IL], water-soluble: Fluoro-Hance from RPI [Dupont NEN, Boston, MA], or equivalent).
7. Hydrogen peroxide (15% solution).
8. Scintillation fluid.

2.3. Double Silver Staining

1. Silver stain kit (Bio-Rad, Hercules, CA).
2. Methanol, ethanol, acetic acid glacial.
3. Solution A: Sodium thiosulfate 436 g/L.
4. Solution B: Sodium chloride 37 g, cupric sulfate 37 g, ammonium hydroxide 850 mL to 1 L of dd H_2O. Store these solutions at room temperature.

3. Methods

3.1. Autoradiography

3.1.1. Wet Gels

When radioactive proteins have to be recovered from the gel, the gel should not be fixed and dried before autoradiography. However, only high-energy isotopes (^{32}P or ^{125}I) should be used.

1. At the end of the electrophoresis, turn off the power supply and disassemble the gel apparatus. With a plastic spatula, pry apart the glass plates, cut one corner of the gel for orientation, carefully remove the gel, place it over two pieces of Whatman paper, and cover with a plastic wrap, avoiding the formation of bubbles or folds.
2. In a dark room, place the gel in direct contact with the X-ray film, and place an intensifying screen over the film. Expose for the appropriate length of time at room temperature or freeze at –70°C.
3. Develop the film using an automatic X-ray processor or a commercially available developer as following: immerse the gel for 5 min in the X-ray developer, rinse in water for 30 s, transfer into the fixer for 5 min, rinse in running water for 10–15 min, and let dry. All the solutions should be at 18–20°C.

   If the exposure of the gel is made at –70°C, allow the cassette to warm to room temperature and wipe off any condensation before opening. Alternatively, develop the film immediately as soon as the cassette is removed from the freezer, before condensation forms. If an additional exposure is needed, allow the cassette to dry completely before reusing.

   Cardboard exposure holders may work better with wet gel, since metal cassettes may compress the gel too tightly. When an intensifying screen is used, the film should be exposed at –70°C. The screen enhances the detection of radioactivity up to 10-fold, but may decrease the resolution.

3.1.2. Dry Gels

Gels containing urea should always be fixed to remove urea crystals. When high resolution is required, gels should always be dried before exposure to film. Gels should also be dried when ^{35}S or other low-energy β-emitters are used as radioactive tracers. For improved sensitivity and resolution, gels containing ^{32}P should also be fixed and dried before autoradiography. (Note: Gel drying is covered in detail in Chapter 34.)

1. At the end of the electrophoresis, turn off the power supply, disassemble the gel apparatus, and carefully pry open the two glass plates with a plastic spatula. (To assure that the gel will adhere to one glass plate only, one of the glass plates should be treated with silicon before casting the gel.)
2. Cut one corner of the gel for orientation, place the gel with the supporting glass plate into a shallow tray containing a volume of fixing solution sufficient to cover the gels, and fix for 30 min (the time necessary for fixation varies according to the thickness of the gel, but longer times do not have deleterious effects).
3. After fixation, carefully remove the plate from the tray, taking care not to float away the gel, or with a pipet connected to a vacuum pump, remove the solution from the tray and carefully from the glass plate. Place a wet piece of Whatman paper over the gel, being careful to avoid formation of bubbles and folding of the gel. Blot dry the Whatman paper with dry paper towels, applying gentle pressure.

4. Flip over the plate and maintain it exactly over the tray of a gel dryer. Carefully begin to detach the gel from one corner of the glass plate. The gel will remain adherent to the Whatman paper and will detach easily from the plate. Carefully cover the gel with plastic wrap, avoiding folding, and then dry the gel in a slab gel vacuum dryer for 1 h at 80°C (*see* Notes 1–5).
5. When the gel is dry, remove the plastic wrap and expose to X-ray film. After an appropriate length of time, develop the film as described in Section 3.1.1., step 3.

To increase the resolution of the radiogaphy, gels in which several bands of similar molecular mass are visualized, should be exposed at room temperature, without an intensifying screen. The use of a screen will result in increased intensity of the radioactive signal, but in decreased sharpness of the image. SDS-PAGE or other gels in which fewer bands need to be resolved may be exposed with an intensifying screen.

3.2. Fluorography

Gels containing weak β-emitters (^{35}S, ^{14}C, and ^{3}H) should be fixed, impregnated with autoradiography enhancers, and dried to reduce the film exposure time necessary for visualization of radioactive bands. SDS-PAGE should also be stained if nonradiolabeled mol-wt markers are used and for quantitative experiments (e.g., immunoprecipitation, detection of cell-free translated products, and so forth) where the radioactivity contained in specific proteins is to be measured in bands cut out from the gel *(8,9)*.

1. Turn off the power supply, remove the gel from the mold, cut a corner for orientation, and place the gel on a tray containing Coomassie blue-staining solution (the volume of the solution should always be adequate to cover the gel, so that it can float freely). Incubate for 45 min at room temperature with gentle shaking.
2. Remove the staining solution, and replace with SDS-PAGE fixing solution (the staining solution can be filtered through filter paper and reused as long as the radioactivity on it remains low, or until the color changes from blue to purple).
3. Incubate overnight at room temperature with gentle shaking, and replace the fixing solution at least once to accelerate the destaining. Gels should be destained until the mol-wt markers are clearly visible and the background is clear (*see* Note 6).
4. Discard the fixing solution in accordance with radioactive liquid waste disposal procedures, and add the autoradiography enhancer.
5. If the enhancer used is based on acetic acid (e.g., En^{3}Hance from Dupont or its equivalent), the gel can be soaked in the enhancer without rinsing, and steps 6–8 should be followed (*see* Note 1).
6. Allow the gel to impregnate with enhancer for 1 h with gentle shaking. Initially, a white precipitate may form on the surface of the gel, but it will disappear within the first 15 min of impregnation. Following impregnation, discard the used enhancer solution (do not mix with waste containing NaOH, $NaHCO_3$, and so forth). Add cold tap water to the gel to precipitate the fluorescent material and incubate the gel in water for 30 min. At this stage, the gel should appear uniformly opaque.
7. After the precipitation step, carefully place the gel over two pieces of wet filter paper, cover with plastic wrap, and dry under heat (60–70°C) and vacuum for 1–2 h on a slab gel dryer.
8. Remove the plastic wrap, tape the gel on a rigid support, and place it against a suitable blue-sensitive X-ray film, with an intensifying screen. Expose at –70°C for an appropriate length

of time. Do not store the gel at room temperature for >48 h before exposure, since evaporation of the fluors may occur, resulting in reduced sensitivity. For a longer period of storage prior to exposure, freeze the gel at –70°C.

9. If water-soluble fluorography solutions are used (e.g., Fluoro-Hance from R.P.I. or equivalent), the gel must be equilibrated in water after destaining, and steps 10 and 11 should be followed.
10. Discard the destaining solution, and wash the gel in distilled water for 30 min at room temperature, with shaking.
11. Discard the water and impregnate the gel with the enhancer for 30 min at room temperature with shaking. Remove the enhancer, place the gel over two wet pieces of filter paper, cover with a plastic wrap, and dry under vacuum with heat (60–80°C) for 2 h (*see* Note 2). Expose the gel as described above. (Flouro-Hance can be reused, but should be discarded as soon as the solution shows sign of discoloration) (*see* Notes 1–6).

    If the gels are not stained with Coomassie blue, after electrophoresis, place the gel in SDS-PAGE fixing solution, and incubate for 45 min at room temperature, with gentle shaking. After incubation, discard the fixing solution and impregnate the gel with enhancer as described (steps 5 or 9).
12. Radioactivity incorporated into specific proteins can be determined by cutting out the radioactive bands from the dried fluorographed gel.
13. Precisely position the film over the gel. With a sharp blade or scalpel, cut out the area of the gel corresponding to the band on the film. Also cut out an area of the gel free from radioactivity immediately below (or above) the radioactive band, for subtraction of background.
14. Place each gel slice into a scintillation vial (detaching the filter paper from the slice is not necessary), add 1 mL of a 15% solution of hydrogen peroxide, and incubate overnight in a water bath at 60°C to digest the gel and release the radioactivity.
15. After incubation, allow the vials to cool down to room temperature, add scintillation fluid, and measure the radioactivity in a scintillation counter. Alternatively, autoradiographic images can be quantitated by densitometric scanning of different exposures of the film (*see* Note 7).

3.3. Double Silver Staining (see Note 9)

1. After electrophoresis, transfer the gel in a glass tray containing 40% methanol and 10% acetic acid, and incubate at room temperature for at least 30 min (longer periods of time have no detrimental effect).
2. Discard this solution, and incubate for 15 min in 10% ethanol, 5% acetic acid.
3. Repeat step 2 one more time.
4. Add Oxidizer (diluted according to the manufacturer's protocol) and incubate 5 min, taking care that the gel is completely submerged in the solution.
5. Rinse the gel twice in distilled water.
6. Incubate 15 min in double-distilled water.
7. Repeat step 6 until the yellow color is completely removed from the gel.
8. Add Silver reagent (diluted according to the manufacturer's protocol), and incubate for 15 min.
9. Wash the gel once in double-distilled water.
10. Add developer (prepared according to the manufacturer's protocol), swirl the gel for 30 s, discard the solution, and wash once in double-distilled water.
11. Repeat step 10 and develop the gel until bands appear and the mol-wt markers become clearly visible.

12. Stop the reaction with 5% acetic acid, incubate 5 min, and wash with double-distilled water.
13. Add 3.5% solution A and 3.5% solution B, and incubate 5–10 min, or until the gel is clear.
14. Incubate two times in 10% acetic acid, 30 min each time.
15. Restain the gel by repeating steps 1–12.
16. Dry the gel in a slab dryer with heat and vacuum, or use a gel rap (Bio-Rad or equivalent), and dry overnight.

4. Notes

1. Gels of high polyacrylamide concentration (>10%) or gradient gels (acrylamide concentration 5–15%) may crack when dried. This problem is reduced by adding glycerol (1–5%) before drying. When acid-based fluorography enhancers are used, the glycerol should be added during the fluors precipitation step in water after removal of the enhancer. When water-soluble enhancers are used, the glycerol is added during equilibration of the gel in water, before addition of the enhancer. If the concentration of glycerol is too high, gels are difficult to dry, and they may stick to the film. Addition of the enhancer does not increase the cracking. We currently use water-based enhancers for our experiments, because they give sharp autoradiography images and good sensitivity, and can be reused for several gels. Fluorography with commercially available enhancers is simpler and less tedious than the traditional method with PPO-DMSO, and the results are as good as, or better than, those obtained with this method.
2. Enhancers are also used to increase the sensitivity of the autoradiograpy of DNA and RNA of agarose, acryalmide, or mixed gels. When enhancers are used, the gels must be dried at the suggested temperature, because excessive heat will cause damage of the fluors crystal of the enhancer and formation of brown spots on the surface of the gel.
3. Cracking may also be owing to the formation of air bubbles between the gel and the rubber cover of the dryer. This is generally caused by a weak vacuum that is insufficient to maintain the gel well adherent to the paper filter and to the dryer's tray. Air bubbles can be eliminated by rolling a pipet over the rubber cover while the vacuum is being applied.
4. Also, excessive stretching of the gel during the transfer to the filter paper may contribute to cracking, particularly for gels that contain high acrylamide concentration. A filter paper should be placed under the gel when removing it from the solution, and the method described in Section 3.1.2. should be used for larger gels.
5. It is always necessary to cover the gel with a plastic wrap to prevent sticking to the cover of the dryer. In addition, the vacuum should never be released during the drying procedure until the gel is completely dry, since this will cause the gel to shatter.
6. Staining of the gels with Coomassie blue before fluorography may quench the effect of the enhancer, particularly when low amounts of radioactivity are used. Therefore, the gels must be destained throughly, until the background is clear and only the protein bands are stained.
7. X-ray films that are not pre-exposed to light respond to radiation in a sigmoidal fashion, because the halide crystals of the emulsion are not fully activated *(10)*. In contrast, in a pre-exposed film, the response becomes linear and proportional to the amount of radioactivity, therefore, allowing precise quantitative measurement of radioactivity by scanning the autoradiography *(11)*. In addition, pre-exposure (preflashing) of the film results in a two- to threefold increase in sensitivity, for levels of radioactivity near the minimum level of detectability *(7)*, enabling autoradiography of gels containing ^{14}C and ^{3}H radioisotopes to be performed at room temperature, instead of –70°C. To pre-expose a film, a stroboscope or a flash of light (<1 ms) from an electronic flash unit can be used, but it is necessary to

reduce the light emission with a Deep Orange Kodak Wratten No. 22, or an Orange Kodak Wratten No. 21 filter, and to diffuse the image of the bulb with two pieces of Whatman No. 1 filter paper placed over the light-emitting lens. The distance between the film and the light source should be determined empirically *(11)*.

8. Autoradiography of SDS-PAGE is a powerful technique that permits detection of very low amounts of protein. However, in some instances, radioactive protein labeling cannot be easily accomplished. For example, metabolic labeling requires active protein synthesis *(12)*; thus, proteins present in body fluids or in tissue biopsy cannot be labeled *(13)*. Moreover, in in vivo animal experiments, it is often difficult to obtain radiolabeled proteins with high specific activity. In this case, silver staining of gels can be used as an alternative method, since it approaches the sensitivity of autoradiography *(14)*. We have developed a double silver-staining technique that is about 10-fold more sensitive than the conventional silver staining. This method has not been previously published in detail, except for figures presented elsewhere *(15)* (*see* Section 3.3.).
9. In silver staining, surface artifacts can be caused by pressure, fingerprints, and surface drying. Gloves should always be worn when handling the gel. A gray precipitate on the gel may be owing to insufficient washing.

Acknowledgment

We thank Peter A. Rice for support and for critical review of the manuscript.

References

1. Thomas, P. (1980) Hybridization of denatured RNA and small DNA fragments to nitrocellulose. *Proc. Natl. Acad. Sci. USA* **77,** 5201–5205.
2. Bonner, W. M. and Laskey, R. A. (1974) A film detection method for Tritium-labeled proteins and nucleic acid. *Eur. J. Biochem.* **46,** 83–88.
3. Garnier, G., Ault, B., Kramer, M., and Colten, H. R. (1992) *cis* and *trans* elements differ among mouse strains with high and low extrahepatic complement factor B gene expression. *J. Exp. Med.* **175,** 471–479.
4. Sanger, F., Nicklen, S., and Coulson, A. R. (1977) DNA sequencing with chain termination inhibitors. *Proc. Natl. Acad. Sci. USA* **74,** 5463–5467.
5. Sambrook, J., Fritsch, E. F., and Maniatis, T. (1989) Autoradiography, in *Molecular Cloning, A Laboratory Manual* (Nolan, C., ed.), Cold Spring Harbor Laboratory, Cold Spring Harbor, NY, pp. E.21–E.24.
6. Swanstrom, R., and Shank, P. R. (1978) X-ray intensifying screens greatly enhance the detection of radioactive isotopes ^{32}P and ^{125}I. *Anal. Biochem.* **86,** 184–192.
7. Bonner, W. M. (1984) Fluorography for the detection of radioactivity in gels. *Methods Enzymol.* **104,** 460–465.
8. Circolo, A., Welgus, H. G., Pierce, F. G., Kramer, J., and Strunk, R. C. (1991) Differential regulation of the expression of proteinases/antiproteinases in human fibroblasts: effects of IL-1 and PDGF. *J. Biol. Chem.* **266,** 12,283–12,288.
9. Garnier, G., Circolo, A., and Colten, H. R. (1995) Translational regulation of murine complement factor B alternative transcripts by upstream AUG codons. *J. Immunol.* **154,** 3275–3282.
10. Laskey, R. A. (1977) Enhanced autoradiographyc detection of ^{32}P and ^{125}I using intensifying screens and hypersensitized films. *FEBS Lett.* **82,** 314–316.
11. Laskey, R. A. and Mills, A. D. (1975) Quantitative film detection of ^{3}H and ^{14}C in polyacrylamide gels by fluorography. *Eur. J. Biochem.* **56,** 335–341.

12. Switzar, R. C. III, Merril, C. R., and Shifrin, S. (1979) A highly sensitive silver stain for detecting proteins and peptides in polyacrylamide gels. *Anal. Biochem.* **98,** 231–237.
13. Irie, S., Sezaki, M., and Kato, Y. (1982) A faithful double stain of proteins in the polyacrylamide gel with Coomassie Blue and silver. *Anal. Biochem.* **126,** 350–354.
14. Berry, M. J. and Samuel, C. E. (1982) Detection of subnanogram amounts of RNA in polyacrylamide gels in the presence and absence of proteins by staining with silver. *Anal. Biochem.* **124,** 180–184.
15. Densen, P., Gulati, S., and Rice, P. A. (1987) Specificity of antibodies against *Neisseria gonorrhoeae* that stimulate neutrophil chemotaxis. *J. Clin. Invest.* **80,** 78–87.

Part III

Blotting and Detection Methods

37

Protein Blotting by Electroblotting

Mark Page and Robin Thorpe

1. Introduction

Identification of proteins separated by gel electrophoresis or electricfocusing is often compounded by the small pore size of the gel, which limits penetration by macromolecular probes. Overcoming this problem can be achieved by blotting the proteins onto an adsorbent porous membrane (usually nitrocellulose or diazotized paper), which gives a mirror image of the gel *(1)*. A variety of reagents can be incubated with the membrane specifically to detect and analyze the protein of interest. Antibodies are widely used as detecting reagents, and the procedure is sometimes called Western blotting. However, protein blotting or immunoblotting is the most descriptive.

Electroblotting is usually preferred for immunoblotting in which proteins are transferred to the membrane support by electrophoresis. A possible exception to this is in transfer from isoelectric focusing gels, where the proteins are at their isoelectric points and uncharged. Therefore, there is considerable delay before the proteins start to migrate in an electric field; also they can migrate in different directions and their rate of transfer can vary. For these reasons, transfer from isoelectric focusing gels by capillary blotting (*see* Chapter 39) may be preferable, particularly if very thin gels are used.

Electroblotting has the advantage that transfer takes only 1–4 h, and lateral diffusion of proteins (which causes diffuse bands) is reduced. Overall, nitrocellulose membranes are recommended. These are efficient protein binders and do not require activation, but are fragile and need careful handling. Nitrocellulose membranes, such as Hybond-C extra, are more robust. If the antigens of interest do not bind efficiently or if the blot is to be reused, then diazotized paper, or possibly nylon membranes, can be used. A suitable electrophoretic transfer chamber and power pack are required; these are available commercially or can be made in a laboratory work shop. The apparatus consists of a tank containing buffer, in which is located a cassette, clamping the gel and membrane tightly together. A current is applied from electrodes situated at either side of the cassette. The buffer is often cooled during transfer to avoid heating effects.

2. Materials

1. Transfer apparatus.
2. Orbital shaker.
3. Nitrocellulose sheet, 0.45-µm pore size, e.g., Hybond-C extra.

From: The Protein Protocols Handbook
Edited by: J. M. Walker Humana Press Inc., Totowa, NJ

4. Filter paper, Whatman 3MM (Maidstone, UK).
5. Plastic box large enough to hold the blot and allow movement on agitation.
6. Transfer buffer: 0.025*M* Tris, 0.052*M* glycine, 20% methanol.
7. Blocking buffer: PBS (0.14*M* NaCl, 2.7 m*M* KCl, 1.5 m*M* KH_2PO_4, 8.1 m*M* Na_2HPO_4) containing 5% dried milk powder.

3. Method

1. Assemble transfer apparatus and fill the tank with transfer buffer (*see* Note 1). If a cooling device is fitted to the apparatus, switch on before it is required to allow the buffer to cool down sufficiently (to ~8–15°C).
2. Cut two pieces of filter paper to the size of the cassette clamp, soak in transfer buffer, and place one on the cathodal side of the cassette on top of a wetted sponge pad (*see* Note 2).
3. Place the gel on the filter paper covering the cathodal side of the cassette (*see* Note 3). Keep the gel wet at all times with transfer buffer. Soak the nitrocellulose sheet (cut to the same size as the gel) in transfer buffer, and place it on the gel, i.e., on the anodal side (*see* Note 4). Avoid trapping air bubbles throughout the process.
4. Place the remaining filter paper over the nitrocellulose, and expel all air bubbles between the nitrocellulose and gel. This is achieved by soaking the gel/nitrocellulose/filter paper assembly liberally with transfer buffer and then pressing with a small hand roller.
5. Finally, place a wetted sponge pad on top of the filter paper, and clamp securely in the cassette. This should be a tight fit (*see* Note 5).
6. Place the cassette in the tank and fit the lid. Recirculate the transfer buffer either by a recirculating pump or a magnetic stirrer.
7. Electrophorese for 1–4 h at 0.5 A (*see* Note 6).
8. Turn off power, remove nitrocellulose sheet, and agitate in 50–200 mL of PBS containing 5% dried milk powder (*see* Note 7).
9. Process nitrocellulose sheet as required (*see* Chapters 44, 45, 48–50, and 97).

4. Notes

1. Methanol prevents polyacrylamide gels, removes SDS from polypeptides, and enhances the binding of proteins to the membrane, but it reduces the efficacy of transfer of larger proteins. It can be omitted from the transfer buffer with no adverse effects, but this has to be established empirically. Methanol is not necessary for non-SDS gels or isoelectric focusing gels.
2. Cassette clamps are normally provided with two sponge pads that fit either side of the gel/membrane/filter paper assembly to fill any dead space to squeeze the gel and membrane tightly together during electrophoresis.
3. Handle the gel and nitrocellulose sheet with gloved hands to prevent contamination with finger-derived proteins.
4. Cut a small piece of the bottom corner of the gel and nitrocellulose to orient the assembly.
5. If the gel and nitrocellulose are not clamped tightly together, then the proteins migrating from the gel can move radially and give a smeared blot.
6. Higher current can be used, but may result in uneven heating effects and blurred or distorted blots. The use of lower currents is not recommended, since transfer efficiency is reduced and poor quality blots are obtained. Overnight transfer can be used, but is not generally recommended. The time required for efficient transfer depends on the acrylamide concentration, gel thickness, gel buffer system, and the molecular size and shape of the proteins. Most proteins will pass through the nitrocellulose sheet if transfer is contin-

ued for too long. Proteins migrate faster from SDS gels (they are coated with SDS and highly charged), and transfer from non-SDS gels takes longer (around 4–5 h).

7. Thirty-minute incubation with blocking protein is sufficient to saturate all the protein binding sites on the blot. Longer times and overnight incubation can be used if this is more convenient. Protein blocked sheets can be stored frozen for long periods. For this, drain excess blocking solution and place in a plastic bag when required, wash with blocking buffer for 10 min, then continue with the next processing steps.

References

1. Towbin, H., Staehelin, T., and Gordon, J. (1979) Electrophoretic transfer of proteins from polyacrylamide gels to nitrocellulose sheets: procedure and some applications. *Proc. Natl. Acad. Sci. USA* **76,** 4350–4354.

38

Protein Blotting by the Semidry Method

Patricia Gravel and Olivier Golaz

1. Introduction

Protein blotting involves the transfer of proteins to an immobilizing membrane. The most widely used blotting method is the electrophoretic transfer of resolved proteins from a polyacrylamide gel to a nitrocellulose or polyvinylidene difluoride (PVDF) sheet, and is often referred to as "Western blotting." Electrophoretic transfer uses the driving force of an electric field to elute proteins from gels and to immobilize them on a matrix. This method is fast, efficient, and maintains the high-resolution of the protein pattern *(1)*. There are currently two main configurations of electroblotting apparatus: tanks of buffer with vertically placed wire (*see* Chapter 37) or plate electrodes, and semidry transfer with flat-plate electrodes.

For semidry blotting, the gel and membrane are sandwiched horizontally between two stacks of buffer-wetted filter papers, which are in direct contact with two closely spaced solid-plate electrodes. The name semidry refers to the limited amount of buffer that is confined to the stacks of filter paper. Semidry blotting requires considerably less buffer than the tank method, the transfer from single gels is simpler to set up, it allows the use of multiple transfer buffers (i.e., different buffers in the cathode and anode electrolyte stacks), and it is reserved for rapid transfers, because the amount of buffer is limited and the use of external cooling is not possible. Nevertheless, both techniques have a high efficacy, and the choice between the two types of transfer is a matter of preference.

Once transferred to a membrane, proteins are more readily and equally accessible to various ligands than they were in the gel. The blot (i.e., the immobilizing matrix containing the transferred proteins) is therefore reacted with different probes, such as antibody for the identification of the corresponding antigen, lectin for the detection of glycoproteins, or ligand for the detection of blotted receptor components, as well as for studies of protein–ligand associations *(2*; and *see* Chapters 48–50 and 97*)*. The blot analysis generally requires small amounts of reagents, the transferred proteins on membrane can be stored for many weeks prior to their use, and the same blot can be used for multiple successive analyses. For reviews on the basic principles involved in performing protein blotting and for an overview of some possible applications, *see* articles by Garfin and Bers *(1)*, Beisiegel *(3)*, Gershoni and Palade *(4)*, and Towbin and Gordon *(5)*.

From: The Protein Protocols Handbook
Edited by: J. M. Walker Humana Press Inc., Totowa, NJ

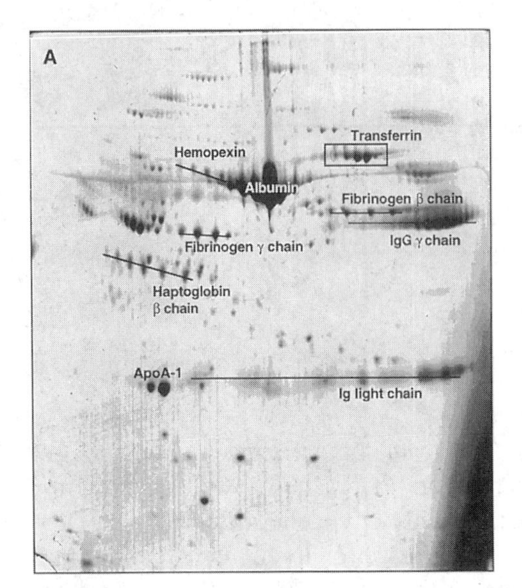

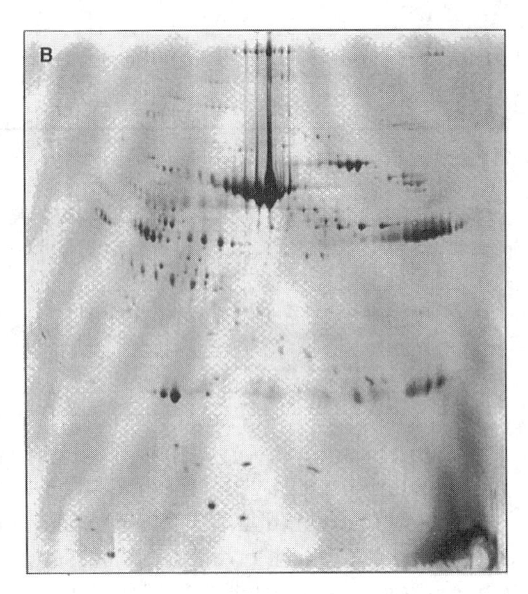

Fig. 1. Plasma proteins separated by 2D PAGE and **(A)** stained with Coomassie brilliant blue R250 or **(B)** transferred to PVDF membrane using the semidry system (2 h, 15 V) with Towbin buffer diluted 1:2 in water and stained with Coomassie blue.

In the following Sections, we describe a protocol for semidry blotting that uses a simple buffer system *(6)*. The efficacy of this method is illustrated in Fig. 1 with human plasma proteins separated by two-dimensional polyacrylamide gel electrophoresis (2D PAGE), transferred on PVDF membrane and stained with Coomassie blue. The blot pattern is compared to the Coomassie blue staining of the same protein sample before transfer from 2D PAGE. The resolution, shape, and abundance of protein spots on membrane are comparable to the 2D polyacrylamide gel pattern. This blotting procedure allows a good and almost complete elution of proteins from the gel and their immobilization on the membrane.

We also review in Section 4. the principal types of transfer matrix, and the different discontinuous or continuous buffers that can be used with this technique.

2. Materials

1. Buffer: The transfer and equilibration buffer is the Towbin buffer diluted 1:2 with distilled water: 12.5 m*M* Tris, 96 m*M* glycine, and 10% (v/v) methanol *(7)* (*see* Note 1 and Table 1).
2. Membranes and filter papers: PVDF (0.2 μm, 200 × 200 mm, Bio-Rad, Hercules, CA) or nitrocellulose (0.45 μm, Schleicher & Shuell, Keene, NH). Filter papers are chromatography papers (grade 3 mm CHr, Whatman, Maidstone, UK). Thicker blotting papers can also be used (Pharmacia-LKB, Uppsala, Sweden, filter paper for blotting, 200 × 250 mm).
3. Electroblotting apparatus: Proteins are transferred with a Trans-Blot SD semidry cell (Bio-Rad). The anode of the apparatus is made of platinum-coated titanium, and the cathode is made of stainless steel. The maximum gel size that can be used with this apparatus is 25 × 18.5 cm.
4. Coomassie brilliant blue staining: Proteins in the 2D gel and on the blot (Fig. 1) are stained with a 0.025% (w/v) solution of Coomassie brilliant blue R250 solubilized in 43% (v/v) methanol and 7% (v/v) acetic acid. The destaining solution contains 30% (v/v) methanol and 7% (v/v) acetic acid.

Table 1
Examples of Transfer Buffers Used for Semidry Blotting of Proteins from Polyacrylamide Gels to Immobilized Matrices

Buffer		pH	Remarks	Refs.
Continuous buffer systems				
Towbin buffer	25 m*M* Tris 192 m*M* glycine 20% (v/v) methanol	8.3	First practical method for the transfer of proteins in an electric field from gel electropherograms to nitrocellulose matrix	*8*
Towbin buffer with addition of SDS	25 m*M* Tris 192 m*M* glycine 20% (v/v) methanol 0.1% (w/v) SDS	8.3	Facilitates the elution of high-mol-wt proteins	*9*
Towbin buffer diluted 1:2 with water	12.5 m*M* Tris 96 m*M* glycine 10% (v/v) methanol	8.3		*7*
Dunn carbonate buffer	10 m*M* $NaHCO_3$ 3 m*M* Na_2CO_3 20% (v/v) methanol	9.9	Carbonate buffer enhances immunochemical recognition and therefore the sensitivity of Western blots by increasing the retention of small proteins and restoration of a more nearly native state for the larger proteins	*10*
Bjerrum and Schafer-Nielsen buffer	40 m*M* Tris 39 m*M* glycine 20% (v/v) methanol	9.2	This buffer and the Dunn carbonate buffer have a higher pH and a lower conductivity than the Towbin buffer and are recommended for semidry transfers	*11*
Buffer without methanol for transferring SDS-protein complexes using nylon membrane	All previous buffers without the addition of methanol	—	Without methanol, the SDS from SDS gels are still complexed with proteins and confer similar negative charge to mass ratio for all proteins; therefore, electrostatic interactions between the nylon matrix (positively charged) and the proteins can occur	*12*

(continued)

Table 1 ***(continued)***

Buffer		pH	Remarks	Refs.
CAPS[a] buffer	10 m*M* CAPS 10% (v/v) methanol	11.0	This buffer without glycine is useful to perform a complete amino acid composition analysis	*13,14*
Discontinuous buffer systems				
Tris-glycine buffer with addition of SDS at the cathodic side and methanol at the anodic side (Svoboda buffer)	Cathodic side 0.1% (w/v) SDS 25 m*M* Tris 192 m*M* glycine (no methanol)	8.3	The presence of 20% methanol at the anodic side facilitates the dissociation of protein–SDS complexes and the protein binding to the negative charge matrix, whereas the presence of 0.1% SDS at the cathodic side increases the rate of transfer of cationic and/or hydrophobic proteins	*15*
	Anodic side 25 m*M* Tris 192 m*M* glycine 20% (v/v) methanol	8.3		
Kyhse-Andersen buffer	Cathodic side 40 m*M* 6-amino-*n*-hexanoic acid 25 m*M* Tris 20% (v/v) methanol	9.4	This system uses isotachophoresis to elute proteins from the gel; SDS, glycine, and SDS–protein complexes from the gel move toward the anode according to their net mobilities in the stated order. At the cathodic side, 6-amino-*n*-hexanoic acid is the terminating ion and thereby carries away the proteins from the polyacrylamide gel	*16*
	Anodic side 300 m*M* Tris 20% (v/v) methanol (in contact with the anode)	10.4		
	25 m*M* Tris 20% (v/v) methanol (in contact with the membrane)	10.4		

Laurière buffer	Cathodic side		This system of buffers introduces a difference of pH that remains stable during the entire electrotransfer time period and an asymmetrical disposition of methanol and SDS on each side of the gel–membrane pair	*17*
	0.4% (w/v) SDS 60 m*M* lactic acid 100 m*M* Tris (in contact with the cathode)	8.4		
	0.1% (v/v) SDS 15 m*M* lactic acid 25 m*M* Tris (in contact with the gel)	8.4		
	Anodic side		Under these conditions, semidry blotting allows quantitative transfer of proteins from SDS gels to membranes almost regardless of their mol-wt and solubility	
	20% (v/v) methanol 60 m*M* lactic acid 20 m*M* Tris (in contact with the membrane)	3.8		
	20% (v/v) methanol 100 m*M* Tris (in contact with the anode)	10.4		

[a]CAPS: 3-[cyclohexylamino]-1-propanesulfonic acid.

All chemicals and methanol are of analytical reagent grade. Metallic contaminants in low-grade methanol normally deposit on the electrodes.

3. Methods

To avoid membrane contamination, always use forceps or wear gloves when handling membranes.

3.1. Preparation for Semidry Blotting

1. Prepare the transfer buffer the day preceding the blotting experiments and store it at 4°C.
2. After the separation of proteins by sodium dodecyl sulfate (SDS-PAGE) or 2D PAGE, the gel is briefly rinsed in distilled water for a few seconds and then equilibrated in transfer buffer for 10 min under gentle agitation (*see* Note 2).
3. During the equilibration, the filter paper and the transfer membrane are cut to the dimensions of the gel. Six pieces of filter paper per gel are needed for each gel/membrane sandwich (or two pieces of thick filter papers).
4. If the hydrophobic PVDF membrane is used, it should be prewetted prior to equilibration in transfer buffer. Immerse the membrane in a small volume of 100% methanol for a few seconds, until the entire membrane is translucent, rinse it in deionized water, and then equilibrate it in transfer buffer for 3–5 min. It is important to keep the membrane wet at all times, since proteins will not bind to the dried PVDF membrane (*see* Note 3).

   For hydrophilic nitrocellulose membrane, wet it in transfer buffer directly, and allow it to soak for approx 5 min (*see* Notes 4 and 5 and Table 2 for the description of the different transfer membranes).
5. If multiple full-size gels are to be transferred at one time, there is a necessity to interleave a dialysis membrane between the gel-membrane pairs to prevent proteins being driven through membranes into subsequent stacks. Cut a piece of dialysis membrane with the appropriate mol-wt cutoff to the dimensions of the gel, and soak it in the transfer buffer (*see* Note 6).

3.2. Assembly of the Semidry Unit

The assembly of a semidry electroblotting apparatus is represented in Fig. 2. Four mini gels can also be transferred at the same time by placing them side-by-side on the anode platform.

1. The filter papers are briefly soaked in transfer buffer for a few seconds. Place a prewetted filter paper onto the anode. Use a pipet or a painter roller to eliminate all air bubbles by pressing firmly all over the area of the paper (*see* Note 7). If a thin filter paper is used, add two more sheets and remove air bubbles between each layer.
2. Place the pre-equilibrated transfer membrane on top of the filter paper.
3. Place the equilibrated gel on top and on the center of the membrane.
4. Place another wetted sheet of filter paper on top of the gel. If a thin filter paper is used, add 2 more sheets. Roll out air bubbles.
5. Place the cathode onto the stack.
6. The blotting unit is then connected to a power supply, and proteins are transferred for 2 h at a constant voltage of 15 V and at room temperature.
7. After protein blotting, the membrane is rinsed (3 × 5 min) with distilled water and is then ready for blot analysis. Following transfer, the first step for blot analysis is to block unoccupied binding sites on the membrane to prevent nonspecific binding of probes, most of

Table 2
Membranes Used for Electroblotting Proteins from Polyacrylamide Gels

Transfer matrix	Charge of the matrix	Mechanism of protein binding[a]	Capacity of protein binding, μg/cm²[b]	Advantages	Disadvantages
Nitrocellulose	Negative	Hydrophobic and electrostatic forces	249	High capacity Low cost	They are mechanically fragile The presence of cellulose acetate in nitrocellulose membranes seems to reduce their capacity to bind proteins; pure cellulose nitrate has the highest binding capacity *(12)* High cost
PVDF	Negative	Hydrophobic forces	172	High capacity Ideal for N-terminal microsequencing, amino acid composition analysis (resists to acidic and organic solvent) Protein transfer efficacy can be visualized without staining (*see* Note 4)	
Modified nylon (made by incorporating tertiary and quaternary amines)	Positive	Electrostatic forces	149	Multiple reprobing possible	Protein staining is difficult because common anionic dyes cannot be used (Coomassie blue, Amido black, colloidal gold, *ponceau S)* Blocking of unoccupied bindng sites is difficult owing to the high charge density of the matrices Most acidic proteins are poorly transferred
Carboxymethylcellulose	Negative	Ionic interactions	2500 for histones (probably owing to the high affinity of histones for anionic groups)	Helpful for the transfer and microsequencing of basic proteins or peptides *(20)*	An elution step is required before microsequencing

[a]Refers to the mechanism of adsorption between the surface of the membrane and structures on the protein.
[b]Ref. *20*.

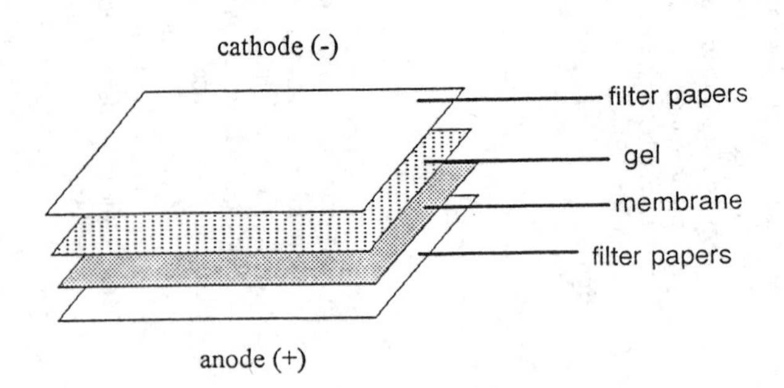

Fig. 2. Assembly of a horizontal electroblotting apparatus.

which are proteins (antibody, lectin). *See* Note 8 and Table 3 for a description of the blocking procedure.

8. The blots can be stored for many weeks prior to their use. *See* Note 9 for storage conditions.

4. Notes

1. Since the introduction of Western blotting in 1979 by Towbin et al. *(8)*, many other buffer systems have been developed in order to improve electrophoretic transfer of proteins. The most common systems used with semidry apparatus are listed in Table 1.

   The two critical factors during transfer are the elution efficacy of proteins out of the gel and the binding capacity of the matrix for proteins. The elution efficacy is mainly determined by the acrylamide concentration of the gel, the ionic strength, the pH of the buffer, and additional constituents of the transfer buffer, such as SDS and methanol. The binding capacity is mainly determined by the character of the membrane, but also by the transfer buffer composition *(4)*.

   Alkaline pH and SDS favor the elution of the proteins from the gel, whereas acidic pH and methanol favor their adsorption on the negatively charged membrane *(17)*. Methanol increases the binding capacity of matrix presumably by exposing hydrophobic protein domains, so that they can interact with the matrix. Also, methanol decreases the elution efficacy by fixing the proteins in the gel and by reducing the gel pore size *(1,4)*. When there is SDS in transfer buffer (up to about 0.1% [w/v]), the proteins are negatively charged and elute efficiently from the gel. Semidry conditions for blotting allow the use of multiple transfer buffers (discontinuous buffer systems) to ensure a faster and better electrotransfer *(16)*. Examples of discontinuous buffer systems are listed in Table 1. However, Bjerrum and Schafer-Nielsen *(11)* showed that there is no advantage in using different buffers in the cathode and anode electrolyte stacks. They found comparable transfer efficiencies for semidry blots performed in continuous and discontinuous buffers. We tested different buffer systems (Towbin buffer, Towbin buffer diluted 1:2 with water, and Laurière buffer) for transferring plasma proteins from 2D PAGE by semidry method, and we similarly found no advantage in using discontinuous buffer systems (unpublished results).

2. Rinsing and equilibrating the gel facilitate the removal of electrophoresis buffer salt and excess of detergent. If salts are still present, the conductivity of the transfer buffer increases, and excessive heat is generated during the transfer. Also, an equilibration period allows the gel to adjust to its final size prior to electroblotting, because the gel shrinks in methanol-containing transfer buffer.

   The duration of equilibration depends on the gel thickness. For a 1.5 mm thick gel, 10 min of equilibration are enough.

Table 3
Blocking Agents for Blots

Blocking agent	Remarks
Proteins[a]	
BSA	Can contain carbohydrate contaminants that can increase background when lectin probes are used *(1,22)*
	For nylon membranes, high concentration of BSA is required because of their very high protein binding capacity; a blocking solution made of 10% BSA in PBS is used for at least 12 h at 45–50°C for satisfactory quenching *(12)*
Nonfat dried milk	Inexpensive and easy to use
	Can contain competing reactants, such as biotin, which can cause a diminution of the signal intensity *(1)*
Gelatin	The use of gelatin as a blocking agent is contested: it may mask blotted proteins *(1),* it leads to strong unspecific reactions with peroxidase-labeled antibodies *(4)*, and when used alone, gelatin is not a satisfactory blocking agent *(5)*
	Fisch skin gelatin gives lower background than mammalian gelatin and does not have to be heated to dissolve (can be used at 4°C) *(23)*
Nonionic detergents[b]	
Tween-20	Simple and effective blocker; widely used at 0.05% for immunoblotting on nitrocellulose or PVDF *(1,24,25)*
	Tween-20 minimizes nonspecific protein–protein and protein–matrix adsorption by disrupting the underlying noncovalent/hydrophobic interactions, while leaving antigen antibody interactions relatively unaffected *(25)*
	However, some studies reported protein and antibody removal from the membrane after Tween-20 incubation *(26)*
NP-40 Triton X-100	These agents carry slight charges and should be avoided since they displace proteins from nitrocellulose at a higher extent than Tween-20 *(26)*
Other	
Polyvinylpyrolidone (PVP-40, 40, 000 M_r)	As adjunct to Tween-20, PVP-40 produces backgrounds lower than those of Tween-20 alone, approximately doubling the signal-to-background ratio (without decreasing specific immunoreactivity in Western blots); should be used at a concentration of 1% (w/v) and added to 0.05% (w/v) Tween-20 *(25)*

[a]Should not be used in excess of 3%.
[b]Should not be used in excess of 0.3–0.5%.

Recently, poly(ethyleneglycol) polymers (PEG 1000–2000) have been used to complete electroblotting, in order to obtain better resolution and enhancement of sensitivity of proteins on membrane. PEG reduces background, raises signal-to-noise ratio and sharpens protein band *(18)*. After PAGE, 30% (w/v) PEG (solubilized in transfer buffer: 12.5 m*M* Tris, 96 m*M* glycine, and 15% [v/v] methanol) is applied to fix proteins reversibly within the gel. The intragel proteins can then be electroblotted directly onto membranes using the same transfer buffer containing PEG.

3. If the PVDF membrane does dry out during use, it can be rewet in methanol. Membranes that contain adsorbed proteins and that have been allowed to dry can also be rewet in methanol. In our experience, we have not seen any difficulty in protein staining, glycoprotein detection, and immunostaining after this rewetting procedure in methanol.
4. Nitrocellulose membrane was the first matrix used in electroblotting *(8)* and is still the support used for most protein-blotting experiments. It has a high binding capacity, the nonspecific protein-binding sites are easily and rapidly blocked, allowing low background staining, and it is not expensive. For blotting with mixtures of proteins, standard nitrocellulose with a pore size of 0.45 µm should be used first. However, membranes having 0.1- and 0.2-µm pore sizes should be tried if some low-mol-wt proteins do not bind efficiently to the standard membrane *(1)*. After staining, the nitrocellulose membranes become transparent simply by impregnating the membrane with concentrated Triton X-114 *(19)*. The blot can thus, be photographed by transillumination or scanned with a densitometer for quantitative analysis. Long-term stability (several months) of transparent stained blots is possible if they are stored at –20°C.

   PVDF membranes are more expensive, but have high mechanical strength, high protein-binding capacity, and are compatible with most commonly used protein stains and immunochemical detection systems. The chemical structure of the membrane offers excellent resistance to acidic and organic solvents. This makes PVDF membrane an appropriate support for N-terminal protein sequencing and amino acid composition analysis. Another interesting advantage of PVDF matrix over the nitrocellulose is the possibility to visualize the protein pattern on the blot without staining. After blotting, the PVDF membrane should be placed on top of a vessel containing distilled water. The immersion of the membrane in water should be avoided. It should be laid down at the surface. The protein spots (or bands) contrast with the remainder of the membrane and can be visualized. To obtain a clear image of the protein pattern, the surface of the membrane should be observed from different angles and under appropriate lighting. This procedure, which was found unintentionally, is easy to perform and allows rapid and good evaluation of the transfer quality.

   Table 2 summarizes the most common matrices that can be used for transferring proteins from polyacrylamide gels.
5. Whatever the membrane used, exceeding its binding capacity tends to reduce the signal eventually obtained on blots. It can be assumed that excess protein, weakly associated with the membrane, may be readily accessible to react with the probe in solution, but the probe–protein complexes formed may then be easily washed off during the further processing of the membrane *(4)*. This situation does not occur if the proteins are initially in good contact with the membrane.

   For 2D PAGE, the best recovery and resolution of proteins are obtained when loading 120 µg of human plasma or platelet proteins. When 400 µg of protein are separated by 2D PAGE and transferred on membrane, the spots are diffused and the basic proteins poorly transferred (not shown). This could be attributed to overloading, which prevents a good separation of proteins and an adequate binding to transfer matrix.
6. It is very difficult to form large stacks of gel-membrane pairs. Even two pairs are often associated with the introduction of air bubbles. We prefer to use a semidry unit to transfer proteins from a single gel only.
7. Air bubbles create points of high resistance, and this results in spot (or band) areas of low efficacy transfer and spot (or band) distortion.
8. The quality or extent of the blocking step determines the level of background interference. It has been recognized that the blocking step may also promote renaturation of epitopes

(21). This latter aspect is particularly important when working with monoclonal antibodies (MAbs) (which often fail to recognize the corresponding antigenic site after electroblotting). Hauri and Bucher *(21)* suggest that MAbs may have individual blocking requirements, probably owing to different degrees of epitope renaturation and/or accessibility of antibody under the various blocking conditions. Some common blocking agents are listed in Table 3.

For immunoblotting on nitrocellulose membrane, we obtain good results by using a blocking solution made of 0.5% (w/v) bovine serum albumin (BSA), 0.2% (w/v) Tween-20, and 5% (w/v) nonfat dried milk in phosphate buffered saline (PBS) (137 m*M* NaCl, 27 m*M* KCl, 10 m*M* Na_2HPO_4, 1.8 m*M* KH_2PO_4, pH 7.4) *(27)*. For lectin blotting on PVDF membrane, we use 0.5% (w/v) Tween-20 in PBS buffer *(6)* (*see also* Chapter 97).

9. We store the blots at 4°C in PBS containing 0.005% (w/v) sodium azide. In order to evaluate the effect of storage on blotted proteins, we stained with colloidal gold a nitrocellulose blot of platelet proteins stored for a period of 4 mo in PBS/azide at 4°C. We observed a protein pattern identical to the same blot stained immediately after Western blotting (not shown). On the other hand, when we dried a nitrocellulose blot at room temperature and stained 24 h later the proteins with colloidal gold, we observed important contaminating spots on the blots.

   Only one point should be kept in mind when storing blots in PBS/azide. Sodium azide inhibits peroxidase activity, and therefore, good washing of the membrane with PBS is necessary before blot analysis using probe labeled with peroxidase.

   As an alternative to blot storage, the gel can be frozen at –80°C. Immediately after electrophoresis, the gel should be rinsed in distilled water for a few seconds and placed in a plastic bag between two precooled glass plates. It is important to precool glass plates at –80°C; otherwise the gel will crack when thawed. These frozen gels can be stored at –80°C for at least 3 mo *(17)*. Before the transfer procedure, the frozen gels are thawed in their plastic bags and then equilibrated as described in Section 3.

Acknowledgments

P. Gravel acknowledges a grant from the Swiss Foundation for Research on Alcohol (SFRA). The authors thank Marianne Gex-Fabry and Claude Walzer for revision of the manuscript.

References

1. Garfin, D. E. and Bers, G. (1982) Basic aspects of protein blotting, in *Protein Blotting* (Baldo, B. A., and Tovey, E. R., eds.), Karger, Basel, Switzerland, pp. 5–42.
2. Gershoni, J. M. (1985) Protein blotting: developments and perspectives. *TIBS* **10,** 103–106.
3. Beisiegel, U. (1986) Protein blotting. *Electrophoresis* **7,** 1–18.
4. Gershoni, J. M. and Palade, G. E. (1983) Protein blotting: principles and applications. *Anal. Biochem.* **131,** 1–15.
5. Towbin, H., and Gordon, J. (1984) Immunoblotting and dot immunobinding: current status and outlook. *J. Immunol. Methods* 7**2,** 313–340.
6. Gravel, P., Golaz, O., Walzer, C., Hochstrasser, D. F., Turler, H., and Balant, L. P. (1994) Analysis of glycoproteins separated by two-dimensional gel electrophoresis using lectin blotting revealed by chemiluminescence. *Anal. Biochem.* **221,** 66–71.
7. Sanchez, J. C., Ravier, F., Pasquali, C., Frutiger, S., Bjellqvist, B., Hochstrasser, D. F., and Hughes, G. J. (1992) Improving the detection of proteins after transfer to polyvinylidene difluoride membranes. *Electrophoresis* **13,** 715–717.

8. Towbin, H., Staehelin, T., and Gordon, J. (1979) Electrophoretic transfer of proteins from polyacrylamide gels to nitrocellulose sheets: procedure and some applications. *Proc. Natl. Acad. Sci. USA* **76,** 4350–4354.
9. Erickson, P. F., Minier, L. N., and Lasher, R. S. (1982) Quantitative electrophoretic transfer of polypeptides from SDS polyacrylamide gels to nitrocellulose sheets: a method for their re-use in immunoautoradiographic detection of antigens. *J. Immunol. Methods* **51,** 241–249.
10. Dunn, S. D. (1986) Effects of the modification of transfer buffer composition and the renaturation of proteins in gels on the recognition of proteins on Western blots by monoclonal antibodies. *Anal. Biochem.* **157,** 144–153.
11. Bjerrum, O. J. and Schafer-Nielsen, C. (1986) Buffer systems and transfer parameters for semidry electroblotting with a horizontal apparatus, in *Electrophoresis 1986* (Dunn, M. J., ed.), VCH, Weinheim, Germany, pp. 315–327.
12. Gershoni, J. M. and Palade, G. E. (1982) Electrophoretic transfer of proteins from sodium dodecyl sulfate-polyacrylamide gels to a positively charged membrane filter. *Anal. Biochem.* **124,** 396–405.
13. Jin, Y. and Cerletti, N. (1992) Western blotting of transforming growth factor b2. Optimization of the electrophoretic transfer. *Appl. Theor. Electrophoresis* **3,** 85–90.
14. Matsudaira, P. J. (1987) Sequence from picomole quantities of proteins electroblotted onto polyvinylidene difluoride membranes. *J. Biol. Chem.* **21,** 10,035–10,038.
15. Svoboda, M., Meuris, S., Robyn, C., and Christophe, J. (1985) Rapid electrotransfer of proteins from polyacrylamide gel to nitrocellulose membrane using surface-conductive glass as anode. *Anal. Biochem.* **151,** 16–23.
16. Kyhse-Andersen, J. J. (1984) Electroblotting of multiple gels: a simple apparatus without buffer tank for rapid transfer of proteins from polyacrylamide to nitrocellulose. *J. Biochem. Biophys. Methods* **10,** 203–209.
17. Laurière, M. (1993) A semidry electroblotting system efficiently transfers both high and low molecular weight proteins separated by SDS-PAGE. *Anal. Biochem.* **212,** 206–211.
18. Zeng, C., Suzuki, Y., and Alpert, E. (1990) Polyethylene glycol significantly enhances the transfer of membrane immunoblotting. *Anal. Biochem.* **189,** 197–201.
19. Vachereau, A. (1989) Transparency of nitrocellulose membranes with Triton X-114. *Electrophoresis* **10,** 524–527.
20. Alimi, E., Martinage, A., Sautière, P., and Chevaillier, P. (1993) Electroblotting proteins onto carboxymethylcellulose membranes for sequencing. *BioTechniques* **15,** 912–917.
21. Hauri, H. P. and Bucher, K. (1986) Immunoblotting with monoclonal antibodies: importance of the blocking solution. *Anal. Biochem.* **159,** 386–389.
22. Rohringer, R. and Holden, D. W. (1985) Protein blotting: detection of proteins with colloidal gold, and of glycoproteins and lectins with biotin-conjugated and enzyme probes. *Anal. Biochem.* **144,** 118–127.
23. Saravis, C. A. (1984) Improved blocking of nonspecific antibody binding sites on nitrocellulose membranes. *Electrophoresis* **5,** 54,55.
24. Batteiger, B., Newhall, W. J., and Jones, R. B. (1982) The use of Tween-20 as a blocking agent in the immunological detection of proteins transferred to nitrocellulose membranes. *J. Immunol. Methods* **55,** 297–307.
25. Haycock, J. W. (1993) Polyvinylpyrrolidone as a blocking agent in immunochemical studies. *Anal. Biochem.* **208,** 397–399.
26. Hoffman, W. L. and Jump, A.A. (1986) Tween-20 removes antibodies and other proteins from nitrocellulose. *J. Immunol. Methods* **94,** 191–196.
27. Gravel, P., Sanchez, J. C., Walzer, C., Golaz, O., Hochstrasser, D. F., Balant, L. P., Hughes, G. J., Garcia-Sevilla, J., and Guimon, J. (1995) Human blood platelet protein map established by two-dimensional polyacrylamide gel electrophoresis. *Electrophoresis* **16,** 1152–1159.

39

Protein Blotting by the Capillary Method

John M. Walker

1. Introduction

The ability to transfer (blot) separated proteins from a polyacrylamide gel matrix onto a sheet of nitrocellullose paper (where the proteins bind to the surface of the paper) has provided a powerful tool for protein analysis. Once immobilized on the surface of the nitrocellulose sheet, a variety of analytical procedures may be carried out on the proteins that otherwise would have proven difficult or impossible in the gel. Such procedures may include hybridization with labeled DNA or RNA probes, detection with antibodies (probably the most commonly used procedure), detection by specific staining procedures, autoradiographic assay, and so forth. The most commonly used, and indeed most efficient, methods for transferring proteins from gels to nitrocellulose (blotting) are by electrophoresis, and these methods are described in Chapters 37 and 38. An alternative method, capillary blotting, is described here. Although this method takes longer than electroblotting methods (it takes overnight) and transfer of proteins from the gel is not as complete as it is for electroblotting (although sufficient protein is transferred for most purposes) the method does have its uses. It is of course ideal for those who only wish to carry out occasional blotting experiments and therefore do not wish to commit themselves to the purchase or the purpose-built apparatus (plus power pack) needed for electroblotting. Second, this method is particularly useful for blotting isoelectric focusing gels where the proteins are at their isoelectric point (i.e., they have zero overall change) and are not easy therefore to transfer by electrophoresis. The method simply involves placing the gel on filter paper soaked in buffer. Buffer is drawn through the gel by capillary action by placing a pad of **dry** absorbent material on top of the gel. As the buffer passes through the gel, it carries the protein bands with it, and these bind to the nitrocellulose sheet that is placed between the top of the gel and the dry absorbent material.

2. Materials

1. Blotting buffer: 20 m*M* Tris, 150 m*M* glycine, 20% methanol, pH 8.3. Dissolve 4.83 g of Tris base, 20.5 g of glycine, and 400 mL of methanol in 2 L of distilled water. The solution is stable for weeks at 4°C.
2. Nitrocellulose paper.
3. Whatman 3MM filter paper

From: The Protein Protocols Handbook
Edited by: J. M. Walker Humana Press Inc., Totowa, NJ

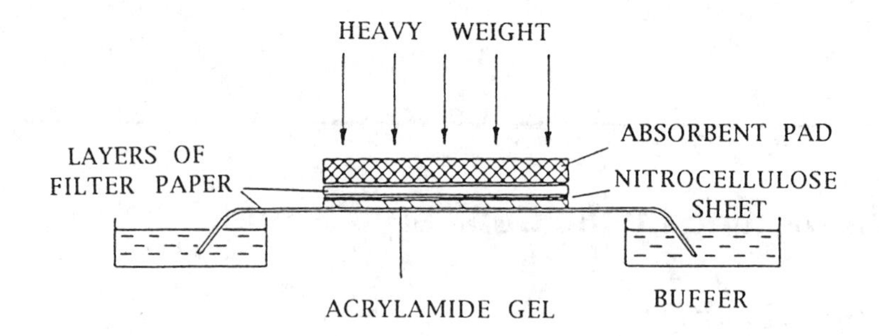

Fig. 1. A typical arrangement for capillary blotting.

4. Absorbent material: e.g., a wad of filter paper or paper towels; this author uses baby diapers.

3. Method

1. Place two sheets of 3MM filter paper on a glass plate, and thoroughly soak them in blotting buffer. The ends should be dipped in reservoirs containing blotting buffer to ensure this filter paper pad remains wet overnight (*see* Fig. 1).
2. Place the gel to be blotted on top of this filter paper bed. Make sure the bed is fully wetted, and that no bubbles are trapped between the gel and filter paper. Thoroughly wet the top of the gel with blotting buffer.
3. Cut a piece of nitrocellulose paper to the size of the gel to be blotted. Wet the paper by dipping it in blotting buffer. **(Care: Use gloves; there are more proteins on your fingers than you are blotting from the gel!)** Then lay the nitrocellulose sheet on the gel surface. Take great care to ensure no air bubbles are trapped (buffer cannot pass through an air bubble).
4. Now dry material must be placed on top of the nitrocellulose. Start with three sheets of 3MM filter paper, cut to the same size as the gel, and then place on top of this a pad of your absorbent material (*see* step 4, Section 2.).
5. Finally, place a heavy weight on top of the absorbent material, e.g., a sheet of thick glass that supports a 2–3 L flask filled with water (*see* Note 1).
6. Allow blotting to take place overnight (preferably for 24 h). The setup may then be dismantled. the nitrocellulose sheet stained for protein to confirm that transfer has been achieved (e.g., with 0.2% Ponceau S in 10% acetic acid), and then the nitrocellulose sheet blocked and probed using any of the methods described in Chapters 44, 45, 48–50, and 97.

4. Notes

1. The only error you can make with this method is to have any of your dry material (e.g., filter paper or absorbance pad) overhanging the gel and making contact with the wet base of filter paper. In this case, buffer preferentially travels around the gel into the absorbant pad, rather than through the gel. When the heavy weight is applied to the setup check that this is not happening (the absorbent pad often "sags" quite easily). If there is overlap, simply trim with scissors. Do not expect the absorbent pad to be particularly wet after an overnight blot. It should be barely damp after an overnight run. If it is soaking wet, then this indicates that buffer has traveled around the gel. However, even if this is the case, it is probably worth proceeding since some protein will have transferred nevertheless and this can probably still be detected with your probe.

40

Protein Blotting of Basic Proteins Resolved on Acid-Urea-Triton-Polyacrylamide Gels

Geneviève P. Delcuve and James R. Davie

1. Introduction

The electrophoretic resolution of histones on acetic acid-urea-Triton (AUT) polyacrylamide gels is the method of choice to separate basic proteins, such as histone variants, modified histone species, and high-mobility group proteins 14 and 17 *(1–6* and *see* Chapters 14 and 15). Basic proteins are resolved in this system on the basis of their size, charge, and hydrophobicity. In previous studies, we analyzed the abundance of ubiquitinated histones by resolving the histones on two-dimensional (AUT into SDS) polyacrylamide gels, followed by their transfer to nitrocellulose membranes, and immunochemical staining of nitrocellulose membranes with an antiubiquitin antibody *(7–9)*. However, transfer of the basic proteins directly from the AUT polyacrylamide gel circumvents the need to run the second-dimension SDS gel and accomplishes the analysis of several histone samples. We have described a method that efficiently transfers basic proteins from AUT polyacrylamide gels to nitrocellulose membranes *(10)*. This method has been used in the immunochemical detection of modified histone, isoforms, and histone H1 subtypes *(6,11–13)*.

To achieve satisfactory transfer of basic proteins from AUT gels to nitrocellulose, polyacrylamide gels are submerged in 50 m*M* acetic acid, and 0.5% SDS (equilibration buffer 1; 2 × 30 min) to displace the Triton X-100, followed by a 30-min incubation in a Tris-SDS buffer (equilibration buffer 2). The transfer buffer is an alkaline transfer buffer (25 m*M* CAPS, pH 10.0, 20% [v/v] methanol). Szewczyk and Kozloff *(14)* reported that alkaline transfer buffers increase the efficiency of transfer of strongly basic proteins from SDS gels to nitrocellulose membranes. We reasoned that this transfer buffer would improve the transfer of histones from AUT gels that had been treated with SDS.

2. Materials

Buffers are made from analytical-grade reagents dissolved in double-distilled water.

1. Equilibration buffer 1: 0.575 mL of glacial acetic acid (50 m*M*), 10 mL of 10% (10 g in 100 mL of water) SDS (0.5%), and water to 200 mL.
2. Equilibration buffer 2: 6.25 mL of 1*M* Tris-HCl (62.5 m*M*), pH 6.8, 23 mL of 10% (w/v) SDS (2.3%), 5 mL of β-mercaptoethanol (5%), and water to 100 mL. 1*M* Tris-HCl: 121 g

From: The Protein Protocols Handbook
Edited by: J. M. Walker Humana Press Inc., Totowa, NJ

of Tris base in 800 mL of water and adjusted to pH 6.8 with hydrochloric acid. The buffer is made up to 1000 mL.
3. Transfer membrane: Nitrocellulose membranes (0.45-μm pore size, Schleicher & Schuell [Keene, NH], BA85) are used.
4. Transfer buffer: 187 mL of Caps (3-[cyclohexylamino]-1-propanesulfonic acid) stock (16X) solution (final concentration, 25 m*M*), 600 mL of methanol (20%), and water to 3 L. Caps stock solution (400 m*M*): 88.5 g of Caps dissolved in 800 mL of water and adjusted to pH 10 with 10*M* NaOH. The buffer is made up to 1000 mL.
5. Electroblotting equipment: Proteins are transferred with the Bio-Rad (Hercules, CA) Trans-Blot transfer cell containing a super cooling coil (Bio-Rad). Cooling is achieved with a Lauda circulating bath.
6. Protein stain: Proteins in the AUT or SDS polyacrylamide gel are stained with Coomassie blue (Serva Blue G). Proteins on the membrane are stained with Indian ink (Osmiroid International Ltd.) (0.01% v/v in TBS-TW). TBS-TW: 20 m*M* Tris-HCl, pH 7.5, 500 m*M* NaCl, and 0.03% (v/v) Tween-20.

3. Methods

3.1. Protein Blotting

1. Nitrocellulose membranes are cut to size of gel at least a day before transfer and stored in water at 4°C.
2. The proteins are electrophoretically resolved on AUT- or SDS-polyacrylamide minislab gels (6 mm long, 10 mm wide, 0.8–1.0 mm thick).
3. Following electrophoresis, the AUT polyacrylamide gel is gently shaken in 100 mL of equilibration buffer 1 for 30 min at room temperature. This solution is poured off, and another 100 mL of equilibration buffer 1 are added. The gel is again shaken for 30 min. The solution is discarded and replaced with equilibration buffer 2. The gel is agitated in this solution for 30 min.
4. Onto the gel holder, place the porous pad that is equilibrated with transfer buffer. Three sheets of 3MM paper soaked in transfer buffer are placed on top of the porous pad. The treated AUT or SDS slab gel is placed onto the 3MM paper sheets. Nitrocellulose membrane is placed carefully on top of the gel, avoiding the trapping of air between the gel and nitrocellulose membrane. One sheet of 3MM paper soaked in transfer buffer is placed on top of the nitrocellulose membrane, followed by the placement of a porous pad that has been equilibrated with transfer buffer.
5. The gel holder is put into the transblot tank with the polyacrylamide gel facing the cathode and the nitrocellulose membrane facing the anode. The tank is filled with transfer buffer. Protein transfer is carried out at 70 V for 2 h and/or at 30 V overnight with cooling at 4°C.
6. Following transfer, the nitrocellulose membrane is placed onto a sheet of 3MM paper and allowed to dry for 30 min at room temperature. The gel is stained with Coomassie blue. The air-dried nitrocellulose membrane is placed between two sheets of 3MM paper. This is wrapped with aluminum foil and baked at 65°C for 30 min. The proteins transferred onto the nitrocellulose membrane may be visualized by staining the nitrocellulose membrane with India ink (*see* Section 3.2.2.).

3.2. Protein Staining

1. The nitrocellulose membrane is agitated in TBS-TW (0.7 mL per cm^2) for 10 min at room temperature in a sealed plastic box. The solution is discarded, and fresh TBS-TW is added. These steps are repeated twice more for a total of four washes of the membrane in the TBS-TW solution.

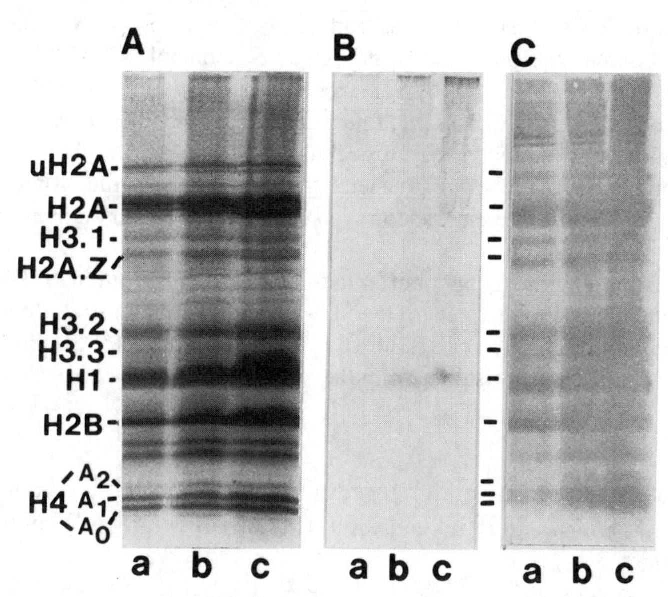

Fig. 1. Electrophoretic transfer of histones from AUT minislab gels. **(A)** Histones (9, 18, and 36 μg in lanes a, b, and c, respectively) isolated from T-47D-5 human breast cancer cells were electrophoretically resolved on AUT minislab gels. The gel was stained with Coomassie blue. **(B)** The Coomassie blue-stained AUT minislab gel pattern of histones remaining after transfer (30 V overnight) to nitrocellulose is shown. **(C)** The India ink-stained nitrocellulose pattern of histones transferred from the AUT minislab gel in B is shown. A_0, A_1, and A_2 correspond to the un-, mono-, and diacetylated species of histone H4, respectively. The ubiquitin adduct of histone H2A is denoted as uH2A. Reprinted with permission from ref. *10* (copyright by the Academic Press).

2. The India ink stain is added to the nitrocellulose membrane (0.56 mL/cm^2) which is agitated for 30 min to 2 h at room temperature.
3. The stain is discarded, and the nitrocellulose membrane is shaken in water for 5 min at room temperature.
4. The nitrocellulose membranes are dried and stored.

4. Notes

1. Nitrocellulose membranes have been used in the majority of our studies. However, these membranes are fragile and must be handled with care. An alternate membrane, which is stronger than nitrocellulose, is PVDF (Bio-Rad). The PVDF membranes are wetted with 100% methanol for 3 min and then equilibrated with transfer buffer for 3 min.
2. The efficiency of transfer of basic proteins (histones) from AUT polyacrylamide slab gels to nitrocellulose membranes is shown in Fig. 1. Most of the histones were efficiently transferred. The efficiency of elution was poorest for histone H1. Figure 1C demonstrates that the histone variants of histone H2A (H2A, H2A.Z) and of histone H3 (H3.1, H3.2, H3.3) and the modified histone species (e.g., ubiquitinated histone H2A, acetylated histone H4) were transferred. Densitometric tracings of the gel patterns before and after transfer demonstrated that >90% of the histones H2A, H2B, H3, and H4, and approx 80% of histone H1 were eluted from the AUT gel.

3. The transfer efficiency of basic proteins from AUT minislab polyacrylamide gels to nitrocellulose membranes was poor when a Tris-glycine-methanol (25 m*M* Tris, 192 m*M* glycine, 20% [v/v] methanol, and 0.1% SDS) transfer buffer was used.
4. With the Bio-Rad Trans-Blot cassette, four minislab gels can be easily accommodated.
5. Following transfer and baking, the nitrocellulose membrane may be stored for several weeks at room temperature before proceeding with immunochemical staining.
6. Leaving the nitrocellulose membrane in water for too long after staining with India ink will result in removal of the stain.
7. We have used this alkaline transfer buffer to transfer histones from SDS slab gels to nitrocellulose membranes. Pretreatment of the SDS slab gel is not required. However, we have found that washing the SDS slab gel in equilibration buffer 2 for 30 min improved the efficiency of elution of the histones from the SDS gel.

Acknowledgments

This work was supported by grants from the Medical Research Council of Canada (MT-9186, MT-12147, MA-12283, PG-12809) and the University of Manitoba Research Development Fund.

References

1. Urban, M. K., Franklin, S. G., and Zweidler, A. (1979) Isolation and characterization of the histone variants in chicken erythrocytes. *Biochemistry* **18,** 3952–3959.
2. Strickland, M., Strickland, W. N., and Von Holt, C. (1981) The occurrence of sperm isohistones H2B in single sea urchins. FEBS *Lett.* **135,** 86–88.
3. Meistrich, M. L., Bucci, L. R., Trostle-Weige, P. K., and Brock, W. A. (1985) Histone variants in rat spermatogonia and primary spermatocytes. *Dev. Biol.* **112,** 230–240.
4. Waterborg, J. H. (1990) Sequence analysis of acetylation and methylation in two histone H3 variants of alfalfa. *J. Biol. Chem.* **265,** 17,157–17,161.
5. Davie, J. R. and Delcuve, G. P. (1991) Characterization and chromatin distribution of the H1 histones and high-mobility-group non-histone chromosomal proteins of trout liver and hepatocellular carcinoma. *Biochem. J.* **280,** 491–497.
6. Li, W., Nagaraja, S., Delcuve, G. P., Hendzel, M. J., and Davie, J. R. (1993) Effects of histone acetylation, ubiquitination and variants on nucleosome stability. *Biochem. J.* **296,** 737–744.
7. Nickel, B. E., Allis, C. D., and Davie, J. R. (1989) Ubiquitinated histone H2B is preferentially located in transcriptionally active chromatin. *Biochemistry* **28,** 958–963.
8. Davie, J. R. and Murphy, L. C. (1990) Level of ubiquitinated histone H2B in chromatin is coupled to ongoing transcription. *Biochemistry* **29,** 4752–4757.
9. Davie, J. R., Lin, R., and Allis, C. D. (1991) Timing of the appearance of ubiquitinated histones in developing new macronuclei of *Tetrahymena thermophila. Biochem. Cell Biol.* **69,** 66–71.
10. Delcuve, G. P. and Davie, J. R. (1992) Western blotting and immunochemical detection of histones electrophoretically resolved on acid-urera-triton- and sodium dodecyl sulfate-polyacrylamide gels. *Anal. Biochem.* **200,** 339–341.
11. Lee, D. Y., Hayes, J. J., Pruss, D., and Wolffe, A. P. (1993) A positive role for histone acetylation in transcription factor access to nucleosomal DNA. *Cell* **72,** 73–84.
12. Davie, J. R. and Murphy, L. C. (1994) Inhibition of transcription selectively reduces the level of ubiquitinated histone H2B in chromatin. *Biochem. Biophys. Res. Commun.* **203,** 344–350.

13. Nagaraja, S., Delcuve, G. P., and Davie, J. R. (1995) Differential compaction of transcriptionally competent and repressed chromatin reconstituted with histone H1 subtypes. *Biochim. Biophys. Acta* **1260,** 207–214.
14. Szewczyk, B. and Kozloff, L. M. (1985) A method for the efficient blotting of strongly basic proteins from sodium dodecyl sulfate-polyacrylamide gels to nitrocellulose. *Anal. Biochem.* **150,** 403–407.

41

Alkaline Phosphatase Labeling of Antibody Using Glutaraldehyde

G. Brian Wisdom

1. Introduction

The chemistry of the homobifunctional reagent. glutaraldehyde, is complex. It reacts with the amino and, to a lesser extent, the thiol groups of proteins, and when two proteins are mixed in its presence, stable conjugates are produced without the formation of Schiff bases. Self-coupling can be a problem, unless the proteins are at appropriate concentrations. In the method *(1)* described, there is usually little self-coupling of the enzyme or the IgG antibody. However, the size of the conjugate is large (>10^6 Daltons), since several molecules of each component are linked. This is the simplest labeling procedure to carry out and, although the yields of enzyme activity and immunoreactivity are small, the conjugates obtained are stable and practical reagents.

Glutaraldehyde has also been used to label antibodies with other enzymes, such as horseradish peroxidase *(2),* but it is necessary to modify the method when the enzyme's chemistry is significantly different.

2. Materals

1. Alkaline phosphatase from bovine intestinal mucosa, 2000 U/mg or greater (with 4-nitrophenyl phosphate as substrate); this is usually supplied at a concentration of 10 mg/mL. If the enzyme is in the presence of ammonium sulfate, Tris, or any other amine, these substances must be removed (*see* Note 1).
2. IgG antibody: This should be the pure IgG fraction or, better, affinity-purified antibody from an antiserum or pure MAb.
3. PBS: 20 m*M* sodium phosphate buffer, pH 7.2, containing 0.15*M* NaCl.
4. Glutaraldehyde.
5. 50 m*M* Tris-HCl buffer, pH 7.5, containing 1 m*M* $MgCl_2$, 0.02% NaN_3, and 2% bovine serum albumin (BSA).

3. Methods

1. Add 0.5 mg of IgG antibody in 100 µL of PBS to 1.5 mg of alkaline phosphatase.
2. Add 5% glutaraldehyde (about 10 µL) to give a final concentration of 0.2% (v/v), and stir the mixture for 2 h at room temperature.
3. Dilute the mixture to 1 mL with PBS and dialyze against PBS (2 L) at 4°C overnight.

From: The Protein Protocols Handbook
Edited by: J. M. Walker Humana Press Inc., Totowa, NJ

4. Dilute the solution to 10 mL with 50 m*M* Tris-HCl buffer, pH 7.5, containing 1 m*M* $MgCl_2$, 0.02% NaN_3, and 2% BSA, and store at 4°C (*see* Notes 2–4).

4. Notes

1. Dialysis of small volumes can be conveniently done in narrow dialysis tubing by placing a short glass tube, sealed at both ends, in the tubing so that the space available to the sample is reduced. Transfer losses are minimized by carrying out the subsequent steps in the same dialysis bag. There are also various microdialysis systems available commercially.
2. The conjugates are stable for several years at 4°C, since the NaN_3 inhibits microbial growth and the BSA minimizes denaturation and adsorption losses. These conjugates should not be frozen.
3. Purification of the conjugates is usually unnecessary; however, if there is evidence of the presence of free antibody, it can be removed by gel filtration in a medium, such as Sepharose CL-6B (Pharmacia Biotech, Uppsala, Sweden) or Bio-Gel A-0.5 m (Bio-Rad, Hercules, CA) with PBS as solvent.
4. The efficacy of the enzyme-labeled antibody may be tested by immobilizing the appropriate antigen on the wells of a microtiter plate or strip, incubating various dilutions of the conjugate for a few hours, washing the wells, adding substrate, and measuring the amount of product formed. This approach may also be used for monitoring conjugate purification in chromatography fractions.

References

1. Engvall, E. and Perlmann, P. (1972) Enzyme-linked immunosorbent assay, ELISA. III. Quantitation of specific antibodies by enzyme-labelled anti-immunoglobulin in antigen-coated tubes. *J. Immunol.* **109,** 129–135.
2. Avrameas, S. and Ternynck, T. (1971) Peroxidase labelled antibody and Fab conjugates with enhanced intracellular penetration. *Immunochemistry* **8,** 1175–1179.

42

β-Galactosidase Labeling of Antibody Using MBS

G. Brian Wisdom

1. Introduction

The heterobifunctional reagent, *m*-maleimidobenzoyl-*N*-hydroxysuccinimide ester (MBS), is of value when one of the components of a conjugate has no free thiol groups, e.g., IgG. In this method *(1)*, the IgG antibody is first modified by allowing the *N*-hydroxysuccinimide ester group of the MBS to react with amino groups in the IgG; the β-galactosidase is then added and the maleimide groups on the modified IgG react with thiol groups in the β-galactosidase to form thioether links. This procedure produces conjugates with molecular weights in the range $0.6–1 \times 10^6$.

The method may also be used for labeling with other enzymes containing nonessential thiol groups. It can also be extended to enzymes lacking free thiols; these groups may be introduced into proteins by various reagents, for example, *N*-succinimidyl 3-(2-pyridyldithio)propionate *(2)*. In addition many other other crosslinking reagents similar to MBS have become available in recent years *(3)*.

2. Materials

1. β-Galactosidase from *Escherichia coli,* 600 U/mg or greater (with 2-nitrophenyl-β-galactopyranoside as substrate) (*see* Note 1).
2. IgG antibody: This should be the pure IgG fraction or, better, affinity–purified antibody from an antiserum or pure MAb.
3. 0.1*M* Sodium phosphate buffer, pH 7.0, containing 50 m*M* NaCl.
4. MBS.
5. Dioxan.
6. Sephadex G25 (Pharmacia Biotech, Uppsala, Sweden) or an equivalent gel-filtration medium.
7. 10 m*M* Sodium phosphate buffer, pH 7.0, containing 50 m*M* NaCl and 10 m*M* $MgCl_2$.
8. 2-Mercaptoethanol.
9. DEAE-Sepharose (Pharmacia Biotech) or an equivalent ion-exchange medium.
10. 10 m*M* Tris-HCl buffer, pH 7.0, containing 10 m*M* $MgCl_2$ and 10 m*M* 2-mercaptoethanol.
11. Item 10 containing 0.5*M* NaCl.
12. Item 10 containing 3% bovine serum albumin (BSA) and 0.6% NaN_3.

3. Methods

1. Dissolve 1.5 mg of IgG in 1.5 mL of 0.1*M* sodium phosphate buffer, pH 7.0, containing 50 m*M* NaCl.

From: The Protein Protocols Handbook
Edited by: J. M. Walker Humana Press Inc., Totowa, NJ

2. Add 0.32 mg of MBS in 15 µL of dioxan, mix, and incubate for 1 h at 30°C.
3. Fractionate the mixture on a column of Sephadex G25 (0.9 × 30 cm) equilibrated with 10 m*M* sodium phosphate buffer, pH 7.0, containing 50 m*M* NaCl and 10 m*M* $MgCl_2$, and elute with the same buffer. Collect 0.5-mL fractions, measure the A_{280}, and pool the fractions in the first peak (about 3 mL in volume).
4. Add 1.5 mg of enzyme, mix, and incubate for 1 h at 30°C.
5. Stop the reaction by adding 1*M* 2-mercaptoethanol to give a final concentration of 10 m*M* (about 30 µL).
6. Fractionate the mixture on a column of DEAE-Sepharose (0.9 × 15 cm) equilibrated with 10 m*M* Tris-HCl buffer, pH 7.0, containing 10 m*M* $MgCl_2$ and 10 m*M* 2-mercaptoethanol. Wash with this buffer (50 mL) followed by the buffer containing 0.5*M* NaCl (50 mL). Collect 3-mL fractions in tubes with 0.1 mL of Tris-HCl buffer, pH 7.0, containing 3% BSA and 0.6% NaN_3. Pool the major peak (this is eluted with 0.5*M* NaCl), and store at 4°C (*see* Notes 2 and 3).

4. Notes

1. The thiol groups of β-galactosidase may become oxidized during storage, thus diminishing the efficacy of the labeling. It is relatively easy to measure these groups using 5.5'-dithiobis(2-nitrobenzoic acid) *(4,5*; and *see* Chapter 82*)*; about 10 thiol groups/enzyme molecule allow the preparation of satisfactory conjugates.
2. The conjugates are stable for a year at 4°C, since the NaN_3 inhibits microbial growth and the BSA minimizes denaturation and adsorption losses.
3. The activity of the conjugate can be checked by the method described in Note 4 of Chapter 41.

References

1. O'Sullivan, M. J., Gnemmi, E., Morris, D., Chieregatti, G., Simmonds, A. D., Simmons, M., Bridges, J. W., and Marks, V. (1979) Comparison of two methods of preparing enzyme-antibody conjugates: application of these conjugates to enzyme immunoassay. *Anal. Biochem.* **100,** 100–108.
2. Carlson, J., Drevin, H., and Axen, R. (1 978) Protein thiolation and reversible protein–protein conjugation, *N*-succinimidyl 3-(2-pyridyldithio)propionate—a new heterobifunctional reagent. *Biochem. J.* **173,** 723–737.
3. *Pierce Catalog and Handbook* (1994) Pierce, Rockford, IL.
4. Ellman, G. L. (1959) Tissue sulfhydryl groups. *Arch. Biochem. Biophys.* **82,** 7–75.
5. Kastenschmidt, L. L., Kastenschmidt, J., and Helmrich, E. (1968) Subunit interactions and their relationship to the allosteric properties of rabbit skeletal muscle phosphorylase b. *Biochemistry* **7,** 3590.

43

Horseradish Peroxidase Labeling of Antibody Using Periodate Oxidation

G. Brian Wisdom

1. Introduction

The most commonly used method *(1)* for labeling IgG antibody molecules with horseradish peroxidase exploits the glycoprotein nature of the enzyme. The saccharide residues are oxidized with sodium periodate to produce aldehyde groups that can react with the amino groups of the IgG molecule, and the Schiff bases formed are then reduced to give a stable conjugate of high molecular weight ($0.5– 1 \times 10^6$). The peroxidase has few free amino groups, so self-coupling is not a significant problem.

This method has been used for other glycoprotein enzymes, such as glucose oxidase *(2)*.

2. Materials

1. Horseradish peroxidase, 1000 U/mg or greater (with 2,2-azinobis[3-ethylbenzthiazoline-6-sulfonic acid] as substrate) (*see* Note 1).
2. IgG antibody: This should be the pure IgG fraction or, better, affinity-purified antibody from an antiserum or pure MAb.
3. 0.1*M* Sodium periodate.
4. 1 m*M* Sodium acetate buffer, pH 4.4.
5. 10 m*M* Sodium carbonate buffer, pH 9.5.
6. 0.2*M* Sodium carbonate buffer, pH 9.5.
7. Sodium borohydride, 4 mg/mL (freshly prepared).
8. Sepharose CL6B (Pharmacia Biotech, Uppsala, Sweden) or a similar gel-filtration medium.
9. PBS: 20 m*M* sodium phosphate buffer, pH 7.2, containing 0.15*M* NaCl.
10. Bovine serum albumin (BSA).

3. Method

1. Dissolve 2 mg of peroxidase in 500 µL of water.
2. Add 100 µL of freshly prepared 0.1*M* sodium periodate, and stir the solution for 20 min at room temperature. (The color changes from orange to green.)
3. Dialyze the modified enzyme against 1 m*M* sodium acetate buffer, pH 4.4 (2 L), overnight at 4°C.
4. Dissolve 4 mg of IgG in 500 µL of 10 m*M* sodium carbonate buffer, pH 9.5.

From: The Protein Protocols Handbook
Edited by: J. M. Walker Humana Press Inc., Totowa, NJ

5. Adjust the pH of the dialyzed enzyme solution to 9.0–9.5 by adding 10 µL of 0.2*M* sodium carbonate buffer, pH 9.5, and immediately add the IgG solution. Stir the mixture for 2 h at room temperature.
6. Add 50 µL of freshly prepared sodium borohydride solution (4 mg/mL), and stir the mixture occasionally over a period of 2 h at 4°C.
7. Fractionate the mixture by gel filtration on a column of Ultrogel AcA34 (1.5 × 85 cm) in PBS. Determine the A_{280} and A_{403} (*see* Note 2).
8. Pool the fractions in the first peak (both A_{280} and A_{403} peaks coincide), add BSA to give a final concentration of 5 mg/mL, and store the conjugate in aliquots at –20°C (*see* Notes 3 and 4).

4. Notes

1. Preparations of horseradish peroxidase may vary in their carbohydrate content, and this can affect the oxidation reaction. Free carbohydrate can be removed by gel filtration. Increasing the sodium periodate concentration to 0.2*M* can also help, but further increases lead to damage to the peroxidase.
2. The absorbance at 403 nm is caused by the peroxidase's heme group. The enzyme is often specified in terms of its RZ value; this is the ratio of A_{403}–A_{280}, and it provides a measure of the heme content and purity of the preparation. Highly purified peroxidase has an RZ of about three. Conjugates with an RZ of 0.4 perform satisfactorily.
3. BSA improves the stability of the conjugate and minimizes losses caused by adsorption. NaN_3 should not be used with peroxidase conjugates, because it inhibits the enzyme. If an antimicrobial agent is required, 0.2% merthiolate should be used.
4. The activity of the conjugate can be checked by the method described in Note 4 of Chapter 41.

References

1. Wilson, M. B. and Nakane, P. P. (1978) Recent developments in the periodate method of conjugating horseradish peroxidase (HRPO) to antibodies, in *Immunofluorescence and Related Staining Techniques* (Knapp, W., Holubar, K., and Wick, G., eds.), Elsevier, North Holland Biomedical, Amsterdam, pp. 215–224.
2. Maseyeff, R., Maiolini, R., Ferrua, B., and Ragimbeau-Gilli, J. (1976) Quantitation of alpha fetoprotein by enzyme immunoassay, in *Protides of the Biological Fluids, Proc. 24th Colloquium* (Peeters, H., ed.), Pergamon, Oxford, pp. 605–612.

44

Protein Staining and Immunodetection Using Colloidal Gold

Susan J. Fowler

1. Introduction

Probes labeled with colloidal gold were originally used as electron dense markers in electron microscopy *(1–3)* and as color markers in light microscopy *(4)*. Their application to immunoblotting was not examined until later *(5–7)*. Gold-labeled antibodies and protein A were demonstrated to be suitable for the visualization of specific antigens on Western blots and dot blots *(5,6)*. When gold-labeled antibodies are used as probes on immunoblots, the antigen-antibody interaction is seen as a pinkish signal owing to the optical characteristics of colloidal gold *(5)*. Used on its own, the sensitivity of immunogold detection is equivalent to indirect peroxidase methods and, hence, only suitable for situations where there are higher levels of antigen. In addition, the signal produced is not permanent. In order to overcome this problem and to allow the technique to be used for more demanding applications, a way of amplifying the signal was subsequently developed using the capacity of gold particles to catalyze the reduction of silver ions *(8)*. This reaction results in the growth of the gold particles by silver deposition. A stable dark brown signal is produced on the blot, and sensitivity is increased 10-fold. The sensitivity achieved using immunogold silver staining (IGSS) is similar to that obtained with alkaline phosphatase using colorimetric detection and several times more sensitive than ^{125}I-labeled antibodies. However, unlike colorimetric detection, the result is stable and not prone to fading, and the chemicals used present no hazards. In addition, the signal-to-noise ratio of IGSS is usually very high, and there are none of the handling or disposal problems that are asssociated with ^{125}I-labeled antibodies.

The binding of the gold to antibodies is via electrostatic adsorbtion. It is influenced by many factors, including particle size, ionic concentration, the amount of protein added, and its molecular weight. Most importantly, it is pH dependent *(9)*. An additional feature is that the binding of the gold does not appear to alter the biological or immunological properties of the protein to which it is attached. The colloidal gold particles used to label antibodies can be produced in different sizes ranging from 1 to 40 nm in diameter. For immunoblotting, Amersham International plc. (Amersham, UK) supplies AuroProbe BLplus antibodies labeled with 10-nm particles and AuroProbe

From: The Protein Protocols Handbook
Edited by: J. M. Walker Humana Press Inc., Totowa, NJ

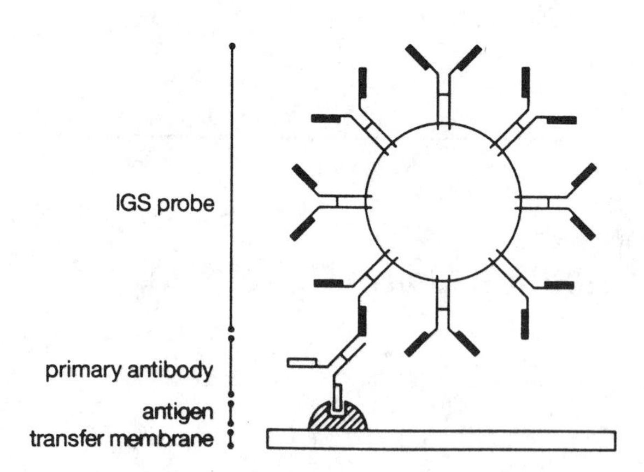

Fig. 1. Principle of the indirect visualization of antigens using immunogold probes. The primary antibodies bind to immobilized antigens on the blot and are in turn recognized by gold-labeled secondary antibodies.The above illustration of the binding pattern of secondary antibodies to gold particles represents the type of conjugate formed with gold particles of 10 nm or larger. The exact configuration adopted by the antibodies on the gold particle is not known.

One antibodies labeled with 1-nm gold particles. In the case of the larger 10-nm gold particles, there will be several antibodies bound to each gold particle, and the probe can be considered to be a gold particle coated with antibodies (*see* Fig. 1). For the small 1-nm particles, each antibody has at least one individual gold particle bound to it. Reducing the size of the gold particles allows increased labeling efficiency and can give greater sensitivity than larger particles *(10)*. This may be owing in part to the larger number of intensifiable gold particles/unit or antigen *(11)*. In addition, when prolonged incubations in AuroProbe One are performed, the nonspecific binding that can sometimes occur with larger gold probes is absent, and potential background problems are avoided. However, when using antibodies labeled with AuroProbe One, the small size of the gold particles means that visualization of the antigen–antibody interaction can only be achieved using silver enhancement.

Silver enhancement was first reported by Danscher *(8),* who used silver lactate and hydroquinone at a pH of 3.5. In the presence of the gold particles that act as catalysts, the silver ions are reduced to metallic silver by the hydroquinone. The silver atoms formed are deposited in layers on the gold surface, resulting in significantly larger particles and a more intense macroscopic signal (*see* Fig. 2). This classical enhancement worked well, but had several disadvantages. The system was sensitive to light and chemical contamination. In addition, the components were not stable and were prone to self-nucleation, a phenomenon whereby the reduction of silver ions occurs spontaneously in solution to form silver particles that can be deposited and lead to high background.

More recently, silver-enhancement reagents have been developed that overcome these problems. IntenSE BL (Amersham International) is light insensitive and has a neutral pH. It also exhibits delayed self-nucleation, which allows a fairly large time margin before it is necessary to stop the reaction (*see* Fig. 3). Using silver enhancement allows sensitivity to be increased 10-fold over immunogold detection.

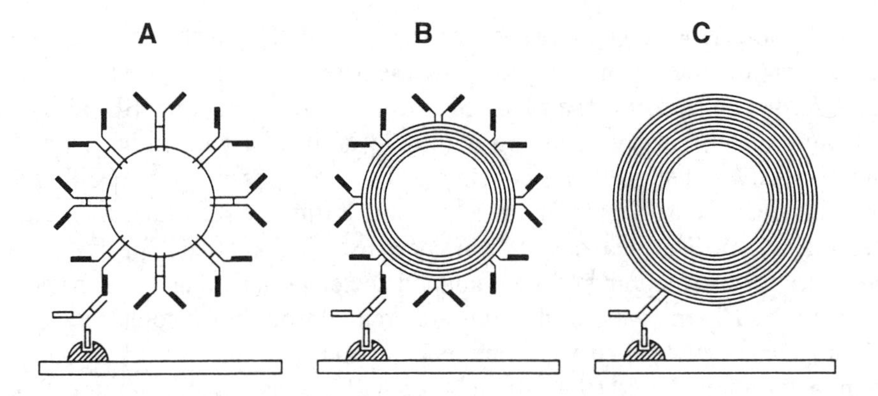

Fig. 2. Schematic representation of the silver-enhancement process. In the initial phase **(A)**, the gold probe attaches to the primary antibody, which is bound to immobilized antigen. During the silver-enhancement process, layers of silver selectively precipitate on the colloidal gold surface **(B)**. The result is a significantly larger particle and a silver surface that generates a more intense macroscopic signal **(C)**.

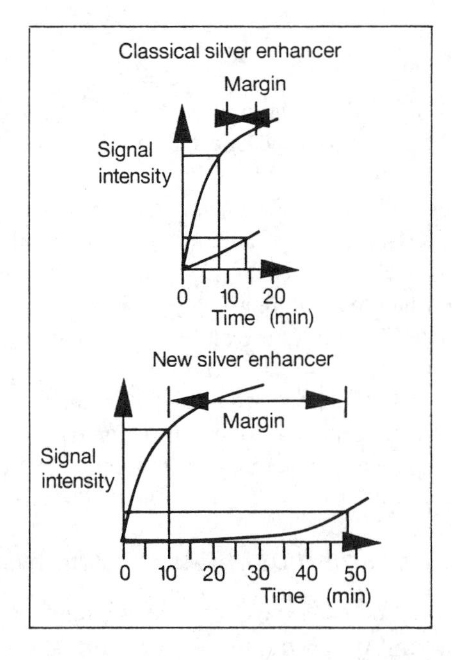

Fig. 3. Silver-enhancement time dependency for both the classical and IntenSE BL silver enhancers.

In addition to the immunological detection of proteins, gold particles can also be used as a general stain for proteins on blots *(12)*. AuroDye forte (Amersham International) is a stabilized colloidal gold solution adjusted to a pH of 3.0. At this pH, the negatively charged gold particles bind very selectively to proteins by hydrophobic and ionic interactions. The proteins thus stained appear as dark red. The sensitivity obtained is comparable to that of silver staining for polyacrylamide gels. Segers and

Rabaey *(13)* found it detected more spots on transfers of 2D gels than silver staining of the gels. For applications demanding very high sensitivity, it is also possible to amplify the signal further by performing a silver-enhancement step using IntenSE BL. An additional feature of total protein staining with gold is that, as with India ink *(14),* the immunoreactivity of the proteins is not altered. Thus, it is possible for specific proteins to be immunodetected with chemiluminescent or colorimetric substrates after total protein staining *(15,16)*. This dual detection has been used with blotted 2D gels *(17),* where it allows very precise mapping of the immunodetected protein against the background of total proteins. Alternatively, if proteins are omitted from block solutions, total protein staining can be performed after immunodetection *(18)*.

In summary, immunogold silver staining is a highly sensitive detection method that provides a permanent record of results. Protocols are simple to use, and provided the silver-enhancement step is carefully monitored, background interference is negligible. Results are obtained without the prolonged autoradiograplhy exposures associated with radioactive methods, and the chemicals used do not present any hazards or disposal problems.

2. Materials

2.1. General

1. Nitrocellulose, nylon, or PVDF membranes: AuroProbe BLplus can be used with any of the three types of membrane; AuroProbe One and AuroDye forte are only compatible with nitrocellulose and PVDF membranes.
2. Phosphate-buffered saline (PBS) pH 7.2, containing 0.02% azide: 8 g NaCl 0.2 g KCl, 1.4 g $Na_2HPO_4 \cdot 2H_2O$, 0.2 g KH_2PO_4, 0.2 g sodium azide. Adjust pH to 7.2, and make up to 1000 mL with distilled water.
3. Gelatin: the gelatin used should be of high-quality if it is to inhibit nonspecific binding of gold probes effectively. IGSS quality is supplied as a component of AuroProbe BLplus and AuroProbe One kits.
4. Analytical-grade chemicals should be used throughout, and water should be distilled and deionized. Where silver enhancement is used, it is important that the glassware and plastic containers used are scrupulously clean and are not contaminated with heavy metals or their salts.

2.2. Immunogold Silver Staining Using AuroProbe BLplus

The buffer system outlined below gives very clean backgrounds without the use of Tween-20. The use of Tween-20 in blocking, incubation, and/or washing can lead to nonspecific binding of gold probes to blotted proteins from certain types of sample, such as whole cultured cells and isolated nuclei extracts.

1. Wash buffer: 0.1% (w/v) bovine serum albumin (BSA) in PBS.
2. Block buffer for nitrocellulose or PVDF membranes: 5% (w/v) BSA in PBS.
3. Block buffer for nylon membranes: 10% (w/v) BSA in PBS.
4. Primary antibody diluent buffer: 1% normal serum v/v (from the same species as that in which the secondary antibody was raised) diluted in wash buffer.
5. Gelatin buffer: 1% (v/v) IGSS-quality gelatin in wash buffer (equivalent to a 1:20 dilution of the gelatin supplied with AuroProbe BLplus).

6. AuroProbe BLplus secondary antibody, 1:100 diluted in gelatin buffer; or biotinylated secondary antibody 2 μg/mL diluted in gelatin buffer, and AuroProbe BLplus streptavidin 1:100 diluted in gelatin buffer.
7. Enhancer solution: ready-to-use component of the IntenSE BL kit.
8. Initiator solution: ready-to-use component of the IntenSE BL kit.

2.3. Immunogold Silver Staining Using AuroProbe One

1. Wash buffer: 0.8% (w/v) BSA, 0.1% (v/v) gelatin in PBS.
2. Block buffer: 4% (w/v) BSA, 0.1% (v/v) gelatin in PBS.
3. Primary and secondary antibody diluent buffer:
   a. For AuroProbe One goat antirabbit and goat antimouse: 1% (v/v) normal goat serum diluted in wash buffer.
   b. For AuroProbe One streptavidin or antibiotin: 1% (v/v) normal serum (from the same species as that in which the biotinylated secondary antibody was raised) diluted in wash buffer.
4. AuroProbe One secondary antibody: 1:200 to 1:400 diluted in antibody diluent buffer, or biotinylated antibody 2 μg/mL diluted in antibody diluent buffer, and AuroProbe One streptavidin or antibiotin 1:200 to 1:400 diluted in antibody diluent buffer.
5. Enhancer solution: ready-to-use component of the IntenSE BL kit.
6. Initiator solution: ready-to-use component of the IntenSE BL kit.

2.4. General Staining of Blotted Proteins Using AuroDye Forte

1. Tween-20: component of AuroDye forte kit (not all brands of Tween-20 give satisfactory results; it is important to use reagent that has been quality controlled for this purpose).
2. Wash and block buffer: 0.3% (v/v) Tween-20 in PBS.

3. Methods

3.1. General

Avoid skin contact with the transfer membrane. Wear gloves throughout the procedure. Handle blots by their edges using clean plastic forceps. Incubations and washes should be carried out under constant agitation. Plastic containers on an orbital shaker are ideal for this purpose. Alternatively, if it is necessary to conserve antibodies, antibody incubations can be performed in cylindrical containers on roller mixers *(19)*.

3.2. Immunogold Silver Staining Using AuroProbe BLplus

1. After transfer of the proteins to this membrane, incubate the blot in block solution for 30 min at 37°C. A shaking water bath is suitable for this incubation. If nylon membranes are used, this period should be extended to overnight. All subsequent steps are performed at room temperature. It is important that there be enough block solution to cover the blot easily.
2. Remove excess block by washing the blot three times for 5 min in wash buffer. As large a volume of wash buffer as possible should be used each time.
3. Prepare a suitable dilution of primary antibody in diluent buffer. If the antibody is purified, 1–2 μg/mL is al suitable concentration. If unpurified antiserum is used, a dilution >1:500 is recommended.
4. Incubate the blot in this solution for 1–2 h.
5. Wash the blot as described in step 2.

6. Prepare a 1:100 dilution of AuroProbe BLplus antibody in gelatin buffer. If the AuroProbe BLplus streptavidin system is being used, the biotinylated second antibody should be diluted to a concentration of 2 μg/mL in gelatin buffer.
7. Incubate the blot in second antibody for 2 h under constant agitation.
8. Wash blot as described in step 2.
9. If using the AuroProbe BLplus streptavidin system, incubate the blot in a 1:100 dilution of streptavidin-gold for 2 h.
10. Wash blot as described in step 2.
11. Wash the gold-stained blot twice for 1 min in distilled water. Do not leave the blot in distilled water for long periods, since this may lead to the release of gold particles from the surface. The result can be reviewed at this stage before going on to perform the silver-enhancement step.
12. Pour equal volumes of enhancer and initiator solutions into a plastic container (100 mL is sufficient for a 10 × 15 cm blot). Immediately add the gold-stained blot, and incubate under constant agitation for 15–40 min. The enhancement procedure can be monitored and interrupted or extended as necessary.
13. Wash the blot three times for 10 min in a large volume of distilled water.
14. Remove blot and leave to air-dry on filter paper (*see* Fig. 4).

3.3. Immunogold Silver Staining with AuroProbe One

1. After transfer of the proteins to the membrane, incubate the blot in block solution for 30 min at 45°C. A shaking water bath is suitable for this incubation. All subsequent incubations are performed at room temperature.
2. Remove excess block by washing the blot three times for 5 min in wash buffer with constant agitation. As large a volume of wash buffer as possible should be used each time.
3. Prepare a suitable dilution of primary antibody in antibody diluent buffer. If the antibody is pure, 1 μg/mL is a suitable concentration. If unpurified antiserum is used, a dilution >1:500 is recommended.
4. Incubate the blot in this solution for 1 h with constant agitation.
5. Wash the blot as described in step 2.
6. Prepare a 1:200 to 1:400 dilution of AuroProbe One secondary antibody in antibody diluent buffer. If using AuroProbe One streptavidin or AuroProbe One antibiotin, dilute the biotinylated secondary antibody 2 μg/mL in antibody diluent buffer.
7. Incubate blot in second antibody for 1 h.
8. Wash blot as described in step 2.
9. If using the AuroProbe One streptavidin or antibiotin, incubate the blot in a 1:200 to 1:400 dilution of streptavidin or antibiotin in antibody diluent buffer for 2 h.
10. Wash blot as described in step 2.
11. Further wash the gold-stained blot twice for 5 min in distilled water. Do not leave the blot in distilled water for prolonged periods, as this may lead to the release of gold particles from the blot. The result can be reviewed at this stage before going on to perform the silver-enhancement step
12. Pour equal volumes of enhancer and initiator solutions into a plastic container (100 mL is sufficient for a 10 × 15 cm blot). Immediately add the gold-stained blot, and incubate under constant agitation for 15–40 min. The enhancement procedure can be monitored and interrupted or extended as necessary.
13. Wash the blot three times for 10 min in a large volume of distilled water.
14. Remove blot, and leave to air-dry on filter paper (*see* Fig. 5).

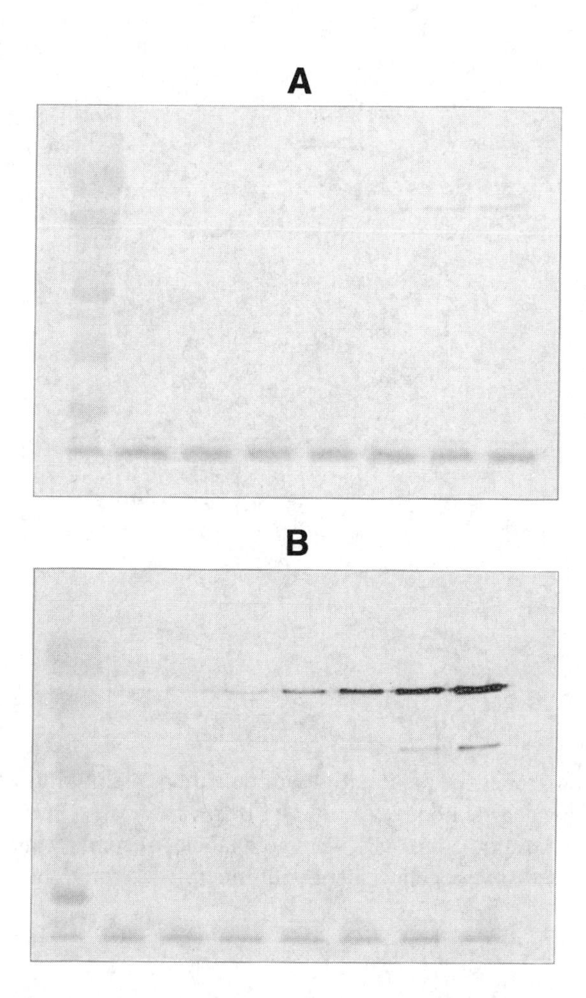

Fig. 4. Detection of bound proteins using colloidal gold-labeled antibodies with **(B)** and without **(A)** silver-enhancement. (A) Doubling dilutions of rat brain homogenate separated by 12% SDS-PAGE and transferred to nitrocellulose membrane, followed by immunodetection with mouse monoclonal anti-B-tubulin (1:1000) and AuroProbe BLplus GAM IgG (1:1000). (B) As for (A), but then subjected to silver-enhancement with IntenSE BL for 20 min.

3.4. General Protein Staining with AuroDye Forte

1. Incubate the blot in an excess of PBS containing 0.3% Tween-20 at 37°C for 30 min. Perform subsequent incubations at room temperature.
2. Further incubate the blot in PBS containing 0.3% Tween-20 three times for 5 min at room temperature.
3. Rinse the blot for 1 min in a large volume of distilled water.
4. Place the blot in AuroDye forte for 2–4 h. The staining can be monitored during this time.
5. When sufficient staining has been obtained, wash the blot in a large volume of distilled water, and leave to air-dry on a piece of filter paper.
6. In cases where extremely high sensitivity or contrast is required, the AuroDye signal can be further amplified with IntenSE BL as described in Section 3.2., steps 12–14 (*see* Fig. 6).

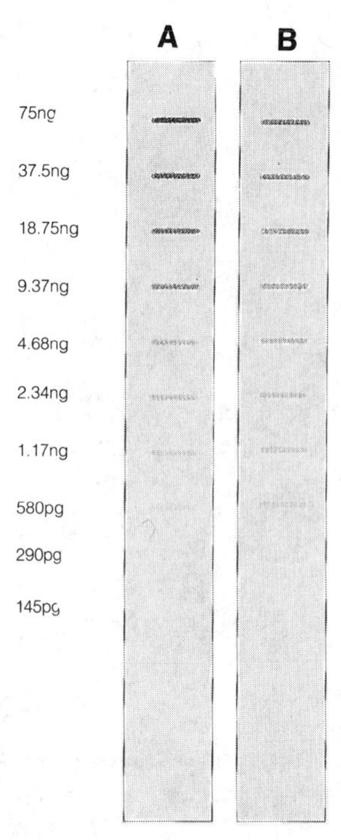

Fig. 5. Effect of gold probe size on sensitivity of detection of slot blots of mouse IgG. **(A)** 10-nm gold-labeled secondary antibody: Auroprobe BLplus GAM IgG (1:100) and silver enhancement with IntenSE BL for 45 min. **(B)** 1-nm gold labeled secondary antibody: AuroProbe One GAM IgG (1:200) and silver enhancement with IntenSE BL for 45 min.

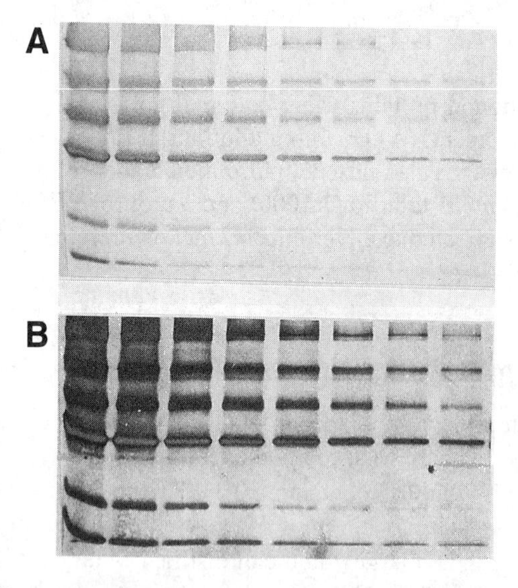

Fig. 6. Comparison of AuroDye forte total protein staining with AuroDye forte staining with silver enhancement. Doubling dilutions of mol-wt-marker proteins: phosphorylase b, BSA, ovalbumin, carbonic anhydrase, trypsin inhibitor, and lysozyme from 200 to 0.375 ng were separated on a 12% gel and transferred to Hybond PVDF membrane (Amersham International, plc.). Both blots were stained with AuroDye after which blot (B) was amplified with IntenSE BL for 15 min.

4. Notes

4.1. Electrophoresis and Electroblotting

1. During electrophoresis, care should be taken to ensure that no extraneous proteins are introduced. Glassware should be thoroughly cleaned, and all solutions should be prepared freshly. Low-ionic-strength transfer buffers are recommended for blotting (i.e., 25 m*M* Tris, 192 m*M* glycine, 20% methanol, pH 8.3).
2. For optimum total protein staining with AuroDye forte, it is important to place a piece of transfer membrane on both sides of the gel. The use of the extra transfer membrane at the cathodic side of the gel, combined with the use of high-quality filter paper, ensures high contrast staining with negligible background. For semidry blotting, the extra piece of transfer membrane is not necessary.

4.2. Immunogold Silver Staining with AuroProbe BLplus and AuroProbe One

3. When using a primary antibody for the first time, it is recommended that its concentration be optimized by performing a dot-blot assay. Antigen dot blots of a suitable concentration are prepared and air-dried. The blocking, washing, and incubation conditions are as outlined in Section 3. A series of primary antibody dilutions are then made, and a dot blot incubated in each. The dilution of the gold-labeled second antibody is kept constant. The primary antibody dilution giving maximum signal with minimum nonspecific binding should be chosen.
4. It is essential to have gelatin in the incubation with the immunogold reagent to prevent nonspecific binding. The source of gelatin is extremely important. If gelatin other than the one supplied in the AuroProbe kits is used, the inclusion of a negative control is essential
5. If desired, the incubations in immunogold reagent can be extended to overnight when concentrations of 1:100 to 1:400 are used. If this is neccessary, AuroProbe One has been shown to give less nonspecific binding on prolonged incubations than AuroProbe BLplus *(10)*.
6. If, after performing the experiment, there is a complete absence of signal, the reactivity of the immunogold reagent with the primary antibody should be checked by performing a dot-blot assay. Prepare a dilution series of primary antibody, e.g., from 250 to 0.5 ng/μL. Spot out 1 μL onto the membrane, and allow to air-dry. Proceed with the appropriate immunogold silver staining using the blocking, washing, and incubation conditions as described in Sections 3.2. and 3.3. (*see* Table 1).

4.3. Silver Enhancement

7. The enhancement reagents are extremely sensitive to the purity of the water used. Low-quality water results in the formation of precipitates that reduce the reactivity of the enhancement reagent and can lead to high backgrounds. In addition, glassware contaminated with heavy metals in elemental form or as heavy metal salts will decrease the performance of the enhancer reagents.
8. The silver enhancement reagents are prepared by mixing the enhancer and initiator solutions in equal quantities. The mixture is only useable over a defined time period (*see* Table 2), so it is important that the components be combined immediately before use. There is no need to shield the silver-enhancer from normal daylight.
9. Both the enhancement time and the stability of the silver-ehancement mixture vary considerably with ambient temperature. A typical enhancement time for most blotting experi-

Table 1
Trouble Shooting Immunogold Silver Staining (IGSS)

Observation	Probable cause	Remedy
Precipitation of silver enhancement mixture before indicated time interval	Glassware: chromic acid was used for rinsing and it was not washed away with HCl	Rinse glassware several times with 0.1*M* HCl and distilled water
	Glassware: traces of metals originating from metal parts, e.g., cleaning brushes, or originating from previous experiments in which metallic compounds were used, e.g., silver staining	Avoid contact with metallic objects; do not use glassware brush with metallic handle to clean glassware; use disposable plastics instead
High background and nonspecific staining	Microprecipitates that are macroscopically invisible produce a high background when the stability time limit is reached	Incubate for a shorter time in silver-enhancement mixture
	Primary antibody is too concentrated	Optimize dilution of primary antibody using dot-blot assay
	Sample is too concentrated	Use more diluted sample
	Wrong type of gelatin or no gelatin used during incubation with AuroProbe	Use the type of gelatin prescribed
No staining	Difficulties regarding the reactivity of the primary antibody with the immunogold reagents	Perform a dot-blot assay (*see* Section 4.3.)
	Inefficient transfer from gel to membrane	Optimize blotting conditions; silver-stain gel to see what remains
	Error in handling: the steps were not performed in the right order or a step was omitted	Repeat the procedure in the right order
	Excessive dilution of primary antibody	Optimize primary antibody dilution using a dot-blot assay
	Nonreactive primary antibody or a primary antibody that was destroyed by inappropriate storage conditions	Use a primary antibody of highest possible antibody quality, and repeat the procedure with a new batch of primary antibody
	The gold probe may have been denatured owing to wrong storage conditions	Repeat procedure with fresh gold probe
Signal too weak	Excessive dilution of primary antibody	Optimize dilution of primary antibody using a dot-blot assay
	Excessive dilution of AuroProbeX reagent or too short an incubation time	Use AuroProbe reagent as recommended in Sections 3.2. and 3.3.
	Silver-enhancement time too short	Rinse the membrane in distilled water, and repeat the silver-enhancement in fresh reagent.

Table 2
Effect of Temperature on Enhancement Time and Enhancement Reagent Stability

Temperature, °C	Typical enhancement time, min	Typical stability time, min
16	27–45	>80
18	22–38	>70
20	20–35	>55
22	18–33	>45
24	16–27	>40

ments of 15–40 min at room temperature (22°C) is recommended. For some applications, it may be necessary to extend the enhancement time, and for others, to shorten it. At room temperature, there is a comfortable safety margin to enable maximum enhancement before there is any danger of self-nucleation of the enhancer reagent occurring.

10. When a very strong amplification signal is desired, it is possible to perform a second silver-enhancement step before self nucleation starts. In this case, the blot is subjected to silver enhancement for 30 min at room temperature (22°C). It is then rinsed with distilled water and immersed in a freshly prepared silver-enhancement solution for another 20–30 min. The increase in signal slows down considerably with time. After enhancement, the blot should be washed in distilled water and dried.

4.4. General Protein Staining with AuroDye Forte

11. Owing to the high sensitivity of AuroDye forte, special care needs to be taken to avoid background staining. It is important to wear gloves when handling gels and blots. Where possible, handle blots by their edges with forceps, since gloves can leave smears. High-quality chemicals should be used throughout.
12. If large amounts of protein are loaded on the gel, when transferred to membrane, they will not only be heavily stained, but will leak off the membrane from saturated sites. Excessive protein leakage will cause an aggregation of the gold particles and destroy the AuroDye forte reagent. The problem is generally more severe with 1D, than 2D gels, where the proteins are more spread out. For 1D gels, it is recommended that protein loads should be equivalent to amounts capable of giving resolvable bands after silver staining. Single bands in 1D gels should not exceed 1000 ng of protein. Molecular-weight standards should be loaded at approx 200 ng/band. In general, the use of lower protein loadings will give better separation, and samples will be conserved.
13. When staining 2D gels, they should be thoroughly washed in several changes of excess transfer buffer after electrophoresis to remove any remaining ampholytes.
14. AuroDye forte is a stabilized gold colloid sol, adjusted to a pH of approx 3. At this low pH, the negatively charged gold particles bind very selectively to proteins by hydrophobic and ionic interactions. This may result in different staining intensities, depending on the isoelectric point of the proteins being stained. However, this has been reported to be a feature of other protein staining methods, such as silver staining and Coomassie blue *(20)*.
15. AuroDye forte is designed for use on nitrocellulose and PVDF membranes. It cannot be used on nylon membranes. where its charge will result in staining of the whole membrane (*see* Table 3).

Table 3
Trouble-Shooting General Protein Staining with AuroDye Forte

Observation	Probable cause	Remedy
AuroDye forte turns purplish during staining	Agglutination of gold particles by proteins released from the blot	Wash blot briefly in excess distilled water; replace AuroDye forte; if possible, use lower protein loads; this phenomenon is less frequently observed with PVDF membrane because it retains proteins better than nitrocellulose
AuroDye forte turns colorless during staining	Adsorption of all the gold particles by excess protein on the blot	Replace AuroDye forte and double the volume used per cm^2 of blot; if possible use lower protein loads
Spotty background	Impurities released from filter paper adsorbed onto blot during transfer	Use high-quality filter paper during electroblotting procedure
High background	Interference by proteinaceous contaminants	Use extra transfer membrane on cathodic side of the gel; use clean Scotch-Brite pads
	Optional silver-enhancement time was too long	Use a shorter silver-enhancement time
	Interference by chemical contaminants	Always use high-quality chemicals
Smears on background	Incorrect handling	Handle blots by their edges using clean plastic forceps; avoid contact with gloves

References

1. Faulk, W. P. and Taylor, G. M. (1971) An immunocolloid method for the electron microscope. *Immunochemistry* **8,** 1081.
2. Romano, E. L., Stolinski, C., and Hughes-Jones, N. C. (1974) An immunoglobulin reagent labelled with colloidal gold for use in electron microscopy. *Immunocytochemistry* **14,** 711–715.
3. Horisberger, M. and Rosset, J. (1977) Colloidal gold, a useful marker for transmission and scanning electron microscopy. *J. Histochem. Cytochem.* **25,** 295–305.
4. Roth, J. (1982) Applications of immunocolloids in light microscopy: preparation of protein A-silver and protein A-gold complexes and their applications for the localization of single and multiple antigens in paraffin sections. *J. Histochem. Cytochem.* **30,** 691–696.
5. Moeremans, M., Daneels, G., Van Dijck, A., Langanger, G., and De Mey, J. (1984) Sensitive visualisation of antigen-antibody reactions in dot and blot immune overlay assays with immunogold and immunogold/silver staining. *J. Immunol. Methods* **74,** 353–360.

6. Brada, D. and Roth, J. (1984) Golden Blot—detection of polyclonal and monocional antibodies bound to antigens on nitrocellulose by protein A–gold complexes. *Anal. Biochem.* **142,** 79–83.
7. Hsu, Y. H. (1984) Immunogold for detection of antigen on nitrocellulose paper. *Anal. Biochem.* **142,** 221–225.
8. Danscher, G. (1981) Histochemical demonstration of heavy metals, a revised version of the sulphide silver method suitable for both light and electron microscopy. *Histochemistry* **71,** 1–16.
9. Geoghegan, W. D. and Ackerman, G. A. (1977) Adsorbtion of horseradish peroxidase, ovomucoid and anti-immunoglobulin to colliodal gold for the indirect detection of concanavilin A, wheat germ agglutinin and goat antihuman immunoglobulin G on cell surfaces at the electron microscopic level: a new method, theory and application. *J. Histochem. Cytochem.* **25,** 1182–1200.
10. Western blotting technical manual: Amersham International plc. 1991, Amersham UK.
11. Moeremans, M., Daneels, G., De Raeymaeker, M., and Leunissen, J. L. M. (1989) AuroProbe One in immunoblotting, in *Aurofile 02,* Janssen Life Sciences, Wantage, UK, pp. 4,5.
12. Moeremans, M., Daneels, G., and De Mey, J. (1985) Sensitive colloid metal (gold or silver) staining of protein blots on nitrocellulose membrane. *Anal. Biochem.* **145,** 315–321.
13. Segers, J. and Rabaey, M. (1985) Sensitive protein stain on nitrocellulose blots. *Protides Biol. Fluids* **33,** 589–591.
14. Glenney, J. (1986) Antibody probing of Western blots which have been stained with India ink. *Anal. Biochem.* **156,** 315–319.
15. Egger, D. and Bienz, K. (1987) Colloidal gold staining and immunoprobing of proteins on the same nitrocellulose blot. *Anal. Biochem.* **166,** 413–417.
16. Egger, D. and Bienz, K. (1992) Colloidal gold staining and immunoprobing on the same Western blot, in *Methods in Molecular Biology, vol. 10, Immunochemical Protocols* (Manson, M., ed.), Humana, Totowa, NJ, pp. 247–253.
17. Schapira, A. H. V. (1992) Colloidal gold staining and immunodetection in 2-D protein mapping, in *Methods in Molecular Biology, vol. 10, Immunochemical Protocols* (Manson, M., ed.), Humana, Totowa, NJ, pp. 255–266.
18. Daneels, I. J., Moeremans, M., De Raemaeker, M., and De Mey, J. (1986) Sequential immunostaining (gold/silver) and complete protein staining (AuroDye) on Western blots. *J. Immunol. Methods* **89,** 89–91.
19. Thomas, N., Jones, C. N., and Thomas, P. L. (1988) Low volume processing of protein blots in rolling drums. *Anal. Biochem.* **170,** 393–396.
20. Jones, A. and Moeremans, M. (1988) Colloidal gold for the detection of proteins on blots and immunoblots, in *Methods in Molecular Biology, vol. 3, New Protein Techniques* (Walker, J. M., ed.), Humana, Totowa, NJ, pp. 441–479.

45

Fluorescent Protein Staining on Nitrocellulose with Subsequent Immunodetection of Antigen

Donald F. Summers and Boguslaw Szewczyk

1. Introduction

The transfer of proteins from gels to nitrocellulose or other immobilizing matrices has become increasingly popular as a powerful tool for the subsequent analysis of proteins. Most frequently, the blotted proteins are analyzed for their antigenic properties by Western blotting.

A frequently encountered problem when performing immunoblotting is to make a direct correlation between the total electrophoretic pattern and the bands detected by reaction with antibodies. A number of stains, such as amido black, India ink, and colloidal gold, have been described for detection of proteins on nitrocellulose *(1–3*, and *see* Chapters 26, 27, 31, 35, and 48*)*. When these stains are used, a duplicate nitrocellulose blot has to be made, which may lead to discrepancies when comparing two patterns. The only stain that binds reversibly to proteins and does not give high background on nitrocellulose is Ponceau S *(4)*. The detection limit of proteins stained with Ponceau S is between 250–500 ng.

Here, we describe a staining method that is based on the coupling of a fluorescent reagent to proteins. The proteins stained in this way are colorless in visible light but can be detected by illuminating blots with long-range UV light. Dichlorotriazynylaminofluorescein (DTAF) is a reagent we have used for coupling fluorochrome to proteins on nitrocellulose. Most of the proteins are detectable at levels of about 50 ng. The coupling of the fluorochrome to proteins does not alter their antigenic properties, and the blots can be subsequently probed with antibodies using one of the many protocols for the immunodetection of blots (*see* refs. *5–9* and Chapters 44, 48–50, 97).

2. Materials

1. Borate-KCl buffer: 50 m*M* borate and 50 m*M* KCl, pH 9.3. It is prepared by mixing 50 mL of 0.1*M* of boric acid and 0.1*M* KCl with 29.3 mL of 0.1*M* NaOH and adding water to 100 mL.
2. Dichlorotriazynylaminofluorescein (DTAF) from Research Organics, Cleveland, OH. It should be stored desiccated at –20°C (*see* Note 1).
3. Blocking solution: 3% nonfat dry milk in TBS.
4. Primary antibody.

From: The Protein Protocols Handbook
Edited by: J. M. Walker Humana Press Inc., Totowa, NJ

5. Secondary antibody, e.g., goat antirabbit IgG–peroxidase conjugated if primary antiserum is from rabbit.
6. 4-Chloro-1-naphthol. The reagent should be discarded when crystals, originally white, turn to grey. It is stable for a few months at –20°C when stored desiccated. It can also be stored as a solution in ethanol at –20°C.
7. Color reagent solution: 60 mg of 4-Chloro-1-naphthol is dissolved in 20 mL of ice-cold 95% ethanol and added to 100 mL of TBS containing 60 µL of 30% H_2O_2. The reagent should be prepared just before using.
8. Nitrocellulose membrane filters (BA83 or BA85) from Schleicher and Schuell, Keene, NH.
9. Whatman 3MM (Maidstone, UK) filter paper.
10. Scotch-Brite pads (size depends on the size of the gel holder).
11. Glass vessels with flat bottom (e.g., Pyrex™ baking dishes). Their dimensions should be slightly greater than the size of the nitrocellulose sheet.
12. Rocker platform.
13. Transfer apparatus (e.g., Trans Blot Cell, Bio-Rad Laboratories, Richmond, CA).
14. Aluminum foil.
15. Long-range UV transilluminator (e.g., Spectroline TL302, Spectronics Corp., Westbury, NY).

3. Methods

3.1. Fluorescent Staining of Proteins

1. Transfer proteins from the gel to nitrocellulose using any of the methods described in Chapters 37–39 (*see also* Note 2).
2. After transfer, wash the nitrocellulose in 50 mL of borate-KCl buffer for 5 min.
3. Immerse the membrane for 10 min in 50 mL of DTAF solution (0.5–2 µg/mL) in borate-KCl buffer. The dish should be wrapped in aluminium foil to avoid photodegradation of DTAF (*see* Notes 3–5).
4. Remove the excess reagent by two washings (100 mL each) in borate-KCl buffer for 5 min.
5. Place the nitrocellulose on a long-range UV transilluminator. The protein bands appear green on a light yellow background (*see* Notes 6 and 7).
6. Mark the bands of interest with a pencil or photograph the blot using a Polaroid camera equipped with an orange filter. The exposures range from 30–120 s (*see* Notes 8 and 9).

3.2. Immunodetection

1. Wash the nitrocellulose sheet that was previously subjected to staining with DTAF in 200 mL of TBS for 5 min.
2. Immerse the sheet in 200 mL of blocking solution and agitate for 1 h at room temperature.
3. Transfer the membrane to a second dish containing 50 mL of primary antibody solution (the antiserum is diluted 1:50 to 1:3000 with blocking solution). Incubate with shaking at room temperature for 1 h.
4. Wash the membrane with 200 mL of blocking solution, 10 min each wash.
5. Transfer the membrane to 50 mL of peroxidase-conjugated secondary antibody solution (diluted as suggested by the manufacturer in blocking solution) and incubate with agitation for 1 h at room temperature. Instead of the system with peroxidase-conjugated secondary antibody, a number of other conjugates described in refs. *5–9* can be also used (*see also* Chapters 48–50).
6. Wash the membrane once in 200 mL of blocking solution, and then twice with 200 mL of TBS alone, 10 min each wash.
7. Prepare the color reagent solution just before using.
8. Incubate the membrane for 5–10 min in color reagent solution. The bands (dark blue in color) should normally be visible within 1–2 min.

9. Transfer the membrane to distilled water for 10 min. Change the water and leave the membrane for an additional 10 min.
10. Dry the membrane in air. Keep it away from light, otherwise, the background will turn yellow.

4. Notes

1. Fluorescein isothiocyanate (FITC) can be used instead of DTAF. The detection limit for staining of proteins with FITC is 2–3 times lower than for staining with DTAF.
2. It is advisable to run protein mol wt standards in one of the electrophoresis lanes. Their position after transfer to the membrane is helpful in the assessment of antigen mol wt and also, their intensity after staining provides information on the quality of transfer.
3. After fluorescent staining, the nitrocellulose sheet can be stored practically indefinitely at 4°C, provided it is wrapped in aluminium foil or otherwise protected from light.
4. Most of the other fluorochromes for protein labeling (e.g., TRi TC-tetramethylrhodamine isothiocyanate) cannot be used for protein staining on nitrocellulose because these fluorescent reagents bind irreversibly to nitrocellulose.
5. Nylon membranes cannot be stained with DTAF because nylon binds the stain. However, Immobilon P membranes (Millipore, [Bedford, MA]), which are composed of polyvinylidine difluoride, are stained equally as well as nitrocellulose.
6. The limit of protein detection with DTAF will vary depending on the intensity of UV radiation. The power of a transilluminator should be at least 100 W. In case of transilluminators with more than one wavelength (e.g., 302 nm and 365 nm), the longer wavelength should be used (absorption maximum for DTAF is 489 nm).
7. Staining before immunodetection permits unambiguous cutting of the membrane into strips for incubation in more than one type of primary antibody (or more than one concentration of the same antibody).
8. When immunodetection is performed with peroxidase-conjugated secondary antibody, the fluorescent protein bands disappear during incubation with the color reagent solution (oxidation with H_2O_2). If immunodetection is done with radioiodinated protein A (*see* Chapter 48), the fluorescence of the proteins remains visible under UV light through all steps of immunoblotting. It is, however, advisable to take a photograph of the membrane immediately after finishing the fluorescent staining because long incubations during immunodetection lead to some quenching of fluorescence.
9. The sensitivity of antigen detection by immunoblotting is not affected by coupling DTAF to proteins if polyclonal antibodies are used. However, this may not always be the case for monoclonal antibodies because, depending on the epitope, DTAF binding may have a more or less drastic effect on the conformation of the antigen, and hence, on its reactivity with antibodies.

References

1. Gershoni, J. M. and Palade, G. E. (1982) Electrophoretic transfer of proteins from sodium dodecyl sulfate-polyacrylamide gels to a positively charged membrane filter. *Anal. Biochem.* **124,** 396–405.
2. Hancock, K. and Tsang, V. (1983) India ink staining of proteins on nitrocellulose paper. *Anal. Biochem.* **133,** 157–162.
3. Moeremans, M., Daneels, G. and de Mey, J. (1985) Sensitive colloidal metal (gold or silver) staining of protein blots on nitrocellulose membranes. *Anal. Biochem.* **145,** 315–321.
4. Salinovich, O. and Montelaro, R. C. (1986) Reversible staining and peptide mapping of proteins transferred to nitrocellulose after separation by sodium dodecyl sulfate-polyacrylamide gel electrophoresis. *Anal. Biochem.* **156,** 341–347.

5. Burnette, W. N. (1981) "Western blotting": electrophoretic transfer of proteins from sodium dodecyl sulfate-polyacrylamide gels to unmodified nitrocellulose and radiographic detection with antibody and radioiodinated protein A. *Anal. Biochem.* **112,** 195–203.
6. Hawkes, R., Niday, E. and Gordon, J. (1982) A dot immunobinding assay for monoclonal and other antibodies. *Anal. Biochem.* **119,** 142–147.
7. Blake, M. S., Johnston, K. H., Russell-Hones, G. J., and Gotschlich, E. C. (1984) A rapid, sensitive method for detection of alkaline phosphatase anti-antibody on Western blots. *Anal. Biochem.* **136,** 175–179.
8. Brada, D. and Roth, J. (1984) "Golden blot"—detection of polyclonal and monoclonal antibodies bound to antigens on nitrocellulose by protein A–gold complexes. *Anal. Biochem.* **142,** 79–83.
9. Davis, L. G., Dibner, M. D., and Batley, J. F. (1986) *Basic Methods in Molecular Biology.* Elsevier, New York.

46

Coupling of Antibodies with Biotin

Rosaria P. Haugland and Wendy W. You

1. Introduction

The avidin–biotin bond is the strongest known biological interaction between a ligand and a protein ($K_d = 1.3 \times 10^{-15} M$ at pH 5) *(1)*. The affinity is so high that the avidin–biotin complex is extremely resistant to any type of denaturing agent *(2)*. Biotin (Fig. 1) is a small, hydrophobic molecule that functions as a coenzyme of carboxylases *(3)*. It is present in all living cells. Avidin is a tetrameric glycoprotein of 66,000–68,000 mol wt, found in egg albumin and in avian tissues. The interaction between avidin and biotin occurs rapidly, and the stability of the complex has prompted its use for *in situ* attachment of labels in a broad variety of applications, including immunoassays, DNA hybridization *(4–6)*, and localization of antigens in cells and tissues *(7)*. Avidin has an isoelectric point of 10.5. Because of its positively charged residues and its oligosaccharide component, consisting mostly of mannose and glucosamine *(8)*, avidin can interact nonspecifically with negative charges on cell surfaces and nucleic acids, or with membrane sugar receptors. At times, this causes background problems in histochemical and cytochemical applications. Streptavidin, a near-neutral, biotin binding protein *(9)* isolated from the culture medium of *Streptomyces avidinii,* is a tetrameric nonglycosylated analog of avidin with a mol wt of about 60,000. Like avidin, each molecule of streptavidin binds four molecules of biotin, with a similar dissociation constant. The two proteins have about 33% sequence homology, and tryptophan residues seem to be involved in their biotin binding sites *(10,11)*. In general, streptavidin gives less background problems than avidin. This protein, however, contains a tripeptide sequence Arg-Tyr-Asp (RYD) that apparently mimics the binding sequence of fibronectin Arg-Gly-Asp (RGD), a universal recognition domain of the extracellular matrix that specifically promotes cell adhesion. Consequently, the streptavidin–cell-surface interaction causes high background in certain applications *(12)*.

As an alternative to both avidin and streptavidin, a chemically modified avidin, NeutraLite™ avidin (NeutraLite is a trademark of Belovo Chemicals, Bastogne, Belgium), has recently become available. NeutraLite avidin consists of chemically deglycosylated avidin, which has been modified to reduce the isoelectric point to a neutral value, without loss of its biotin binding properties and without significant change in the lysines available for derivatization *(13)*. (Fluorescent derivatives and

From: The Protein Protocols Handbook
Edited by: *J. M. Walker Humana Press Inc., Totowa, NJ*

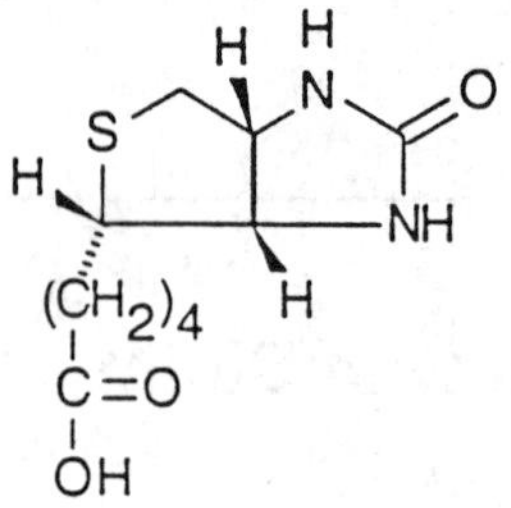

Biotin MW 244.31

Fig. 1. Structure of biotin.

enzyme conjugates of NeutraLite avidin, as well as the unlabeled protein, are available from Molecular Probes [Eugene, OR].)

As shown in Fig. 1, biotin is a relatively small and hydrophobic molecule. The addition to the carboxyl group of biotin of one (*X*) or two (*XX*) aminohexanoic acid "spacers" greatly enhances the efficiency of formation of the complex between the biotinylated antibody (or other biotinylated protein) and the avidin–probe conjugate, where the probe can be a fluorochrome or an enzyme *(14,15)*. Each of these 7- or 14-atom spacer arms has been shown to improve the ability of biotin derivatives to interact with the binding cleft of avidin. The comparison between streptavidin binding activity of proteins biotinylated with biotin-*X* or biotin-*XX* (labeled with same number of moles of biotin/mol of protein) has been performed in our laboratory (Fig. 2). No difference was found between the avidin or streptavidin–horse radish peroxidase conjugates in their ability to bind biotin-*X* or biotin-*XX*. However, biotin-*XX* gave consistently higher titers in enzyme-linked immunosorbent (ELISA) assays, using biotinylated goat antimouse IgG (GAM), bovine serum albumin (BSA), or protein A (results with avidin and with protein A are not presented here). Even nonroutine conjugations performed in our laboratory have consistently yielded excellent results using biotin-*XX*.

Biotin, biotin-*X*, and biotin-*XX* have all been derivatized for conjugation to amines or thiols of proteins and aldehyde groups of glycoproteins or other polymers. The simplest and most popular biotinylation method is to label the ε-amino groups of lysine residues with a succinimidyl ester of biotin. Easy-to-use biotinylation kits are commercially available that facilitate the biotinylation of 1–2 mg of protein or oligonucleotides *(16)*. One kit for biotinylating smaller amounts of protein (0.1–3 mg) utilizes biotin-*XX* sulfosuccinimidyl ester *(17)*. This compound is water-soluble and allows for the efficient labeling of dilute protein samples. Another kit uses biotin-*X* 2,4-dinitrophenyl-X-lysine succinimidyl ester (DNP-biocytin) as the biotinylating reagent. DNP-biocytin was developed by Molecular Devices (Menlo Park, CA) for their patented Threshold-Immunoligand System *(18)*. DNP-biocytin permits the direct measurement of the degree of biotinylation of the reaction product by using the molar extinction coefficient of DNP (15,000M^{-1} cm^{-1} at 364 nm). Conjugates of DNP-biocytin can be probed separately or simultaneously using either anti-DNP antibodies or avidin/streptavidin; this flexibility is useful when combining techniques such as fluorescence and electron microscopy. Biotin iodoacetamide or maleimide, which could biotinylate the reduced

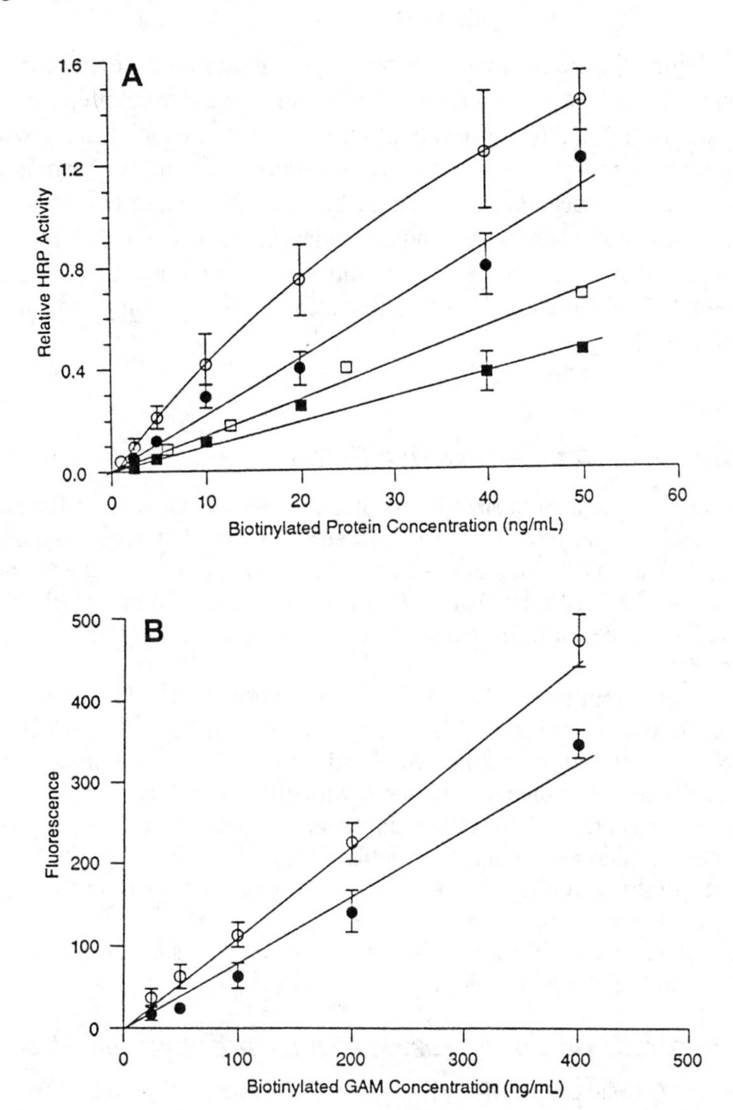

Fig. 2. (A) ELISA-type assay comparing the binding capacity of BSA and GAM biotinylated with biotin-*X* or biotin-*XX*. The assay was developed using streptavidin-HRP conjugate (0.2 μg/mL) and *o*-phenylenediamine dihydro-chloride (OPD). The number of biotin/mol was: 4.0 biotin-*X*/GAM (●), 4.4 biotin-*XX*/GAM (○), 6.7 biotin-*X*/BSA (■), and 6.2 biotin-*XX*/BSA (□). Error bars on some data points have been omitted for clarity. **(B)** Similar assay using GAM biotinylated with biotin-*X* (●) or biotin-*XX* (○). The assay was developed with streptavidin–R-phycoerythin conjugate (25 μg/mL using a Millipore CytoFluor™ fluorescence microtiter plate reader).

sulfhydryls located at the hinge region of antibodies, is not usually used for this purpose. More examples in the literature describe biotinylation of antibodies with biotin hydrazide at the carbohydrate prosthetic group, located in the Fc portion of the molecule, relatively removed from the binding site. Conjugation of carbohydrates with hydrazides requires the oxidation of two adjacent hydroxyls to aldehydes and optional stabilization of the reaction with cyanoborohydride *(19)*.

Because of its strength, the interaction between avidin and biotin cannot be used for preparing matrices for affinity column purification, unless columns prepared with avidin monomers are used *(20)*. The biotin analog, iminobiotin, which has a lower affinity for avidin, can be used for this purpose *(21,22)*. Iminobiotin in reactive form is commercially available, and the procedure for its conjugation is identical to that used for biotin. Detailed, practical protocols for biotinylating antibodies at the lysine or at the carbohydrate site, and a method to determine the degree of biotinylation are described in detail in this chapter (*see* Notes 1–10 for review of factors that affect optimal conjugation and yield of biotinylated antibodies).

2. Materials

2.1. Conjugation with Amine-Reactive Biotin

1. Reaction buffer: 1*M* sodium bicarbonate, stable for about 2 wk when refrigerated. Dissolve 8.3 g of $NaHCO_3$ in 100 mL of distilled water. The pH will be about 8.3. Dilute 1:10 before using to obtain a 0.1*M* solution. Alternate reaction buffer: 0.1*M* sodium phosphate, pH 7.8. Dissolve 12.7 g Na_2HPO_4 and 1.43 g NaH_2PO_4 in 800 mL of distilled water. Adjust pH to 7.8 if necessary. Bring the volume to 1000 mL. This buffer is stable for 2 mo when refrigerated.
2. Anhydrous dimethylformamide (DMF) or dimethyl sulfoxide (DMSO).
3. Phosphate-buffered saline (PBS): Dissolve 1.19 g of K_2HPO_4, 0.43 g of $KH_2PO_4{\cdot}H_2O$ and 8.8 g NaCl in 800 mL of distilled water, adjust the pH to 7.2 if necessary or to the desired pH, and bring the volume to 1000 mL with distilled water.
4. Disposable desalting columns or a gel-filtration column: Amicon GH-25 and Sephadex G-25 or the equivalent, equilibrated with PBS or buffer of choice.
5. Good-quality dialysis tubing as an alternative to the gel-filtration column when derivatizing small quantities of antibody.
6. Biotin, biotin-*X* or biotin-*XX* succinimidyl ester: As with all succinimidyl esters, these compounds should be stored well desiccated in the freezer.

2.2. Conjugation with Biotin Hydrazide at the Carbohydrate Site

1. Reaction buffer: 0.1*M* acetate buffer, pH 6.0. Dilute 5.8 mL acetic acid in 800 mL distilled water. Bring the pH to 6.0 with 5*M* NaOH and the volume to 1000 mL. The buffer is stable for several months when refrigerated.
2. 20 m*M* Sodium metaperiodate: Dissolve 43 mg of $NaIO_4$ in 10 mL of reaction buffer, protecting from light. Use fresh.
3. Biotin-*X* hydrazide or biotin-*XX* hydrazide.
4. DMSO.
5. Optional: 100 m*M* sodium cyanoborohydride, freshly prepared. Dissolve 6.3 mg of $NaBH_3CN$ in 10 mL of 0.1 m*M* NaOH.

2.3. Determination of the Degree of Biotinylation

1. 10 m*M* 4'Hydroxyazobenzene-2-carboxylic acid (HABA) in 10 m*M* NaOH.
2. 50 m*M* Sodium phosphate and 150 m*M* NaCl, pH 6.0. Dissolve 0.85 g of Na_2HPO_4 and 6.07 g of NaH_2PO_4 in 800 mL of distilled water. Add 88 g of NaCl. Bring the pH to 6.0 if necessary and the volume to 1000 mL.
3. 0.5 mg/mL Avidin in 50 m*M* sodium phosphate and 150 m*M* NaCl, pH 6.0.
4. 0.25 m*M* Biotin in 50 m*M* sodium phosphate, and 150 m*M* NaCl, pH 6.0.

3. Methods

3.1. Conjugation with Amine-Reactive Biotin

1. Dissolve the antibody, if lyophilized, at approx 5–15 mg/mL in either of the two reaction buffers described in Section 2.1. If the antibody to be conjugated is already in solution in 10–20 m*M* PBS, without azide, the pH necessary for the reaction can be obtained by adding 1/10 vol of 1*M* sodium bicarbonate. IgM should be conjugated in PBS, pH 7.2 (*see* Note 3).
2. Calculate the amount of a 10 mg/mL biotin succinimidyl ester solution (biotin-NHS) needed to conjugate the desired quantity of antibody at the chosen biotin/antibody molar ratio, according to the following formula:

$$\text{(mL of 10 mg/mL biotin-SE)} = \{[(\text{mg antibody} \times 0.1)/\text{mol wt of antibody}] \times R \times \text{mol wt of biotin-SE})\} \tag{1}$$

where R = molar incubation ratio of biotin/protein. For example, using 5 mg of IgG and a 10:1 molar incubation ratio of biotin-*XX*-SE, Eq. (1) yields:

$$\text{(mL of 10 mg/mL biotin-}XX\text{-SE)} = \{[(5 \times 0.1)/145{,}000] \times (10 \times 568)\} = 0.02 \text{ mL} \tag{2}$$

3. Weigh 3 mg or more of the biotin-SE of choice, and dissolve it in 0.3 mL or more of DMF or DMSO to obtain a 10 mg/mL solution. **It is essential that this solution be prepared immediately before starting the reaction,** since the succinimidyl esters or any amine-reactive reagents hydrolyze quickly in solution. Any remaining solution should be discarded.
4. While stirring, slowly add the amount of 10 mg/mL solution, calculated in step 2, to the antibody prepared in step 1, mixing thoroughly.
5. Incubate this reaction mixture at room temperature for 1 h with gentle stirring or shaking.
6. The antibody conjugate can be purified on a gel-filtration column or by dialysis. When working with a few milligrams of dilute antibody solution, care should be taken not to dilute the antibody further. In this case, dialysis is a very simple and effective method to eliminate unreacted biotin. A few mL of antibody solution can be effectively dialyzed in the cold against 1 L of buffer with three to four changes. Small amounts of concentrated antibody can be purified on a prepackaged desalting column equilibrated with the preferred buffer, following the manufacturer's directions. Five or more milligrams of antibody can be purified on a gel-filtration column. The dimensions of the column will have to be proportional to the volume and concentration of the antibody. For example, for 5–10 mg of antibody in 1 mL solution, a column with a bed volume of 10 × 300 mm will be adequate. To avoid denaturation, dilute solutions of biotinylated antibodies should be stabilized by adding BSA at a final concentration of 0.1–1%.

3.2. Conjugation with Biotin Hydrazide at the Carbohydrate Site

1. It is essential that the entire following procedure be carried out with the sample completely protected from light (*see* Note 9).
2. Dissolve antibody (if lyophilized) or dialyze solution of antibody to obtain a 2–10 mg/mL solution in the reaction buffer described in Section 2.1., item 1. Keep at 4°C.
3. Add an equal volume of cold metaperiodate solution. Incubate the reaction mixture at 4°C for 2 h in the dark.
4. Dialyze overnight against the same buffer protecting from light, or, if the antibody is concentrated, desalt on a column equilibrated with the same buffer. This step removes the iodate and formaldehyde produced during oxidation.

5. Dissolve 10 mg of the biotin hydrazide of choice in 0.25 mL of DMSO to obtain a 40 mg/mL solution, warming if needed. This will yield a 107 m*M* solution of biotin-*X* hydrazide or an 80 m*M* solution of biotin-*XX* hydrazide. These solutions are stable for a few weeks.
6. Calculate the amount of biotin hydrazide solution needed to obtain a final concentration of approx 5 m*M*, and add it to the oxidized antibody. When using biotin-*X* hydrazide, 1 vol of hydrazide should be added to 20 vol of antibody solution. When using biotin-*XX* hydrazide, 1 vol of hydrazide should be added to 15 vol of antibody solution.
7. Incubate for 2 h at room temperature with gentle stirring.
8. This step is optional. The biotin hydrazone–antibody conjugate formed in this reaction (steps 6 and 7) is considered by some researchers to be relatively unstable. To reduce the conjugate to a more stable, substituted hydrazide, treat the conjugate with sodium cyanoborohydride at a final concentration of 5 m*M* by adding a 1/20 vol of a 100-m*M* stock solution. Incubate for 2 h at 4°C (*see* Note 5).
9. Purify the conjugate by any of the methods described for biotinylating antibodies at the amine site (*see* Section 3.1., step 6).

3.3. Determination of the Degree of Biotinylation

The dye HABA interacts with avidin yielding a complex with an absorption maximum at 500 nm. Biotin, because of its higher affinity, displaces HABA, causing a decrease in absorbance at 500 nm proportional to the amount of biotin present in the assay.

1. To prepare a standard curve, add 0.25 mL of HABA reagent to 10 mL of avidin solution. Incubate 10 min at room temperature and record the absorbance at 500 nm of 1 mL avidin–HABA complex with 0.1 mL buffer, pH 6.0. Distribute 1 mL of the avidin–HABA complex into six test tubes. Add to each the biotin solution in a range of 0.005–0.10 mL. Bring the final volume to 1.10 mL with pH 6.0 buffer, and record the absorbance at 500 nm of each concentration point. Plot a standard curve with the nanomoles of biotin vs the decrease in absorbance at 500 nm. An example of a standard curve is illustrated in Fig. 3.
2. To measure the degree of biotinylation of the sample, add an aliquot of biotinylated antibody of known concentration to 1 mL of avidin–HABA complex. For example, add 0.05–0.1 mL of biotinylated antibody at 1 mg/mL to 1 mL of avidin–HABA mixture. Bring the volume to 1.10 mL, if necessary, incubate for 10 min, and measure the decrease in absorbance at 500 nm.
3. Deduct from the standard curve the nanomoles of biotin corresponding to the observed change in absorbance. The ratio between nanomoles of biotin and nanomoles of antibody used to displace HABA represents the degree of biotinylation, as seen from the following equation:

$$[(\text{nmol biotin} \times 145{,}000 \times 10^{-6})/(\text{mg/mL antibody} \times 0.1\ \text{mL})] = (\text{mol of biotin/mol of antibody}) \quad (3)$$

where 145,000 represents the mol wt of the antibody and 0.1 mL is the volume of 1 mg/mL of biotinylated antibody sample.

4. Notes

4.1. Factors that Influence the Biotinylation Reaction

1. Protein concentration: As in any chemical reaction, the concentration of the reagents is a major factor in determining the rate and the efficiency of the coupling. Antibodies at a concentration of 5–20 mg/mL will give better results; however, it is often difficult to have such concentrations or even such quantities available for conjugation. Nevertheless, the

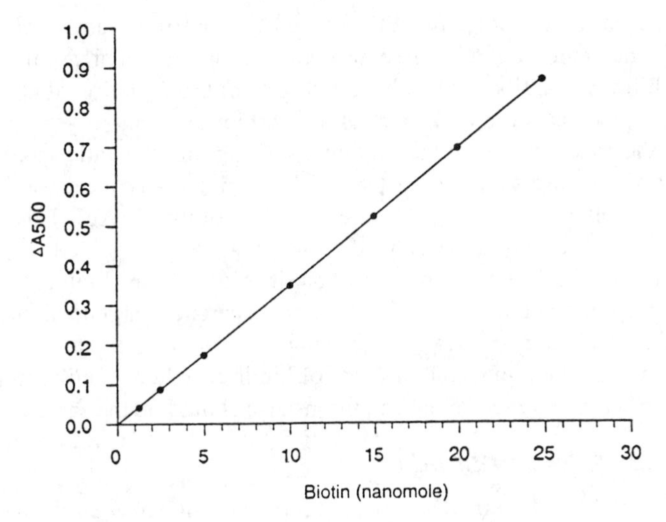

Fig. 3. Examples of standard curve for biotin assay with avidin-HABA reagent, obtained as described in Section 3.3.

antibody should be as concentrated as possible. In the case of solutions of antibody <2–3 mg/mL, the molar ratio of biotinylating reagent (or of both the oxidizing and biotinylating reagent, in the case of labeling the carbohydrate region) should be increased. It is also essential that the antibody solutions do not contain gelatin or BSA, which are often added to stabilize dilute solutions of antibodies. These proteins, generally present at a 1% concentration, will also react with biotinylating reagents.

2. pH: The reactivity of amines increases at basic pH. Unfortunately, so does the rate of hydrolysis of succinimidyl esters. We have found that the best pH for biotinylation of the ε-amino groups of lysines is 7.5–8.3. IgM antibodies, which denature at basic pH, can be biotinylated at pH 7.2 by increasing the molar ratio of the biotinylating reagent to antibody to at least 20. The optimum pH for oxidation and conjugation with hydrazides is 5.5–6.0.
3. Buffer: Bicarbonate or phosphate buffers are suitable for biotinylation. Organic buffers, such as Tris which contain amines, should be avoided, because they react with amino-labeling reagents or interfere with the reaction between aldehydes and hydrazides. However, HEPES and EPPS, which contain tertiary amines, are suitable. Antibodies dissolved in 10–20 m*M* PBS can be readily prepared for conjugation at the lysine site by adding 1/10–1/5 of the volume of 1*M* sodium bicarbonate. As noted, because IgM antibodies are unstable in basic solution, biotinylation at the ε-amino group of lysines should be attempted in PBS or equivalent buffer at pH 7.2. Reactions of antibodies with periodate and biotin hydrazide can be performed in PBS at pH 7 or in acetate buffer, pH 6.0 (*see* Section 2.2).
4. Temperature: Biotinylations at the amino group sites are run at room temperature, at the carbohydrate site at 0–4°C.
5. Time: Succinimidyl ester derivatives will react with a protein within 1 h. Periodate oxidation will require 2 h at pH 6.0. Reaction with biotin hydrazide can be performed in a few hours. Stabilization with cyanoboro-hydride requires <2 h.
6. Desired degree of biotinylation and stability of the conjugate: Reaction of an antibody with biotin does not significantly alter the size or charge of the molecule. However, because of the size of avidin or its analogs (mol wt = 60,000–68,000), an increase in the number of biotins per antibody will not necessarily increase the number of avidins capable of react-

ing with one antibody molecule. Because biotin, biotin-*X*, and biotin-*XX* are very hydrophobic molecules, a high degree of biotinylation might increase the background or might destabilize the antibody. To obtain a degree of biotinylation of about 3–7 biotins/IgG, generally a molar ratio of 15 mol of amino biotinylating reagent/mol of protein is used. When the concentration of the antibody is <3 mg/mL, this ratio should be increased. The amount of increase should be determined experimentally, because the reactivity of the lysines available for conjugation varies for each antibody (Ab). This could become a significant factor, especially at low antibody concentrations.

The succinimidyl esters or hydrazides of biotin, biotin-*X*, and biotin-*XX* exhibit similar degrees of reactivity, and the choice is up to the researcher. In general, the longer spacer arm in biotin-*XX* should be advantageous (Fig. 2). The overall stability of biotinylated MAbs derivatized with a moderate number of biotin should be similar to the stability of the native antibody, and the storage conditions also should be the same.

4.2. Factors that Affect Antibodies

7. Most Abs can withstand biotinylation with minimal change in activity and stability, especially if the degree of biotinylation is about 3–6 biotins/mol.
8. Biotin or any of its longer chain derivatives do not contribute to the absorbance of the antibody at 280 nm. Consequently, the concentration of the antibody can be measured by using $A^{1\%}_{1\ cm} = 14$ at 280 nm.
9. It is essential that the entire procedure for biotinylation of antibodies at the carbohydrate site (Section 3.2.) be performed in the dark, protected from light.
10. It should be noted that dry milk, serum, and other biological fluids contain biotin and, consequently, they should not be used as blocking agents in systems where blocking is required.

References

1. Green, N. M. (1963) Avidin. 3. The nature of the biotin binding site. *Biochem. J.* **89,** 599–609.
2. Green, N. M. (1963) Avidin. 4. Stability at extremes of pH and dissociation into subunits by guanidine hydrochloride. *Biochem. J.* **89,** 609–620.
3. Knappe, J. (1970) Mechanism of biotin action. *Annu. Rev. Biochem.* **39,** 757–776.
4. Wilchek, M. and Bayer, E. A. (1988) The avidin–biotin complex in bioanalytical applications. *Anal. Biochem.* **171,** 1–32.
5. Wilchek, M. and Bayer, E. A. (1990) Avidin–biotin technology, in *Methods in Enzymology,* vol. 184, Academic, New York, pp. 213–217.
6. Levi, M., Sparvoli, E., Sgorbati, S., and Chiantante, D. (1990) Biotin–streptavidin immunofluorescent detection of DNA replication in root meristems through Brd Urd incorporation: cytological and microfluorimetric applications. *Physiol. Plantarum* **79,** 231–235.
7. Armstrong, R., Friedrich, V. L., Jr., Holmes, K. V., and Dubois-Dalcq, M. (1990) *In vitro* analysis of the oligodendrocyte lineage in mice during demyelination and remyelination. *J. Cell Biol.* **111,** 1183–1195.
8. Bruch, R. C. and White, H. B. III (1982) Compositional and structural heterogeneity of avidin glycopeptides. *Biochemistry* **21,** 5334–5341.
9. Hiller, Y., Gershoni, J. M., Bayer, E. A., and Wilchek, M. (1987) Biotin binding to avidin: oligosaccharide side chain not required for ligand association. *Biochem. J.* **248,** 167–171.
10. Green, N. M. (1975) Avidin, in *Advances in Protein Chemistry,* vol. 29 (Anfinsen, C. B., Edsall, J. T., and Richards, F. M., eds.), Academic, New York, pp. 85–133.
11. Chaiet, L. and Wolf, F. J. (1964) The properties of streptavidin, a biotin-binding protein produced by *Streptomycetes. Arch. Biochem. Biophys.* **106,** 1–5.

12. Alon, R., Bayer, E. A., and Wilcheck, M. (1990) Streptavidin contains an Ryd sequence which mimics the RGD receptor domain of fibronectin. *Biochem. Biophys. Res. Commun.* **170,** 1236–1241.
13. Wilchek, M. and Bayer, E. A. (1993) Avidin–biotin immobilization systems, in *Immobilized Macromolecules: Application Potentials* (Sleytr, U. B., ed.), Springer-Verlag, New York, pp 51–60.
14. Gretch, D. R., Suter, M., and Stinski, M. F. (1987) The use of biotinylated monoclonal antibodies and streptavidin affinity chomatography to isolate Herpes virus hydrophobic proteins or glycoproteins. *Anal. Biochem.* **163,** 270–277.
15. Hnatowich, D. J., Virzi, F., and Rusckowski, M. (1987) Investigations of avidin and biotin for imaging applications. *J. Nucl. Med.* **28,** 1294–1302.
16. Haugland, R. P. (1996) Biotin derivatives, in *Handbook of Fluorescent Probes and Research Chemicals,* 6th ed. (Spence, M., ed.), Molecular Probes, Inc., Eugene, OR, Chapter 4.
17. LaRochelle, W. J. and Froehner, S. C. (1986) Determination of the tissue distributions and relative concentrations of the postsynaptic 43-kDa protein and the acetylcholine receptor in *Torpedo. J. Biol. Chem.* **261,** 5270–5274.
18. Briggs, J. and Panfili, P. R. (1991) Quantitation of DNA and protein impurities in biopharmaceuticals. *Anal. Chem.* **63,** 850–859.
19. Wong, S. S. (1991) Reactive groups of proteins and their modifying agents, in *Chemistry of Protein Conjugation and Crosslinking,* CRC, Boston, MA, pp. 27–29.
20. Kohanski, R. A. and Lane, M. D. (1985) Receptor affinity chomatography. *Ann. NY Acad. Sci.* **447,** 373–385.
21. Orr, G. A. (1981) The use of the 2-iminobiotin-avidin interaction for the selective retrieval of labeled plasma membrane components. *J. Biol. Chem.* **256,** 761–766.
22. Hoffmann, K., Wood, S. W., Brinton, C. C., Montibeller, J. A., and Finn, F. M. (1980) Iminobiotin affinity columns and their application to retrieval of streptavidin. *Proc. Natl. Acad. Sci.* **77,** 4666–4668.

47

Preparation of Avidin Conjugates

Rosaria P. Haugland and Mahesh K. Bhalgat

1. Introduction

The high-affinity avidin-biotin system has found applications in different fields of biotechnology, including immunoassays, histochemistry, affinity chromatography, and drug delivery, to name a few. A brief description of avidin and avidin-like molecules, streptavidin, deglycosylated avidin, and NeutraLite avidin, is presented in the previous chapter (Chapter 46). With four biotin binding sites per molecule, the avidin family of proteins is capable of forming tight complexes with one or more biotinylated compounds *(1)*. Typically, the avidin-biotin system is used to prepare signal-amplifying "sandwich" complexes between specificity reagents (e.g., antibodies) and detection reagents (e.g., fluorophores, enzymes, and so on). The specificity and detection reagents are independently conjugated, one with avidin and the other with biotin, or both with biotin, providing synthetic flexibility *(2)*.

Avidin conjugates of a wide range of fluorophores, phycobiliproteins, secondary antibodies, microspheres, ferritin, and enzymes commonly used in immunochemistry are available at reasonable prices, making their small scale preparation impractical and not cost effective (*see* Note 1). However, conjugations of avidin to specific antibodies, to uncommon enzymes, and to other proteins and peptides are often performed on-site. A general protocol for the conjugation of avidin to enzymes, antibodies, and other proteins is described in this chapter.

Avidin conjugates of oligodeoxynucleotides are hybrid molecules that not only provide multiple biotin binding sites, but can also be targeted to complimentary DNA or RNA sequences, by annealing interactions. Such conjugates are useful for the construction of macromolecular assemblies with a wide variety of constituents *(3)*. The protocol outlined in Section 3.1. can be modified (*see* Note 2) for the conjugation of oligonucleotides to avidin.

Streptavidin conjugates are also being evaluated for use in drug delivery systems. A two-step imaging and treatment protocol has been developed that involves injection of a suitably prepared tumor-specific monoclonal antibody, followed by a second reagent that carries an imaging or therapeutic agent, capable of binding to the tumor-targeted antibody *(4)*. Owing to complications associated with the injection of radiolabeled biotin *(5)*, conjugation of the imaging or therapeutic agent to streptavidin is being considered instead. A protocol for radioiodination of streptavidin using IODO-BEADS *(6)* is

From: The Protein Protocols Handbook
Edited by: J. M. Walker Humana Press Inc., Totowa, NJ

described in Section 3.2. Some other methods that have been developed include the iodogen method *(7,8; see also* Chapter 115*)*, the Bolton-Hunter reagent method *(9; see also* Chapter 114*)*, and a few that do not involve direct iodination of tyrosine residues *(10–13)*. Streptavidin-drug conjugates are also candidates for therapeutic agents. Synthesis of a streptavidin-drug conjugate involves making a chemically reactive form of the drug followed by its conjugation to streptavidin. The synthetic methodology thus depends on the structure of the specific drug to be conjugated *(14–16)*.

The avidin-biotin interaction can also be exploited for affinity chromatography; however, there are limitations to this application. For example, a biotinylated protein captured on an avidin affinity matrix would likely be denatured by the severe conditions required to separate the high affinity avidin-biotin complex. On the other hand, an avidin affinity matrix may find utility in the removal of undesired biotinylated moieties from a mixture or for the purification of compounds derivatized with 2-iminobiotin. The biotin derivative 2-iminobiotin has reduced affinity for avidin, and its moderate binding to avidin at pH 9.0 is greatly diminished at pH 4.5 *(17)*. Another approach to reducing the affinity of the interaction is to denature avidin to its monomeric subunits. The monomeric subunits have greatly reduced affinity for biotin *(18)*. We describe here a protocol for preparing native *(19)* and monomeric avidin matrices *(20)*. Recently, modified streptavidins, hybrids of native and engineered subunits with lower binding constants, have been prepared that may also be suitable for affinity matrices *(21)*.

2. Materials

2.1. Conjugation with Antibodies and Enzymes

1. Avidin (mol wt = 66,000).
2. Antibody, enzyme, peptide, protein, or thiolated oligonucleotide to be conjugated to avidin.
3. Succinimidyl 3-(2-pyridyldithio)propionate (SPDP; mol wt = 312.36) (*see* Note 3).
4. Succinimidyl *trans*-4-(N-maleimidylmethyl)cyclohexane-1-carboxylate (SMCC; mol wt = 334.33).
5. Dithiothreitol (DTT; mol wt = 154.24).
6. *Tris* (2-carboxyethyl)phosphine (TCEP; mol wt = 286.7).
7. N-ethylmaleimide (NEM; mol wt = 125.13).
8. Anhydrous dimethylsulfoxide (DMSO) or anhydrous dimethylformamide (DMF).
9. 0.1*M* Phosphate buffer: 0.1*M* sodium phosphate, 0.1*M* NaCl at pH 7.5. Dissolve 92 g of Na_2HPO_4, 21 g of $NaH_2PO_4 \cdot H_2O$ and 46.7 g of NaCl in approx 3.5 L of distilled water and adjust the pH to 7.5 with 5*M* NaOH. Dilute to 8 L. Store refrigerated.
10. 1*M* Sodium bicarbonate (*see* Note 4). Dissolve 8.4 g in 90 mL of distilled water and adjust the volume to 100 mL. A freshly prepared solution has a pH of 8.3–8.5.
11. Molecular exclusion matrix with properties suitable for purification of the specific conjugate. Sephadex G-200 (Pharmacia Biotech, Uppsala, Sweden), Bio-Gel A-0.5 m or Bio-Gel A-1.5 m (Bio-Rad Laboratories, Hercules, CA) are useful for relatively small to large conjugates, respectively.
12. Sephadex G-25 (Pharmacia Biotech) or other equivalent matrix.

2.2. Radioiodination Using IODO-BEADS

1. Streptavidin (mol wt = 60,000).
2. $Na^{131}I$ or $Na^{125}I$, as desired.

3. IODO-BEADS (Pierce Chemical, Rockford, IL).
4. Phosphate buffered saline (PBS), pH 7.2: Dissolve 1.19 g of K_2HPO_4, 0.43 g of KH_2PO_4 and 9 g of NaCl in 900 mL of distilled water. Adjust the pH to 7.2 and dilute to 1 L with distilled water.
5. Saline solution: 9 g of NaCl dissolved in 1 L of distilled water.
6. 0.1% Bovine serum albumin (BSA) solution in saline: 0.1 g of BSA dissolved in 100 mL of saline solution.
7. Trichloroacetic acid (TCA), 10% w/v solution in saline: Dissolve 1 g TCA in 10 mL of saline solution.
8. Bio-Gel P-6DG Gel (Bio-Rad).

2.3. Avidin Affinity Matrix

1. 50–100 mg of avidin.
2. Sodium borohydride.
3. 1,4-butanediol-diglycidylether.
4. Succinic anhydride.
5. 6*M* Guanidine · HCl in 0.2*M* KCl/HCl, pH 1.5: Dissolve 1.5 g of KCl in 50 mL of distilled water. Add 57.3 g of guanidine · HCl with stirring. Adjust the pH to 1.5 with 1*M* HCl. Adjust the volume to 100 mL with distilled water.
6. 0.2*M* Glycine · HCl pH 2.0: Dissolve 22.3 g of glycine · HCl in 900 mL of distilled water. Adjust the pH to 2.0 with 6*M* HCl and the volume to 1 L with distilled water.
7. PBS: *see* Section 2.2., item 4.
8. 0.2*M* Sodium carbonate pH 9.5: Dissolve 1.7 g of sodium bicarbonate in 80 mL of distilled water. Adjust the pH to 9.5 with 1*M* NaOH and the volume to 100 mL with distilled water.
9. 0.2*M* Sodium phosphate, pH 7.5: Weigh 12 g of Na_2HPO_4 and 2.5 g of $NaH_2PO_4{\cdot}H_2O$ and dissolve in 900 mL of distilled water. Adjust the pH to 7.5 with 5*M* NaOH and the volume to 1 L with distilled water.
10. 20 m*M* Sodium phosphate, 0.5*M* NaCl, 0.02% sodium azide pH 7.5: Dilute 100 mL of the buffer described in item 9 to 900 mL with distilled water. Add 28 g of NaCl and 200 mg of sodium azide. Adjust pH if necessary, and dilute to 1 L with distilled water.
11. Sepharose 6B (Pharmacia Biotech) or other 6% crosslinked agarose gel.

3. Methods

3.1. Conjugation with Antibodies and Enzymes

3.1.1. Avidin Thiolation

An easy-to-use, protein-to-protein crosslinking kit is now commercially available (Molecular Probes, Eugene, OR). This kit allows predominantly 1:1 conjugate formation between two proteins (0.2–3.0 mg) through the formation of a stable thioether bond *(22)*, with minimal generation of aggregates. A similar protocol is described here for conjugation of 5 mg avidin to antibodies or enzymes. Modifications of the procedure for conjugation of avidin to thiolated oligonucleotides and peptides are described in Notes 2 and 5, respectively. Although the protocol described in this section uses avidin for conjugation, it can be applied for the preparation of conjugates using either avidin, streptavidin, deglycosylated avidin or NeutraLite avidin.

1. Dissolve 5 mg of avidin (76 nmol) in 0.5 mL of 0.1*M* phosphate buffer to obtain a concentration of 10 mg/mL.

2. Weigh 3 mg of SPDP and dissolve in 0.3 mL of DMSO to obtain a 10 mg/mL solution. This solution must be prepared **fresh** immediately before using. Vortex or sonicate to ensure that the reagent is completely dissolved.
3. Slowly add 12 µL (380 nmol) of the SPDP solution (*see* Note 3) to the stirred solution of avidin. Stir for 1 h at room temperature.
4. Purify the thiolated avidin on a 7 × 250 mm size exclusion column, such as Sephadex G-25 equilibrated in 0.1*M* phosphate buffer.
5. Determine the degree of thiolation (optional):
   a. Prepare a 100-m*M* solution of DTT by dissolving 7.7 mg of the reagent in 0.5 mL of distilled water.
   b. Transfer the equivalent of 0.3–0.4 mg of thiolated avidin (absorbance at 280 nm of a 1.0 mg/mL avidin solution = 1.54) and dilute to 1.0 mL using 0.1*M* phosphate buffer. Record the absorbance at 280 nm and at 343 nm.
   c. Add 50 µL of DTT solution. Mix well, incubate for 3–5 min at room temperature and record the absorbance at 343 nm.
   d. Using the extinction coefficient at 343 nm of 8.08×10^3/cm/M (23), calculate the amount of pyridine-2-thione liberated during the reduction, which is equivalent to the number of thiols introduced on avidin, using the following equation along with the appropriate extinction coefficient shown in Table 1:

$$\text{Number of thiols/avidin} = [\Delta A_{343}/(8.08 \times 10^3)] \times [E^{M}_{avidin}/(A_{280}\text{-}0.63\Delta A_{343})] \quad (1)$$

   where ΔA_{343} = change in absorbance at 343 nm; E^{M}_{avidin} = molar extinction coefficient; and $0.63\Delta A_{343}$ = correction for the absorbance of pyridyldithiopropionate at 280 nm *(23)*.
6. Equation (1) allows the determination of the average number of moles of enzyme or antibody that can be conjugated with each of avidin (*see* Note 6). For a 1:1 protein–avidin conjugate, avidin should be modified with 1.2–1.5 thiols/mole. Thiolated avidin prepared by the above procedure can be stored in the presence of 2 m*M* sodium azide at 4°C for 4–6 wk.

3.1.2. Maleimide Derivatization of the Antibody or Enzyme

In this step, which should be completed prior to the deprotection of thiolated avidin, some of the amino groups from the antibody or enzyme are transformed into maleimide groups by reacting with a bifunctional crosslinker, SMCC (*see* Note 7).

1. Dissolve or, if already in solution, dialyze the protein in 0.1*M* phosphate buffer to obtain a concentration of 2–10 mg/mL. If the protein is an antibody, 11 mg are required to obtain an amount equimolar to 5 mg of avidin (*see* Note 6).
2. Prepare a **fresh** solution of SMCC by dissolving 5 mg in 500 µL of dry DMSO to obtain a 10 mg/mL solution. Vortex or sonicate to ensure that the reagent is completely dissolved.
3. While stirring, add an appropriate amount of SMCC solution to the protein solution to obtain a molar ratio of SMCC-to-protein of approx 10. (If 11 mg of an antibody is the protein used, 30 µL of SMCC solution are required.)
4. Continue stirring at room temperature for 1 h.
5. Dialyze the solution in 2 L of 0.1*M* phosphate buffer at 4°C for 24 h, with four buffer changes using a membrane with a suitable molecular weight cut off.

3.1.3. Deprotection of the Avidin Thiol Groups

This procedure is carried out immediately before reacting thiolated avidin with the maleimide derivative of the antibody or enzyme prepared in Section 3.1.2.

Table 1
Molar Extinction Coefficients at 280 nm and Molecular Weights of Avidin and Avidin-Like Proteins

Protein	Molecular weight	E^{M}_{avidin}/cm/M
Avidin	66,000	101,640
Deglycosylated avidin/neutraLite avidin	60,000	101,640
Streptavidin	60,000	180,000

1. Dissolve 3 mg of TCEP in 0.3 mL of 0.1*M* phosphate buffer.
2. Add 11 µL of TCEP solution to the thiolated avidin solution. Incubate for 15 min at room temperature.

3.1.4. Formation and Purification of the Conjugate

1. Add the thiolated avidin-TCEP mixture dropwise to the dialyzed maleimide derivatized protein solution with stirring. Continue stirring for 1 h at room temperature, followed by stirring overnight at 4°C.
2. Stop the conjugation reaction by capping residual sulfhydryls with the addition of NEM at a final concentration of 50 µ*M*. Dissolve 6 mg of NEM in 1 mL DMSO and dilute 1:1000 in the conjugate reaction mixture. Incubate for 30 min at room temperature or overnight at 4°C (*see* Note 8). The conjugate is now ready for final purification.
3. Concentrate the avidin-protein conjugate mixture to 1–2 mL in a Centricon-30 (Amicon, Beverly, MA) or equivalent centrifuge tube concentrator.
4. Pack appropriate size columns (e.g., 10 × 60 mm for approx 15 mg of final conjugate) with a degassed matrix suitable for the isolation of the conjugate from unconjugated reagents. If the protein conjugated is an antibody, a matrix such as Bio-Gel A-0.5 m is suitable. For other proteins, Sephadex G-200 or a similar column support may be appropriate, depending on the size of the protein-avidin conjugate.
5. Collect 0.5–1 mL fractions. The first protein peak to elute contains the conjugate; however, the first or second fraction may contain some aggregates. Analyze each fraction absorbing at 280 nm for biotin binding and assay it for the antibody or enzyme activity. HPLC may also be performed for further purification, if necessary.

3.2. Radioiodination Using IODO-BEADS

The radioiodination procedure (*see* Note 9) described here uses IODO-BEADS, which contain the sodium salt of N-chloro-benzenesulfonamide immobilized on nonporous, polystyrene beads. Immobilization of the oxidizing agent allows for easy separation of the latter from the reaction mixture. This method also prevents the use of reducing agents.

1. Wash 6–8 IODO-BEADS twice with 5 mL of PBS. Dry the beads by rolling them on a clean filter paper.
2. Add 500 µL of PBS to the supplier's vial containing 8–10 mCi of carrier free $Na^{125}I$ or $Na^{131}I$. Place the beads in the same vial and gently mix the contents by swirling. Allow the mixture to sit for 5 min at room temperature with the vial capped.
3. Dissolve or dilute streptavidin in PBS to obtain a final concentration of 1 mg/mL. Add 500 µL of streptavidin solution to the vial containing sodium iodide. Cap the vial immediately and mix the contents thoroughly. Incubate for 20–25 min at room temperature, with occasional swirling (*see* Note 10).

4. Carefully remove and save the liquid from the reaction vessel; this is the radioiodinated streptavidin solution. Wash the beads by adding 500 µL of PBS to the reaction vial. Remove the wash solution and add it to the radioiodinated streptavidin.
5. For purification, load the reaction mixture onto a 9 × 200 mm Bio-Gel P-6DG column packed in PBS (0.1% BSA may be added as a carrier to the PBS to reduce loss of streptavidin by adsorption to the column). Elute the column with PBS and collect 0.5-mL fractions. The first set of radioactive fractions (as determined by counting in a γ-ray counter) contains radioiodinated streptavidin, while the unreacted radioiodine elutes in the later fractions. Pool the radioiodinated streptavidin fractions.
6. Assessment of protein-associated activity with trichloroacetic acid precipitation:
   a. Dilute a small volume of the pooled radiolabeled streptavidin with saline solution such that 50 µL of the diluted solution has 10^4–10^6 cpm.
   b. Add 50 µL of the diluted streptavidin solution to a 12 × 75 mm glass tube, followed by 500 µL of a 0.1% BSA solution in saline.
   c. For precipitating the proteins, add 500 µL of 10% (w/v) TCA solution in saline.
   d. Incubate the solution for 30 min at room temperature and count the radioactivity of the solution for 10 min ("Total Counts").
   e. Centrifuge the tube at 500 g for 10 min and carefully discard the supernatant in a radioactive waste container.
   f. Resuspend the pellet in 1 mL of saline and count its radioactivity for 10 min ("Bound Counts").
   g. The percentage of radioactivity bound to streptavidin is determined using the following equation:

$$[(\text{Bound counts})/(\text{Total counts})] \times 100 = \%\ \text{of radioactivity bound to streptavidin} \quad (2)$$

3.3. Avidin Affinity Matrices

3.3.1. Native Avidin Affinity Matrix

1. Wash 10 mL of sedimented 6% crosslinked agarose with distilled water on a glass or Buchner filter and remove excess water by suction.
2. Dissolve 14 mg of $NaBH_4$ in 7 mL of 1*M* NaOH. Add this solution along with 7 mL of 1,4-butanediol-diglycidylether to the washed agarose, with mixing. Allow the reaction to proceed for 10 h or more at room temperature with gentle stirring.
3. Extensively wash the activated gel with distilled water on a supporting filter. The washed gel can be stored in water at 4°C, for up to 10 d.
4. Dissolve 50–100 mg of avidin in 10–20 mL of 0.2*M* sodium carbonate pH 9.5 and suspend the sedimented activated agarose gel in the same buffer to obtain a workable slurry.
5. Slowly drip the agarose slurry into the stirred protein solution and allow the binding to take place at room temperature for 2 d with continuous gentle mixing.
6. Wash the avidin-agarose mixture in PBS until the filtrate shows no absorbance at 280 nm. Store at 4°C in the presence of 0.02% sodium azide.

3.3.2. Monomeric Avidin Affinity Matrix

1. Filter the avidin-agarose matrix (from Section 3.3.1., step 6) on a glass or Buchner filter (or pack in a column) and wash four times with 2 vol of 6*M* guanidine · HCl in 0.2*M* KCl, pH 1.5, to dissociate the tetrameric avidin.
2. Thoroughly wash the gel with 0.2*M* potassium phosphate, pH 7.5, and suspend in 10 mL of the same buffer.
3. Add 3 mg of solid succinic anhydride to succinylate the monomeric avidin and incubate for 1 h at room temperature with gentle stirring.

4. Wash the gel with 0.2*M* potassium phosphate, pH 7.5, pack in a column, and saturate the binding sites by running through three volumes of 1 m*M* biotin dissolved in the same buffer.
5. Remove biotin from the low affinity binding sites by washing the column with 0.2*M* glycine · HCl, pH 2.0.
6. Store the column equilibrated in 20 m*M* sodium phosphate, 0.5*M* NaCl, 0.02% sodium azide, pH 7.5. The column is now ready to use.
7. Load the column with the mixture to be purified. Elute any unbound protein by adding 20 m*M* sodium phosphate, 0.5*M* NaCl, pH 7.5. Add biotin to the same buffer to obtain a final concentration of 0.8 m*M* to elute the biotinylated compound.
8. Regenerate the column after each run by washing with 0.2*M* glycine · HCl, pH 2.0.

4. Notes

1. A detailed procedure for the conjugation of fluorophores to antibodies has been recently published *(24)*. This protocol can be modified for conjugation of fluorophores to avidin or avidin-related proteins by using a dye-to-avidin molar ratio of 5–8:1.
2. The conjugation reaction for oligonucleotides synthesized with a disulfide containing a protecting group should be performed under nitrogen or argon. Deprotect the disulfide of the oligonucleotide using DTT. Add 1 mg of DTT to 140 µL of a 6 µ*M* oligonucleotide (21–33 mer) solution in 0.1*M* phosphate buffer containing 5 m*M* ethylenediamine-tetraacetic acid. Stir the solution at 37°C for 0.5 h. Purify the reaction mixture using a disposable desalting column. Combine the oligonucleotide-containing fractions with thiolated avidin prepared as described in Section 3.1.1. It should be noted that, in this case, conjugation occurs through the formation of a disulfide bond instead of a thioether bond. Disulfides are sensitive to reducing agents; however, they make reasonably stable conjugates, useful in most applications *(25)*. Purify the conjugate as outlined in Section 3.1.4.
3. Using a molar ratio of SPDP to avidin of 5 yields 1–2 protected sulfhydryls per molecule of avidin. This range of thiols per mole is found to produce the best yield of a 1:1 conjugate.
4. Buffer and pH: The entire procedure for preparation of conjugates through thioether bonds can be performed at pH 7.5. (**Note:** Organic buffers containing amines, such as Tris, are unsuitable.) Antibodies or enzymes in PBS can be prepared for reaction with SMCC by adding 1/10 volume of 1*M* sodium bicarbonate solution. This step eliminates dialysis and consequent dilution of the protein. Presence of azide at concentrations above 0.1% may interfere with the reaction of the protein with SMCC or of avidin with SPDP. IgM antibodies denature above pH 7.2. They can, however, be conjugated in PBS at pH 7.0 by increasing the molar ratio of maleimide to antibody.
5. Peptides (20–25 amino acids) containing a single cysteine can also be conjugated to thiolated avidin by modifying the procedure described in Section 3.1. and performing the reaction under argon or nitrogen *(26)*. Peptide-avidin conjugate formation described here also involves the formation of a disulfide bond. For conjugation with 5 mg avidin, dissolve 1.6 mg of a lyophilized cysteine-containing peptide in 900 µL water/methanol (2:1 v/v) using 50 m*M* NaOH (a few microliters at a time) to improve solubility. Immediately prior to use, cleave any cystine-bridged homodimer that may be present by the addition of TCEP solution (10 mg/mL in 0.1*M* phosphate buffer) to obtain a TCEP to peptide ratio of 3. Incubate for 15 min at room temperature. Purify the peptide-TCEP mixture using a disposable desalting column. Combine the peptide-containing fractions with thiolated avidin prepared as described in Section 3.1.1. Purify the conjugate as described in Section 3.1.4.

6. Avidin and antibody or enzyme concentration: The concentration of avidin as well as that of the protein to be conjugated should be 2–10 mg/mL. The crosslinking efficiency and, consequently, the yield of the conjugate decreases at lower concentrations of the thiolated avidin and maleimide-derivatized protein. To obtain 1:1 conjugates, equimolar concentrations of avidin and the protein are desirable. However, most methods of conjugation will generate conjugates of different sizes, following the Poisson distribution. The size range obtained with the method described here is much narrower because the number of proteins reacting with each mole of avidin can be regulated by the degree of thiolation of avidin.
7. It is essential that the procedure described in Section 3.1.2. be performed approx 24 h before the procedure described in Section 3.1.3., because the deprotected thiolated avidin and the maleimide derivative of the protein are unstable. Purification of the maleimide-derivatized protein by size exclusion chromatography can be performed more rapidly than dialysis; however, the former leads to dilution of the protein and a decrease in the yield of the conjugate.
8. If the molecule being conjugated to avidin is β-galactosidase or other free thiol-containing oligonucleotide or protein, NEM treatment is not performed.
9. Radioiodination of streptavidin uses procedures similar to those used for stable nuclides. However, some distinct differences remain, since radioiodinations are performed in dilute solutions. Also, the radioiodination mixture contains minor impurities formed during the preparation and purification of the radionuclide. Thus, optimization of reaction parameters is essential for performing radioiodination. This reaction is carried out in small volumes; it is therefore essential to ensure adequate mixing at the outset of the reaction. Inadequate mixing is often responsible for poor radioiodination yield.
10. Specific activity using the method described in Section 3.2. is usually in the range of 10–50 mCi/mg and the protein-bound radioactivity obtained is >95%. Higher specific activity can be achieved by increasing the reaction time of step 3 in Section 3.2. by using more beads, or by increasing the amount of radioiodine. However, one must bear in mind that at longer incubation times, the risk of damage to streptavidin is greater.
11. Storage and stability of avidin conjugates: Most avidin conjugates can be stored at 4°C or –20°C after lyophilization. Because of the variation in antibody structure, there is no general rule on the best method to store avidin-antibody conjugates, and the best conditions are determined experimentally. Aliquoting in small amounts and freezing is generally satisfactory. Radiolabeled streptavidin is aliquoted (~100 mL/tube) and stored at 4°C or –20°C until use.

References

1. Green, N. M. (1975) Avidin, in *Advances in Protein Chemistry*, vol. 29 (Anfinsen, C. M., Edsall, J. T., and Richards, F. M., eds.), Academic, New York, 85–133.
2. Bayer, E. A. and Wilchek, M. (1980) The use of the avidin-biotin complex as a tool in molecular biology. *Meth. Biochem. Anal.* **26,** 1–45.
3. Niemeyer, C. M., Sano, T., Smith, C. L., and Cantor, C. R. (1994) Oligonucleotide-directed self-assembly of proteins: semisynthetic DNA-streptavidin hybrid molecules as connectors for the generation of macroscopic arrays and the construction of supramolecular bioconjugates. *Nucleic Acids Res.* **22,** 5330–5339.
4. Paganelli, G., Belloni, C., Magnani, P., Zito, F., Pasini, A., Sassi, I., Meroni, M., Mariani, M., Vignali, M., Siccardi, A. G., and Fazio, F. (1992) Two-step tumor targetting in ovarian cancer patients using biotinylated monoclonal antibodies and radioactive streptavidin. *Eur. J. Nucl. Med.* **19,** 322–329.

5. van Osdol, W. W., Sung, C., Dedrick, R. L., and Weinstein, J. N. (1993) A distributed pharmacokinetic model of two-step imaging and treatment protocols: application to streptavidin-conjugated monoclonal antibodies and radiolabeled biotin. *J. Nucl. Med.* **34,** 1552–1564.
6. Markwell, M. A. K. (1982) A new solid-state reagent to iodinate proteins. I. Conditions for the efficient labeling of antiserum. *Anal. Biochem.* **125,** 427–432.
7. Salacinski, P. R. P., McLean, C., Sykes, J. E. C., Clement-Jones, V. V., and Lowry, P. J. (1981) Iodination of proteins, glycoproteins, and peptides using a solid-phase oxidizing agent, 1,3,4,6,-tetrachloro-3α,6α-diphenyl glycoluril (iodogen). *Anal. Biochem.* **117,** 136–146.
8. Mock, D. M. (1990) Sequential solid-phase assay for biotin based on ^{125}I-labeled avidin. *Methods Enzymol.* **184,** 224–233.
9. Bolton, A. E. and Hunter, W. M. (1973) The labeling of proteins to high specific radioactivity by conjugation to an ^{125}I-containing acylating agent. Applications to the radioimmunoassay. *Biochem. J.* **133,** 529–539.
10. Vaidyanathan, G., Affleck, D. J., and Zalutsky, M. R. (1993) Radioiodination of proteins using N-succinimidyl 4-hydroxy-3-iodobenzoate. *Bioconjugate Chem.* **4,** 78–84.
11. Vaidyanathan, G. and Zalutsky, M. R. (1990) Radioiodination of antibodies via N-succinimidyl 2,4-dimethoxy-3-(trialkylstannyl)benzoates. *Bioconjugate Chem.* **1,** 387–393.
12. Hylarides, M. D., Wilbur, D. S., Reed, M. W., Hadley, S. W., Schroeder, J. R., and Grant, L. M. (1991) Preparation and *in vivo* evaluation of an N-(p-[^{125}I]iodophenethyl)maleimide-antibody conjugate. *Bioconjugate Chem.* **2,** 435–440.
13. Arano, Y., Wakisaka, K., Ohmomo, Y., Uezono, T., Mukai, T., Motonari, H., Shiono, H., Sakahara, H., Konishi, J., Tanaka, C., and Yokoyama, A. (1994) Maleimidoethyl 3-(tri-*n*-butylstannyl)hippurate: A useful radioiodination reagent for protein radiopharmaceuticals to enhance target selective radioactivity localization. *J. Med. Chem.* **37,** 2609–2618.
14. Willner, D., Trail, P. A., Hofstead, S. J., Dalton King, H., Lasch, S. J., Braslawsky, G. R., Greenfield, R. S., Kaneko, T., and Firestone, R. A. (1993) (6-Maleimidocaproyl)hydrazone of doxorubicin—A new derivative for the preparation of immunoconjugates of doxorubicin. *Bioconjugate Chem.* **4,** 521–527.
15. Arnold Jr., L. J. (1985) Polylysine-drug conjugates. *Methods Enzymol.* **112,** 270–285.
16. Pietersz, G. A. and McKenzie, I. F. (1992) Antibody conjugates for the treatment of cancer. *Immunol. Rev.* **129,** 57–80.
17. Orr, G. A. (1981) The use of the 2-iminobiotin-avidin interaction for the selective retrieval of labeled plasma membrane components. *J. Biol. Chem.* **256,** 761–766.
18. Dimroth, P. (1986) Preparation, characterization, and reconstitution of oxaloacetate decarboxylase from *Klebsiella aerogenes*, a sodium pump. *Methods Enzymol.* **125,** 530–540.
19. Dean, P. D. G., Johnson, W. S., and Middle, F. S. (1985) Activation procedures, in *Affinity Chromatography. A Practical Approach,* IRL, Washington, DC, pp. 34,35.
20. Kohanski, R. A. and Lane, D. (1990) Monovalent avidin affinity columns. *Methods Enzymol.* **184,** 194–220.
21. Chilkoti, A., Schwartz, B. L., Smith, R. D., Long, C. J., and Stayton, P. S. (1995) Engineered chimeric streptavidin tetramers as novel tools for bioseparations and drug delivery. *Bio/Technology* **13,** 1198–1204.
22. Wong, S. S. (1991) Reactive groups of proteins and their modifying agents, in *Chemistry of Protein Conjugation and Crosslinking*, CRC, Boston, MA, pp. 7–48.
23. Carlsson, J., Drevin, H., and Axen, R. (1978) Protein thiolation and reversible protein–protein conjugation. N-Succinimidyl 3-(2-pyridyldithio)propionate, a new heterobifunctional reagent. *Biochem. J.* **173,** 723–737.

24. Haugland, R. P. (1995) Coupling of monoclonal antibodies with fluorophores. In *Methods in Molecular Biology*, vol. 45 (Davis, W. C., ed.), Humana, Totowa, NJ, pp. 205–221.
25. Kronick, M. N. and Grossman, P. D. (1983) Immunoassay techniques with fluorescent phycobiliprotein conjugates. *Clin. Chem.* **29,** 1582–1586.
26. Bongartz, J.-P., Aubertin, A.-M., Milhaud, P. G., and Lebleu, B. (1994) Improved biological activity of antisense oligonucleotides conjugated to a fusogenic peptide. *Nucleic Acids Res.* **22,** 4681–4688.

48

Detection of Polypeptides on Blots Using Secondary Antibodies or Protein A

Nicholas J. Kruger

1. Introduction

Immunoblotting provides a simple and effective method for identifying specific antigens in a complex mixture of proteins. Initially, the constituent polypeptides are separated using sodium dodecyl sulfate-polyacrylamide gel electrophoresis (SDS-PAGE), or a similar technique, and then are transferred either electrophoretically or by diffusion onto a nitrocellulose filter. Once immobilized on a sheet of nitrocellulose, specific polypeptides can be identified using antibodies that bind to antigens retained on the filter and subsequent visualization of the resulting antibody–antigen complex. This chapter describes conditions suitable for binding antibodies to immobilized proteins and methods for locating these antibody–antigen complexes using appropriately labeled ligands. These methods are based on those of Blake et al. *(1)*, Burnette *(2)*, and Towbin et al. *(3)*.

Although there are several different techniques for visualizing antibodies bound to nitrocellulose, most exploit only two different types of ligand. One is protein A, which is either radiolabeled or conjugated with a marker enzyme. The other ligand is an antibody raised against IgG from the species used to generate the primary antibody. Usually, this secondary antibody is either conjugated with a marker enzyme or linked to biotin. In the later instance, the biotinylated antibody is subsequently detected using avidin linked to a marker enzyme.

Detection systems based on protein A are both convenient and sensitive. Protein A specifically binds the Fc region of immunoglobulin G (IgG) from many mammals *(4)*. Thus, this compound provides a general reagent for detecting antibodies from several sources. Using this ligand, as little as 0.1 ng of protein may be detected, though the precise amount will vary with the specific antibody titer *(5)*. The principal disadvantage of protein A is that it fails to bind effectively to major IgG subclasses from several experimentally important sources, such as rat, goat, and sheep (*see* Table 1). For antibodies raised in such animals, a similar method using protein G derivatives may be suitable (*see* Note 1). Alternatively, antibody bound to the nitrocellulose filter may be detected using a second antibody raised against IgG (or other class of immunoglobulin) from the species used to generate the primary antibody. The advantage of such secondary antibody systems is that they bind only to antibodies from an individual

From: The Protein Protocols Handbook
Edited by: *J. M. Walker Humana Press Inc., Totowa, NJ*

Table 1
Binding of Protein A to Immunoglobulins from Various Species

Species	Serum IgG level[a], mg/mL	Affinity[b] Strong	Weak	Unreactive
Rabbit	5	IgG		
Human	12	IgG1[c],2,4 IgA2 IgM (some)		IgG3
Guinea pig	6	IgG1,2[c]		
Mouse	7	IgG2a,2b,3	IgG1[c] IgM	
Pig	18	IgG IgM (some) IgA (some)		
Goat	–	IgG2	IgG1	
Sheep	–	IgG2		IgG1
Dog	9	IgG IgM (some) IgA (some)		
Rat	16	IgG1,2c	IgG2b	IgG2a
Cow	20	IgG2		IgG1

[a]The values for serum IgG levels are approximate, and significant variation may occur among individuals.

[b]Immunoglobulins failing to bind protein A at pH 8.0 are described as unreactive. Immunoglobulins that bind protein A at pH 8.0, but not at pH 6.0 are considered to have a weak affinity. High affinity indicates binding below pH 6.0. Data from Tijssen *(4)* and references therein.

[c]Denotes the major IgG subclass.

species. When combined with different marker enzymes, the specificity of secondary antibodies may be exploited to identify multiple polypeptides on a single nitrocellulose membrane *(4)*.

The marker enzymes most commonly used for detection are alkaline phosphatase and horseradish peroxidase. Both enzymes can be linked efficiently to other proteins, such as antibodies, protein A, and avidin, without interfering with the function of the latter proteins or inactivating the enzyme. Moreover, a broad range of synthetic substrates have been developed for each of these enzymes. Enzyme activity is normally visualized by incubating the membrane with an appropriate chromogenic substrate, which is converted to a colored, insoluble product. The latter precipitates onto the membrane in the area of enzyme activity, thus identifying the site of the antibody–antigen complex (*see* Note 2).

Both antigens and antisera can be screened efficiently using immunoblotting. Probing of a crude extract after fractionation by SDS-PAGE indicates the specificity of an antiserum. The identity of the antigen can be confirmed using a complementary technique, such as immunoprecipitation of enzyme activity. This information is essential if the antibodies are to be used reliably. Once characterized, an antiserum may be used to identify antigenically related proteins in other extracts using the same technique. Examples of the potential of immunoblotting have been described by Towbin and Gordon *(6)*.

2. Materials

1. Electrophoretic blotting system, such as Trans-Blot, supplied by Bio-Rad (Hercules, CA).
2. Nitrocellulose paper: 0.45-µm pore size.
3. Protein A derivatives.
   a. Alkaline phosphatase-conjugated protein A obtained from Sigma (St. Louis, MO). Dissolve 0.1 mg in 1 mL of 50% (v/v) glycerol in water. Store at –20°C.
   b. Horseradish peroxidase-conjugated protein A obtained from Sigma. Dissolve 0.1 mg in 1 mL of 50% (v/v) glycerol in water. Store at –20°C.
   c. ^{125}I-labeled protein A, SA 30 mCi·mg^{-1}. Affinity-purified protein A, suitable for blotting, is available commercially (*see* Note 3). ^{125}I **emits γ-radiation. Check the procedures for safe handling and disposal of this radioisotope.**
4. Secondary antibody: A wide range of both alkaline phosphatase- and horseradish peroxidase-conjugated antibodies are available commercially. They are usually supplied as an aqueous solution containing protein stabilizers. The solution should be stored under the conditions recommended by the supplier. **Ensure that the enzyme-linked antibody is against IgG of the species in which the primary antibody was raised.**
5. Washing solutions: Phosphate-buffered saline (PBS): Make 2 L containing 10 m*M* NaH_2PO_4, 150 m*M* NaCl adjusted to pH 7.2, using NaOH. This solution is stable and may be stored at 4°C. It is susceptible to microbial contamination, however, and is usually made as required.

   The other washing solutions are made by dissolving the appropriate weight of bovine serum albumin (BSA) or Triton X-100 in PBS. Dissolve BSA by rocking the mixture gently in a large, sealed bottle to avoid excessive foaming. The "blocking" and "antibody" solutions containing 8% albumin may be stored at –20°C and reused several times. Microbial contamination can be limited by filter-sterilizing these solutions after use or by adding 0.05% (w/v) NaN_3 (but *see* Note 4). Other solutions are made as required and discarded after use.
6. Alkaline phosphatase substrate mixture:
   a. Diethanolamine buffer: Make up 100 m*M* diethanolamine and adjust to pH 9.8 using HCl. This buffer is usually made up as required, but may be stored at 4°C if care is taken to avoid microbial contamination.
   b. 1*M* $MgCl_2$: This can be stored at 4°C.

   Combine 200 µL of 1*M* $MgCl_2$, 5 mg nitroblue tetrazolium, 2.5 mg 5-bromo-4-chloroindolyl phosphate (disodium salt—*see* Note 5). Adjust the volume to 50 mL using 100 m*M* diethanolamine buffer. Make up this reaction mixture as required, and protect from the light before use.
7. Horseradish peroxidase substrate mixture.
   a. Make up 50 m*M* acetic acid, and adjust to pH 5.0 using NaOH. This buffer is usually made up as required, but may be stored at 4°C if care is taken to avoid microbial contamination.
   b. Diaminobenzidine stock solution of 1 mg/mL dissolved in acetone. Store in the dark at –20°C. **Caution: Diaminobenzidine is potentially carcinogenic; handle with care.**
   c. Hydrogen peroxide at a concentration of 30% (v/v). This compound decomposes, even when stored at 4°C. The precise concentration of the stock solution can be determined by measuring its absorbance at 240 nm. The molar extinction coefficient for H_2O_2 is 43.6M^{-1}·cm^{-1} at this wavelength (*see* Note 6).

   Combine 50 mL of acetate buffer, 2 mL of diaminobenzidine stock solution, and 30 µL of hydrogen peroxide immediately before use. Mix gently and avoid vigorous shaking to prevent unwanted oxidation of the substrate. Protect the solution from the light.

8. Protein staining solutions: These are stable at room temperature for several weeks and may be reused.
   a. Amido black stain (100 mL): 0.1% (w/v) Amido black in 25% (v/v) propan-2-ol, 10% (v/v) acetic acid.
   b. Ponceau S stain (100 mL): 0.2% (w/v) Ponceau S, 10% (w/v) acetic acid.
   c. Ponceau S destain (400 mL): distilled water.

3. Methods

3.1. Immunodetection of Polypeptides

1. Following SDS-PAGE (*see* Chapter 11), electroblot the polypeptides from the gel onto nitrocellulose at 50 V for 3 h using a Bio-Rad Trans-Blot apparatus, or at 100 V for 1 h using a Bio-Rad Mini Trans-Blot system (*see* Chapters 37–39).
2. After blotting, transfer the nitrocellulose filters individually to plastic trays for the subsequent incubations. Ensure that the nitrocellulose surface that was closest to the gel is uppermost. Do not allow the filter to dry out, since this often increases nonspecific binding and results in heavy, uneven backgrounds. The nitrocellulose filter should be handled sparingly to prevent contamination by grease or foreign proteins. Always wear disposable plastic gloves, and only touch the sides of the filter.
3. If desired, stain the blot for total protein using Ponceau S as described in Section 3.3.2. (*see* Note 7).
4. Rinse the nitrocellulose briefly with 100 mL of PBS. Then incubate the blot at room temperature with the following solutions, shaking gently (*see* Note 8).
   a. 50 mL of PBS/8% BSA for 30 min. This blocks the remaining protein binding sites on the nitrocellulose (*see* Note 9).
   b. 50 mL of PBS/8% BSA containing 50–500 μL of antiserum for 2–16 h (*see* Note 10).
   c. Wash the nitrocellulose at least five times, each time using 100 mL of PBS for 15 min to remove unbound antibodies.
   d. 50 mL of PBS/4% BSA containing an appropriate ligand for 2 h (*see* Note 11). This is likely to be one of the following:
      i. 1 μCi ^{125}I-labeled protein A.
      ii. 5 μg enzyme-conjugated protein A.
      iii. Enzyme-conjugated secondary antibody at the manufacturers' recommended dilution (normally between 1:1000 and 1:10,000).
   e. Wash the nitrocellulose at least five times, each time using 100 mL PBS/1% Triton X-100 for 5 min to remove unbound protein A or secondary antibody.

   To ensure effective washing of the filter, pour off each solution from the same corner of the tray, and replace the lid in the same orientation.

3.2. Visualization of Antigen–Antibody Complex

3.2.1. Alkaline Phosphatase-Conjugated Ligand

In this method, the enzyme hydrolyzes 5-bromo-4-chloroindolyl phosphate to the corresponding indoxyl compound. The product tautomerizes to a ketone, oxidizes, and then dimerizes to form an insoluble blue indigo that is deposited on the filter. Hydrogen ions released during the dimerization reduce nitroblue tetrazolium to the corresponding diformazan. The latter compound is an insoluble intense purple compound that is deposited alongside the indigo, enhancing the initial signal.

1. Briefly rinse the filter twice, each time using 50 mL of diethanolamine buffer.

2. Incubate the filter with 50 mL of alkaline phosphatase substrate mixture until the blue-purple products appear, usually after 5–30 min.
3. Prevent further color development by removing the substrate mixture and washing the filter three times, each time in 100 mL of distilled water. Finally, dry the filter thoroughly before storing (*see* Note 12).

3.2.2. Horseradish Peroxidase-Conjugated Ligand

Peroxidase catalyzes the transfer of hydrogen from a wide range of hydrogen donors to H_2O_2, and it is usually measured indirectly by the oxidation of the second substrate. In this method, soluble 3,3'-diaminobenzidine is converted to a red-brown insoluble complex that is deposited on the filter. The sensitivity of this technique may be increased up to 100-fold by intensifying the diaminobenzidine-based products using a combination of cobalt and nickel salts that produce a dense black precipitate (*see* Note 13).

1. Briefly rinse the filter twice, each time using 50 mL of sodium acetate buffer.
2. Incubate the filter with 50 mL of horseradish peroxidase substrate mixture until the red-brown insoluble products accumulate. Reaction times longer than about 30 min are unlikely to be effective owing to substrate inactivation of peroxidase (*see* Note 6).
3. When sufficient color has developed, remove the substrate mixture and wash the filter three times with 100 mL of distilled water. Then dry the filter, and store it in the dark (*see* Note 12).

3.2.3. ^{125}I-Labeled Protein A

1. If desired, stain the blot for total protein as described below.
2. Allow the filter to dry. Do not use excessive heat, since nitrocellulose is potentially explosive when dry.
3. Mark the nitrocellulose with radioactive ink to allow alignment with exposed and developed X-ray film.
4. Fluorograph the blot using suitable X-ray film and intensifying screens. Expose the film at –70°C for 6–72 h, depending on the intensity of the signal.

3.3. Staining of Total Protein

Either of the following stains is suitable for visualizing polypeptides after transfer onto nitrocellulose. Each can detect bands containing about 1 µg of protein. Coomassie blue is unsuitable for nitrocellulose membranes, since generally it produces heavy background staining.

3.3.1. Amido Black

Incubate the filter for 2–5 s in 100 mL of stain solution. Transfer immediately to 100 mL of destain solution, and wash with several changes to remove excess dye. Unacceptably dark backgrounds are produced by longer incubation times in the stain solution.

3.3.2. Ponceau S

Incubate the filter with 100 mL of Ponceau S stain solution for 30 min. Wash excess dye off the filter by rinsing in several changes of distilled water. The proteins may be destained by washing the filter in PBS (*see* Note 7).

4. Notes

1. Protein G, a cell-wall component of group G streptococci, binds to the Fc region of IgG from a wider range of species than does protein A *(7)*. Therefore, antibodies that react

poorly with protein A, particularly those from sheep, cow, and horse, may be detected by a similar method using protein G derivatives. In principle, the latter are more versatile. At present, however, the limited availability of suitable protein G conjugates means that, where suitable, protein A derivatives remain preferable.

2. Several visualization systems have been developed for both alkaline phosphatase and horseradish peroxidase. Alternatives to the assay systems provided in this chapter are described by Tijssen *(4)*. Currently, the greatest sensitivity is provided by luminescent detection systems that have been developed recently by several biochemical companies (*see* Chapter 50). The alkaline phosphatase system is based on the light emission that occurs during the hydrolysis of AMPPD (3-[2'-spiroadamantane]-4-methoxy-4-[3'-phosphoryloxy]-phenyl-1,2-dioxetane). The mechanism involves the enzyme-catalyzed formation of the dioxetane anion, followed by fragmentation of the anion to adamantone and the excited state of methyl *meta*-oxybenzoate. This latter anion is the source of light emission. The peroxidase detection system relies on light-emitted during the oxidation of luminol (3-aminophthalhydrazine) by peroxide radicals. The latter are formed during the enzyme-catalyzed reduction of hydrogen peroxide or other suitable substrates.
3. Iodination of protein A using Bolton and Hunter reagent labels the ε-NH_2 group of lysine, which apparently is not involved directly in the binding of protein A to the Fc region of IgG. This method is preferable to others, such as those using chloramine T or iodogen, which label tyrosine. The only tyrosine residues in protein A are associated with Fc binding sites, and their iodination may reduce the affinity of protein A for IgG *(8)*.

   ^{35}S-labeled protein A may be substituted for iodinated protein A in the protocol described in this chapter by those researchers not wishing to handle ^{125}I. ^{35}S-labeled protein A has the additional advantage of producing a far sharper image on X-ray film. However, this radioisotope requires longer exposure times.
4. Many workers include up to 0.05% sodium azide in the antibody and washing buffers to prevent microbial contamination. However, azide inhibits horseradish peroxidase. Therefore, do not use buffers containing azide when using this enzyme.
5. In the original description of this protocol, 5-bromo-4-chloroindolyl phosphate was made up as a stock solution in dimethylformamide. However, this is not necessary if the disodium salt is used since this compound dissolves readily in aqueous buffers.
6. Urea peroxide may be used instead of hydrogen peroxide as a substrate for peroxidase. The problems of instability, enzyme inactivation, and possibility of caustic burns associated with hydrogen peroxide are eliminated by using urea peroxide. A 10% (w/v) stock solution of urea peroxide is stable for several months and is used at a final concentration of 0.1% in the peroxidase substrate mixture.
7. If desired, the nitrocellulose filter may be stained with Ponceau S immediately after electroblotting. This staining apparently does not affect the subsequent immunodetection of polypeptides, if the filter is thoroughly destained using PBS before incubation with the antiserum. In addition to confirming that the polypeptides have been transferred successfully onto the filter, initial staining allows tracks from gels to be separated precisely and probed individually. This is useful when screening several antisera.
8. Nonspecific binding is a common problem in immunoblotting. Several factors are important in reducing the resulting background.

   First, the filter is washed in the presence of an "inert" protein to block the unoccupied binding sites. BSA is the most commonly used protein, but others, such as fetal calf serum, hemoglobin, gelatin, and nonfat dried milk, have been used successfully. Economically, the latter two alternatives are particularly attractive.

   The quality of protein used for blocking is important, since minor contaminants may interfere with either antigen–antibody interactions or the binding of protein A to IgG.

These contaminants may vary between preparations and can be sufficient to inhibit completely the detection of specific polypeptides. Routinely we use BSA (fraction V) from Sigma (St. Louis, MO) (product no. A 4503), but no doubt albumin from other sources is equally effective. The suitability of individual batches of protein should be checked using antisera known to react well on immunoblots.

Second, the background may be reduced further by including nonionic detergents in the appropriate solutions. These presumably decrease the hydrophobic interactions between antibodies and the nitrocellulose filter. Tween-20, Triton X-100, and Nonidet P-40 at concentrations of 0.1–1.0% have been used. In my experience, such detergents may supplement the blocking agents described above, but cannot substitute for these proteins. In addition, these detergents sometimes remove proteins from nitrocellulose (*see* Note 9).

Third, the nitrocellulose must be washed effectively to limit nonspecific binding. For this, the volumes of the washing solutions should be sufficient to flow gently over the surface of the filter during shaking. The method described in this chapter is suitable for 12×7 cm filters incubated in 14×9 cm trays. If the size of the filter is significantly different, the volumes of the washing solutions should be adjusted accordingly.

Finally, reducing the incubation temperature to 4°C may greatly decrease the extent of nonspecific background binding *(9)*.

9. Protein desorption from the membrane during the blocking step and subsequent incubations can result in the loss of antigen and decrease the sensitivity of detection *(9,10)*. In some instances, this problem may be reduced by incubating the membrane in 0.1*M* phosphate buffer (pH 2.0) for 30 min and then rinsing in PBS prior to treatment with the blocking agent. Such acid treatment is particularly effective when using nondenaturing gel blots, or SDS-PAGE blots transferred onto polyvinylidene difluoride rather than nitrocellulose membrane *(11)*.

   Alternatively, polyvinyl alcohol may be used as a blocking agent *(12)*. In comparative tests, PBS containing 1 μg/mL polyvinyl alcohol produced lower background staining than other commonly used blocking agents. Moreover, the blocking effect of polyvinyl alcohol is virtually instantaneous, allowing the incubation time to be reduced to 1 min and decreasing the opportunity for loss of protein from the membrane *(12)*.
10. The exact amount of antibody to use will depend largely on its titer. Generally, it is better to begin by using a small amount of antiserum. Excessive quantities of serum tend to increase the background rather than improve the sensitivity of the technique. Nonspecific binding can often be reduced by decreasing the amount of antibody used to probe the filter.
11. Deciding which form of detection system to use is largely a personal choice. ^{125}I-labeled protein A is extremely sensitive. This method has the advantage of allowing the polypeptide recognized by the antibody to be precisely identified by aligning the fluorograph with the original filter after staining for total protein. However, many researchers prefer not to work with this radioactive isotope. Comparison between the two enzymic detection systems is difficult because the reported sensitivity limits of both systems vary considerably, and most studies use different antigens, different primary antibodies, and different protocols. Despite these uncertainties, alkaline phosphatase is generally considered more sensitive than horseradish peroxidase. For routine work, I prefer to use alkaline phosphatase-conjugated protein A.
12. The products of the peroxidase reaction are susceptible to photobleaching and fading. Consequently, the developed filters should be stored in the dark, and the results photographed as soon as possible. The products of the phosphatase reaction are reportedly stable in the light. However, the author treats such filters in the same way–just in case!
13. To increase sensitivity of the diaminobenzidine-based staining protocol, replace the standard substrate mixture with the following intensifying solution *(13)*. Dissolve 100 mg

diaminobenzidine in 100 mL of 200 m*M* phosphate buffer (pH 7.3). To this solution, add, dropwise and with constant stirring, 5 mL of 1% (w/v) cobalt chloride followed by 4 mL of 1% (w/v) nickel ammonium sulfate. Finally, add 60 µL of 30% (v/v) hydrogen peroxide just before use.

14. Particular care should be taken when attempting to detect antigens on nitrocellulose using MAb. Certain cell lines may produce antibodies that recognize epitopes that are denatured by detergent. Such "conformational" antibodies may not bind to the antigen after SDS-PAGE.
15. Even before immunization, serum may contain antibodies, particularly against bacterial proteins. These antibodies may recognize proteins in an extract bound to the nitrocellulose filter. Therefore, when characterizing an antiserum, control filters should be incubated with an equal amount of preimmune serum to check whether such pre-existing antibodies interfere in the immunodetection of specific proteins.
16. Quantitation of specific antigens using this technique is difficult and must be accompanied by adequate evidence that the amount of product or radioactivity bound to the filter is directly related to the amount of antigen in the initial extract. This is important, since polypeptides may vary in the extent to which they are eluted from the gel and retained by the nitrocellulose (*see* Note 9). Additionally, in some tissues, proteins may interfere with the binding of antigen to the filter. Therefore, the reliability of the technique should be checked for each extract.

    Perhaps the best evidence is provided by determining the recovery of a known amount of pure antigen. For this, duplicate samples are prepared, and to one is added a known amount of antigen comparable to that already present in the extract. Ideally, the pure antigen should be identical to that in the extract. The recovery is calculated by comparing the antigen measured in the original and supplemented samples. Such evidence is preferable to that obtained from only measuring known amounts of pure antigen. The latter indicates the detection limits of the assay, but does not test for possible interference by other components in the extract.

    The other major problem in quantifying the level of antigen on immunoblots derives from the technical problems associated with relating densitometric measurements from photographs or fluorographs to the amount of antibody bound to the filter. A combined radiochemical-color method has been developed that circumvents these problems *(14)*. The technique involves challenging the filter sequentially with alkaline phosphatase-conjugated secondary antibody and ^{125}I-labeled protein A (which binds to the secondary antibody). The color reaction derived from the enzyme conjugate is used to localize the antibody–antigen complex. The appropriate region of the filter is then excised, and the radioactivity derived from the protein A associated with the band is measured to provide a direct estimate of the amount of antigen.

References

1. Blake, M. S., Johnson, K. H., Russell-Jones, G. J., and Gotschlich, E. C. (1984) A rapid, sensitive method for detection of alkaline phosphatase-conjugated anti-antibodies on Western blots. *Anal. Biochem.* **136,** 175–179.
2. Burnette, W. N. (1981) "Western blotting": Electrophoretic transfer of proteins from sodium dodecyl sulfate-polyacrylamide gels to unmodified nitrocellulose and radiographic detection with antibody and radioiodinated protein A. *Anal. Biochem.* **112,** 195–203.
3. Towbin, H., Staehelin, T., and Gordon, J. (1979) Electrophoretic transfer of proteins from polyacrylamide gels to nitrocellulose sheets: Procedure and some applications. *Proc. Natl. Acad. Sci. USA* **76,** 4350–4354.

4. Tijssen, P. (1985) *Practice and Theory of Enzyme Immunoassays.* Elsevier, Amsterdam.
5. Vaessen, R. T. M. J., Kreide, J., and Groot, G. S. P. (1981) Protein transfer to nitrocellulose filters. *FEBS Lett.* **124,** 193–196.
6. Towbin, H. and Gordon, J. (1984) Immunoblotting and dot immunobinding-current status and outlook. *J. Immunol. Methods* **72,** 313–340.
7. Akerstrom, B., Brodin, T., Reis, K., and Bjorck, L. (1985) Protein G: A powerful tool for binding and detection of monoclonal and polyclonal antibodies. *J. Immunol.* **135,** 2589–2592.
8. Langone, J. J. (1980) ^{125}I-labeled protein A: reactivity with IgG and use as a tracer in radioimmunoassay, in *Methods in Enzymology,* vol. 70 (Vunakis, H. V. and Langone, J. J., eds.) Academic, New York, pp. 356–375.
9. Thean, E. T. and Toh, B. H. (1989) Western immunoblotting: temperature-dependent reduction in background staining. *Anal. Biochem.* **177,** 256–258.
10. DenHollander, N. and Befus, D. (1989) Loss of antigens from immunoblotting membranes. *J. Immunol. Methods* **122,** 129–135.
11. Hoffman, W. L., Jump, A. A., and Ruggles, A. O. (1994) Soaking nitrocellulose blots in acidic buffers improves the detection of bound antibodies without loss of biological activity. *Anal. Biochem.* **217,** 153–155.
12. Miranda, P. V., Brandelli, A., and Tezon, J. G. (1993) Instantaneous blocking for immunoblots. *Anal. Biochem.* **209,** 376,377.
13. Adams, J. C. (1981) Heavy metal intensification of DAB-based HRP reaction product. *J. Histochem. Cytochem.* **29,** 775.
14. Esmaeli-Azad, B. and Feinstein, S.C. (1991) A general radiochemical-color method for quantitation of immunoblots. *Anal. Biochem.* **199,** 275–278.

49

Detection of Proteins on Blots Using Avidin- or Streptavidin-Biotin

William J. LaRochelle

1. Introduction

Since the initial publication by Towbin and coworkers *(1)* on the preparation of replicas of sodium dodecyl sulfate(SDS)-polyacrylamide gel patterns, commonly called protein blots, the technique of transferring proteins from inaccessible gel matrices to accessible solid supports, such as nitrocellulose or nylon membranes, has become widely utilized *(2,3)*.

The detection of proteins on blots has ranged from the specific visualization of an individual protein of interest to the general staining of total protein. Specific proteins are detected with probes, such as antibodies or toxins, that are either directly radiolabeled or conjugated to an enzyme (and *see* Chapters 44, 48–50, and 97 and refs. *4* and *5*). Alternatively, bound and unlabeled antibody or toxin is detected by a secondary affinity probe similarly conjugated. Total protein detection is based on physical staining methods, such as Coomassie blue *(6)*, amido black *(7)*, India ink *(8)*, or silver-enhanced copper staining *(9)*. Other approaches often require derivatization of protein with hapten followed by antihapten antibody and labeled secondary antibody or protein A *(10)*.

Here, we exploit the high-affinity and well-characterized interactions of biotin with either avidin or streptavidin. Initially, proteins are resolved by SDS-PAGE, transferred to nitrocellulose paper, and the amino groups covalently derivatized *(11)* with sulfosuccinimidobiotin. Depending on the sensitivity required, either of two techniques that are illustrated in Fig. 1 are used to stain the proteins, which appear as dark bands against an essentially white background *(12,13)*. The first method utilizes avidin or streptavidin conjugated to horseradish peroxidase and detects <25 ng of protein in a single band. The second technique, although slightly more lengthy, requires streptavidin amplification with antistreptavidin antisera followed by a second antibody horseradish peroxidase conjugate and detects <5 ng of protein/band.

The methods described here *(12–14)* permit direct comparison of stained replicas with a duplicate blot that has been probed with antibody or ligand. Our approach is rapid and possesses greater sensitivity than the commonly used dyes. Moreover, our detection scheme is less costly and time-consuming than the use of the metal stains, which however possess greater sensitivity. Problems associated with gel shrinkage on

From: The Protein Protocols Handbook
Edited by: J. M. Walker Humana Press Inc., Totowa, NJ

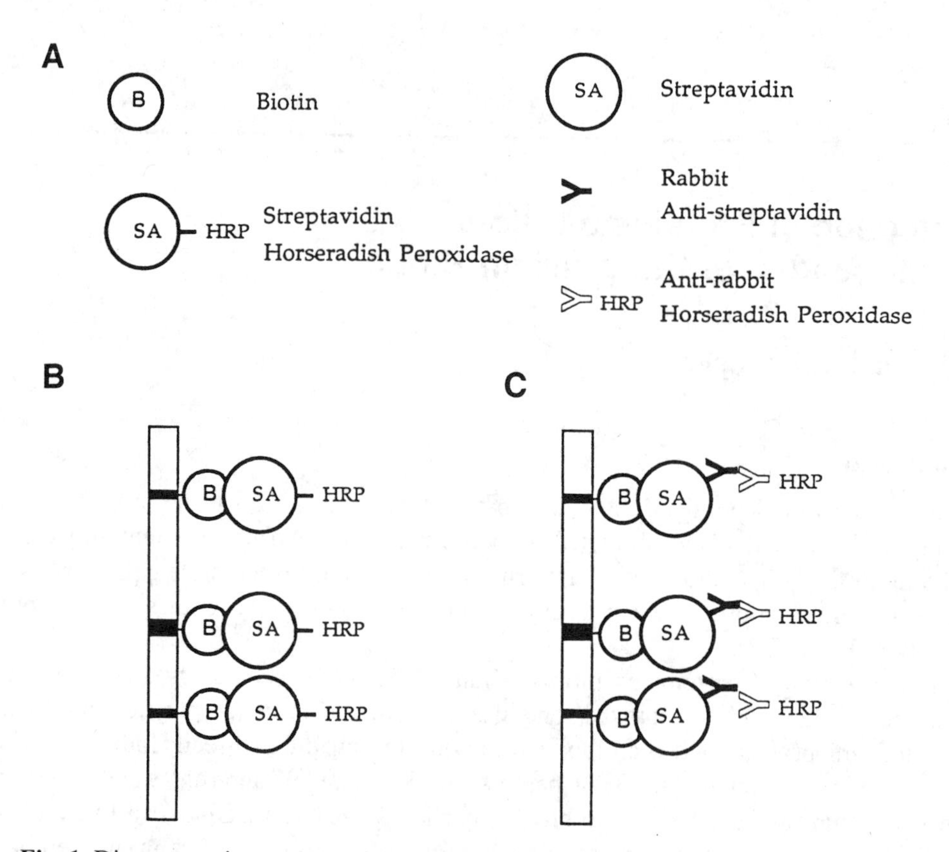

Fig. 1. Diagrammatic representation of streptavidin or amplified antistreptavidin staining of nitrocellulose replicas. Electrophoretically transferred proteins were biotinylated on a nitrocellulose replica depicted here as a strip blot. **(A)** B, Biotin; SA, streptavidin; HRP, horseradish peroxidase. **(B)** Schematic diagram of streptavidin staining method (Section 3.1.). Total protein, shown as individual dark bands, was biotinylated and detected with streptavidin conjugated to horseradish peroxidase followed by the α-chloronaphthol color reaction. **(C)** Schematic diagram of amplified antistreptavidin staining method (Section 3.2.). After electrophoretic transfer and biotinylation, streptavidin binding to biotinylated proteins was detected with rabbit antistreptavidin followed by goat antirabbit antibody conjugated to horseradish peroxidase as described in Section 3.2.

drying or altered electrophoretic mobility of proteins cause by staining or derivatization prior to electrophoresis are avoided. Since our initial study *(12)*, this approach has also proven useful in detection and labeling of DNA on membrane supports *(15)*.

2. Materials

2.1. Avidin or Streptavidin Horseradish Peroxidase Staining

1. Distilled, deionized water.
2. Plastic trays rather than glass are preferred.
3. Nitrocellulose membranes, 0.45 μm (Bio-Rad, Hercules, CA).
4. Sulfosuccinimidobiotin (Pierce Chemical Company, Rockford, IL) solution: 10 m*M* in 100 m*M* sodium bicarbonate, pH 8.0.

5. Avidin conjugated to horseradish peroxidase (Cappel Laboratories, Malvern, PA).
6. Streptavidin conjugated to horseradish peroxidase (Gibco-BRL, Gaithersburg, MD).
7. PBS: 10 m*M* sodium phosphate, 150 m*M* sodium chloride, pH 7.4.
8. Block solution: PBS containing 5% v/v newborn calf serum (Gibco-BRL) and 3% w/v bovine serum albumin (BSA) (Fraction V, 98–99%, Sigma, St. Louis, MO).
9. Wash solution: PBS containing 0.05% w/v Tween-80 (Sigma).
10. PBS containing 0.05% w/v Tween-80 and 1.0% w/v BSA.
11. 1.0*M* Glycine-HCl, pH 6.5.
12. α-Chloro-naphthol solution: PBS containing 0.6 mg/mL α-chloronaphthol (4-chloro-1-naphthol, Sigma) and 0.01% hydrogen peroxide.

2.2. Streptavidin/Antistreptavidin Amplified Staining

1. Materials listed in Section 2.1., excluding avidin or streptavidin-HRP.
2. Streptavidin (Gibco-BRL, Gaithersburg, MD).
3. Antistreptavidin (Sigma or Zymed Laboratories [South San Francisco, CA]).
4. Goat antirabbit immunoglobulin conjugated to horseradish peroxidase (Cappel Laboratories).

3. Methods

3.1. Avidin or Streptavidin Horseradish Peroxidase Staining

1. After SDS-PAGE, transfer proteins to nitrocellulose membranes (*see* Notes 1 and 2, and Chapters 37–40). Rinse the nitrocellulose replicas three times and soak in 100 m*M* sodium bicarbonate, pH 8.0, for 5 min (0.25–0.50 mL/cm^2 nitrocellulose). Typically, a 10-mL volume is used for a minigel replica. The solution volume should permit the filter to move freely on agitation. Carry out all incubation and washing reactions using an orbital shaker or rocker platform at ambient temperature.
2. Transfer and submerge the replicas in the same volume of freshly prepared sulfosuccinimidobiotin solution for 45 min (*see* Notes 3 and 4).
3. Add 1*M* glycine-HCl, pH 6.5, to a final concentration of 1 m*M* for approx 5 min to quench the derivatization reaction.
4. Wash the filters three times (5 min each wash) with PBS to remove free sulfosuccinimidobiotin. Incubate filters for 30 min with block solution.
5. Incubate the replicas for 1 h with either avidin conjugated to horseradish peroxidase (5 μg/mL) or streptavidin conjugated to horseradish peroxidase (1 μg/mL) diluted in PBS containing 1% BSA, 0.05% Tween-80 (*see* Note 5).
6. Wash filters three times for 15 min each time with the same volume of wash solution. Protein bands are visualized by immersing the replicas in α-chloronaphthol solution (*see* Note 6).
7. After allowing sufficient time for color development (usually 30 min), rinse the replicas with distilled water, and dry between two sheets of dialysis membrane. The replicas may also be dried, and stored in cellophane for future use.

3.2. Streptavidin/Antistreptavidin Amplified Staining

1. If desired, an alternative procedure to amplify fivefold the detection method described above is utilized. First, biotinylate and block the nitrocellulose filters as described in Section 3.1., steps 1–4 of the avidin- or streptavidin-staining protocol.
2. Incubate blots with streptavidin (1 μg/mL) for 1 h in PBS containing 1% BSA, 0.05% Tween-80. Wash the replicas three times for 15 min each time with wash solution.

3. Next, dilute affinity-purified rabbit antistreptavidin IgG (0.5 μg/mL) in PBS containing 1% BSA, 0.05% Tween-80. Add solution to replica for 4 h or overnight if convenient. Wash replicas three times for 15 min each time with wash solution.
4. Incubate replicas with goat antirabbit immunoglobulin conjugated to horseradish peroxidase diluted (4 μg/mL) in PBS containing 1% BSA, 0.05% Tween-80 for 4 h. Wash replicas three times for 15 min each time with wash solution.
5. Protein bands are visualized by immersing the replicas in α-chloronaphthol solution for approx 30 min (*see* Note 6). The replicas are rinsed with distilled water and dried between two sheets of dialysis membrane.

4. Notes

1. This method is more sensitive for nitrocellulose membranes than for proteins transferred to Biodyne membranes by approx 10-fold owing to the higher background staining of Biodyne membranes.
2. Because of the sensitivity of the staining, care should be taken to avoid protein contamination of replicas with fingertips, and so forth. Gloves or forceps should be used.
3. This method derivatizes the free amino groups of proteins bound to filters. Amine containing compounds, such as Tris or glycine buffers, will compete for biotinylation with the sulfosuccinimidobiotin. Blot transfer buffers that contain free amino groups, such as Tris or glycine, may be used, but must be thoroughly removed by soaking and rinsing as indicated in Section 3.1., step 1.
4. The staining is highly dependent on the sulfosuccinimidobiotin concentration. Sulfosuccinimidobiotin concentrations of >10 μ*M* have resulted in a dramatic decrease of protein staining intensity *(12)*. In some instances, it may be necessary to determine empirically the optimal concentration of sulfosuccinimidobiotin to use for staining particular proteins.
5. Little or no differences were observed when avidin conjugated to horseradish peroxidase was substituted for streptavidin conjugated to horseradish peroxidase. However, for some applications, streptavidin may present fewer problems owing to a neutral isoelectric point and apparent lack of glycosylation.
6. Sodium azide will inhibit horseradish peroxidase and, accordingly, must be removed before addition of the enzyme solution.
7. All reagent concentrations were determined empirically and were chosen to give maximum staining sensitivity. In principle, our procedure can be used in double-label experiments in which all proteins on the replica are biotinylated and the same blot is then probed with radioactive antibody or protein A.

Acknowledgments

The author thanks Stanley C. Froehner for helpful advice, discussions, and continued encouragement. This work was supported by grants to SCF from the NIH (NS-14781) and the Muscular Dystrophy Association.

References

1. Towbin, H. E., Staehelin, T., and Gordon, J. (1979) Electrophoretic transfer of proteins from polyacrylamide gels to nitrocellulose sheets: procedure and some applications. *Proc. Natl. Acad. Sci. USA* **76,** 4350–4354.
2. Gershoni, J. M. and Palade, G. E. (1983) Electrophoretic transfer of proteins from sodium dodecyl sulfate-polyacrylamide gels to a positively charged membrane filter. *Anal. Biochem.* **124,** 396–405.

3. Bers, G. and Garfin, D. (1985) Protein and nucleic acid blotting and immunobiochemical detection. *Biotechniques* **3,** 276–288.
4. Moremans, M., Daneels, G., Van Dijck, A., Langanger, G., and De Mey, J. (1984) Sensitive visualization of antigen-antibody reactions in dot and blot immune overlay assays with immunogold and immunogold/silver staining. *J. Immunol. Methods* **74,** 353–360.
5. Hsu, Y-. H. (1984) Immunogold for detection of antigen on nitrocellulose paper. *Anal. Biochem.* **142,** 221–225.
6. Burnette, W. N. (1981) Western blotting: electrophoretic transfer of proteins from sodium dodecyl sulfate-polyacrylamide gels to unmodified nitrocellulose and radiographic detection with antibody and radioiodinated protein A. *Anal. Biochem.* **112,** 195–203.
7. Harper, D. R., Liu, K.-M., and Kangro, H. O. (1986) The effect of staining on the immunoreactivity of nitrocellulose bound proteins. *Anal. Biochem.* **157,** 270–274.
8. Hancock, K. and Tsang, V. C. W. (1983) India ink staining of proteins on nitrocellulose paper. *Anal. Biochem.* **133,** 157–162.
9. Root, D. D. and Wang, K. (1993) Silver-enhanced copper staining of protein blots. *Anal. Biochem.* **209,** 15–19.
10. Wojtkowiak, Z., Briggs, R. C., and Hnilica, L. S. (1983) A sensitive method for staining proteins transferred to nitrocellulose paper. *Anal. Biochem.* **129,** 486–489.
11. Bayer, E. A., Wilchek, M., and Skutelsky, E. (1976) Affinity cytochemistry: the localization of lectin and antibody receptors on erythrocytes via the avidin-biotin complex. *FEBS Lett.* **68,** 240–244.
12. LaRochelle, W. J. and Froehner, S. C. (1986) Immunochemical detection of proteins biotinylated on nitrocellulose replicas. *J. Immunol. Methods* **92,** 65–71.
13. LaRochelle, W. J. and Froehner, S. C. (1990) Staining of proteins on nitrocellulose replicas, in *Methods in Enzymology*, vol. 184 (Wilchek, M. and Bayer, E., eds.), Academic, San Diego, CA, pp. 433–436.
14. Bio-Rad Biotin-Blot Protein Detection Kit Instruction Manual (1989) Bio-Rad Laboratories Lit. No. 171., Richmond, CA, pp. 1–11.
15. Didenko, V. V. (1993) Biotinylation of DNA on membrane supports: a procedure for preparation and easy control of labeling of nonradioactive single-stranded nucleic acid probes. *Anal. Biochem.* **213,** 75–78.

50

Detection of Proteins on Blots Using Chemiluminescent Systems

Graeme R. Guy

1. Introduction

The detection of membrane-bound antibodies is normally carried out using either radioactive labeled secondary antibody (or protein A) or enzyme conjugates of an appropriate secondary antibody followed by an enzyme-catalyzed color reaction. These detection systems, however, may be either hazardous, time-consuming, or lacking in sensitivity. In addition, colors tend to fade with time and exposure to light.

Enhanced chemiluminsecent methods for the detection of proteins on Western blots have become popular in recent years. In this light-emitting nonradioactive method for the detection of immobilized specific antigens, antibodies are employed that have horseradish peroxidase (HRP) either conjugated directly to a primary antibody or to a secondary antibody that interacts with the primary antibody. The chemical luminescence relies on the properties of cyclic diacylhydrazides, such as luminol, that are oxidized in alkaline conditions when catalyzed by hydrogen peroxide and HRP *(1)*. Immediately following oxidation, luminol is in an excited state, which then decays to ground state via a light-emitting pathway. Various phenols are added to the reaction to enhance the luminescence by increasing the light output by approx 1000-fold and by extending the time of light emission *(2)*. The light produced by this reaction peaks in 5–20 min and decays slowly with a half-life of 60 min. Blue-light-sensitive X-ray film is used to record the resultant image.

1.1. Advantages of Enhanced Chemiluminescent Detection

Previous methods for detecting proteins on blots have relied on different systems for visualizing the antibody bound to the antigen that is fixed on the blot matrix. Such systems include: dyes conjugated to the secondary antibody or enzymes conjugated to the antibody that catalyze a colorimetric reaction, gold or silver particles attached to antibodies, and radioactive tracers attached to secondary antibodies. Some methods have drawbacks in the intensity or stability of the signal output, whereas others, like radioactivity, produce a strong and quantifiable permanent record, but have associated safety and disposal problems and require special handling facilities during use. The advent of chemiluminescent techniques seems to overcome many problems associated with other systems and offers a number of advantages:

From: The Protein Protocols Handbook
Edited by: J. M. Walker Humana Press Inc., Totowa, NJ

High sensitivity: This method can detect 1 pg of antigen and is at least 10× more sensitive than dye-conjugated systems and around 2–5× more sensitive than radioactive methods.
High resolution: A high-contrast, easily quantifiable signal is generated.
Speed: The transfer of the light output to a completed image is around 3 min.
Multiple images: The signal can be duplicated with identical or varying exposures giving several copies of the image. The resultant image is easy to quantify on a densitometer.
Conservative use of antibodies: Small amounts of antibodies or low-affinity antibodies can give good results.
Versatility: Detection of Western blotted proteins from all types of gels: one-dimensional, two-dimensional, and agarose gels are possible.
Nondestructive: The Western blot can be reprobed many times with the same or different antibodies.
Nonradioactive: Chemiluminescence protocols are safe and produce no radioactive waste for disposal.

2. Materials

2.1. Solutions

1. 10 n*M* luminol (Boehringer Mannheim, Mannheim, Germany) in 20 m*M* Tris-HCl, pH 8.5.
2. 5 n*M* 4-iodine-phenol (Aldrich, Milwaukee, WI) in 20 m*M* Tris-HCl buffer, pH 8.5.
3. 20 n*M* aqueous H_2O_2 solution.
4. Phosphate buffered saline (PBS): 11.5 g disodium hydrogen orthophosphate (anhydrous) (80 m*M*), 2.96 g sodium dihydrogen orthophosphate (20 m*M*), sodium chloride (100 m*M*). Dilute to 1000 mL with distilled water and adjust the pH to 7.5.
5. PBS/Tween and PBS/T/BSA: Dilute the required volume of Tween-20 in the above PBS; 0.1% of Tween is suitable for most blotting work, but weaker affinity antibodies may require less or no amounts of the detergent. For most blocking protocols and incubation with the primary and secondary antibodies, 1% of bovine serum albumin (BSA) is added to the PBS/Tween. Low-fat dried milk (5%) can be used instead of BSA in the primary blocking bath, but it cannot be used in any streptavidin-HRP incubations because of the endogenous biotin content of milk.

All buffers should be stored in a refrigerator (2–8°C). PBS and PBS/T will keep for 2–3 mo. PBS/T/BSA should be replaced each week because of the likelihood of bacterial contamination. Sodium azide interferes with the chemiluminescent reaction and should not be used as a bactericide.

2.2. Proprietary Kitsets

Several manufacturers produce kits for chemiluminescence detection of proteins on blots. Some of these are (all names of kits are trademarks of the respective companies).

ECL (Amersham, Bucks, UK)
BM Chemiluminescence (Boehringer Mannheim, Germany)
Photoblot (Gibco, Gaithersburg, MD)
Immuno-Lite (Bio-Rad, Hercules, CA)
Renaissance (Dupont NEN, Boston, MA)

2.3. Equipment

1. Membrane, e.g., Immobilon P (Millipore, Bedford, MA) or Nitrocellulose, e.g., Hyperbond-ECL (Amersham).

2. X-ray film cassette.
3. X-ray film, e.g., Kodak X-Omat AR, Amersham Hyperfilms-ECL or Fuji RX.
4. Timer.
5. Developer, water bath, and fixer in tanks or in an X-ray film processing machine.
6. Transparent plastic bag or container and Saran Wrap.
7. Bag sealer.
8. Reciprocal shaker.
9. Forceps with rounded edges.

3. Methods

3.1. Blotting and Probing

The following is a standard protocol, but the solutions, times, and temperatures should be optimized for the user's experimental system. Constant agitation should be used during each step.

1. After the separation of protein mixtures on electrophoretic systems, they are transferred to a membrane. Similar results are obtained for nitrocellulose or polyvinyl membranes. The blots may be used immediately or air-dried and stored in a desiccator at 2–8°C. Blots should be manipulated at all times with forceps or handled only while wearing rubber gloves.
2. The membrane should be blocked for an hour at room temperature or overnight at 2–8°C in PBS/T/BSA.
3. The membrane should then be briefly rinsed with PBS/T (2 × 5 min).
4. The primary antibody can be diluted in the blocking solution and incubated for 1 h at room temperature. The antibody should be diluted as recommended by the manufacturer, or dilution experiments should be done to determine the optimum working strength. MAb blot well at 250 ng/mL.
5. The membrane should be rinsed in a relatively large volume of PBS/T for 5–10 min, and this should be repeated 3–5 times.
6. The secondary antibody, which is conjugated to HRP, is diluted in the blocking solution PBS/T/BSA also. The dilution again should be optimized for individual experimental conditions or according to manufacturer's protocols. The aim is to get the maximum signal with the minimum background. (The author uses Sigma [St. Louis, MO] antimouse IgG peroxidase conjugate [catalog # A-4416] or Sigma antirabbit IgG peroxidase conjugate [catalog # A-4914] diluted at 1:1000).
7. The above washing protocol (step 5) is repeated, after which the wet blot is drained and agitated in a solution of the chemiluminescence chemicals for 1 min at room temperature.
8. The excess reagent is drained off, and the blot is placed under cling film or in a clear plastic "sandwich" and fixed into an X-ray cassette with adhesive tape. Multiple exposures are made onto X-ray film in a darkroom. Optimum exposure occurs from 5 s to 5 min. Most proprietary X-ray film is suitable, but some have a high inherent background. We find Fuji RX film gives a very clear background that renders the resultant image suitable for reproducing in publications. Specialized X-ray films, such as Hyperfilm, ECL, exhibit a linear response to the light produced from enhanced chemiluminescence, which enables direct quantification. The range over which the film response is linear can be extended by preflashing, which makes the measurements of low amounts of protein more accurate. Preflashing can be performed by using a modified flash unit, such as the Sensitize unit marketed by Amersham. The flash duration should be in the region of 1 ms.
9. The wet membrane can be stored in a refrigerator for several days before further treatment. The membrane can be reprobed to clarify or confirm results. If the same antibody is

to be reused, the membrane can be washed in PBS/T for 2 × 10 min at room temperature before the described protocol is repeated.

10. For reprobing, both the primary and secondary antibody can be removed from membranes with the stripping buffer. Incubation is at 50°C for 30 min with occasional agitation. Wash the membrane in PBS/T for 2 × 10 min, and the complete immunodetection procedure described in Section 3.1., steps 2–8 is carried out. Blots can be stripped 5–10 times to detect different antigens.

3.2. Quantification of Proteins on Enhanced Chemiluminescence Western blots

Several films, such as Hyperfilm ECL (Amersham), exhibit a linear response to the light produced from enhanced chemiluminescence. The range over which the film response is linear can be extended by preflashing the film prior to exposure, which makes quantification of proteins, especially at lower levels, more accurate. The sample to be quantified can be diluted to five different dilutions within a range of not more than one order of magnitude. Antigens in the resultant Western blot are detected using standard protocols and then exposed to preflashed film. For quantification to be accurate, it is important that the light produced is in the linear range of the film. To achieve this, the film is exposed for several lengths of time, and the one that only just shows the lowest concentration of the antigen standard is selected. The film can be scanned using a densitometer and a graph of peak area against protein concentration plotted. The concentration of the protein being quantified can be read off this graph after taking dilutions into account.

The major use for ECL detection on Western blots is for:

1. Detection of single antigens on one-or two-dimensional Western blots.
2. The detection of antigens on multiple proteins (such as proteins containing phosphorylated tyrosine) on one- or two-dimensional Western blots. Examples of each are given in Figs. 1 and 2.

4. Notes

1. There are many things that can cause problems in the above protocol or the gel electrophoretic and blotting protocols that precede the antigen detection step. A full diagnostic system is beyond the scope of this chapter, but an abbreviated troubleshooting guide is appended below. Failures can be minimized by paying attention to the following three points:
   a. Optimize: Kits using the chemiluminescence system for detecting antigen levels on Western blots are readily available. These kits vary slightly in their recommendations, but the chosen kit should be characterized for a particular project taking into account the manufacturer's instructions and troubleshooting guides. The most frequent mistake made is a failure to spend some time to optimize the system and produce standards or use controls.
   b. Standardize: A standard sample of the antigen being detected should be prepared and run on each gel and blot to assess the methodology and to make sure that the detection system is working on each occasion. For example, if phosphotyrosine blots are being run, a cell extract that gives an even distribution of tyrosine-phosphorylated proteins should be prepared, standardized, and run with each set of experiments.
   c. Minimize: Our laboratory has found that the best results are achieved if the blots are kept on a miniscale, which keeps the antigens in a compact state. We estimate that

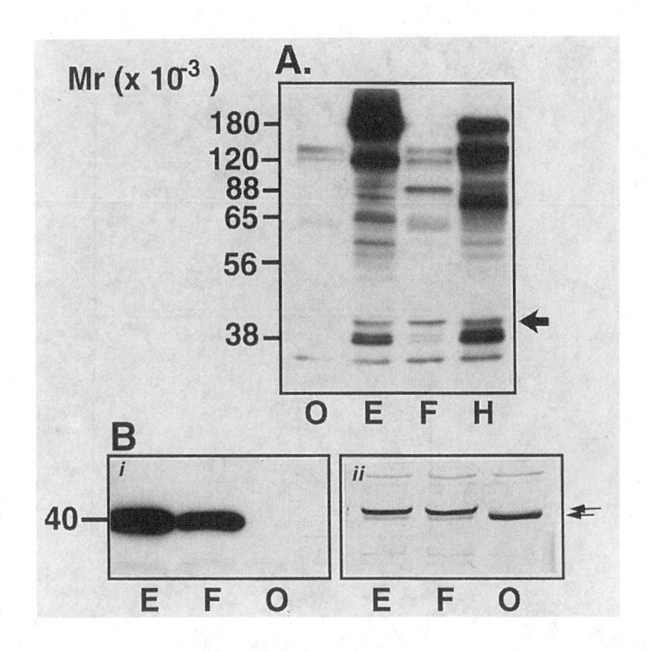

Fig. 1. (A) Primary human MRC-5 fibroblasts were stimulated with various agonists (O = control [no ligand added]), E = epidermal growth factor (50 ng/mL), F = fibroblast derived growth factor (10 ng/mL), and H = hydrogen peroxide (50 m*M*) for 5 min prior to the cells' being solubilized and separated on a 10% SDS-PAGE gel on a Bio-Rad minigel apparatus. The gel was blotted onto a PVDF (immobilon) membrane, and the subsequent blocking and immunodetection as described in the text were performed. The primary antibody was anti-phosphotyrosine PY-20, diluted 1:1000 (Transduction Laboratories, Lexington, KY) conjugated to HRP. The arrow marks the location of the extracellular-regulated kinase 2 (Erk-2) protein. This figure illustrates the clarity of the system for revealing the proteins that are tyrosine-phosphorylated following treatment with mitogens (EGF or FGF) or with a tyrosine phosphatase inhibitor (H_2O_2). Tyrosine phosphorylation is an important early stage in signal transduction pathways. Many proteins that are modified in this way are present in low abundance, and alternate means of detection would involve the use of relatively large amounts of radioactive phosphate. The blot can be developed for multiple times over the next 15 min to obtain an optimal image. **(B**, part i**)** The Erk-2 protein was immunoprecipitated from cellular extracts of MRC-5 cells using a MAb diluted 1:1000 (#E162220, Transduction Laboratories), the resultant extract was subjected to electrophoresis and blotting *(as above),* the blot was probed for antiphosphotyrosine using the PY-20 antibody, and the location was determined using the enhanced chemiluminescent technique. (B, part ii) A similar blot to that shown in (Bi and A) was stripped and reprobed with a pan-Erk antibody (#E17120, Transduction Laboratories), diluted 1:1000, that picked up all members of the Erk family of kinases. A secondary antimouse antibody (# A4416 antimouse IgG [whole molecule] from Sigma, St. Louis, MO) that had HRP conjugated to it was used to locate the primary antibody, and thus the location of Erk-2 was revealed by the chemiluminescent system. This also illustrates the clarity of the system for revealing minor details. In (B, part ii) there is a shift in molecular weight of Erk-2 in mitogen stimulated cells and this increase in molecular weight has been shown to be dependent on tyrosine phosphorylation. It can be seen in the immunoprecipitated sample of Erk-2 (B, part i) that only the mitogen stimulated samples are tyrosine-phosphorylated.

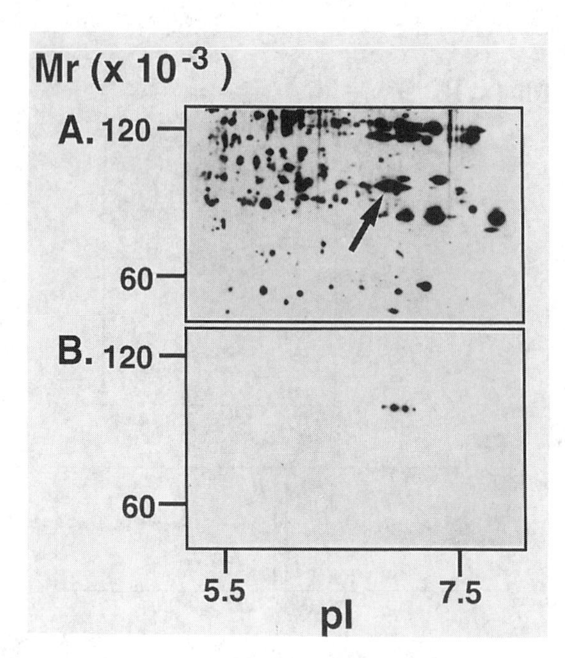

Fig. 2. The chemiluminescent system of detection can also be used on blots derived from two-dimensional gels. **(A)** Shows the distribution of tyrosine-phosphorylated proteins on a Western blot derived from MRC-5 cells that had been treated with epidermal growth factor (EGF) for 5 min prior to extraction. The antiphosphotyrosine antibody PY-20 conjugated to HRP was used at a dilution of 1:1000, and the above protocol was used to reveal the phosphotyrosine containing proteins. Such sensitivity is not possible with other detection systems and could open up new avenues in signal transduction research by detecting low-abundance tyrosine-phosphorylated proteins that cannot be detected on one-dimensional gels. **(B)** The blot used in (A) was stripped and reprobed with anticortactin MAb (#P13320, Transduction Laboratories), diluted 1:1000 in combination with a secondary antimouse antibody conjugated to HRP after which the location of the antibodies was revealed by chemiluminsecence. The location of cortactin in the parent blot is indicated by an arrow. The same blot can be stripped and reprobed 5–10 times, and the tyrosine containing proteins for which antibodies are available can be mapped and characterized in various signal transduction pathways.

moving from a mini to a maxi electrophoresis gel results in a spreading or dilution of the protein by 5- to 10-fold, which means loading more protein to larger gels with the possibility of overloading and the problems that that brings with lack of resolution.

2. If no signal appears, check the following:
   a. Has the protein been transferred properly to the membrane during blotting? The blot should be stained with Ponceau S solution and the gel should be silver stained to ascertain the efficiency of the transfer.
   b. Does the primary antibody detect denatured proteins on blots? Most manufacturers will describe the optimum binding conditions for the antibody.
   c. Check the affinity of the primary antibody, and optimize the antibody concentration.
   d. Check the activity of the HRP conjugated to the secondary antibody. This can be done by dotting various dilutions of the HRP conjugate onto a membrane and developing the image.

   e. Check detection reagent. Make sure solutions are freshly prepared and that there is sufficient solution in the bag (or container), and that the times of incubation and those between incubation and developing the image are correct.
3. If signals are weak:
   a. Optimize the concentrations of antibodies.
   b. Prolong the incubation time with primary antibody to overnight at +4°C.
   c. Prolong the incubation of the secondary antibody to 3 h.
   d. Prolong the detection time.
   e. Shorten the washing times, using washing buffer without Tween-20.
   f. Incubate the primary and secondary antibodies in buffer without blocking reagent.
   g. Increase the amount of protein applied to the gel.
   h. Check for efficient blotting (*see* Note 2a).
4. If background is too high:
   a. Use clean equipment, freshly prepared buffers, and new membranes.
   b. Increase washing times.
   c. Increase the detergent concentration in washing buffer.
   d. Block overnight.
   e. Avoid touching the membrane; use gloves and blunt-ended forceps.
   f. Use different membrane types.
5. Protein mol-wt-markers: Several manufacturers (e.g., Amersham) market a mixture of different proteins labeled with biotin for use in Western blotting following electrophoresis on sodium dodecyl sulfate-polyacrylamide gel electrophoresis (SDS-PAGE) gels. The resultant blot is incubated with streptavidin horseradish peroxidase followed by detection with the enhanced chemiluminescence blotting system, and a ladder of bands can provide internal mol-wt reference markers.
6. Time: The time required for detecting proteins by the above chemiluminescent protocol is around 5 h. The actual manipulation time is minimal, and other procedures can be carried out at the same time. This can be reduced to 3 h if the primary antibody has HRP conjugated directly to it. The wash times can also be shortened.

References

1. Gunderman, K. D. and McCapra, F. (1987) *Chemiluminescence in Organic Chemistry.* Springer-Verlag, Berlin, Germany, pp. 77–108.
2. Thorpe, G. H. G., Kricka, L. J., Moseley, M. R., and Whitehead, T. P. (1985) Phenols as enhancers of the chemiluminescent horse radish peroxidase-luminol-hydrogen peroxide reaction: application in luminescence-monitored enzyme immunoassays. *Clin. Chem.* **31,** 1335–1341.

Part IV

Chemical Modification of Proteins and Peptide Production and Purification

51

Carboxymethylation of Cysteine Using Iodoacetamide/Iodoacetic Acid

Alastair Aitken and Michèle Learmonth

1. Introduction

If cysteine or cystine are identified in a protein, they require modification if they are to be quantified. Thiol groups may be blocked by a variety of reagents, including iodoacetic acid and iodoacetamide. Iodoacetate produces the *S*-carboxymethyl derivative of cysteine, effectively introducing new negative charges into the protein. Where such a charge difference is undesirable, iodoacetamide may be used to produce *S*-carboxyamidomethylcysteine (on acid hydrolysis, as for amino acid analysis, this yields *S*-carboxymethylcysteine). The charge difference between these two derivatives has been utilized in a method to quantify the number of cysteine residues in a protein (*[1]*, *see* Chapter 83).

Carboxymethylation may be carried out without prior reduction in order to modify only those cysteine residues that are not involved in disulfide bridges. If the protein is to be analyzed using gas-phase protein microsequencing, the derivatizing agent of choice is commonly vinylpyridine, since this produces a well-separated PTH derivative (PTH-*S*-pyridylethylcysteine) (*see* Chapter 54).

2. Materials

1. Denaturing buffer: 6*M* guanidinium hydrochloride (or 8*M* deionized urea) in 0.6*M* Tris-HCl, pH 8.6.
2. 4 m*M* dithiothreitol, freshly prepared in distilled water.
3. β-Mercaptoethanol.
4. SDS.
5. Oxygen-free nitrogen source.
6. Iodoacetic acid: 500 m*M* in distilled water, pH adjusted to 8.5 with NaOH. This is light-sensitive, and although the solution may be stored in the dark at –20°C, it is better to use freshly prepared.
7. Iodoacetamide: 500 m*M*, freshly prepared in distilled water.
8. Ammonium bicarbonate: 5 and 50 m*M* in distilled water.
9. 0.1% (v/v) TFA, HPLC-grade.
10. Acetonitrile, far-UV HPLC-grade (e.g., Romil, Cambridge, UK).
11. Microdialysis kit (e.g., supplied by Pierce, Chester, UK).
12. HPLC apparatus.

From: The Protein Protocols Handbook
Edited by: J. M. Walker Humana Press Inc., Totowa, NJ

3. Method

3.1. Large Scale Reaction

1. Dissolve protein (up to 30 mg) in 3 mL 6*M* guanidium hydrochoride (or 8*M* deionized urea), 0.6*M* Tris-HCl, pH 8.6.
2. Add 30 µL β-mercaptoethanol (or 5 µL 4*M* dithiothreitol) and incubate under N_2 for 3 h at room temperature. Some proteins may need the presence of 1% SDS with an incubation time of up to 18 h at 30–40°C in order to completely denature and expose buried thiol groups.
3. Add 0.3 mL colorless 500 m*M* iodoacetate solution (*see* Note 1) or 500 m*M* iodoacetamide solution, with stirring, and incubate in the dark for 30 min at 37°C. Proceed to Section 3.3.

3.2 Small Scale Reaction

1. Dissolve the protein (1–10 µg) in 50 µL of denaturing buffer in an Eppendorf tube and flush with N_2.
2. Add an equal volume of 4 m*M* DTT solution to give a final concentration of 2 m*M*.
3. Wrap the tube in aluminum foil and then add 40 µL iodoacetic acid (or iodoacetamide) solution dropwise with stirring. Keep under N_2, and incubate in the dark for 30 min at 37°C. (*See* Note 2). Proceed to Section 3.3.

3.3. Purification

The removal of excess reagents may be carried out in a variety of ways.

1. Desalting on a 20 × 1 cm Sephadex G10 or Biogel P10 column, eluting with 5 m*M* ammonium bicarbonate, pH 7.8.
2. Dialysis or microdialysis using 10 mL to 1 L of 50 m*M* ammonium bicarbonate, pH 7.8, as dialysis buffer with two changes of buffer for a few hours each.
3. By HPLC using a C_4 or C_8 matrix, such as an Aquapore RP-300 column, equilibrated with 0.1% (v/v) TFA and eluting with an acetonitrile/0.1% TFA gradient.

4. Notes

1. The iodoacetic acid used must be colorless. A yellow color indicates the presence of iodine. This will rapidly oxidize thiol groups, preventing alkylation, and may also modify tyrosine residues. It is possible to recrystallize from hexane.
2. Reductive alkylation may also be carried out using iodo-[^{14}C]-acetic acid. The radiolabeled material should be diluted with carrier iodoacetic acid to ensure an excess of this reagent over total thiol groups.

Reference

1. Creighton, T. E. (1980) Counting integral numbers of amino acid residues per polypeptide chain. *Nature* **284,** 487,488.

52

Performic Acid Oxidation

Alastair Aitken and Michèle Learmonth

1. Introduction

Where it is necessary to carry out quantitative amino acid analysis of cysteine or cystine residues, carboxymethylation (Chapter 51) or pyridethylation (Chapter 54) are not the methods of choice since it may be difficult to assess the completeness of the reactions. It is often preferable to oxidize the thiol or disulfide groups with performic acid to give cysteic acid. This also has the effect of converting methionine residues to methionine sulfone. Tyrosine residues may be protected from reaction by the use of phenol in the reaction mixture.

Performic acid oxidation is also used in the determination of disulfide linkages by diagonal electrophoresis (*see* Chapter 81).

2. Materials

1. 30% (v/v) Hydrogen peroxide (care: strong oxidizing agent).
2. 88% (v/v) Formic acid.
3. Phenol (care).
4. 1 m*M* Mercaptoethanol.

3. Method

1. Add 0.5 mL 30% (v/v) H_2O_2 to 4.5 mL of 88% formic acid containing 25 mg phenol.
2. After at least 30 min at room temperature, cool to 0°C.
3. Add this reagent (performic acid, HCOOOH) to the dry protein sample (or protein in a small aqueous volume) at 0°C to give a final protein concentration of 1%.
4. After 16–18 h, dilute the solution with an equal volume of water and dialyze against two changes of 100 vol of water at 4°C (for a few hours with each solution), and finally against 100 vol of 1 m*M* mercaptoethanol at 4°C.
5. The derivatized protein can then be lyophilized or dried in a hydrolysis tube ready for acid cleavage and subsequent amino acid analysis.

Elution positions for separations of derivatized amino acids by ion-exchange and reverse-phase chromatography are shown in Table 1.

From: The Protein Protocols Handbook
Edited by: J. M. Walker Humana Press Inc., Totowa, NJ

Table 1
Elution Order of Amino Acids and Derivatives from Ion-Exchange Analyzers

1. *O*-Phosphoserine	23. *S*-Ethylcysteine
2. *O*-Phosphothreonine	24. Glucosamine
3. Cysteic acid	25. Mannosamine
4. Glucosaminic acid	26. Valine
5. Methionine sulfoxides	27. Cysteine
6. Hydroxyproline	28. Methionine
7. CM-Cysteine	29. α-Methylmethionine
8. Aspartic acid	30. Isoleucine
9. Methionine sulfone	31. Leucine
10. α-Methyl aspartic acid	32. Norleucine
11. Threonin	33. Tyrosine
12. Serine	34. Phenylalanine
13. Asparagine	35. Ammonia
14. Glutamine	36. Hydroxylysine
15. α-Methyl serine	37. Lysine
16. Homoserine	38. l-Methylhistidine
17. Glutamic acid	39. Histidine
18. α-Methyl glutamic acid	40. 3-Methylhistidine
19. Proline	41. Tryptophan
20. *S*-methylcysteine	42. Pyridylethyl-cystein
21. Glycine	43. Homocysteine thiolactone
22. Alanine	44. Arginine

The precise elution order may vary slightly depending on the exact instrument used and the precise temperature, molarity, and pH of buffers.

53

Succinylation of Proteins

Alastair Aitken and Michèle Learmonth

1. Introduction

Modification of lysine with dicarboxylic anhydrides, such as succinic anhydride, prevents the subsequent cleavage of lysyl peptide bonds with trypsin. In addition, modified proteins act as better substrates for proteases. Succinic anhydride reacts with the ε-amino group of lysine and the N-terminal α-amino group of proteins, in their nonprotonated forms, converting them from basic to acidic groups *(1)*. Thus, one effect of succinylation is to alter the net charge of the protein by up to two charge units. This is an effect that has been exploited for the counting of integral numbers of lysine residues within a protein *(2)*. Succinic anhydride has been reported to react with sulfhydryl groups. Therefore it is advisable to modify any cysteines within the protein (*see* Chapters 51 and 54) prior to succinylation. The reaction occurs between pH 7.0 and 9.0. The reaction is carried out using an approx 50-fold excess of the anhydride over native or carboxymethylated protein.

2. Materials

1. 8*M* deionized urea (*see* Note 1).
2. Distilled water containing up to 0.1*M* NaCl or 0.2*M* sodium borate buffer, pH 8.5.
3. Succinic anhydride.

3. Method

1. 5 mg of the protein should be dissolved in 5 mL of buffer (*see* Note 2).
2. A pH electrode should be placed within the solution to allow monitoring of the pH. The solution should be continuously stirred.
3. The solid succinic anhydride should be added in 0.5-mg portions over a period of 15 min to 1 h to give a 50-fold excess. The pH should be adjusted back to 7.0 with 1*M* sodium hydroxide after each addition.
4. The reaction should be allowed to proceed for at least 30 min after the last addition.
5. The modified protein may be separated from the side products of the reaction by dialysis (for a few hours in each solution) against three changes of 1 L of 50 m*M* ammonium carbonate buffer, or by gel-filtration (e.g., Sephadex G25).

From: The Protein Protocols Handbook
Edited by: J. M. Walker Humana Press Inc., Totowa, NJ

4. Notes

1. Deionization of urea: Urea should be deionized before use to remove cyanates, which may react with amino and thiol groups. Immediately before use, the urea solution should be filtered through mixed-bed Dowex or Amberlite resin in a filter flask.
2. The choice of buffer is dependent on the solubility of the protein. Distilled water may be used, or it may be necessary to use 0.1*M* NaCl or 0.2*M* sodium borate buffer in the presence of 8*M* deionized urea.
3. Other dicarboxylic anhydrides may be used. Particularly commonly used is citraconic anhydride *(3)*. This reagent has the advantage of reversibility, being readily removed from the protein by low-pH treatment (e.g., 10 m*M* HCl, pH 2.0, for 2 h or 5% formic acid, 8 h at room temperature). When used in conjunction with trypsin, the proteolytic reaction may be stopped with acid, which will also remove the lysine blocking groups.

References

1. Klapper, M. H. and Klotz, I. M. (1972) Acylation with dicarboxylic acid anhydrides. *Methods Enzymol.* **25,** 531–536.
2. Hollecker, M. and Creighton, T. E. (1980) Counting integral numbers of amino groups per polypeptide chain. *FEBS Lett.* **119,** 187–189.
3. Yarwood, A. (1989) Manual methods of protein sequencing, in *Protein Sequencing—a Practical Approach* (Findlay J. B. C. and Geisow M. J., eds.), IRL, Oxford, pp. 119–145.

54

Pyridylethylation of Cysteine Residues

Malcolm Ward

1. Introduction

To help maintain their three-dimensional structure, many proteins contain disulfide bridges between cysteine residues. Cysteine residues can cause problems during Edman sequence analysis, and quantification of cysteine and cystine by amino acid analysis is difficult since these residues are unstable during acid hydrolysis.

Chemical modification of cysteine residues can enhance the solubility of the protein, enable more effective enzymatic digestion with proteases, such as trypsin, and facilitate quantification by amino acid analysis.

Oxidation with performic acid can be used to convert cysteine and cystine to cysteic acid *(1)* (and *see* Chapter 52). This, however, can lead to other nondesirable side reactions, such as oxidation of methionine residues and destruction of tryptophan residues. Alkylation with iodoacetic acid has been used extensively, since this enables the addition of negative charges to the protein. The use of iodoacetic acid containing ^{14}C provides a means of incorporating a radioactive label into the polypeptide chain (*see* Chapter 51) *(2)*.

The method described in this chapter is an effective alternative, where reduction and alkylation can be achieved in one step. 4-Vinylpyridine is used to convert cysteine residues to *S*-pyridylethyl derivatives. The *S*-pyridylethyl group is a strong chromophore at λ254 nm, which facilitates the detection of cysteine containing peptides as well as aiding the identification of cysteine residues during Edman sequencing (*see* Note 1).

2. Materials (*see* Note 2)

1. Denaturing buffer: 0.1*M* Tris-HCl, pH 8.5, 6*M* guanidine hydrochloride.
2. 4-Vinylpyridine: Store at –20°C.
3. 2-Mercaptoethanol.

3. Method

1. Dissolve 10–50 µg protein/peptide in the denaturing buffer (1 mL).
2. Add 2-mercaptoethanol (5 µL) and 4-vinyl-pyridine (2 µL) to the sample tube and shake. Since both reagents are extremely volatile and toxic, all experimental work should be carried out in a fumehood wearing appropriate safety clothing.

From: The Protein Protocols Handbook
Edited by: J. M. Walker Humana Press Inc., Totowa, NJ

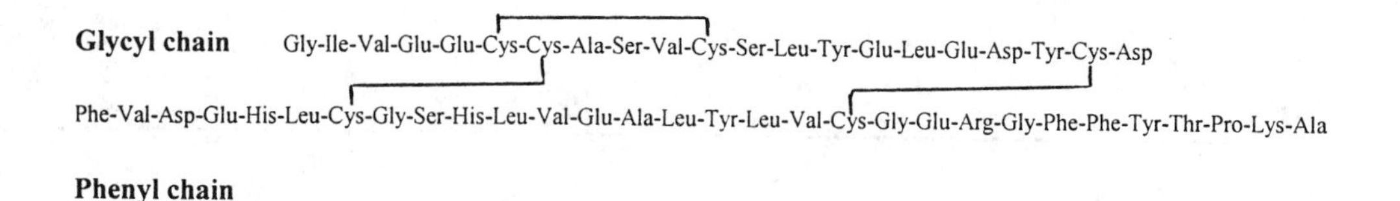

Fig. 1. Amino acid sequence of insulin showing position of disulfide bonds.

3. Blow nitrogen gas over the reaction mixture to expel any oxygen.
4. Seal the tubes with a screw cap.
5. Allow the reaction to proceed at 37°C for 30 min.
6. After this time, an aliquot of the reaction mixture may be taken and analyzed by mass spectrometry. Bovine insulin is shown as an example. This protein contains two interchain bridges and one intrachain bridge (Fig. 1). The matrix-assisted laser desorption ionization (MALDI) mass spectrum of reduced and alkylated insulin shows molecular ions at m/z 2763 and m/z 3612 corresponding to the two fully alkylated peptide chains (Fig. 2) (*see* Note 3). There are no other signals present, indicating that no side reactions have occurred.
7. The sample can now be loaded onto a hydrophobic column for either *N*-terminal sequence analysis, using the HPG1000A protein sequencer, or *in situ* enzymatic or chemical digestion (*see* Chapter 69).

4. Notes

1. The PTH-pyridylethylcysteine derivative can be readily assigned during Edman sequence analysis. The relative elution position on the HPLC system of the Applied Biosystems gas-phase sequencer is between PTH-valine and diphenylthiourea. On the Hewlett Packard G1000A system, PTH-pyridylethylcysteine elutes at 14.2 min just before PTH-methionine.
2. All reagents should be of the highest quality available.
3. The pyridylethylation reaction as described allows for fast, effective derivatization of cysteine residues. The reaction is efficient, giving a high yield of fully alkylated product with no side products.
4. The concept of monitoring chemical reactions by mass spectrometry is not new, yet the sensitivity and speed of MALDI provide a means of examining reaction products to enable the controlled use of reagents that have previously proven troublesome. A recent publication by Vestling et al. *(3)* describes the controlled use of BNPS Skatole, a reagent that to date has been seldom used owing to side reactions. The reactivity of such a reagent may be more widely used in the future now that the progress of the reaction can be easily monitored. See Chapter 91 for further uses of MALDI.

References

1. Glazer, A. N., Delange, R. J., and Sigman, D. S. (1975) Chemical characterization of proteins and their derivatives, in *Chemical Modifications of Proteins* (Work, T. S and Work. E., eds.), North Holland, Amsterdam, 21–24.
2. Allen, G. (1989) Sequencing of proteins and peptides, in *Laboratory Techniques in Biochemistry and Molecular Biology* (Burdon, R. H. and Van Knippenberg, P. H., eds.), Elsevier, Amsterdam.
3. Vestling, M. M., Kelly, M. A., and Fenselau, C. (1994) Optimisation by mass spectrometry of a tryptophan-specific protein cleavage reaction. *Rapid Commun. Mass Spectrom.* **8,** 786–790.

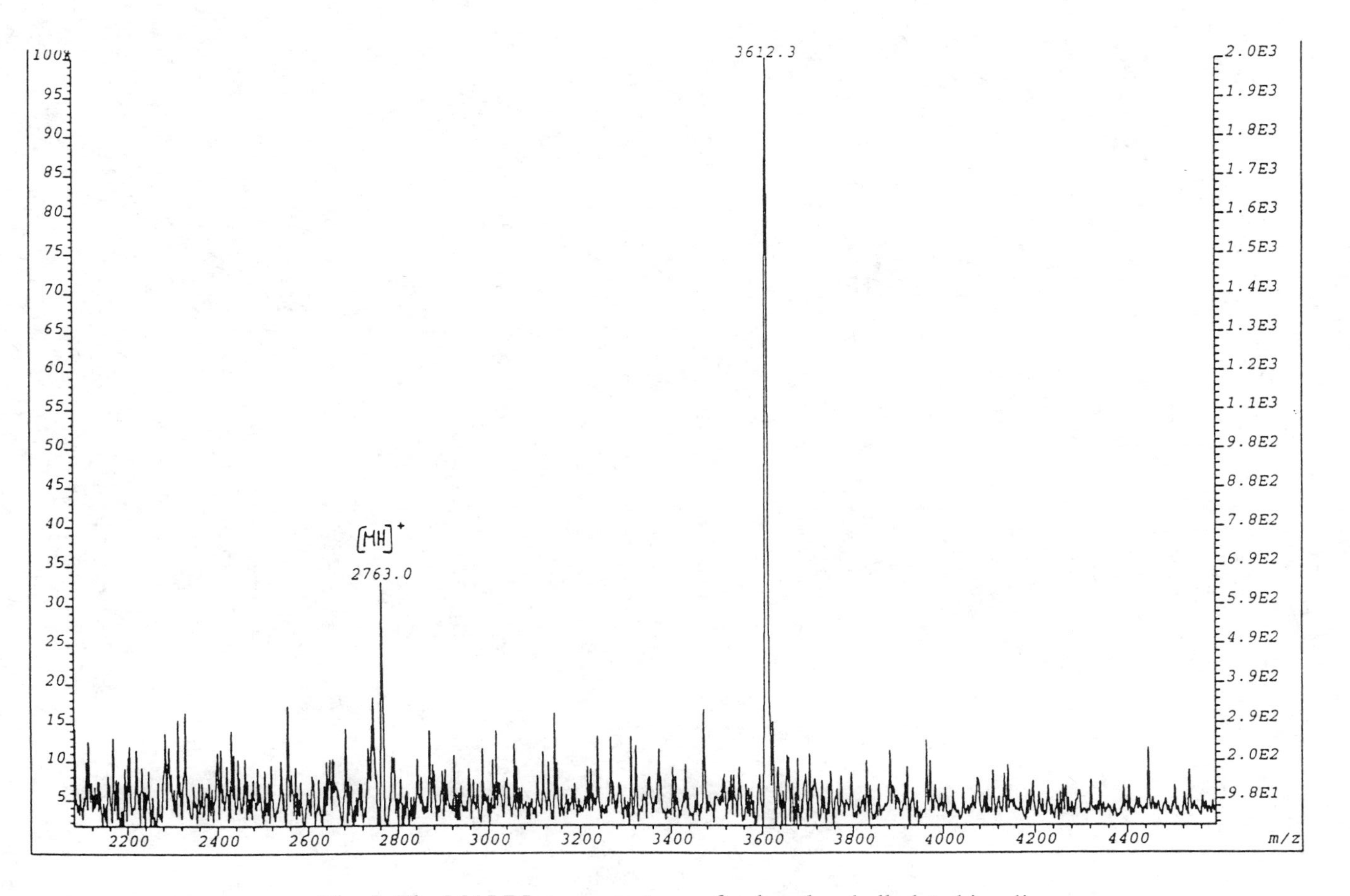

Fig. 2. The MALDI mass spectrum of reduced and alkylated insulin.

55

Side-Chain Selective Chemical Modifications of Proteins

Dan S. Tawfik

1. Introduction

Chemical modifications of proteins may be performed simply and rapidly, and can readily provide preliminary data regarding the role of particular amino acids in a given protein. Even in the era of molecular biology, when site-directed mutagenesis has become so popular, selective chemical modifications are often useful. Many reviews and books cover these aspects of protein chemistry; only a few are cited here *(1–4)*. In particular, the book by Means and Feeney *(1)*, although more than 20 years old, is an excellent introduction and a practical guide to this field.

The modifications discussed in the following chapters are side-chain selective—namely, under appropriate conditions, the reagents mentioned below (and additional reagents mentioned in refs. *1–4*) react specifically with a single type of amino acid side chain. Hence, loss of activity (enzymatic, binding, or other biological activity) following treatment of the protein with such a modifying reagent is considered to be an indication for the role of that side chain in the active site of the protein.

Data obtained by side-chain modifications must be analyzed carefully (as for data obtained by genetic site-directed mutagenesis). Loss of activity on treatment with a reagent might be the result of conformational changes or other changes that occur far from the active site. Some of the reagents, in particular when applied in large excess or under inappropriate conditions may react with more than one type of side chains or may even disrupt the overall fold of the protein. In general, the type of modifications that alter the size of a particular residue, but not its charge, are preferred (*see* Chapters 59 and 60). In addition, the reactivity of a certain type of side chain in a protein varies by several orders of magnitude owing to interactions with neighboring groups that affect the accessibility and reactivity. For example, the pK_a of the carboxylate side chain of aspartic acid is generally around 4.5; however, interactions with other side chains may increase the pK_a by more than 3 U. This change will have a major effect on the reactivity of such a carboxylate group toward the modifying reagent. Dramatic pK_a changes are often found in active sites; thus, certain residues might be particularly hard to modify, thereby forcing conditions that result in a nonspecific loss of activity.

There are a number of quite simple experiments that may strongly support the results obtained by chemical modifications:

From: The Protein Protocols Handbook
Edited by: J. M. Walker Humana Press Inc., Totowa, NJ

1. A simple control that allows the modification to be ascribed to the active site of a protein is to demonstrate protection (i.e., lack of modification) in the presence of a specific ligand to that site, e.g., a hapten for an antibody, a substrate or an inhibitor for an enzyme (for an example, *see* Chapter 56).
2. The extent of modification is primarily determined by the molar excess of the modifying reagent, but also by other conditions, such as pH, temperature, and reaction time. Reliable and reproducible results are generally obtained only after a wide range of reagent concentrations are applied under different reaction conditions. Following these modifications, one should determine not only the remaining biological activity, but also, when possible, the number of modified side chains (details for each reaction are provided in the following chapters). These data may allow one to assess whether the modification is indeed site specific (e.g., loss of activity is the result of modification of one or two amino acids of the type modified) or if loss of activity is owing to a complete disruption of the protein (e.g., the result of modification of a large number of amino acids).
3. Some of the modifications are reversible; for example, histidine side chains modified by diethyl pyrocarbonate can be recovered by a short treatment with hydroxylamine. Recovery of the activity of the modified protein following this treatment will demonstrate the specificity of the modification (*see* Chapter 57). Additional reversible modifications are described in Chapters 60 and 62.
4. Demonstrating pseudo first-order kinetics for the inactivation of the protein may indicate that the modification proceeds like an ordinary bimolecular reaction and not via the formation of a binding complex (as with affinity labelers or suicide inhibitors).

In the following chapters, I have provided basic protocols for the specific modifications of different side chains; these, or very similar, protocols can be applied with other reagents as well (*see* refs. *1–4*). The first protocol for the nitration of tyrosine side chains with tetranitromethane (Chapter 56) is written for a specific protein (an antidinitrophenyl [DNP] antibody) and provides as many experimental details as possible. Hence, it is recommended to read Chapter 56 (including Section 4.) before applying any of the other modifications described in Chapters 57–62.

1.1. General Notes

1. To avoid misleading results and waste of time and materials, it is important to be familiar with the chemistry of each of the reagents (*see* refs. *1–4* and references therein). A detailed mechanistic discussion is beyond the scope of this book, though examples for typical problems or side reactions are given in each chapter.
2. Many of the reagents and solvents described below are harmful. These modifications should be performed in a well-ventilated hood.
3. Examine the buffer you intend to use with your protein in light of the modification reaction. For example, while amidating carboxylate groups (*see* Chapter 59), the use of an acetate buffer, or of any other buffer that contains carboxylate groups or other nucleophiles (e.g., Tris buffer), should obviously be avoided. Likewise, the use of certain organic solvents should be avoided (e.g., acetone with 2-hydroxy-5-nitrobenzylbromide, *see* Chapter 61).
4. Most of the reagents described in the following chapters are reactive with water; however, in some cases, quenching the excess of unreacted reagent is required (for an example, *see* Chapter 56).
5. Chemically modified proteins are often unstable. Likewise, some of the modifications are removed even under mild conditions. This must be taken into consideration while the

protein is purified and its activity is being determined. Hence, when possible, it is best to determine the results of the modification reaction (i.e., the remaining biological activity and the number of modified residues) immediately after the reaction and to avoid further manipulations of the protein (e.g., dialysis or gel filtration).

References

1. Means, G. E. and Feeney, R. E. (1971) *Chemical Modifications of Proteins.* Holden-Day, San Francisco, CA.
2. Hirs, C. H. N. and Timasheff, S. N. (eds.) (1972) Enzyme Structure B. *Methods Enzymology* **25.**
3. Lundblad, R. L. and Noyes, C. M. (1984) *Chemical Reagents for Protein Modifications,* vols. 1 and 2. CRC, Boca Raton, FL.
4. Feeney, R. E. (1987) Chemical modification of proteins: comments and perspectives. *Int. J. Pept. Protein Res.* **27,** 145–161.

56

Nitration of Tyrosines

Dan S. Tawfik

1. Introduction

Tetranitromethane (TNM) reacts with the phenolic side chain of tyrosine under relatively mild conditions to give 3-nitrotyrosine *(1)*. The enclosed protocol was developed for an anti-DNP antibody, but can be used with any other antibody or protein. The major side reaction is oxidation of thiols, although under more extreme conditions, tryptophans might be oxidized as well. The main factor that controls the reactivity of the protein tyrosyl side chains toward nitration by TNM is pH, since the reactive species is the phenolate ion; thus, increasing the pH will usually enhance the rate of modification.

The number of nitrated tyrosines can be determined spectrophotometrically during the reaction. In addition, the stability of nitrotyrosine allows the specific site of modification to be determined by amino acid analysis of cleaved fragments of the protein. Finally, 3-nitrotyrosine has a much lower pK_a than tyrosine; examining the pH-activity profile of the nitrated protein (for example, the binding of an antibody to its hapten) can be readily exploited to demonstrate the specificity of the modification to the active site (*see* Section 3.2. and ref. *2*).

2. Materials

1. TBS 8.0: 0.05*M* Tris-HCl, pH 8.0, 0.15*M* NaCl.
2. Monoclonal or polyclonal antibody preparation (1–2 mg/mL; purified by protein A affinity chromatography.
3. Acetonitrile.
4. TNM (store in aliquots at –20°C). **Please note**—TNM should be handled with care. Preparation of aliquots and of stock solutions should be done in a ventilated hood!
5. 2-Mercaptoethanol.
6. PBS: 0.01*M* phosphate buffer, pH 7.4, 0.15*M* NaCl.
8. Dialysis tubes: 10,000 mol-wt cutoff.

3. Method

3.1. Nitration of Tyrosines with TNM

1. Dialyze the antibody against TBS 8.0 (4 h at 4°C). Determine the protein concentration in the sample by measuring the absorbance at 280 nm (for IgG, ε = 1.45/cm/mg·mL), and adjust it with TBS 8.0 to OD 1.09 or 0.75 mg/mL (= 5 μM antibody = 10 μM sites).

From: The Protein Protocols Handbook
Edited by: J. M. Walker Humana Press Inc., Totowa, NJ

2. Prepare a set of TNM solutions in acetonitrile—0, 2.1, 10.5, 42, 105, and 420 m*M* (corresponding to 21 times the final reagent concentration or to 0, 10, 50, 200, 500, and 2000 molar ratios of TNM/antibody site).
3. Add 5 µL of each of the TNM solutions to 100-µL aliquots of the cold antibody solution in Eppendorf tubes immersed in an ice bath. Incubate, with occasional stirring, for 1.5 h at 4°C and then for 30 min at room temperature.
4. Quench the reaction by adding 2-mercaptoethanol, 2 µL, to the samples containing 0- to 200-fold excess of TNM, and 10 µL to the samples containing 500- to 2000-fold excess of TNM. Incubate for 15 min at room temperature.
5. Dilute the samples with TBS to a total volume of 500 µL and dialyze them against TBS or PBS (at least twice; each round for 4 h at 4°C).
6. Determine the activity of the various antibody samples by ELISA (or any other immunoassay) at increasing antibody dilutions (e.g., 1:50 up to 1:50,000 in PBS).
7. Determine the number of 3-nitrotyrosines/antibody molecule by measuring the OD at 428 nm ($\varepsilon = 4100$ *M*/cm at pH ≥8.5; *see* Note 2).
8. For modification in the presence of the hapten—dinitrophenol (DNPOH): Incubate the antibody with 1 m*M* DNPOH for 30–60 min at 4°C. Proceed with the addition of TNM, quenching, and dialysis as described above (*see* Note 3).

   Determine the number of 3-nitrotyrosines by measuring the OD at 428 nm (and compare it to the number of tyrosines modified under the same conditions in the absence of the hapten).

3.2. pH-Dependency of Binding of Nitrated Antibodies

Nitration of the phenolic group of tyrosine induces a dramatic shift in the pK_a of this residue. The pK_a of tyrosine is normally around 10; thus the hydroxyl group is mostly protonated under, pH ≤9.5. In contrast, the pK_a of 3-nitrotyrosine is around 7.0 *(1)*; hence, loss of activity of the nitrated protein is often the result of deprotonation of the hydroxyl group at pH above 7.0. In such cases, activity could be recovered at pH <6.0, where the hydroxyl of the nitrated tyrosine regains its proton. This was recently demonstrated with several antibodies (including an anti-DNP antibody) in which a pH dependency of binding was observed after site-specific modification of tyrosine with TNM *(2)*. Recovery and loss of binding of these antibodies to the corresponding haptens (at pH < 6.0 and at pH > 8.0, respectively) were ascribed to the protonation and deprotonation of the hydroxyl group of a 3-nitrotyrosine side chain at their binding sites.

This approach can be utilized to determine the role of the modified tyrosine residue at the binding site; it may also find use in a variety of applications in which controlled modulation of binding under mild conditions is required (e.g., affinity chromatography, cell sorting, or immunosensors).

1. Nitrate the antibody as described above (Section 3.1., steps 1–5).
2. Perform a series of dilutions of the nitrated antibody (1:50–1:50,000; *see* Note 5) in 50 m*M* MES saline buffer, pH 5.8, and in TBS, pH 9.0. Determine the binding activity of the diluted antibody at pH 5.8 and 9.0 by ELISA on microtiter plates coated with DNP-BSA (*see* Note 6).

4. Notes

1. In the first modification experiment of a protein, a wide range of TNM concentrations should be applied, e.g., 0–10,000 molar excess. If loss of activity is not observed, it is

recommended to try again at higher pH (e.g., at pH 9.0) and with longer incubations at room temperature.

2. Although proteins hardly absorb at this 428 nm, a sample of the same concentration of the unmodified protein should be used as blank. Relatively high quantities of a protein are required for the determination of low modification ratios; for example, a single 3-nitrotyrosine per site (i.e., two per antibody molecule) would give an $OD_{428\ nm}$ of ~0.11 at antibody concentration of 2 mg/mL (13.3 μ*M*).
3. Demonstrating specificity by modifying the protein in the presence of an active site-specific ligand should be done under the mildest conditions that cause full loss of activity; these conditions (e.g., excess of TNM and, pH) should be determined in a preliminary experiment. In some cases, dialysis with 6*M* urea (or a similar reagent) is required to release the ligand from the protein to allow the determination of its activity. In any case, control samples (without the addition of TNM) containing the protein alone and the protein incubated with the ligand should be included for comparison of the remaining activity after modification.
4. Nitrated proteins (and other chemically modified proteins as well) are often unstable. Therefore, measure the residual activity soon after modification, and avoid freezing and defrosting of the samples. In those cases where the activity assay can be performed at low protein concentrations (e.g., measuring the binding activity of an antibody by ELISA), dialysis that follows the quenching can be avoided. Protein stability can be improved by adding an equal volume of a bovine serum albumin solution (BSA; 10 mg/mL) after the addition of the 2-mercaptoethanol.
5. In order to be able to observe pH-dependent binding, the nitration should be performed under mild modification conditions (e.g., 200 molar excess of TNM at pH 8.3); under more extreme conditions, e.g., 1000 molar excess of TNM, the antibody is irreversibly inactivated, i.e., hapten binding is not recovered at pH 5.8 *(2)*.
6. Conditions of the binding assay (e.g., ELISA) should be optimized in order to identify clearly the pH dependency of binding; in particular, the concentration of the immobilized ligand (DNP in this example) must be low enough so that the differences in binding affinities at pH 5.8 vs 9.0 may be observed. Thus, antigen carrying low ratios of DNP (3–10 DNPs/molecule of BSA) should be used, and its concentration for the coating of the ELISA microtiter plates should be titrated.

References

1. Riordan, J. F., Sokolovsky, M., and Valee, B. L. (1966) Tetranitromethane: a reagent for the nitration of tyrosine and tyrosyl residues in proteins. *J. Am. Chem. Soc.* **88,** 4104,4105.
2. Tawfik, D. S., Chap, R., Eshhar, Z., and Green, B. S. (1994) pH "On-off" switching of antibody-hapten binding obtained by site-specific chemical modification of tyrosine. *Protein Eng.* **7,** 431–434.

57

Ethoxyformylation of Histidine

Dan S. Tawfik

1. Introduction

Diethyl pyrocarbonate (DEP) reacts with various nucleophiles (amines, alcohols, thiols, imidazoles, or guanido groups) to yield the respective ethoxyformyl derivatives. At low pH (generally under 6.0), the reaction is quite selective for histidine, since the main side reaction with the ε-amino group of lysine proceeds very slowly (*see*, for example, ref. *1*). Still, side reactions even with hydroxyl groups (e.g., of serine or tyrosine) were observed *(2)*. The fact that the ethoxyformyl group can be removed from the imidazole side chain by mild treatment with hydroxylamine can be exploited to ascribe the modification to a histidine residue. In addition, ethoxyformylation of histidines is characterized by an increase in absorbance at 242 nm, which is also used to determine the number of modified histidines *(1,2)*.

2. Materials

1. DEP (*see* Note 1).
2. Acetonitrile.
3. Protein for modification (~5 µ*M*) diluted in 0.1*M* sodium acetate buffer, pH 5.0.
4. 1*M* Hydroxylamine, pH 7.0.

3. Method

1. Prepare a series of **fresh** DEP solutions in acetonitrile (1–30 m*M*) (*see* Note 2).
2. Add 5-µL aliquots of each of the DEP solutions to 95-µL aliquots of the protein solution (*see* Note 3).
3. Incubate for 15–60 min.
4. Determine the activity of the modified protein.
5. To assay the recovery of the ethoxyformylated histidine residues:
   a. Add to a solution of the modified protein 1/10 volume of 1*M* hydroxylamine, pH 7.0, and incubate for 10 min.
   b. Dilute with acetate buffer, and dialyze extensively against a buffer suitable for the protein being studied.
   c. Determine the activity of the protein.

From: The Protein Protocols Handbook
Edited by: J. M. Walker Humana Press Inc., Totowa, NJ

4. Notes

1. Concentration of commercial DEP is often lower than indicated owing to hydrolysis. The concentration of the sample after dilution with an organic solvent can be readily determined by adding an aliquot of an imidazole solution (1–10 m*M* in phosphate, pH 7.0) and measuring the increase in absorbance at 230 nm after 5 min ($\varepsilon = 3000M^{-1}$/cm).
2. The final acetonitrile concentration in the protein reaction mixture should be <5%—*see* Chapter 56, Section 3.1. for typical dilutions and reaction volumes.
3. A molar excess of 10–300 of DEP is usually sufficient; however, with some proteins, higher concentrations of the reagent might be needed. Likewise, if no modification is observed at pH 5.0, the reaction can be performed at higher pH. In such cases, it is recommended to ensure that loss of activity is indeed owing to the ethoxyformylation of histidine (*see* Note 4).
4. Measuring the differential UV spectra during modification with DEP is useful not only to determine the number of modified histidines, but also to eliminate the possibility that residues other than histidines were modified. The ethoxyformylation of histidine side chains by DEP should result in an increase in the absorbance of the protein at 242 nm ($\varepsilon = 3200M^{-1}$/cm). Likewise, restoration of the activity of the modified protein after treatment with hydroxylamine should be accompanied by a parallel decrease in the absorbance at 242 nm. Ethoxyformylation of the hydroxyl of tyrosine would increase the absorbance at ~280 nm, whereas similar modification of serine or threonine does not cause a significant change in absorbance at this range.
5. The ethoxyformyl product is quite unstable. Because the reagent is rapidly hydrolyzed (to give ethanol and carbonate), there is hardly a need to purify the protein after modification. In any case, dialysis even in neutral buffers may result in a significant removal of the ethoxyformyl group. Hence, purification by exclusion chromatography (e.g., on Sephadex G-25) is preferred.

References

1. Dominicini, P., Tancini, B., and Voltattorni, C. B. (1985) Chemical modification of pig kidney 3,4-dihydroxyphenylalanine decarboxylase with diethyl pyrocarbonate. *J. Biol. Chem.* **260,** 10,583–10,589.
2. Melchior, W. B., Jr. and Fahrney, D. (1970) Ethoxyformylation of proteins. Reaction of ethoxyformic anhydride with α-chymotrypsin, pepsin and pancreatic ribonuclease at pH 4. *Biochemistry* **9,** 251–258.

58

Modification of Arginine Side Chains with *p*-Hydroxyphenylglyoxal

Dan S. Tawfik

1. Introduction

A variety of dicarbonyl compounds, including phenylglyoxal, 2,3-butanendione, and 1,2-cyclohexanedione, selectively modify the guanidine group of arginine *(1,2)*. The main advantage of *p*-hydroxyphenylglyoxal is in the ability to determine the number of modified arginines spectrophotometrically. This reagent is also reactive at mildly alkaline, pH (usually 8.0–9.0) and yields a single product that is relatively stable *(1)*.

2. Materials

1. *p*-Hydroxyphenylglyoxal.
2. 1*M* NaOH.
3. Protein for modification (~10 μ*M*) diluted in 0.1*M* sodium pyrophosphate buffer, pH 9.0.
4. Sephadex G-25 column.

3. Method

1. Prepare a 100-m*M* solution of *p*-hydroxyphenylglyoxal in water, and adjust the pH of the solution with 1*M* NaOH to pH 9.0.
2. Prepare a series of dilutions (5–50 m*M*; *see* Note 2) of the above solution in 0.1*M* sodium pyrophosphate buffer, pH 9.0.
3. Add 10-μL aliquots of the *p*-hydroxyphenylglyoxal solutions to 90-μL aliquots of the protein solution. Check the pH and, if necessary, adjust it back to 9.0.
4. Incubate for 60–180 min in the dark.
5. Pass the sample through a Sephadex G-25 column. Elute with deionized water or with an appropriate buffer (*see* Note 4).
6. Determine the activity of the modified protein.
7. Determine the number of modified arginines by measuring the absorbance of the purified protein (*see* Note 4) at 340 nm (at pH 9.0, $\varepsilon = 18{,}300 M^{-1}/cm$).

4. Notes

1. An optimal rate and selectivity of modification are generally obtained at pH 8.0–9.0; however, in some proteins, a higher pH might be required to modify a particular arginine residue.

From: The Protein Protocols Handbook
Edited by: J. M. Walker Humana Press Inc., Totowa, NJ

2. A molar excess of *p*-hydroxyphenylglyoxal in the range of 50–500 is usually sufficient for a first trial.
3. The absorbance of *p*-hydroxyphenylglyoxal modified arginines changes with the pH. Maximal absorbance is observed at 340 nm at pH ≥ 9.0 ($\varepsilon = 18{,}300M^{-1}$/cm) *(1)*.
4. Prolonged dialysis in neutral or mildly alkaline buffers may cause a significant release of the modifying group. Purification of the protein to determine the extent of modification should therefore be performed by exclusion chromatography *(1)*.

References

1. Yamasaki, R. B., Vega, A., and Feeney, R. E. (1980) Modification of available arginine residues in proteins by *p*-hydroxyphenylglyoxal. *Anal. Biochem.* **109,** 32–40.
2. Rogers, T. B., Børresen, T., and Feeney, R. E. (1978) Chemical modification of the arginines in transferring *Biochemistry* **17,** 1105–1109.

59

Amidation of Carboxyl Groups

Dan S. Tawfik

1. Introduction

Several reactions have been described for the modification of the carboxylic side chains of aspartic and glutamic acid *(1,2)*; of these, amidation, using an amine and a water-soluble coupling carbodiimide reagent, is most often applied to proteins. The advantages of this approach are the stability of the modification (an amide bond) and the ability to achieve some site specificity (namely, to modify selectively a particular carboxylic side chain) by using different carbodiimide reagents and by variations in the structure (e.g., size, charge, hydrophobicity) of the amine *(3)*.

The methyl or ethyl esters of glycine are commonly used as the amine nucleophile. Hydrolysis of these groups (by a short treatment with hydroxylamine [0.1*M*, pH 8.0, 5–30 min] or a base [0.1*M* carbonate, pH 10.8, 2–6 h]) affords a free carboxyl group (e.g., Protein-COOH is converted into Protein-CO-$NHCH_2COOH$). Hence, a mild modification that affects only the size, but not the charge of the aspartyl or glutaryl side chains of the protein is achieved. Determination of the number of modified carboxylate groups can be performed only by labeling the amine group, e.g., by using a radiolabeled glycine ethyl ester (which is commercially available). In principle, different amines can be used to achieve selectivity or to assist the identification of the modified residues. It is important, however, to ensure that the reaction with these amines is rapid enough and yields a single product *(3)*.

Several side reactions (e.g., with tyrosines and cysteines) may occur mainly at neutral or mildly basic pH; most of these can be ruled out by demonstrating that the activity of the modified protein is not regained after treatment with hydroxylamine.

2. Materials

1. 1-Ethyl-3-(3-dimethylaminopropyl)carbodiimide (EDC) (*see* Note 1).
2. Glycine ethyl ester.
3. Protein for modification (~5 μg/mL) diluted in 0.1*M* MES buffer, pH 5.5.
4. 0.1*M* acetate buffer, pH 5.0.

3. Method

1. Add the glycine ethyl ester to the protein solution to give a final concentration of up to 50 m*M* (*see* Note 2); check the pH and, if necessary, adjust it back to pH 5.5 (*see* Note 3).

From: The Protein Protocols Handbook
Edited by: J. M. Walker Humana Press Inc., Totowa, NJ

2. Add EDC to a final concentration of 0.5–10 m*M*, and incubate for 1–6 h (*see* Note 4).
3. Add 1 vol of acetate buffer to quench the reaction.
4. Dialyze against an appropriate buffer or pass the sample through Sephadex G-25 column.
5. Determine the activity of the modified protein.

4. Notes

1. EDC is most commonly used. Several other water-soluble carbodiimide reagents are available (e.g., 1-cyclohexyl-3-[2-morpholinoethyl]carbodiimide) and can be used for this reaction; these reagents, however, are usually derived from more bulky side chains than EDC and may therefore have more limited accessibility to certain carboxylic residues of the protein.
2. Crosslinking of the protein in the presence of the coupling reagent is avoided by using relatively dilute protein solutions (≤0.5 mg/mL) and a large excess of the amine nucleophile (e.g., glycine ethyl ester).
3. The reaction is usually performed at acidic pH (4.5–5.5) to minimize side reactions. However, the modification of particular carboxylate side chains may require higher pH (*see* Chapter 55).
4. The reaction is usually run at ambient temperature; nevertheless, lowering the temperature to 4°C may eliminate the appearance of certain side products.

References

1. Means, G. E. and Feeney, R. E. (1971) *Chemical Modifications of Proteins.* Holden-Day, San Francisco.
2. Lundblad, R. L. and Noyes, C. M. (1984) *Chemical Reagents for Protein Modifications,* vols. 1 and 2, CRC, Boca Raton, FL.
3. Hoare, D. G. and Koshland, D. E., Jr. (1967) A method for quantitative modification and estimation of carboxylic acid groups in proteins. *J. Biol. Chem.* **242,** 2447–2453.

60

Amidination of Lysine Side Chains

Dan S. Tawfik

1. Introduction

Perhaps the largest variety of modifications available is that for the ε-amino group of lysine *(1–4)*. The amino side chain can be acylated (using, for example, acetic anhydride) or alkylated by trinitrobenzenesulfonic acid (TNBS); these reactions alter both the size and the charge of the amino group. Other modifications, using anhydrides of dicarboxylic acids (e.g., succinic anhydride; *see* Chapter 53) replace the positively charged amino group with a negatively charged carboxyl group. Amidinations *(5,6)* and reductive alkylations (*see* ref. *7*) offer an opportunity to modify the structure of the ε-amino group of lysines, while maintaining the positive charge. Modifications that usually do not disrupt the overall structure of the protein are preferred particularly in those cases when one wishes to identify the specific role of lysine in the active site of the protein being studied.

Amidination is performed by reacting the protein with imidoesters, such as methyl or ethyl acetimidate, at basic pH. The reaction proceeds solely with amino groups to give mainly the positively charged acetimidine derivative, which is stable under acidic and mildly basic pH. Side products can be avoided by maintaining the pH above 9.5 throughout the reaction (*see* Note 1 and ref. *6*). The modification can be removed at a higher pH (≥11.0) and in the presence of amine nucleophiles (e.g., ammonia) *(5,6)*.

A major drawback of this modification is that the number of amidinated lysines cannot be readily determined. However, it is possible to take advantage of the fact that the amidine group is not reactive with amine-modifying reagents, like TNBS, and thereby to determine indirectly the number of the remaining unmodified lysine residues after the reaction *(8)*.

2. Materials

1. Methyl acetimidate hydrochloride.
2. 0.1*M* and 1*M* NaOH.
3. Protein for modification.
5. 0.1*M* borate buffer, pH 9.5.

3. Method

1. Dissolve the protein (1–2 mg/mL) in 0.1*M* borate buffer, pH 9.5; check the pH and, if necessary, adjust it back to pH 9.5 using 0.1*M* NaOH.

From: The Protein Protocols Handbook
Edited by: *J. M. Walker Humana Press Inc., Totowa, NJ*

2. Dissolve 110 mg of methyl acetimidate hydrochloride in approx 1.1 mL of 1*M* NaOH (~0.9*M*); check the pH, and if necessary, adjust it to pH ~10.0 with 1*M* NaOH (*see* Note 1).
3. Add an aliquot of the methyl acetimidate solution to the protein solution, and check the pH again (*see* Note 1).
4. Incubate for 40 min.
5. Dialyze the sample against an appropriate buffer (pH < 8.5) or filter on a Sephadex G-25 column.
6. Determine the activity of the protein.

4. Notes

1. The amidination reaction proceeds with almost no side products only at pH above 9.5; at lower pH, the side reactions proceed very rapidly. Hence, it is important to add the methyl acetimidine solution to the buffered protein solution without causing a change of pH *(6)*. The methyl acetimidine is purchased as the hydrochloride salt, which is neutralized by dissolving it in 1*M* NaOH (*see* Section 3., step 2). The pH of the resulting solution should be ~10; if necessary, the pH may be adjusted before the addition to the protein solution with 1*M* NaOH or 1*M* HCl. Since acetimidates are rapidly hydrolyzed at basic pH, the entire process should be performed very rapidly. It is therefore recommended in a preliminary experiment to dissolve the methyl acetimidine hydrochloride, and determine the exact amount of 1*M* NaOH that yields a solution of pH 10.0. The same process is then repeated with a freshly prepared methyl acetimidine solution that is rapidly added to the protein.
2. The acetimidyl group may be removed by treatment of the modified protein with an ammonium acetate buffer prepared by adding concentrated ammonium hydroxide solution to acetic acid to a pH of 11.3. (**Caution!** Preparation of this buffer must be done carefully and in a well-ventilated chemical hood.)
3. Amidination is obviously unsuitable for the modification of proteins that are sensitive to basic pH. Reductive alkylation, using formaldehyde and sodium cyanoborohydride (*see* ref. *7*), can be performed at neutral pH and is recommended for the modification of such proteins.

References

1. Means, G. E. and Feeney, R. E. (1971) *Chemical Modifications of Proteins.* Holden-Day, San Francisco.
2. Hirs, C. H. N. and Timasheff, S. N. (eds.) (1972) *Enzyme Structure B. Methods Enzymology,* vol. 25.
3. Lundblad, R. L. and Noyes, C. M. (1984) *Chemical Reagents for Protein Modifications,* vols. 1 and 2. CRC, Boca Raton, FL.
4. Feeney, R. E. (1987) Chemical modification of proteins: comments and perspectives. *Int. J. Pept. Protein Res.* **27,** 145–161.
5. Hunter, M. J. and Ludwig, M. L. (1962) The reaction of imidoesters with small proteins and related small molecules. *J. Am. Chem. Soc.* **84,** 3491–3504.
6. Wallace, C. J. A. and Harris, D. E. (1984) The preparation of fully *N*-ε-acetimidylated cytochrome c. *Biochem. J.* **217,** 589–594.
7. Jentoft, N. and Dearborn, D. G. (1979) Labeling of proteins by reductive methylation using sodium cyanoborohydride. *J. Biol. Chem.* **254,** 4359–4365.
8. Fields, R. (1972) The rapid determination of amino groups with TNBS. *Methods Enzymol.* **25,** 464–468.

61

Modification of Tryptophan with 2-Hydroxy-5-Nitrobenzylbromide

Dan S. Tawfik

1. Introduction

2-Hydroxy-5-nitrobenzylbromide, Koshland's reagent *(1),* reacts rapidly and under mild conditions with tryptophan residues. At low pH (<7.5), this reagent exhibits a marked selectivity for tryptophan; under more basic pH or at higher reagent concentrations, cysteine, tyrosine, and even lysine residues can be modified as well. The reaction is extremely rapid either with the protein or with water. The reagent is relatively insoluble in water; it is therefore necessary first to dissolve it in an organic solvent (e.g., dioxane) and then to add it to the protein solution in buffer. Unlike most other modifying reagents, the final organic solvent concentration in the reaction mixture is relatively high (5–15%). Determination of the number of modified tryptophans is achieved spectrophotometrically.

2. Materials

1. 2-Hydroxy-5-nitrobenzylbromide.
2. Dioxane (water-free).
3. Protein for modification: 1 mg/mL in 0.1*M* phosphate buffer, pH 7.0.
4. 1*M* NaOH.
5. A Sephadex G-25 column.

3. Method

1. Prepare a **fresh** solution of 2-hydroxy-5-nitrobenzylbromide (200 m*M*) in dioxane (keep the solution **in the dark**).
2. Dilute the above solution in dioxane to 10 times the final reagent concentration.
3. Add 10 µL of the 2-hydroxy-5-nitrobenzylbromide solution to a 90-µL aliquot of the protein solution and shake for 2 min (e.g., on a Vortex).
4. Centrifuge (for few minutes at >12,000*g;* e.g., bench-centrifuge at 13,500 rpm) if a precipitate forms (*see* Note 3).
5. Filter on a Sephadex G-25, and then dialyze against an appropriate buffer (*see* Note 3).
6. Determine the number of modified tryptophans by measuring the absorbance of the purified protein at 410 nm at pH ≥ 10.0 ($\varepsilon = 18{,}450 M^{-1}$/cm).
7. Determine the activity of the protein.

From: The Protein Protocols Handbook
Edited by: J. M. Walker Humana Press Inc., Totowa, NJ

4. Notes

1. At neutral and slightly acidic pH the reaction is generally selective. At higher pH, the major side reaction is the benzylation of sulfhydryl groups; to prevent the modification of free cysteine residues, the protein can be carboxymethylated (*see* Chapter 56) prior to the modification with 2-hydroxy-5-nitrobenzylbromide.
2. Avoid the use of acetone, methanol, or ethanol as organic cosolvents for this reagent. Water-free dioxane is commercially available or can be prepared by drying dioxane over sodium hydroxide pellets.
3. The reaction of 2-hydroxy-5-nitrobenzylbromide is accompanied by the release of hydrobromic acid. The pH of the reaction should be maintained, if necessary, by the subsequent addition of small aliquots of 1*M* NaOH solution.
4. At high reagent concentrations, a precipitate of 2-hydroxy-5-nitrobenzyl alcohol is sometimes observed and can be removed by centrifugation. Purification of the modified protein from **all** the remaining alcohol product is sometimes difficult and may require extensive dialysis in addition to gel filtration.

Reference

1. Horton, H. R. and Koshland, D. E., Jr. (1972) Modification of proteins with active benzyl halides. *Methods Enzymol.* **25,** 468–482.

62

Modification of Sulfhydryl Groups with DTNB

Dan S. Tawfik

1. Introduction

5-5'-Dithiobis(2-nitrobenzoic acid) (DTNB or Ellman's reagent; ref. *1*) reacts with the free sulfhydryl aide chain of cysteine to form an *S—S* bond between the protein and a thio-nitrobenzoic acid (TNB) residue. The modification is generally rapid and selective. The main advantage of DTNB over alternative reagents (e.g., *N*-ethylmaleimide or iodoacetamide; *see* Chapter 51) is in the selectivity of this reagent and in the ability to follow the course of the reaction spectrophotometrically. The reaction is usually performed at pH 7.0–8.0, and the modification is stable under oxidative conditions. The TNB group can be released from modified protein by treatment with reagents that are routinely used to reduce *S—S* bonds, e.g., mercaptoethanol, or by potassium cyanide *(2)*.

2. Materials

1. DTNB.
2. 0.1*M* Tris-HCl, pH 8.0.
3. Protein for modification: ~5 μ*M* in 0.1*M* Tris-HCl, pH 8.0.

3. Method

1. Prepare **fresh** solutions of DTNB (0.5–5 m*M*; *see* Note 1) in 0.1*M* Tris-HCl buffer, pH 8.0.
2. Add 20-μL aliquots of the DTNB solution to 180 μL of the protein solution, and incubate for 30 min.
3. Determine the number of modified cysteines by measuring the absorbance of the released TNB anion at 412 nm ($\varepsilon = 14{,}150 M^{-1}/\text{cm}$).
4. Dialyze against an appropriate buffer (*see* Note 3).
5. Determine the activity of the protein.

4. Notes

1. A molar excess of 10–100 of DTNB and an incubation time of 30 min is usually sufficient with most proteins. The modification of a particular cysteine residue may require a higher DTNB concentration, a longer reaction time, or a higher pH.

From: The Protein Protocols Handbook
Edited by: J. M. Walker Humana Press Inc., Totowa, NJ

2. The extent of modification is determined by measuring the absorbance of the **released** TNB anion and, hence, can be followed during the reaction and in the presence of an excess of unreacted DTNB.
3. Dialysis of the reaction mixture is not always necessary; in many cases, the activity of the protein (enzymatic or binding) can be determined directly after the reaction (for an example, *see* ref. *2*).
4. Release of the TNB modification can be achieved by treatment of the modified protein (preferably after dialysis) with thiols (e.g., mercaptoethanol) or by potassium cyanide (20 m*M* final concentration; 10–60 min). The release of the TNB anion can be followed spectrophotometrically at 412 nm *(2)*.

References

1. Ellman, G. L. (1959) Tissue sulfhydryl groups. *Arch. Biochem. Biophys.* **82,** 70–77.
2. Fujioka, M., Takata, Y., Konishi, K., and Ogawa, H. (1987) Function and reactivity of sulfhydryl groups of rat liver glycine methyltransferase. *Biochemistry* **26,** 5696–5702.

63

Chemical Cleavage of Proteins at Methionyl Residues

Bryan John Smith

1. Introduction

One of the most commonly used methods for proteolysis uses cyanogen bromide to cleave the bond to the C-terminal side of methionyl residues. The reaction is highly specific with few side reactions and a typical yield of 90–100%. It is also relatively simple and adaptable to large or small scale. Since methionine is one of the least abundant amino acids, cleavage at that residue tends to generate relatively few peptides of size up to 10,000–20,000 Dalton. Such cleavage may be useful to confirm estimates of methionine content by amino acid analysis, which has a tendency to be somewhat inaccurate for this residue *(1)*. Cyanogen bromide may be used for the purposes of peptide mapping and preparation of peptides for amino acid sequencing. In addition the method has been used to map the binding sites of antibodies *(2)* or ligands *(3)*, whereas in combination with molecular engineering techniques, it has been used to cleave at specific points in order to generate functionally distinct domains (from hirudin, by Wallace et al. *[4])* or proteins of interest from fusion proteins *(5)*.

2. Materials

1. 0.4*M* Ammonium bicarbonate solution in distilled water: Stable for weeks in refrigerated stoppered bottle.
2. 2-Mercaptoethanol: Stable for months in dark, stoppered, refrigerated bottle.
3. Formic acid, minimum assay 98%, Aristar-grade.
4. Cyanogen bromide: Stable for months in dry, dark, refrigerated storage. Warm to room temperature before opening. Use only white crystals, not yellow ones. **Beware of the toxic nature of this reagent.**
5. Sodium hypochlorite solution (domestic bleach).
6. Equipment includes a nitrogen supply, fume hood, and suitably sized and capped tubes (e.g., Eppendorf microcentrifuge tubes).

3. Methods

3.1. Reduction

1. Dissolve the polypeptide in water to between 1 and 5 mg/mL, in a suitable tube. Add 1 vol of ammonium bicarbonate solution, and add 2-mercaptoethanol to between 1 and 5% (v/v).

From: The Protein Protocols Handbook
Edited by: J. M. Walker Humana Press Inc., Totowa, NJ

2. Blow nitrogen over the solution to displace oxygen, seal the tube, and incubate at room temperature for approx 18 h.

3.2. Cleavage

1. Dry down the sample under vacuum, warming if necessary to help drive off all of the bicarbonate. Any remaining ammonium bicarbonate will form a nonvolatile salt on subsequent reaction with formic acid.
2. Redissolve the dried sample in formic acid (98%) to 1–5 mg/mL. Add water to make the acid 70% (v/v) finally.
3. Add excess white crystalline cyanogen bromide to the sample solution, to between 2- and 100-fold molar excess over methionyl residues. Practically, this amounts to approximately equal weights of protein and cyanogen bromide. To very small amounts of protein, add one small crystal of reagent. Carry out this stage in the fume hood.
4. Seal the tube, and incubate at room temperature for 24 h.
5. Terminate the reaction by drying down under vacuum. Store samples at –10°C or use immediately.
6. Immediately after use, decontaminate spatulas, tubes, and so on, that have contacted cyanogen bromide by immersion in hypochlorite solution (bleach) until effervescence stops (a few minutes).

4. Notes

1. The mechanism of the action of cyanogen on methionine-containing peptides is shown in Fig. 1. For further details, *see* the review by Fontana and Gross *(6)*. Methionine sulfoxide does not take part in this reaction. An acid environment is required to protonate basic groups, and so prevent reaction there and maintain a high degree of specificity. Met-Ser and Met-Thr bonds may give <100% yields of cleavage because of the involvement of the β-hydroxyl groups of seryl and threonyl residues in alternative reactions, which do not result in cleavage *(6)*. Morrison et al. *(7)*, however, have found that use of 70% v/v trifluoroacetic acid instead of formic acid (*see* Note 3) improved the yield of cleavage of a Met-Ser bond in apolipoprotein A1.
2. Although the specificity of this reaction is excellent, some side reactions may occur. This is particularly so if colored (yellow or orange) cyanogen bromide crystals are used, when destruction of tyrosyl and tryptophanyl residues may occur.

   The acid conditions employed for the reaction may lead to small degrees of deamidation of glutamine and asparagine side chains (which occurs below pH 3) and cleavage of acid-labile bonds, e.g., Asp-Pro. A small amount of oxidation of cysteine to cysteic acid may occur if these residues have not previously been reduced and carboxymethylated.
3. Acid conditions are required for the reaction to occur—the pH of 70% formic acid is about 1. Formic acid is commonly used, because it is a good protein solvent and denaturant. However, it may damage tryptophan and tyrosine residues *(7)*, and furthermore the advent of mass spectrometric methods applicable to peptides has shown that incubation of polypeptides in formic acid can result in formylation and so increase in mass *(8)*. To avoid this, acids other than formic acid may be used. Trifluoroacetic acid (TFA, obtainable in HPLC-grade purity) may be used in preference to formic acid in concentrations of 50–100% (v/v), the rate of cleavage of horse myoglobin, for example, being greater with 50% TFA (approx pH 0.5). The rate of cleavage in 50% TFA may be somewhat slower than in 70% formic acid, but similar reaction times of hours, up to 24 h, will provide satisfactory results. Caprioli et al. *(9)* and Andrews et al. *(10)* have illustrated the use of 60 and 70% TFA (respectively) for cyanogen bromide cleavage of proteins. Acetic acid (50–

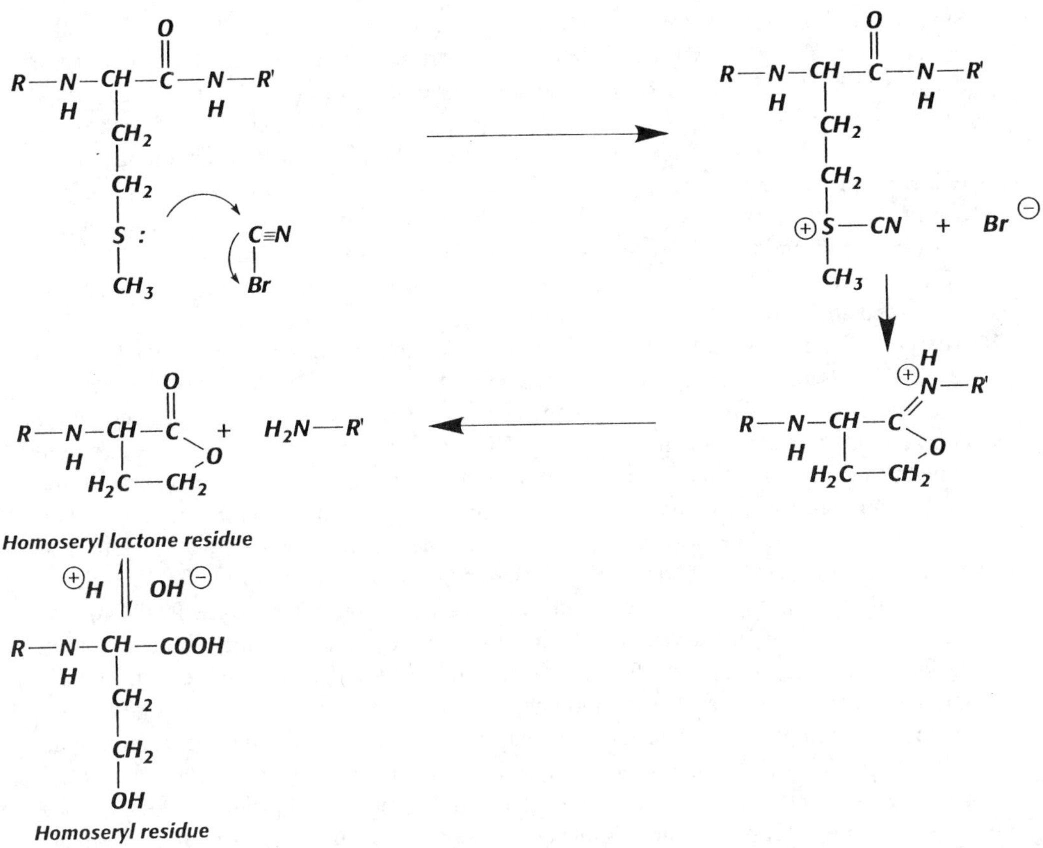

Fig. 1. Mechanism of cleavage of Met-X bonds by cyanogen bromide.

100% v/v) may be used as an alternative, but reaction is somewhat slower than in TFA. Alternatively, 0.1*M* HCl has been used *(8)*.

4. Incomplete cleavage, generating combinations of (otherwise) potentially cleaved peptides, may be advantangeous for ordering peptides within a protein sequence. Mass spectrometric methods are suitable for this type of analysis *(8)*. Such partial cleavage may be achieved by reducing the duration of reaction, even to <1 h *(8)*.
5. Frequently, the protein of interest is impure in a preparation containing other proteins. Polyacrylamide gel electrophoresis is a popular means by which to resolve such mixtures. Proteins in gel slices may be subjected to treatment with cyanogen bromide *(11)* as follows: the piece of gel containing the band of interest is cut out, lyophilized, and then exposed to vapor from a solution of cyanogen bromide in TFA for 24 h, at room temperature in the dark. The vapor is generated from a solution of 20 mg cyanogen bromide in 1 mL of 50% v/v TFA by causing it to boil by placing it under reduced pressure, in a sealed container together with the sample. The gel piece is then lyophilized again, and the peptides in it analyzed by PAGE.
6. Alternatively, proteins that have been transferred from polyacrylamide gel to polyvinylidene difluoride membrane may be cleaved *in situ,* as described by Stone et al. *(12)*. The protein band (of a few micrograms) is first cut from the membrane on the minimum size of polyvinylidene difluoride (since excess membrane reduces final yield). The dried membrane is then wetted with about 50 µL of cyanogen bromide solution in 70% formic acid—

Stone et al. *(12)* report the use of cyanogen bromide applied in the ratio of about 70 μg/1 g protein. Note that although polyvinylidene difluoride does not wet directly in water, it does do so in 70% formic acid, or in the alternative of 50 or 70% v/v TFA. Cleavage is achieved by incubation at room temperature in the dark for 24 h. Oxidation of methionine during electrophoresis and blotting was not found to be a significant problem in causing reduction in cleavage yield, this being about 100% in the case of myoglobin *(12)*. The peptides generated by cleavage may be extracted for further analysis, first in the solution of cyanogen bromide in 70% formic acid, second in 100 μL of acetonitrile (40% v/v, 37°C, 3 h), and third in 100 μL of TFA (0.05% v/v in 40% acetonitrile, 50°C). All extracts are pooled and dried under vacuum before any subsequent analysis.

7. Analysis of protein samples in automated peptide sequencing may sometimes yield no result. The alternative causes (lack of sample or *N*-terminal blockage) may be tested by cleavage at methionyl residues by cyanogen bromide. Generation of new sequence(s) indicates blockage of the original *N*-terminus. The method is similar if the sample has been applied to a glass fiber disk, or to a piece of polyvinylidene difluoride membrane in the sequencing cartridge. The cartridge containing the filter and/or membrane is removed from the sequencer, and the filter and/or membrane saturated with a fresh solution of cyanogen bromide in 70% formic acid. The cartridge is wrapped in sealing film in order to prevent drying out, and then incubated in the dark at room temperature for 24 h. The sample is then dried under vacuum, replaced in the sequencer, and the sequence started again. Yields tend to be poorer than in the standard method described in Section 3. If the sample contains more than one methionine, more than one new *N*-terminus is generated, leading to a complex of sequences. This may be simplified by subsequent reaction with orthophthaladehyde, which blocks all *N*-termini except those bearing prolyl residues *(13)*. For success with this approach, prior knowledge of the location of proline in the sequence is required, the reaction with orthophthaladehyde being conducted at the appropriate cycle of sequencing.
8. As described in Note 1 above, the peptide to the *N*-terminal side of the point of cleavage has at its C-terminus a homoserine or homoserine lactone residue. The lactone derivative of methionine can be coupled selectively and in good yield *(14)* to solid supports of the amino type, e.g., 3-amino propyl glass. This is a useful technique for sequencing peptides on solid supports. The peptide from the C-terminus of the cleaved protein will, of course, not end in homoserine lactone (unless the C-terminal residue was methionine!) and so cannot be so readily coupled.
9. The reagents used are removed by lyophilization, unless salt has formed following failure to remove all of the ammonium bicarbonate.

   The products of cleavage may be fractionated by the various forms of electrophoresis and chromatography currently available. If analyzed by reverse-phase HPLC, the reaction mixture may be applied to the column directly without lyophilization. Since methionyl residues are among the less common residues, peptides resulting from cleavage at Met-X may be large, so in HPLC, use of wide-pore column materials may be advisable (e.g., 30-μm pore size reverse-phase columns, using gradients of acetonitrile in 0.1% v/v TFA in water).

References

1. Strydom, D. J., Tarr, G. E., Pan, Y.-C. E., and Paxton, R. J. (1992) Collaborative trial analysers of ABRF-91AAA, in *Techniques in Protein Chemistry III* (Angeletti, R. H., ed.), Academic, San Diego, CA, pp. 261–274.
2. Malouf, N. N., McMahon, D., Oakeley, A. E., and Anderson, P. A. W. (1992) A cardiac troprin T epitope conserved across phyla. *J. Biol. Chem.* **267,** 9269–9274.

3. Vickers, P. J., Adam, M., Charleson, S., Coppolino, M. G., Evans, J. F., and Mancini, J. A. (1992) Identification of amino acid residues of 5-lipoxygenase-activating protein essential for the binding of leucotrine biosynthesis inhibitors. *Mol. Pharmacol.* **42,** 94–102.
4. Wallace, D. S., Hofsteenge, J., and Store, S. R. (1990) Use of fragments of hirudin to investigate thrombin-hirudin interaction. *Eur. J. Biochem.* **188,** 61–66.
5. Callaway, J. E., Lai, J., Haselbeck, B., Baltaian, M., Bonnesen, S. P., Weickman, J., Wilcox, G., and Leis, P. (1993) Antimicrobial Agents Chemother. **17,** 1614–1619.
6. Fontana, A. and Gross, E. (1986) Fragmentation of polypeptides by chemical methods, in *Practical Protein Chemistry A Handbook* (Darbre, A., ed.), Wiley, Chichester, pp. 67–120.
7. Morrison, J. R., Fidge, N. H., and Greo, B. (1990) Studies on the formation, separation, and characterization of cyanogen bromide fragments of human A1 apolipoprotein. *Anal. Biochem.* **186,** 145–152.
8. Beavis, R. C. and Chait, B. T. (1990) Rapid, sensitive analysis of protein mixtures by mass spectrometry. *Proc. Natl. Acad. Sci. USA* **87,** 6873–6877.
9. Caprioli, R. M., Whaley, B., Mock, K. K., and Cottrell, J. S. (1991) Sequence-ordered peptide mapping by time-course analysis of protease digests using laser description mass spectrometry, in *Techniques in Protein Chemistry II* (Angeletti, R. M., ed.), Academic, San Diego, CA, pp. 497–510.
10. Andrews, P. C., Allen, M. M., Vestal, M. L., and Nelson, R. W. (1992) Large scale protein mapping using infrequent cleavage reagents, LD TOF MS, and ES MS, in *Techniques in Protein Chemistry II* (Angeletti, R. M., ed.), Academic, San Diego, CA, pp. 515–523.
11. Wang, M. B., Boulter, D., and Gatehouse, J. A. (1994) Characterisation and sequencing of cDNA clone encoding the phlorem protein pp2 of Cucurbita pepo. *Plant Mol. Biol.* **24,** 159–170.
12. Stone, K. L., McNulty, D. E., LoPresti, M. L., Crawford, J. M., DeAngelis, R., and Williams, K. R. (1992) Elution and internal amino acid sequencing of PVDF blotted proteins, in *Techniques in Protein Chemistry III* (Angeletti, R. M., ed.), Academic, San Diego, CA, pp. 23–34.
13. Wadsworth, C. L., Knowth, M. W., Burrus, L. W., Olivi, B. B., and Niece, R. L. (1992) Reusing PVDF electroblotted protein samples after N-terminal sequencing to obtain unique internal amino acid sequence, in*Techniques in Protein Chemistry III* (Angeletti, R. M., ed.), Academic, San Diego, CA, pp. 61–68.
14. Horn, M. and Laursen, R. A. (1973) Solid-phase Edman degradation. Attachment of carboxyl-terminal homoserine peptides to an insoluble resin. *FEBS Lett.* **36,** 285–288.

64

Chemical Cleavage of Proteins at Tryptophan Residues

Bryan John Smith

1. Introduction

Tryptophan is represented in the genetic code by a single codon and has proven useful in cloning exercises in providing an unambiguous oligonucleotide sequence as part of a probe or primer. It is also one of the less abundant amino acids found in polypeptides, and cleavage of bonds involving tryptophan generates large peptides. Although there are no proteases showing specificity for tryptophanyl residues, various chemical methods have been devised for cleavage of the bond to the C-terminal side of tryptophan. These are summarized in Table 1. Some show relatively poor yields of cleavage, and/or result in modification of other residues (such as irreversible oxidation of methionine to its sulfone), or cleavage of other bonds (such as those to the C-terminal side of tyrosine or histidine). The method described below is one of the better ones, involving the use of cyanogen bromide. Cleavage of bonds to the C-terminal side of methionyl residues (*see* Chapter 63) is prevented by prior reversible oxidation to methionine sulfoxide. Cleavage by use of *N*-bromosuccinimide or *N*-chlorosuccinimide remains a popular method, however, despite some chance of alternative reactions (*see* Table 1).

Apart from its use in peptide mapping and generation of peptides for peptide sequencing, cleavage at tryptophan residues has also been used to generate peptides used to map the binding site of an antibody *(1)* or the sites of phosphorylation *(2)*. Another application has been generation of a recombinant protein from a fusion protein *(3)*: tryptophan was engineered at the end of a β-galactosidase leader peptide, adjacent to the *N*-terminus of phospholipase A2. Reaction with *N*-chlorosuccinimide allowed subsequent purification of the enzyme without leader peptide.

2. Materials

1. Oxidizing solution: Mix together 30 vol glacial acetic acid, 15 vol 9*M* HCl, and 4 vol dimethylsulfoxide. Use best-grade reagents. Though each of the constituents is stable separately, mix and use the oxidizing solution when fresh.
2. Ammonium hydroxide (15*M*).
3. Cyanogen bromide solution in formic acid (60% v/v): Make 6 mL formic acid (minimum assay 98%, Aristar-grade) to 10 mL with distilled water. Add white crystalline cyanogen bromide to a concentration of 0.3 g/mL. Use when fresh.

From: The Protein Protocols Handbook
Edited by: J. M. Walker Humana Press Inc., Totowa, NJ

Table 1
A Summary of Methods for Cleavage of TRP-X Bonds in Polypeptides

Brief method details[a]	Comments	Example ref.
1. Incubation in molar excess of cyanogen bromide:	*See text* for further detail	*4*
a. Incubation in glacial acetic acid: 9*M* HCl (2/1 [v/v]) with DMSO, room temperature, 30 min	Yields and specificity excellent; Met oxidized to sulfoxide by step a.	
b. Neutralization		
c. Incubation with cyanogen bromide in acid (e.g., 60% formic acid) 4°C, 30 h in the dark		
2. Incubation in very large (up to 10,000-fold) molar excess of cyanogen bromide over Met, in heptafluorobutyric acid/88% HCOOH (1/1 [v/v]), room temperature, 24 h, in the dark	To inhibit Met-X cleavage, Met is photoxidized irreversibly to Met sulfone; yield poor.	*7*
3. a. Incubate protein for 30 min, room temperature, in: phenol (21.6 mg/mL) in glacial acetic acid:12*M* HCl: DMSO::24:12:1, v/v/v	Cys and Met oxidized; some deamidation and cleavage of bonds around Asp may occur; fresh (colorless) HBr required.	*8*
b. Add 0.1 vol of 48% HBr, 0.03 vol DMSO, and incubate for 30 min, room temperature	Trp converted to dioxindolyl alanyl lactone. Protein containing dioxindolylalanine derivative(s) but remaining uncleaved may be cleaved to improve yield to ~80% as follows: Incubate in 10% acetic acid, 60°C 10–15 h.	

Method	Comments	Reference
4. Incubation in BNPS-skatole[b] (100-fold molar excess over Trp) in 50% acetic acid (v/v) room temperature, 48 h, in the dark	Reagent is unstable—fresh reagent required to minimize side reactions; some reactions with Tyr may occur, but addition of free Tyr to reaction mixture minimizes this; Met and Cys may be oxidized, yields up to 60%.	*9*
	Alternative conditions for rapid reaction: neat acetic acid 47°C 15–60 min; adaptable for cleavage of proteins bound to glass fiber or PVDF.	*10*
5. Incubation in *N*-bromosuccinimide (threefold molar excess over His, Trp, and Tyr) pH 3.0–4.0 (e.g., pyridine-acetic acid, pH 3.3) 1 h, 100°C	Also less rapid cleavage at His–X and Tyr–X; yields moderate to poor; Trp converted to lactone derivative.	*11*
6. Incubation in *N*-chlorosuccinimide (10-fold molar excess over (protein) in 27.5% acetic acid, 4.68*M* urea; room temperature, 30 min; stopped by addition of *N*-acetyl-L-methionine	Yield approx 50%; oxidation of methionine (to sulfone) and cysteine (to cysteic acid), especially in higher NCS concentrations; Tyr and His not modified or cleaved (cf. NBS—method 5); adaptable for cleavage of proteins in gels.	*12*, *13*
7. Incubation in 80% acetic acid containing 4*M* guanidine; HCl, 13 mg/mL iodosobenzoic acid; 20 µL/mL *p*-cresol, for 20 h, room temperature, in the dark	Specificity and yields good; the *p*-cresol is used to prevent cleavage at Tyr; Trp converted to lactone derivative.	*14*

[a] All methods cleave bond to the C-terminal side of residue.
[b] BNPS-skatole = 3-bromo-3-methyl-2-(2'-nitrophenylsulfenyl)-indolenine.
[c] PVDF = polyvinylidene difluoride.

Store cyanogen bromide refrigerated in the dry and dark, where it is stable for months. Use only white crystals. **Beware of the toxic nature of this reagent.**

4. Sodium hypocholorite solution (domestic bleach).
5. Equipment includes a fume hood and suitably sized capped tubes (e.g., Eppendorf microcentrifuge tubes).

3. Methods

1. Oxidation: Dissolve the sample to approx 0.5 nmol/μL in oxidizing solution (e.g., 2–3 nmol in 4.9 μL oxidizing solution). Incubate at 4°C for 2 h.
2. Partial neutralization: To the cold sample, add 0.9 vol of ice cold NH_4OH (e.g., 4.4 μL of NH_4OH to 4.9 μL oxidized sample solution). Make this addition carefully so as to maintain a low temperature.
3. Cleavage: Add 8 vol of cyanogen bromide solution. Incubate at 4°C for 30 h in the dark. Carry out this step in a fume hood.
4. To terminate the reaction, lyophilize the sample (all reagents are volatile).
5. Decontaminate equipment, such as spatulas, that have contacted cyanogen bromide, by immersion in bleach until the effervescence stops (a few minutes).

4. Notes

1. The method described is that of Huang et al. *(4)*. Although full details of the mechanism of this reaction are not clear, it is apparent that tryptophanyl residues are converted to oxindolylalanyl residues in the oxidation step, and the bond to the C-terminal side at each of these is readily cleaved in excellent yield (approaching 100% in ref. *4*) by the subsequent cyanogen bromide treatment. The result is seemingly unaffected by the nature of the residues surrounding the cleavage site.

   During the oxidation step, methionyl residues are protected by conversion to sulfoxides, bonds at these residues not being cleaved by the cyanogen bromide treatment. Cysteinyl residues will also suffer oxidation if they have not been reduced and alkylated beforehand (*see* Chapter 51). The peptide to the C-terminal side of the cleavage point has a free N-terminus, so it is suitable for sequencing.
2. Methionyl sulfoxide residues in the peptides produced may be converted back to the methionyl residues by the action, in aqueous solution, of thiols (e.g., dithiothreitol, as described in ref. *3*).
3. The acid conditions used for oxidation and cleavage reactions seem to cause little deamidation *(4)*, but one side reaction that can occur is hydrolysis of acid-labile bonds. The use of low temperature minimizes this problem. If a greater degree of such acid hydrolysis is acceptable, speedier and warmer alternatives to the reaction conditions described above can be used as follows:
   a. Oxidation at room temperature for 30 min, but cool to 4°C before neutralization.
   b. Cleavage at room temperature for 12–15 h.
4. As alternatives to the volatile base NH_4OH, other bases may be used (e.g., the nonvolatile potassium hydroxide or Tris base).
5. In connection with cleavage of methionyl-X bonds by cyanogen bromide, it has been noted that use of 70% (v/v) formic acid can cause formylation of the polypeptide (seen as an increase in molecular mass *[5]*) and damage to tryptophan and tyrosine (evidenced by spectral changes *[6]*). As an alternative to 70% formic acid, 5*M* acetic acid may be used. Possibly, as in the use of cyanogen bromide in cleaving methionyl-X bonds, 70% (v/v) trifluoroacetic acid may prove an acceptable alternative, although this has not been reported.

6. Samples eluted from sodium dodecylsulfate (SDS) gels may be treated as described, but for good yields of cleavage, Huang et al. *(4)* recommend that the sample solutions are acidified to pH 1.5 before lyophilization in preparation for dissolution in the oxidizing solution. Any SDS present may help to solubilize the substrate and, in small amounts at least, does not interfere with the reaction. However, nonionic detergents that are phenolic or contain unsaturated hydrocarbon chains (e.g., Triton, Nonidet P-40), and reducing agents are to be avoided.
7. The method is suitable for large-scale protein cleavage, this requiring simple scaling up. Huang et al. *(4)* made two points, however:
   a. The neutralization reaction generates heat. Since this might lead to protein or peptide aggregation, cooling is important at this stage. Ensure that the reagents are cold and are mixed together slowly and with cooling. A transient precipitate is seen at this stage. If the precipitate is insoluble, addition of SDS may solubilize it (but will not interfere with the subsequent treatment).
   b. The neutralization reaction generates gases. Allow for this when choosing a reaction vessel.
8. At the end of the reaction. all reagents may be removed by lyophilization and the peptide mixture analyzed, for instance, by polyacrylamide gel electrophoresis or by reverse-phase HPLC. Peptides generated may tend to be large, ranging up to a size on the order of 10,000 Dalton or more.
9. Note that all reactions are done in one reaction vial, eliminating transfer of sample between vessels, and so minimizing peptide losses that can occur in such exercises.
10. Alternative methods for cleavage of tryptophanyl-X bonds are outlined in Table 1. The method (4) that employs *N*-chlorosuccinimide is the most specific, but shows only about 50% yield.
11. Both BNPS-skatole and *N*-chlorosuccinimide methods (*see nos.* 4 and 6 of Table 1) have been adapted to cleave small amounts (micrograms or less) of proteins on solid supports or in gels. Thus, proteins bound to glass fiber (as used in automated peptide sequences) may be cleaved by wetting the glass fiber with 1 μg/mL BNPS-skatole in 70%, v/v, acetic acid, followed by incubation in the dark for 1 h at 47°C. After drying, sequencing may proceed as normal. Alternatively, protein blotted to polyvinylidene difluoride membrane may be similarly treated, and resulting peptides eluted for further analysis *(10)*. Alternatively, protein in slices of polyacrylamide gel (following SDS-PAGE) may be cleaved by soaking for 30 min in *N*-chlorosuccinimide, 0.015*M*, in urea (0.5 g/mL) in 50%, v/v, acetic acid. Following washing, peptides may be electrophoresed to generate peptide maps *(13)*.

References

1. Kilic, F. and Ball, E. H. (1991) Partial cleavage mapping of the cytoskeletal protein vinculin. Antibody and talin binding sites. *J. Biol. Chem.* **266,** 8734–8740.
2. Litchfield, D. W., Lozeman, F. J., Cicirelli, M. F., Harrylock, M., Ericsson, L. H., Piening, C. J., and Krebs, E. G. (1991) Phosphorylation of the beta subunit of casein kinase II in human A431 cells. Identification of the autophosphorylation site and a site phosphorylated by p34cdc2. *J. Biol. Chem.* **266,** 20,380–20,389.
3. Tseng, A., Buchta, R., Goodman, A. E., Loughman, M., Cairns. D., Seilhammer, J., Johnson, L., Inglis, A. S., and Scott, K. F. (1991) A strategy for obtaining active mammalian enzyme from a fusion protein expressed in bacteria using phospholipase A2 as a model. *Protein Exp. Purif.* **2,** 127–135.

4. Huang, H. V., Bond, M. W., Hunkapillar, M. W., and Hood, L. E. (1983) Cleavage at tryptophanyl residues with dimethyl sulfoxide-hydrochloric acid and cyanogen bromide. *Methods Enzymol.* **91,** 318–324.
5. Beavis, R. C. and Chait, B. T. (1990) Rapid, sensitive analysis of protein mixtures by mass spectrometry. *Proc. Natl. Acad. Sci. USA* **87,** 6873–6877.
6. Morrison, J. R., Fidge, N. H., and Grego, B. (1990) Studies on the formation, separation, and characterization of cyanogen bromide fragments of human A1 apolipoprotein. *Anal. Biochem.* **186,** 145–152.
7. Ozols, J. and Gerard, C. (1977) Covalent structure of the membranous segment of horse cytochrome b5. Chemical cleavage of the native hemprotein. *J. Biol. Chem.* **252,** 8549–8553.
8. Savige, W. E. and Fontana, A. (1977) Cleavage of the tryptophanyl peptide bond by dimethyl sulfoxide-hydrobromic acid. *Methods Enzymol.* **47,** 459–469.
9. Fontana, A. (1972) Modification of tryptophan with BNPS skatole (2-(2-nitro-phenylsulfenyl)-3-methyl-3'-bromoindolenine). *Methods Enzymol.* **25,** 419–423.
10. Crimmins, D. L., McCourt, D. W., Thoma, R. S., Scott, M. G., Macke, K., and Schwartz, B. D. (1990) In situ cleavage of proteins immobilized to glass-fibre and polyvinylidene difluoride membranes: cleavage at tryptophan residues with 2-(2'-nitropheylsulfenyl)-3-methyl-3'-bromoindolenine to obtain internal amino acid sequence. *Anal. Biochem.* **187,** 27–38.
11. Ramachandran, L. K. and Witkop, B. (1976) *N*-Bromosuccinimide cleavage of peptides. *Methods Enzymol.* **11,** 283–299.
12. Lischwe, M. A. and Sung, M. T. (1977) Use of *N*-chlorosuccinimide/urea for the selective cleavage of tryptophanyl peptide bonds in proteins. *J. Biol. Chem.* **252,** 4976–4980.
13. Lischwe, M. A. and Ochs, D. (1982) A new method for partial peptide mapping using *N*-chlorosuccinimide/urea and peptide silver staining in sodium dodecyl sulfatepolyacrylamide gels. *Anal. Biochem.* **127,** 453–457.
14. Fontana, A., Dalzoppo, D., Grandi, C., and Zambonin, M. (1983) Cleavage at tryptophan with o-iodosobenzoic acid. *Methods Enzymol.* **91,** 311–318.

65

Chemical Cleavage of Proteins at Aspartyl Residues

Bryan John Smith

1. Introduction

Some methods for chemically cleaving proteins, such as those described in Chapters 63 and 64, are fairly specific for a particular residue, show good yields, and generate usefully large peptides (since reaction occurs at relatively rare amino acid residues). There may be cases, however, when such peptides (or indeed proteins) lacking these rarer residues need to be further fragmented, and in such instances, cleavage at the more common aspartyl residue may prove useful. The method described below (and in ref. *1*) for cleavage to the C-terminal side of aspartyl residues is best limited to smaller polypeptides rather than larger proteins because yields are <100% and somewhat variable according to sequence. Partial cleavage of aspartyl-X bonds in a larger protein leads to a very complex set of peptides, which may be difficult to analyze. Partial hydrolysis of smaller peptides yields simpler mixtures which may even be preferable to complete fragmentation for some purposes, such as mass spectrometry, whereby the series of overlapping peptides that are the product of partial hydrolysis may be used to order the peptides in the protein sequence (by analogy with partial cleavage at methionyl-X bonds by cyanogen bromide *[2]* and Chapter 63).

2. Materials

1. Dilute hydrochloric acid (approx 0.013*M*) pH 2. Dilute 220 µL of constant boiling (6*M*) HCl to 100 mL with distilled water.
2. Pyrex glass hydrolysis tubes.
3. Equipment includes a blowtorch suitable for sealing the hydrolysis tubes, a vacuum line, and an oven for incubation of samples at 108°C.

3. Method

1. Dissolve the protein or peptide in the dilute acid to a concentration of 1–2 mg/mL in a hydrolysis tube.
2. Seal the hydrolysis tube under vacuum, i.e., with the hydrolysis (sample) tube connected to a vacuum line. Using a suitably hot flame, draw out and finally seal the neck of the tube.
3. Incubate at 108°C for 2 h.
4. To terminate the reaction, cool and open the hydrolysis tube, dilute the sample with water, and lyophilize.

From: The Protein Protocols Handbook
Edited by: *J. M. Walker Humana Press Inc., Totowa, NJ*

4. Notes

1. The bond most readily cleaved in dilute acid is the Asp-X bond, by the mechanism outlined in Fig. 1A. The bond X-Asp may also be cleaved, in lesser yields (*see* Fig. 1B). Thus, either of the peptides resulting from any one cleavage may keep the aspartyl residue at the point of cleavage, or neither might, if free aspartic acid is generated by a double cleavage event. Any of these peptides is suitable for sequencing.
2. The amino acid sequence of the protein can affect the lability of the affected bond. Thus, the aspartyl-prolyl bond is particularly labile in acid conditions (*see* Note 3). Again, ionic interaction between the aspartic acid side chains and basic residue side chains elsewhere in the molecule can adversely affect the rate of cleavage at the labile bond. Such problems as these make prediction of cleavage points somewhat difficult, particularly if the protein is folded up (e.g., a native protein). The method may well prove suitable for use in cleaving small proteins or peptides, in which case such intramolecular interactions are less likely. Nevertheless, yields are <100%—up to about 70% have been reported *(1)*.
3. As noted above, the aspartyl-prolyl bond is particularly acid-labile, and the following conditions have been proposed in order to promote cleavage of this particular bond *(3)*: dissolution of the sample in guanidine · HCl (7*M*) in dilute acid (e.g., acetic acid, 10% v/v, adjusted to pH 2.5 by pyridine); incubation at moderate temperature (e.g., 37°C) for prolonged periods (e.g., 24 h); and termination by lyophilization. Inclusion of guanidine · HCl, intended to denature the protein, may still fail to render all aspartyl-prolyl bonds sensitive to cleavage.
4. The conditions of low pH can be expected to cause a number of side reactions: cleavage at glutamyl residues; deamidation of (and possibly some subsequent cleavage at) glutaminyl and asparaginyl residues; partial destruction of tryptophan; cyclization of *N*-terminal glutaminyl residues to residues of pyrrolidone carboxylic acid; and α-β-shift at aspartyl residues. The last two changes create a blockage to Edman degradation. The short reaction time of 2 h is intended to minimize these side reactions. A small degree of loss of formyl or acetyl groups from N-termini *(1)* is another possible side reaction, but is not generally recognized as a significant problem.
5. A polypeptide substrate that is insoluble in cold dilute HCl may dissolve during the incubation at 108°C. Formic acid is a good protein denaturant and solvent, and may be used instead of HCl as follows: Dissolve the sample in formic acid (minimum assay 98%, Aristar-grade), and then dilute 50-fold to pH 2.0; proceed as in method for HCl. Note, however, that incubation of protein in formic acid may result in formylation (increased molecular mass *[2]*) and damage to tryptophan and tyrosine residues (altered spectral properties *[4]*).
6. The comments above concerning the effect of the amino acid sequence and of the environment around potentially labile bonds, and the various side reactions that can occur, indicate that the consequences of incubation of a protein in dilute acid are difficult to predict—they are best investigated empirically by monitoring production of peptides by electrophoresis or HPLC.
7. The method described has the benefit of simplicity. It is carried out in a single reaction vessel, with reagents being removed by lyophilization at the end of reaction. Thus, sample handling and losses incurred during this are minimized. This makes it suitable for subnanomolar quantities of protein, though the method may be scaled up for larger amounts also.
8. Note that bonds involving aspartyl residues may also be cleaved by commercially available enzymes: endoproteinase Asp-N hydrolyzes the bond to the *N*-terminal side of an aspartyl residue, but also of a cysteinyl residue; Glu-C cleaves the bond to the C-terminal side of glutamyl and aspartyl residues.

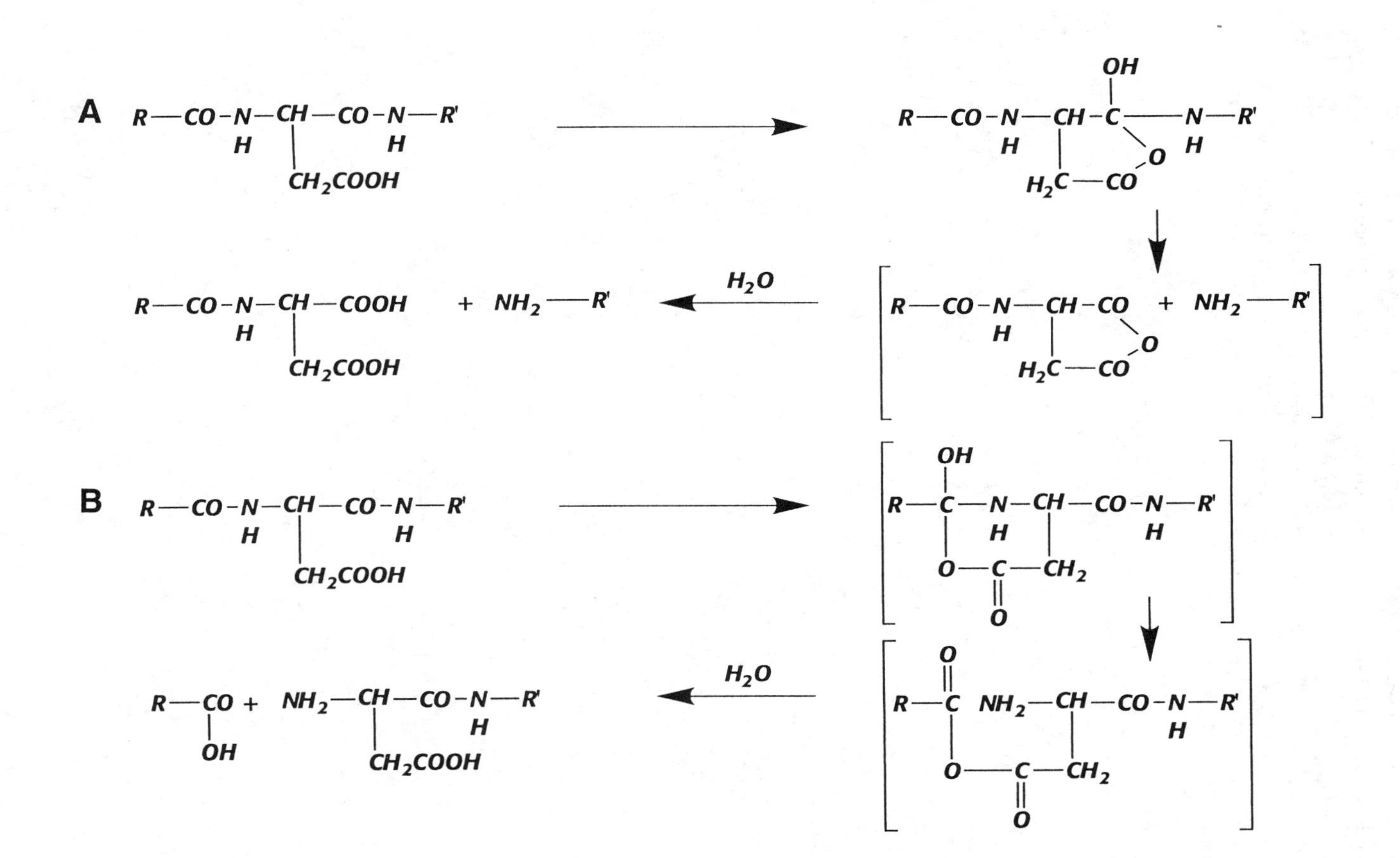

Fig. 1. Mechanisms of the cleavage of bonds to the COOH side (A) and to the NH_2 side (B) of aspartyl residues in dilute acid.

References

1. Ingris, A. S. (1983) Cleavage at aspartic acid. *Methods Enzymol.* **91,** 324–332.
2. Beavis, R. C. and Chait, B. T. (1990) Rapid, sensitive analysis of protein mixtures by mass spectrometry. *Proc. Natl. Acad. Sci. USA* **87,** 6873–6877.
3. Landon, M. (1977) Cleavage at aspartyl-prolyl bonds. *Methods Enzymol.* **47,** 132–145.
4. Morrison, J. R., Fiolge, N. H., and Grego, B. (1990) Studies on the formation, separation and characterization of cyanogen bromide fragments of human A1 apolipoprotein. *Anal. Biochem.* **186,** 145–152.

66

Chemical Cleavage of Proteins at Cysteinyl Residues

Bryan John Smith

1. Introduction

Cysteine is a significant amino acid residue in that it can form a disulfide bridge with another cysteine (to form cystine). Such disulfide bridges are important determinants of protein structure. No known endoproteinase shows specificity solely for cysteinyl or cystinyl residues, although endoproteinase Asp-N is able to hydrolyze bonds to the *N*-terminal side of aspartyl or cysteinyl residues. Modification of the former *(1)* can generate specificity for cysteinyl residues. Again, as discussed by Aitken *(2),* modification of cysteinyl to 2-aminoethylcysteinyl residues makes the bond to the C-terminal side susceptible to cleavage by trypsin or the bond to the *N*-terminal side sensitive to Lys-N. However, specific cleavage of bonds to the *N*-terminal side of cysteinyl residues may be achieved in good yield by chemical means *(3).* The cleavage generates a peptide blocked at its *N*-terminus as the cysteinyl residue is converted to an iminothiazolidinyl residue, but peptide sequencing can be carried out after conversion of this residue to an alanyl residue *(4).*

Cysteine is one of the less frequent amino acids in proteins, and cleavage at cysteinyl residues tends to generate relatively large peptides. Thus, the method described below has been used to generate separate domains of troponin C for the purpose of studying Ca and peptide binding *(5),* and peptides from vinculin to study talin- and anti-vinculin antibody binding *(6).*

2. Materials

1. Modification buffer: 0.2*M* Tris acetate, pH 8.0, 6*M* guanidine/HCl, 10 m*M* dithiothreitol. Use Analar-grade reagents and HPLC-grade water.
2. 2-Nitro-5-thiocyanobenzoate (NTCB): Commercially available (Sigma, Poole, UK) as yellowish powder. Contact with skin, eyes, and so forth, may cause short-term irritation. Long-term effects are unknown, so handle with care (protective clothing). Sweep up spillages. Store at 0–5°C.
3. NaOH: Sodium hydroxide solution, sufficiently concentrated to allow convenient alteration of reaction pH. For example, 2*M* in HPLC-grade water.
4. Deblocking buffer: 50 m*M* Tris-HCl, pH 7.0.
5. Raney nickel-activated catalyst: Commercially available (e.g., from Sigma as 50% slurry in water, pH >9.0). Wash in deblocking buffer prior to use. A supply of N_2 gas is also required for use with the Raney nickel.

From: The Protein Protocols Handbook
Edited by: *J. M. Walker Humana Press Inc., Totowa, NJ*

3. Method

1. Dissolve the polypeptide to a suitable concentration (e.g., 2 mg/mL) in the modification buffer (pH 8.0). To reduce disulfides in the dithiothreitol, incubate at 37°C for 1–2 h.
2. Add NTCB to 10-fold excess over sulfydryl groups in polypeptide and buffer. Incubate at 37°C for 20 min.
3. To cleave the modified polypeptide, adjust to pH 9.0 by addition of NaOH solution. Incubate at 37°C for 16 h or longer.
4. Dialyze against water. Alternatively, submit to gel filtration or reverse-phase HPLC to separate salts and peptides. Lyophilize peptides.
5. If it is necessary to convert the newly formed iminothiazolidinyl *N*-terminal residue to an alanyl group, dissolve the sample to, for example, 0.5 mg/mL in deblocking buffer (pH 7.0) and add to Raney nickel (10-fold excess [w/w], over polypeptide), and incubate at 50°C for 7 h under an atmosphere of nitrogen. Cool and centrifuge briefly to pellet the Raney nickel. Store supernatant at –20°C or further analyze as required.

4. Notes

1. The reactions involved in the above are illustrated in Fig. 1. The method described is basically that used by Swenson and Fredrickson *(5),* an adaptation of that of Jacobson et al. (*7; see also* ref. *3*). The principal difference is that the earlier method *(3,7)* describes desalting (by gel filtration or dialysis) at the end of the modification step (Section 3., step 2), followed by lyophilization and redissolution in a pH 9.0 buffer to achieve cleavage. Simple adjustment of pH as described in Section 3., step 3 has the advantages of speed and avoiding the danger of sample loss on desalting.
2. Swenson and Fredrickson *(5)* describe cleavage (Section 3., step 3) at 37°C for 6 h, but report yields of 60–80%. Other references recommend longer incubations of 12 h or 16 h at 37°C to obtain better yields *(3,7,8).*
3. A slightly modified procedure is described in ref. *8:*
   a. 1 mg/mL sample in borate buffer (20 m*M*) pH 8.0, urea (6*M*) mixed with NTCB, added as 0.1*M* solution in 33% v/v dimethylformamide, at the rate of 40 µL/mL sample solution. Incubation was at 25°C for 1 h.
   b. Cleavage was by adjusting to pH 9.0 with NaOH and incubation at 55°C for 3 h.
   c. Reaction was quenched by addition of 2-mercaptoethanol to 80-fold excess over NCTB.
4. If the sample contains no intramolecular or intermolecular disulfide bonds, the dithiothreitol content of the modification buffer may be lessened, at 1 m*M*.
5. If blockage of the *N*-terminal residue of the newly generated peptide(s) to the C-terminal side of the cleavage point(s) is not a problem (i.e., if sequencing is not required, e.g., ref. *5*), then Section 3., step 5, may be omitted.
6. Reaction with Raney nickel (Section 3., step 5) converts methionyl residues to β-aminobutyryl residues.
7. Although Raney nickel is available commercially, Otieno *(4)* has reported that a more efficient catalyst may be obtained by the method he described, starting from Raney nickel-aluminum alloy. This is reacted with NaOH, washed, deionized, and washed again (under H_2 gas).
8. Treatment of protein with Raney nickel without prior treatment with NTCB causes desulfurization of methioninyl residues (to give β-aminobutyryl residues) and of cysteinyl and cysteinyl residues (to alanyl residues). Otieno *(4)* has suggested that this modification might be used to study dependence of protein function on Met and Cys content.

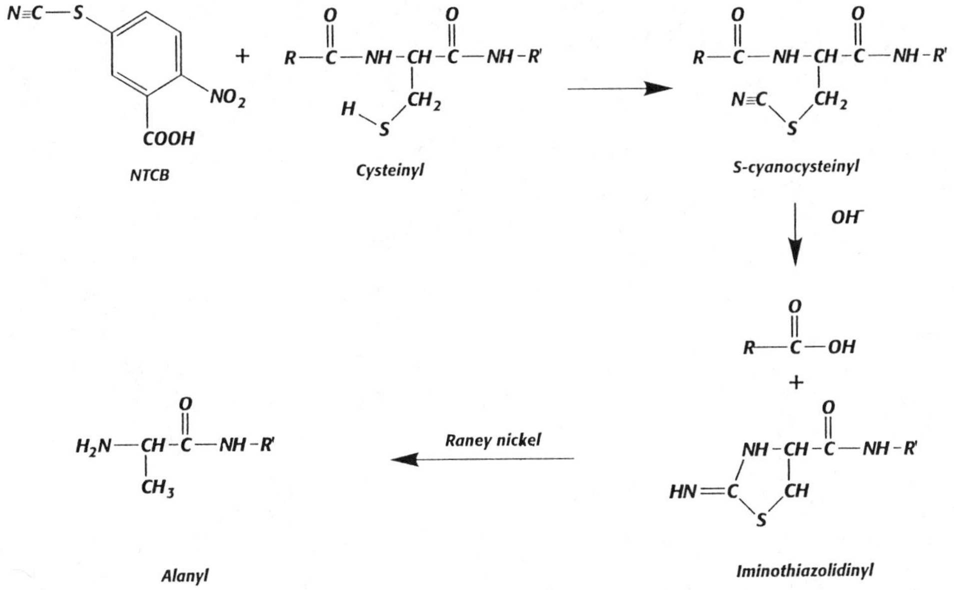

NTCB + Cysteinyl → S-cyanocysteinyl

OH^-

Iminothiazolidinyl

Raney nickel

Alanyl

Fig. 1. Reactions in modification of, and cleavage at, cysteinyl residues by NTCB, and subsequent generation of alanyl *N*-terminal residue.

References

1. Wilson, K. J., Fischer, S., and Yuau, P. M. (1989) Specific enzymatic cleavage at cystine/cysteine residues. The use of Asp-N endoproteinase, in *Methods in Protein Sequence Analysis* (Wittman-Liebold, B., ed.), Springer-Verlag, Berlin, pp. 310–314.
2. Aitken, A. (1994) Analysis of cysteine residues and disulfide bonds, in *Methods in Molecular Biology, vol. 32, Basic Protein and Peptide Protocols* (Walker, J. M., ed.), Humana, Totowa, NJ, pp. 351–360.
3. Stark, G. R. (1977) Cleavage at cysteine after cyanylation. *Methods Enzymol.* **47,** 129–132.
4. Otieno, S. (1978) Generation of a free α-amino group by Raney nickel after 2-nitrothiocyanobenzoic acid cleavage at cysteine residues: applications to automated sequencing. *Biochemistry* **17,** 5468–5474.
5. Swenson, C. A. and Fredrickson, R. S. (1992) Interaction of troponin C and troponin C fragments with troponin I and the troponin I inhibitory peptide, *Biochemistry* **31,** 3420–3427.
6. Kilic, F. and Ball, E. H. (1991) Partial cleavage mapping of the cytoskeletal protein vinculin. Antibody and talin binding sites. *J. Biol. Chem.* **266,** 8734–8740.
7. Jacobson, G. R., Schaffer, M. H., Stark, G., and Vanaman, T. C. (1973) Specific chemical cleavage in high yield at the amino peptide bonds of cysteine and cystine residues. *J. Biol. Chem.* **248,** 6583–6591.
8. Peyser, Y. M., Muhlrod, A., and Werber, M. M. (1990) Tryptophan-130 is the most reactive tryptophan residue in rabbit skeletal myosin subfragment-1. *FEBS Lett.* **259,** 346–348.

67

Chemical Cleavage of Proteins at Asparaginyl-Glycyl Peptide Bonds

Bryan John Smith

1. Introduciton

Reaction with hydroxylamine has been used to cleave DNA and to deacylate proteins (at neutral pH—*see* Note 8). At alkaline pH, however, hydroxylamine may be used to cleave the asparaginyl-glycyl bond. This cleavage tends to generate large peptides, since this pairing of relatively common residues is relatively uncommon, representing about 0.25% of amino acid pairs, according to Bornstein and Balian *(1)*. Apart from generation of peptides for sequencing purposes (e.g., ref. *2*), cleavage of Asn-Gly bonds by hydroxylamine has been used to generate peptides for use in mapping ligand binding sites (e.g., ref. *3*) and phosphorylation sites (e.g., ref. *4*). The method has also been used to cleave a fusion protein at the point of fusion of the constituent polypeptides, although it was noted that formation of hydroxamates may occur *(5)*, as may minor cleavage reactions *(3)*.

2. Materials

1. Cleavage buffer: 2*M* hydroxylamine.HCl, 2*M* guanidine.HCl, 0.2*M* K_2CO_3, pH 9.0. Use Analar-grade reagents and HPLC-grade water.
   Beware of the mutagenic, toxic, and irritant properties of hydroxylamine. Wear protective clothing. Clear wet spillages with absorbent material or clear dry spillages with a shovel, and store material in containers prior to disposal.
2. Stopping solution: Trifluoroacetic acid, 2% (v/v) in water (HPLC-grade).

3. Method

1. Dissolve the protein sample directly in the cleavage buffer to give a concentration in the range of 0.1–5 mg/mL. Alternatively, if the protein is in aqueous solution already, add 10 vol of the cleavage buffer (i.e., sufficient buffer to maintain pH 9.0 and high concentration of guanidine.HCl and hydroxylamine). Use a stoppered container (Eppendorf tube or similar) with small head space, so that the sample does not dry out during the following incubation.
2. Incubate the sample (in stoppered vial) at 45°C for 4 h.
3. To stop reaction, cool and acidify by addition of 3 vol of stopping solution. Store frozen (–20°C) or analyze immediately.

From: The Protein Protocols Handbook
Edited by: J. M. Walker Humana Press Inc., Totowa, NJ

4. Notes

1. The reaction involved in this cleavage is illustrated in Fig. 1 and is described in more detail in ref. *1,* with the proposed role of the succinimide being confirmed by Blodgett et al. *(6)*. Note that, strictly speaking, the reaction of hydroxylamine is with the cyclic imide, which derives from the Asn-Gly pair. Asp-Gly cannot form this succinimide, so that bond is resistant to cleavage by hydroxylamine. The succinimide residue is involved in spontaneous asparagine deamidation and aspartate racemization and isomerization, for it can rapidly hydrolyze in neutral or alkaline conditions to aspartyl-glycyl and isoaspartyl-glycyl (or α aspartyl-glycyl and β aspartyl-glycyl). Stable succinimides have recently been identified in proteins, the succinimidyl version being slightly more basic (by 1 net negative charge) than the aspartate version, which forms after incubation in neutral pH *(7)*. Bornstein and Balian *(1)* have reported an Asn-Gly cleavage yield of about 80%, but yields are somewhat dependent on the sequence of the protein.
2. Inclusion of guanidine.HCl as a denaturant seems to be a factor in improving yields. Kwong and Harris *(7)* reported that omission of guanidine.HCl eliminated Asn-Gly cleavage (though allowed cleavage to the C-terminal side of succinimides, with yields of about 50%, and so allowed identification of such sites). Nevertheless, the literature has examples of the use of buffers lacking guanidine.HCl, for instance *(3,5)*. Both of these examples report use of a Tris-HCl buffer of approx pH 9.0, with ref. *5* including 1 m*M* EDTA and ethanol (10% v/v). Yet other examples *(1,2,8)* describe the use of more concentrated guanidine.HCl, at 6*M,* the cleavage buffer being prepared by titrating a solution of guanidine.HCl (6*M,* finally) and hydroxylamine.HCl (2*M,* finally) to pH 9.0 by addition of a solution of lithium hydroxide (4.5 *M*). Note that preparation of this lithium hydroxide solution may generate insoluble carbonates, which can be removed by filtration. Other reaction conditions were as described in Section 3.
3. As when making peptides by other cleavage methods, it may be advisable, prior to the above operations, to reduce disulfide bonds and alkylate cysteinyl residues (*see* Chapter 51). This denatures the substrate and prevents formation of interpeptide disulfide bonds. Niles and Christen *(8)* describe alkylation and subsequent cleavage by hydroxylamine on a few microliters scale.
4. The hydroxylamine cleavage method has been adapted to cleave proteins in polyacrylamide gel pieces *(9)*. The cleavage buffer was 2*M* hydroxylamine.HCl, 6*M* guanidine.HCl, in 15 m*M* Tris titrated to pH 9.3 by addition of 4.5*M* lithium hydroxide solution. Pieces of gel that had been washed in 5% methanol to remove SDS and then dried *in vacuo* were submerged in the cleavage solution (50–200 µL per 3 µL of gel), and incubated at 45°C for 3 h. Analysis by electrophoresis on a second SDS gel then followed. Peptides of about 10,000 Dalton or less tended to be lost during washing steps, and about 10% of sample remained bound to the treated gel piece. Recoveries, in the second (analytical) SDS gel, were reported to be approx 60%, and cleavage yield was about 25%.
5. In addition to cleavage at Asn-Gly, treatment with hydroxylamine may generate other, lower-yielding, cleavages. Thus, Bornstein and Balian *(1)* mention cleavage of Asn-Leu, Asn-Met, and Asn-Ala, whereas Hiller et al. *(3)* report cleavage of Asn-Gln, Asp-Lys, Gln-Pro, and Asn-Asp. Prolonged reaction times tend to generate more of such cleavages.
6. Treatment of protein with hydroxylamine may generate hydroxamates of asparagine and glutamine, these modifications producing more acidic variants of the protein *(5)*.
7. After the cleavage reaction has been stopped by acidification, the sample may be loaded directly onto reverse-phase HPLC or gel filtration for analysis/peptide preparation. Alternatives are dialysis and PAGE. The reaction itself may be stopped not by acidification, but by mixing with SDS-PAGE sample solvent and immediate electrophoresis *(3)*.

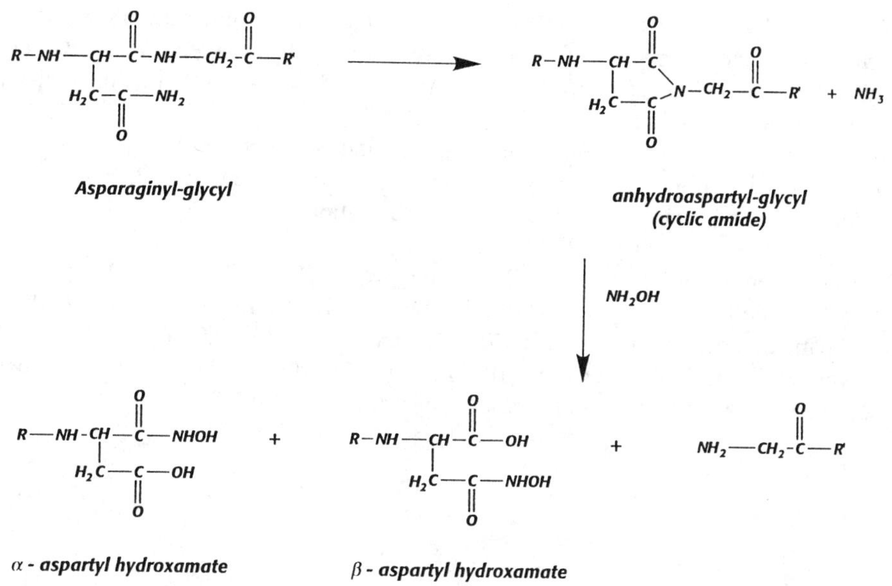

Asparaginyl-glycyl

anhydroaspartyl-glycyl
(cyclic amide)

NH_2OH

α - aspartyl hydroxamate

β - aspartyl hydroxamate

Fig. 1. Illustration of reactions leading to cleavage of Asn-Gly bonds by hydroxylamine.

8. In approximately neutral pH conditions, reaction of protein with hydroxylamine may cause esterolysis, and so may be a useful method in studying posttranslational modification of proteins. Thus, incubation in 1*M* hydroxylamine, pH 7.0, 37°C, for up to 4 h cleaved carboxylate ester-type ADP-ribose-protein bonds (on histones H2A and H2B) and arginine-ADP-ribose bonds (in histones H3 and H4) *(10)*. Again, Weimbs and Stoffel *(11)* identified sites of fatty acid-acylated cysteine residues by reaction with 0.4*M* hydroxylamine at pH 7.4, such that the fatty acids were released as hydroxamates. Omary and Trowbridge *(12)* adapted the method to release [^{3}H]-palmitate from transferrin receptor in polyacrylamide gel pieces, soaking these for 2 h in 1*M* hydroxylamine. HCl titrated to pH 6.6 by addition of sodium hydroxide.

References

1. Bornstein, P. and Balian, G. (1977) Cleavage at Asn-Gly bonds with hydroxlyamine. *Methods Enzymol.* **47,** 132–145.
2. Arselin, G., Gandar, J. G., Guérin, B., and Velours, J. (1991) Isolation and complete amino acid sequence of the mitochondral ATP synthase ε-subunit of the yeast *Saccharomyces cerevisiae. J. Biol. Chem.* **266,** 723–727.
3. Hiller, Y., Bayer, E. A., and Wilchek, M. (1991) Studies on the biotin-binding site of avidin. Minimised fragments that bind biotin. *Biochem. J.* **278,** 573–585.
4. Hoeck, W. and Groner, B. (1990) Hormone-dependent phosphorylation of the glucocorticoid receptor occurs mainly in the amino-terminal transactivation domain. *J. Biol. Chem.* **265,** 5403–5408.
5. Canova-Davis, E., Eng, M., Mukku, V., Reifsnyder, D. H., Olson, C. V., and Ling, V. T. (1992) Chemical heterogeneity as a result of hydroxylamine cleavage of a fusion protein of human insulin-like growth factor I. *Biochem. J.* **278,** 207–213.
6. Blodgett, J. K., Londin, G. M., and Collins, K. D. (1985) Specific cleavage of peptides containing an aspartic acid (beta-hydroxamic) residue. *J. Am. Chem. Soc.* **107,** 4305–4313.

7. Kwong, M. Y. and Harris, R. J. (1994) Identification of succinimide sites in proteins by *N*-terminal sequence analysis after alkaline hydroxylamine cleavage. *Protein Sci.* **3,** 147–149.
8. Niles, E. G. and Christen, L. (1993) Identification of the vaccinia virus mRNA guanyl-transferase active site lysine. *J. Biol. Chem.* **268,** 24,986–24,989.
9. Saris, C. J. M., van Eenbergen, J., Jenks, B. G., and Bloemers, H. P. J. (1983) Hydroxylamine cleavage of proteins in polyacrylamide gels. *Anal. Biochem.* **132,** 54–67.
10. Golderer, G. and Gröbner, P. (1991) ADP-ribosylation of core histones and their acetylated subspecies. *Biochem. J.* **277,** 607–610.
11. Weimbs, T. and Stoffel, W. (1992) Proteolipid protein (PLP) of CNS myelin: positions of free, disulfide-bonded and fatty acid thioester-linked cysteine residues and implications for the membrane topology of PLP. *Biochemistry* **31,** 12,289–12,296.
12. Omary, M. B. and Trowbridge I. S. (1981) Covalent binding of fatty acid to the transferrin receptor in cultured human cells. *J. Biol. Chem.* **256,** 4715–4718.

68

Preparation of Peptides for Microsequencing from Proteins in Polyacrylamide Gels

Robin J. Philp

1. Introduction

Polyacrylamide gel electrophoresis (PAGE) is undoubtedly one of the most effective methods for the separation of proteins and is used extensively in the course of protein purification. Usually the separation is carried out in one dimension only and separates according to the size of the protein, i.e., sodium dodecyl sulfate-PAGE (SDS-PAGE) *(1)*. However, to enhance further the separation of complex mixtures, such as whole-cell lysates, two-dimensional PAGE (2D PAGE) will provide an extremely powerful separation of proteins *(2)*.

It is clearly advantageous to be able to identify proteins from either of these separations, and, to this end, many techniques have been developed to do so. Indeed, several groups are routinely using 2D PAGE in the development of databases of particular cell lines by the systematic identification of the separated proteins *(3,4)*. Previous approaches have been based on the removal of the protein from the gel matrix by passive elution *(5)*, electroelution *(6)*, or by electrophoretic transfer to a membrane, such as nitrocellulose *(7,8)* or polyvinylidene difluoride (PVDF) *(9)*, and have been recently reviewed *(10)*. These methods have resulted in the ability to carry out microsequence analysis of the proteins either directly by N-terminal sequence analysis, or by first fragmenting the protein and sequencing internal peptides.

One such method that was developed *(11,12)* and later modified *(13)*, allows complete proteolytic digestion of a protein within the polyacrylamide matrix followed by passive extraction of the peptides. The advantages of this method are:

1. High recoveries of peptides are achieved as they elute through the pores more easily than the intact protein.
2. The procedure can be carried out in 1D or 2D gel pieces.
3. The gels can be "wet," i.e., just after running and staining or having been dried on a gel drier (this has the added advantage where gels have to be accumulated to pool multiple bands or spots when protein levels are extremely low).
4. This circumvents the problem of N-terminal blockage.
5. This allows choice of peptides for sequencing after HPLC separation, e.g., aromatic amino acid containing peptides.

From: The Protein Protocols Handbook
Edited by: J. M. Walker Humana Press Inc., Totowa, NJ

The essence of this method is the ability to shrink the gel piece prior to adding the enzyme solution. This allows the enzyme to be drawn into the gel matrix as it reswells, permitting an efficient digestion in a small volume. The resulting peptides are then easily eluted from within the gel as the smaller mol-wt fragments will passively elute with a second shrinking step after the digestion has occurred. Finally, a high-resolution reverse-phase HPLC separation of the peptides is carried out prior to microsequencing. Sensitivity using this method for preparation can be extremely high, and sequence data can be obtained on as little as 15 pmol of a given protein loaded onto the gel.

Although this is an ideal step in the identification of a protein by microsequencing using automated Edman chemistry, it is also compatible with the mass-spectrometry based methods of identification *(14–16)*.

2. Materials

2.1. Buffers

1. Stock 1*M* ammonium bicarbonate, pH 9.0: Dissolve 7.906 g ammonium bicarbonate (Sigma, St. Louis, MO) in about 90 mL of Milli-Q water (Millipore Corp., Bedford, MA). Adjust pH to 9.0 with ammonium hydroxide (Aristar, BDH), and make up to a final volume of 100 mL. Filter through 0.22-μ Minisart filter (Sartorius AG, Germany).
2. 200 m*M* ammonium bicarbonate, pH 9.0, 50% acetonitrile: To a 50-mL measuring cylinder, add 10 mL of the 1*M* ammonium bicarbonate stock, and make up to 25 mL with Milli-Q water. Premeasure 25 mL of acetonitrile (HPLC-grade, J. T. Baker), and add to the dilute ammonium bicarbonate.
3. 200 m*M* ammonium bicarbonate, pH 9.0: To a 50-mL measuring cylinder add 10 mL of the stock 1*M* ammonium bicarbonate solution, and make up to 50 mL with Milli-Q water.
4. 60% acetonitrile, 0.1% Trifluoroacetic acid (TFA): To a 100-mL measuring cylinder, add 1 mL of a 10% solution of TFA (Sequencer-grade, Pierce, Rockford, IL), and make up to 40 mL with Milli-Q water. Premeasure 60 mL of acetonitrile (HPLC-grade), and add to the TFA solution.

2.2. Enzyme Solutions

1. Trypsin (modified trypsin, Promega, WI) is diluted with the supplied dilution buffer to a concentration of 0.1 mg/mL and frozen in 10-μL aliquots at –20°C.
2. Endoproteinase LysC (Sequencing-grade, Boehringer Mannheim, Mannheim, Germany) is diluted to a concentration of 0.1 mg/mL with Milli-Q water and stored in aliquots as with trypsin.

2.3. Reverse-Phase HPLC

To achieve high-sensitivity separation with low amounts of protein, attention to the configuration and optimization of the HPLC is important. However, it is not intended to cover this in any detail here, so the reader is advised to consult literature covering this topic if they are not already familiar with the area (*see,* for example, ref. *16*).

2.3.1. Mobile Phases

1. Solvent A: 0.1% TFA in Milli-Q water: Add 5 mL of a 10% solution of TFA (sequencer-grade, Pierce) to a 500-mL measuring cylinder, and make up to 500 mL with Milli-Q water. Filter through 0.22-μm membrane filter and degas with a helium sparge.

2. Solvent B: 0.08% TFA in 80% acetonitrile. Add 4 mL of a 10% solution of TFA to a 500-mL cylinder, and make up to 100 mL with Milli-Q water. Premeasure 400 mL of HPLC-grade acetonitrile, and add to the TFA solution. Degas with a helium sparge.
3. Silica based reverse-phase column with a pore size of 300 Å, e.g., Vydac C18 Protein/Peptide column (Vydac, Hesperia, CA), 1-mm internal diameter with a length of 100–250 mm.

3. Method

1. Assuming that a gel has already been run and stained/destained appropriately, e.g., Coomassie brilliant blue R250, cut out the band(s) or, in the case of 2D gels, spot(s) to be treated.
2. Incubate the gel piece(s) in the ammonium bicarbonate/acetonitrile buffer using about three times the volume of the gel. After 30 min, discard the buffer and replace with an equal volume. Repeat one more time.
3. Discard the buffer and dry the gel piece(s) with a gentle stream of nitrogen or argon (2–3 min).
4. Place the dry gel piece(s) onto a clean surface (a clean glass electrophoresis plate is ideal).
5. Using a pipet, add 1–2 µL of the enzyme solution (trypsin or LysC) to each gel piece and allow to soak in. Add a second aliquot of the enzyme and, again, allow to soak in.
6. Put gel piece(s) back into the Eppendorf tube, and add 5-µL aliquots of the 200 m*M* ammonium bicarbonate buffer, allowing each to soak into the gel. Finally, add enough of the same buffer to just cover the gel.
7. Incubate the tube at 37°C for 4–18 h.
8. After the required incubation time, transfer the supernatant to a clean Eppendorf tube, and retain.
9. Using a clean scalpel blade, cut up the gel piece(s) very finely and return them to the tube.
10. Using about four times the volume of the gel piece(s), add some 60% acetonitrile, 0.1% TFA, and shake for 20 min.
11. Remove the supernatant and add to the previously removed aliquot. To the gel piece(s), add the same volume again of the acetonitrile/TFA and shake. After a further 20 min, add this supernatant to the pool and dry down using a Speedvac (Savant) or lyophilize.
12. Redissolve residue in an appropriate volume of Milli-Q water to inject onto the HPLC system.
13. Peptides can be collected from the HPLC for sequence analysis.

4. Notes

1. Omit any fixing step and stain for a reduced time to minimize fixing the protein in the gel. Since the pH of the digestion needs to be between 8.0 and 9.0, it is helpful to reduce the acetic acid in the destain step or indeed omit it altogether. A 10 or 20% methanol solution is adequate to destain the gel enough to determine the band or spot of interest. If the gel has been stained and destained using normal conditions, then extend the ammonium bicarbonate/acetonitrile step to ensure that the pH is suitable. When cutting out the band or spot, it is advisable to cut as closely to the stained area as possible. This reduces any SDS that may still be present in the gel as well as lowering the gel-associated artifacts that coelute from the gel and are seen in the early part of the HPLC separation.
2. If a "wet" gel is being used, this step will result in a noticeable decrease in the size of the gel piece(s) as well as some further destaining. In the case of gels that have been dried, there will be some reswelling seen, but it will not extend to the original size. Any 3MM filter paper and cellophane used in the gel drying procedure will float away from the gel piece(s) and it is important to remove as much of this as possible (*see also* Note 12).

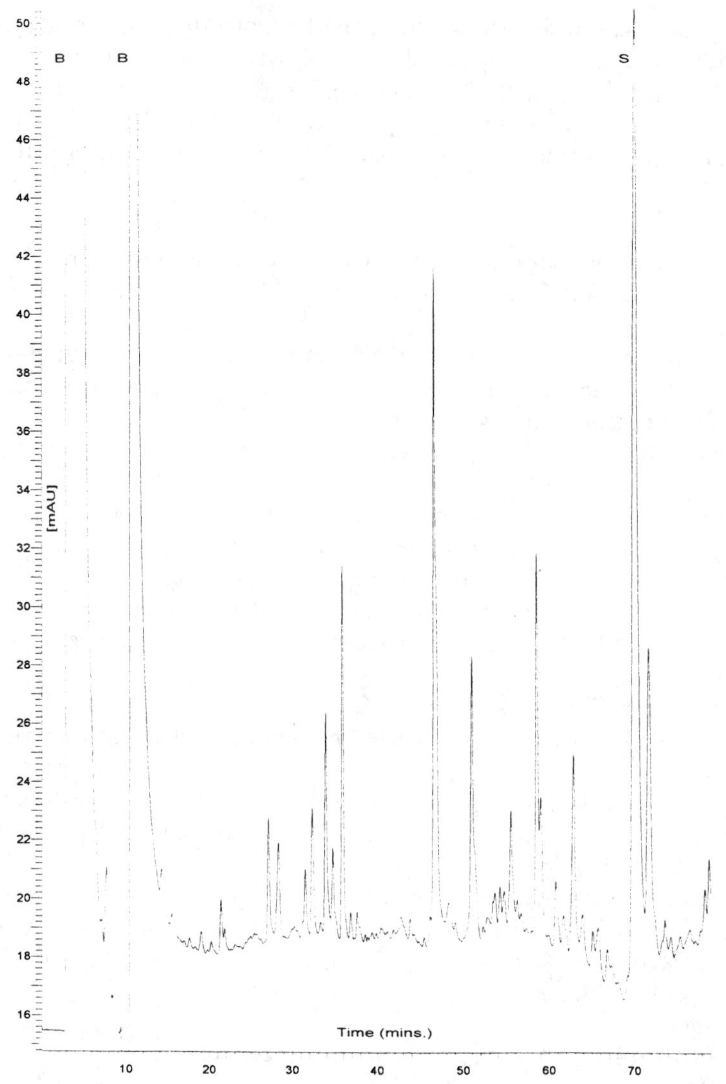

Fig. 1. HPLC separation of peptides following an in-gel digestion of approx 1.5 μg (35 pmol) of a 42-kDa protein. Two faintly stained bands were excised from the gel and digested in-gel with endoproteinase LysC. Subsequent peptides were separated on a 1 × 250 mm Vydac C18 column with a gradient of acetonitrile in 0.08% TFA from 2 to 60% over 80 min at a flowrate of 50 μL/min. The peaks labeled "B" are from the gel extract and include the extraction buffer, residual SDS, and any gel-derived artifacts. The peak labeled "S" is the eluting Coomassie blue stain. The peptide eluting at 36.1 min was sequenced on an Applied Biosystems Procise sequencing system (ABI, Foster City) giving an initial yield of 9.4 pmol and the following sequence: MEISYHGHSIVAWRE (first 15 cycles).

3. It is not essential to dry the gel piece(s) totally at this step, but enough so that the gel shows a "sticky" consistency.
4. If only a single gel piece is used, this step can be carried out directly in the Eppendorf tube.
5. It is advisable to keep the enzyme-to-substrate ratio as high as possible, especially in the case of trypsin that does autodigest. The modified trypsin from Promega reduces this problem, but occasionally autodigested trypsin peptides have been sequenced! The peptides

from a trypsin digest do tend to be rather short owing to the multiple cleavage sites, although quite satisfactory for most cases. Using endoproteinase LysC results in longer fragments being obtained, but possibly at the expense of not eluting these with such high efficiency. The enzymes trypsin and endoproteinase LysC have given best results in the author's experience. Other enzymes, such as Staphylococcal V8 protease, have been tried, but with limited success.

6. After the digestion has proceeded for 30 min or so, check to see that the gel piece(s) is still submerged. Ocassionally, the gel is still reswelling and the level of buffer drops, exposing the gel.
7. Four hours is usually long enough for digestion to take place. However, if a very high ratio of enzyme-to-substrate was used, then an overnight incubation is suggested.
8. If the level of stain is high enough, the supernatant is often blue since the stain tends to diffuse from the gel as the digest proceeds.
9. This step improves the extraction of peptides from the gel during the next addition.
10. As this incubation proceeds, the gel pieces will be seen to shrink.
11. If necessary, a third elution step can be included.
12. It is important to include a hard spin in a bench-top centrifuge prior to injecting the sample. There is usually some insoluble debris that pellets during this spin. This is probably from fibers of 3MM paper if a dried gel was used in the preparation as well as some small pieces of acrylamide, which fragments during the procedure. Failure to clear this may cause a problem in the HPLC separation and can lead to a blockage of the column, resulting in high back-pressure.

    The gradient chosen to elute the peptides from the column should be such that all peaks are well resolved. To achieve this, a shallow increase in the B buffer is essential. It is also recommended, after injection, to hold the gradient at the initial condition until the large void volume peak containing any buffer salts has eluted. About 5 min is usually sufficient, but the more gel pieces that are used, the larger the initial peak is. A linear gradient from 0 to 60% solvent B over 80 min is sufficient to achieve a succesful separation. Eluting peptides should be monitored at between 205 and 220 nm for the peptide bond and at 280 nm for peptides containing aromatic amino acids (especially tryptophan). Other wavelengths may be used as appropriate. Coomassie blue stain, which is extracted from the gel with the peptides, elutes late in the gradient and has a strong absorbance at both the wavelengths mentioned (*see* Fig. 1).

Acknowledgment

This procedure was derived from the original method of Jorge Rosenfeld et al. *(13)*

References

1. Laemmli, U. K. (1970) Cleavage of structural proteins during the assembly of the head of bacteriophage T4. *Nature* **227,** 680–685.
2. O'Farrell, P. H. (1975) High resolution two-dimensional electrophoresis of proteins. *J. Biol. Chem.* **250,** 4007–4021.
3. Celis, J. E., Gesser, B., Rasmussen, H. H., Madsen, P., Leffers, H., Dejgaard, K., Honore, B., Olsen, E., Ratz, G., et al. (1990) Comprehensive two-dimensional gel protein databases offer a global approach to the analysis of human cells: the transformed amnion cells (AMA) master database and its link to genome DNA sequence analysis. *Electrophoresis* **11,** 989–1071.
4. Garrels, J. and Franza, B. (1989) The REF52 protein database: methods of database construction and analysis using the QUEST system and characterization of protein patterns from proliferating and quiescent REF52 cells. *J. Biol. Chem* **264,** 5283–5298.

5. Ward, D. L. and Simpson, R. J. (1991) Micropreparative protein isolation from polyacrylamide gels following detection by high-resolution dynamic imaging: application to microsequencing. *Pept. Res.* **4,** 187–193.
6. Ward, D. L., Reid, G. E., Moritz, R. L., and Simpson, R. J. (1990) Strategies for internal amino acid sequence analysis of proteins separated by polyacrylamide gel electrophoresis. *J. Chromatogr.* **519,** 199–216.
7. Towbin, H., Staehelin, T., and Gordon, J. (1979) Electrophoretic transfer of proteins from acrylamide gels to nitrocellulose sheets: procedure and some applications. *Proc. Natl. Acad. Sci. USA* **76,** 4350–4354.
8. Aebersold, R. H., Leavitt, J., Saavedra, R. A., Hood, L. E., and Kent, S. B. (1987) Internal amino acid sequence analysis of proteins separated by one- or two-dimensional gel electrophoresis after *in situ* protease digestion on nitrocellulose. *Proc. Natl. Acad. Sci USA* **84,** 6970–6974.
9. Matsudaira, P. (1987) Sequence from picomole quantities of proteins electroblotted onto polyvinylidene difluoride membranes. *J. Biol. Chem.* **262,** 10,035–10,038.
10. Patterson, S. D. (1994) From electrophoretically separated protein to identification: strategies for sequence and mass analysis. *Anal. Biochem.* **221,** 1–15.
11. Kawasaki, H., Emori, Y., and Suzuki, K. (1990) Production and separation of peptides from proteins stained with Coomassie brilliant blue R-250 after separation by sodium dodecyl sulfate-polyacrylamide gel electrophoresis. *Anal. Biochem.* **191,** 332–336.
12. Eckerskorn, C. and Lottspiech, F. (1989) Internal amino acid sequence analysis of proteins separated by gel electrophoresis after tryptic digestion in polyacrylamide matrix. *Chromatographia* **28,** 92–94.
13. Rosenfeld, J., Capdevielle, J., Guillemot, J. C., and Ferrara, P. (1992) In-gel digestion of proteins for internal sequence analysis after one- or two-dimensional gel electrophoresis. *Anal. Biochem.* **203,** 173–179.
14. Pappin, D. J. C., Hojrup, P., and Bleasby, A. J. (1993) Rapid identification of proteins by peptide-mass fingerprinting. *Curr. Biol.* **6,** 327–332.
15. Henzel, W. J., Billeci, T. M., Stults, J. T., Wong, S., Grimley, C., and Watanabe, C. (1993) Identifying proteins from two-dimensional gels by molecular mass searching of peptide fragments in protein sequence databases. *Proc. Natl. Acad. Sci. USA* **90,** 5011–5015.
16. Elicone, C., Lui, M., Geromanos, S., Erdjument-Bromage, H., and Tempst, P. (1994) Microbore reversed-phase high-performance liquid chromatographic purification of peptides for combined chemical sequencing—laser description mass spectrometric analysis. *J. Chromatogr. A.* **676,** 121–137.

69

In Situ Chemical and Enzymatic Digestions of Proteins Immobilized on Miniature Hydrophobic Columns

Malcolm Ward

1. Introduction

Protein and peptide samples that require sequencing are, more often than not, received in biological buffers containing free amines, salts, and denaturants. Tris, guanidinium chloride, and urea are commonly used when purifying proteins, and these can interfere with the Edman degradation if present during sequencing. Additional sample manipulation, for example, dialysis, precipitation, or desalting by HPLC, is usually required to make the sample matrix compatible with the sequencing chemistries.

The Hewlett Packard G1000A Protein Sequencing system enables proteins in complex matrices to be sequenced directly by performing Edman chemistry on samples retained on a unique support *(1)*. Proteins are loaded onto the hydrophobic half of a two-part miniature column via a sample loading device (Fig. 1). The inorganic salts and buffers are removed using aqueous washes facilitating the sequence analysis of the retained material.

However, since many proteins are blocked at their *N*-terminus and will not undergo the Edman reaction, it is often necessary to cleave the protein into fragments to obtain internal sequence information. This chapter will illustrate the use of chemical (Fig. 2) and enzymatic (Fig. 3) digestions of proteins that have been immobilized on miniature hydrophobic columns.

Chemical cleavage of proteins with cyanogen bromide (CNBr) is a widely used process for the generation of a relatively small number of large peptide fragments (*see* Chapter 63). The reaction conditions can be adapted to allow *in situ* digestion of proteins bound to the hydrophobic support *(2)*. The methods for performing *in situ* enzymatic digestions on the Hewlett Packard sequencing columns were first reported by Burkhart *(3)*. Initially, digests were performed using trypsin, endoproteinase Lys-C, and endoproteinase Asp-N. A more recent publication *(4)* presents additional data for the use of enzymes, such as chymotrypsin, papain, and subtilisin.

Coupling of the miniature column to an HPLC column is achieved by using a column adaptor. This allows for the separation of peptides produced by chemical or enzymatic digests without the need for buffer exchange or lyophilization.

From: The Protein Protocols Handbook
Edited by: J. M. Walker Humana Press Inc., Totowa, NJ

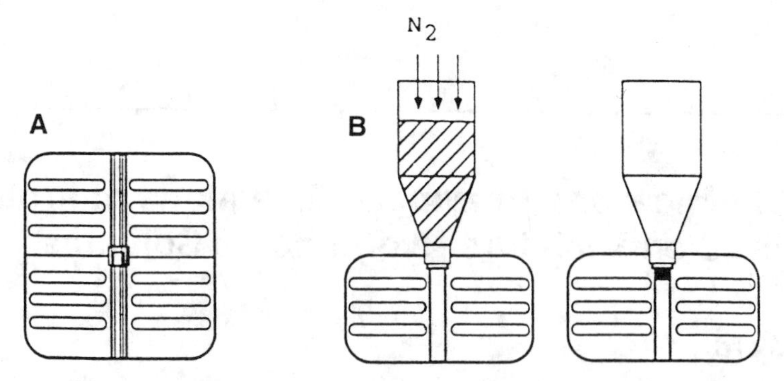

Fig. 1. (A) Miniature biphasic column assembly consisting of an upper hydrophobic half and a lower hydrophilic half. **(B)** Protein solutions are loaded onto the hydrophobic half of the column via a 5-mL plastic funnel under nitrogen pressure using the sample preparation station.

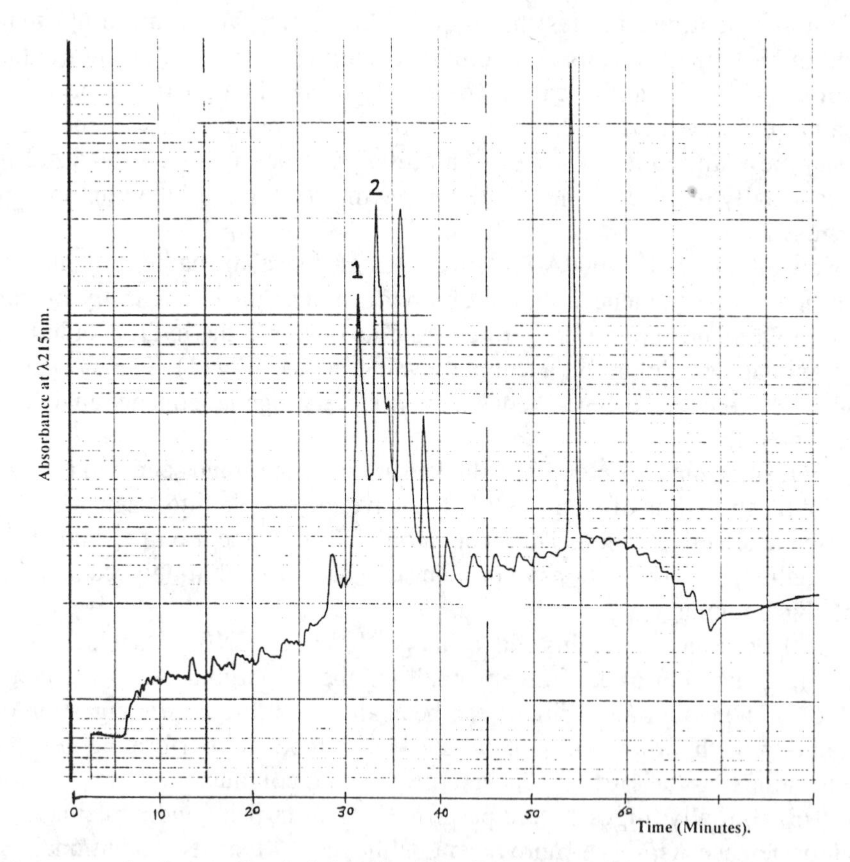

Fig. 2. An example of *in situ* chemical digestion using CNBr. A protein known to inhibit rubber biosynthesis was isolated from *Hevea brasiliensis* serum. The *N*-terminus of the protein was shown to be blocked to the Edman reaction. Internal sequence data were obtained by loading the sample onto a hydrophobic column. All salts and buffers were removed by washing with 2 mL 2% TFA. A CNBr digest was performed using 0.25*M* CNBr in 70% formic acid for 1 h at 60°C. After this time, the column was placed in line to a 2.1-mm Aquapore RP-300 C18 reversed-phase column via the adaptor. Peptide fragments were separated by gradient elution at a flowrate of 200 μL/min beginning at 20% B and rising to 60% B in 40 min, where solvent A = 0.1% TFA/H_2O and solvent B = 0.1% TFA/MeCN. Detection was at λ215 nm 0.500 AUFS. Peptide fragments were collected and sequenced. Peaks 1 and 2 gave sequences beginning VDFHL and AQFPD, respectively. Sufficient sequence information was obtained to design probes for cloning experiments.

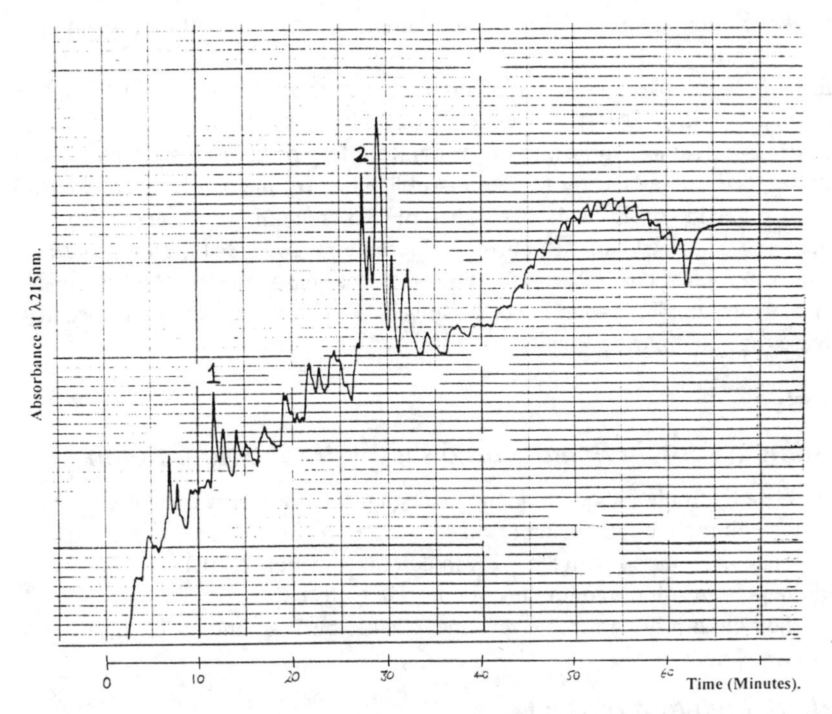

Fig. 3. An example of *in situ* enzymatic digestion using trypsin. An *in situ* tryptic digest was performed on a hydrophobic column to yield internal sequence data for the protein Aspartate amino transferase isolated from *Thermus aquaticus*. The digestion buffer consisted of 5 μg sequence-grade trypsin/400 μL of 0.1*M* Tris-HCl, pH 8.5 containing 20% acetonitrile. The digest was incubated for 5 h at 37°C. After this time, the column was placed in line to a 2.1 mm Aquapore RP–300C18 reversed-phase column via the adaptor. Peptide fragments were separated by gradient elution at a flowrate of 200 μL/min beginning at 20% B and rising to 60% B in 40 min, where solvent A = 0.1% TFA/H_2O and solvent B = 0.1% TFA/MeCN. Detection was at λ215 nm 0.300 AUFS. Peptide fragments were collected and sequenced. Peaks 1 and 2 gave sequences beginning DAGGV and ALVVN, respectively.

Chemical and enzymatic modifications of proteins can be performed *in situ* to yield similar or complementary results to in vitro experiments. This alternative approach may offer information with regard to protein structure, such aspects as conformation, folding, or hydrophobicity *(5)*.

In situ digestion can provide a simple procedure for handling low amounts of problematic proteins. The advantages of reversed-phase chromatography can be fully utilized while avoiding many sample handling stages. Thus, sample losses are minimized.

2. Materials

2.1. Equipment

1. Miniature sequencing columns (Hewlett Packard, Palo Alto, CA).
2. Sample loading station G1001A (Hewlett Packard).

2.2. Chemicals

1. Methanol (HPLC-grade reagent).
2. 2% Trifluoroacetic acid (TFA) in distilled water.

3. Pyridylethylation solution: 20 m*M* dithiothreitol (DTT) and 50 m*M* 4-vinylpyridine in 6*M* guanidinium-HCl.
4. 0.25*M* CNBr in 70% formic acid. This is prepared by diluting a stock solution of 5*M* CNBr in acetonitrile (Aldrich Chemical Co., Milwaukee, WI) with 70% formic acid.

   CNBr is extremely toxic and appropriate safety precautions should be adhered to at all times (for details, refer to Sigma-Aldrich library of chemical safety data 2,938C).
5. Protease digestion buffer: 0.1*M* Tris-HCl, pH 8.5 containing 20% acetonitrile.
6. Sequencing-grade trypsin (Boehringer Mannheim 1047-841), endoproteinase Lys-C (Boehringer Mannheim 1047-825), and endoproteinase Asp-N (Boehringer Mannheim 1054-589) are supplied lyophilized. Dissolve 2–5 μg of the required protease in 400 μL of protease digestion buffer.

3. Methods

3.1. Loading a Protein Sample onto the Hydrophobic Column

1. Assemble a hydrophobic column, and place in the sample prep station.
2. Rinse the column with methanol (1 mL).
3. Equilibriate the column with 2% TFA in deionized water (1 mL).
4. Load the protein solution onto the column (*see* Note 1).
5. Wash the bound protein sample to remove salts, buffers, and so forth, using deionized water (*see* Note 2).

3.2. Pyridylethylation of Cysteine

1. Load the protein sample onto a hydrophobic column as described in Section 3.1., steps 1–4.
2. Transfer 400 μL of the pyridylethylation solution to the reservoir funnel.
3. Allow the solution to drip through the column leaving a small volume in the funnel.
4. Incubate for 15 min at room temperature.
5. Wash the column with water (500 μL) and 20% acetonitrile (1 mL) to remove buffer and excess reagents.

3.3. CNBr Digestion

1. Load the protein sample onto a hydrophobic column as described in Section 3.1., steps 1–4.
2. Reduce and alkylate the protein if required as described in Section 3.2.
3. Transfer 400 μL of the CNBr solution to the reservoir funnel.
4. Allow the solution to drip through the column leaving a small volume in the funnel.
5. Incubate for 4 h at room temperature (*see* Note 4).
6. Wash the column with water (500 μL) and 20% acetonitrile (1 mL) to remove buffer and excess reagents.

3.4. Digestion Using Proteases

1. Load the protein sample onto a hydrophobic as described in Section 3.1., steps 1–4.
2. Reduce and alkylate the protein if required as described in Section 3.2.
3. Allow 400 μL of the buffered protease solution to drip through the column, leaving a small volume in the funnel.
4. Remove the column and attached reservoir funnel from the sample prep station, and incubate at 37°C overnight.

3.5. Elution of Peptides

1. Peptides produced can either be batch-eluted using 80% acetonitrile (400 μL) and collected in an Eppendorf tube for subsequent analysis by mass spectrometry, or the column

can be coupled to another HPLC column via the column adaptor, which enables direct separation of the peptides without any further sample manipulation. Peptides can be collected for Edman sequence analysis or analyzed directly by LC-MS techniques.

4. Notes

1. Proteins will not bind to the column if loaded in >1% SDS or Triton X100. If detergent is present, the sample should be diluted to approx 0.1% detergent before being applied to the column.
2. Three consecutive 1-mL washes of deionized water are usually sufficient to eliminate any salts or detergents present in the sample solution. The bed volume of the miniature column is 30 µL; therefore, 3 mL are equivalent to 100 column volumes.
3. Take care not to allow the column to run dry at any stage, since this may cause sample loses.
4. It is possible to use a shorter digestion time of 1 h if the reaction is carried out at an elevated temperature of 60°C. Carefully blow helium gas across the top of the funnel to create an inert atmosphere. Then seal both the top of the funnel and the bottom of the column with parafilm. All experimental work should be performed in a fumehood, following the appropriate safety precautions.

References

1. Slattery, T. K. and Harkins, R. N. (1993) Analysis of complex protein mixtures on the HP-G1000 A sequencer, in *Techniques in Protein Chemistry IV* (Angeletti, R., ed.), pp. 443–452.
2. Miller, C. G. (1993) Sequence Analysis of an On-column CNBr Digestion of a Lipase. Hewlett Packard Application Note 93–8.
3. Burkhart, W. (1993) *In situ* proteolytic digestions of proteins bound to the Hewlett Packard hydrophobic sequencing column, *Techniques in Protein Chemistry IV* (Angeletti, R., ed.), Academic, Sandiego, CA, pp. 399–406
4. Burkhart, W., Moyer, M. B., and Kassel, D. B. (1994) Additional proteases for *in situ* digestions on the Hewlett Packard sequencing column and its coupling to capillary LC and LC/MS 7th Symposium of the Protein Society, Boston, MA (Poster presentation).
5. Miller, C. G. (1992) 6th Symposium of the Protein Society, San Diego, CA (Poster presentation).

70

Enzymatic Digestion of Membrane-Bound Proteins for Peptide Mapping and Internal Sequence Analysis

Joseph Fernandez and Sheenah M. Mische

1. Introduction

Enzymatic digestion of membrane-bound proteins is a sensitive procedure for obtaining internal sequence data of proteins that either have a blocked amino terminus or require two or more stretches of sequence data for DNA cloning or confirmation of protein identification. Since the final step of protein purification is usually SDS-PAGE, electroblotting to either PVDF or nitrocellulose is the simplest and most common procedure for recovering protein free of contaminants (SDS, acrylamide, and so forth) with a high yield. The first report for enzymatic digestion of a nitrocellulose-bound protein for internal sequence analysis was by Aebersold et al. in 1987, with a more detailed procedure later reported by Tempst et al. in 1990 *(1,2)*. Basically, these procedures first treated the nitrocellulose-bound protein with PVP-40 (polyvinyl pyrrolidone, M_r 40,000) to prevent enzyme adsorption to any remaining nonspecific protein binding sites on the membrane, washed extensively to remove excess PVP-40, and the sample was enzymatically digested at 37°C overnight. Attempts with PVDF-bound protein using the above procedures *(3,4)* give poor results and generally require >25 µg of protein. PVDF is preferred over nitrocellulose because it can be used for a variety of other structural analysis procedures, such as amino-terminal sequence analysis and amino acid analysis. In addition, peptide recovery from PVDF-bound protein is higher, particularly from higher retention PVDF (ProBlott, Westran, Immobilon P^{sq}). Finally, PVDF-bound protein can be stored dry as opposed to nitrocellulose, which must remain wet during storage and work up to prevent losses during digestion.

Enzymatic digestion of both PVDF- and nitrocellulose-bound protein in the presence of 1% hydrogenated Triton X-100 (RTX-100) buffers as listed in Table 1 was first performed after treating the protein band with PVP-40 *(5)*. Unfortunately, the RTX-100 buffer also removes PVP-40 from the membrane, which can interfere with subsequent reverse-phase HPLC. Further studies *(6,7)* demonstrate that treatment with PVP-40 is unnecessary when RTX-100 is used in the digestion buffer. It appears that RTX-100 acts as both a blocking reagent and a strong elution reagent.

PVDF-bound proteins are visualized by staining and subsequently excised from the blot. Protein bands are immersed in hydrogenated Triton X-100 (RTX-100), which

From: The Protein Protocols Handbook
Edited by: J. M. Walker Humana Press Inc., Totowa, NJ

Table 1
Digestion Buffers Recipes for Various Enzymes

Enzyme	Digestion buffer	Recipe (using RTX-100)[a]	Recipe (using OGP)[c]	Comments
Trypsin or Lys-C	1% Detergent/ 10% acetonitrile/ 100 m*M* Tris, pH 8.0	100 µL 10% RTX-100 stock, 100 µL acetonitrile, 300 µL HPLC-grade water, and 500 µL 200 m*M* Tris stock[b]	10 mg OGP, 100 µL acetonitrile, 400 µL HPLC-grade water, and 500 µL 200 m*M* Tris stock[b]	Detergent prevents enzyme adsorption to membrane and increases recovery of peptides
Glu-C	1% Detergent/ 100 m*M* Tris, pH 8.0	100 µL 10% RTX-100 stock, 400 µL HPLC-grade water, and 500 µL 200 m*M* Tris stock[b]	10 mg OGP, 500 µL HPLC-grade water, and 500 µL 200 m*M* Tris stock[b]	Acetonitrile decreases digestion efficiency of Glu-C
Clostripain	1% Detergent/ 10% acetonitrile/ 2 m*M* DTT/ 1 m*M* $CaCl_2$/ 100 m*M* Tris, pH 8.0	100 µL 10% RTX-100 stock, 100 µL acetonitrile, 45 µL 45 m*M* DTT, 10 µL 100 m*M* $CaCl_2$, 245 µL HPLC-grade water, and 500 µL 200 m*M* Tris stock[b]	10 mg OGP, 100 µL acetonitrile, 45 µL 45 m*M* DTT, 10 µL 100 m*M* $CaCl_2$, 345 µL HPLC-grade water, and 500 µL 200 m*M* Tris stock[b]	DTT and $CaCl_2$ are necessary for Clostripain activity

[a]Hydrogenated Triton X-100 (RTX-100) as described in Tiller et al. *(13)*.
[b]200 m*M* Tris stock (pH 8.0) is made up as follows: 157.6 mg Tris-HCl and 121.1 mg trizma base to a final volume of 10 mL with HPLC-grade water.
[c]Octyl glucopyranoside can be substituted for RTX-100.

acts as both a reagent for peptide extraction and a blocking reagent for preventing enzyme adsorption to the membrane during digestion. Remaining cystine bonds are reduced with dithiotreitol (DTT) and carboxyamidomethylated with iodoacetamide. After incubation with the enzyme of choice, the peptides are recovered in the digestion buffer. Further washes of the membrane remove the remaining peptides, which can be analyzed by microbore HPLC. Purified peptides can then be subjected to automated sequence analysis.

Recently, additional studies have been reported that have enhanced this procedure. Best et al. *(8)* have reported that a second aliquot of enzyme several hours later improves the yield of peptides. Reduction and alkylation of cysteine is possible directly in the digestion buffer, allowing identification of cysteine during sequence analysis *(9)*. Finally, octyl- or decylglucopyranoside can be substituted for RTX-100 in order to obtain cleaner mass spectrometric analysis of the digestion mixture *(10)*.

2. Materials

The key to success with this procedure is cleanliness. Use of clean buffers, tubes, and staining/destaining solutions, as well as using only hydrogenated Triton X-100 as opposed to the nonhydrogenated form, greatly reduces contaminant peaks obscured during reverse-phase HPLC. A corresponding blank piece of membrane must always be analyzed at the same time as a sample is digested, as a negative control. All solutions should be prepared with either HPLC-grade water or double-glass-distilled water that has been filtered through an activated charcoal filter, and passed through a 0.22-μm filter *(11)*.

2.1. Preparation of the Membrane-Bound Sample

Protein should be analyzed by SDS-PAGE or 2D-IEF using standard laboratory techniques. Electrophoretic transfer of proteins to the membrane should be performed in a full immersion tank rather than a semidry transfer system to avoid sample loss and obtain efficient transfer *(12)*. PVDF membranes with higher protein binding capacity such as Immobilon P^{sq} (Millipore, Bedford, MA), Problott (Applied Biosystems, Foster City, CA), and Westran (Bio-Rad, Hercules, CA), are preferred owing to greater protein recovery on the blot, although all types of PVDF and nitrocellulose can be used with this procedure. The following stains are compatible with the technique: Ponceau S, Amido black, india ink, and chromatographically pure Coomassie brilliant blue with a dye content >90%. **A blank region of the membrane should be excised to serve as a negative control.**

2.2. Enzymatic Digestion Buffers

Digestion buffer should be made as described in Table 1. Make up 1 mL of buffer at a time and store at –20°C for up to 1 wk. Hydrogenated Triton X-100 (RTX-100) (protein grade, cat. # 648464, Calbiochem, LaJolla, CA) is purchased as a 10% solution, which should be stored at –20°C. Note: Only hydrogenated Triton X-100 should be used since UV-absorbing contaminants are present in ordinary Triton X-100, making identification of peptides on subsequent HPLC impossible (*see* Fig. 1). Alternately, octyl glucopyranoside (OGP) (Ultrol-grade, Calbiochem) or decyl glucopyranoside (DGP) (Ultrol-grade, Calbiochem) can be substituted for RTX-100 as described in Table 1.

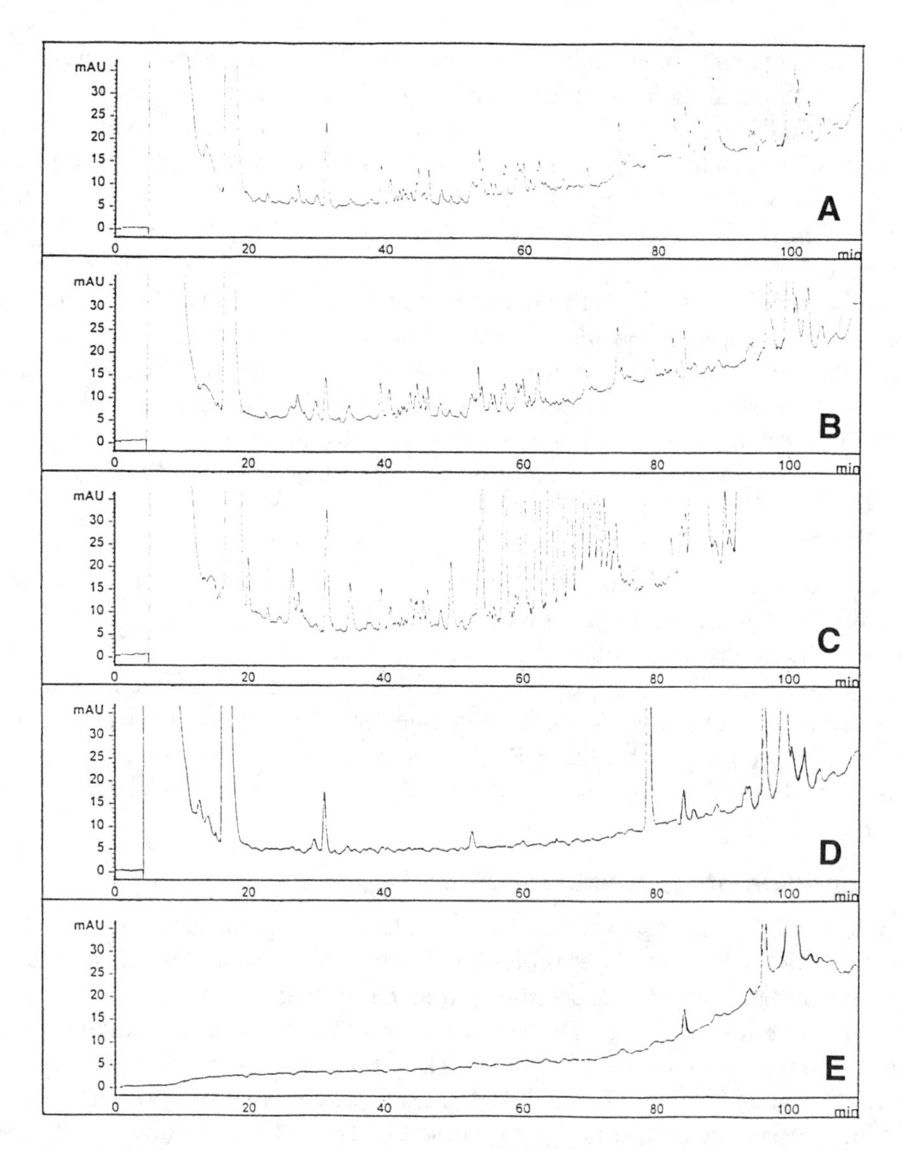

Fig. 1. HPLC analysis of PVDF-bound human transferrin digested with trypsin **(A–C)**. Four micrograms of human transferrin was analyzed by SDS-PAGE, electroblotted to Immobilon P[sq], stained with either amido black (B,C) or Coomassie blue (A), and digested with trypsin as described in the Methods section. Chromatograms A and B were digested using hydrogenated Triton X-100, and chromatogram C was digested with nonhydrogenated Triton X-100. Chromatogram **D** is a negative control digestion (amido black stain) indicating which peaks are contaminants. Chromatogram **E** is a gradient blank indicating which peaks arise from the HPLC or HPLC buffers. Chromatography conditions were using a VYDAC C18 reverse phase column (2.1 mm id × 250 mm) as described elsewhere *(5)*.

2.3. Reduction and Carboxyamidomethylation

1. 45 m*M* DTT: bring 3.5 mg DTT (Ultrol-grade, Calbiochem) up in 500 µL HPLC-grade water. This can be stored at –20°C for up to 3 mo.
2. 100 m*M* iodoacetamide solution: bring 9.25 mg iodoacetamide (reagent-grade) up in 500 µL HPLC-grade water. This solution must be made fresh just prior to use. Dry DTT and iodoacetamide should be stored at 4°C or –20°C.

2.4. Enzyme Solutions and Inhibitors

Enzymes should be stored as small aliquots at –20°C, and made up as 0.1 µg/µL solutions immediately before use. These aliquots can be stored for at least 1 mo at –20°C without significant loss of enzymatic activity.

1. Trypsin (25 µg, sequencing-grade, Boehringer Mannheim, Indianapolis, IN): Solubilize trypsin in 25 µL of 0.01% trifluoroacetic acid (TFA), and let stand ~10 min. Aliquot 5 µL (5 µg) quantities to clean microcentrifuge tubes, dry in a SpeedVac, and store at –20°C. Reconstitute the dry enzyme in 50 µL 0.01% TFA for a 0.1 µg/µL working solution, which is good for 1 d. **Trypsin cleaves at arginine and lysine residues.**
2. Endoproteinase Lys-C (3.57 mg, Wako Pure Chemicals, Osaka, Japan): Solubilize enzyme in 1000 µL HPLC-grade water, and let stand ~10 min. Make nine 100-µL quantities to clean tubes, and store at –20°C. Disperse remaining 100 µL into 20 × 5 µL aliquots (17.85 µg each), and store at –20°C. When needed, take one 5-µL aliquot and add 173 µL of HPLC-grade water to establish a 0.1 µg/µL working solution, which can be used for up to 1 wk if stored at –20°C between uses. When 5-µL aliquots are used up, disperse another 100-µL aliquot. **Endoptoteinase Lys-C cleaves only at lysine residues.**
3. Endoproteinase Glu-C (50 µg, sequencing-grade, Boehringer Mannheim) : Solubilize in 100 µL of HPLC-grade water, and let stand ~10 min. Aliquot 10-µL (5 µg) quantities to clean tubes, dry in a SpeedVac, and store at –20°C. Reconstitute the enzyme in 50 µL HPLC-grade water for a 0.1 µg/µL solution, which is good for only 1 d. **Under these conditions, endoproteinase Glu-C cleaves predominantly at glutamic acid residues, but can sometimes cleave at aspartic acid residues.**
4. Clostripain (20 µg, sequencing-grade, Promega, Madison, WI): Solubilize enzyme in 200 µL of manufacturer's supplied buffer for a concentration of 0.1 µg/µL. Enzyme solution can be stored at –20°C for 1 mo. **Clostripain cleaves at arginine residues only.**
5. 1% Diisopropyl fluorophosphate (DFP) solution: DFP is a dangerous neurotoxin that must be handled with double gloves in a chemical hood. Please follow all precautions listed with this chemical. Add 10 µL of DFP to 990 µL absolute ethanol in a capped microcentrifuge tube. Store at –20°C.

3. Method

3.1. Preparation of the Membrane-Bound Sample

Protein should be electrophoresed in one or two dimensions, followed by electrophoretic transfer to the membrane and visualization of the bands according to the following suggestions.

1. Electroblotting of proteins will be most efficient with a tank transfer system, rather than a semidry system. Concentrate as much protein into a lane as possible; however, if protein resolution is a concern, up to 5 cm^2 of membrane can be combined for digestion (*see* Notes 3 and 5).
2. Stain PVDF membrane with either Ponceau S, Amido black, india ink, or chromatographically pure Coomassie brilliant blue with a dye content of 90%. Destain the blot until the background is clean enough to visualize the stained protein. Complete destaining of the blot is unnecessary, but at least three washes (~5 min) with distilled water should be done to remove excess acetic acid, which is used during destaining (*see* Note 5).
3. Excise protein band(s), and place into a clean 1.5-mL microcentrifuge tube. In addition, excise a blank region of the membrane approximately the same size as the protein blot to serve as a negative control (*see* Note 4).
4. Air dry PVDF-bound protein dry at room temperature and store at –20°C or 4°C.

3.2. Digestion of the Membrane-Bound Protein

NOTE: Gloves should be worn during all steps to avoid contamination of sample with skin keratin.

1. Place ~100 µL of HPLC-grade water onto a clean glass plate, and submerge the membrane-bound protein into the water. Transfer the wet membrane to a dry region of the plate and with a clean razor blade, cut the membrane first lengthwise into 1-mm wide strips, and then perpendicular so that the membrane pieces are 1 × 1 mm. Treat the negative control under the same conditions as the sample. Keeping the membrane wet will simplify manipulation of the sample as well as minimize static charge, which could cause PVDF to "jump" off the plate. The 1 × 1 mm pieces of membrane will settle to the bottom of the tube and require less digestion buffer to immerse the membrane completely (*see* Note 7).
2. Slide the cut membrane onto the forceps with the razor blade, and return it to a clean 1.5-mL microcentrifuge tube. Use the cleanest tubes possible to minimize contamination during peptide mapping. Surprisingly, many UV-absorbing contaminants can be found in microcentrifuge tubes, and this appears to vary with supplier and lot number. Tubes can be cleaned by rinsing with 1 mL of HPLC-grade methanol followed by 2 rinses of 1 mL HPLC-grade water prior to adding the protein band.
3. Add 50 µL of the appropriate digestion buffer (Table 1), and vortex thoroughly for 10–20 s. Optionally, add 50 µL digestion buffer to an empty microcentrifuge tube to serve as a digestion blank for HPLC analysis. The amount of digestion buffer can be increased or decreased depending on the amount of membrane; however, the best results will be obtained with a minimum amount of digestion buffer. PVDF membrane will float in the solution at first, but will submerge after a short while, depending on the type and amount of PVDF .
4. Add 5 µL of 45 m*M* DTT, vortex thoroughly for 10–20 s, seal the tube cap with parafilm, and incubate at 55°C for 30 min. DTT will reduce any remaining cystine bonds. DTT should be of the highest grade to reduce uv absorbing contaminants that might interfere with subsequent peptide mapping.
5. Allow the sample to cool to room temperature. Add 5 µL of 100 m*M* iodoacetamide, vortex thoroughly for 10–20 s, and incubate at room temperature for 30 min in the dark. Iodoacetamide alkylates cysteine residues to generate carboxyamidomethyl cysteine, allowing identification of cysteine during sequence analysis. Allowing the sample to cool prior to adding iodoacetamide and incubating at room temperature are necessary to avoid side reaction to other amino acids.
6. Add enough of the required enzyme solution to obtain an estimated enzyme to substrate ratio of 1:10 (w/w) and vortex thoroughly for 10–20 s. Incubate the sample (including digestion buffer blank) at 37°C for 22–24 h. The amount of protein (substrate) can be estimated by comparison of staining intensity to that of known quantities of stained standard proteins. The 1:10 ratio is a general guideline. Ratios of 1:2 through 1:50 can be used without loss of enzyme efficiency or peptide recovery. An enzyme should be selected that would likely produce peptides of >10 amino acids long. Amino acid analysis of the protein would be informative for estimating the number of cleavage sites. A second aliquot of enzyme can be added after 4–6 h (*see* Note 6).

3.3. Extraction of the Peptides

1. After digestion, vortex the sample for 5–10 s, sonicate for 5 min by holding in a sonicating water bath, spin in a centrifuge (~1800*g*) for 2 min, and transfer the supernatant to a separate vial that will be used directly for HPLC analysis.

2. Add a fresh 50 μL of digestion buffer to the sample, repeat step 11, and pool the supernatant with the original buffer supernatant.
3. Add 100 μL 0.1 % TFA to the sample, and repeat step 1. The total volume for injection onto the HPLC is 200 μL. Most of the peptides (~80%) are recovered in the original digestion buffer; however, these additional washes will ensure maximum recovery of peptides from the membrane.
4. Terminate the enzymatic reaction by either analyzing immediately by HPLC or adding 2 μL of the DFP solution.

   CAUTION: DFP is a dangerous neurotoxin and must be handled with double gloves under a chemical fume hood. Please follow all precautions listed for this chemical.

3.4. Analysis of Samples by Reverse-Phase HPLC and Storage of Peptide Fractions

1. Prior to reverse-phase HPLC, inspect the pooled supernatants for small pieces of membrane or particles that could clog the HPLC tubing. If membrane or particles are observed, either remove the membrane with a clean probe (such as thin tweezers, a thin wire, thin pipet tip, and so forth), or spin in a centrifuge for 2 min and transfer the sample to a clean vial. A precolumn filter will help increase the life of HPLC columns, which frequently have problems with clogged frits.
2. Sample is ready to be fractionated by HPLC (*see* Chapter 72). Fractions can be collected in capless 1.5-mL plastic tubes, capped, and stored at –20°C until sequenced (*see* Note 9).

4. Notes

1. This procedure is generally applicable to proteins that need to have their primary structure determined and offers a simple method for obtaining internal sequence data in addition to amino terminal sequence analysis data. The procedure is highly reproducible and is suitable to peptide mapping by reverse-phase HPLC. Proteins 12–300 kDa have been successfully digested with this procedure with the average size around 100 kDa. Types of proteins analyzed by this technique include DNA binding, cystolic, peripheral, and integral-membrane proteins, including glycosylated and phosphorylated species. The limits of the procedure appear to be dependent on the sensitivity of both the HPLC used for peptide isolation and the protein sequencer.
2. There are several clear advantages of this procedure over existing methods. First, it is applicable to PVDF (especially high-retention PVDF membranes), which is the preferred membrane owing to higher recovery of peptides after digestion, as well as being applicable to other structural analysis. The earlier procedures *(1,2)* have not been successful with PVDF. Second, because of the RTX-100 buffer, recovery of peptides from nitrocellulose is higher than earlier nitrocellulose procedures *(5)*. Third, the procedure is a one-step procedure and does not require pretreatment of the protein band with PVP-40. Fourth, since the procedure does not require all the washes that the PVP-40 procedures do, there is less chance of protein washout. Fifth, the time required is considerably less than with the other procedures. Overall, the protocol described here is the simplest and quickest method to obtain quantitative recovery of peptides.
3. The largest source of sample loss is generally not the digestion itself, but rather electroblotting of the sample. Protein electroblotting should be performed with the following considerations. Use PVDF (preferably a higher binding type, i.e., ProBlott, Immobilon P^{sq}) rather than nitrocellulose, since peptide recovery after digestion is usually higher with PVDF *(5,8)*. If nitrocellulose must be used, e.g., protein is already bound to nitrocellulse before digestion is required, never allow the membrane to dry out since this

will decrease yields. Always electroblot protein using a transfer tank system, since yields from semidry systems are not as high *(12)*. Using stains such as Ponceau S, Amido black, or chromatographically pure Coomassie brilliant blue, with a dye content >90% will increase detection of peptide fragments during reverse-phase HPLC.

Note: Most commercial sources of Coomassie brilliant blue are extremely dirty and should be avoided. Only chromatographically pure Coomassie brilliant blue with a dye content of 90% appears suitable for this procedure.

4. The greatest source of failure in obtaining internal sequence data is not enough protein on the blot, which results in the failure to detect peptide during HPLC analysis. An indication of insufficient protein is that either the intensity of the stain is weak, i.e., cannot be seen with Amido black even though observable with india ink (about 10-fold more sensitive), or possibly detectable by radioactivity or immunostaining, but not by protein stain. Amino acid analysis, amino-terminal sequence analysis, or at the very least, comparison with stained standard proteins on the blot should be performed to help determine if enough material is present. When <10 μg of protein is present, the most problematic item is misidentification of peptides on reverse-phase HPLC owing to artifact peaks and contaminants. Although elimination of every contaminant is usually impossible, there are several strategic points and steps that can be taken to help alleviate these contaminants. A negative control of a blank region of the membrane blot (preferably from a blank lane) that is approximately the same size as the protein band will help to identify contaminants present that are associated with the membrane and digestion buffer (*see* Fig. 1D). The blank membrane should have gone through the same preparation steps as the sample, including electroblotting and staining, and should be analyzed by HPLC immediately before or after the sample. A positive control (membrane-bound standard protein) is generally unnecessary, but should be performed if the activity of the enzyme is in question or a new lot number of enzyme is to be used.
5. Major sources of contaminants are stains used to visualize the protein, the microcentrifuge tubes used for digestion, reagents used during digestion and extraction of peptides, and the HPLC itself. Stains are the greatest source of contaminants, and Coomassie brilliant blue in particular is a problem (*see* Fig. 1A). Amido black and Ponceau S are generally the cleanest, whereas chromatographically pure Coomassie brilliant blue with a dye content <90% does appear to generate less contaminants than most other commercially available Coomassie brilliant blue stains. Surprisingly, microcentrifuge tubes can produce significant artifact peaks, which seem to vary with supplier and lot number. A digestion blank of just the microcentrifuge tube should be done, since some contaminants only appear after incubation in the RTX-100 buffer. The major concern with the digestion buffer is the hydrogenated Triton X-100 (*see* Fig. 1C). Additional late-eluting peaks may be observed with certain lots of RTX-100, whereas other lots are completely free of UV-absorbing contaminants. HPLC-grade water or water prepared as described by Atherton *(11)* should be used for all solution preparation. An HPLC blank (gradient run with no injection) should always be performed to determine what peaks are related to the HPLC (*see* Fig. 1E).
6. The key to success of the procedure and quantitative recovery of peptides from both PVDF and nitrocellulose membranes is the use of hydrogenated Triton X-100 (RTX-100) in the buffer. This should be purchased from Calbiochem as a 10% stock solution, protein-grade (cat. # 648464). Figure 1C demonstrates why hydrogenated Triton X-100 should be used, since nonhydrogenated Triton X-100 has several strong UV-absorbing contaminants *(13)*. RTX-100 acts as a block to prevent enzyme adsorption to the membrane as well as a strong elution reagent of peptides *(6,7)*. In addition, RTX-100 does not inhibit enzyme activity or interfere with peak resolution during HPLC, as do ionic detergents, such as SDS *(5)*. The concentration of RTX-100 can be decreased to 0.1% *(7)*. However, with a

large amount of membrane, there could be a decrease in peptide recovery. Optionally, octyl- or decylglucopyranoside can be substituted for RTX-100 with no loss in peptide recovery *(10)*.

7. The membrane should be cut into 1 × 1 mm pieces while keeping it wet to avoid static charge buildup. The 1 x 1 mm pieces allow using the minimum volume of digestion buffer to cover the membrane. The volume of the digestion buffer should be enough to cover the membrane (about 50 µL), but can be increased or decreased depending on the amount of membrane present. The enzyme solution should be selected based on additional knowledge of the protein, such as amino acid composition or whether the protein is basic or acidic. If the protein is a complete unknown, endoproteinase Lys-C or Glu-C would be a good choice. The enzyme-to-substrate ratio should be about 1:10; however, if the exact amount of protein is unknown, ratios of 1:2 through 1:50 will not affect the quantitative recovery of peptides. After digestion, most of the peptides are recovered in the original buffer (about 80%), and the additional washes are performed to ensure maximum peptide recovery. Microbore reverse-phase HPLC is the best isolation procedure for peptides.
8. As mentioned earlier, previous procedures *(1–3,5)* require pretreatment with PVP-40 to prevent enzyme adsorption to the membrane. RTX-100 is essential for quantitative recovery of peptides from the membrane; however, RTX-100 also strips PVP-40 from the membrane, resulting in a broad, large UV-absorbing contaminant that can interfere with peptide identification. The PVP-40 contaminant is not dependent on the age or lot number of PVP-40, and making fresh solutions did not help as previously suggested *(4)*. This appears to be more of a problem with nitrocellulose and higher binding PVDF (ProBlott and Immobilon P^{sq}) than lower binding PVDF (Immobilon P), and also is dependent on the amount of membrane. The PVP-40 contaminant also appears to elute earlier in the chromatogram as the HPLC column ages, becoming more of a nuisance in visualizing peptides. Therefore, using PVP-40 to prevent enzyme adsorbtion to the membrane should be avoided.
9. There are a few considerations that should be addressed regarding peptide mapping by HPLC using the protocol described here. A precolumn filter (Upchurch Scientific) must be used to prevent small membrane particles from reaching the HPLC column, thus decreasing its life. Inspection of the pooled supernatants for visible pieces of PVDF can prevent clogs in the microbore tubing, and can be removed either with a probe, or by spinning in a centrifuge and transferring the sample to a clean vial.

   Peptide mapping by reverse-phase HPLC after digestion of the membrane-bound protein should result in several peaks on the HPLC. Representative peptide maps from trypsin digestions of human transferrin bound to PVDF and stained with Coomassie blue or amido black (Immobilon P^{sq}) is shown in Figs. 1A and B. Peptide maps should be reproducible under identical digestion and HPLC conditions. In addition, the peptide maps from proteins digested on membranes are comparable if not identical to those digested in solution, indicating that the same number of peptides are recovered from the membrane as from in solution. The average peptide recovery is generally 40–70 % based on the amount analyzed by SDS-PAGE, and 70–100 % based on the amount bound to PVDF as determined by amino acid analysis or radioactivity counting. The recovery of peptides from the membrane appears to be quantitative, and the greatest loss of sample tends to be in the electroblotting.
10. The entire procedure can be done in approx 24 h plus the time required for peptide mapping by reverse-phase HPLC. Cutting the membrane takes about 10 minutes, reduction with DTT takes 30 min, carboxyamidomethylation take another 30 min, digestion at 37°C takes 22–24 h, and extraction of the peptides requires about 20 min.

References

1. Aebersold, R. H., Leavitt, J., Saavedra, R. A., Hood, L. E., and Kent, S. B. (1987) Internal amino acid sequence analysis of proteins separated by one- or two-dimensional gel electrophoresis after *in situ* protease digestion on nitrocellulose. *Proc. Natl. Acad. Sci. USA* **84,** 6970–6974.
2. Tempst, P., Link, A. J., Riviere, L. R., Fleming, M., and Elicone, C., (1990) Internal sequence analysis of proteins separated on polyacrylamide gels at the submicrogram level: improved methods, applications and gene cloning strategies. *Electrophoresis* **11,** 537–553.
3. Bauw, G., Van Damme, J., Puype, M., Vandekerckhove, J., Gesser, B., Ratz, G. P., Lauridsen, J. B., and Celis, J. E. (1989) Protein-electroblotting and -microsequencing strategies in generating protein data bases from two-dimensional gels. *Proc. Natl. Acad. Sci. USA* **86,** 7701–7705.
4. Aebersold, R. (1993) Internal amino acid sequence analysis of proteins after *in situ* protease digestion on nitrocellulose, in *A Practical Guide to Protein and Peptide Purification for Microsequencing*, 2nd ed. (Matsudaira, P., ed.), Academic, New York, pp. 105–154.
5. Fernandez, J., DeMott, M., Atherton, D., and Mische, S. M. (1992) Internal protein sequence analysis: enzymatic digestion for less than 10 μg of protein bound to polyvinylidene difluoride or nitrocellulose membranes. *Anal. Biochem.* **201,** 255–264.
6. Fernandez, J., Andrews, L., and Mische, S. M. (1994) An improved procedure for enzymatic digestion of polyvinylidene difluoride-bound proteins for internal sequence analysis. *Anal. Biochem.* **218,** 112–117.
7. Fernandez, J., Andrews, L., and Mische, S. M. (1994) A one-step enzymatic digestion procedure for PVDF-bound proteins that does not require PVP-40, in *Techniques in Protein Chemistry V* (Crabb, J., ed.), Academic Press, San Diego, pp. 215–222.
8. Best, S., Reim, D. F., Mozdzanowski, J., and Speicher, D. W., (1994) High sensitivity sequence analysis using *in-situ* proteolysis on high retention PVDF membranes and a biphasic reaction column sequencer, in *Techniques in Protein Chemistry V* (Crabb, J., ed.), Academic, New York, pp. 205–213.
9. Atherton, D., Fernandez, J., and Mische, S. M. (1993) Identification of cysteine residues at the 10 pmol level by carboxamidomethylation of protein bound to sequencer membrane supports. *Anal. Biochem.* **212,** 98–105.
10. Kirchner, M., Fernandez, J., Agashakey, A., Gharahdaghi, F., and Mische, S. M. (1996) in *Techniques in Protein Chemistry VII* (Marshak, O., ed.), Academic, New York (in press).
11. Atherton, D., (1989) Successful PTC amino acid analysis at the picomole level, in *Techniques in Protein Chemistry* (Hugli, T., ed.) Academic, New York, pp. 273–283.
12. Mozdzanowski, J. and Speicher, D. W. (1990) Quantitative electrotransfer of proteins from polyacrylamide gels onto PVDF membranes, in *Current Research in Protein Chemistry: Techniques, Structure, and Function* (Villafranca, J., ed.), Academic, New York, pp. 87–94.
13. Tiller, G. E., Mueller, T. J., Dockter, M. E., and Struve, W. G. (1984) Hydrogenation of Triton X-100 Eliminates Its Fluorescence and Ultraviolet Light Absorbance while Preserving Its Detergent Properties. *Anal. Biochem.* **141,** 262–266.

71

Enzymatic Digestion of Proteins in Solution and in SDS Polyacrylamide Gels

Kathryn L. Stone and Kenneth R. Williams

1. Introduction

Although most prokaryotic proteins have free NH_2-termini and therefore can be directly sequenced, most eukaryotic proteins have blocked NH_2-termini which precludes Edman degradation *(1,2)*. In these instances, one of the most direct approaches to obtaining partial amino acid sequences is via enzymatic or chemical cleavage followed by peptide fractionation and sequencing. Although several different approaches may be taken to cleave proteins, one of the most common is to digest the protein enzymatically with a relatively specific protease such as trypsin or lysyl endopeptidase. Since final purification is often dependent on SDS-PAGE, cleavage procedures that can either be carried out in the polyacrylamide gel matrix *(3,4)* or that may be used on samples that have been blotted from SDS polyacrylamide gels onto PVDF *(5)* or nitrocellulose *(5,6)* membranes are extremely useful. In both instances, the proteins are usually stained with Coomassie blue or Ponceau S prior to excision, and proteolytic digestion and the resulting peptides are separated by reverse-phase HPLC. Relatively straightforward solution and in-gel digestion procedures that have been used extensively in the Keck Foundation Biotechnology Resource Laboratory at Yale University will be described in this chapter, whereas a procedure suitable for *in situ* digestion of SDS-PAGE blotted proteins is described in Chapter 70.

2. Materials

2.1. Enzymatic Digestion of Proteins

1. Enzymatic digestion of proteins is usually accomplished using either sequencing grade, modified trypsin (from Promega [Madison, WI] or Boehringer Mannheim [Indianapolis, IN]) or lysyl endopeptidase (Achromobacter Protease I from *Achromobacter lyticus*) from Wako Pure Chemical Industries, Ltd. (Osaka, Japan). Occasionally, chymotrypsin or endoproteinase Glu-C from Boehringer Mannheim may also be used. All enzyme stocks are divided into 100 μL aliquots and stored at –20°C.
2. Sequencing-grade chymotrypsin and modified trypsin (Boehringer Mannheim): prepare by dissolving the dried 100-μg aliquots (obtained from the manufacturer) in 1 mL 1 m*M* HCl to make a 0.1 mg/mL stock solution that appears to be stable for at least 6 mo at –20°C.

From: The Protein Protocols Handbook
Edited by: J. M. Walker Humana Press Inc., Totowa, NJ

3. Sequencing grade, modified trypsin (Promega): dissolve the 20-µg aliquot (obtained from the manufacturer) in 200 µL 1 m*M* HCl to make a 0.1 mg/mL stock solution that appears to be stable for at least 6 mo at –20°C.
4. Endoproteinase Glu-C: dissolve the 50-µg aliquot from the manufacturer in 500 µL 50 m*M* NH_4HCO_3. According to the manufacturer, the dissolved enzyme is stable for 1 mo at –20°C.
5. Lysyl endopeptidase: dissolve 2.2 mg as purchased in 2.2 mL of 2 m*M* Tris-HCl, pH 8.0, to make a 1 mg/mL stock, which according to the manufacturer is stable for at least 2 yr when stored at –20°C. More dilute solutions are made by adding 10 µ L of this 1 mg/mL stock solution to 90 µL 2 m*M* Tris-HCl, pH 8.0, for a 0.1 mg/mL stock.
6. Pepsin (Sigma Chemical Co., St. Louis, MO): dissolve in 5% formic acid (Baker) at a concentration of 0.1 mg/mL.
7. Digestion buffer for in-solution digestion: 8*M* urea, 0.4*M* NH_4HCO_3. Prepare by dissolving 4.8 g Pierce Sequanal-Grade urea and 0.316 g Baker ammonium bicarbonate in H_2O to make a final volume of 10 mL.
8. 50% CH_3CN/0.1*M* Tris-HCl, pH 8.0: prepare by diluting 5 mL 1.0*M* Tris-HCl, pH 8.0 (12.1 g Tris-HCl dissolved in 100 mL H_2O with the pH adjusted to 8.0 with HCl) with 20 mL H_2O and 25 mL 100% acetonitrile.
9. In gel digestion buffer: 0.1*M* Tris-HCl, pH 8.0/0.1% Tween 20. Prepared by adding 1 mL 1.0*M* Tris-HCl, pH 8.0, and 10 µL polyoxyethylene-sorbitan monolaurate (Tween 20 from Sigma Chemical Co.) to 9.0 mL H_2O.
10. Cysteine modification buffer: 0.1*M* Tris-HCl, pH 8.0/60% CH_3CN is prepared by diluting 5.0 mL 1.0*M* Tris-HCl, pH 8.0 with 15 mL H_2O plus 30 mL 100% CH_3CN.
11. 45 m*M* DTT (Pierce Chemical Co.) solution for protein reduction: dissolve 69 mg DTT in 10 mL H_2O.
12. 100 m*M* iodoacetic acid (IAA) (Pierce Chemical Co.) solution for alkylation: dissolve 185.9 mg IAA in 10 mL H_2O.
13. 200 m*M* methyl 4-nitrobenzene sulfonate (Aldrich Chemical Co.) solution for cysteine modification: is made by dissolving 0.0434 g methyl 4-nitrobenzene sulfonate in 100% CH_3CN. This solution is made immediately before use and is not stored.
14. 0.1% TFA/60% CH_3CN solution for peptide extraction from the gel slices: add 50 µL 100% trifluoracetic acid (TFA) to 20 mL H_2O and 30 mL 100% CH_3CN.

3. Methods

3.1. Enzymatic Digestion of Proteins

3.1.1. Digestion of Proteins in Solution

3.1.1.1. Sample Preparation for in Solution Digestion

Proper sample preparation is critical both in avoiding sample loss and in ensuring successful digestion (*see* Note 1). If the sample contains a sufficiently low level of salt and glycerol (such that the concentration of salt and glycerol in the final digest will be less than the equivalent of 1*M* NaCl and 15% glycerol, respectively), it may simply be reduced to dryness in a SpeedVac prior to carrying out the digest in the tube in which it was dried. Samples containing higher levels of salts, glycerol, and/or detergents, such as SDS, may be precipitated using either trichloroacetic acid (TCA) or acetone in order to remove the salts, glycerol, and detergents. To TCA-precipitate the protein, 1/9th vol of 100% TCA is added to the sample prior to incubating on ice for 30 min. The sample is then centrifuged (10,000*g*/15 min) and the supernatant is removed. Residual TCA is then removed by suspending the pellet in 50 μ L cold acetone. After vortexing and centrifuging, the supernatant is removed with a pipet. The air-dried pellet is then digested as described below. To acetone precipitate the protein, it is often necessary first to reduce the salt concentration by dialysis vs 0.05% SDS, 5 m*M* NH_4HCO_3. At this point, the volume is reduced to <50 μL in a SpeedVac, and 9 vol cold acetone are added followed by a 1 h incubation at –20°C. The sample is then centrifuged and the pellet washed (as described above) with 50 μ L cold acetone. When large amounts of SDS are present, the acetone wash should be repeated once or twice more to remove excess detergent. The recommended minimum protein concentration during either precipitation is 100 μg/mL, and for TCA precipitation, the glycerol concentration should be below 15%.

3.1.1.2. Digestion of Proteins in Solution with Trypsin, Lysyl Endopeptidase, Chymotrypsin, and Endoproteinase Glu-C

After the salts and detergents have been removed/minimized, the sample is ready for digestion. Typically, trypsin is the enzyme of choice because of its relatively high cleavage specificity (the COOH-terminal side of lysine and arginine) and its ability to digest insoluble substrates. That is, proteins that are only partially soluble in the digest buffer will often cleave with trypsin as evidenced by rapid clearing of the solution. Cleavage occurs more slowly when there is an acidic residue following the lysine or arginine and not at all when a lysine-proline or arginine-proline linkage is present. Since lysyl endopeptidase only cleaves on the COOH-terminal side of lysine, it provides longer peptides than trypsin. Based on the average occurrence of lysine (5.7%) and arginine (5.4%) in proteins in the Protein Identification Resource Data Base, the average length of a tryptic and lysyl endopeptidase peptide is about 9 and 18 residues, respectively. However, since lysyl endopeptidase does not generally cleave insoluble substrates and since we have occasionally encountered proteins that will digest with trypsin, but not with lysyl endopeptidase, we generally use trypsin unless there is sufficient protein to carry out two digests (if needed), in which case we would initially use lysyl endopeptidase.

Since chymotrypsin readily cleaves on the COOH-terminal side of tryptophan, tyrosine, and phenylalanine, and generally gives partial cleavage after leucine, methionine, and several other amino acids, its specificity is too broad to be of general use. However, it is occasionally used to redigest larger peptides or proteins that fail to digest with trypsin. Although endoproteinase Glu-C is relatively specific in that it cleaves after glutamic acid in either ammonium bicarbonate (pH 8.0) or ammonium acetate (pH 4.0) buffers and after both aspartic acid and glutamic acid in phosphate buffers (pH 7.8), it also does not generally cleave insoluble substrates, and the HPLC profiles we have obtained with this enzyme usually suggest relatively incomplete cleavage and the resulting generation of overlapping peptides in lower yield than might often be obtained with trypsin.

Although this brief survey of proteolytic enzymes is far from complete, it does cover most of the enzymes that are frequently used. The enzymatic digestion protocol outlined below may be used with any of these four enzymes, providing the appropriate buffer changes are made in the case of Glu-C.

1. Dissolve the dried or precipitated protein in 20 μL 8*M* urea, 0.4*M* NH_4HCO_3 and then remove a 10–15% aliquot for acid hydrolysis/ion-exchange amino acid analysis. If the

analysis indicates there is sufficient protein to digest (*see* Note 2), proceed with step 2; otherwise additional protein should be prepared to pool with the sample.

2. Check the pH of the sample by spotting 1–2 µL on pH paper. If necessary, adjust the pH to between 7.5 and 8.5.
3. Add 5 µL 45 m*M* DTT and incubate at 50°C for 15 min to reduce the protein. (*See* Note 3.)
4. After cooling to room temperature, alkylate the protein by adding 5 µL 100 m*M* IAA and incubating at room temperature for 15 min. (*See* Note 3.)
5. Dilute the digestion buffer with H_2O so the final digestion will be carried out in 2*M* urea, 0.1*M* NH_4HCO_3.
6. Add the enzyme in a 1/25, enzyme/protein (wt/wt) ratio. (*See* Note 4.)
7. Incubate at 37°C for 24 h.
8. Stop the digest by freezing, acidifying the sample with TFA, or by injecting onto a reverse-phase HPLC system.

3.1.1.3. Digestion of Proteins in Solution with Pepsin

Although the very broad specificity of pepsin hinders its routine use for comparative peptide mapping, its low pH optimum enables it to cleave proteins that might otherwise be intransigent. It is also applicable for digesting relatively small peptides and, particularly, for studies directed at identifying disulfide bonds (*see* Note 5). Although pepsin cleaves preferentially between adjacent aromatic or leucine residues, it has been shown to cleave at either the NH_2- or COOH-terminal side of any amino acid, except proline. A typical digestion procedure follows:

1. Dissolve the dried protein in 100 µL 5% formic acid.
2. Add pepsin at a 1:50, enzyme:protein, (wt:wt) ratio.
3. Incubate the sample at room temperature for 1–24 hours with the time of incubation being dependent on the desired extent of digestion.
4. Dry the digest in a SpeedVac prior to dissolving in 0.05% TFA and immediately injecting onto a reverse-phase HPLC system.

3.1.2. Digestion of Proteins in SDS Polyacrylamide Gels

3.1.2.1. Sample Preparation for In Gel Digestion

As in the case of samples destined for in-solution digests, care must be exercised in preparing samples for SDS-PAGE so that sample losses are minimized (*see* Note 6) and so that the final ratio of protein/gel matrix is as high as possible (*see* Note 7). Although prior carboxymethylation does not appear to be essential with in-gel digests, the presence of greater than 10–20% carbohydrate (by weight) often appears to hinder cleavage significantly (*see* Note 3). Samples that have been purified by SDS-PAGE can be digested directly in the gel matrix, thereby eliminating the need for electroelution or electroblotting of the intact protein from the gel. SDS PAGE-separated proteins destined for in-gel digestion should be stained with 0.1% Coomassie blue in 50% methanol, 10% acetic acid for 1 hour prior to destaining with 50% methanol, 10% acetic acid for a minimum of 2 h (*see* Note 8). Alternatively, when sufficient sample is available and the band of interest is well separated, a guide lane in the gel can be stained and the protein of interest excised from the gel using this guide. However, in this instance, the gel still must be exposed to 50% methanol, 10% acetic acid for 3 h (to ensure adequate removal of SDS) prior to excising the protein band of interest.

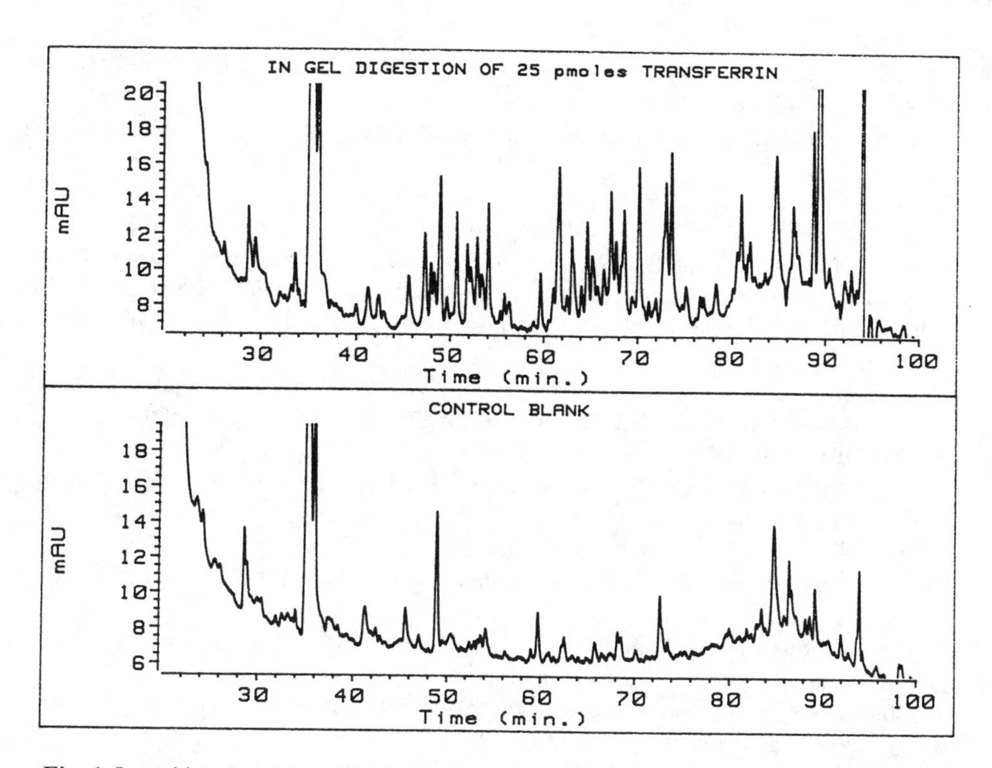

Fig. 1. In-gel lysyl endopeptidase digestion of transferrin. Following SDS-PAGE of 25 pmol of transferrin in a 12.5% polyacrylamide gel, the gel was stained and destained as described in Section 3. The protein band was then excised, along with a control blank, and digested in the gel with lysyl endopeptidase as described in Section 3. Peptides were chromatographed on a Vydac C-18, 2.1 × 250 mm reverse-phase column that was eluted at a flow rate of 150 μL/min as described in Chapter 72. A comparison of the transferrin digest (top profile) with the control digest (bottom profile), which was carried out on a blank section of gel, indicates that the digest proceeded well.

The % polyacrylamide gel that is used is determined by the size of the protein. Proteins >100 kDa are typically electrophoresed in 7–10% polyacrylamide gels, whereas smaller proteins are electrophoresed in 10–17.5% polyacrylamide.

Although the following procedure, which is a modification of the Rosenfeld et al. *(4)* procedure as reported in Williams and Stone *(3)*, is capable of succeeding with as little as 25 pmol protein (e.g., *see* Fig. 1), the minimum amount of protein recommended for in gel digestion is 50–100 pmol at a minimum protein to gel density (after staining/destaining) of 0.05 μg/mm^3 (*see* Notes 7 and 9). The protein of interest, along with a blank section of gel that serves as a control, are excised using a razor blade and tweezers. To prevent going ahead with insufficient protein, we recommend that a 10–15% section of the gel band of interest be subjected to hydrolysis/ion exchange amino acid analysis to quantitate the amount of protein remaining (*see* Note 10).

3.1.2.2 In-Gel Digestion of Proteins with Trypsin and Lysyl Endopeptidase

1. Determine the approximate volume of gel to be digested (length × width × thickness).
2. Cut the gel band(s) containing the protein of interest into approx 1 × 2 mm pieces, and place in an Eppendorf tube. Repeat for the "blank" section of gel.

3. Add 150 µL (or more, if necessary to cover the gel pieces) 50% CH_3CN/0.1M Tris-HCl, pH 8.0, to the gel pieces.
4. Wash for 15 min at room temperature on a rocker table.
5. Remove wash.
6. Semidry the washed gel pieces in a SpeedVac to approx 25–50% of original volume.
7. Make up the enzyme solution by diluting 5 µL 0.1 mg/mL enzyme stock solution with 10 µL 0.1M Tris-HCl, pH 8.0, 0.1% Tween 20 for every 15 mm^3 gel that is to be digested.
8. Rehydrate gel pieces with the enzyme solution from step 7, which should be equal in volume to that of the gel pieces and should provide a final enzyme ratio of about 0.5 µg/15-mm^3 gel volume.
9. If the gel pieces are not totally immersed in the enzyme solution, add an additional volume of 0.1M Tris-HCl, pH 8.0, 0.1% Tween 20 to submerge the gel pieces totally.
10. Incubate at 37°C for 24 h.
11. After incubation, add a volume of 0.1M Tris-HCl, pH 8.0/60% CH_3CN equal to the total volume of buffer added in steps 8 and 9.
12. Estimate the total volume of the sample plus gel.
13. Calculate the volume of 45 mM DTT needed to give a final (DTT) of about 1 mM in the sample.
14. Add the above volume of 45 mM DTT and then incubate at 50°C for 20 min.
15. Remove samples from the incubator; cool to room temperature and add an equal volume of 100 mM IAA or 200 mM methyl 4-nitrobenzene sulfonate (MNS) (*see* Note 11) as the volume of DTT added in step 14.
16. For IAA alkylation, incubate at room temperature in the dark for 20 min; for MNS treatment, incubate at 37°C for 40 min.
17. Extract peptides by adding at least 100 µL 0.1% TFA, 60% CH_3CN (or, if it is greater, a volume equal to the gel volume estimated in step 2), and shake on a rocker table at room temperature for at least 40 min.
18. Sonicate for 5 min in a water bath sonicator.
19. Remove and save the supernatant, which contains the released peptides, and repeat steps 17 and 18.
20. SpeedVac dry the combined washes from step 19.
21. Redissolve the dried samples in 80 µL H_2O and bring to a final volume of 110 µL.
22. Filter through a Millipore Ultrafree-MC 0.22-µm filter unit.
23. The sample is now ready for reverse-phase HPLC peptide separation as described in Chapter 71.

4. Notes

1. In many instances, large losses occur during the final purification steps when the protein concentrations are invariably lower. Hence, although ultrafiltration or dialysis of a 5 mg/mL crude solution of a partially purified enzyme may lead to nearly 100% recovery of activity, similar treatment of a 25 µg/ml solution of the purified protein might well lead to significant, if not total loss of activity owing to nonspecific adsorption. Similarly, the effectiveness of organic and acid-precipitation procedures often decreases substantially as the final protein concentration is decreased below about 100 µg/mL. Whenever possible, therefore, the final purification step should be arranged such that the resulting protein solution is as concentrated as possible and, ideally, can simply be dried in a SpeedVac prior to enzymatic digestion. In this regard, it should be noted that a final NaCl concentration of 1M does not significantly affect the extent of trypsin digestion *(7)*. When it is necessary to carry out an organic or acid precipitation to remove salts or detergents, the

protein should first be dried in a SpeedVac (in the 1.5-mL tube in which it will ultimately be digested) prior to redissolving or suspending in a minimum volume of water and then adding the acetone or TCA. In this way, the protein concentration will be as high as possible during the precipitation and losses will be minimized. Two common contaminants that are extremely deleterious to enzymatic cleavage are detergents (as little as 0.005% SDS will noticeably decrease the rate of tryptic digests carried out in the presence of 2*M* urea *[7]*) and ampholines. Since detergent removal is often associated with protein precipitation, and since many detergents (such as SDS) form large micelles, which cannot be effectively dialyzed, it is usually preferable to extract the detergent from the protein (that has been dried in the tube in which it will be digested) rather than to dialyze it away from the protein.

In the case of ampholines, our experience is that even prolonged dialysis extending over several days with a 15,000-Dalton cutoff membrane is not sufficient to decrease the ampholine concentration to a level that permits efficient trypsin digestion. Rather, the only effective methods we have found for complete removal of ampholines are TCA precipitation, hydrophobic or reverse-phase chromatography, or SDS-PAGE followed by staining and destaining.

2. One of the most common causes of "failed" digests is that the amount of protein being subjected to digestion has been overestimated. Often this is because of the inaccuracy of dye binding and colorimetric assays. For this reason, we recommend that an aliquot of the sample be taken for hydrolysis and amino acid analysis prior to digestion. The aliquot for amino acid analysis should be taken either immediately prior to drying the sample in the tube in which it will be digested or after redissolving the sample in 8*M* urea, 0.4*M* NH_4HCO_3. Although up to 10 µL of 8*M* urea is compatible with acid hydrolysis/ion-exchange amino acid analysis, this amount of urea may not be well tolerated by PTC amino acid analysis. Hence, in the latter case, the amino acid analysis could be carried out prior to drying and redissolving the sample in urea. Although it is possible to succeed with less material, to ensure a high probability of success, we recommend that a minimum of 50–100 pmol protein be digested. Typically, 10–15% of this sample would be taken for amino acid analysis. In the case of a 25-kDa protein, the latter would correspond to only 0.125–0.188 µg protein being analyzed. When such small amounts of protein are being analyzed, it is important to control for the ever-present background of free amino acids that are in buffers, dialysis tubing, plastic tubes and tips, and so forth. If sufficient protein is available, aliquots should be analyzed both before (to determine the free amino acid concentration) and after hydrolysis. Alternatively, an equal volume of sample buffer should be hydrolyzed and analyzed and this concentration of amino acids should then be subtracted from the sample analysis.
3. Since many native proteins are resistant to enzymatic cleavage, it is usually best to denature the protein prior to digestion. Although some proteins may be irreversibly denatured by heating in 8*M* urea (as described in the above protocol), this treatment is not sufficient to denature transferrin. In this instance, prior carboxymethylation, which irreversibly modifies cysteine residues, brings about a marked improvement in the resulting tryptic peptide map *(7)*. Another advantage of carboxymethylating the protein is that this procedure enables cysteine residues to be identified during amino acid sequencing. Cysteine residues have to be modified in some manner prior to sequencing to enable their unambiguous identification. Under the conditions that are described in Section 3., the excess dithiothreitol and iodoacetic acid do not interfere with subsequent digestion. Although carboxymethylated proteins are usually relatively insoluble, the 2*M* urea that is present throughout the digest is frequently sufficient to maintain their solubility. However, even in those instances where the carboxymethylated protein precipitates following dilution of

the 8*M* urea to 2*M*, trypsin and chymotrypsin will usually still provide complete digestion. Often, the latter is evidenced by clearing of the solution within a few minutes of adding the enzyme.

If carboxymethylation is insufficient to bring about complete denaturation of the substrate, an alternative approach is to cleave the substrate with cyanogen bromide (1000-fold molar excess over methionine, 24 h at room temperature in 70% formic acid). The resulting peptides can then either be separated by SDS-PAGE (since they usually do not separate well by reverse-phase HPLC) or, preferably, they can be enzymatically digested with trypsin or lysyl endopeptidase and then separated by reverse phase HPLC. If this approach fails, the protein may be digested with pepsin, which, as described above, is carried out under very acidic conditions or can be subjected to partial acid cleavage *(8)*. However, the disadvantage of these later two approaches is that they produce an extremely complex mixture of overlapping peptides. Finally, extensive glycosylation (i.e., typically >10–20% by weight) can also hinder enzymatic cleavage. In these instances, it is usually best to remove the carbohydrate prior to beginning the digest. In the case of in gel digests, this may often be best carried out immediately prior to SDS-PAGE, which thus prevents loss (owing to insolubility) of the deglycosylated protein and effectively removes the added glycosidases.

4. Every effort should be made to use as a high substrate and enzyme concentration as possible to maximize the extent of cleavage. Although the traditional 1:25, weight:weight ratio of enzyme to substrate provides excellent results with milligram amounts of protein, it will often fail to provide complete digestion with low microgram amounts of protein. For instance, using the procedures outlined above, this weight:weight ratio is insufficient to provide complete digestion when the substrate concentration falls below about 20 μg/mL *(7)*. The only reasonable alternative to purifying additional protein is either to decrease the final digestion volume below the 80-μL value used above or to compensate for the low substrate concentration by increasing the enzyme concentration. The only danger in doing this, of course, is the increasing risk that some peptides may be isolated that are autolysis products of the enzyme. Assuming that only enzymes, such as trypsin, chymotrypsin, lysyl endopeptidase, and Protease V8 are used, whose sequences are known, it is usually better to risk sequencing a peptide obtained from the enzyme (which can be quickly identified via a data base search) than it is to risk incomplete digestion of the substrate.

   Often, protease autolysis products can be identified by comparative HPLC peptide mapping of an enzyme (i.e., no substrate) control that has been incubated in the same manner as the sample and by subjecting candidate HPLC peptide peaks to matrix-assisted laser desorption mass spectrometry prior to sequencing (*see* Chapter 91). The latter can be extremely beneficial both in identifying (via their mass) expected protease autolysis products and in ascertaining the purity of candidate peptide peaks prior to sequencing. To promote more extensive digestion, we have sometimes used enzyme:substrate mole ratios that approach unity. If there is any doubt concerning the appropriate enzyme concentration to use with a particular substrate concentration, it is usually well worth the effort to carry out a control study (using a similar concentration and size standard protein) where the extent of digestion (as judged by the resulting HPLC profile) is determined as a function of enzyme concentration.
5. One approach to identifying disulfide-linked peptides is to comparatively HPLC peptide map a digest that has been reduced/carboxymethylated vs one that has only been carboxymethylated (thus leaving disulfide-linked peptides intact). In this instance, pepsin offers an advantage in that the digest can be carried out under acidic conditions where disulfide interchange is less likely to occur.

6. As in the case of in-solution digests, care must be exercised to guard against sample loss during final purification. Whenever possible, SDS (0.05%) should simply be added to the sample prior to drying in a SpeedVac and subjecting to SDS-PAGE. Oftentimes, however, if the latter procedure is followed, the final salt concentration in the sample will be too high (i.e., >1*M*) to enable it to be directly subjected to SDS-PAGE. In this instance, the sample may either be concentrated in a SpeedVac and then precipitated with TCA (as described above) or it may first be dialyzed to lower the salt concentration. If dialysis is required, the dialysis tubing should be rinsed with 0.05% SDS prior to adding the sample, which should also be made 0.05% in SDS. After dialysis versus a few m*M* NH_4HCO_3 containing 0.05% SDS, the sample may be concentrated in a SpeedVac and then subjected to SDS PAGE (note that samples destined for SDS PAGE may contain several % SDS). Another approach that works extremely well is to use an SDS polyacrylamide gel containing a funnel-shaped well that allows samples to be loaded in volumes as large as 300 μ L*(9)*.
7. In general, the sample should be run in as few SDS-PAGE lanes as possible to maximize the substrate concentration and to minimize the total gel volume present during the digest. Whenever possible, a 0.5–0.75-mm thick gel should be used and at least 0.5–0.75 μg of the protein of interest should be run in each gel lane so the density of the protein band is at least 0.05 $\mu g/mm^3$. As shown in Table 1, the in-gel procedure has an average success rate of close to 98%. The least amount of an unknown protein that we have successfully digested and sequenced was 5 pmol of a protein that was submitted at a comparatively high density of 0.1 $\mu g/mm_3$. We believe the latter factor contributed to the success of this sample.

   Although several enzymes (i.e., trypsin, chymotrypsin, lysyl endopeptidase, and endoproteinase GluC) may be used with the in gel procedure, nearly all our experience has been with trypsin. In general, we recommend using 0.5 μg enzyme/15 mm^3 of gel with the only caveat being that we use a corresponding lower amount of enzyme if the mole ratio of protease/substrate protein would exceed unity.
8. Since high concentrations of Coomassie blue interfere with digestion, it is best to use the lowest Coomassie blue concentration possible and to stain for the minimum time necessary to visualize the bands of interest. In addition, the gel should be well destained so that the background is close to clear.
9. As shown in Table 1, the in-gel digestion procedure outlined in this chapter appears to have a success rate of nearly 98% with unknown proteins. Surprisingly, this success rate does not appear to vary significantly over the range of protein extending from an average of about 37–323 pmol. Obviously, however, the quality of the resulting sequencing data is improved by going to the higher levels, and this is probably reflected by the increased number of positively called residues/peptide sequenced that was observed at the >200 pmol level. One critical fact that has so far not been noted is that 71% of the proteins on which the data in Table 1 are based were identified via data base searching of the first peptide sequence obtained. Hence, all internal peptide sequences obtained should be immediately searched against all available databases to determine if the protein that has been digested is unique.
10. Although most estimates of protein amounts are based on relative staining intensities, our data suggest there is a 5–10-fold range in the relative staining intensity of different proteins. Obviously, when working in the 50–100 pmol level, such a 5–10-fold range could well mean the difference between success and failure. Hence, we routinely subject an aliquot of the SDS-PAGE gel (usually 10–15% based on the length of the band) to hydrolysis and ion-exchange amino acid analysis prior to proceeding with the digest. As these analyses will often contain <0.5 μg protein, it is important that a "blank" section of

Table 1
Summary of 88 In-Gel Digests Carried Out in the W. M. Keck Foundation Biotechnology Resource Laboratory at Yale University

Parameter	<50 pmol	51–100 pmol	101–200 pmol	>200 pmol	Total
Number of proteins digested	7	19	34	28	88
Average mass of protein, kDa	67	53	56	63	58
Average amount of protein digested, pmol[a]	37	77	143	323	177
Average density of protein band, µg/mm^3 gel volume	0.117	0.178	0.255	0.489	0.302
Average amount of protease added, µg	0.943	1.11	1.48	3.30	
Total number of peptides sequenced	14	31	74	53	172
Peptides sequenced that provided >6 positively called residues, %	80	77	85	87	84
Average initial yield, %[b]	21.1	8.1	9.0	12.3	10.8
Median initial yield, %[b]	10.2	3.2	6.1	10.0	7.1
Average number of positively called residues/peptide sequenced	12.5	11.0	12.6	14.5	12.9
Overall average background, µg[c]	—	—	—	—	0.09
Overall digest success rate[d]	100	94.7	100	96.4	97.7

[a]Data on the amount of protein digested and on the density of the protein band are based on hydrolysis/amino acid analysis of 10–15% of the stained band.

[b]Calculated as the yield of Pth-amino acid sequencing at the first or second cycle compared to the amount of protein that was digested. Since initial yields of purified peptides directly applied onto the sequencer are usually about 50%, the actual % recovery of peptides from the above digests is probably twice the % initial yields that are given.

[c]Based on hydrolysis/amino acid analysis of a 10–15% aliquot of a piece of SDS polyacrylamide gel that is equal in volume to that of the sample band and that is from an area of the sample gel that should not contain protein.

[d]As judged by the ability to sequence a sufficient number of internal residues to either identify the protein via database searches or to allow oligonucleotide probes/primers to be synthesized to enable cDNA cloning studies to proceed. The total number of positively identified residues for those proteins that were scored as a success and that were not identified via database searches varied (at the request of the submitting investigator) from 9 to 56 residues.

gel, which is about the same size as that containing the sample, be hydrolyzed and analyzed as a control to correct for the background level of free amino acids that are usually present in polyacrylamide gels. Based on samples taken from about 20 different gels submitted by users of the W. M. Keck Foundation Biotechnology Resource Laboratory, the background ranges up to about 0.2 μg and averages about 0.09 μg in these 10–15% aliquots. Although amino acid analyses on gel slices are complicated by this background level of free amino acids and by the fact that some amino acids (i.e., glycine, histidine, methionine, and arginine) usually cannot be quantitated following hydrolysis of gel slices, these estimates are still considerably more accurate than are estimates based on relative staining intensities. In those instances where amino acid analysis indicates <50 pmol protein, the stained band may be stored frozen while additional material is purified.

11. Modification of cysteine residues with MNS provides the advantage that the resulting *S*-methyl cysteine phenylthiohydantoin derivative elutes in a favorable position (i.e., between PTH-Tyr and PTH-Pro) using an Applied Biosystems/Perkin Elmer PTH C-18 column. Either the Applied Biosystems Premix or sodium acetate buffer system can be used, with a linear gradient extending from 10 to 38% buffer B over 27 min.

References

1. Brown, J. L. and Roberts, W. K. (1976) Evidence that approximately eighty percent of the soluble proteins from Ehrlich Ascites cells are amino terminally acetylated. *J. Biol. Chem.* **251,** 1009–1014.
2. Driessen, H. P., DeJong, W. W., Tesser, G. I., and Bloemendal, H. (1985) The mechanism of *N*-terminal acetylation of proteins in *Critical Reviews in Biochemistry* vol. 18 (Fasman, G. D., ed.), CRC, Boca Raton, FL, pp. 281–325.
3. Williams. K. R. and Stone, K. L. (1995) In gel digestion of SDS PAGE-separated protein: observations from internal sequencing of 25 proteins in *Techniques in Protein Chemistry VI* (Crabb, J. W., ed.), Academic, San Diego, pp. 143–152.
4. Rosenfeld, J., Capdevielle, J., Guillemot, J., and Ferrara, P. (1992) In gel digestions of proteins for internal sequence analysis after 1 or 2 dimensional gel electrophoresis. *Anal. Biochem.* **203,** 173–179.
5. Fernandez, J., DeMott, M., Atherton, D., and Mische, S. M. (1992) Internal protein sequence analysis: enzymatic digestion for less than 10 μg of protein bound to polyvinylidene difluoride or nitrocellulose membranes. *Anal. Biochem.* **201,** 255–264.
6. Aebersold, R. H., Leavitt, J., Saavedra, R. A., Hood, L. E., and Kent, S. B. (1987) Internal amino acid sequence analysis of proteins separated by one- or two-dimensional gel electrophoresis after *in situ* protease digestion on nitrocellulose. *Proc. Natl. Acad. Sci. USA* **84,** 6970–6974.
7. Stone, K. L., LoPresti, M. B., and Williams, K. R. (1990) Enzymatic digestion of proteins and HPLC peptide isolation in the subnanomole range in *Laboratory Methodology in Biochemistry: Amino Acid Analysis and Protein Sequencing* (Fini, C., Floridi, A., Finelli, V., and Wittman-Liebold, B., eds.), CRC, Boca Raton, FL, pp. 181–205.
8. Sun, Y., Zhou, Z., and Smith, D. (1989) Location of disulfide bonds in proteins by partial acid hydrolysis and mass spectrometry in *Techniques in Protein Chemistry* (Hugli, T., ed.), Academic, San Diego, pp. 176–185.
9. Lombard-Platet, G. and Jalinot, P. (1993) Funnel-well SDS-PAGE: a rapid technique for obtaining sufficient quantities of low-abundance proteins for internal sequence analysis. *BioTechniques* **15,** 669–672.

72

Reverse-Phase HPLC Separation of Enzymatic Digests of Proteins

Kathryn L. Stone and Kenneth R. Williams

1. Introduction

The ability of reverse-phase HPLC to resolve complex mixtures of peptides within a few hours' time in a volatile solvent makes it the current method of choice for fractionating enzymatic digests of proteins. In general, we find that peptides that are less than about 30 residues in length usually separate based on their content of hydrophobic amino acids and that their relative elution positions can be reasonably accurately predicted from published retention coefficients *(1,2)*. Since proteins often retain some degree of folding under the conditions used for reverse-phase HPLC, the more relevant parameter in this instance is probably surface rather than total hydrophobicity. Although larger peptides and proteins may be separated on HPLC, sometimes their tight binding, slow kinetics of release, propensity to aggregate, and relative insolubility in the usual acetonitrile/0.05% trifluoroacetic acid mobile phase results in broad peaks and/or carryover to successive chromatograms. In our experience, these problems are seldom seen with peptides that are less than about 30 residues in length, which thus makes reverse-phase HPLC an ideal method for fractionating tryptic and lysyl endopeptidase digests of proteins. Although it is sometimes possible to improve a particular separation by lessening the gradient slope in that region of the chromatogram, generally, enzymatic digests from a wide variety of proteins can be reasonably well fractionated using a single gradient that might extend over 1–2 h. Another advantage of reverse-phase HPLC is its excellent reproducibility which greatly facilitates using comparative HPLC peptide mapping to detect subtle alterations between otherwise identical proteins. Applications of this approach might include identifying point mutations as well as sites of chemical and posttranslational modification and demonstrating precursor/product relationships. Finally, since peptides are isolated from reverse-phase HPLC in aqueous mixtures of acetonitrile and 0.05% TFA, they are ideally suited for subsequent analysis by matrix-assisted laser desorption mass spectrometry (MALDI-MS, *see* Chapter 91) and automated Edman sequencing.

From: The Protein Protocols Handbook
Edited by: J. M. Walker Humana Press Inc., Totowa, NJ

2. Materials

1. HPLC system: Digests of 25 pmol to 10 nmol amounts of proteins may be fractionated on a Hewlett Packard 1090M or comparable HPLC system (*see* Note 1 for suggestions on evaluating HPLC systems and for a general discussion of important parameters that affect HPLC reproducibility, resolution, and sensitivity) capable of generating reproducible gradients in the flowrate range extending at least from about 75 to 500 μL/min. The HP 1090M HPLC used in this study was equipped with an optical bench upgrade (HP #79891A), a 1.7-μL high-pressure microflow cell with a 0.6-cm path length, a diode array detector, a Waters Chromatography static mixer, and a 250-μL injection loop. The detector outlet was connected to an Isco Foxy fraction collector and Isco Model 2150 Peak Separator to permit collection by peaks into 1.5-mL Sarstedt capless Eppendorf tubes that were positioned in the tops of 13 × 100 mm test tubes. We have found that by applying a varying resistance in parallel to the input detector signal that enters the Model 2150 Peak Separator, it is possible to improve peak detection significantly in the <500 pmol range. This can be easily accomplished via a small "black box" that is equipped with a four-position switch and that contains four different resistors labeled and configured as follows: 50 pmol/2,000 Ω, 100 pmol/1000 Ω, 250 pmol/470 Ω, 500 pmol/220 Ω. With this arrangement, best peak detection is obtained with the following settings on the 2150 Peak Separator: input = 10 mV, peak duration = 1 min, slope sensitivity = high. To minimize the "transit time" during which a given peak is traveling from the flow cell to the fraction collector, these two components are connected with 91 cm of 75-μ id fused silica capillary tubing (*see* Note 2). With this configuration, the dead volume is equal to about 7.5 μL, which corresponds to a peak delay of about 6 s at a flowrate of 75 μL/min, and the drop size is sufficiently small that even extremely small peak volumes can be accurately fractionated.
2. A 5-μ particle size, 2–2.1 × 250 mm C18 column is recommended for fractionating 25–250 pmol amounts of digests, whereas 250 pmol to 10 nmol amounts may be fractionated on 3.9–4.6 mm id columns. Although many commercially available columns would undoubtedly be satisfactory (*see* Note 3), the two columns that are currently being used in our laboratory are the 300-Å Vydac (cat. # 218TP52 for the 2.1-mm column, Separations Group, Hesperia, CA) and the 120 Å YMC column (cat. #MCQ-112 for the 2.0-mm column, YMC, Inc., Wilmington, NC).
3. pH 2.0 Buffer system: *see* Notes 4 and 5.
   Buffer A: 0.06% trifluoroacetic acid (TFA) (3 mL 20% TFA/H_2O/L final volume HPLC-grade H_2O).
   Buffer B: 0.052% TFA/80% acetonitrile (2.7 mL 20% TFA/H_2O, 800 mL CH_3CN [HPLC-grade], HPLC-grade H_2O to a final volume of 1000 mL).
4. pH 6.0 Buffer system: *see* Note 5.
   Buffer A: 5 m*M* potassium phosphate, pH 6.0 (10 mL 0.5*M* KH_2PO_4 in a total volume of 1000 mL HPLC-grade H_2O).
   Buffer B: 80% (v/v) CH_3CN (200 mL HPLC-grade H_2O and 800 mL acetonitrile [HPLC-grade]).
5. Peptide dilution buffer: 2*M* urea, 0.1*M* NH_4HCO_3. Dissolve 1.2 g Pierce Sequanal Grade urea and 79 mg Baker NH_4HCO_3 in a final volume of 10 mL HPLC-grade H_2O. This solution should be made up at least weekly and stored at –20°C.
6. 0.02% (v/v) Tween 20 solution for peptide dilution: Add 2 μL polyoxyethylene-sorbitan monolaurate (Tween 20 from Sigma Chemical Co., St. Louis, MO) to 10 mL HPLC-grade H_2O.

3. Methods

3.1. HPLC Separation of Peptides

The TFA acetonitrile buffer system described in Section 2, step 3 is an almost universal reverse-phase solvent system owing to its low-UV absorbance, high resolution, and excellent peptide solubilizing properties. The gradient we generally use is:

Time	% B
0–60 min	2–37.5%
60–90 min	37.5–75%
90–105 min	75–98%

In the case of extremely complex digests (i.e., tryptic digests of proteins that are above about 100 kDa), the gradient times may be doubled. In general, we recommend using the lowest flowrates consistent with near-optimum resolution for the column diameter that is being used *(3,4)*. Hence, we recommend a flowrate of 75 μL/min for 2.0–2.1 mm columns and a flowrate of 0.5 mL/min for 3.9–4.6 mm columns. Immediately following their collection, all fractions are tightly capped (to prevent evaporation of the acetonitrile—*see* Note 6) and are then stored in plastic boxes (USA/Scientific Plastics, Part #2350-5000) at 5°C. With the reduced flowrate of 75 μL/min, the average peak detected fraction volume is about 50 μL, which is sufficiently small that, if necessary, the entire fraction can be directly spotted onto support disks used for automated Edman degradation. To prevent adsorptive peptide losses onto the plastic tubes, fractions should not be concentrated prior to further analysis and, after spotting the peptide sample, the empty tube should be rinsed with 50 μL 100% TFA, which is then overlaid on top of the sample. If <100% of the sample is to be sequenced, we recommend that the fraction that is to be saved be transferred to a second tube, so that the tube in which the sample was collected can be rinsed in the same manner as described with 100% TFA. Since one of the important applications of reverse-phase HPLC is comparative peptide mapping, we have included below (*see* Note 7) a brief discussion of the use of this approach to identifying subtle structural modifications between otherwise identical proteins.

3.2. HPLC Repurification of Peptides

Peptides whose absorbance profile and/or MALDI-MS spectrum (*see* Chapter 91) indicate they are insufficiently pure for amino acid sequencing may be further purified by chromatography on a second (different) C18 column developed with the same mobile phase and gradient as was used for the initial separation. Because of their differing selectivity (*see* Note 8), we recommend that (when necessary) peptides that are initially separated on a Vydac C-18 column be further purified by injection onto a YMC C-18 column that is eluted with the same mobile phase and gradient as was used for the initial separation. Peptides destined for repurification are mixed with 20 μL 0.02% Tween 20 and then a volume of 2*M* urea, 0.1*M* NH_4HCO_3 such that the volume of Tween 20 and 2*M* urea, 0.1*M* NH_4HCO_3 is equal to or greater than the volume of 0.05% TFA, acetonitrile in which the fraction was originally isolated (*see* Note 9). In this way, the acetonitrile concentration is diluted by at least 50%, which, in our experience, is sufficient to permit peptide binding to the second C-18 column.

In those few instances where the sequential use of Vydac and YMC C-18 columns fails to bring about sufficient purification, the sample may be further purified by chromatography at pH 6.0 on either of these columns (*see* Note 10). Again, the same gradient is used with the only difference being the change in mobile phase.

4. Notes

1. Although general suggestions for selecting suitable HPLC systems may be found in Stone et al. *(5)*, three factors that critically impact on peptide HPLC and that will be briefly discussed are reproducibility, resolution, and sensitivity.

   Although reproducibility will not have a significant impact on the success of a single analytical HPLC separation, comparative peptide mapping requires that successive chromatograms of digests of the same protein be sufficiently similar that they can be overlaid onto one another with little or no detectable differences. In general, the latter requires that average peak retention times not vary by more than about ±0.20% *(5)*. Assuming the digests were carried out under identical conditions, problems with regard to reproducibility often relate to the inability of the HPLC pumps to deliver accurate flowrates at the extremes of the gradient range. That is, to accurately deliver a 99% buffer A/1% buffer B composition at an overall flowrate of 75 μL/min requires that pump B be able to accurately pump at a flowrate of only 0.75 μL/min. The latter is well beyond the capabilities of many conventional HPLC systems. Although reproducibility can be improved somewhat by restricting the gradient range to 2–98%, as opposed to 0–100% buffer B, the reproducibility of each HPLC system will be inherently limited in this regard by the ability of its pumps to deliver low flowrates accurately. Obviously, some HPLC systems that provide reproducible chromatograms at an overall flowrate of 0.5 mL/min might be unable to do so at 75 μL/min *(5)*. Similarly, minor check valve, piston seal, and injection valve leaks that go unnoticed at 0.5 mL/min might well account for reproducibility problems at 75 μL/min.

   The ability of HPLC to discriminate between chemically similar peptides and to resolve adequately a reasonable number of peptides from a high-mol-wt protein, which might well produce 100 or more tryptic peptides, is critically dependent on resolution, which, in turn, depends on a large number of parameters, including the flowrate, gradient time, column packing, and dimensions as well as the mobile phase *(3–5)*. Studies with tryptic digests of transferrin suggest that, within reasonable limits, gradient time is a more important determinant of resolution than is gradient volume. In general, a total gradient time of ~100 min seems to represent a reasonable compromise between optimizing resolution and maintaining reasonable gradient times *(3)*. As mentioned in Section 3.1., optimal flowrates depend on the inner diameter of the column being used, which is dictated primarily by the amount of protein that has been digested. In general, we find that amounts of protein digests in the 25–250 pmol range are best chromatographed at ~75 μL/min on 2.0–2.1 mm id columns, whereas larger amounts are best chromatographed at ~0.5 mL/min on 3.9–4.6 mm id columns. Amounts of digests that are below ~25 pmol can probably be best chromatographed on 1-mm id columns developed at flowrates near 25 μL/min. Unless precautions are taken to minimize dead volumes, significant problems may, however, be encountered in terms of automated peak detection/collection and postcolumn mixing as flowrates are lowered much below 0.15 mL/min *(5)*. Typically, the use of flowrates in the 25–75 μL/min range require that fused silica tubing be used between the detector and the fraction collector, and that a low volume flow cell (i.e., 1–2 μL) be substituted for the standard flow cell in the UV detector. Several commercially available C-18, reverse-phase supports provide high resolution, including (but certainly not limited to)

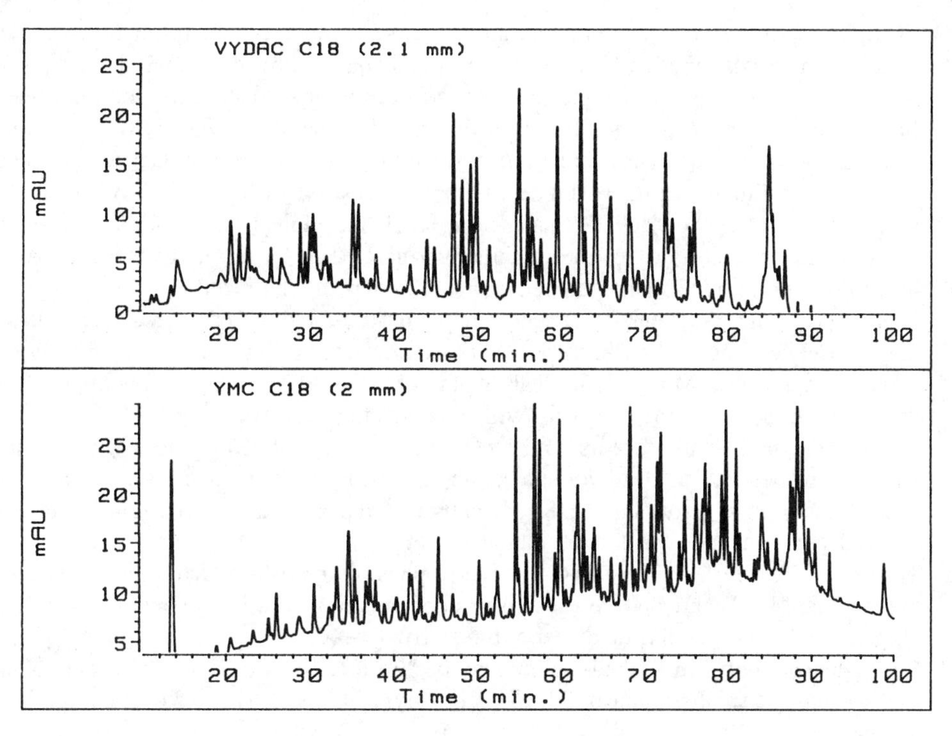

Fig. 1. Reverse-phase HPLC of 50-pmol aliquots of a large scale, in solution digest of carboxamidomethylated transferrin *(5)* that was eluted with the pH 2.0 mobile phase at a flow rate of 75 µL/min as described in Section 3. The top chromatogram was obtained on a 2.1 mm × 25 cm Vydac-18 column (5-µ particle size, 300-Å pore size), whereas the bottom chromatogram was obtained on a YMC C-18 column (5-µ particle size, 120-Å pore size). In both instances, detection was at 210 nm.

Alltech Macrosphere, Reliasil, Vydac (Fig. 1), Waters' Delta Pak *(3–5)* and YMC (Fig. 1). Since peptide resolution appears to be directly related to column length *(3–5)*, whenever possible, the 25-cm versions of these columns should be used. One caveat with regard to the latter is that under the conditions tested, we found that a 15-cm Delta-Pak C-18 column provided similar resolution as that obtained on a 25-cm Vydac C-18 column *(3)*. Although the low-UV absorbance, high resolution, and excellent solubilizing properties of the 0.05% TFA/acetonitrile, pH 2.0, buffer system have made it the almost universal mobile phase for reverse-phase HPLC, there are occasions when a different mobile phase might be advantageous. Hence, the differing selectivity of the 5 m*M,* pH 6.0, phosphate system *(5)* makes this a valuable mobile phase for detecting posttranslational modifications (such as deamidation) that may be more difficult to detect at the lower pH of the TFA system (where ionization of side-chain carboxyl groups would be suppressed). In addition, as noted in Section 3.2., changing the mobile phase provides another approach for further purifying peptides that were originally isolated in the TFA system. In our experience, however, the pH 2.0 mobile phase provides somewhat better resolution than the higher pH mobile phase *(5)*. Hence, we recommend using the pH 2.0 system for the initial separation.

The sensitivity of detection of HPLC is dictated primarily by the volume in which each peak is eluted. Although sensitivity can be increased by simply decreasing the flowrate (while maintaining a constant gradient time program) eventually, the linear flow velocity on the column will be reduced to such an extent that optimal resolution will be lost *(3–5)*. At this point, the column diameter needs to be decreased so that a more optimal linear flow velocity can be maintained at a lower flowrate. In general, the sensitivity of detection is increased as the wavelength is decreased with the practical limit in 0.05% TFA being about 210 nm. Finally, an important determinant of sensitivity (that is often overlooked) is the path length of the flow cell. For instance, an HP1090 equipped with a 0.6-cm path length cell provides (at the same flowrate) a threefold increase in sensitivity over that afforded by a Michrom UMA System equipped with a 0.2-cm path length cell.

2. If the transit time (i.e., peak delay) between the detector and the fraction collector is too long, closely eluting peaks will be pooled together. The reason is that if a second peak is detected by the Isco Model 2150 Peak Separator while it is "counting down" the peak delay for the first peak, the two peaks will be pooled together. Although our experience is this phenomenon seldom occurs with a peak delay of 6 s (at a flowrate of 75 µL/min), which corresponds to a "dead volume" of about 7.5 µL, it often occurs if the peak delay exceeds about 15 s. To our knowledge, no commercial peak detector is currently available that can simultaneously track more than one peak.
3. The procedure we use to evaluate C-18 reverse-phase columns is to determine the relative number of peaks that are detected at a given slope sensitivity during the fractionation of an aliquot of a large-scale tryptic digest of transferrin *(3–5)*.
4. For high-sensitivity work, the baseline may be "balanced" (after running the first blank run) by adding a small volume of 20% TFA (i.e., typically 10 to 100 µL) to either buffer A or B as needed *(5)*.
5. Because filtering HPLC solvents may result in their contamination *(6)*, we recommend they be made with HPLC-grade water and acetonitrile, and that they not be filtered prior to use.
6. Provided the fractions are tightly capped within a few hours of collection (to prevent loss of acetonitrile owing to evaporation), the acetonitrile is extremely effective at preventing microbial growth and peptide loss owing to adsorption. Under these conditions, we have often successfully sequenced peptide fractions that have been stored for longer than a year.
7. Provided that samples of both the modified and unmodified protein are available, comparative HPLC peptide mapping provides an extremely facile means of rapidly identifying peptides that contain posttranslational modifications. In the case of proteins that have been expressed in *E. coli*, the latter can often serve as the unmodified control, since relatively few posttranslational modifications occur in this organism. Certainly, the first attempt at comparative HPLC peptide mapping should be with enzymes, such as trypsin or lysyl endopeptidase, that have high specificity, and the digests should be separated using acetonitrile gradients in 0.05% TFA. Although elution position (as detected by absorbance at 210 nm) provides a sensitive criterion to detect subtle alterations in structure, the value of comparative HPLC peptide mapping can be further enhanced by multiwavelength monitoring and, especially, by on-line or off-line mass spectrometry of the resulting peptide -fractions (*see* Chapter 91, which details the off-line use of MALDI-MS for analyzing peptides). If comparative peptide mapping fails to reveal any significant changes, it is often worthwhile running the same digest in the pH 6.0 phosphate-buffered system. At this higher pH, some changes, such as deamidation of asparagine and glutamine, produce a larger effect on elution position than at pH 2.0, where ionization of the side-chain carboxyl groups would be suppressed. Another possible reason for failing to detect differences on comparative HPLC is that the peptide(s) containing the modifications are either too hydrophilic to bind or too hydrophobic to elute from reverse-phase supports. Hence,

in addition to trying a different HPLC solvent system, another approach that may be taken to expand the capabilities of comparative HPLC peptide mapping is to try a different proteolytic enzyme, such as chymotrypsin or Protease V8. Finally, the failure to observe any difference on comparative HPLC peptide mapping may result from loss of the post-translational modification during either the cleavage or the subsequent HPLC. Assignment of disulfide bonds is one example where this can be a problem in that disulfide interchange may occur during enzymatic cleavage, which is typically carried out at pH 8.0. This problem can be addressed by either going to shorter digestion times *(7)* or by carrying out the cleavage under acidic conditions, where disulfide interchange is less likely to occur. For this reason, pepsin (which is active in 5% formic acid) digests are often used for isolating disulfide bonded peptides. Providing that the control sample is reduced, comparative HPLC peptide mapping can be used to identify disulfide-linked peptides rapidly. If the sequence of the protein of interest is known, then comparison of the MALDI-MS spectra obtained before and after reduction of the disulfide-linked peptide can be used to identify the two peptides that are disulfide-bonded.

8. The differing selectivity of the YMC and Vydac C-18 columns that is evident in Fig. 1 explains why peptides that are initially separated on a Vydac C-18 column usually may be purified further by chromatography on a YMC C-18 column that is eluted with the same mobile phase and gradient as was used for the initial separation.
9. The purpose of the 2*M* urea and Tween 20 is to minimize adsorptive losses on diluting the peptide fraction (which is accompanied by a 50% decrease in the acetonitrile concentration). We have found that this amount of Tween 20 has no effect on the subsequent HPLC separation and that provided the urea is made up (at least weekly) in NH_4HCO, no detectable NH_2-terminal blocking occurs as the result of cyanate formation.
10. The advantage of increasing the pH from 2.0, which is the approximate pH of the usual 0.05% TFA mobile phase, to pH 6.0 is that (as mentioned above) this change accentuates the separation of peptides based on their content of acidic amino acid residues. That is, since the PK_a of the acidic side chains of aspartic and glutamic acid is about 4, increasing the pH of the mobile phase from 2.0 to 6.0 results in ionization of their COOH side chains. The increased charge that accompanies ionization greatly decreases retention of peptides on reverse-phase supports *(1)*.

References

1. Guo, D., Mant, C. T., Taneja, A. K., Parker, J. M. R., and Hodges, R. S. (1986) Prediction of peptide retention times in reversed-phase high-performance liquid chromatography. I. Determination of retention coefficients of amino acid residues of model synthetic peptides. *J. Chromatog.* **359,** 499–517.
2. Guo, D., Mant, C. T., Taneja, A. K., and Hodges, R. S. (1986) Prediction of peptide retention times in reversed-phase high-performance liquid chromatography. II. Correlation of observed and predicted peptide retention times and factors influencing the retention times of peptides. *J. Chromatog.* **359,** 519–532.
3. Stone, K. L., LoPresti, M. B., Crawford, J. M., DeAngelis, R., and Williams, K. R. (1991) Reversed-phase HPLC separation of sub-nanomole amounts of peptides obtained from enzymatic digests, in *High-Performance Liquid Chromatography of Peptides and Proteins: Separation, Analysis, and Conformation* (Mant, C. T. and Hodges, R. S., eds.), CRC, Boca Raton, FL, pp. 669–677.
4. Stone, K. L., LoPresti, M. B., and Williams, K. R. (1990) Enzymatic digestion of proteins and HPLC peptide isolation in the subnanomole range, in *Laboratory Methodology in Biochemistry* (Fini, C., Floridi, A., Finelli, V. N., and Wittman-Liebold, B., eds.), CRC, Boca Raton, FL, pp. 181–205.

5. Stone, K. L., Elliott, J. I., Peterson, G., McMurray, W., and Williams, K. R. (1990) Reversed-phase high-performance chromatography for fractionation of enzymatic digests and chemical cleavage products of proteins, in *Methods in Enzymology*, vol. 193, Academic, New York, pp. 389–412.
6. Stone, K. L. and Williams, K. R. (1986) High-performance liquid chromatographic peptide mapping and amino acid analysis in the sub-nanomole range. *J. Chromatog.* **359,** 203–212.
7. Glocker, M. O., Arbogast, B., Schreurs, J., and Deinzer, M. L. (1993) Assignment of the inter- and intramolecular disulfide linkages in recombinant human macrophage colony stimulating factor using fast atom bombardment mass spectrometry. *Biochemistry* **32,** 482–488.

Part V

Protein/Peptide Characterization

73

Peptide Mapping by Two-Dimensional Thin-Layer Electrophoresis–Thin-Layer Chromatography

Ralph C. Judd

1. Introduction

The principle behind peptide mapping is straightforward: If two proteins have the same primary structures, then cleavage of each protein with a specific protease or chemical cleavage reagent will yield identical peptide fragments. However, if the proteins have different primary structures, and then the cleavage will generate unrelated peptides. The similarity or dissimilarity of the proteins' primary structure is reflected in the similarity or dissimilarity of the peptide fragments. Separation of peptides by 2D thin-layer electrophoresis–thin-layer chromatography (2D TLE-TLC) results in very high resolution of the peptides, making subtle comparisons possible. There are four phases to the 2D TLE-TLC peptide mapping process:

1. Identification and purification of the proteins to be compared;
2. Radiolabeling of the proteins, and thus the peptide fragments, to minimize the quantity of protein required;
3. Cleavage of the proteins with specific endopeptidic reagents, either chemical or enzymatic; and
4. Separation and visualization of the peptide fragments for comparison.

Each step can be accomplished in different ways depending on the amount of protein available, the technologies available, and the needs of the researcher. Basic procedures that have proven reliable are presented for each phase. Because peptide mapping is empirical by nature, reaction times, reagent concentrations, and amounts of proteins and peptides may need to be altered to accommodate different research requirements.

2. Materials

2.1. Sodium Dodecyl Sulfate-Polyacrylamide Gel Electrophoresis (SDS-PAGE)

See Chapter 11 for reagents and procedures for SDS-PAGE.

2.2. Electroblotting

1. Blotting chamber with cooling coil (e.g., Transblot chamber, Bio-Rad, Inc., Hercules, CA, or equivalent).

From: The Protein Protocols Handbook
Edited by: J. M. Walker Humana Press Inc., Totowa, NJ

2. Power pack (e.g., EC 420, EC Apparatus Inc. [St. Petersburg, FL], or equivalent).
3. Nitrocellulose paper (NCP).
4. Polyvinylidene difluoride (PVDF) nylon membrane.
5. Ponceau S: 1–2 mL/100 mL H_2O.
6. Naphthol blue black (NBB): 1% in H_2O.
7. India ink (Pelikan, Hannover, Germany).
8. 0.05% Tween-20 in phosphate buffered saline (PBS), pH 7.4.
9. 20 m*M* phosphate buffer, pH 8.0: 89 mL 0.2*M* Na_2HPO_4 stock, 11 mL 0.2*M* NaH_2PO_4 stock in 900 mL H_2O.

2.3. Radiolabeling

1. γ-Radiation detector.
2. Speed-Vac concentrator (Savant Inst. Inc., Farmingdale, NY) or any other drying system, such as heat lamps, warm air, and so forth, will suffice.
3. Carrier-free ^{125}I (*see* Notes 1 and 2).
4. 1,3,4,6-tetrachloro-3a,6α-glycouril (Iodogen). Iodogen-tubes are prepared by placing 10 µL of chloroform containing 1 mg/mL Iodogen in the bottom of 1.5-mL polypropylene microfuge tubes and allowing to air-dry. Iodogen tubes can be stored at –20°C for up to 6 mo.
5. PBS, pH 7.4 (any dilute, neutral buffer should work).
6. Dowex 1-X-8, 20–50 mesh, anion-exchange resin (#451421, Bio-Rad).
7. Twenty-four-well disposable microtiter plate.
8. Sephadex G-25 or G-50 (Pharmacia, Piscataway, NJ).
9. 15% Methanol in H_2O.
10. XAR-5 film (Kodak, Rochester, NY, or equivalent).
11. Lightening Plus intensifying screens (DuPont, Wilmington, DE, or equivalent).

2.4. Protein Cleavage

1. Enzymatic and chemical cleavage reagents and appropriate buffers as described in Chapter 71.
2. 88 or 70% Formic acid.
3. 50 m*M* NH_4HCO_3 adjusted to the appropriate pH with sodium hydroxide.
4. 50% Glacial acetic acid in H_2O.
5. Glacial acetic acid added to H_2O to bring pH to 3.0.

2.5. Peptide Separation

2.5.1. 2D TLE-TLC

1. Forma 2095 refrigerated cooling bath (Forma Scientific, Marietta, OH) or equivalent.
2. Immersion TLE chamber (e.g., Savant TLE 20 electrophoresis chamber, or equivalent).
3. 1200-V Power pack.
4. Chromatography chambers.
5. "Varsol" (EC123, Savant, or equivalent).
6. 0.1-mm Mylar-backed cellulose sheets (E. Merck, MCB Reagents, Gibbstown, NJ, or equivalent).
7. TLE buffer: 2 L H_2O, 100 mL glacial acetic acid, 10 mL pyridine.
8. TLC buffer: 260 mL *n*-butanol, 200 mL pyridine, 160 mL H_2O, 40 mL glacial acetic acid.
9. H_2O containing Tyr, Ile, and Asp (1 mg/mL). These are 2D TLC-TLE amino acid markers.
10. 1% Methyl green in H_2O (w/v).
11. Laboratory sprayer.

12. 0.25% Ninhydrin in acetone.
13. XAR-5 film (Kodak or equivalent).
14. Lightening Plus intensifying screens (DuPont or equivalent).

3. Methods

3.1. Protein Punfication

3.1.1. SDS-PAGE

Any protein purification procedure that results in 95–100% purity is suitable for peptide mapping. For analytical purposes, the discontinuous buffer, SDS-PAGE procedure is the best choice *(1,2)*. SDS-PAGE provides apparent molecular-mass information, and the ability to probe SDS-PAGE-separated proteins by immunoblotting, helps ensure that the proper proteins are being studied.

If the proteins to be compared are abundant (>100 μg), peptide fragments can be visualized following 2D TLE-TLC by ninhydrin staining. Much smaller amounts (<0.05 μg) can be visualized by autoradiography if the proteins are extrinsically labeled with ^{125}I. Resolution of peptides increases as the amount of each peptide decreases. For these reasons, it is highly recommended that radiolabeled proteins be used *(1,3–5)*. Proteins separated in SDS-PAGE gels can be labeled and cleaved directly in gel slices (*3–7* and *see* Chapter 71), but labeling and cleavage are much more efficient if the proteins are first electroblotted to NCP *(4)*. Proteins can also be intrinsically or extrinsically labeled before SDS-PAGE separation *(5)*. It is strongly recommended that even highly pure proteins be separated in SDS-PAGE gels and transferred to NCP because of the ease of labeling and cleavage using this system. Blotted proteins can be readily located by staining with Ponceau S in water (preferred), NBB in water, or India ink-0.05% Tween-20-PBS *(8)*. Proteins of interest can then be excised, labeled using ^{125}I, and cleaved directly on the NCP *(4)*. The peptides are released into the supernatant and can then be separated using 2D TLE-TLC *(4)*.

A single SDS-PAGE separation is often adequate to purify proteins for peptide mapping. Occasionally, a second separation may be required. Alternately, 2D isoelectric focusing–SDS-PAGE (Chapters 20–22) can be used to purify proteins. If ^{125}I-labeling is used, a single protein band from a single lane of a 24-tooth comb is ample material for numerous separations of peptide fragments. Again, labeling and cleavage are greatly facilitated by electroblotting the protein to NCP.

3.1.1.1. Single SDS-PAGE Separation

1. Samples to be compared can be separated in individual lanes of an SDS-PAGE gel; "preparative" gels, where each sample is loaded over the entire stacking gel *(4)*, may also be used.
2. After electrophoresis, fixation, Coomassie brilliant blue (CBB) staining, and destaining (*see* Chapter 11), excise the protein bands of interest for use in the "gel slice" methods described below. The preferred method is to electroblot the protein to NCP, at 20 V constant current, 0.6 A for 16 h in degassed 20 m*M* phosphate buffer, pH 8.0 *(9)* (*see* Chapters 37–39).
3. To stain the proteins on NCP, shake the NCP in Ponceau S for 15 min, then destain with H_2O, or shake in 0.1% NBB in H_2O for 1 h, and then destain with H_2O. If the proteins cannot be located by using these stains, place the NCP in 100 mL of 0.05% Tween-20 PBS, and mix for 1 h. Then add three drops of India ink and mix for another hour. Protein bands will be black and the background white (*see* Note 3).

4. Excise the protein band from the NCP (a 1 × 5-mm band is more than ample), and place the excised strip in a 1.5-mL microfuge tube. Wash with H_2O until no stain is released into the supernatant. The protein is now ready for labeling and cleavage (*see* Note 4).

3.1.1.2. Double SDS-PAGE Separation

1. Separate the samples in individual lanes of an SDS-PAGE gel or in "preparative" SDS-PAGE gels. Fix, stain with CBB, and then destain (*see* Chapter 11).
2. Excise the protein bands of interest. Soak the bands in 50% ethanol–50% stacking buffer (1*M* Tris-HCl, pH 6.8) for 30 min to shrink the gel strip to facilitate loading onto a second SDS-PAGE gel.
3. Push the excised band into contact with the stacking gel of a second SDS-PAGE gel of a different acrylamide concentration (generally use high concentration in the first gel and lower concentration in the second gel).
4. Separate proteins in a second gel (CBB runs just behind the dye front). Stain or electroblot the proteins as in Section 3.1.1. The protein is now ready for labeling and cleavage.

3.2. Protein Labeling

Proteins can be intrinsically labeled by growing organisms in the presence of a uniform mixture of ^{14}C-amino acids *(5)*, but this is quite expensive. Intrinsic labeling with individual amino acids, such as ^{35}S-Met or ^{35}S-Cys, will not work, since many peptide fragments will not be labeled. Iodination with ^{125}I is inexpensive and reproducible. Iodinated peptides are readily visualized by autoradiography *(1,3,4)*. Comparative cleavages of a 40,000-Dalton protein intrinsically labeled with ^{14}C-amino acids extrinsically labeled with ^{125}I showed that 61 of 66 α-chymotryptic peptides were labeled with ^{125}I, whereas all 22 *Staphylococcus aureus* V8 protease-generated peptides were labeled with ^{125}I, demonstrating the effectiveness of radioiodination *(10)*. This demonstrates that tyrosine (Tyr) is not the only amino acid labeled using this procedure.

Iodination mediated by chloramine-T (CT) *(11)* produces extremely high specific activities, but the procedure requires an extra step to remove the CT and can cleave some proteins at tryptophan residues *(12)*. This can be beneficial since it is specific and increases the number of peptides, thus increasing the sensitivity of the procedure (*see* ref. *13* for peptide maps of CT- vs Iodogen-labeled proteins). Unfortunately, small peptides generated by CT cleavage, followed by a second enzymatic or chemical cleavage, can be lost during the removal of the CT and unbound ^{125}I.

The 1,3,4,6-tetrachloro-3α,6α-glycouril (Iodogen) *(13)* procedure, where the oxidizing agent is bound to the reaction vessel, does not damage the protein and produces high specific activities. Aspiration of the reaction mixture stops the iodination and separates the protein from the oxidant in a single step. For these reasons, Iodogen-mediated labeling is the preferred method for radioiodination. (Radioemission of ^{125}I will be expressed as counts per minute [cpm]. This assumes a detector efficiency of 70%. If detector efficiency varies, multiply the cpm presented here by 1.43 to determine decays per minute (dpm), and then multiply the dpm by the efficiency of your detector.)

3.2.1. NCP Strip (Preferred Method)

1. Put the protein-containing NCP strip in an Iodogen-coated (10 μg) microfuge tube.
2. Add 50–100 μL PBS, pH 7.4 (any dilute, neutral buffer should work) and 50–100 μCi ^{125}I (as NaI, carrier-free, 25 μCi/μL) (*see* Notes 1 and 2).

3. Incubate at room temperature for 1 h. Aspirate the supernatant. **(Caution: supernatant is radioactive.)**
4. Place the NCP strip in a fresh microfuge tube and wash three to five times with 1.5 mL H_2O (radioactivity released should stabilize at <10,000 cpm/wash).
5. The protein on the NCP strip is now ready for cleavage (*see* Note 5).

3.2.2. Gel Slice

1. Dry the gel slice containing protein using a Speed-Vac concentrator or other drying system, such as heat lamps, warm air, and so forth.
2. Put the slice in an Iodogen-coated (10-μg) microfuge tube.
3. Add 100 μL of PBS, pH 7.4 (any dilute, neutral buffer should work) plus 50–100 μCi ^{125}I (as NaI, carrier-free, 25 μCi/μL) (*see* Notes 1 and 2).
4. Incubate at room temperature for 1 h. Aspirate the supernatant. **(Caution: supernatant is radioactive.)**
5. Remove the gel slice and soak for 0.5 to 1 h in 1.5 mL of H_2O. Repeat three times.
6. Place 0.5 g Dowex 1-X-8, 20–50 mesh, anion-exchange resin and 1.5 mL of 15% methanol in H_2O in the wells of a 24-well microtiter plate.
7. Add the iodinated gel slice to a well with anion-exchange resin and incubate at room temperature for 16 h. The resin binds unreacted iodine, becoming **extremely** radioactive.
8. Remove the gel slice from the resin, and soak it in 1.5 mL of H_2O. Repeat several times, and dry the gel slice. The protein is now ready for cleavage (*see* Note 6).

3.2.3. Lyophilized/Soluble Protein

1. Suspend up to 1 mg/mL of protein in 100–200 μL of PBS, pH 7.4 (any dilute, neutral buffer should work), in a 1.5-mL microfuge tube containing 10 μg Iodogen.
2. Add 100–200 μCi ^{125}I (as NaI, carrier-free, 25 μCi/μL) (*see* Notes 1 and 2).
3. Incubate on ice for 1 h.
4. Remove the protein-containing supernatant and separate the protein from salts and unbound iodine by the following methods:
   a. (Preferred method) Separate on a Sephadex G-25 or G-50 desalting column using H_2O as the eluant and lyophilize.
   b. (Relatively easy) Solubilize the sample in 2X sample buffer (10–20 μg/lane), and separate in an SDS-PAGE gel. Stop the electrophoresis before the ion front reaches the bottom of the gel and cut the gel just above the dye front. Unbound iodine will be in this portion of the gel. Either fix, stain, and destain the gel to locate the protein band, or electroblot onto NCP and locate the protein by Ponceau S. NBB, or India ink staining (*see* Section 3.1.1., step 3). Excise the protein band from the gel or NCP.
   c. (Excellent if available) Separate the protein using reverse-phase or molecular-exclusion HPLC columns, and then dialyze and lyophilize.
   d. (Least preferred) Dialysis, followed by lyophilization, can be used, but it produces excessive radioactive liquid waste.
5. The protein is now ready for cleavage.

3.3. Protein Cleavage

The use of cleavage reagents (e.g., α-chymotrypsin) or combinations of cleavage reagents, which generate many fragments, tends to accentuate differences in primary structure, whereas cleavage reagents that produce small numbers of fragments (e.g., V8 protease, thermolysin, CNBr, BNPS-skatole) emphasize similarities in primary structure. Enzymatic reagents are often easiest to use, safest, and most reliable, but they can interfere with results, since they are themselves proteins. Chemical reagents are also easy to use and reliable, but can be toxic, requiring careful handling (*see* Note 7). Several practical cleavage reagents are presented in Table 1.

Volatile buffers must be used when using 2D TLE-TLC peptide separation, since this system is negatively affected by salts. For formate and CNBr cleavages, the acid is diluted in H_2O to 88 or 70%, respectively. BNPS-skatole works well in 50% glacial acetic acid–50% H_2O. Ammonium bicarbonate (50 m*M*), adjusted to the appropriate pH with sodium hydroxide, is excellent for enzymes requiring weak base environments (trypsin, α-chymotrypsin, thermolysin, V8 protease). The acid peptidase, Pepsin A, is active in H_2O adjusted to pH 3.0 with glacial acetic acid (*see also* Chapter 71).

3.3.1. Protein on NCP Strip

1. Put the NCP strip containing the radiolabeled protein in a 1.5-mL microfuge tube, and measure the radioemission using a γ-radiation detector.
2. Add 90 μL of the appropriate buffer and 10 μL of chemical or enzymatic cleavage reagent in buffer (1 mg/mL) to the NCP strip.
3. Incubate with shaking at 37°C for 4 h (for enzymes) or at room temperature for 24–48 h in dark under nitrogen (for chemical reagents).
4. Aspirate the peptide-containing supernatant, and count the NCP strip and supernatant. Enzymes should release 60–70% of counts in the slice into the supernatant; CNBr should release >80%, BNPS-skatole rarely releases more than 50% (*see* Notes 8 and 9).
5. Completely dry the supernatant in a Speed-Vac, and wash the sample at least four times by adding 50 μL of H_2O, vortexing, and redrying in a Speed-Vac. Alternate drying systems will work.
6. The sample is now ready for peptide separation (*see* Note 10).

3.3.2. Protein in Gel Slice

1. Put the dry gel slice containing the radiolabeled protein in a 1.5-mL microfuge tube and measure the radioemission using a γ-radiation detector.
2. Add 10 μL of cleavage reagent in buffer (1 mg/mL) directly to the dry gel slice. Allow slice to absorb cleavage reagent, and then add 90 μL of appropriate buffer.
3. Continue as from step 3, Section 3.3.1. Release of peptides into the supernatant will be less efficient than with the NCP strip.
4. The sample is now ready for peptide separation (*see* Note 10).

3.3.3. Lyophilized/SolubleProteins

1. Rehydrate the lyophilized radiolabeled proteins in the appropriate buffer at 1 mg/mL (less concentrated samples can be used successfully).
2. Add up to 25 μL of the appropriate cleavage reagent (1 mg/mL) to 25 μL of suspended protein.
3. Continue as from step 3, Section 3.3.1., except there is no strip to count.
4. The sample is now ready for peptide separation. Be aware that the sample will usually contain uncleared protein along with the peptide fragments (*see* Note 10).

3.4. 2D TLE-TLC Peptide Separation

It is strongly recommended that iodinated samples be used in the 2D TLE-TLC system. The technique described is precise enough that peptide maps can be overlaid to facilitate comparisons. Flat-bed electrophoresis can be used, but systems that cool by immersion of the thin-layer sheet in an inert coolant, such as "varsol" (such as the Savant TLE 20 electrophoresis chamber or equivalent), yield superior results. Cooling

Table 1
Cleavage Reagents

Reagent	Site of cleavage	Buffer
Chemical		
Cyanogen bromide[a,d]	Carboxy side of Met	70% formate in H_2O
BNPS-skatole[a] (-bromo-methyl-2-(nitrophenylmercapto)-3H-indole)	Carboxy side of Trp	50% H_2O-50% glacial acetic acid
Formic acid	Between ASP and Pro	88% in H_2O
Chloramine T[b]	Carboxy side of Trp	H_2O
Enzymatic[c]		
α-chymotrypsin	Carboxy side of Tyr, Trp, Phe, Leu	50 m*M* NH_4HCO_3, pH 8.5
Pepsin A	Amino side of Phe>Leu	Acetate-H_2O, pH 3.0
Thermolysin	Carboxy side of Leu>Phe	50 m*M* NH_4HCO_3, pH 7.85
Trypsin	Carboxy side of Arg, Lys	50 m*M* NH_4HCO_3, pH 8.5
V8 protease	Carboxy side of Glu, Asp, *or* carboxy side of Glu	50 m*M* NH_4HCO_3, pH 8.5 pH 6–7 (H_2O)

[a]Cyanogen bromide and BNPS-skatole are used at 1 mg/mL. Room temperature incubation should proceed for 24–48 h under nitrogen in the dark.
[b]Chloramine T is used at 10 mg/mL in H_2O.
[c]All enzymes are used at 1 mg/mL in the appropriate buffer.
[d]**Caution:** CNBr is extremely toxic—handle with care in chemical hood.

should be supplied by as large a refrigerated bath as possible, such as the Forma 2095 refrigerated cooling bath. Extra cooling coils, made by bending 1/4 in. aluminum tubing, are helpful. Peptides migrate based on charge, which is a function of pH, in an electric field. The buffer pH is a function of temperature. Therefore, maintenance of the running buffer temperature is crucial. Inconsistent cooling results in inconsistent peptide migration. For best results, use only 0.1 mm Mylar backed cellulose sheets of E. Merck. Run two or three peptide maps per 20 × 20-cm sheet. If necessary, increased resolution can be obtained by running one sample/sheet and increasing running times (*see* Note 11).

1. Set the cooling bath at 8.5°C to keep the electrophoresis tank at 10–13.5°C. The temperature of the cooling tank should not increase more than 1.5°C during a run.
2. Rehydrate the peptide sample to 10^5 cpm/μL in H_2O containing Tyr, Ile, and Asp (1 mg/mL) as amino acid markers.
3. For two samples/run: draw a line down the center of the back of the sheet with a laboratory marker parallel to the machine lines (they can be subtle, but always electrophorese parallel to these lines). Mark two spots 2.5 cm from the end of the sheet and 1.5 cm from center line on the back of the sheet to indicate where to load samples. When the sheet is turned with the cellulose facing up, the marks will show through. For three samples/run: draw two lines of the back of the sheet (parallel to machine lines) 6.7 and 13.4 cm from left edge of sheet and mark three spots, each 8 cm from the end of the plate and 1 cm to the right of the left edge and each line (*see* Note 12).

4. Use a graduated, 1–5 μL capillary pipette to spot 2 μL (~2×10^5 cpm) if two samples are used, or 1.5 μL (~1.5×10^5 cpm) if three samples are used, 0.5 μL at a time (dry spot with hair-dryer each time) to one mark on the sheet. Repeat for each sample on the other mark(s). More sample can be run, but resolution will decrease. To verify proper electrophoresis, spot 1 μL of 1% methyl green on the center line. The methyl green should migrate rapidly toward the cathode in a straight line. Veering indicates a problem.
5. Spray the plate with TLE buffer using a laboratory sprayer. Do not over wet. Remove any standing buffer with one paper towel. Always blot TLE plates in exactly the same manner.
6. Place in the electrophoresis chamber with the samples toward the anode. Run the electrophoresis at 1200 V (about 20 W and 20 mA) for 45 min (two samples/run) or 31 min (three samples/run).
7. Remove the sheet from the chamber, and immediately dry with a hair-dryer. The "varsol" will dry first, and then the buffer. Cut along the lines on the back of the sheet. Score the cellulose 0.5 cm down from top edge of each piece (bottom is the edge closest to the sample) to form a moat.
8. Place the sheets in the chromatography chamber so that chromatography can proceed perpendicularly to the electrophoresis. The TLC buffer should be about 0.5-cm deep. Chromatograph until the buffer reaches the moat. Remove and dry with a hair-dryer (best done in hood).
9. Spray the sheet with 0.25% ninhydrin in acetone (do not saturate) and dry with a hair-dryer to locate amino acid markers. Ninhydrin can also be used to locate peptides if larger amounts of sample are separated (10–100 μg). Be sure to run enzyme controls to distinguish sample from enzyme. Markers should migrate identically in all separations.
10. Overlay the sheets with X-ray film, place Lightening Plus intensifying screen over film, and place in cassette. Expose for 16–24 h at –70°C or expose film without a screen for about 4 d at room temperature. Develop the film (*see* Note 13).

Figure 1 is presented to demonstrate the separation of peptides generated by cleavage with trypin by 2D TLE-TLC. These peptide maps indicate that the porin protein (POR) of *Neisseria gonorrhoeae* is structurally unrelated to the 44-kDa proteins, whereas the 44-kDa protein from the sarkosyl insoluble (membrane) extract (44-kDa Mem.) is structurally indistinguishable from the 44-kDa protein from the periplasmic extract (44-kDa Peri.). Note the high resolution of the peptide fragments using this technique.

4. Notes

1. Regardless of the labeling procedure, never use ^{125}I that is over one half-life (60 d) old. Do not increase the amount of older ^{125}I to bring up the activity; it does not work.
2. Never use carrier-free ^{125}I in acid buffers. The iodine becomes volatile and could be inhaled.
3. Ponceau S can be completely removed with H_2O, but NBB or India ink cannot.
4. Do not compare proteins stained with Ponceau S with those stained with NBB or India ink. Use the same staining procedure for all proteins to be compared.
5. It is common to have between 3×10^6 and 6×10^6 cpm for a strip 1 × 5 mm. This provides enough material to run 15–30 peptide maps/strip.
6. It is common to have between 2×10^6 and 4×10^6 cpm in a gel slice (1 × 5 mm). This provides enough material to run 10–20 peptide maps/slice.
7. Chemical cleavage reagents are preferred when peptides are not radiolabeled since enzymes cleave themselves, resulting in confusing data. **CNBr is extremely toxic—handle with great care in a chemical hood.**

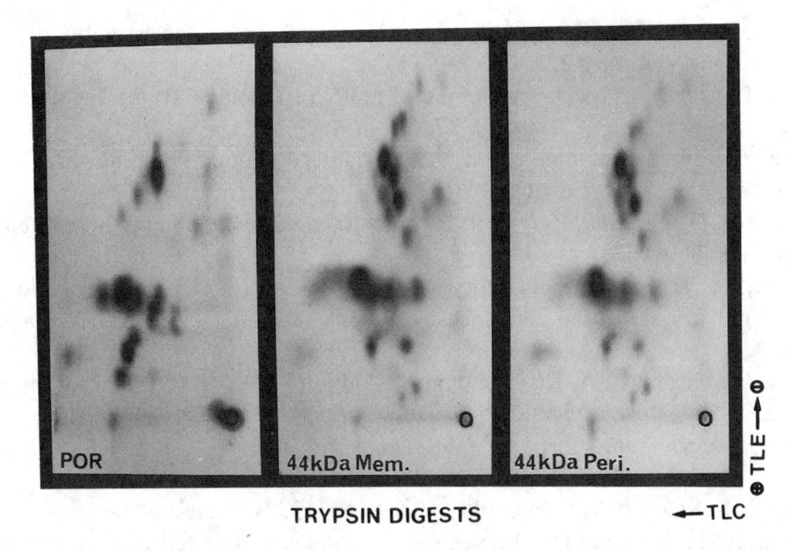

Fig. 1. Example of peptides separated by 2D TLE-TLC. Proteins were radiolabeled on NCP strips as described in Section 3.2.1. and cleaved with trypsin as described in Section 3.3.1. The peptides (1×10^5 cpm) were spotted on a thin-layer cellulose sheet and subjected to 2D TLE-TLC as described in Section 3.4. The origin (O) is at the lower right of each map. TLE-direction of thin-layer electrophoresis; TLC—direction of thin-layer chromatography. The ^{125}I-labeled peptides were visualized by autoradiography.

8. Repeated digestions will release about the same percentage of counts. Generally, one or two digestions are adequate. Only 1×10^5 cpm are necessary to produce a peptide map.
9. If >95% of counts are released in the first enzyme digestion, there may be excess, unbound iodine left in sample. This could cause serious problems, since the enzyme may become labeled. The resultant peptide maps will all be identical maps of the enzyme and not your sample.
10. Peptides prepared in this manner can also be separated by high-performance liquid chromatography (*see* Chapter 75).
11. CNBr should not be used to generate peptides for use in 2D TLE-TLC, since it produces very hydrophobic peptides, which tend to compress at the top of the chromatogram. CNBr is excellent for generating peptides to be separated by SDS-PAGE or HPLC.
12. Always spot sample to be compared the same distance from the anode (positive) terminal, since peptides migrate more rapidly close to the anode and more slowly far from the anode.
13. Migration of peptides should be consistent enough to overlay peptide maps directly for comparisons. Coordinates of amino acid markers and peptides can be determined and used to compare migration. The labeling procedures described here are precise enough to use emission intensities as a criterion for comparison.

Acknowledgments

The author thanks Pam Gannon for her assistance and the Public Health Service, NIH, NIAID (grant RO1 AI21236) and UM Research Program for their continued support.

References

1. Judd, R. C. (1986) Evidence for N-terminal exposure of the PIA subclass of protein I of *Neisseria gonorrhoeae. Infect. Immunol.* **54,** 408–414.

2. Laemmli, U. K. (1970) Cleavage of structural proteins during the assembly of the head of bacteriophage T4. *Nature* **227,** 680–695.
3. Judd, R. C. (1982) I^{125}-peptide mapping of protein III isolated from four strains of *Neisseria gonorrhoeae. Infect. Immunol.* **37,** 622–631.
4. Judd, R. C. (1987) Radioiodination and ^{125}I-labeled peptide mapping on nitrocellulose membranes. *Anal. Biochem.* **160,** 306–315.
5. Caldwell, H. D. and Judd, R. C. (1982) Structural analysis of chlamydial proteins. *Infect. Immunol.* **38,** 960–968.
6. Swanson, J. (1979) Studies on gonococcus infection. XVIII. ^{125}I-labeled peptide mapping of the major protein of the gonococcal cell wall outer membrane. *Infect. Immunol.* **23,** 799–810.
7. Elder, J. H., Pickett, R. A. III, Hampton, J., and Lerner, R. A. (1977) Radioiodination of proteins in single polyacrylamide gel slices. *J. Biol. Chem.* **252,** 6510–6515.
8. Hancock, K. and Tsang, V. C. W. (1983) India ink staining of protein on nitrocellulose paper. *Anal. Biochem.* **133,** 157–162.
9. Batteiger, B., Newhall, W. J., V., and Jones, R. B. (1982) The use of Tween-20 as a blocking agent in the immunological detection of proteins transferred to nitrocellulose membranes. *J. Immunol. Methods* **55,** 297–307.
10. Judd, R. C. and Caldwell, H. D. (1985) Comparison of ^{125}I- and ^{14}C-radiolabeled peptides of the major outer membrane protein of *Chlamydia trachomatis* strain L2/434 separated by high-performance liquid chromatography. *J. Liq. Chromatogr.* **8,** 1109–1120.
11. Greenwood, F. C., Hunter, W. M., and Glover, J. S. (1963) The preparation of ^{131}H-labeled human growth hormone of high specific radioactivity. *Biochem. J.* **89,** 114–123.
12. Alexander, N. M. (1973) Oxidation and oxidative cleavage of tryptophanyl peptide bonds during iodination. *Biochem. Biophys. Res. Commun.* **54,** 614–621.
13. Markwell, M. A. K. and Fox, C. F. (1978) Surface-specific iodination of membrane proteins of viruses and eukaryotic cells using 1,2,3,6-tetrachloro-3α, 6α-diphenylglycoluril. *Biochemistry* **112,** 278–281.

74

Peptide Mapping by Sodium Dodecyl Sulfate-Polyacrylamide Gel Electrophoresis

Ralph C. Judd

1. Introduction

The comparison of the primary structure of proteins is an important facet in the characterization of families of proteins from the same organism, similar proteins from different organisms, and cloned gene products. There are many methods available to establish the sequence similarities of proteins. A relatively uncomplicated approach is to compare the peptide fragments of proteins generated by enzymatic or chemical cleavage, i.e., peptide mapping. The similarity or dissimilarity of the resultant peptides reflects the similarity or dissimilarity of the parent proteins.

One reliable method of peptide mapping is to separate peptide fragments by sodium dodecyl sulfate-polyacrylamide gel electrophoresis (SDS-PAGE). Comparison of the separation patterns reveals the structural relationship of the proteins. Moderate separation of peptides can be accomplished using this procedure. The technique, first described by Cleveland et al. *(1),* is simple, inexpensive, requires no special equipment, and can be combined with Western blotting to locate epitopes (i.e., epitope mapping *[1–3]*) or blotting to nylon membranes for microsequencing *(4)*. Microgram amounts of peptide fragments can be visualized by in-gel staining, making this system fairly sensitive. Sensitivity can be greatly enhanced by radiolabeling (*see* Chapter 73).

2. Materials

2.1. SDS-PAGE

1. SDS-PAGE gel apparatus and power pack (e.g., EC 500, EC Apparatus, Inc., St. Petersburg, FL, or equivalent).
2. SDS-PAGE Solubilization buffer: 2 mL 10% SDS (w/v) in H_2O, 1.0 mL glycerol, 0.625 mL 1*M* Tris-HCl, pH 6.8, 6 mL H_2O, bromophenol blue to color.
3. Enzyme buffer for Cleveland et al. *(1)* "in-gel" digestion: 1% SDS, 1 m*M* EDTA, 1% glycerol, 0.1*M* Tris-HCl, pH 6.8.
4. All buffers and acrylamide solutions necessary for running SDS-PAGE (*see* Chapter 11).
5. Ethanol: 1*M* Tris-HCl, pH 6.8 (50:50; [v/v]).
6. Fixer/destainer: 7% acetic acid, 25% isopropanol in H_2O (v/v/v).
7. Coomassie brilliant blue: 1% in fixer/destainer (w/v).
8. Mol-wt markers, e.g., low-mol-wt kit (Bio-Rad, Hercules, CA, or equivalent) or peptide mol-wt markers (Pharmacia, Piscataway, NJ, or equivalent).

From: The Protein Protocols Handbook
Edited by: J. M. Walker Humana Press Inc., Totowa, NJ

2.2. Electroblotting (for Epitope or Sequence Analyses)

1. Blotting chamber with cooling coil (e.g., Transblot chamber, Bio-Rad, Inc. or equivalent).
2. Power pack (e.g., EC 420, EC Apparatus, Inc. or equivalent).
3. Nitrocellulose paper (NCP).
4. Polyvinylidene difluoride (PVDF) nylon membrane.
5. Ponceau S: 1–2 mL/100 mL H_2O.
6. 1% Naphthol blue black (NBB) in H_2O (w/v).
7. 0.05% Tween-20 in phosphate-buffered saline (PBS), pH 7.4.
8. India ink, three drops in 0.05% Tween-20 in PBS, pH 7.4.
9. 20 m*M* phosphate buffer, pH 8.0: 89 mL 0.2*M* Na_2HPO_4 stock, 11 mL 0.2*M* NaH_2PO_4 stock in 900 mL H_2O.

3. Methods

3.1. Protein Purification

Purification of the proteins to be compared is the first step in peptide mapping. Any protein purification procedure that results in 95–100% purity is suitable for peptide mapping. For analytical purposes, the discontinuous buffer, SDS-PAGE procedure is often the best choice (Chapter 11 and refs. *2* and *5*). The advantages of SDS-PAGE are: Its resolving power generally can bring proteins to adequate purity in one separation, whereas a second SDS-PAGE separation almost always provides the required purity for even the most difficult proteins; the simple reliability of the procedure; both soluble and insoluble proteins can be purified; apparent molecular-mass information; and the ability to probe SDS-PAGE-separated proteins by immunoblotting help ensure that the proper proteins are being studied. If even greater separation is required, 2D isoelectric focusing-SDS-PAGE can be used (*see* Chapters 20–22).

Proteins separated in SDS-PAGE gels can be cleaved directly in gel slices *(6)*. However, cleavage is more efficient if the proteins are first electroblotted to NCP *(3)*. If required, proteins can also be intrinsically or extrinsically labeled before SDS-PAGE separation (Chapters 19 and 73 and ref. *6*).

3.2. Protein Cleavage

Cleavages of purified proteins can be accomplished in the stacking gel with the resultant peptides separated directly in the separating gel, or they can be performed prior to loading the gel (preferred). Several lanes, with increasing incubation time or increasing concentration of enzyme in each lane, should be run to determine optimal proteolysis conditions. Standard Laemmli SDS-PAGE *(5)* system is able to resolve peptide fragments >3000 Dalton. Smaller peptides are best separated in the tricine gel system of Schagger and von Jagow *(7)* (*see also* Chapter 16). The methods described in Chapter 74, Section 3.3. can be used to generate peptides. Cleavage technique varies slightly depending on the form of the purified proteins to be compared.

3.2.1. In-Gel Cleavage/Separation

3.2.1.1. Lyophilized/Soluble Protein

1. Boil the purified protein (~1 mg/mL) for 5 min in enzyme buffer for Cleveland "in-gel" digestion.
2. Load 10–30 µL (10–30 µg) of each protein to be compared into three separate wells of an SDS-PAGE gel (*see* Note 1).

3. Overlay samples of each protein with 0.005, 0.05, and 0.5 μg enzyme in 10–20 μL of Cleveland "in-gel" digestion buffer to separate wells. V8 protease (endoproteinase Glu-C) works very well in this system (*see* Note 2). SDS hinders the activity of trypsin, α-chymotrypsin, and themolysin, so cleavage with these enzymes may be very slow. In-gel cleavage with chemical reagents is not generally recommended, since they can be inefficient in neutral, oxygenated environments. Gently fill wells and top chamber with running buffer.
4. Subject the samples to electrophoresis until the dye reaches the bottom of the stacking gel. Turn off the power, and incubate for 2 h at 37°C to allow the enzyme to digest the protein partially. Following incubation, continue electrophoresis until the dye reaches the bottom of the gel. Fix, stain, and destain, or electroblot onto NCP for immunoanalysis (*see* Note 3).

3.2.1.2. Protein In-Gel-Slice

Peptides of proteins purified by SI)S-PAGE usually retain adequate SDS to migrate into a second gel without further treatment.

1. Run SDS-PAGE gels, fix, stain with Coomassie brilliant blue, and destain.
2. Excise protein bands to be compared with a razor blade.
3. Soak excised gel slices containing proteins to be compared in ethanol–1*M* Tris-HCl, pH 6.8, for 30 min to shrink the gel, making loading on the second gel easier. Place the gel slices into the wells of a second gel.
4. Continue from step 3, section 3.2.1.1.

3.2.1.3. Protein on NCP Strip

1. Run SDS-PAGE gels, and electroblot to NCP.
2. Stain blot with Ponceau S, NBB, or block with 0.05% Tween-20 in PBS for 30 min, and then add three drops India ink (*see* Note 4).
3. Excise protein bands to be compared with a razor blade.
4. Push NCP strips to bottom of the wells of a second SDS-PAGE gel.
5. Continue from step 3, Section 3.2.1.1.

Once the separation conditions, protein concentrations, and enzyme concentrations have been established, a single digestion lane for each sample can be used for comparative purposes.

3.2.2. Protein Cleavage Followed by SDS-PAGE

It is often preferable to cleave the proteins to be compared prior to loading onto an SDS-PAGE gel for separation. This generally gives more complete, reproducible cleavage and allows for the use of chemical cleavage reagents not suitable for in-gel cleavages.

3.2.2.1. Lyophilized/Soluble Protein

1. Rehydrate the lyophilized proteins in the appropriate buffer at 1 mg/mL (less concentrated samples can be used successfully). Soluble proteins may need to be dialyzed against the proper buffer.
2. Place 10–30 μL (10–30 μg) of each protein to be compared in 1.5-mL microfuge tubes.
3. Add up to 25 μL of the appropriate chemical cleavage reagent (1 mg/mL) to suspended protein. For enzymes, use only as much enzyme as needed to achieve complete digestion (1:50 enzyme to sample maximum) (*see* Note 5).
4. Incubate with shaking at 37°C for 4 h (for enzymes) or at room temperature for 24–48 h in dark under nitrogen (for chemical reagents).
5. Add equal volume of 2X SDS-PAGE solubilization buffer and boil for 5 min (*see* Note 6).
6. Load entire sample on SDS-PAGE gel and proceed with electrophoresis.

3.2.2.2. Protein In-Gel Slice

1. Run SDS-PAGE gels, fix, stain with Coomassie brilliant blue, and destain.
2. Excise protein bands to be compared with a razor blade.
3. Dry gel slices containing proteins using a Speed-Vac concentrator or other drying system, such as heat lamps, warm air, and so forth.
4. Put the dry gel slice containing the protein in a 1.5-mL microfuge tube.
5. Add up to 10 μL of the appropriate chemical cleavage reagent (1 mg/mL) directly to the gel slice protein, and then add 90 μL of appropriate buffer. For enzymes, use only as much enzyme as needed to achieve complete digestion (1:50 enzyme to sample maximum) (*see* Note 5).
6. Incubate with shaking at 37°C for 4 h (for enzymes) or at room temperature for 24–48 h in dark under nitrogen (for chemical reagents).
7. Aspirate peptide-containing supernatant.
8. Completely dry-down the supernatant in a Speed-Vac, and wash the sample several times by adding 50 μL of H_2O, vortexing, and redrying in a Speed-Vac. Alternate drying systems will work.
9. Add 10–20 μL SDS-PAGE solubilizing solution to samples and boil for 5 min (*see* Note 6).
10. Load samples onto SDS-PAGE gel and proceed with electrophoresis.

3.2.2.3. Protein on NCP Strip

1. Run SDS-PAGE gels, and electroblot to NCP.
2. Stain blot with Ponceau S, NBB, or block with 0.05% Tween-20 in PBS for 30 min, and then add three drops India ink (*see* Note 3).
3. Excise protein bands to be compared with a razor blade.
4. Put the NCP strip containing the protein in a 1.5-mL microfuge tube.
5. Continue from step 5, Section 3.2.2.1.

Figure 1 is presented to demonstrate the separation of peptides generated by cleavage with BNPS-skatole in an SDS-PAGE gel. These peptide maps indicate that the porin protein (POR) was structurally unrelated to the 44-kDa proteins, whereas the 44-kDa proteins from a sarkosyl insoluble (membrane) extract (44-kDa Mem.) and a periplasmic extract (44-kDa Peri.) appeared to have similar primary structures.

4. Notes

1. Between 5×10^4 cpm and 10^5 cpm should be loaded in each lane if radioiodinated samples are to be used. Autoradiography should be performed on unfixed gels. Fixation and staining may wash out small peptides. Place gel in a plastic bag and overlay with XAR-5 film, place in a cassette with a Lightening Plus intensifying screen, and expose for 16 h at –70°C.
2. Control lanes containing only enzyme must be run to distinguish enzyme bands from sample bands.
3. PVDF nylon membrane is preferable to NCP when blotting small peptides.
4. Do not compare proteins stained with Ponceau S with those stained with NBB or India ink. Use the same staining procedure for all proteins to be compared.
5. To ensure complete digestion, it is advisable to incubate the samples for increasing periods of time or to digest with increasing amounts of enzyme. Once optimal conditions are established, a single incubation time and enzyme concentration can be used.
6. It is often advisable to solubilize protein in SDS prior to cleavage, since some peptides do not bind SDS well.

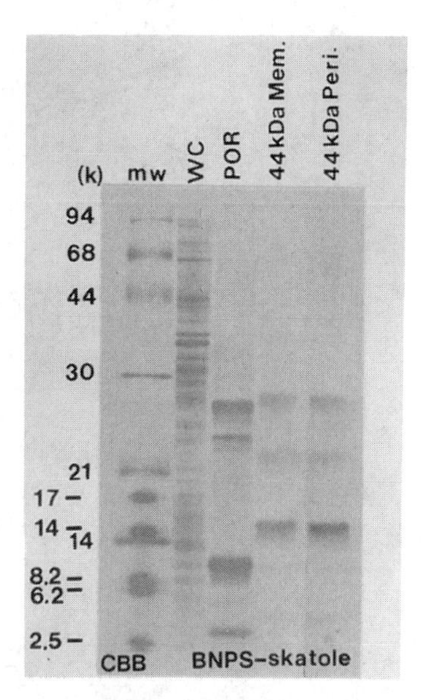

Fig. 1. Example of peptides separated in an SDS-PAGE gel. Whole cells (WC), a sarkosyl insoluble pellet, and a periplasmic extract of *Neisseria gonorrhoeae* were separated in "preparative" 15% SDS-PAGE gels and blotted to NCP as described in Chapter 73, Section 3.1.1. The 37,000-Dalton major outer membrane protein (POR) and two 44,000-Dalton (44-kDa) proteins, one isolated from a sarkosyl insoluble (membrane) extract (44-kDa Mem.), and the other isolated from a periplasmic extract (44-kDa Peri.), were located on the NCP by Ponceau S staining, excised, and cleaved with BNPS-skatole as described in Chapter 73, Section 3.3.1. Approximately 30 μg of peptides of each protein were solubilized and separated in an SDS-PAGE gel along with whole-cells (WC), Bio-Rad low-mol-wt markers, and Pharmacia peptide mol-wt markers (mw) (expressed in thousands of Dalton [k]). The gel was stained with Coomassie brilliant blue (CBB) to visualize peptides.

Acknowledgments

The author thanks Pam Gannon for her assistance and the Public Health Service, NIH, NIAID (grant RO1 AI21236), and UM Research Program for their continued support.

References

1. Cleveland, D. W., Fischer, S. G., Kirschner, M. W., and Laemmli, U. K. (1977) Peptide mapping by limited proteolysis in sodium dodecyl sulfate and analysis by gel electrophoresis. *J. Biol. Chem.* **252,** 1102–1106.
2. Judd, R. C. (1986) Evidence for *N*-terminal exposure of the PIA subclass of protein I of *Neisseria gonorrhoeae. Infect. Immunol.* **54,** 408–414.
3. Judd, R. C. (1987) Radioiodination and ^{125}I-labeled peptide mapping on nitrocellulose membranes. *Anal. Biochem.* **160,** 306–315.
4. Moos, M., Jr. and Nguyen, N. Y. (1988) Reproducible high-yield sequencing of proteins electrophoretically separated and transferred to an inert support. *J. Biol. Chem.* **263,** 6005–6008.
5. Laemmli, U. K. (1970) Cleavage of structural proteins during the assembly of the head of bacteriophage T4. *Nature* **227,** 680–695.
6. Caldwell, H. D. and Judd, R. C. (1982) Structural analysis of chlamydial proteins. *Infect. Immunol.* **38,** 960–968.
7. Schagger, H. and von Jagow, G. (1987) Tricine-sodium dodecyl sulfate-polyacrylamide gel electrophoresis for the separation of proteins in the range from 1 to 100 kDa. *Anal. Biochem.* **166,** 368–397.

75

Peptide Mapping by High-Performance Liquid Chromatography

Ralph C. Judd

1. Introduction

Peptide mapping is a convenient method for comparing the primary structures of proteins in the absence of sequence data. There are many techniques for specifically cleaving peptides with enzymes or chemical cleavage reagents (*see* Chapters 63–72).

Peptides generated by specific endopeptidic cleavage must be separated and visualized if comparisons are to be made. A convenient separation system is reverse-phase high-performance liquid chromatography (HPLC). The precision of this technique allows for rigorous comparison of primary structure with the added benefit that peptides can be recovered for further analysis. The availability of extremely sensitive in-line UV and γ-detectors makes it possible to visualize extremely small amounts of material.

2. Materials

2.1. Protein Purification, Radiolabeling, and Cleavage

See Chapter 73, Section 2. for materials needed to purify, radiolabel, and cleave proteins to be compared (*see* Note 1).

2.2. HPLC

1. HPLC capable of generating binary gradients.
2. *Preferred:* In-line UV detector and in-line γ-radiation detector (e.g., Model 170, Beckman [Fullerton, CA], or equivalent). *Alternate:* Manual UV and γ-radiation detectors.
3. Fraction collector.
4. Computing integrator or strip chart recorder.
5. Reverse-phase C_{18} column (P/N 27324 S/N, Millipore, Bedford, MA).
6. 0.005% Trifluoroacetic acid (TFA).
7. HPLC-grade methanol.
8. H_2O containing Phe, Trp, and Tyr (1 mg/mL). These are HPLC amino acid markers.

3. Method

HPLC separation of peptides can be used for structural comparisons, but its main advantage is the ability to recover peptides for further studies *(1–4)*. If radioiodinated

From: The Protein Protocols Handbook
Edited by: *J. M. Walker Humana Press Inc., Totowa, NJ*

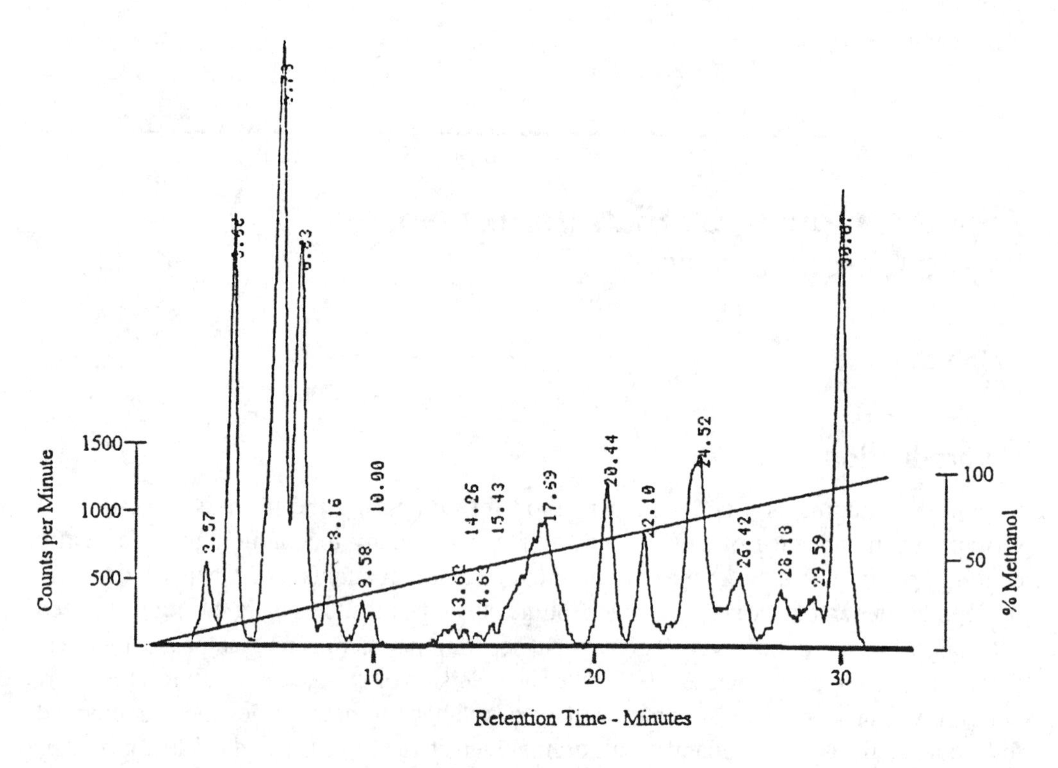

Fig. 1. Example of peptides separated by HPLC. A 37,000-Dalton membrane protein of *Neisseria gonorrhoeae* was radiolabeled and cleaved with thermolysin on an NCP strip as described in Chapter 73, Section 3.3.1. The peptides (2×10^5 cpm) were injected into a mobile phase of 0.05% TFA-H_2O (flow rate of 1 mL/min) and separated over a 35-min linear gradient to 100% methanol as described in Section 3. The ^{125}I-labeled peptides were visualized using an in-line γ-radiation detector linked to a computing integrator.

peptides are to be separated, the iodogen method of radiolabeling should be used (*see* Chapter 73, Sections 2.3. and 3.2.), since chloramine-T-mediated labeling results in considerable "noise" in HPLC chromatograms. Most peptides can be separated by reverse-phase chromatography using a C_{18} column and a linear gradient of H_2O-0.005% TFA to 100% methanol *(2)*. Different gradients using these solvents or other solvents, such as acetonitrile–0.005% TFA *(1,2,4),* isopropanol, and so forth, may be needed to achieve adequate separation. An in-line γ-radiation detector (e.g., Beckman Model 170) is helpful, but fractions can be collected and counted. Ten to 100 µg of sample should be used if UV absorbance (280_{nm}) is to be used to detect peptides. The sensitivity of UV detection roughly parallels that of Coomassie brilliant blue (CBB) staining in gels. Precision of repeated separation should be ± 0.005 min retention time for all peaks, allowing for direct comparisons of elution profiles of different samples (*see* Note 1).

1. Rehydrate the peptides in about 0.1 mL of H_2O containing 1 mg/mL of Phe, Trp, and Tyr (internal amino acid markers to verify consistent separations).
2. Inject between 1.5×10^5 and 1×10^6 cpm/separation into H_2O–0.005% TFA mobile phase running at 1 mL/min. A linear gradient (0.05% TFA to 100% methanol) over 0.5–1 h

should yield adequate preliminary separation of peptides (but *see* Note 2). Time and gradient profile will vary with the number of peptides to be separated and the nature of the peptides.

3. Monitor amino acid marker elution, or peptide elution if using UV absorbance, using an in-line UV detector at 280_{nm}. Monitor radiolabeled peptide elution using an in-line g-radiation detector. Alternatively, fractions can be collected, and the marker and peptide elution times monitored manually. Peaks can be collected, washed, dried, and reseparated by SDS-PAGE or 2D TLE-TLC.

4. Notes

1. CNBr and BNPS-skatole are excellent cleavage reagents to generate peptides to be separated by HPLC (*see* Chapter 73 for details).
2. Figure 1 demonstrates an HPLC separation of peptides generated by cleavage of a 37,000-Dalton membrane protein of *Neisseria gonorrhoeae* with themolysin. Note that there appears to be several peptides eluting in a single diffuse peak in the 17-min region of the chromatogram. Compression of peptides in this manner is relatively common. The gradient must be modified in this region to resolve these peptides adequately. Common modifications include changing the gradient slope, addition of a third solvent, such as acetonitrile, alteration of flow rate, or extending the time of separation.

Acknowledgments

The author thanks Pam Gannon for her assistance and the Public Health Service, NIH, NIAID (grant RO1 AI21236) and UM Research Program for their continued support.

References

1. Judd, R. C. (1983) ^{125}I-labeled peptide mapping and high-performance liquid chromatography ^{125}I-peptide separation of protein I of four strains of *Neisseria gonorrhoeae. J. Liq. Chromatogr.* **6,** 1421–1439.
2. Judd, R. C. (1987) Radioiodination and ^{125}I-labeled peptide mapping on nitrocellulose membranes. *Anal. Biochem.* **160,** 306–315.
3. Judd, R. C. and Caldwell, H. D. (1985) Comparison of ^{125}I- and ^{14}C-radio-labeled peptides of the major outer membrane protein of *Chlamydia trachomatis* strain L2/434 separated by high-performance liquid chromatography. *J. Liq. Chromatogr.* **8,** 1109–1120.
4. Judd, R. C. and Caldwell, H. D. (1985) Identification and isolation of surface-exposed portions of the major outer membrane protein of *Chlamydia trachomatis* by 2D peptide mapping and high-performance liquid chromatography. *J. Liq. Chromatogr.* **8,** 1559–1571.

76

Production of Protein Hydrolysates Using Enzymes

John M. Walker and Patricia J. Sweeney

1. Introduction

Traditionally, protein hydrolysates for amino acid analysis are produced by hydrolysis in 6*N* HCl. However, this method has the disadvantage that tryptophan is totally destroyed, serine and threonine partially (5–10%) destroyed, and most importantly, asparagine and glutamine are hydrolyzed to the corresponding acids. Digestion of the protein/peptide with enzymes to produce protein hydrolysate overcomes these problems, and is particularly useful when the concentration of asparagine and glutamine is required. For peptides less than about 35 residues in size, complete digestion can be achieved by digestion with aminopeptidase M and prolidase. For larger polypeptides and proteins, an initial digestion with the nonspecific protease Pronase is required, followed by treatment with aminopeptidase M and prolidase. Since it is important that all enzymes have maximum activity, the following sections will discuss the general characteristics of these enzymes.

1.1. Pronase

Pronase (EC 3.4.24.4) is the name given to a group of proteolytic enzymes that are produced in the culture supernatant of *Streptomyces griseus* K-1 *(1–3)*. Pronase is known to contain at least ten proteolytic components: five serine-type proteases, two Zn^{2+} endopeptidases, two Zn^{2+}-leucine aminopeptidases, and one Zn^{2+} carboxypeptidase *(4,5)*. Pronase therefore has very broad specificity, hence its use in cases where extensive or complete degradation of protein is required. The enzyme has optimal activity at pH 7.0–8.0. However, individual components are reported to retain activity over a much wider pH range *(6–9)*. The neutral components are stable in the pH range 5.0–9.0 in the presence of calcium, and have optimal activity at pH 7.0–8.0. The alkaline components are stable in the pH range 3.0–9.0 in the presence of calcium, and have optimal activity at pH 9.0–10.0 *(4)*. The aminopeptidase and carboxypeptidase components are stable at pH 5.0–8.0 in the presence of calcium *(9)*. Calcium ion dependence for the stability of some of the components (mainly exopeptidases) was one of the earliest observations made of Pronase *(2)*. Pronase is therefore normally used in the presence of 5–20 m*M* calcium. The addition of excess EDTA results in the irreversible loss of 70% of proteolytic activity *(10)*. Two peptidase components are inactivated by EDTA, but activity is restored by the addition of Co^{2+} or Ca^{2+}. One of these components, the

From: The Protein Protocols Handbook
Edited by: J. M. Walker Humana Press Inc., Totowa, NJ

leucine aminopeptidase, is heat stable up to 70°C. All other components of Pronase lose 90% of their activity at this temperature *(5)*. The leucine aminopeptidase is not inactivated by 9*M* urea, but is labile on dialysis against distilled water *(2)*. Some of the other components of Pronase are also reported to be stable in 8*M* urea *(2)*, and one of the serine proteases retains activity in 6*M* guanidinium chloride *(11)*. Pronase retains activity in 1% SDS (w/v) and 1% Triton (w/v) *(12)*.

Among the alkaline proteases, there are at least three that are inhibited by diisopropyl phosphofluoridate (DFP) *(10)*. In general, the neutral proteinases are inhibited by EDTA, and the alkaline proteinases are inhibited by DFP *(4)*. No single enzyme inhibitor will inhibit all the proteolytic activity in a Pronase sample.

1.2. Aminopeptidase M

Aminopeptidase M (EC 3.4.11.2), a zinc-containing metalloprotease, from swine kidney microsomes *(13–16)* removes amino acids sequentially from the N-terminus of peptides and proteins. The enzyme cleaves N-terminal residues from all peptides having a free α-amino or α-imino group. However, in peptides containing an X-Pro sequence, where X is a bulky hydrophobic residue (Leu, Tyr, Trp, Met sulfone), or in the case of an N-blocked amino acid, cleavage does not occur. It is for this reason that prolidase is used in conjunction with aminopeptidase M to produce total hydrolysis of peptides. The enzyme is stable at pH 7.0 at temperatures up to 65°C, and is stable between pH 3.5 and 11.0 at room temperature for at least 3 h *(15)*. It is not affected by sulfhydryl reagents, has no requirements for divalent metal ions, is stable in the presence of trypsin, and is active in 6*M* urea. It is not inhibited by PMSF, DFP, or PCMB. It is, however, irreversibly denatured by alcohols and acetone, and 0.5*M* guanidinium chloride, but cannot be precipitated by trichloroacetic acid *(15)*. It is inhibited by 1,10-phenanthroline (10M) *(16)*.

Alternative names for the enzyme are amino acid arylamidase, microsomal alanyl aminopeptidase, and α-aminoacyl peptide hydrolase.

1.3. Prolidase

Prolidase (EC 3.4.13.9) is highly specific, and cleaves dipeptides with a prolyl or hydroxyprolyl residue in the carboxyl-terminal position *(17,18)*. It has no activity with tripeptides *(19)*. The rate of release is inversely proportional to the size of the amino-terminal residue *(19)*. The enzyme's activity depends on the nature of the amino acids bound to the imino acid. For optimal activity, amino acid side chains must be as small as possible and apolar to avoid steric competition with the enzyme receptor site. The enzyme has the best affinity for alanyl proline and glycyl proline. The enzyme has optimal activity at pH 6.0–8.0, but it is normally used at pH 7.8–8.0 *(20)*. Manganous ions are essential for optimal catalytic activity. The enzyme is inhibited by 4-chloromercuribenzoic acid, iodoacetamide, EDTA, fluoride, and citrate. However, if Mn^{2+} is added before iodoacetamide, no inhibition is observed *(21)*.

Alternative names for the enzyme are imidodipeptidase, proline dipeptidase, amino acyl L-proline hydrolyase, and peptidase D.

2. Materials

1. Buffer: 0.05*M* ammonium bicarbonate, pH 8.0 (no pH adjustment needed) or 0.2*M* sodium phosphate, pH 7.0 (*see* Note 1).

2. Pronase: The enzyme is stable at 4°C for at least 6 mo and is usually stored as a stock solution of 5–20 mg/mL in water at –20°C.
3. Aminopeptidase M: The lyophilized enzyme is stable for several years at –20°C. A working solution can be prepared by dissolving about 0.25 mg of protein in 1 mL of deionized water to give a solution of approx 6 U of activity/mL. This solution can be aliquoted and stored frozen for several months at –20°C.
4. Prolidase: The lyophilized enzyme is stable for many months when stored at –20°C and is stable for several weeks at 4°C if stored in the presence of 2 m*M* $MnCl_2$ and 2 m*M* β-mercaptoethanol *(18)*.

3. Methods

3.1. Digestion of Proteins (22)

1. Dissolve 0.2-μmol of protein in 0.2 mL of 0.05*M* ammonium bicarbonate buffer, pH 8.0, or 0.2*M* sodium phosphate, pH 7.0 (*see* Note 1).
2. Add Pronase to 1% (w/w), and incubate at 37°C for 24 h.
3. Add aminopeptidase M at 4% (w/w), and incubate at 37°C for a further 18 h.
4. Since in many cases the X-Pro- bond is not completely cleaved by these enzymes, to ensure complete cleavage of proline-containing polypeptides, the aminopeptidase M digest should be finally treated with 1 μg of prolidase for 2 h at 37°C.
5. The sample can now be lyophilized and is ready for amino acid analysis (*see* Note 2).

3.2. Digestion of Peptides (22)

This procedure is appropriate for polypeptides less than about 35 residues in length. For larger polypeptides, use the procedure described in Section 3.1.

1. Dissolve the polypeptides (1 nmol) in 24 μL of 0.2*M* sodium phosphate buffer, pH 7.0, or 0.05*M* ammonium bicarbonate buffer, pH 8.0 (*see* Note 1).
2. Add 1 μg of aminopeptidase M (1 μL), and incubate at 37°C.
3. For peptides containing 2–10 residues, 8 h are sufficient for complete digestion. For larger peptides (11–35 residues), a further addition of enzyme after 8 h is needed, followed by a further 16-h incubation.
4. To ensure complete cleavage at proline residues, finally treat the digest with 1 μg of prolidase for 2 h at 37°C.
5. The sample can now be lyophilized and is ready for amino acid analysis (*see* Note 2).

4. Notes

1. Sodium phosphate buffer should be used if ammonia interferes with the amino acid analysis.
2. When using two enzymes or more, there is often an increase in the background amino acids owing to hydrolysis of each enzyme. It is therefore important to carry out a digestion blank to correct for these background amino acids.

References

1. Hiramatsu, A. and Ouchi, T. (1963) On the proteolytic enzymes from the commercial protease preparation of *Streptomyces griseus* (Pronase P). *J. Biochem. (Tokyo)* **54(4),** 462–464.
2. Narahashi, Y. and Yanagita, M. (1967) Studies on proteolytic enzymes (Pronase) of *Streptomyces griseus* K-1. Nature and properties of the proteolytic enzyme system. *J. Biochem. (Tokyo)* **62(6),** 633–641.
3. Wählby, S. and Engström, L. (1968) Studies on *Streptomyces griseus* protease. Amino acid sequence around the reactive serine residue of DFP-sensitive components with esterase activity. *Biochim. Biophys. Acta* **151,** 402–408.

4. Narahashi, Y., Shibuya, K., and Yanagita, M. (1968) Studies on proteolytic enzymes (Pronase) of *Streptomyces griseus* K-1. Separation of exo- and endopeptidases of Pronase. *J. Biochem.* **64(4),** 427–437.
5. Yamskov, I. A., Tichonova, T. V., and Davankov, V. A. (1986) Pronase-catalysed hydrolysis of amino acid amides. *Enzyme Microb. Technol.* **8,** 241–244.
6. Gertler, A. and Trop, M. (1971) The elastase-like enzymes from *Streptomyces griseus* (Pronase). Isolation and partial characterization. *Eur. J. Biochem.* **19,** 90–96.
7. Wählby, S. (1969) Studies on *Streptomyces griseus* protease. Purification of two DFP-reactin enzymes. *Biochim. Biophys. Acta* **185,** 178–185.
8. Yoshida, N., Tsuruyama, S., Nagata, K., Hirayama, K., Noda, K., and Makisumi, S. (1988) Purification and characterisation of an acidic amino acid specific endopeptidase of *Streptomyces griseus* obtained from a commercial preparation (Pronase). *J. Biochem.* **104,** 451–456.
9. Narahashi, Y. (1970) Pronase. *Methods Enzymol.* **19,** 651–664.
10. Awad, W. M., Soto, A. R., Siegei, S., Skiba, W. E., Bernstrom, G. G., and Ochoa, M. S. (1972) The proteolytic enzymes of the K-1 strain of *Streptomyces griseus* obtained from a commercial preparation (Pronase). Purification of four serine endopeptidases. *J. Biol. Chem.* **257,** 4144–4154.
11. Siegel, S., Brady, A. H., and Awad, W. M. (1972) Proteolytic enzymes of the K-1 strain of *Streptomyces griseus* obtained from a commercial preparation (Pronase). Activity of a serine enzyme in 6*M* guanidinium chloride. *J. Biol. Chem.* **247,** 4155–4159.
12. Chang, C. N., Model, P., and Blobel, G. (1979) Membrane biogenesis: cotranslational integration of the bacteriophage F1 coat protein into an *Escherichia coli* membrane fraction. *Proc. Natl. Acad. Sci. USA* **76(3),** 1251–1255.
13. Pfleiderer, G., Celliers, P. G., Stanulovic, M., Wachsmuth, E. D., Determann, H., and Braunitzer, G. (1964) Eizenschafter und analytische anwendung der aminopeptidase aus nierenpartikceln. *Biochem. Z.* **340,** 552–564.
14. Pfleiderer, G. and Celliers, P. G. (1963) Isolation of an aminopeptidase in kidney tissue. *Biochem. Z.* **339,** 186–189.
15. Wachsmuth, E. D., Fritze, I., and Pfleiderer, G. (1966) An aminopeptidase occurring in pig kidney. An improved method of preparation. Physical and enzymic properties. *Biochemistry* **5(1),** 169–174.
16. Pfleiderer, G. (1970) Particle found aminaopeptidase from pig kidney. *Methods Enzymol.* 19, 514–521.
17. Sjöstrom, H., Noren, O., and Jossefsson, L. (1973) Purification and specificity of pig intestinal prolidase. *Biochim. Biophys. Acta* **327,** 457–470.
18. Manao, G., Nassi, P., Cappugi, G., Camici, G., and Ramponi, G. (1972) Swine kidney prolidase: Assay isolation procedure and molecular properties. *Physiol. Chem. Phys.* **4,** 75–87.
19. Endo, F., Tanoue, A., Nakai, H., Hata, A., Indo, Y., Titani, K., and Matsuela, I. (1989) Primary structure and gene localization of human prolidase. *J. Biol. Chem.* **264(8),** 4476–4481.
20. Myara, I., Charpentier, C., and Lemonnier, A. (1982) Optimal conditions for prolidase assay by proline colourimetric determination: application to iminodipeptiduria. *Clin. Chim. Acta* **125,** 193–205.
21. Myara, I., Charpentier, C., and Lemonnier, A. (1984) Minireview: Prolidase and prolidase deficiency. *Life Sci.* **34,** 1985–1998.
22. Jones, B. N. (1986) Amino acid analysis by *o*-phthaldialdehyde precolumn derivitization and reverse-phase HPLC, in *Methods of Protein Microcharacterization* (Shively, J. E., ed.), Humana, Clifton, NJ, pp. 127–145.

77

Amino Acid Analysis Using Precolumn Derivatization with 6-Aminoquinolyl-*n*-Hydroxysuccinimidyl Carbamate

Malcolm Ward

1. Introduction

Amino acid analysis is one of the major analytical techniques used in the bioanalytical environment. It is the most reliable method for protein/peptide quantitation and can provide an extra dimension of structural information when used in conjunction with Edman sequencing and mass spectrometry.

The Steine and Moore method of amino acid analysis, developed during the 1950s, was based on ion-exchange chromatography followed by postcolumn derivatization using ninhydrin *(1)*. Higher sensitivity can be achieved by the use of precolumn derivatization using reagents, such as *ortho*-phthalialdyhyde (OPA) *(2)* or the Edman coupling reagent phenylisothiocyanate (PITC) *(3,4)*.

The method described in this chapter was introduced by the Waters chromatography division of the Millipore Corporation (Bedford, MA) in 1993. The technique involves the precolumn derivatization of amino acids using the reagent-6-aminoquinolyl-*n*-hydroxysuccinimidyl carbamate (AQC), which provides a fluorescent tag *(5)*. Amino acid derivatives are then separated by reversed-phase chromatography and detected by fluorescence at EMλ395 nm.

The reaction of AQC with primary and secondary amines produces highly stable ureas which fluoresce strongly at EMλ395 nm. The reaction is very rapid and is essentially complete within seconds (*see* Fig. 1). Excess reagent is slowly hydrolyzed to yield 6-aminoquinoline (AMQ), *N*-hydroxysuccinimide (NBS), and carbon dioxide (*see* Fig. 2).

These biproducts do not fluoresce strongly at EMλ395 nm (AMQ is <1% relative to an equivalent amount of amino acid derivative). This allows for the direct injection of the derivatized mixture.

2. Materials

2.1. Apparatus

Protein samples are hydrolyzed using a Picotag work station. This comprises an oven, a vacuum manifold, and a cold trap. A nitrogen gas line and a vacuum pump are also required.

From: The Protein Protocols Handbook
Edited by: J. M. Walker Humana Press Inc., Totowa, NJ

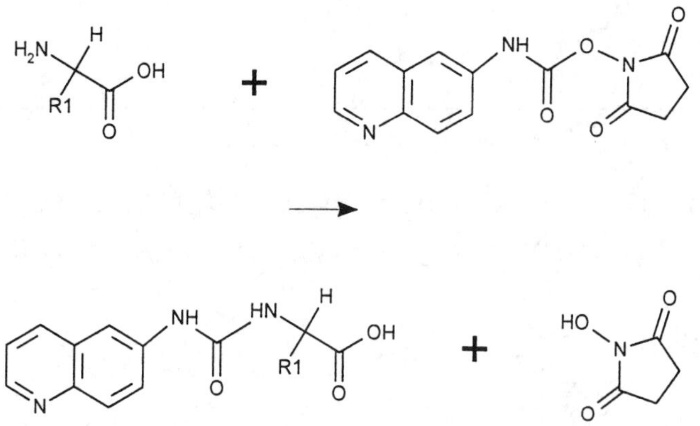

Fig. 1. The reaction of AQC with 1 and 2° amines.

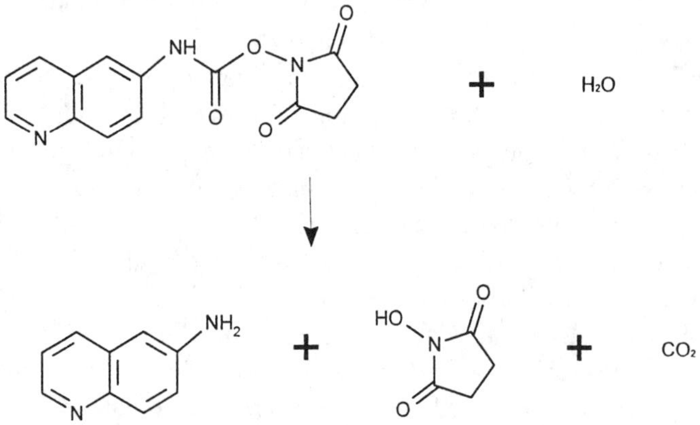

Fig. 2. The hydrolysis of AQC.

Sample tubes are inserted into a vacuum vial containing 6*M* HCl and fitted with an air-tight screw cap. A valve on the screw cap allows the vial to be sealed following evacuation using the vacuum pump and cold trap.

Hydrolysis is performed by the HCl vapor produced from heating the vial in the built-in oven. After hydrolysis, the HCl is removed under vacuum allowing derivatization of the amino acids present.

Separation and quantification of the derivatized amino acids are achieved using a reversed-phase high-performance liquid chromatography system. The system comprises two pumps, a gradient controller, an automated sample injector, a C18 column and heater unit, a fluorescence detector, and an integrator.

2.2. Chemicals (See Note 1)

1. Constant boiling HCl (Pierce Chemical Co., Rockford, IL).
2. AccQFluor reagent in acetonitrile (3 mg/mL) (available from Millipore).
3. 0.2*M* sodium borate buffer, pH 8.8.

Table 1
Gradient Table for Separation of Amino Acid Derivatives

Time, min	Flow rate, mL/min	%A	%B	%C	Curve
Initial	1.0	100	0	0	[a]
0.5	1.0	99	1	0	11
18.0	1.0	95	5	0	6
19.0	1.0	91	9	0	6
29.5	1.0	82	18	0	6
33.0	1.0	82	18	0	6
33.1	1.0	0	60	40	11
36.0	1.0	100	0	0	11
65.0	1.0	0	60	40	11
100	0.0	0	60	40	6

[a]The Waters gradient controller uses a curve parameter. A value of 6 is linear, whereas a value of 11 maintains the initial segment conditions.

4. Amino acid calibration standard (Sigma, St. Louis, MO): Dilute 40 μL Sigma H standard with 960 μL water. This produces a solution containing 17 amino acids at 100 pmol/μL (except cysteine 50 pmol/μL).
5. Eluent A: 140 m*M* sodium acetate containing 17 m*M* triethylamine, pH 5.02. Prepare by diluting 100 mL of concentrated Eluent A (Millipore WAT052890; store at 4°C) with 100 mL of Milli-Q water.
6. Eluent B: 60% acetonitrile 40% water. Degas by sonicating under vacuum for 30 s.
7. Eluent C: Milli-Q water.

3. Methods

3.1. Hydrolysis

1. Dry an aliquot containing 0.2–5 μg of sample (*see* Note 2) in a Pyrex tube (6 × 50 mm).
2. Add constant boiling HCl containing one crystal of phenol (approx 0.5 mg) to the bottom of the vial (*see* Note 3).
3. Seal the vial under vacuum.
4. Hydrolyze at 110°C for 24 h (*see* Note 4).
5. Cool the vial, and dry under vacuum.

3.2. Derivatization

1. Add an aliquot (20 μL) of 20 m*M* HCl to the sample tube, and mix thoroughly by vortexing.
2. Add Borate buffer solution (60 μL) and vortex.
3. Add AQC reagent (20 μL) and vortex. At this stage, the sample may be transferred to an autosampler tube and capped with a silicone-lined septum.
4. Heat the derivatization mixture at 50°C for 10 min.

3.3. Chromatography

1. Inject a 5-μL aliquot of the standard (50 p*M* each amino acid), and separate the components using the gradient shown in Table 1. The chromatogram resulting from the separation of a standard mixture of 17 amino acids after derivatization with AQC (*see* Fig. 3) shows that all the derivatives have eluted within 40 min.

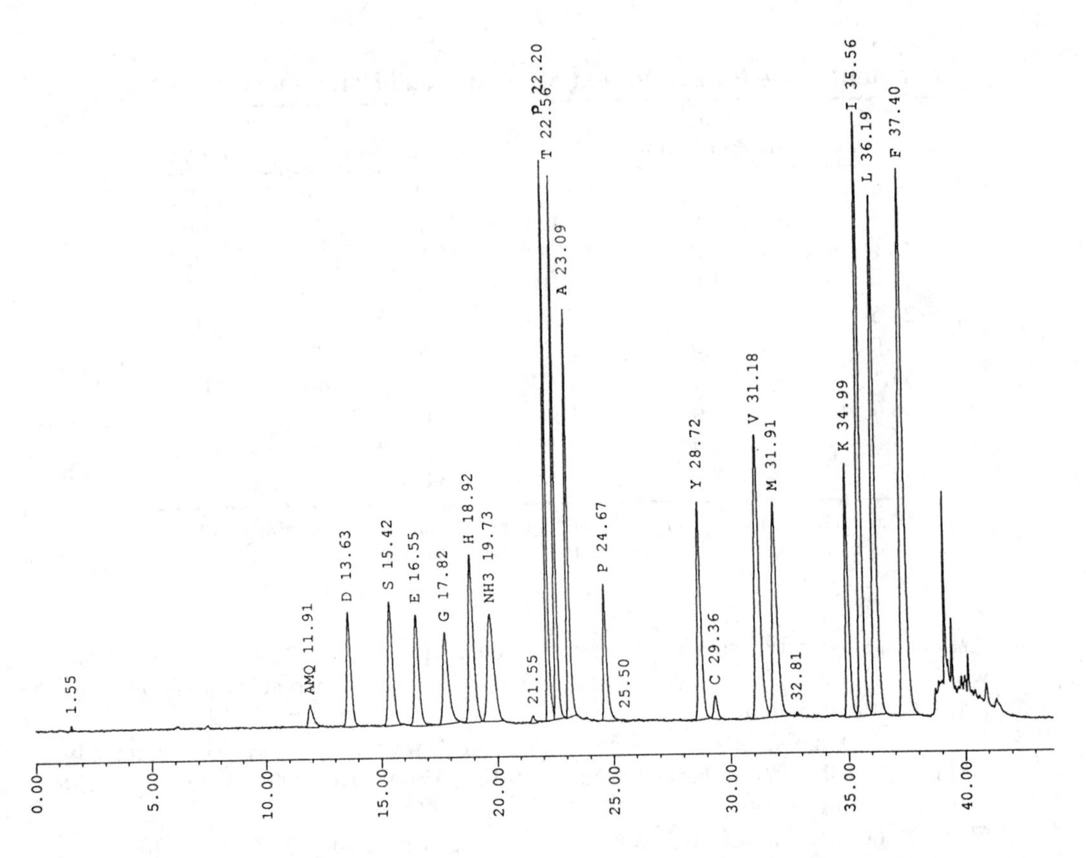

Fig. 3. Chromatogram resulting from the separation of a standard mixture of 17 amino acids after derivatization with AQC as described in Section 3.1. Each derivative is detected by flouresecence at EMλ395 nm, and identified using the one-letter code for that amino acid.

2. Inject a suitable aliquot of the unknown amino acid mixture, and separate the components using the gradient as above. For each component, record the fluorescence at EMλ395 nm.
3. After the separation, peaks can be integrated, and the amino acid composition of the unknown is determined by comparison with peak areas given by the standard amino acid mixture.

4. Notes

1. The highest quality reagents should be used, particularly in cases were high sensitivity is required. Check reagent quality using control blank derivatizations. If possible, include controls in every batch of analyses.
2. Protein samples should be free of salts, amine buffers, and detergents. Reversed-phase HPLC using microbore column technology provides a suitable means of sample preparation. Material may be collected directly into hydrolysis tubes and dried down ready for hydrolysis.
3. Phenol is added to prevent the destruction of the aromatic amino acids.
4. Rapid hydrolysis may be carried out at 150°C for 1 h *(6)*. However, more satisfactory results are obtained using the lower temperature and longer time described in Section 3.1.

Acid hydrolysis causes asparagine and glutamine to be converted to aspartic acid and glutamic acid, respectively. Serine and threonine are slowly destroyed (serine suffers 10% loss over 22 h at 110°C). Therefore, accurate quantification of these amino acids requires the hydrolysis of replicate samples over a time-course of 24, 48, and 72 h. The amount of each amino acid can then be extrapolated to zero time. Some loss of sulfur containing amino acids must also be expected. Oxidation with performic acid may be used to convert methionine to methionine sulfone and cysteine and cystine to cysteic acid *(7)*. Alternatively, the pyridylethylation reaction as described in Chapter 60, may be used to convert cysteine and cystine to more stable *S*-pyridylethyl derivatives. Tryptophan is totally destroyed by acid hydrolysis, but may be analyzed by hydrolysis with methanesulfonic acid *(8)*.

5. Avoid the use of gloves containing talc. These can cause background contamination affecting the quantification of amino acids, such as serine and glycine.
6. All glasswear should be pyrolyzed at 500°C overnight to ensure it is sufficiently clean.

References

1. Spackman, D. H., Steine, W. H., and Moore, S. (1958) Automatic recording apparatus for the use in chromatography of amino acids. *Anal. Chem.* **30,** 1190–1206.
2. Lindroth, P. and Mopper, K. (1979) High-performance liquid chromatographic determination of subpicomole amounts of amino acids by precolumn derivatization with *o*-phthalaldialdehyde. *Anal. Chem.* **51,** 1667–1674.
3. Hendrikson, R. L. and Meridith, S. C. (1984) Amino acid analysis by reverse-phase high-performance liquid chromatography: precolumn derivatization with phenylisothiocyanate. *Anal. Biochem.* **136,** 65–74.
4. Cohen, S. and Strydom, D. J. (1988) Amino acid analysis utilising phenylisothiocyanate derivatives. *Anal. Biochem.* **174,** 1–16.
5. Cohen, S., DeAntonis, K., and Michaud, D. P. (1993) Compositional protein analysis using 6-aminoquinolyl-*n*-hydroxysuccinimidyl carbamate, a novel derivatising agent, in *Techniques in Protein Chemistry IV* (Angeletti. R., ed.), Academic, San Diego, CA, pp. 289–306.
6. Bidlingmeyer, B. A., Cohen, S. A., and Tarvin, T. L. (1984) Rapid analysis of amino acids using precolumn derivatization. *J. Chromatogr.* **336,** 93–104.
7. Glazer, A. N., Delange, R. J., and Sigman, D. S. (1975) Chemical charactorization of proteins and their derivatives, in *Chemical Modifications of Proteins* (Work, T. S. and Work, E., eds.), Elsevier, North Holland, Amsterdam, pp. 21–24.
8. Hendrickson, R. L., Mora, R., and Maraganore, J. M. (1987) A practical guide to the general application of PTC-amino acid analysis, in *Proteins: Structure and Function* (L'Italien, J. J., ed.), Plenum, New York, pp. 187–195.

78

Amino Acid Analysis Using Precolumn Derivatization with Phenylisothiocyanate

G. Brent Irvine

1. Introduction

The method described in this chapter is based on derivatization of amino acids, produced by hydrolysis of peptides or proteins, with phenylisothiocyanate. This forms phenylthiocarbamyl amino acids, which are then separated by reversed-phase high-performance liquid chromatography and quantified from their UV absorbance at 254 nm.

Quantitative amino acid analysis, based on separation by ion-exchange chromatography followed by postcolumn derivatization using ninhydrin for detection, was developed during the 1950s *(1)* and remained the predominant method for 20 years. With the advent of reversed-phase high-performance liquid chromatography, however, rapid separation of amino acid derivatives became possible. Precolumn derivatization also avoids dilution of peaks and so increases sensitivity. Methods involving derivatization with fluorogenic reagents, such as dansyl chloride *(2)* and *o*-phthaldialdehyde *(3),* were the first to be developed, and these enabled detection of <1 pmol of an amino acid. These methods have some disadvantages, however, including instability of derivatives, reagent interference, and lack of reaction with secondary amino acids. Derivatization using phenylisothiocyanate, the reagent used in the first step of the Edman method for determining the sequence of proteins, avoids many of these problems. The reaction, shown in Fig. 1, is rapid and quantitative with both primary and secondary amino acids.

The products are relatively stable, and excess reagent, being volatile, is easily removed. Sensitivity, at about the level of 1 pmol, is more than adequate, since it is difficult to reduce background contamination below this level. Quantitative analysis of phenylthiocarbamyl amino acids was first described by Knoop and coworkers *(4)*. Full details of the application of this method to the analysis of protein hydrolysates were published in 1984 *(5)*. A similar procedure was published *(6)* by employees of the Waters Chromatography Division of Millipore Corporation, and the equipment that they developed is commercially available as the Waters Pico-Tag® system (Millipore, Milford, MA). The method described in this chapter was carried out using Waters' equipment, but equivalent instrumentation could be used. A review *(7)* describes the application of the method to the analysis of free amino acids in physiological fluids, amino acids in foodstuffs, and unusual amino acids.

From: The Protein Protocols Handbook
Edited by: J. M. Walker Humana Press Inc., Totowa, NJ

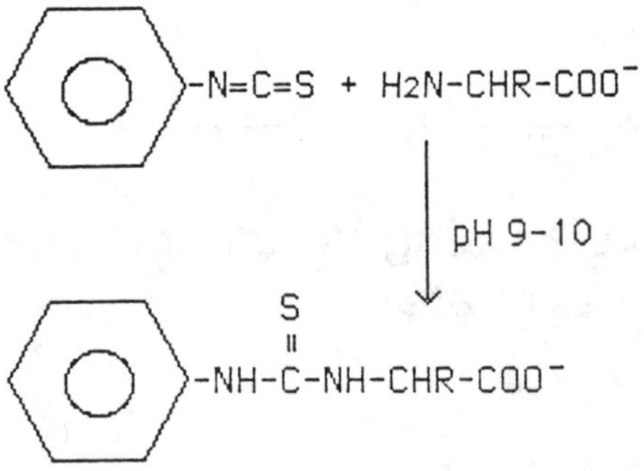

$$C_6H_5\text{-N=C=S} + H_2N\text{-CHR-COO}^-$$

pH 9–10

$$C_6H_5\text{-NH-C(=S)-NH-CHR-COO}^-$$

Fig. 1. Reaction of phenylisothiocyanate with an amino acid to form a phenylthiocarbamyl amino acid derivative.

2. Materials

2.1. Apparatus

Two pieces of equipment are required. The first is for hydrolysis of proteins and for derivatization of the resulting amino acids. This can be carried out using the Pico-Tag work station, which comprises an oven that can accommodate up to four vacuum vials, and a vacuum/purge manifold with vacuum gage and a cold trap. Also required are a nitrogen line and a vacuum pump. Up to 12 tubes (6 × 50 mm, Waters), each containing a sample of protein for hydrolysis, can be inserted into a vacuum vial. The vial can be sealed with an air-tight plastic screw cap that has a drilled-through Teflon™ valve fitted with a slider control that can be pushed to seal or release the vacuum. To the bottom of the vacuum vial is added HCl. The top is screwed on and inserted into the vacuum manifold, and the vial is evacuated using a vacuum pump and cold trap. The valve is then sealed, and the vial is removed and heated in the built-in oven, causing hydrolysis of protein by HCl in the vapor phase. The vial has a bulge in the middle to prevent condensing HCl from running down the inside into the sample tubes. After removal of HCl under vacuum, phenylthiocarbamyl amino acids are formed by addition of phenylisothiocyanate. Excess reagent is then removed under vacuum.

The second piece of equipment is a reversed-phase high-performance liquid chromatography system for separation and quantification of the phenylthiocarbamyl amino acids. This requires two pumps, a gradient controller, an automated injector, a C18 column and column heater, a UV detector set at 254 nm, and an integrator.

2.2. Chemicals (see Note 1)

2.2.1. Hydrolysis and Derivatization

1. HCl/phenol: add melted crystalline phenol (10 µL) to hydrochloric acid (constant boiling at 760 mm) (1.0 mL) (Pierce, Rockford, IL).
2. Redry solution: ethanol:water:triethylamine (2:2:1 by vol).
3. Derivatization solution: ethanol:water:triethylamine:phenylisothiocyanate (7:1:1:1 by vol). Vortex and allow to stand for 5 min before using. Use within 2 h. Phenylisothiocyanate (Pierce): Store at –20°C under nitrogen. After opening an ampule, it can be divided into aliquots that should be resealed under nitrogen. It is important to allow the container to come to room temperature before opening, since this reagent is sensitive to moisture. Each aliquot should be used within 3 wk of opening.

4. Amino Acid Standard H (Pierce): This contains a solution of 17 amino acids (2.5 m*M* each, except cystine, which is 1.25 m*M*) in 0.1*N* HCl.

2.2.2. Chromatography

1. Sample buffer: Dissolve anhydrous sodium dihydrogen phosphate (0.71 g) in water (1 L). Adjust the pH to 7.4 with 1% (v/v) orthophosphoric acid. Add 52.6 mL acetonitrile. Filter through a 0.22-µm filter (Millipore type GV).
2. Eluent A: Dissolve sodium acetate trihydrate (19.0 g) in water (1 L). Add triethylamine (0.5 mL). Adjust the pH to 5.7 with acetic acid (*see* Note 2). Add acetonitrile (63.8 mL). Add 1.07 mL of a solution of ethylenediaminetetra-acetic acid dipotassium salt (100 mg) in water (100 mL). Filter through a 0.22-µm filter (Millipore type GV).
3. Eluent B: Mix acetonitrile (600 mL) with water (400 mL). Filter through a 0.22-µm filter (Millipore type GV).

3. Methods

3.1. Hydrolysis and Derivatization

1. To a pyrex tube (6 × 50 mm), add an aliquot (10 µL) of a solution (2.5 m*M*) of peptide or protein in a volatile solvent, such as methanol or water (*see* Note 3). Volumes in excess of 50 µL should not be used, since solutions may "bump" out of the tube. Place these tubes into the vacuum vial.
2. Dry under vacuum on the work station until the pressure has fallen to 65 mtorr. Drying times using the work station depend on the efficiency of the vacuum pump. With a good pump, this step should take <1 h.
3. To the bottom of the vial (not into the tubes), add HCl/phenol (200 µL).
4. Flush the vial with oxygen-free nitrogen gas, and then evacuate to 1–2 torr. Repeat this twice, sealing the vial after the third evacuation step.
5. Place the vial in the heating block of the work station at 110°C for 22 h (*see* Note 4).
6. Remove the vial, and allow it to cool. Remove excess HCl from the outside of tubes by wiping with a tissue.
7. Add to the vial two tubes, each containing an aliquot (10 µL) of Amino Acid Standard H. Dry under vacuum on the work station until the pressure has fallen to 65 mtorr (about 1 h).
8. Clean out the cold trap to remove any traces of acid that might react with cyanate in the next stages.
9. To each tube, add redry solution (20 µL) and vortex.
10. Dry under vacuum on the work station (about 30 min).
11. To each tube, add derivatization solution (20 µL) and vortex. Leave at room temperature for 20 min.
12. Dry under vacuum on the work station. After the pressure has dropped to 65 mtorr (about 2 h), leave for a further 10 min to ensure complete removal of phenylisothiocyanate.

3.2. Chromatography

1. To each tube, add sample buffer (200 µL) and vortex. The relatively large volume added here facilitates the next (filtration) step. Only a small proportion of the filtered sample is analyzed. If sample quantity is limited, *see* Note 5.
2. Filter the samples through a 0.45-µm filter (Millipore type HV). Use a 1-mL plastic syringe and needle to withdraw samples. Then place the filter over the syringe and add a fresh needle to ensure that the sample is placed at the bottom of the tube from which it will be injected for chromatography. This should be carried out as soon as possible, and certainly within a few hours if samples are left at room temperature (*see* Note 6).
3. Place the Pico-Tag column in the heating module at 39.0°C, and commence flow of Eluent B (in which the column is stored) at a rate of 1.0 mL/min.

Table 1
Gradient Table

Time, min	Flow, mL/min	%A	%B	Curve no.[a]
Initial	1.0	100	0	*
10.0	1.0	54	46	5
10.5	1.0	0	100	6
11.5	1.0	0	100	6
12.0	1.5	0	100	6
12.5	1.5	100	0	6
20.0	1.5	100	0	6
20.5	1.0	100	0	6

[a]The column headed "Curve No." refers to the setting used on the Waters gradient controller. Number 6 is linear, whereas number 5 is a shallow convex curve (the rate of change is greater in the earlier part of the period).

4. Run a linear gradient from 100% Eluent B to 100% Eluent A during 2 min. Allow the column to equilibrate for 30 min in the latter mobile phase.
5. Inject an aliquot (5 µL) of standard using the gradient shown in Table 1. Run time should be set at 21.0 min. The use of an automatic injector, such as the WISP (Waters), is highly recommended (*see* Note 7).
6. The second and third injections should be of the same solution (standard) as the first. Invariably, poor chromatography is obtained with the first injection, but good separation should be achieved with the second and third injections. These chromatograms should be checked to ensure that this is the case before proceeding with further samples.
7. Inject an aliquot (5 µL) of each unknown sample, recording the absorbance at 254 nm.
8. After the run, peaks are integrated, and the amino acid composition of the unknown is determined by comparison with peak areas given by the Amino Acid Standard H.

4. Notes

1. The highest quality of reagents should be used. Suitable water can be obtained using a Milli-Q purification system (Millipore, Bedford, MA) fed by a supply of tap water that has been distilled once. For other chemicals, suppliers of some suitable grades are suggested in Section 2. This is particularly important if high sensitivity is required (*see* Note 5).
2. Other workers often adjust the pH of Eluent A to 6.35 *(6)* or 6.40 *(7)* rather than the lower pH of 5.7 described above. At this lower pH, the resolution of Asp and Glu is increased.
3. Protein samples should be free of salts, amines, and detergents, although it has been reported that salts at concentrations up to 4*M* do not affect derivatization or separation *(8)*.
4. Rapid hydrolysis may be carried out at 150°C for 1 h *(6)*. However, more satisfactory results are obtained using the lower temperature and longer time described in Section 3.1. Acid hydrolysis of proteins causes conversion of asparagine and glutamine to aspartic acid and glutamic acid, respectively. Serine, threonine, and, to a lesser extent, tyrosine are slowly destroyed (serine suffers about 10% loss during 22 h at 110°C; threonine and tyrosine rather less than this). Accurate quantification of these amino acids requires hydrolysis of replicate samples for 24, 48, and 72 h and extrapolation back to zero time. The longer hydrolysis times can also give more accurate values for alanine, valine, and isoleu-

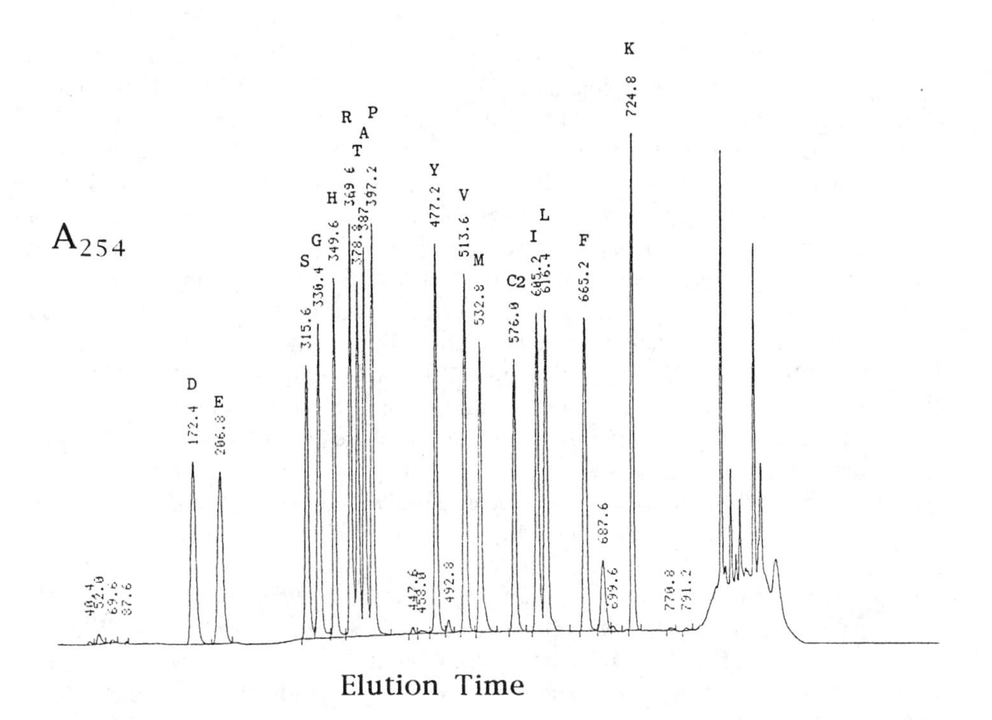

Fig. 2. Separation of a standard mixture of 17 amino acids after derivatization with phenylisothiocyanate. The peak owing to each phenylthiocarbamyl amino acid is identified using the one-letter code for that amino acid, but with C2 being cystine rather than cysteine. The number above each peak is the elution time in seconds. The absorbance at 254 nm of the largest peak (K) was 0.14.

Amino acid standard H (10 µL) (Pierce) was treated as described in Section 3.1. The phenylthiocarbamyl derivatives were dissolved in sample buffer (200 µL), and an aliquot (5 µL, containing 625 pmol of each amino acid derivative) was subjected to chromatography as described in Section 3.2. Injections were made using a WISP 712 with two Model 510 pumps and an Automated Gradient Controller onto a Pico-Tag column (3.9 × 150 mm) equilibrated at 39.0°C in a Column Heater controlled by a Temperature Control Module. Peaks were measured by absorbance at 254 nm using a 441 Absorbance Detector (all above equipment from Waters) linked to a Trio Chromatography Computing Integrator (Trivector Inc., West Chester, PA). It can be seen that all the amino acid derivatives have been eluted by 12.1 min. The peaks eluting between F and K and between 13 and 15 min are the result of side reactions of phenylisothiocyanate with nonamino acid material.

cine, since hydrolysis of peptide bonds between certain aliphatic amino acids is incomplete after 24 h. Methionine, cysteine, and cystine are partially destroyed by acid hydrolysis with HCl, and for accurate determination, require conversion to more stable derivatives. Tryptophan is totally destroyed by acid hydrolysis with HCl, but may be analyzed after hydrolysis with methanesulfonic acid *(7)*.

After hydrolysis, in order to prevent condensing HCl from running down the side of the vial into the tubes, keep the tubes upright. With fewer than 12 samples, add blank tubes for support.

5. The amount of each amino acid present in the chromatogram shown in Fig. 2 is 625 pmol. This value is very much higher than the sensitivity limit of the method (about 1 pmol).

However, if adequate quantities of protein are available, working at this sensitivity will avoid problems owing to background contamination of samples. If limited quantities of protein are available, say 200 pmol (2 µg of a protein of mol wt 10,000), the procedures described above will need to be modified by redissolving dried, derivatized material in less sample buffer and/or injecting a larger volume (maximum 25 µL). This will ensure that more of the sample goes onto the column. A corresponding adjustment in the treatment of standards will also be required. However, if it is necessary to carry out analyses in the low pmol range, special precautions must be taken, since background contamination (especially of serine and glycine, where it can reach several pmol per sample) is a common problem. These precautions include: pyrolysing glassware at 500°C overnight or steeping in sulfuric acid (250 mL) containing sodium nitrate (1 g); handling with clean gloves (nontalc) and forceps; including a control blank that has been subjected to hydrolysis. Of course, it is good practice to use these procedures in any case, even for lower sensitivity analyses.

6. Phenylthiocarbamyl amino acids in solution at neutral pH are relatively stable, with <10% loss of the least stable derivatives (Leu, Ile) during 10 h at room temperature. Losses are much reduced in the cold, with <5% loss during 48 h at 4°C.
7. The use of automatic injectors, apart from their labor-saving function, gives constant injection volumes and constant intervals between injections (when using the WISP, this interval is actually about 21.7 min for run time set at 21.0 min). An identical interval between injections is an important criterion for obtaining reproducible retention times for each amino acid in different chromatograms, since the column is not given sufficient time to reequilibrate in Eluent A.

Acknowledgment

Thanks are owed to Adrienne Healy for expert technical assistance.

References

1. Spackman, D. H., Stein, W. H., and Moore, S. (1958) Automatic recording apparatus for use in the chromatography of amino acids *Anal. Chem.* **30,** 1190–1206.
2. Hsu, K-T. and Currie, B. L. (1978) High-performance liquid chromatography of Dns-amino acids and application to peptide hydrolysates. *J. Chromatogr.* **166,** 555–561.
3. Lindroth, P. and Mopper, K. (1979) High performance liquid chromatographic determination of subpicomole amounts of amino acids by precolumn fluorescence derivatization with *o*-phthaldialdehyde. *Anal. Chem.* **51,** 1667–1674.
4. Knoop, D. R., Morgan, E. T., Tarr, G. E., and Coon, M. J. (1982) Purification and characterization of a unique isoenzyme of cytochrome P-450 from liver microsomes of ethanol-treated rabbits. *J. Biol. Chem.* **257,** 8472–8480.
5. Hendrikson, R. L. and Meridith, S. C. (1984) Amino acid analysis by reverse-phase high-performance liquid chromatography: precolumn derivatization with phenylisothiocyanate. *Anal. Biochem.* **136,** 65–74.
6. Bidlingmeyer, B. A., Cohen, S. A., and Tarvin, T. L. (1984) Rapid analysis of amino acids using pre-column derivatization. *J. Chromatogr.* **336,** 93–104.
7. Cohen, S. A. and Strydom, D. J. (1988) Amino acid analysis utilizing phenylisothiocyanate derivatives. *Anal. Biochem.* **174,** 1–16.
8. Hendrikson, R. L., Mora, R., and Maraganore, J. M. (1987) A practical guide to the general application of PTC-amino acid analysis, in *Proteins: Structure and Function* (L'Italien, J. J., ed.), Plenum, New York, pp. 187–195.

79

Molecular-Weight Estimation for Native Proteins Using Size-Exclusion High-Performance Liquid Chromatography

G. Brent Irvine

1. Introduction

The chromatographic separation of proteins from small molecules on the basis of size was first described by Porath and Flodin, who called the process "gel filtration" *(1)*. Moore applied a similar principle to the separation of polymers on crosslinked polystyrene gels in organic solvents, but named this "gel permeation chromatography" *(2)*. Both terms came to be used by manufacturers of such supports for the separation of proteins, leading to some confusion. The name size-exclusion chromatography is more descriptive of the principle on which separation is based and has largely replaced the older names *(3)*.

The original matrices were based on crosslinked polysaccharides, but these are too compressible for use in high-performance liquid chromatography (HPLC). Currently, the most efficient columns are based on surface-modified silica of particle size about 5 μm. These columns (about 1 × 30 cm) have many thousands of theoretical plates. They can be operated at flow rates of about 1 mL/min, giving run times of about 12 min, 10–100 times faster than conventional chromatography on soft gels. The improved peak sharpness and speed have led to a resurgence of interest in the technique.

As well as being a standard chromatographic mode for the purification of proteins, size-exclusion chromatography can be used for estimation of mol wts. For polymers of the same shape, plots of log mol wt against K_d (*see* Note 1) give straight lines within the range $0.1 < K_d < 0.8$ *(3)*. This is true only for ideal size-exclusion chromatography, in which the support does not interact with solute molecules (*see* Note 2). In any case, it must be borne in mind that it is the size, rather than the mol wt, of a solute molecule that determines its elution volume. Hence, calibration curves prepared with globular proteins as standards cannot be used for the assignment of mol wts to proteins with different shapes, such as the rod-like protein, myosin. It has been found that the most reliable measurements of mol wt by size-exclusion HPLC are obtained under denaturing conditions, when all proteins have the same random-coil structure. Disulfide bonds must be reduced, usually with dithiothreitol, in a buffer that destroys secondary and

From: The Protein Protocols Handbook
Edited by: J. M. Walker Humana Press Inc., Totowa, NJ

tertiary structure. Buffers containing guanidine hydrochloride *(4,5)* or sodium dodecylsulfate (SDS) *(6)* have been used for this purpose.

However, the use of denaturants has many drawbacks, which are described in Note 3. In any case, polyacrylamide gel electrophoresis in sodium dodecylsulfate-containing buffers is widely used for determining the mol wt of protein subunits. This technique can also accommodate multiple samples in the same run, although each run takes longer than for size exclusion HPLC.

The method described below is for the estimation of mol wts of native proteins. The abilities to measure bioactivity and to recover native protein in high yield make this an important method, even though a number of very basic, acidic or hydrophobic proteins will undergo nonideal size exclusion under these conditions.

2. Materials

2.1. Apparatus

A high-performance liquid chromatography system for isocratic elution is required. This comprises a pump, an injector, a size-exclusion column and guard column (*see* Note 4), a UV detector, and a data recorder. There are many size-exclusion columns based on surface-modified silica on the market. The results described below were obtained using a Zorbax Bioseries GF-250 column (0.94 × 25 cm), of particle size 4 µm, sold by Dupont, Wilmington, DE. This column can withstand very high back pressures (up to 380 bar or 5500 psi) and can be run at flow rates up to 2 mL/min with little loss of resolution *(7)*. It has a mol-wt exclusion limit for globular proteins of several hundred thousand. The exclusion limit for the related GF-450 column is about a million.

2.2. Chemicals

1. 0.2*M* Disodium hydrogen orthophosphate (Na_2HPO_4).
2. 0.2*M* Sodium dihydrogen orthophosphate (NaH_2PO_4).
3. 0.2*M* Sodium phosphate buffer, pH 7.0: Mix 610 mL of 0.2*M* Na_2HPO_4 with 390 mL of 0.2*M* NaH_2PO_4. Filter through a 0.22-µm filter (Millipore type GV).
4. Solutions of standard proteins: Dissolve each protein in 0.2*M* sodium phosphate buffer, pH 7.0, at a concentration of about 0.5 mg/mL. Filter the samples through a 0.45-µm filter (Millipore-type HV). Proteins suitable for use as standards are listed in Table 1 (*see* Note 5).
5. Blue dextran, average mol wt 2,000,000 (Sigma), 1 mg/mL in 0.2*M* sodium phosphate buffer, pH 7.0. Filter through a 0.45-µm filter (Millipore-type HV).
6. Glycine, 10 mg/mL in 0.2*M* sodium phosphate buffer, pH 7.0. Filter through a 0.45-µm filter (Millipore type HV).

3. Method

1. Allow the column to equlibrate in the 0.2*M* sodium phosphate buffer, pH 7.0, at a flow rate of 1 mL/min (*see* Note 6) until the absorbance at 214 nm is constant.
2. Inject a solution (20 µL) of a very large molecule, such as blue dextran (*see* Note 7), to determine V_o. Repeat with 20 µL of water (negative absorbance peak) or a solution of a small molecule, such as glycine, to determine V_t. Definitions of V_o, V_e, V_t, and K_d are given in Note 1.
3. Inject a solution (20 µL) of one of the standard proteins, and determine its elution volume, V_e, from the time at which the absorbance peak is at a maximum. Repeat this procedure until all the standards have been injected. A chromatogram showing the separation of

Table 1
Protein Standards

Protein[a]	Mol Wt
Thyroglobulin	669,000
Apoferritin	443,000
β-Amylase	200,000
Immunoglobulin G	160,000
Alcohol dehydrogenase	150,000
Bovine serum albumin	66,000
Ovalbumin	42,700
β-Lactoglobulin	36,800
Carbonic anhydrase	29,000
Trypsinogen	24,000
Soybean trypsin inhibitor	20,100
Myoglobin	16,900
Ribonuclease A	13,690
Insulin	5,900
Glucagon	3,550

[a]All proteins listed were obtained from Sigma, Poole, England.

seven solutes during a single run on a Zorbax Bio-series GF-250 column is shown in Fig. 1. Calculate K_d (*see* Note 1) for each protein, and plot log mol wt against K_d. A typical plot is shown in Fig. 2.

4. Inject a solution (20 μL) of the protein of unknown mol wt and measure its elution volume, V_e, from the absorbance profile. If the sample contains more than one protein and the peaks cannot be assigned with certainty, collect fractions and assay each fraction for the relevant activity.
5. Calculate K_d for the unknown protein and use the calibration plot to obtain an estimate of its mol wt.

4. Notes

1. The support used in size-exclusion chromatography consists of particles containing pores. The molecular size of a solute molecule determines the degree to which it can penetrate these pores. Molecules that are wholly excluded from the packing emerge from the column first, at the void volume, V_o. This represents the volume in the interstitial space (outside the support particles) and is determined by chromatography of very large molecues, such as blue dextran or DNA. Molecules that can enter the pores freely have full access to an additional space, the internal pore volume, V_i. Such molecules emerge at V_t, the total volume available to the mobile phase, which can be determined from the elution volume of small molecules. Hence $V_t = V_o + V_i$. A solute molecule that is partially restricted from the pores will emerge with elution volume, V_e, between the two extremes, V_o and V_t. The distribution coefficient, K_d, for such a molecule represents the fraction of V_i available to it for diffusion. Hence:

$$V_e = V_o + K_d \cdot V_i \quad (1)$$

and

$$K_d = (V_e - V_o)/V_i = (V_e - V_o)/(V_t - V_o) \quad (2)$$

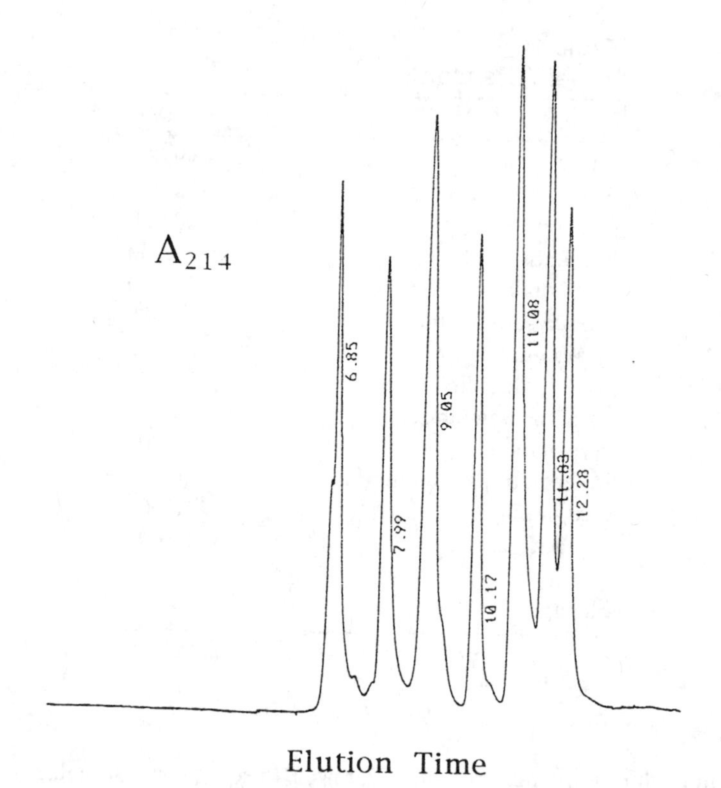

Fig. 1. Separation of a mixture of seven solutes on a Zorbax Bio-series GF-250 column. Twenty microliters of a mixture containing about 1.5 µg of each protein were injected. The solutes were, in order of elution, thyroglobulin, alcohol dehydrogenase, ovalbumin, myoglobin, insulin, glucagon, and sodium azide. The number beside each peak is the elution time in minutes. The absorbance of the highest peak, insulin, was 0.105. The equipment was a Model 501 Pump, a 441 Absorbance Detector operating at 214 nm, a 746 Data Module, all from Waters (Millipore, Milford, MA) and a Rheodyne model 7125 injector (Rheodyne, Cotati, CA) with 20 µL loop. The column was a Zorbax Bio-Series GF-250 with guard column (Dupont, Wilmington, DE). The flow rate was 1 mL/min, and the chart speed was 1 cm/min. The attenuation setting on the Data Module was 128.

2. Silica to which a hydrophilic phase, such as a diol, has been bonded still contains underivatized silanol groups. Above pH 3, these are largely anionic and will interact with ionic solutes, leading to nonideal size-exclusion chromatography. Depending on the value of its isoelectric point, a protein can be cationic or anionic at pH 7. Proteins that are positively charged will undergo ion exchange, causing them to be retarded. Conversely, anionic proteins will experience electrostatic repulsion from the pores, referred to as "ion exclusion," and will be eluted earlier than expected on the basis of size alone. When size-exclusion chromatography is carried out at a low pH, the opposite behavior is found, with highly cationic proteins being eluted early and anionic ones being retarded. To explain this behavior, it has been suggested that at pH 2, the column may have a net positive charge *(8)*. In order to reduce ionic interactions, it is necessary to use a mobile phase of high ionic strength. On the other hand, as ionic strength increases, this promotes the formation of hydrophobic interactions. To minimize both ionic and hydrophobic interactions, the mobile phase should have an ionic strength between 0.2 and 0.5*M (9)*.

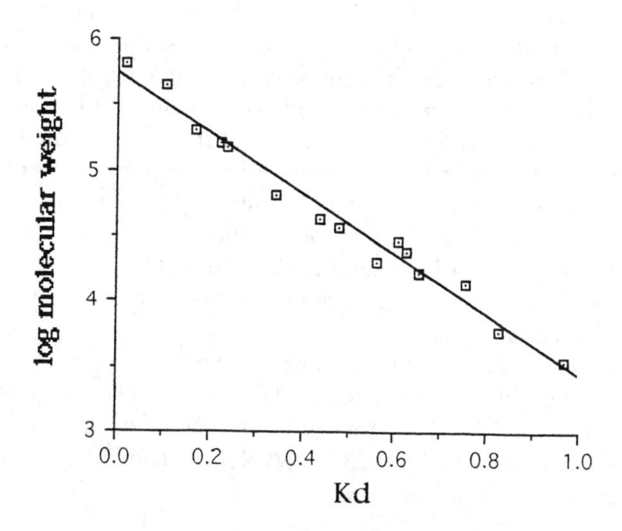

Fig. 2. Plot of log mol wt vs K_d for the proteins listed in Table 1. Chromatography was carried out as described in Fig. 1. V_o was determined to be 6.76 mL from the elution peak of blue dextran. V_t was determined from the elution peak of glycine and from the negative peak given by injecting water, both of which gave a value of 11.99 mL. The regression line ($y = 5.751 - 2.321x$, $r^2 = 0.977$) was computed using all the points shown.

3. Problems arising when size-exclusion chromatography is carried out under denaturing conditions include:
   a. For a particular column, the mol-wt range in which separation occurs is reduced. This is because the radius of gyration, and hence the hydrodynamic size, of a molecule increases when it changes from a sphere to a random coil. For example, the separation range of a TSK G3000SW column operating with denatured proteins is 2000–70,000, compared to 10,000–500,000 for native proteins *(5,10)*. Of course, this may actually be an advantage when working with small proteins or peptides.
   b. Proteins are broken down into their constituent subunits and polypeptide chains, so that the mol wt of the intact protein is not obtained.
   c. Bioactivity is usually destroyed or reduced, and it is not usually possible to monitor enzyme activity. This can be a serious disadvantage when trying to identify a protein in an impure preparation.
   d. The denaturants usually absorb light in the far UV range, so that monitoring the absorbance in the most sensitive region for proteins (200–220 nm) is no longer possible.
   e. Manufacturers often advise that once a column has been exposed to a mobile phase containing denaturants, it should be dedicated to applications using that mobile phase, since the properties of the column may be irreversibly altered. In addition, the denaturant, especially if it is sodium dodecylsulfate, may be difficult to remove completely.
   f. Since these mobiles phases have high viscosities, flow rates may have to be reduced to avoid high back pressures.
   g. High concentrations of salts, especially those containing halide ions, can adversely affect pumps and stainless steel.
4. Most manufacturers sell guard columns appropriate for use with their size-exclusion columns. In order to protect the expensive size-exclusion column it is strongly recommended that a guard column be used.

5. Even in the presence of high-ionic-strength buffers, several proteins show nonideal behavior and are thus unsuitable as standards. For example, the basic proteins cytochrome c (p*I* ≈ 10) and lysozyme (p*I* ≈ 11) have K_d values >1.0, under the conditions described for Fig. 1, because ion-exchange interactions are not totally suppressed. On the other hand, the very acidic protein pepsin (p*I* ≈ 1) emerges earlier than expected on the basis of size, because of ion exclusion. One should be aware that such behavior may also occur when interpreting results for proteins of unknown p*I*.
6. Columns are often stored in 0.02% sodium azide to prevent bacterial growth. When changing mobile phase, some manufacturers recommend that the flow rate should not be greater than half the maximum flow rate.
7. Although V_o is most commonly measured using blue dextran, Himmel and Squire suggested that it is not a suitable marker for the TSK G3000SW column, because of tailing under nondenaturing conditions, and measured V_o using glutamic dehydrogenase from bovine liver (Sigma Type II; mol wt 998,000) *(11)*. Calf thymus DNA is also a commonly used marker for V_o.

References

1. Porath, J. and Flodin, P. (1959) Gel filtration: a method for desalting and group separation. *Nature* **183,** 1657–1659.
2. Moore, J. C. (1964) Gel permeation chromatography. I. A new method for molecular weight distribution of high polymers. *J. Polymer Sci.* **A2,** 835–843.
3. Gooding, K. M. and Regnier, F. E. (1990) Size exclusion chromatography, in *HPLC of Biological Macromolecules* (Gooding, K. M. and Regnier, F. E., eds.), Marcel Dekker, New York, pp. 47–75.
4. Ui, N. (1979) Rapid estimation of molecular weights of protein polypeptide chains using high-pressure liquid chromatography in 6*M* guanidine hydrochloride. *Anal. Biochem.* **97,** 65–71.
5. Kato, Y., Komiya, K., Sasaki, H., and Hashimoto, T. (1980) High-speed gel filtration of proteins in 6*M* guanidine hydrochloride on TSK-GEL SW columns. *J. Chromatogr.* **193,** 458–463.
6. Josic. D., Baumann, H., and Reutter, W. (1984) Size-exclusion high-performance liquid chromatography and sodium dodecyl sulfate-polyacrylamide gel electrophoresis of proteins: a comparison. *Anal. Biochem.* **142,** 473–479.
7. Anspach, B., Gierlich, H. U., and Unger, K. K. (1988) Comparative study of Zorbax Bio Series GF 250 and GF 450 and TSK-GEL 3000 SW and SWXL columns in size-exclusion chromatography of proteins. *J. Chromatogr.* **443,** 45–54.
8. Irvine, G. B. (1987) High-performance size-exclusion chromatography of polypeptides on a TSK G2000SW column in acidic mobile phases. *J. Chromatogr.* **404,** 215–222.
9. Regnier, F. E. (1983) High performance liquid chromatography of proteins. *Methods Enzymol.* **91,** 137–192.
10. Kato, Y., Komiya, K., Sasaki, H., and Hashimoto, T. (1980) Separation range and separation efficiency in high-speed gel filtration on TSK-GEL SW columns. *J. Chromatogr.* **190,** 297–303.
11. Himmel, M. E. and Squire, P. G. (1981) High pressure gel permeation chromatography of native proteins on TSK-SW columns. *Int. J. Peptide Protein Res.* **17,** 365–373.

80

Detection of Disulfide-Linked Peptides by HPLC

Alastair Aitken and Michèle Learmonth

1. Introduction

Classical techniques for determining disulfide bond patterns usually require the fragmentation of proteins into peptides under low-pH conditions to prevent disulphide exchange. Pepsin or cyanogen bromide are particularly useful (*see* Chapter 63). Diagonal techniques to identify disulphide-linked peptides were developed by Brown and Hartley (*see* Chapter 81). A modern micromethod employing reverse-phase HPLC is described here.

2. Materials

1. 1*M* dithiothreitol (DTT, good-quality, e.g., Calbiochem, Nottingham, UK).
2. 100 m*M* Tris-HCl, pH 8.5.
3. 4-Vinylpyridine (Aldrich).
4. 95% Ethanol.
5. Isopropanol.
6. 1*M* triethylamine-acetic acid, pH 10.0.
7. Tri-*n*-butyl-phosphine: 1% in isopropanol.
8. HPLC system.
9. Vydac C_4, C_8, or C_{18} reverse phase HPLC columns.

3. Method *(1,2)*

1. Alkylate the protein (1–10 mg in 2–5 mL buffer) without reduction to prevent possible disulphide exchange by dissolving in 100 m*M* Tris-HCl, pH 8.5, and adding 1 μL of 4-vinylpyridine (*see* Note 1). Incubate for 1 h at room temperature and desalt by HPLC or precipitate with 95% ice-cold ethanol followed by bench centrifugation. The pellet obtained after the latter treatment may be difficult to redissolve and may require addition of 10-fold concentrated acid (HCl, formic or acetic acid) before digestion at low pH. It may be sufficient to resuspend the pellet with acid using a sonic bath if necessary, and then commence the digestion. Vortex the suspension during the initial period until the solution clarifies.
2. Fragment the protein under conditions of low pH (*see* Note 2) and subject the peptides from half the digest to reverse-phase HPLC. Vydac C_4, C_8, or C_{18} columns give particularly good resolution depending on the size range of fragments produced. Typical separation conditions are: column equilibrated with 0.1% (v/v) aqueous TFA , elution with an

From: The Protein Protocols Handbook
Edited by: J. M. Walker Humana Press Inc., Totowa, NJ

acetonitrile/0.1% TFA gradient. A combination of different cleavages, both chemical and enzymatic, may be required if peptide fragments of interest remain large after one digestion method.

3. To the other half of the digest (dried and resuspended in 10 µL of isopropanol) add 5 µL of 1*M* triethylamine-acetic acid, pH 10.0, 5 µL of tri-*n*-butyl-phosphine (1% in isopropanol) and 5 µL of 4-vinylpyridine. Incubate for 30 min at 37°C, and dry *in vacuo*, resuspending in 30 µL of isopropanol twice. This procedure cleaves the disulfides and modifies the resultant—SH groups.
4. Run the reduced and alkylated sample on the same column, under identical conditions on reverse-phase HPLC. Cysteine-linked peptides are identified by the differences between elution of peaks from reduced and unreduced samples. Collection of the alkylated peptides (which can be identified by rechromatography with detection at 254 nm) and a combination of sequence analysis and mass spectrometry (*see* Chapter 84) will allow disulfide assignments to be made.

4. Notes

1. The iodoacetic acid used must be colorless. A yellow color indicates the presence of iodine, this will rapidly oxidize thiol groups, preventing alkylation and may also modify tyrosine residues. It is possible to recrystallize from hexane. Reductive alkylation may also be carried out using iodo-[^{14}C]-acetic acid or iodoacetamide (*see* Chapter 51). The radiolabeled material should be diluted to the desired specific activity before use with carrier iodoacetic acid or iodoacetamide to ensure an excess of this reagent over total thiol groups.
2. Fragmentation of proteins into peptides under low-pH conditions to prevent disulfide exchange is important. Pepsin, Glu-C, or cyanogen bromide (CNBr) are particularly useful (*see* Chapters 63 and 71). Typical conditions for pepsin are 25°C for 1–2 h at pH 2.0–3.0 (10 m*M* HCl or 5% acetic or formic acid) with an enzyme:substrate ratio of about 1:50. Endoproteinase Glu-C has a pH optimum of 4.0 as well as an optimum at pH 8.0. Digestion at the acid pH (typically 37°C overnight in ammonium acetate at pH 4.0 with an enzyme:substrate ratio of about 1:50) will also help minimize disulfide exchange. CNBr digestion in guanidinium HCl (6*M*)/0.1–0.2*M* HCl may be more suitable acid medium owing to the inherent redox potential of formic acid, which is the most commonly used protein solvent. When analyzing proteins that contain multiple disulfide bonds, it may be appropriate to carry out an initial chemical cleavage (CNBr is particularly useful) followed by a suitable proteolytic digestion. The initial acid chemical treatment will cause sufficient denaturation and unfolding as well as peptide bond cleavage to assist the complete digestion by the protease. If a protein has two adjacent cysteine residues, this peptide bond will not be readily cleaved by specific endopeptidases. For example, this problem was overcome during mass spectrometric analysis of the disulfide bonds in insulin by using a combination of an acid proteinase (pepsin) and carboxypeptidase A as well as Edman degradation *(3)*.

References

1. Friedman, M., Zahnley, J. C., and Wagner, J. R. (1980) Estimation of the disulfide content of trypsin inihibitors as *S*-β-(2-pyridylethyl)-L- cysteine. *Anal. Biochem.* **106,** 27–34.
2. Amons, R. (1987) Vapor-phase modification of sulfhydryl groups in proteins *FEBS Lett.* **212,** 68–72.
3. Toren, P., Smith, D., Chance, R., and Hoffman, J. (1988) Determination of interchain crosslinkages in insulin B-chain dimers by fast atom bombardment mass spectrometry. *Anal. Biochem.* **169,** 287–299.

81

Diagonal Electrophoresis for Detecting Disulfide Bridges

Alastair Aitken and Michèle Learmonth

1. Introduction

Methods for identifying disulfide bridges have routinely employed "diagonal" procedures using two-dimensional paper or thin-layer electrophoresis. This essentially utilizes the difference in electrophoretic mobility of peptides containing either cysteine or cystine in a disulfide link, before and after oxidation with performic acid. It was first described by Brown and Hartley *(1)*. Peptides unaltered by the performic acid oxidation have the same mobility in both dimensions and, therefore, lie on a diagonal. After oxidation, peptides that contain cysteine or were previously covalently linked produce one or two spots off the diagonal, respectively. This method has also been adapted for HPLC methodology and is discussed in Chapter 80.

First the protein has to be fragmented into suitable peptides containing individual cysteine residues. It is preferable to carry out cleavages at low pH to prevent possible disulfide bond exchange. In this respect, pepsin (active at pH 3.0) and cyanogen bromide (CNBr) are particularly useful reagents. Proteases with active-site thiols should be avoided (e.g., papain, bromelain).

Before the advent of HPLC, paper electrophoresis was the most commonly used method for peptide separation *(2)*. Laboratories with a history of involvement with protein characterization are likely to have retained the equipment, but it is no longer commercially available. Although it is possible to make a simple electrophoresis tank in house *(3),* thin-layer electrophoresis equipment is still commercially available, and it is advisable that this be used, owing to the safety implications.

For visualization of the peptides, ninhydrin is the classical amino group stain. However, if amino acid analysis or sequencing is to be carried out, fluorescamine is the reagent of choice.

2. Materials

2.1. Equipment

1. Electrophoresis tank.
2. Flat bed electrophoresis apparatus (e.g., Hunter thin layer peptide mapping system, Orme, Manchester, UK).

From: The Protein Protocols Handbook
Edited by: J. M. Walker Humana Press Inc., Totowa, NJ

3. Whatman (Maidstone, UK) 3MM and Whatman No. 1 paper.
4. Cellulose TLC plates (Machery Camlab [Cambridge, UK] Nagel or Merck [Poole, UK]).

2.2. Reagents

1. Electrophoresis buffers (*see* Note 1): Commonly used volatile buffers are:
   pH 2.1 acetic acid/formic acid/water 15/5/80, v/v/v
   pH 3.5 pyridine/acetic acid/water 5/50/945 v/v/v
   pH 6.5 pyridine/acetic acid/water 25/1/225, v/v/v
2. Nonmiscible organic solvents: toluene, for use with pH 6.5 buffer: white spirit for use with pH 2.1 and 3.5 buffers.
3. Formic acid. **Care!**
4. 30% w/v Hydrogen peroxide. **Care!**
5. Fluorescamine: 1 mg/100 mL in acetone.
6. Marker dyes: 2% orange G, 1% acid fuschin, and 1% xylene cyanol dissolved in appropriate electrophoresis buffer.
7. 1% (v/v) Triethylamine in acetone.

3. Methods

3.1. First-Dimension Electrophoresis

3.1.1. Paper Electrophoresis

1. Dissolve the peptide digest from 0.1–0.3 μmol protein in 20–25 μL electrophoresis buffer.
2. Place the electrophoresis sheet on a clean glass sheet (use Whatman No. 1 for analytical work, Whatman 3MM for preparative work). Support the origin at which the sample is to be applied on glass rods. Where paper is used, multiple samples can be run side by side. Individual strips can then be cut out for running in the second dimension.
3. Apply the sample slowly without allowing to dry (covering an area of about 2 × 1 cm) perpendicular to the direction intended for electrophoresis. NB: for pH 6.5 buffer, apply near the center of the sheet, for acidic buffers, apply nearer the anode end.
4. Apply a small volume of marker dyes (2% orange G, 1% acid fuschin, 1% xylene cyanol, in electrophoresis buffer) on the origin and additionally in a position that will not overlap the peptides after the second dimension.
5. Once the sample is applied, wet the sheet with electrophoresis buffer slowly and uniformly on either side of the origin so that the sample concentrates in a thin line. Remove excess buffer from the rest of the sheet with blotting paper.
6. Place the wet sheet in the electrophoresis tank previously set up with electrophoresis buffer covering bottom electrode. An immiscible organic solvent (toluene where pH 6.5 buffer is used, white spirit for acidic buffers) used to fill the tank to the top. Upper trough filled with electrophoresis buffer. **Care!**
7. Electrophorese at 3 kV for about 1 h, with cooling if necessary. The progress of the electrophoresis can be monitored by the movement of the marker dyes (*see* Note 1).
8. Dry sheet in a well-ventilated place overnight, at room temperature, secured from a glass rod with Bulldog clips.

3.1.2. Thin-Layer Electrophoresis

1. Dissolve the peptide digest from 0.1–0.3 μmol protein in at least 10 μL electrophoresis buffer.
2. Mark the sample and dye origins on the cellulose side of a TLC plate with a cross using an extrasoft blunt-ended pencil, or on the reverse side with a permanent marker.

3. Spot the sample on the origin. This can be done using a micropipet fitted with a disposable capillary tip. To keep the spot small, apply 0.5–1 µL spots and dry between applications.
4. Apply 0.5 µL marker dye to the dye origin.
5. Set up electrophoresis apparatus with electrophoresis buffer in both buffer tanks. Prepare electrophoresis wicks from Whatman 3MM paper, wet with buffer, and place in buffer tanks.
6. Prepare a blotter from a double sheet of Whatman 3MM, with holes cut at the positions of the sample and marker origins. Wet with buffer and place over TLC plate. Ensure concentration at the origins by pressing lightly around the holes.
7. Place TLC plate in apparatus.
8. Electrophorese at 1.5 kV for about 30–60 min.
9. Dry plate in a well-ventilated place at room temperature, overnight.

3.2. Performic Acid Oxidation (see Note 2)

1. Prepare performic acid by mixing 19 mL formic acid with 1 mL 30% (w/v) hydrogen peroxide. The reaction is spontaneous. **Care!**
2. Place the dry electrophoresis sheet/plate in a container where it can be supported without touching the sides.
3. Place the performic acid in a shallow dish inside the container. Close the container and leave to oxidize for 2–3 h. (Note the marker dyes change from blue to green.)
4. Dry sheet thoroughly at room temperature overnight.

3.3. Second Dimension Electrophoresis

3.3.1. Paper Electrophoresis

1. To prepare for the second dimension, individual strips from the first dimension can be machine zigzag stitched onto a second sheet. The overlap of the second sheet should be carefully excised with a razor blade/scalpel.
2. Wet the sheet with electrophoresis buffer, applying the buffer along both sides of the sample, thus concentrating the peptides in a straight line.
3. Repeat electrophoresis, at right angles to the original direction.
4. Thoroughly dry sheet, as before.

3.3.2. Thin-Layer Electrophoresis

1. Wet TLC plate with electrophoresis buffer using two sheets of prewetted Whatman 3 MM paper on either side of sample line.
2. Repeat electrophoresis at right angles to the original direction.
3. Thoroughly dry plate as before.

3.4. Visualization

Peptide spots can be seen after reaction with fluorescamine.

1. The reaction should be carried out under alkaline conditions. The sheet or plate should be dipped in a solution of triethylamine (1% [v/v]) in acetone. This should be carried out at least twice if the electrophoresis buffer employed was acidic. Dry the sheet well.
2. Dip the sheet in a solution of fluorescamine in acetone (1 mg/100 mL).
3. Allow most of the acetone to evaporate.
4. View the map under a UV lamp at 300–365 nm (**Care:** Goggles must be worn). Peptides and amino-containing compounds fluoresce. Encircle all fluorescent spots with a soft pencil.

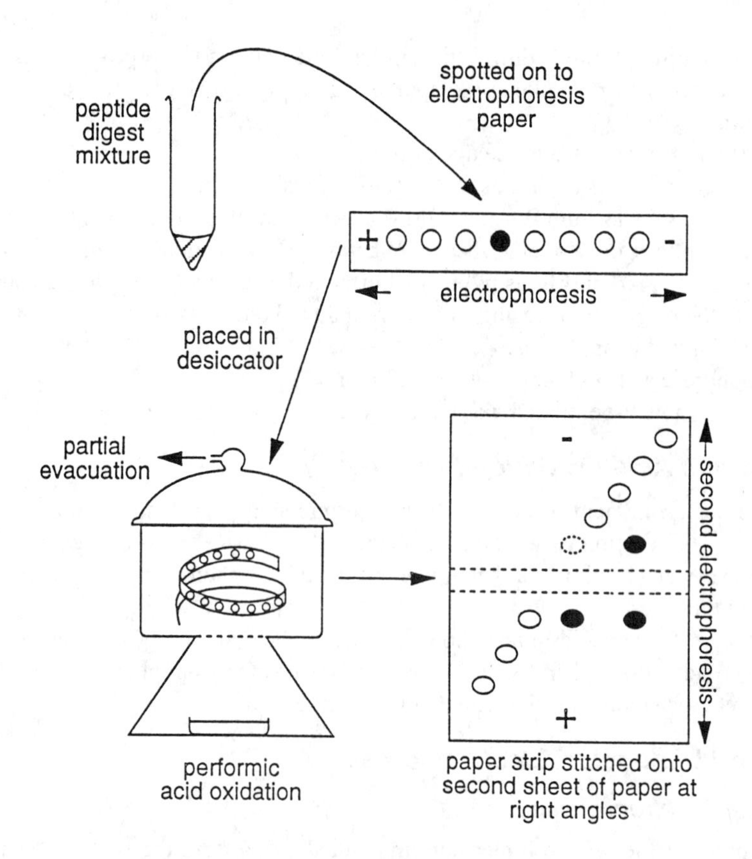

Fig. 1. Diagonal electrophoresis for identification and purification of peptides containing cysteine or disulfide bonds. This figure shows that peptides unaltered by the performic acid treatment (open circles) have the same mobility in both dimensions and therefore lie on a diagonal. Peptides that contain cysteine or were previously covalently linked (closed circles) produce one or two spots respectively, that lie off the diagonal after oxidation.

5. Interpretation of diagonal maps: Figure 1 shows that peptides unaltered by the performic acid treatment have the same mobility in both dimensions and therefore lie on a diagonal. Peptides that contain cysteine or were previously covalently linked produce one or two spots, respectively, that lie off the diagonal after oxidation.

3.5. Elution

Peptides may be eluted from the paper using, for example, 0.1*M* NH_3 or 25% acetic acid. Peptides may be extracted from TLC plates by scraping off the spot into an Eppendorf tube containing elution buffer. This should be vortexed for 5 min and then centrifuged for 5 min. The cellulose can then be re-extracted once or twice with the same buffer to ensure optimal recovery of peptide.

4. Notes

1. The buffer of choice for the initial analysis is pH 6.5. However, if the cysteine residues have already been blocked with iodoacetate (*see* Chapter 51), the pH 3.5 buffer is very useful, since peptides containing these residues lie slightly off the diagonal, being slightly

more acidic in the second dimension after the performic acid oxidation.
2. The movement of the marker dyes will enable progress of the electrophoresis to be followed to ensure that the samples do not run off the end of the paper.
3. It is important to exclude halide ions rigorously, since these are readily oxidized to halogens, which will react with histidine, tyrosine, and phenylalanine residues in the protein.

References

1. Brown, J. R., and Hartley, B. S. (1966) Location of disulphide bridges by diagonal paper electrophoresis *Biochem. J.* **101,** 214–228.
2. Michl, H. (1951) Paper electrophoresis at potential differences of 50 volts per centimetre. *Monatschr. Chem.* **82,** 489–493.
3. Creighton, T. E. (1983) Disulphide bonds between cysteine residues, in *Protein Structure—a Practical Approach* (Rickwood, D. and Hames B. D., eds.), IRL, Oxford, UK, pp. 155–167.

82

Estimation of Disulfide Bonds Using Ellman's Reagent

Alastair Aitken and Michèle Learmonth

1. Introduction

Ellman's reagent 5,5'-dithiobis(2-nitrobenzoic acid) (DTNB) was first introduced in 1959 for the estimation of free thiol groups *(1)*. The procedure is based on the reaction of the thiol with DTNB to give the mixed disulfide and 2-nitro-5-thiobenzoic acid (TNB) which is quantified by the absorbance of the anion (TNB^{2-}) at 412 nm.

The reagent has been widely used for the quantitation of thiols in peptides and proteins. It has also been used to assay disulfides present after blocking any free thiols (e.g., by carboxymethylation) and reducing the disulfides prior to reaction with the reagent *(2,3)*. It is also commonly used to check the efficiency of conjugation of sulfhydryl-containing peptides to carrier proteins in the production of antibodies.

2. Materials

1. Reaction buffer: 0.1*M* phosphate buffer, pH 8.0.
2. Denaturing buffer: 6*M* guanidinium chloride, 0.1*M* Na_2HPO_4, pH 8.0 (*see* Note 1).
3. Ellman's solution: 10 m*M* (4 mg/mL) DTNB (Pierce, Chester, UK) in 0.1*M* phosphate buffer, pH 8.0 (*see* Note 2).
4. Dithiothreitol (DTT) (Boerhinger or Calbiochem) solution: 200 m*M* in distilled water.

3. Methods

3.1. Analysis of Free Thiols

1. It may be necessary to expose thiol groups, which may be buried in the interior of the protein. The sample may therefore be dissolved in reaction buffer or denaturing buffer. A solution of known concentration should be prepared with a reference mixture without protein. Sufficient protein should be used to ensure at least one thiol per protein molecule can be detected; in practice, at least 2 nmol of protein (in 100 µL) are usually required.
2. Sample and reference cuvets containing 3 mL of the reaction buffer or denaturing buffer should be prepared and should be read at 412 nm. The absorbance should be adjusted to zero (A_{buffer}).
3. Add 100 µL of buffer to the reference cuvet.
4. Add 100 µL of Ellman's solution to the sample cuvet. Record the absorbance (A_{DTNB}).
5. Add 100 µL of protein solution to the reference cuvet.

From: The Protein Protocols Handbook
Edited by: J. M. Walker Humana Press Inc., Totowa, NJ

6. Finally, add 100 μL protein solution to the sample cuvet, and after thorough mixing, record the absorbance until there is no further increase. This may take a few minutes. Record the final reading (A_{final}).
7. The concentration of thiols present may be calculated from the molar absorbance of the TNB anion. (*See* Note 3.)

$$\Delta A_{412} = E_{412}TNB^{2-}[RSH] \quad (1)$$

Where $\Delta A_{412} = A_{final} - (3.1/3.2)(A_{DTNB} - A_{buffer})$
and $E_{412}\ TNB^{2-} = 1.415 \times 10^4\ cm^{-1}M^{-1}$.

If using denaturing buffer, use the value $E_{412}\ TNB^{2-} = 1.37 \times 10^4\ cm^{-1}\ M^{-1}$.

3.2. Analysis of Disulfide Thiols

1. Sample should be carboxymethylated (see Chapter 51) or pyridethylated (*see* Chapter 54) without prior reduction. This will derivatize any free thiols in the sample, but will leave intact any disulfide bonds.
2. The sample (at least 2 nmol of protein in 100 μL, is usually required) should be dissolved in 6*M* guanidinium HCl, 0.1*M* Tris-HCl pH 8.0 or denaturing buffer, under a nitrogen atmosphere.
3. Add freshly prepared DTT solution to give a final concentration of 10–100 m*M*. Carry out reduction for 1–2 h at room temperature.
4. Remove sample from excess DTT by dialysis for a few hours each time, with two changes of a few hundred mL of the reaction buffer or denaturing buffer (*see* Section 3.1.). Alternatively, gel filtration into the same buffer may be carried out.
5. Analysis of newly exposed disulfide thiols can thus be carried out as described in Section 3.1.

4. Notes

1. It is not recommended to use urea in place of guanidinium HCl, since this can readily degrade to form cyanates, which will react with thiol groups.
2. Unless newly purchased, it is usually recommended to recrystallize DTNB from aqueous ethanol.
3. Standard protocols for use of Ellman's reagent often give $E_{412}\ TNB^{2-} = 1.36 \times 10^4\ cm^{-1}M^{-1}$. A more recent examination of the chemistry of the reagent indicates that these are more suitable values *(4)*, and these have been used in this chapter.

References

1. Ellman, G. L. (1959) Tissue sulfhydryl groups. *Arch. Biochem. Biophys.* **82,** 70–77.
2. Zahler, W. L. and Cleland, W. W. (1968) A specific and sensitive assay for disulfides. *J. Biol. Chem.* **243,** 716–719.
3. Anderson, W. L. and Wetlaufer, D. B. (1975) A new method for disulfide analysis of peptides. *Anal. Biochem.* **67,** 493–502.
4. Riddles P. W., Blakeley, R. L., and Zerner, B. (1983) Reassessment of Ellman's reagent. *Methods Enzymol.* **91,** 49–60.

83

Quantitation of Cysteine Residues and Disulfide Bonds by Electrophoresis

Alastair Aitken and Michèle Learmonth

1. Introduction

Amino acid analysis quantifies the molar ratios of amino acids per mole of protein. This generally gives a nonintegral result, yet clearly there are integral numbers of the amino acids in each protein. A method was developed by Creighton *(1)* to count integral numbers of amino acid residues, and it is particularly useful for the determination of cysteine residues. Sulfhydryl and disulfide groups are of great structural, functional, and biological importance in protein molecules. For example, the Cys sulfhydryl is essential for the catalytic activity of some enzymes (e.g., thiol proteases) and the interconversion of Cys SH to Cystine S—S is directly involved in the activity of protein disulfide isomerase *(2)*. The conformation of many proteins is stabilized by the presence of disulfide bonds *(3)*, and the formation of disulfide bonds is an important posttranslational modification of secretory proteins *(4)*.

Creighton's method exploits the charge differences introduced by specific chemical modifications of cysteine. A similar method was first used in the study of immunoglobulins by Feinstein in 1966 *(5)*. Cys residues may be reacted with iodoacetic acid, which introduces acidic carboxymethyl ($^{-}O_2CCH_2S$—) groups, or with iodoacetamide, which introduces the neutral carboxyamidomethyl (H_2NCOCH_2S—) groups. The reaction with either reagent is essentially irreversible, thereby producing a stable product for analysis. Using a varying ratio of iodoacetamide/iodoacetate, these acidic and neutral agents will compete for the available cysteines, and a spectrum of fully modified protein molecules having 0,1,2, ... n acidic carboxymethyl residues per molecule is produced (where n is the number of cysteine residues in the protein). These species will have, correspondingly, n, $n-1$, $n-2$, ... 0 neutral carboxyamidomethyl groups. These species may then be separated by electrophoresis, isoelectric focusing, or by a combination of both *(1,6,7)*. The examples of the analysis of the cysteine residues in bovine pancreatic trypsin inhibitor and ovotransferrin are shown in Fig. 1.

Creighton used a low-pH discontinuous system *(1)*. Takahasi and Hirose recommend a high-pH system *(6)*, whereas Stan-Lotter and Bragg used the Laemmli electrophoresis system followed by isoelectric focusing *(7)*. It may therefore be necessary to carry out preliminary experiments to find the best separation conditions for the protein under analysis. The commonly used methods are given below.

From: The Protein Protocols Handbook
Edited by: J. M. Walker Humana Press Inc., Totowa, NJ

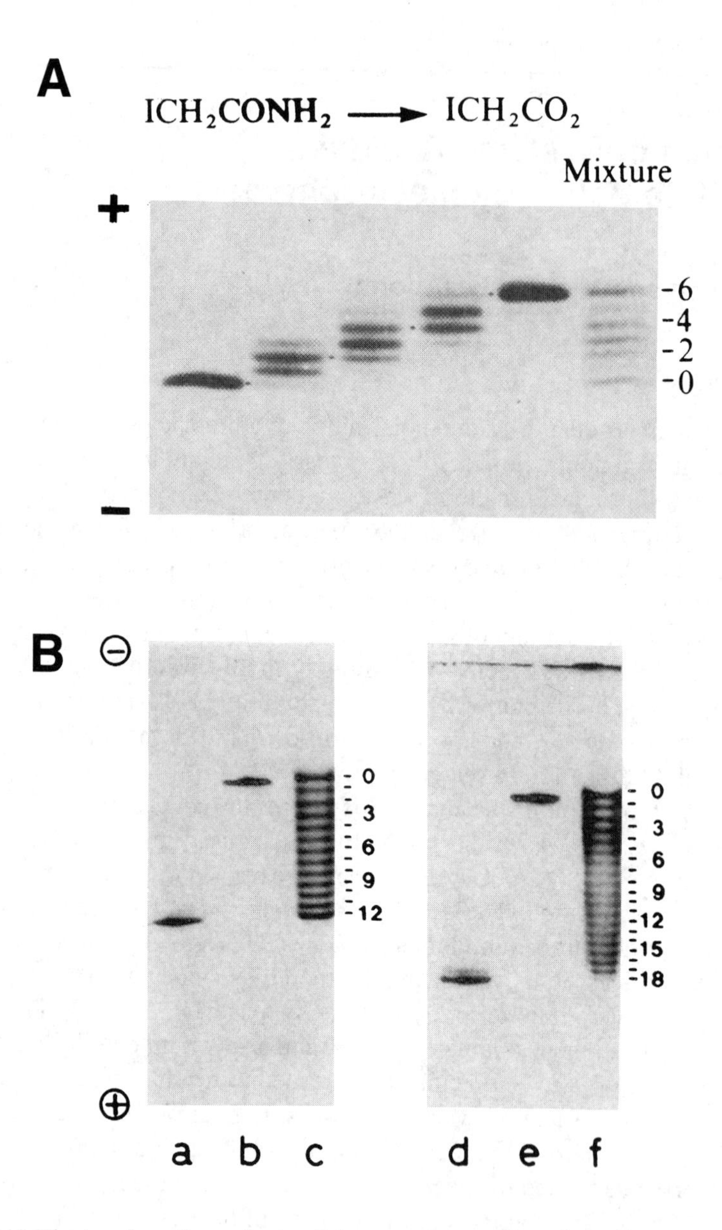

Fig. 1. (A) Electrophoretic analysis of the cysteine residues in bovine pancreatic trypsin inhibitor *(1)* with six cysteine residues, run with the low-pH system. Lanes 1–5 contain (respectively) samples reacted with neutral iodoacetamide; 1:1; 1:3; 1:9 ratios of neutral to acidic reagent; acidic iodoacetate. Lane 6 contains a mixture of equal portions of the samples in lanes 1–5. **(B)** Electrophoretic analysis of the cysteine residues in the N-terminal (lanes a–c) and the C-terminal (lanes d–f) domains of ovotransferrin *(6)* run with the high pH system. The subunits contain 13 and 19 cysteines, respectively. Lanes a and d contain samples alkylated with iodoacetic acid. Lanes b and e contain samples alkylated with iodoacetamide. Lanes c and e contain mixtures of the samples alkylated with different ratios of neutral to acidic reagent.

In order to ensure that all thiol groups are chemically equivalent, the reactions must be carried out in denaturing (in the presence of urea) and reducing (in the presence of dithiothreitol, DTT) conditions. The electrophoretic separation must also be carried out with the unfolded protein (i.e., in the presence of urea) in order that the modification has the same effect irrespective of where it is in the polypeptide chain.

The original method has been modified into a two-stage process to allow for the quantification of both sulfhydryl groups and disulfide bonds (*see* Notes 1 and 2) *(6,8)*. The principle of the method has also been adapted to counting the numbers of lysine residues after progressive modification of the ε-amino groups with succinic anhydride, which converts this basic group to a carboxylic acid-containing moiety.

2. Materials

2.1. Reaction Solutions

1. 1*M* Tris-HCl, pH 8.0.
2. 0.1*M* EDTA, pH 7.0.
3. 1*M* DTT (good-quality, e.g., Calbiochem, Nottingham, UK).
4. 8*M* urea (BDH [Poole, UK], Aristar-grade, *see* Note 2).
5. Solution A: 0.25*M* iodoacetamide, 0.25*M* Tris-HCl, pH 8.0.
6. Solution B: 0.25*M* iodoacetic acid, prepared in 0.25*M* Tris-HCl, pH readjusted to 8.0 with 1*M* KOH.

2.2. Solutions for Electrophoretic Analysis in the Low-pH System (pH 4.0) (9)

1. 30% acrylamide solution containing 30 g acrylamide, 0.8 g bis-acrylamide **(extreme caution: work in fume hood)**, made up to 100 mL with distilled water.
2. 10% acrylamide solution containing 10 g acrylamide **(extreme caution: work in fume hood)** and 0.25 g bis-acrylamide made up to 100 mL with distilled water.
3. Low-pH buffer (eight times concentrated stock) for separating gel; 12.8 mL glacial acetic acid, 1 mL *N,N,N'N'*-tetramethylethylenediamine (TEMED), 1*M* KOH (approx 35 mL) to pH 4.0, made up to 100 mL with distilled water.
4. Low pH buffer (8 times concentrated) for stacking gel; 4.3 mL glacial acetic acid, 0.46 mL TEMED, 1*M* KOH to pH 5.0, to 100 mL with distilled water.
5. 4 mg riboflavin/100 mL water.
6. Low-pH buffer for electrode buffer; dissolve 14.2 g β-alanine in ~800 mL water then adjust to pH 4.0 with acetic acid. Make up to a final volume of 1 L with distilled water.
7. Tracking dye solution (five times concentrated); 20 mg methyl green, 5 mL water, and 5 g glycerol.

2.3. Gel Solution Recipes for Low-pH Electrophoresis (pH 4.0) (see Note 3)

1. 30 mL separating gel (10% acrylamide, photopolymerized with riboflavin) is made up as follows: 10 mL 30% acrylamide stock, 4 mL pH 4.0 buffer stock, 3 mL riboflavin stock, 14.7 g urea, water (approx 2.5 mL) to 30 mL . Degas on a water vacuum pump (to remove oxygen which inhibits polymerization).
2. 8 mL stacking gel (2.5% acrylamide, photopolymerized with riboflavin) is made up with: 2 mL 10% acrylamide stock, 1 mL pH 5.0 buffer stock, 1 mL riboflavin stock, 3.9 g urea, water (approx 1.2 mL) to 8 mL. Degas.

2.4. Electrophoresis Buffers for High-pH Separation (pH 8.9)

1. 30% acrylamide solution containing 30 g acrylamide, 0.8 g bis-acrylamide **(extreme caution: work in fume hood)**, made up to 100 mL with distilled water.
2. 10% acrylamide solution containing 10 g acrylamide **(extreme caution: work in fume hood)** and 0.25 g bis-acrylamide made up to 100 mL with distilled water.
3. High-pH buffer (four times concentrated stock) for separating gel; 18.2 g Tris base (in ~40 mL water), 0.23 mL TEMED, 1*M* HCl to pH 8.9, made up to 100 mL with distilled water.
4. High-pH buffer (four times concentrated) for stacking gel; 5.7 g Tris base (in ~40 mL water), 0.46 mL TEMED, 1*M* H_3PO_4 to pH 6.9, made up to 100 mL with distilled water.
5. 4 mg riboflavin in 100 mL water
6. 10% ammonium persulfate solution (consisting of 0.1 g ammonium persulfate in 1 mL water).
7. High-pH buffer for electrode buffer; 3 g Tris base, 14.4 g glycine, distilled water to 1 L.
8. Tracking dye solution (five times concentrated): 1 mL 0.1% Bromophenol blue, 4 mL water, and 5 g glycerol.

2.5. Gel Solution Recipes for High-pH Electrophoresis (see Note 3)

1. 30 mL separating gel (7.5% acrylamide, polymerized with ammonium persulfate) is made up with: 7.5 mL 30% acrylamide stock, 7.5 mL pH 8.9 buffer stock, 0.2 mL 10% ammonium persulfate (add immediately before casting), 14.7 g urea, water (approx 4.5 mL) to 30 mL. Degas.
2. 8 mL stacking gel (2.5% acrylamide, photopolymerized with riboflavin) is made up with: 2 mL 10% acrylamide stock, 1 mL pH 6.9 buffer stock, 1 mL riboflavin stock solution, 3.9 g urea, water (approx 1.2 mL) to 8 mL. Degas.

3. Methods

3.1. Reduction and Denaturation

1. To a 0.2-mg aliquot of lyophilized protein add 10 µL of each of the solutions containing 1*M* Tris-HCl, pH 8.0, 0.1*M* EDTA, and 1*M* DTT (*see* Note 4).
2. Add 1 mL of the 8*M* urea solution (*see* Note 2).
3. Mix and incubate at 37°C for at least 30 min.

3.2. Reaction

1. Freshly prepare the following solutions using solutions A and B listed in Section 2.
   a. Mix 50 µL of solution A with 50 µL solution B (to give solution C).
   b. Mix 50 µL of solution A with 150 µL solution B (to give solution D).
   c. Mix 50 µL of solution A with 450 µL of solution B (to give solution E).
2. Label six Eppendorf tubes 1–6.
3. Add 10 µL of solutions A, B, C, D, and E to tubes 1–5. Reserve tube 6.
4. Add 40 µL of denatured, reduced protein solution prepared as in Section 3.1. to each of tubes 1–5.
5. Gently mix each tube and leave at room temperature for 15 min. Thereafter, store on ice.
6. After the 15 min incubation period, place 10 µL aliquots from each of tubes 1–5 into tube 6. Mix.

The samples are now ready for analysis (*see* Note 5).

3.3. Electrophoretic Analysis

1. 50 µL aliquots of each sample, Labeled 1–6, mixed with 12 µL of appropriate tracking dye solution, are loaded onto successive lanes of a polymerized high- or low-pH gel, set up in a suitable slab gel electrophoresis apparatus.
2. Low-pH buffer system: Electrophoresis is carried out toward the negative electrode, using a current of 5–20 mA for each gel, overnight at 8°C.
3. High-pH buffer system *(10)*; Electrophoresis is carried out toward the positive electrode at 10–20 mA per gel (or 100–180V) for 3–4 h.
4. Electrophoresis is stopped when the tracking dye reaches bottom of the gel.
5. Proteins are visualized using conventional stains e.g., Coomassie blue (Pierce, Chester, UK), silver staining (*see* Chapters 11 and 35).

4. Notes

1. The method of Takahashi and Hirose *(6)* can be used to categorize the half-cystines in a native protein as:
   a. Disulfide bonded;
   b. Reactive sulfhydryls; and
   c. Nonreactive sulfhydryls.

   In the first step, the protein sulfhydryls are alkylated with iodoacetic acid in the presence and absence of 8*M* urea. In the second step, the disulfide bonded sulfhydryls are fully reduced and reacted with iodoacetamide. The method described above is then used to give a ladder of half-cystines so that the number of introduced carboxymethyl groups can be quantified.
2. Urea is an unstable compound; it degrades to give cyanates which may react with protein amino and thiol groups. For this reason, the highest grade of urea should always be used, and solutions should be prepared immediately before use.
3. Electrophoresis in gels containing higher or lower percent acrylamide may have to be employed depending on the molecular weight of the particular protein being studied.
4. Where protein is already in solution, it is important to note that the pH should be adjusted to around 8.0, and the DTT and urea concentrations should be made at least 10 m*M* and 8*M,* respectively.
5. Other ratios of iodoacetic acid to iodoacetamide may need to be used if more than about eight cysteine residues are expected, since a sufficiently intense band corresponding to every component in the complete range of charged species may not be visible. A greater ratio of iodoacetic acid should be used if the more acidic species are too faint (and vice versa).

References

1. Creighton, T. E. (1980) Counting integral numbers of amino acid residues per polypeptide chain. *Nature* **284,** 487,488
2. Freedman R. B., Hirst, T. R., and Tuite, M. F. (1994) Protein disulphide isomerase: building bridges in protein folding. *Trends Biochem. Sci.* **19,** 331–336
3. Creighton, T. E. (1989) Disulphide bonds between cysteine residues, in *Protein Structure—a Practical Approach* (Creighton, T. E., ed.), IRL, Oxford, pp. 155–167.
4. Freedman, R. B. (1984) Native disulphide bond formation in protein biosynthesis; evidence for the role of protein disulphide isomerase. *Trends Biochem. Sci.* **9,** 438–441.
5. Feinstein, A. (1966) Use of charged thiol reagents in interpreting the electrophoretic patterns of immune globulin chains and fragments. *Nature* **210,** 135–137.

6. Takahashi, N. and Hirose, M. (1990) Determination of sulfhydryl groups and disulfide bonds in a protein by polyacrylamide gel electrophoresis. *Anal. Biochem.* **188,** 359–365.
7. Stan-Lotter, H. and Bragg, P. (1985) Electrophoretic determination of sulfhydryl groups and its application to complex protein samples, *in vitro* protein synthesis mixtures and cross-linked proteins. *Biochem. Cell Biol.* **64,** 154–160.
8. Hirose, M., Takahashi, N., Oe, H., and Doi, E. (1988) Analyses of intramolecular disulfide bonds in proteins by polyacrylamide gel electrophoresis following two-step alkylation. *Anal. Biochem.* **168,** 193–201.
9. Reisfield, R. A., Lewis, U. J., and Williams, D. E. (1962) Disk electrophoresis of basic proteins and peptides on polyacrylamide gels. *Nature* **195,** 281–283.
10. Davis, B. J. (1964) Disk electrophoresis II method and application to human serum proteins. *Ann. N.Y Acad. Sci.* **21,** 404–427.

84

Detection of Disulfide-Linked Peptides by Mass Spectrometry

Alastair Aitken and Michèle Learmonth

1. Introduction

Mass spectrometry is playing a rapidly increasing role in protein chemistry and sequencing (*see* Chapter 91) and is particularly useful in determining sites of co- and posttranslational modification *(1,2)*, and application in locating disulfide bonds is no exception. This technique can of course readily analyze peptide mixtures. Therefore, it is not always necessary to isolate the constituent peptides. A combination of microsequencing and mass spectrometry techniques is now commonly employed for complete covalent structure determination. On-line electrospray mass spectrometry (ESMS) coupled to capillary electrophoresis or high-performance liquid chromatography (HPLC) has proven particularly valuable in the identification of modified peptides *(3)*. Recent developments in ESMS sources permit on-line microbore HPLC using matrices, such as 10-µm Poros resins slurry-packed into columns <0.25 mm in diameter. Polypeptides can be separated on gradients of 5–75% acetonitrile over 2 min in formic or acetic acid (0.1%) *(4)*.

Some fragmentation may be obtained in ESMS with capillary-skimmer source fragmentation, when analyzing pure peptides, but more sequence information has been generally obtained using tandem mass spectrometry after collision-induced dissociation *(5)*. However, an ion trap quadrupole MS (called "LCQ") has just been launched by Finnegan-MAT, which permits sequence information to be readily obtained. Not only can MS-MS analysis be carried out, but owing to the high efficiency of each stage, further fragmentation of selected ions may be carried out to MS^n. The charge state of peptide ions is readily determined by a "zoom-scan" technique, which resolves the isotopic envelopes of multiply charged peptide ions. The instrument still allows accurate mol-wt determination to over 100,000 Dalton at 0.01% mass accuracy.

Selective detection of modified peptides is possible on ESMS. For example, phosphopeptides can be identified by production of phosphate-specific fragment ions of 63 Dalton (PO_2^-) and 79 Dalton (PO_3^-) by collision-induced dissociation during negative ion HPLC-ESMS. This technique of selective detection of posttranslational modifications through collision-induced formation of low-mass fragment ions that serve as characteristic and selective markers for the modification of interest has been extended to identify other modifications such as glycosylation, sulfation, and acylation *(6)*.

From: The Protein Protocols Handbook
Edited by: J. M. Walker Humana Press Inc., Totowa, NJ

2. Materials

Materials for proteolytic and chemical cleavage of proteins are described in Chapters 63–67 and 71.

3. Method

For Detection Of Disulfide-Linked Peptides By Mass Spectrometry.

1. Peptides can be generated by any suitable proteolytic or chemical method that minimize disulfide exchange (e.g., acid pH, *see* Note 1). Partial acid hydrolysis, although nonspecific, has been successfully used in a number of instances. If the peptides are then analyzed by standard techniques on fast-atom bombardment (FAB) MS *(7)*. The use of thioglycerol and mixtures of related compounds in the matrix should be avoided for obvious reasons. Despite this, even with glycerol alone as the matrix compound, for example, disulfide bonds will be partially reduced during the analysis, and peaks corresponding to the individual components of the disulfide-linked peptides will be observed. Control samples with the above reagents (employing the same matrix compounds) are essential to avoid misleading results owing to additional matrix-derived peaks (*see* Note 2).
2. The peptide mixture is incubated with reducing agents, such as mercaptoethanol and DTT, and reanalyzed as before. Peptides that were disulfide linked disappear from the spectrum and reappear at the appropriate positions for the individual components. For example, in the positive ion mode the mass (M) of disulfide-linked peptides (of individual masses A and B) will be detected as the pseudomolecular ion at $(M + H)^+$, and after reduction, this will be replaced by two additional peaks for each disulfide bond in the polypeptide at masses $(A + H)^+$ and $(B + H)^+$. Remember that $A + B = M + 2$, since reduction of the disulfide bond will be accompanied by a consistent increase in mass owing to the conversion of cystine to two cysteine residues, i.e., —S—S → —SH + HS— and peptides containing an intramolecular disulfide bond will appear at 2 amu higher. Such peptides, if they are in the reduced state, can normally be readily reoxidized to form intramolecular disulfide bond by bubbling a stream of air through a solution of the peptide for a few minutes (*see* Fig. 1 and Note 3).
3. Combined with computer programs that will predict the cleavage position of any particular proteinase or chemical reagent, simple knowledge of the mass of the fragment will, in most instances, give unequivocal answers as to which segments of the polypeptide chain are disulfide-linked. If necessary, one cycle of Edman degradation can be carried out on the peptide mixture, and the FAB MS analysis repeated. The shift in mass(es) that correlates with loss of specific residues will confirm the assignment. Recent developments in techniques, such as ladder sequencing, are proving extremely useful (*see* Note 4).

4. Notes

1. Fragmentation of proteins into peptides under low-pH conditions is important (e.g., with pepsin, Glu-C, or cyanogen bromide[CBNr]) to prevent disulfide exchange. Analysis can also be carried out by electrospray mass spectrometers, which will give an accurate mol wt up to 80–100,000 Dalton (and in favorable cases up to 150,000 Dalton has been claimed). The increased mass of 2 Dalton for each disulfide bond will, in all probability, be too small to obtain an accurate estimate for polypeptide of mass greater than ca. 10,000 (accuracy obtainable is >0.01%). It is unlikely that the lower resolution obtained with laser desorption time-of-flight mass spectrometers (ca. 0.1% up to 20,000 Dalton) would permit a meaningful analysis.

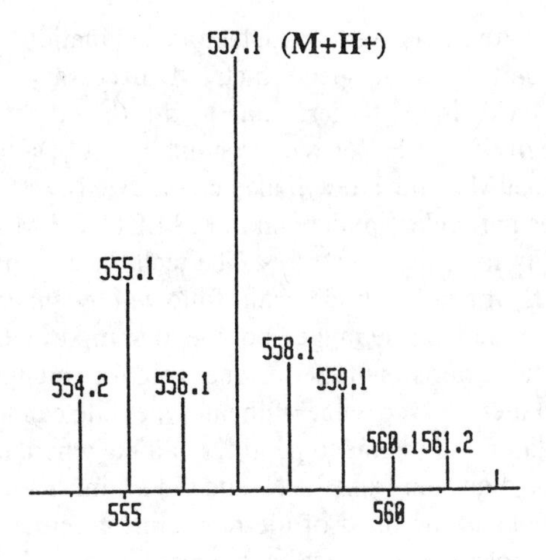

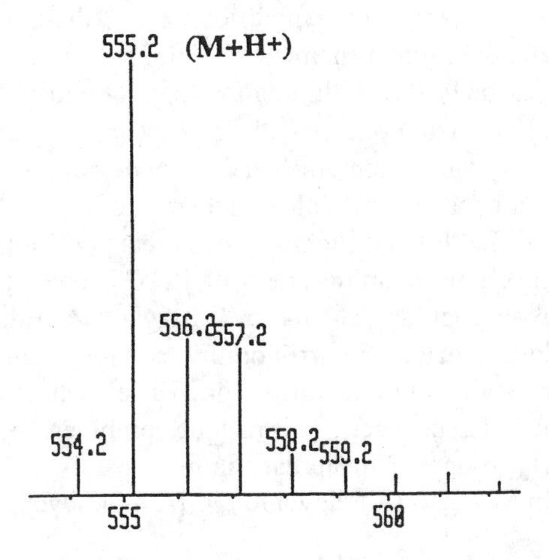

Fig. 1. FAB mass spectrum of a peptide containing an internal disulfide bond. The analysis of the peptide Cys-Ser-Gly-Ser-Thr-Cys was done on a VG 70-250 SE mass spectrometer with a glycerol/1% TFA matrix, using a cesium ion source. Both spectra were run in the positive ion mode *(3)*. Upper panel shows the linear peptide, in reducing conditions. Lower panel shows the peptide oxidized with a stream of air. Pseudomolecular ions $(M + H)^+$ are 557.1 and 555.2 Dalton (2 Dalton lower,) respectively.

2. Mass analysis by ESMS *(3,8)* and MALDITOF *(9)* is affected, seriously in some cases, by the presence of particular salts, buffers, and detergents. In some cases, using nonionic saccharide detergents such as *n*-dodecyl-β-D-glucopyranoside, improvements in signal-to-noise ratios of peptides and proteins were observed *(9)*. The effect on ESMS sensitivity of different buffer salts, detergents, and tolerance to acid type may vary widely with the instrument and particularly with the ionization source. Nonionic saccharide detergents, such as *n*-dodecyl-β-D-glucopyranoside at the 0.01% level gave the best signals in ESMS, accompanied by a shift of the ions to higher charge states (in contrast to the effect predicted by droplet surface tension). Critical micelle concentration is not a good predictor of how well a surfactant will perform *(8)*.

3. The difference of 2 Dalton may allow satisfactory estimation of the number of intramolecular disulfide bonds by mass spectrometry. If necessary a larger mass difference may be generated by oxidation with performic acid (*10*, and *see* Chapter 52). This will cause a mass increase of 48 Dalton for each cysteine and 49 Dalton for each half-cystine residue. (Remember that Met and Trp will also be oxidized).
4. Ladder sequencing has particular application in MALDI TOF MS, which has high sensitivity and greater ability to analyze mixtures. The technique involves the generation of a set of nested fragments of a polypeptide chain followed by analysis of the mass of each component. Each component in the ragged polypeptide mixture differs from the next by loss of a mass that is characteristic of the residue weight (which may involve a modified side chain). In this manner, the sequence of the polypeptide can be read from the masses obtained in MS. The ladder of degraded peptides can be generated by Edman chemistry *(11)* or by exopeptidase digestion from the N- and C-termini. This is essentially a subtractive technique (one looks at the mass of the remaining fragment after each cycle). For example, when a phosphoserine residue is encountered, a loss of 167.1 Dalton is observed in place of 87.1 for a serine residue . This technique therefore avoids one of the major problems of analyzing posttranslational modifications. Although the majority of modifications are stable during the Edman chemistry, *O*- or *S*-linked esters, for example (which are very numerous), may be lost by β-elimination (e.g., *O*-phosphate) during the cleavage step to form the anilinothioazolidone or undergo *O*- (or *S*- in the case of palmitoylated cysteine) to *N*-acyl shifts, which block further Edman degradation.

   Exopeptidase digestion may be difficult, and the rate of release of amino acid may vary greatly. The use of modified Edman chemistry has great possibilities *(12)*. The modification consists of carrying out the coupling step with PITC in the presence of a small amount of phenylisocyanate, which acts as a chain-terminating agent. A development of this technique involves the addition of volatile trifluorethylisothiocyanate (TFEITC) to the reaction tube to which a fresh aliquot of peptide is added after each cycle. This avoids steps to remove excess reagent and byproducts. This may be combined with subsequent modification of the terminal NH_2 group with quaternary ammonium alkyl NHS esters, which allows increased sensitivity in MALDI-TOF down to the 1-pmol. level.

Acknowledgment

We thank Steve Howell (NIMR) for the mass spectrometry analysis.

References

1. Burlingame, A. L., Boyd, R. K., and Gaskell, S. J. (1994) Mass spectrometry. *Anal. Chem.* **66,** 634R–683R.
2. Mann, M. and Wilm, M. (1995) Electrospray mass spectrometry for protein characterisation. *Trends Biochem. Sci.* **20,** 219–224.
3. Kay, I. and Mallet, A. I. (1993) Use of an on-line liquid chromatography trapping column for the purification of protein samples prior to electrospray mass spectrometric analysis *Rapid Commun. Mass Specrom.* **7,** 744–746.
4. Aitken, A., Howell, S., Jones, D., Madrazo, J., and Patel, Y. (1995) 14-3-3 α and δ are the phosphorylated forms of Raf-activating 14-3-3 β and ζ. *In vivo* stoichiometric phosphorylation in brain at a Ser-Pro-Glu-Lys motif. *J. Biol. Chem.* **270,** 5706–5709.
5. Hunt, D. F., Yates, J. R., Shabanowitz, J., Winston, S., and Hauer, C. R. (1986) Protein sequencing by tandem mass spectrometry. *Proc. Natl. Acad. Sci. USA* **83,** 6233–6237.
6. Bean, M. F., Annan, R. S., Hemling, M. E., Mentzer, M., Huddleston, M. J., and Carr, S. A. (1994) LC-MS methods for selective detection of posttranslational modifications in pro-

teins, in *Techniques in Protein Chemistry VI* (Crabb, J. W., ed.), Academic, New York, pp. 107–116.

7. Morris, R. H. and Pucci, P. (1985). A new method for rapid assignment of S—S bridges in proteins. *Biochem. Biophys. Res. Commun.* **126,** 1122–1128.
8. Loo, R., Dales, N., and Andrews, P. C. (1994) Surfactant effects on protein structure examined by electrospray ionisation mass spectrometry. *Protein Sci.* **3,** 1975–1983.
9. Vorm, O., Chait, B. T., and Roepstorff, P. (1993) Mass spectrometry of protein samples containing detergents. Proceedings of the 41st ASMS Conference on Mass Spectrometry and Allied Topics, pp. 621,622.
10. Sun, Y. and Smith, D. L. (1988) Identification of disulfide-containing peptides by performic acid oxidation and mass spectrometry. *Anal. Biochem.* **172,** 130–138.
11. Chait, B. T., Wang, R., Beavis, R. C., and Kent, S. B. H. (1993) Protein ladder sequencing. *Science* **262,** 89–92.
12. Bartlet-Jones, M., Jeffery, W. A. Hansen, H. F., and Pappin, D. J. C. (1994) Peptide ladder sequencing by mass spectrometry using a novel, volatile degradation reagent. *Rapid Commun. Mass Spectrom.* **8,** 737–742.

85

Analyzing Protein Phosphorylation

John Colyer

1. Introduction

Protein phosphorylation is a ubiquitous modification used by eukaryotic cells to alter the function of enzymes, ion channels, and other proteins in response to extracellular stimuli, or mechanical or metabolic change within the cell. In many instances, phosphorylation results in a change in the catalytic activity of the phosphoprotein, which influences one particular aspect of cellular physiology, thereby allowing the cell to respond to the initiating stimulus. A number of different residues within a protein can be modified by phosphorylation. Serine, threonine, and tyrosine residues can be phosphorylated on the side chain hydroxyl group (*o*-phosphoamino acids), whereas others become phosphorylated on nitrogen atoms (*N*-phosphoamino acids, lysine, histidine, and arginine). The former group are involved in dynamic "regulatory" functions and have been studied extensively *(1)*, whereas the latter group may perform both structural/catalytic roles and signaling functions, the study of which has occurred more recently *(2)*. The disparity in our understanding of the role of *o*- and *N*-phosphoamino acids is in part a consequence of the acid lability of *N*-phosphoamino acids, which leads to their destruction during the analysis of many phosphorylation experiments.

In terms of the process of studying an individual phosphoprotein, a number of key issues can be identified. First, one must demonstrate that phosphorylation of the protein takes place; then define the number of sites within the primary sequence that can be phosphorylated and by which protein kinase; identify the individual residue(s) phosphorylated; the functional implication of phosphorylation of each site; and describe the use of each site of phosphorylation in vivo. This chapter aims to describe the conduct of an experiment performed to identify a protein as a phosphoprotein. In the case of oligomeric enzymes, it will identify the subunit(s) phosphorylated by a particular kinase. The determination of the stoichiometry of phosphorylation is also described, which provides the first information concerning the number of phosphorylation sites within a polypeptide. These procedures are most straightforward if one has access to the purified protein kinase of interest and the protein substrate. In the case of the kinase, this can be served in many instances by a number of commercial sources, but the approach may be limited by the availability of sufficient pure protein substrate. If this is the case, phosphorylation of a particular target as part of a complex mixture of proteins (e.g., whole-cell extract) can be performed. Under these conditions, identification

From: The Protein Protocols Handbook
Edited by: J. M. Walker Humana Press Inc., Totowa, NJ

of the protein of interest will require exploitation of a unique electrophoretic property of the protein *(3)* or require purification of the protein by immunoprecipitation or other comparable affinity-interaction means prior to electrophoresis. The identification of the protein as a phosphoprotein can thereby be achieved, although analysis of the stoichiometry of phosphorylation in this way is inadvisable. In each case, the experimental procedure has a common design: an in vitro phosphorylation reaction is followed by separation of the phosphoproteins by SDS-PAGE and subsequent identification by autoradiography. The incorporation of labeled phosphate can be determined by excising the phosphoprotein band from the dried gel, scintillation counting this gel piece, and converting ^{32}P cpm into molar terms from the knowledge of the specific activity of the initial ATP stock, and the amount of protein substrate analyzed.

2. Materials

1. Purified and partially purified multifunctional protein kinases can be obtained from several commercial sources. The availability of a number of enzymes is illustrated in Table 1. The list is not exhaustive, and inclusion in the table does not constitute endorsement of the product:
2. Phosphorylation buffer for the catalytic subunit of protein kinase A (c-PKA): 50 m*M* Histidine-KOH, pH 7.0, 5 m*M* $MgSO_4$, 5 m*M* NaF, 100 n*M* c-pKA, 100 μ*M* ATP.
   a. Histidine-KOH, pH 7.0 is prepared as a concentrated stock, 200 m*M* stored at 4°C for 1 mo, or –20°C for >12 mo. (Warm to 30°C for 30 min to dissolve histidine following storage at –20°C.)
   b. $MgSO_4$, EGTA, NaF: all prepared as 100-m*M* stock, stable at 4°C >12 mo.
   c. ATP, nonradioactive: 20- or 100-m*M* stock (pH corrected to 7.0 with KOH) stable at –20°C for >12 mo. Aliquot to avoid repeat freeze-thaw cycles.
   d. c-pKA, M_r 39,000 *(4)*, sources Table 1: stable at –70°C for ~12 mo. Avoid dilute solutions—enzyme tends to aggregate and inactivate.
3. γ-^{32}P-ATP, ICN Pharmaceuticals: dispense into small aliquots and store –20°C. Avoid freeze to thaw cycles. $T_{1/2}$ ~14 d; discard 1 mo after reference date.
4. SDS-PAGE sample buffer, (double-strength): 125 m*M* Tris-HCl, pH 6.8, 20% glycerol, 2% SDS, 10% 2-mercaptoethanol, 0.01% bromophenol blue, stable at –20°C >12 mo. Aliquot and avoid freeze-thaw cycles.
5. Filter paper: Whatman 3MM.
6. X-ray film, X-ray cassettes, intensifying screens, developer and fixative: X-Ograph Ltd.; developer and fixative are stored at room temperature, are reusable, and are stable for ~2 wks.
7. Scintillation fluid: Emulsifier-safe, Packard.

3. Method

3.1. Phosphorylation Reaction Using Purified Protein Substrates

1. In a designated radioactive area with appropriate acrylic screening prepare a stock of radioactive ATP. The addition of 50 μCi [γ-^{32}P]-ATP to a 0.5-mL solution of 1 m*M* ATP will produce a suitable experimental ATP stock of 220 cpm/pmol (*see* Note 1). Warm to 37°C.
2. Incubate the purified protein of interest (0.1–1.0 mg/mL) in the phosphorylation assay medium lacking ATP. Allow the sample to warm to 37°C for 2 min (*see* Note 2). A number of control samples should be set up in parallel. One control should contain target protein, but no exogenous kinase, and another, kinase but no target protein.

Table 1
Commercial Source of Multifunctional Protein Kinases

Kinase	Source
Protein kinase A	*a,c,d,e*
Protein kinase C	*a,b,c,d*
Calmodulin kinase II	*e*
Casein kinase II	*a,b,d*
CDC2 kinase	*d*
src kinase	*d*

[a]Boehringer Mannheim.
[b]Calbiochem-Novabiochem.
[c]Sigma Chemical Co.
[d]TCS Biologicals Ltd.
[e]PhosphoProtein Research.

3. Start the phosphorylation reaction by the addition of ATP, containing [γ-^{32}P]-ATP (as defined in step 1 above). Cap the tube and vortex briefly. Follow the phosphorylation as a function of time by removing aliquots of the reaction at specific points in time, every 20 s for the first minute, and then at 60-s intervals for the next 4 min.
4. Terminate the reaction by mixing the sample with an equal volume of double-strength SDS-PAGE sample buffer at room temperature. Dispense the phosphorylation sample into an Eppendorf tube containing an equal volume of double-strength sample buffer, cap the tube, and vortex briefly. Store these samples at room temperature, behind appropriate screens, until all samples have been collected.
5. Incubate all samples for 30 min at 37°C. Perform SDS-PAGE in a gel of suitable acrylamide composition, loading a minimum of 5 µg pure target protein/lane (details of electrophoresis in Chapter 11). Allow the dye front to migrate off the bottom of the gel (depositing most of the radioactivity into the electrode buffer), stain with Coomassie brilliant blue, and destain the gel (*see* Chapter 11).
6. Mount the gel on filter paper and cover with clingfilm. The filter paper should be wet with unused destain solution prior to contact with the SDS-PAGE gel. Lower the gel onto the wet filter paper slowly, and flatten to remove air bubbles. Cover with clingfilm, and dry using a vacuum-assisted gel drier for 2 h at 90°C.
7. Once dry, an X-ray film should be placed in contact with the gel for a protracted period to image the location of phosphoproteins. **This procedure must be performed in the dark,** although light emitted by dark room safety lamps is permitted. An X-ray film is first exposed to a conditioning flash of light from a flash gun. Hold a single piece of X-ray film and the flash gun 75 cm apart. Set the flash gun to the minimum power output, and discharge a single flash directly onto the film (*see* Note 3). Place the film on top of a clean intensifying screen within an X-ray cassette. Take the dried SDS-PAGE gel, still sandwiched between filter paper and clingfilm, and place it gel side down onto the X-ray film. Do not allow the gel to move once in contact with the film, and use adhesive tape to secure the contact. With a permanent marker pen, draw distinctive markings from the filter paper backing of the gel onto the X-ray film to facilitate orientation of film and gel once autoradiography is complete (*see* Note 4). Close the X-ray cassette, label the cassette with experimental details, including the current date and time, and store at –70°C.

8. After 16 h of exposure, the autoradiograph can be developed. Remove the cassette from the –70°C freezer, and allow at least 30 min for it to thaw. Once in the dark room, with safety light illumination only, the cassette should be opened, and SDS-PAGE gel removed from the X-ray film. The film should be placed in 2 L developer and agitated for 4 min at room temperature, in the dark. Using plastic forceps, the film is removed from developer, rinsed in water, and then agitated for a further 90 s in 2 L of fixative (room temperature, dark). At the end of 90 s, the autoradiograph is no longer light-sensitive, and normal lighting can be resumed. Wash the autoradiograph extensively, for 10 min in a constant flow of water, and allow to air-dry.
9. Identification of phosphorylated polypeptides can be performed by superimposition of autoradiograph on the SDS-PAGE gel (*see* Note 5). A uniform, almost transparent background should be achieved, with phosphorylated proteins identified as black bands on the autoradiograph of variable intensities (depending on the level of phosphorylation), but regular width and shape (*see* Note 6). Exposure times can be altered in the light of results obtained, and repeated autoradiographs of various durations performed on a single gel (*see* Note 7).

3.2. Phosphorylation of Components in Complex Protein Mixtures

1. The phosphorylation experiment is performed largely as detailed in Section 3.1. with the following modifications. A protein concentration of 1–5 mg/mL is recommended supplemented with 100 n*M* purified c-pKA and 10 μ*M* adenosine cyclic 3,5-monophosphate, and with 1 μ*M* microcystin-LR for additional Ser/Thr phosphatase control.
2. Perform a phosphorylation time-course experiment and process as described in Section 3.1. If the identity of a particular protein cannot be gauged from a peculiar electrophoretic feature (e.g., dissociation of oligomer to monomer on boiling; *3*), then an affinity purification step must be introduced prior to electrophoresis.
3. In this case, phosphorylation will be terminated by placing the sample on ice. Immunoprecipitation of the protein of interest should be performed as detailed in Chapter 87, Section 3.3.) taking care to solubilize membrane proteins effectively if they are of interest. Immunoprecipitates should be processed as described in Section 3.1., step 5 onward (*see* Note 8).

3.3. Determination of Phosphorylation Stoichiometry

1. The specific activity of the ATP (cpm/pmol) needs to be determined empirically. At the time of the phosphorylation experiment, dilute a sample of the experimental ATP (1 m*M* containing 100 μCi/mL [γ-^{32}P]-ATP, as described in Section 3.1., step 1) to 1 μ*M* by serial dilution in water. Dispense triplicate 10-μL aliquots of the 1-μ*M* ATP (10 pmol) into separate scintillation vials, and add 4.6 mL scintillation fluid to each. Cap and label 10 pmol ATP (*see* Note 9).
2. Perform steps 1–9 of Section 3.1. To excise the phosphorylated protein bands from the dry SDS-PAGE gel, identify the location by superimposition of gel and autoradiograph. With a marker pen, outline the autoradiographic limits of the phosphoprotein on the gel. Overlay again to confirm the accuracy of demarkation. Mark similar-sized areas of gel that do not contain phosphoproteins to determine background [^{32}P]. Excise these gel pieces with scissors, remove the clingfilm, and place the acrylamide piece and filter paper support in a scintillation vial. Add 4.6 mL scintillation fluid, and cap the vial.
3. Scintillation count each vial for 5 min or longer, using a program defined for ^{32}P radionucleotides. Minimal quenching of ^{32}P occurs under these conditions.

4. To calculate the pmol phosphate incorporated/μg protein, subtract the background radioactivity (cpm) from experimental data (cpm) to obtain phosphate incorporation into the protein sample (in units of cpm). Convert this to pmol incorporation/μg using the formula:

$$\text{Phosphorylation (pmol/}\mu\text{g protein)} = [\text{protein phosphorylation (cpm)}/\text{SA of ATP (cpm/pmol)} \times \mu\text{g protein}] \quad (1)$$

   With a knowledge of the molecular weight of the polypeptide, a molar phosphorylation stoichiometry can be calculated from these data using the formula:

$$\text{Phosphorylation (mol/mol protein)} = [\text{phosphorylation (pmol/}\mu\text{g protein)} \times \text{molecular weight}/10^6] \quad (2)$$

5. Experimental conditions that result in maximal protein phosphorylation will have to be optimized. Parameters worth considering include alteration of the pH of the reaction, extension of the time-course of phosphorylation (up to several hours), addition of extra protein kinase during the phosphorylation process, and addition of extra ATP throughout the time-course.

4. Notes

1. The specific activity of ATP must be tailored to the experiment intended. Phosphorylation and autoradiography of proteins require ≥200 cpm/pmol, studies that require analysis beyond this point (e.g., phosphoamino acid identification) require ~2000 cpm/pmol, while phosphorylation of peptides requires ~20–50 cpm/pmol.
2. The stability of proteins at low concentration is sometimes an issue; inclusion of an irrelevant protein, but not a phosphoprotein (e.g., bovine serum albumin) at 1 mg/mL (final) is recommended. The phosphorylation example used to illustrate this method (c-pKA) displays catalytic activity in the absence of signaling molecules. In other instances this is not so. Therefore, relevant activators should be included as dictated by the kinase (e.g., Ca^{2+}, calmodulin, acidic phospholipids, and so forth).
3. Film developed at this stage will exhibit very slight discoloration compared to unexposed film.
4. Phosphorescent labels (Sigma-Techware) can be used to label an SDS-PAGE gel prior to autoradiography. It will also facilitate superimposition of gel and autoradiograph. These can also highlight the position of mol-wt markers, an image of which will be captured on the X-ray film.
5. Protein kinases invariably autophosphorylate. This can be identified clearly in control samples lacking phosphorylation target (Section 3.1., step 2).
6. Autoradiographs sometimes have a high background signal. Uniform black coloration over the whole film, extending beyond the area exposed to the gel is indicative of illumination of the X-ray film. Discoloration of part of the film is indicative of light entering the X-ray cassette. Examine the cassette carefully, particularly the corners that are prone to damage by rough handling.
7. Phosphoimage technology represents an alternative to autoradiography, it has the advantage of collecting the image quickly. Exposure times are reduced by an order of magnitude. However, in my experience, this benefit is at the expense of the quality of the image, which is granular.
8. The time required for immunoprecipitation or similar procedure should be kept to a minimum to limit the dephosphorylation of proteins by endogenous phosphatase enzymes. A cocktail of phosphatase inhibitors should be included for the same reason.
9. The hydrolysis of ATP to ADP and Pi occurs at a low rate in the absence of any enzyme. The extent of hydrolysis of [γ-^{32}P]-ATP can affect the determination of phosphorylation

stoichiometry, since it will result in the overestimation of the specific activity of ATP if a correction is not made. Quantification of the purity of ATP is quoted in the product specification from suppliers. Only fresh [γ-^{32}P]-ATP should be used in these procedures or the degree of hydrolysis confirmed by thin-layer chromatography *(6)*.

References

1. Krebs, E. G. (1994) The growth of research on protein phosphorylation. *Trends Biochem. Sci.* **19,** 439.
2. Swanson, R. V., Alex, L. A., and Simon, M. I. (1994) Histidine and aspartate phosphorylation: two component systems and the limits of homology. *Trends Biochem. Sci.* **19,** 485–490.
3. Drago, G. A. and Colyer, J. (1994) Discrimination between two sites of phosphorylation on adjacent amino acids by phosphorylation site-specific antibodies to phospholamban. *J. Biol. Chem.* **269,** 25,073–25,077.
4. Peters, K. A., Demaille, J. G., and Fischer, E. H. (1977) Adenosine 3':5'-monophosphate dependent protein kinase from bovine heart. Characterisation of the catalytic subunit. *Biochemistry* **16,** 5691–5697.
5. Otto, J. J. and Lee, S. W. (1993) Immunoprecipitation methods. *Methods Cell Biol.* **37,** 119–127.
6. Bochner, B. R. and Ames, B. N. (1982) Complete analysis of cellular nucleotides by two dimensional thin layer chromatography. *J. Biol. Chem.* **257,** 9759–9769.

86

Identification of Proteins Modified by Protein (D-Aspartyl/L-Isoaspartyl) Carboxyl Methyltransferase

Darin J. Weber and Philip N. McFadden

1. Introduction

The several classes of *S*-adenosylmethionine-dependent protein methyltransferases are distinguishable by the type of amino acid they modify in a substrate protein. The protein carboxyl methyltransferases constitute the subclass of enzymes that incorporate a methyl group into a methyl ester linkage with the carboxyl groups of proteins. Of these, protein (D-aspartyl/L-isoaspartyl) carboxyl methyltransferase, EC 2.1.1.77 (PCM) specifically methyl esterifies aspartyl residues that through age-dependent alterations are in either the D-aspartyl or the L-isoaspartyl configuration *(1,2)*. There are two major reasons for wishing to know the identity of protein substrates for PCM. First, the proteins that are methylated by PCM in the living cell, most of which have not yet been identified, are facets in the age-dependent metabolism of cells. Second, the fact that PCM can methylate many proteins in vitro, including products of overexpression systems, can be taken as evidence of spontaneous damage that has occurred in these proteins since the time of their translation.

The biggest hurdle in identification of substrates for PCM arises from the extreme base-lability of the incorporated methyl esters, which typically hydrolyze in a few hours or less at neutral pH. Thus, many standard biochemical techniques for separating and characterizing proteins are not usefully applied to the identification of these methylated proteins. In particular, the electrophoresis of proteins by the most commonly employed techniques of sodium dodecyl sulfate polyacrylamide gel electrophoresis (SDS-PAGE) results in a complete loss of methyl esters incorporated by PCM, owing to the alkaline pH of the buffers employed. Consequently, a series of systems employing polyacrylamide gel electrophoresis at acidic pH have been utilized in efforts to identify the substrates of PCM. A pH 2.4 SDS system *(3)* using a continuous sodium phosphate buffering system has received the most attention *(1,3–14)*. The main drawback of this system is that it produces broad electrophoretic bands. Acidic discontinuous gel systems using cationic detergents, *(15)*, have proven useful in certain situations *(16–21)* and can be recommended if the cationic detergent is compatible with other procedures that might be utilized by the investigator (e.g., immunoblotting, protein sequencing). Recently, we have developed an electrophoresis system that employs SDS

From: The Protein Protocols Handbook
Edited by: J. M. Walker Humana Press Inc., Totowa, NJ

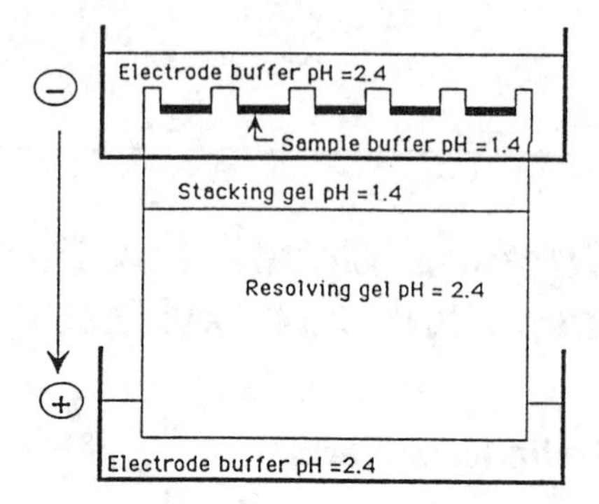

Fig. 1. Schematic of acidic discontinuous gel system. The system employs a pH 1.4 stacking gel on top of a pH 2.4 resolving gel with chloride as the leading ion and acetate as the trailing ion to stack the proteins tightly. The presence of the anionic detergent SDS allows the separation of proteins on the basis of molecular weight. The low pH preserves labile protein methyl esters, and so allows the identification of age-altered substrates of PCM.

and an acidic discontinuous buffering system (Fig. 1). This procedure results in sharp electrophoretic bands and would be a good choice for investigators wishing to adhere to SDS as the anionic detergent. This system is described below, and examples of its ability to resolve proteins are provided.

2. Materials

2.1. Equipment

1. Slab gel electrophoresis unit: We have used the mini-gel electrophoresis units from Idea Scientific (Minneapolis, MN) (10 × 10 × 0.1 cm) with great success, as well as the Sturdier large-format gel system from Hoeffer Scientific Instruments (San Francisco, CA) (16 × 18 × 0.15 cm).
2. Electrophoresis power supply, constant current.
3. X-ray film and photo darkroom.
4. Scintillation counter.

2.2. Reagents

1. 40% (w/v) Acrylamide stock solution containing 37:1 ratio of acrylamide to *N,N'*-methylene-bis-acrylamide (Bio-Rad, Richmond, CA).
2. Resolving gel buffer: 0.1*M* Na H_2PO_4 (Sigma, St. Louis, MO), 2.0 % SDS (United States Biochemical, Cleveland, OH [USB], ultrapure), 6*M* urea [USB], ultrapure) pH 2.4, with HCl.
3. Modified Clark and Lubs buffer (C & L buffer): Add 25.0 mL of 0.2*M* NaCl to 26 mL of 0.2*M* HCl, and bring to a final volume of 0.1 L. The buffer pH should be ~1.4 *(22)* (*see* Note 1).
4. Stacking gel buffer (2X): 2.0% (w/v) SDS, 6.0*M* urea and C & L buffer such that the C & L buffer makes up 66% (v/v) of the total volume with the remaining 34% consisting of water and other buffer components. Buffer will be 0.033*M* in NaCl. Readjust pH to 1.4 with HCl.

5. Sample solubilization buffer (2X): 2.0% (w/v) SDS, 6.0*M* urea, 10% glycerol ([USB], ultrapure), 0.01% pyronin Y dye (Sigma), and C & L 33% (v/v) of the total volume, with the remaining 67% consisting of water and other buffer components. Buffer will be 0.0165*M* in NaCl. Readjust pH to 1.4 with HCl.
6. Electrode buffer (1X): 0.03*M* Na H_2PO_4, 0.1% SDS, 0.2*M* acetate, pH 2.4 with HCl.
7. Gel polymerization catalysts: 0.06% $FeSO_4$, 1.0% H_2O_2, 1.0% ascorbic acid, prepared fresh in separate containers.
8. Colloidal Coomassie G-250 protein stain stock solution: 125 g ammonium sulfate, 25 mL 86% phosphoric acid, 1.25 g Coomassie brilliant blue G-250 (Sigma), deionized water to 1.0 L. The dye will precipitate, so shake well immediately before use. Stable indefinitely at room temperature *(11)*.
9. Destain solution: 10% (v/v) acetic acid.
10. Fluorography solution: 1.0*M* sodium salicylate brought to pH 6.0 with acetic acid *(23)*.
11. X-ray film: Kodak X-Omat AR or equivalent.

2.3. Gel Recipes

2.3.1. 12% Acrylamide Resolving Gel, pH 2.4

The following volumes are sufficient to prepare one 7.5 × 10.5 × 0.15 cm slab gel:

2X Resolving gel buffer	3.75 mL
40% Acrylamide (37:1)	2.25 mL
0.06% $FeSO_4$	0.06 mL
1.0% Ascorbic acid	0.06 mL
0.3% H_2O_2	0.06 mL

2.3.2. 4.0% Stacking Gel, pH 1.4

The following volumes are sufficient to prepare one 3.0 × 10.5 × 0.15 cm stacking gel:

2X Stacking gel buffer	3.75 mL
40% Acrylamide (37:1)	0.75 mL
0.06% $FeSO_4$	0.06 mL
1.0% Ascorbic acid	0.06 mL
0.3% H_2O_2	0.06 mL

3. Methods

3.1. Sample Solubilization

1. An equal volume of 2X solubilization buffer is added and the samples are heated in a 95°C heating block for no more than 30 s (*see* Note 2).
2. To separate samples under reducing conditions, it is critical to add any reducing agent before addition of the solubilization buffer. Up to 10 μg/gel protein band can be resolved with this system; total sample loaded in a single well should not exceed about 100 μg.

3.2. Gel Preparation

3.2.1. Resolving Gel

1. Add all the components listed under Section 2.3.1., except the 0.3% H_2O_2, together in a small Erlenmeyer side-arm flask.
2. Degas the solution for at least 5.0 min using an in-house vacuum.
3. Assemble together gel plates and spacers that have been scrupulously cleaned and dried.

4. Using a pen, make a mark 3.0 cm from the top of the gel plates to denote the space left for the stacking gel.
5. To the degassed gel solution, add the H_2O_2 catalyst. Gently mix solution by pipeting the solution in and out several times. Avoid introducing air bubbles into the solution.
6. Quickly pipet the acrylamide solution between the glass gel plates to the mark denoting 3.0 cm from the top of the gel plates.
7. Carefully overlay the acrylamide solution with about a 2.0-mm layer of water-saturated butanol using a Pasteur pipet, so that the interface will be flat on polymerization.
8. Allow the gel to polymerize at room temperature until a distinct gel–butanol interface is visible.
9. After polymerization is complete, pour off the overlay, and gently rinse the top of the gel with deionized water. Invert the gel on paper towels to blot away any remaining water between the gel plates.

3.2.2. Stacking Gel

1. Add all the components listed under Section 2.3.2., except the 0.3% H_2O_2, together in a small Erlenmeyer sidearm flask.
2. Degas the solution for at least 5.0 min using in-house vacuum.
3. Soak the well-forming combs in the H_2O_2 catalyst solution while preparing the stacking gel.
4. To the degassed gel solution, add the H_2O_2 catalyst. Gently mix solution by pipeting the solution in and out several times. Avoid introducing air bubbles into the solution.
5. Quickly pipet the acrylamide solution between the glass gel plates to the very top of the gel plates.
6. Remove the combs from the H_2O_2 catalyst solution, shake off some of the excess solution, and insert the comb in between the gel plates by angling the comb with one hand and guiding the comb between the plates with the other hand.
7. Ensure the comb is level relative to the top of the resolving gel and that no bubbles are trapped under the comb.
8. After the stacking gel has completely polymerized, carefully remove the comb. Remove any unpolymerized acrylamide from each well by rinsing with deionized water (*see* Note 3).

3.3. Electrophoresis

1. Assemble the gel in the electrophoresis unit.
2. Add sufficient electrode buffer to cover the electrodes in both the upper and lower reservoirs.
3. Remove any air bubbles trapped under the bottom edge of the plates with a bent 25-gage needle and syringe containing electrode buffer.
4. Rinse each well with electrode buffer immediately before adding samples.
5. Load all wells with 40–60 µL of protein samples; load 1X solubilization buffer into any empty wells.
6. Run the gels at 15 mA, constant current, at room temperature for 4–6 h, or until proteins of interest have been adequately resolved (*see* Note 4).
7. Fix the gel in 12% (v/v) TCA with gentle shaking for 30 min.
8. Pour off fixative, and mix 1 part methanol with 4 parts of colloidal Coomassie G-250 stock solution. Slowly shake gel with staining solution for ~12 h (*see* Note 5).
9. Pour off staining solution. Any nonspecific background staining of the gel can be removed by soaking the gel in 10% (v/v) acetic acid. Little destaining of protein bands occurs even after prolonged times in 10% acetic acid.

3.4. Detection of Radioactively Methylated Proteins

Several protocols exist for radioactively methylating proteins with PCM *(2,16)*. Following electrophoresis, gel bands containing radioactively labeled proteins can be detected by fluorography or scintillation counting of gel slices.

3.4.1. Fluorography of Gels

1. If gels have been stained, they are destained using 10% methanol/7% acetic acid to decolorize the gel bands.
2. Expose the gel to fluorography solution for 30 min at room temperature with gentle shaking.
3. The gels are then placed on a piece of filter paper and dried under vacuum without heat for 3 h.
4. In a dark room, X-ray film (Kodak X-Omat AR) is preflashed twice at a distance of 15 cm with a camera flash unit fitted with white filter paper (3M) to act as a diffuser.
5. The gel is placed in direct contact with the film and taped in place. For future alignment, puncture holes in an asymmetric pattern in a noncrucial area of the gel-film sandwich. A 25-gage needle is useful for this purpose.
6. After sealing in a film cassette, the cassette is wrapped with aluminum foil, and exposure takes place at –70°C for several weeks.
7. After exposure, remove the cassette from –70°C, and allow to warm to room temperature. Develop film in darkroom.
8. An example of this technique is shown in Fig. 2.

3.4.2. Scintillation Counting Radioactive Methanol Evolved by Base Hydrolysis of Protein Methyl Esters in Gel Slices

1. Stained gels are soaked in 10% acetic acid containing 3% (v/v) glycerol for 1.0 h, placed on a piece of filter paper, and dried under vacuum without heat for 3.0 h. The glycerol keeps the gel from cracking and keeps it pliable for the steps described below.
2. Using a ruler and fine-tip marking pen, a grid is drawn directly on the surface of the dried gel.
3. A sharp scalpel is then used to cut out uniform slices precisely from each gel lane. Alternatively, selected bands can be individually excised from the gel.
4. 4.0 mL of scintillation fluid are added to each 20-mL scintillation vial. A glass 1-dram vial is then placed inside the scintillation vial, carefully avoiding spilling scintillation fluid into the 1-dram vial.
5. Each dried gel slice is then placed in the inner dram vial of a scintillation vial.
6. After all the gel slices have been placed in a separate inner 1-dram vial, 0.3 mL of 0.2*M* sodium hydroxide is added to each inner vial. The scintillation vial is then immediately tightly capped and allowed to sit undisturbed for at least 3 h. Several hours are required for any volatile methanol that has formed by methyl ester hydrolysis to equilibrate and partition into the organic scintillation fluid, where it can then be detected by the scintillation counter.
7. Controls for measuring the efficiency of equilibration are performed by using a ^{14}C methanol standard, which is added to an inner 1-dram vial containing a nonradioactive gel slice and base hydrolysis solution. This is then placed inside a scintillation vial and allowed to equilibrate along with the other samples. An equal aliquot of the ^{14}C methanol standard is mixed directly with the scintillation fluid.
8. Figures 3 and 4 show examples of gels that have been sliced and counted in a scintillation counter under the conditions just described.

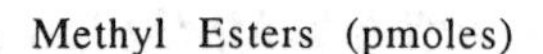

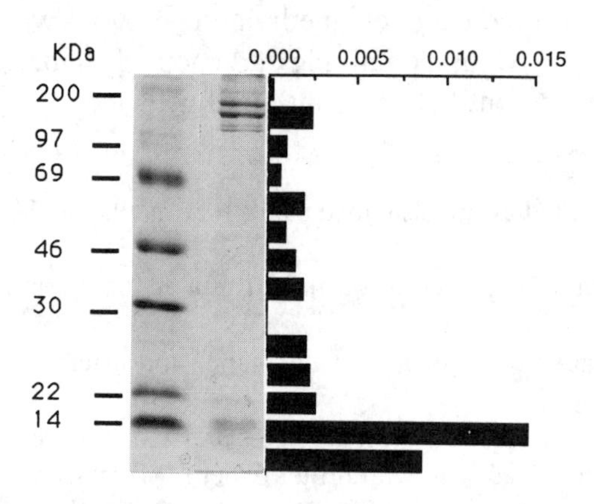

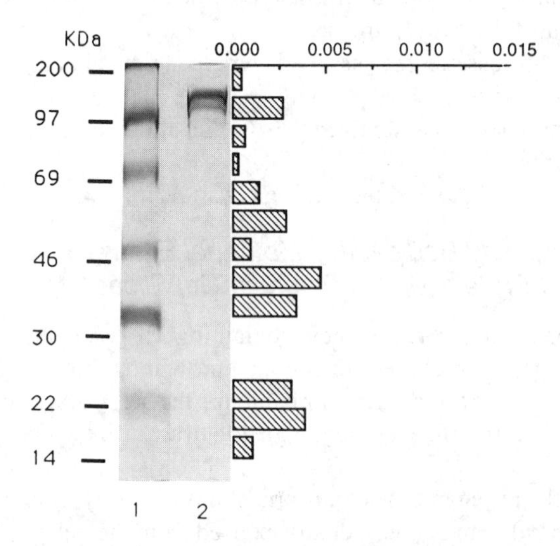

Fig. 2. Comparison of discontinuous acid gel with continuous acid gel. The following experiment tested for the presence of age-altered proteins in a commercial preparation of collagenase. The collagenase preparation (Sigma, type IV) was methylated with purified rabbit erythrocyte PCM and ^{3}H-AdoMet. Aliquots from the same methylation reaction were then resolved on (top) 12% discontinuous acid gel, described in text, or (bottom) 12% continuous acid gel system prepared according to the method of Fairbanks and Auruch *(3)*. Lane 1: Rainbow mol-wt markers (Amersham); Lane 2: 18 µg of methylated collagenase were loaded on each gel. Following electrophoresis and staining, 0.5-cm gel slices were treated with base to detect radioactivity as described under Section 3. Both gel systems are capable of preventing the loss of methyl esters from protein samples, but the discontinuous system provides much higher resolution of individual polypeptide bands.

4. Notes

1. The buffering system employed in the stacking gel and sample solubilization buffer is based on a modification of the C & L buffering system. NaCl is employed rather than KCl of the original system, because K^+ ions cause SDS to precipitate out of solution.

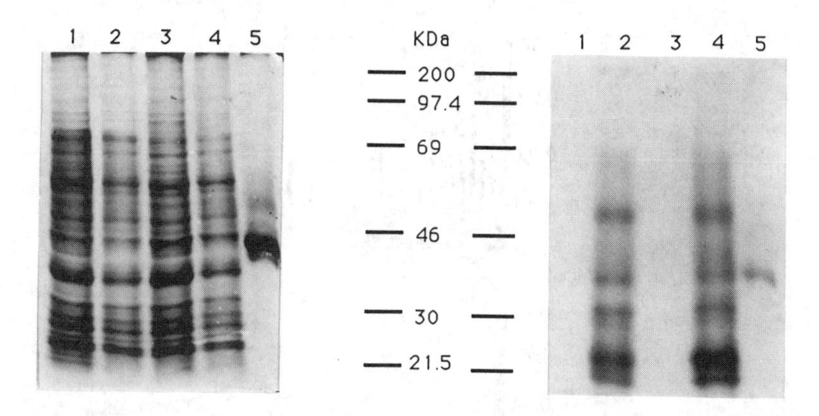

Fig. 3. Coomassie staining and autoradiography of complex protein mixtures by acidic discontinuous SDS gel electrophoresis. The following experiment was performed to measure the varieties of age-altered proteins in a cell cytoplasm. (A) Coomasie G-250 stained acidic discontinuous gel. (B) Autoradiogram of same gel. Lane 1: Cytoplasmic proteins, following incubation cells with ^{3}H-*S*-adenosyl-methione (AdoMet). Lane 2: Cytoplasmic proteins, following incubation of PC12 cytoplasm with ^{3}H-AdoMet. Lane 3: Cytoplasmic proteins, following incubation of intact PC12 cells with ^{3}H-AdoMet and purified rabbit PCM. Lane 4: Cytoplasmic proteins, following incubation of lysed PC12 cells with ^{3}H-AdoMet and purified rabbit PCM. Lane 5: Positive control; 20 μg of ovalbumin methylated with ^{3}H-AdoMet and purified rabbit PCM. Cells, lysates, and subfraction incubated with IU PCM, 300 pmol Adomet in final volume of 50 μL 0.2*M* citrate, pH 6.0, 20 min at 37°C.

2. On occasion, the solubilization buffer will contain precipitates. These can be brought back into solution by briefly heating the buffer at 37°C . The solubilization buffer is stable for at least 2 wk at room temperature; by storing the buffer in small aliqouts at –20°C, it is stable indefinitely.
3. Gels can be stored for up to two weeks at 4°C by wrapping them in damp paper towels and sealing tightly with plastic wrap.
4. Since the dye front is a poor indicator of protein migration, use of prestained mol-wt markers, such as the colored Rainbow markers from Amersham, allows the progress of protein separation to be monitored by simply identifying the colored bands, which are coded according to molecular weight.
5. It is essential that all fixing and staining steps occur at acidic pH. The colloidal Coomassie G-250 procedure described under Section 3. has several advantages: it is acidic, simple to perform, and has higher sensitivity than other dye-based staining methods, including those using Coomassie R-250. Additionally, nonspecific background staining is very low, so only minimal destaining is necessary to visualize protein bands.
6. Avoid using higher glycerol concentrations or prolonged incubation of the gel in this solution. Otherwise, the gel will be sticky after drying and contract sharply away from the paper backing on cutting.

References

1. Aswad, D. W. and Deight, E. A. (1983) Endogenous substrates for protein carboxyl methyltransferase in cytosolic fractions of bovine brain. *J. Neurochem.* **31,** 1702–1709.

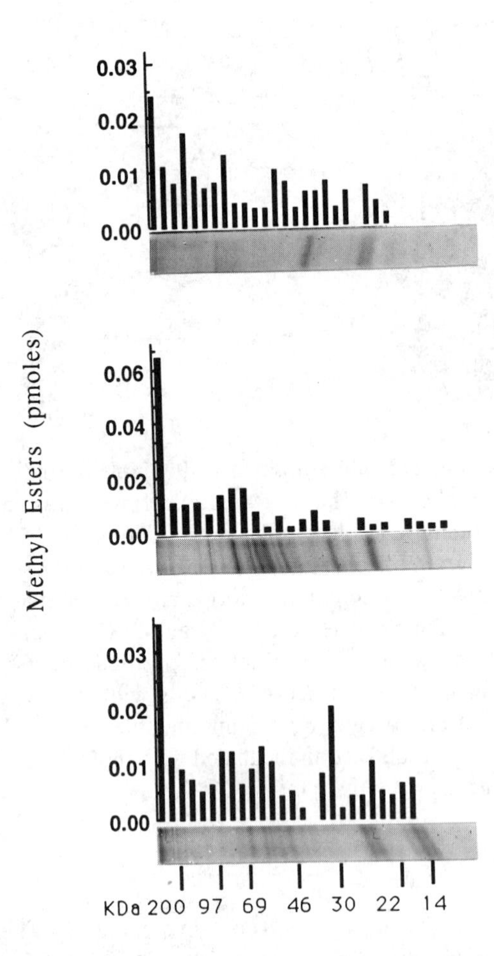

Fig. 4. Coomassie staining and radioactive methyl ester determination in gel slices of electrophoresed proteins from diseased human brain tissue. Extracts prepared from homogenates of Alzheimer's diseased brain (obtained from the Department of Pathology, Oregon Health Sciences University) were methylated in vitro with purified rabbit erythrocyte PCM and ^{3}H-AdoMet. Methylated proteins were then separated on 12% discontinuous, acidic gels, and radioactivity in each gel slice was quantified with scintillation counting as described under Section 3. Top: Distribution of methyl acceptor proteins in tissue protein that was insoluble in the nonionic detergent Triton X-100. Middle: Distribution of methyl acceptor proteins in tissue proteins that was soluble in an aqueous homogenization buffer. Bottom: Distribution of methyl acceptor proteins in crude homogenates of Alzheimer's diseased brains. Incubation conditions are similar to those described in Figure 3.

2. Lou, L. L. and Clarke, S. (1987) Enzymatic methylation of band 3 anion transporter in intact human erythrocytes. *Biochemistry* **26,** 52–59.
3. Fairbanks, G. and Avruch J. (1973) Four gel systems for electrophoretic fractionation of membrane proteins using ionic detergents. *J. Supramol. Struct.* **1,** 66–75.
4. Barber, J. R. and Clarke, S. (1984) Inhibition of protein carboxyl methylation by *S*-adenosyl-L-homocysteine in intact erythrocytes. *J. Biol. Chem.* **259(11),** 7115–7122.
5. Bower, V. E. and Bates, R. G. (1955) pH Values of the Clark and Lubs buffer solutions at 25°C. *J. Res. Natl. Bureau Stand.* **55(4),** 197–200.

6. Gingras, D., Menard, P., and Beliveau, R. (1991) Protein carboxyl methylation in kidney brush-border membranes. *Biochim. Biophys. Acta.* **1066,** 261–267.
7. Johnson, B. A., Najbauer, J., and Aswad, D. W. (1993) Accumulation of substrates for protein L-isoaspartyl methyltransferase in adenosine dialdehyde-treated PC12 cells. *J. Biol. Chem.* **268(9),** 6174–6181.
8. Johnson, B. A., Freitag, N. E., and Aswad, D. W. (1985) Protein carboxyl methyltransferase selectively modifies an atypical form of calmodulin. *J. Biol. Chem.* **260(20),** 10,913–10,916.
9. Lowenson, J. D. and Clarke, S. (1995) Recognition of isomerized and racemized aspartyl residues in peptides by the protein L-isoaspartate (D-aspartate) *O*-methyltransferase, in *Deamidation and Isoaspartate Formation in Peptides and Proteins.* (Aswad, D. W., ed.), CRC, Boca Raton, pp. 47–64.
10. McFadden, P. N., Horwitz, J., and Clarke, S. (1983) Protein carboxyl methytransferase from cow eye lens. *Biochem. Biophys. Res. Comm.* **113(2),** 418–424.
11. Neuhoff, V, Stamm, R., Pardowitz, I., Arold, N., Ehrhardt, W., and Taube, D. (1988) Essential problems in quantification of proteins following colloidal staining with Coomassie brilliant blue dyes in polyacrylamide gels, and their solutions. *Electrophoresis* **9,** 255–262.
12. O'Conner, C. M. and Clarke, S. (1985) Analysis of erythrocyte protein methyl esters by two-dimensional gel electrophoresis under acidic separating conditions. *Anal. Biochem.* **148,** 79–86.
13. O'Conner, C. M. and Clarke, S. (1984) Carboxyl methylation of cytosolic proteins in intact human erythrocytes. *J. Biol. Chem.* **259(4),** 2570–2578.
14. Sellinger, O. Z. and Wolfson, M. F. (1991) Carboxyl methylation affects the proteolysis of myelin basic protein by staphylococcus aureus V8 proteinase. *Biochim. Biophys. Acta.* **1080,** 110–118.
15. MacFarlane, D. E. (1984) Inhibitors of cyclic nucleotides phosphodiesterases inhibit protein carboxyl methylation in intact blood platelets. *J. Biol. Chem.* **259(2),** 1357–1362.
16. Aswad, D. W. (1995) Methods for analysis of deamidation and isoaspartate formation in peptides, in *Deamidation and Isoaspartate Formation in Peptides and Proteins* (Aswad, D. W., ed.), CRC, Boca Raton, pp. 7–30.
17. Freitag, C. and Clarke, S. (1981) Reversible methylation of cytoskeletal and membrane proteins in intact human erythrocytes. *J. Biol. Chem.* **256(12),** 6102–6108.
18. Gingras, D., Boivin, D., and Beliveau, R. (1994) Asymmetrical distribution of L-isoaspartyl protein carboxyl methyltransferases in the plasma membranes of rat kidney cortex. *Biochem. J.* **297,** 145–150.
19. O'Conner, C. M., Aswad, D. W., and Clarke, S. (1984) Mammalian brain and erythrocyte carboxyl methyltranserases are similar enzymes that recognize both D-aspartyl and L-isoaspartyl residues in structurally altered protein substrates. *Proc. Natl. Acad. Sci. USA* **81,** 7757–7761.
20. O'Conner, C. M. and Clarke, S. (1983) Methylation of erythrocyte membrane proteins at extracellular and intracellular D-aspartyl sites in vitro. *J. Biol. Chem.* **258(13),** 8485–8492.
21. Ohta, K., Seo, N., Yoshida, T., Hiraga, K., and Tuboi, S. (1987) Tubulin and high molecular weight microtubule-associated proteins as endogenous substrates for protein carboxyl methyltransferase in brain. *Biochemie* **69,** 1227–1234.
22. Barber, J. R. and Clarke, S. (1983) Membrane protein carboxyl methylation increase with human erythrocyte age. *J. Biol. Chem.* **258(2),** 1189–1196.
23. Chamberlain, J. P. (1979) Fluorographic detection of radioactivity in polyacrylamide gels with the water soluble fluor, sodium salicylate. *Anal. Biochem.* **98,** 132.

87

Analysis of Protein Palmitoylation

Morag A. Grassie and Graeme Milligan

1. Introduction

The incorporation of many membrane proteins into the lipid environment is based on sequences of largely hydrophobic amino acids that can form membrane-spanning domains. However, a number of other proteins are membrane-associated, but do not display such hydrophobic elements within their primary sequence. Membrane association in these cases is often provided by covalent attachment, either cotranslationally or posttranslationally, of lipid groups to the polypeptide chain. Acylation of proteins by either addition of C14:0 myristic acid to an N-terminal glycine residue or addition of C16:0 palmitic acid by thioester linkage to cysteine residues, in a variety of positions within the primary sequence, has been recorded for a wide range of proteins. Palmitoylation of proteins is not restricted to thioester linkage and may occur also through oxyester linkages to serine and threonine residues. Furthermore, thioester linkage of fatty acyl groups to proteins is not restricted to palmitate. Longer chain fatty acids, such as stearic acid (C18:0) and arachidonic acid (C20:4), have also been detected. Artificial peptide studies have provided evidence to support the concept that attachment of palmitate to a protein can provide sufficient binding energy to anchor a protein to a lipid bilayer, but that attachment of myristate is insufficient, in isolation, to achieve this.

Mammalian proteins that have been demonstrated to be palmitoylated include a range of G protein-coupled receptors and G protein α subunits, members of the Src family of nonreceptor tyrosine kinases, growth cone-associated protein GAP 43, endothelial nitric oxide synthase, spectrin, and glutamic acid decarboxylase. Since many of these proteins play central roles in information transfer across the plasma membrane of cells, there has been considerable interest in examining both the steady-state palmitoylation status of these proteins and, because the thioester linkage is labile, the possibility that it may be a dynamic, regulated process *(1–7)*. Palmitoylation thus provides a means to provide membrane anchorage for many proteins and, as such, can allow effective concentration of an enzyme or other regulatory protein at the two-dimensional surface of the membrane. Turnover of the protein-associated palmitate may regulate membrane association of polypeptides and, thus, their functions.

There has been considerable pharmaceutical interest in the development of small-mol-wt inhibitors of the enzyme farnesyl transferase, since attachment of the farnesyl

From: The Protein Protocols Handbook
Edited by: J. M. Walker Humana Press Inc., Totowa, NJ

group to the protooncogene p21^{ras} is integral both for its membrane association and transforming activities. Whether there will be similar interest in molecules able to interfere with protein palmitoylation is more difficult to ascertain as the wide range of proteins modified by palmitate is likely to limit specificity of such effects. It is true, however, that compounds able to interfere with protein palmitoylation are available *(8,9)*. These may prove to be useful experimental reagents in a wide range of studies designed to explore further the role of protein palmitoylation.

This chapter will describe methodology to determine if a protein expressed in a cell maintained in tissue culture is in fact modified by the addition of palmitate. Specific conditions are taken from our own experiences in analysis of the palmitoylation of G protein α subunits *(10–12)*, but should have universal relevance to studies of protein palmitoylation.

2. Materials

1. (9,10-^{3}H [*N*])palmitic acid (Dupont/NEN or Amersham International) (*see* Note 1) and Trans[^{35}*S*]-label (ICN Biomedicals, Inc.).
2. Growth medium for fibroblast derived cell lines: DMEM containing 5% newborn calf serum (NCS), 20 m*M* glutamine, 100 U/mL penicillin, and 100 mg/mL streptomycin (Life Technologies).
3. [^{3}H]palmitate labeling medium: as for growth medium with NCS replaced with 5% dialyzed NCS (*see* item 5 below), 5 m*M* Na pyruvate, and 150 μCi/mL (9,10-^{3}H [*N*]) palmitic acid (*see* Note 2). (9,10-^{3}H [*N*])palmitic acid is usually supplied at 1 μCi/mL in ethanol (*see* Note 3). Dry under N_2 in a glass tube to remove ethanol, and then redissolve in labeling medium to give a final concentration of 150 μCi/mL of medium
4. [^{35}*S*] methionine/cysteine labeling medium: 1 part growth medium, 3 parts DMEM lacking methionine and cysteine (Life Technologies) supplemented with 50 μCi/mL Trans[^{35}*S*]-label (ICN Biomedicals, Inc.) (*see* Note 4).
5. Dialyzed NCS: Prepare in dialysis tubing that has been boiled twice in 10 m*M* EDTA for 10 min (Note 5); 50 mL of NCS are dialyzed against 2 L of Earle's salts (6.8 g NaCl, 0.1 g KCl, 0.2 g $MgSO_4 \cdot 7H_2O$, 0.14 g NaH_2PO_4, 1.0 g glucose) over a period of 12–36 h with three changes of buffer. Remove serum from dialysis tubing and filter-sterilize before storing at –20°C in 2 mL aliquots until required.
6. Phosphate-buffered saline (PBS): 0.2 g KCl, 0.2 g KH_2PO_4, 8 g NaCl, 1.14 g $NaHPO_4$ (anhydrous) to 1000 mL with H_2O (pH should be in range of 7.0–7.4).
7. 1 and 1.33% (w/v) SDS.
8. TE buffer: 10 m*M* Tris-HCl, 0.1 m*M* EDTA, pH 7.5.
9. Solubilization buffer: 1% Triton X-100, 10 m*M* EDTA, 100 m*M* $Na_2H_2PO_4$, 10 m*M* NaF, 100 μ*M* Na_3VO_4, 50 m*M* HEPES, pH 7.2.
10. Immunoprecipitation wash buffer: 1% Triton X-100, 0.5% SDS, 100 m*M* NaCl, 100 m*M* NaF, 50 m*M* NaH_2PO_4, 50 m*M* HEPES, pH 7.2.
11. Gel solutions for 10% gels:
    a. Acrylamide: 30 g acrylamide, 0.8 g bis-acrylamide to 100 mL with H_2O.
    b. Buffer 1: 18.17 g Tris, 4 mL 10% SDS (pH 8.8) to 100 mL with H_2O.
    c. Buffer 2: 6g Tris, 4 mL 10% SDS (pH 6.8) to 100 mL with H_2O.
    d. 50% (v/v) Glycerol.
    e. 10% (w/v) Ammonium persulfate (APS).
    f. TEMED.
    g. 0.1% (w/v) SDS.

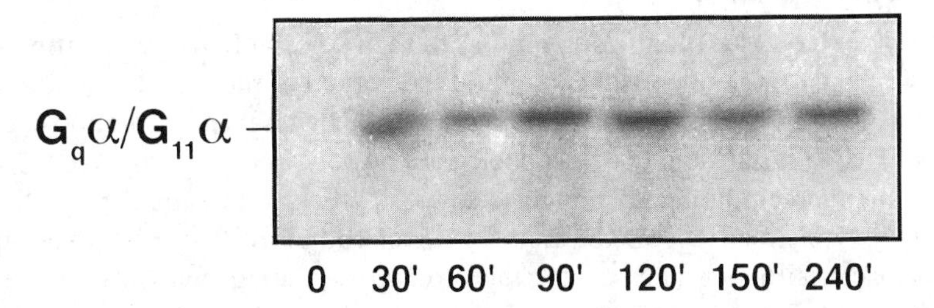

Fig. 1. Incorporation of [^{3}H] palmitate into the α subunits of the G proteins G_q and G_{11}. Rat 1 fibroblasts were metabolically labeled with (9,10-^{3}H [N])palmitic acid (150 μCi/mL) for the times indicated (in minutes) as described in Section 3.1. The cells were harvested and the α subunits of the G proteins G_q and G_{11} were immunoprecipitated as in Section 3.3. using an antipeptide antiserum directed against the C-teminal decapeptide, which is conserved between these two polypeptides *(15)*. Following SDS-PAGE, the gel was treated as in Section 3.5. and exposed to X-ray film for 6 wk.

h. Laemmli sample buffer: 3 g urea, 0.5 g SDS, 0.6 g DTT (*see* Note 6), 0.5 mL 1*M* Tris-HCl pH 8.0, to 10 mL with H_2O.
i. Running buffer: 28.9 g glycine, 6 g Tris, 2 g SDS to 2 L with H_2O.

12. Fixing solution: 25% propan-2-ol, 65% H_2O, 10% acetic acid.
13. Amplify (Amersham International plc, UK) (*see* Note 7).
14. 1*M* hydroxylamine adjusted to pH 8.0 with KOH.

3. Methods

3.1. Cell Culture and Metabolic Labeling

1. Seed equal numbers of cells to be analyzed into six-well tissue-culture dishes using 1.5–2 mL growth medium/well (or in 100 mm tissue-culture dishes with 10 mL of growth medium if cell fractionation is to be carried out). Incubate at 37°C in an atmosphere of 5% CO_2 until cells are 90% confluent. Remove growth medium and replace with 1 mL of (9,10-^{3}H [*N*])palmitic acid-labeling medium for 2 h at 37°C (*see* Fig. 1).
2. Parallel control experiments using Trans[^{35}*S*]-label (50 μCi/mL) are performed. However, addition of [^{35}*S*] methionine/cysteine labeling medium should occur when cell are 60–80% confluent (*see* Note 8), and the cells labeled over a period of 18 h (*see* Note 9).

3.2. Cell Harvesting and Sample Solubilization

1. At the end of the labeling period, remove the labeling medium, and add 200 μL of 1% (w/v) SDS/well. Scrape the monolayer of cells into the SDS solution, and transfer to a 2 mL screw-top plastic tube.
2. Heat to 80°C for 20 min (*see* Note 10) in a heating block to denature proteins. If the samples have a stringy consistency after this stage, pass through a 20–25 gage needle and reboil for 10 min.
3. Remove the samples from the heating block, and allow to cool for 2 min at room temperature. Pulse each tube briefly at high speed in a microfuge to bring the contents to the bottom of the tube. The samples can either now be frozen at –20°C until required or processed immediately.

3.3. Immunoprecipitation

1. Add 800 µL of ice-cold solubilization buffer to each tube, and mix by inverting. Pulse the samples in a microfuge to collect contents at the bottom of the tubes.
2. Retain fractions of these samples to allow analysis of total incorporation into cellular protein of [^{3}H]palmitate and the [^{35}S]labeled amino acids.
3. Before immunoprecipitating the protein of interest, preclear the sample by adding 100 µL of protein A-Sepharose (Sigma, St. Louis, MO) or 100 µL of Pansorbin (a cheaper alternative of bacterial membranes containing protein A, [Calbiochem]) (*see* Note 11), and mix at 4°C for 1–2 h on a rotating wheel. (Ensure caps are firmly closed before rotating.)
4. Centrifuge samples for 2 min at maximum speed in a microfuge to pellet the protein A or Pansorbin. Transfer the precleared (*see* Note 12) supernatants to fresh screw-top tubes.
5. To the precleared samples, add 5–15 µL of protein-specific antibody (volume will vary depending on the antibody used) and 100 µL of protein A-Sepharose. Ensure caps are firmly closed and rotate at 4°C as before for 2–5 h (*see* Note 13).
6. Spin samples for 2 min at maximum speed in a microfuge to pellet the immunocomplex. Remove the supernatant and resuspend the pellet in 1 mL of immunoprecipitation wash buffer. Invert the tube 10 times (do not vortex). (The supernatant can be retained to analyze the efficiency of immunoprecipitation if required.)
7. Repeat step 6.
8. Centrifuge samples for 2 min at maximum speed in a microfuge and discard supernatant. Resuspend agarose-immunocomplex pellet in 40 µL of Laemmli sample buffer (*see* Note 6).
9. Heat samples to 80°C (*see* Note 10) for 5 min and then centrifuge for 2 min at maximum speed in a microfuge. Analyze the samples by SDS-PAGE by loading an equal volume of each sample (e.g., 55 µL) on a 10% (w/v) acrylamide gel (*see* Section 3.4.).

3.4. SDS-PAGE Analysis

Recipes given below are for one 16 × 18 cm 10% acrylamide gel. Resolving gel (lower): 8.2 mL H_2O, 6 mL buffer 1.8 mL acrylamide, 1.6 mL 50% glycerol, 90 µL APS, 8 µL TEMED.

1. Add all reagents in the order given and mix thoroughly. Carefully pour into prepared gel plates.
2. Very carefully overlay gel mixture with approx 1 mL of 0.1% (w/v) SDS, and leave gel to polymerize.
3. Once gel has polymerized, pour off SDS.
4. Prepare stacker gel mixture as indicated below and mix thoroughly. Stacker gel mixture: 9.75 mL H_2O, 3.75 mL buffer 2, 1.5 mL acrylamide, 150 mL APS, 8 mL TEMED.
5. Pour stacker gel on top of resolving gel, and place well-forming comb in top of gel, ensuring no air bubbles are trapped under the comb. Leave to polymerize.
6. Once gel has polymerized remove the comb, place the gel in the gel tank containing enough running buffer in the base to cover the bottom edge of the gel, and add the remaining running buffer to the top.
7. Load the prepared samples in the preformed wells using a Hamilton syringe.
8. Run the gel overnight (approx 16 h) at 12 mA and 60 V until the dye front reaches the bottom of the gel plates.

3.5. Enhancement of [^{3}H] Fatty acid Signal from Gel

In order the increase the effectiveness of detection of the weak β-particle signal emitted by [^{3}H]palmitic acid, the gel is treated with Amplify (Amersham International) (*see* Note 7) according to the manufacturer's instructions.

1. Fix proteins in gel using fixing solution for 30 min.
2. Pour off fixing solution (**caution—this may contain radioactivity**), and soak gel in Amplify with agitation for 15–30 min. Wash.
3. Remove gel from solution, and dry under vacuum at 60–80°C.
4. Expose the gel to X-ray film (Hyperfilm-MP, Amersham International, or equivalent) at –70°C for an appropriate time (*see* Note 14) before developing.

3.6. The Nature of the Linkage Between [^{3}H] Palmitate and Protein

Hydrolysis of the thioester bond between palmitate and cysteine can be achieved by treatment with near neutral hydroxylamine *(9,13)*. Such an approach is amenable to samples following SDS-PAGE resolution and provides clear information on the chemical nature of the linkage. Hydrolysis by hydroxylamine of *O*-esters requires strongly alkaline (pH >10.0) conditions, and amide linkages are stable to treatment with this agent.

1. Following immunoprecipitation, resolution, and fixing of the protein in SDS-PAGE, lanes of the gels are exposed to 1*M* hydroxylamine pH 7.4, or 1*M* Tris-HCl 7.4, for 1 h at 25°C (*see* Note 15).
2. The gels are then washed with H_2O (2 15-min washes) and prepared for fluorography as described in Section 3.5.

3.7. Metabolic Interconversion Between [^{3}H] Fatty Acids

Metabolic interconversion between fatty acids is a well-appreciated and established phenomenon. As such, it is vital in experiments designed to demonstrate that a polypeptide is truly a target for palmitoylation that the [^{3}H] fatty acid incorporated following labeling of cells with [^{3}H] palmitate is shown actually to be palmitate. An excellent and detailed description of the strategies that may be used to examine the identity of protein-linked radiolabeled fatty acids, and the limitations of such analyses, has recently appeared *(14)* and readers are referred to this for details.

4. Notes

1. [^{3}H]-labeled palmitic acid available from commercial sources is not at a sufficiently high concentration that it can be directly added to cells for labeling experiments. This can be concentrated by evaporation of the solvent and dissolution in either ethanol or dimethylsulfoxide.
2. Since the cellular pool of palmitate is large, radiolabeling experiments have to use relatively high amounts of [^{3}H]palmitate (0.1–1.0 μCi/mL) to obtain sufficient incorporation into a protein of interest, such that detection of the [^{3}H]radiolabeled polypeptide can be achieved in a reasonable time frame. Although it would theoretically be possible to use [^{14}C]palmititc acid, which is also available commercially, a combination of low specific activity and concerns about degradation by β-oxidation removing the radiolabel (particularly if labeled in the 1 position) restricts its usefulness.
3. When preparing [^{3}H] palmitic acid (a) ensure radiolabel is dried in a glass tube to minimize loss of material by adsorption, and (b) when resuspending [^{3}H] palmitic acid, take great care to ensure all material is recovered from the sides of the glass tube and fully redissolved in the labeling medium. This can be confirmed by counting the amount of radioactivity present in a small proportion of the labeling medium and comparing it to the amount of radioactivity added originally. Generally, addition of [^{3}H]palmitate in metabolic labeling studies is regulated such that the addition of vehicle is limited to 1% (v/v) or less.
4. **Caution: [^{35}*S*] is volatile and therefore stocks should be opened in a fume hood.**
5. Dialysis tubing can be stored at this point in 20% ethanol until required.

6. When preparing samples for SDS-PAGE, care must be taken with addition of nucleophilic reducing agents, such as dithiothreitol (DTT) and 2-mercaptoethanol, since these can cause the cleavage of thioester linkages. We limit the concentration of DTT in the sample buffer to 20 m*M*. This may not be a universal problem, occurring only with certain proteins.
7. Other commercially available solutions for fluorography, or indeed methods based on salicylate or 2,5 diphenyloxazole (PPO) may be substituted.
8. The confluency of cells required for [^{35}S]metabolic labeling will varying depending on the speed of growth of the cell line used. For rapidly growing cells, such as fibroblasts, add radiolabel at 60% confluency; for slow-growing cells, e.g., of neuronal derivation, add radiolabel when cells are approaching 80–85% confluency.
9. Subsequent immunoprecipitation (*see* Section 3.3.) of [^{35}S] labeled protein can thus provide controls for immunoprecipitation efficiency and confirm that the lack of immunoprecipitation of a [^{3}H]palmitate containing polypeptide was not owing to lack of immunoprecipitation of the relevant polypeptide by the antiserum.
10. It is recommended that heating is not prolonged and does not exceed 80°C.
11. We have found Pansorbin to be a good and cheap alternative to protein A-Sepharose, especially for preclearing; however, the use of protein A-Sepharose is recommended for the immunoprecipitation reaction itself, since the use of Pansorbin has been found to give increased nonspecific background for some antibodies.
12. Preclearing samples before immunoprecipitation removes material that binds nonspecifically to protein A, thus reducing the background levels in the final sample. This is especially useful for [^{35}S] radiolabeled material where it may be advisable to preclear for the maximum 2 h.
13 The length of time of immunoprecipitation should be determined empirically for each antibody used, in order to minimize the amount of nonspecific material present in the final sample. For most antibodies with high titer, the shorter the incubation, the less nonspecific material immunoprecipitated.
14. Time will obviously depend on the levels of expression of the protein and the amount of [^{3}H]palmitate used. For cell lines that have not been transfected to express high levels of a particular protein, 30–40 days of exposure is not an unusual period of time.
15. Stability of incorporation of the [^{3}H]radiolabel to treatment with Tris-HCl, but removal by hydroxylamine under these conditions can be taken to reflect linkage via a thioester bond.

References

1. Milligan, G., Parenti, M., and Magee, A. I. (1995) The dynamic role of palmitoylation in signal transduction. *Trends Biochem. Sci.* **20,** 181–186.
2. Wedegaertner, P. B. and Bourne, H. R. (1994) Activation and depalmitoylation of $G_{s\alpha}$. *Cell* **77,** 1063–1070.
3. Degtyarev, M. Y., Spiegel, A. M., and Jones, T. L. Z. (1993) Increased palmitoylation of the G_s protein α subunit after activation by the β-adrenergic receptor or cholera toxin. *J. Biol. Chem.* **268,** 23,769–23,772.
4. Mouillac, B., Caron, M., Bonin, H., Dennis, M., and Bouvier, M. (1992) Agonist-modulated palmitoylation of β_2-adrenergic receptor in Sf9 cells. *J. Biol. Chem.* **267,** 21,733–21,737.
5. Robinson, L. J., Busconi, L., and Michel, T. (1995) Agonist-modulated palmitoylation of endothelial nitric oxide synthase. *J. Biol. Chem.* **270,** 995–998.
6. Stoffel. R. H., Randall, R. R., Premont, R. T., Lefkowitz, R. J., and Inglese, J. (1994) Palmitoylation of G protein-coupled receptor kinase, GRK6. Lipid modification diversity in the GRK family. *J. Biol. Chem.* **269,** 27,791–27,794.

7. Mumby, S. M., Kleuss, C., and Gilman, A. G. (1994) Receptor regulation of G-protein palmitoylation. *Proc. Natl. Acad. Sci. USA* **91,** 2800–2804.
8. Hess, D. T., Patterson, S. I., Smith, D. S., and Skene, J. H. (1993) Neuronal growth cone collapse and inhibition of protein fatty acylation by nitric oxide. *Nature* **366,** 562–565.
9. Patterson, S. I. and Skene, J. H. P. (1995) Inhibition of dynamic protein palmitoylation in intact cells with tunicamycin. *Meth. Enzymol.* **250,** 284–300.
10. Parenti, M., Vigano, M. A., Newman, C. M. H., Milligan, G., and Magee, A. I. (1993) A novel N-terminal motif for palmitoylation of G-protein α subunits. *Biochem. J.* **291,** 349–353.
11. Grassie, M. A., McCallum, J. F., Guzzi, F., Magee, A. I., Milligan, G., and Parenti, M. (1994) The palmitoylation status of the G-protein $G_o1\alpha$ regulates its avidity of interaction with the plasma membrane. *Biochem. J.* **302,** 913–920.
12. McCallum, J. F., Wise, A., Grassie, M. A., Magee, A. I., Guzzi, F., Parenti, M., and Milligan, G. (1995) The role of palmitoylation of the guanine nucleotide binding protein $G_{11}\alpha$ in defining interaction with the plasma membrane. *Biochem. J.* **310,** 1021–1027.
13. Magee, A. I., Wooton, J., and De Bony, J. (1995) Detecting radiolabeled lipid-modified proteins in polyacrylamide gels. *Methods Enzymol.* **250,** 330–336.
14. Linder, M. A., Kleuss, C., and Mumby, S. M. (1995) Palmitoylation of G-protein α subunits. *Methods Enzymol.* **250,** 314–330.
15. Mitchell, F. M., Mullaney, I., Godfrey, P. P., Arkinstall, S. J., Wakelam, M. J. O., and Milligan, G. (1991) *FEBS Lett.* **287,** 171–174.

88

Removal of Pyroglutamic Acid Residues from the N-Terminus of Peptides and Proteins

John M. Walker and Patricia J. Sweeney

1. Introduction

In both peptides and proteins, N-terminal glutamine residues can readily cyclize to the pyroglutamyl derivative (Fig. 1). This can occur during peptide and protein purification (it is uncertain whether the N-terminal pyroglutamyl residues of a number of naturally occurring peptides and proteins are genuine posttranslational modifications or were introduced by cyclization of N-terminal glutamine during purification). This cyclized derivative does not have a free amino group, and therefore, the peptide or protein is not amenable to sequence determination, unless the pyroglutamyl derivative is removed. This can be achieved by using the enzyme pyroglutamate aminopeptidase, a thiol exoprotease that cleaves N-terminal pyroglutamyl residues (pyrrolidone carboxylic acid) from peptides and proteins *(1–5)*. The enzyme was first purified from *Pseudomonas fluoresens (6)*, but nowadays, the calf liver enzyme is used, and it is this enzyme that we describe here.

The enzyme cleaves N-terminal pyroglutamyl residues from peptides and proteins, but not if the following residue is proline. The enzyme has highest specificity when the pyroglutamate residue is linked to alanine *(7)*. The enzyme is active at pH 7.0–9.0, but is normally used at pH 8.0 *(8)*.

Alternative names for the enzyme are L-pyroglutamyl peptide hydrolase, 5-oxoprolyl-peptidase, pyrrolidonyl peptidase, pyroglutamate aminopeptidase, pyroglutamyl peptidase, and pyroglutamase.

Note: This enzyme is now classified as EC 3.4.19.3. In earlier literature, the enzyme was classified as EC 3.4.11.8.

2. Materials

1. Pyroglutamate aminopeptidase: The lyophilized enzyme is stable at 4°C for months. In solution, the enzyme is generally unstable. Its stability is enhanced by sucrose and EDTA in some commercial preparations. It may be reconstituted in solutions containing 5 m*M* DTT and 10 m*M* EDTA, and stored at –20°C, or it can be used for a maximum of 1 wk, if stored at 4°C. Podell and Abraham *(9)* have found the enzyme to be extremely unstable above room temperature, and it was found that deblocking did not occur at 37°C. It was found that an initial incubation at 4°C followed by a second incubation at room tempera-

From: The Protein Protocols Handbook
Edited by: J. M. Walker Humana Press Inc., Totowa, NJ

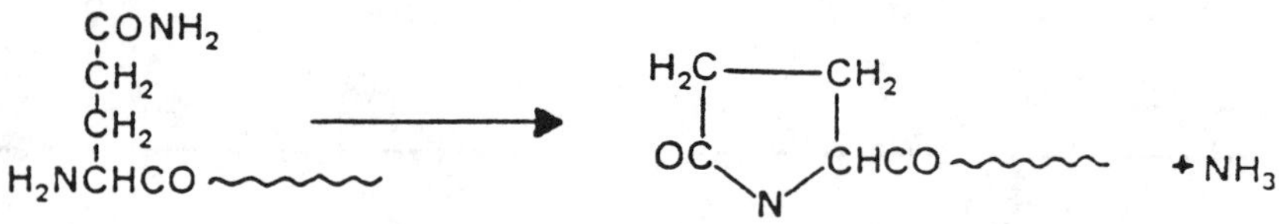

$$H_2NCH(CH_2CH_2CONH_2)CO\sim\sim \longrightarrow \text{(cyclic) } H_2C{-}CH_2,\ OC{-}N{-}CHCO\sim\sim + NH_3$$

Fig. 1. The cyclization of N-terminal glutamine to pyroglutamic (pyrrolidone carboxylic) acid.

ture (*see* Section 3.) was necessary to ensure maximum enzyme-stabilizing conditions. Air oxidation causes severe reduction in enzyme activity, and therefore, incubations involving the enzyme should be carried out under nitrogen.

2. Deblocking buffer: 0.1*M* sodium phosphate buffer, pH 8.0, 5 m*M* DTT, 10 m*M* EDTA, 5% glycerol (*see* Note 1).
3. Screw-top vial.
4. Nitrogen cylinder.

3. Method

This method is based on that described by Podell and Abraham *(9)*.

1. Dissolve the sample (1 mg) in 1 mL of deblocking buffer, and place the solution in a screw-top vial.
2. Add 2.5 µg of enzyme, flush with nitrogen, and seal the vial. Incubate at 4°C for 9 h with occasional mixing.
3. Add a further 2.5 µg of enzyme, and incubate under nitrogen at room temperature for a further 14 h.
4. Since the purpose of deblocking is invariably to render the protein amenable to sequence analysis, the protein can be desalted either by dialysis against 0.05*M* acetic acid and then lyophilized, or by passage through an HPLC gel-filtration column in 0.1% aqueous TFA.

4. Note

1. Podell and Abraham *(9)* investigated factors affecting the stability of the enzyme and developed a deblocking buffer that is compatible with the use of the enzyme. The enzyme requires a thiol compound for activation and is inactivated by thiol-blocking compounds, such as iodoacetamide, and divalent metal ions, such as Hg^{2+}. Activity can be restored by short incubations with mercaptoethanol. Dithiothreitol and EDTA are included in the deblocking buffer to overcome inactivation *(1)*.

References

1. Doolittle, R. F. (1970) Pyrrolidone carboxyl peptidase. *Methods Enzymol.* **19,** 555–569.
2. Szewczuk, A. and Kwiatkowska, J. (1970) Pyrrolidonyl peptidase in animal, plant and human tissues. Occurrence and some properties of the enzyme. *Eur. J. Biochem.* **15,** 92–96.
3. Mudge, A. W. and Fellows, R. E. (1973) Bovine pituitary pyrrolidone carboxyl peptidase. *Endocrinology* **93,** 1428–1434.
4. Browne, P. and O'Cuinn, G. (1983) An evaluation of the role of a pyroglutamyl peptidase, a post-proline cleaving enzyme and a post-proline dipeptidyl aminopeptidase, each purified from the soluble fraction of a guinea-pig brain, in the degradation of thyroliberin in vitro. *Eur. J. Biochem.* **137,** 75–87.

5. O'Connor, B. and O'Cuinn, G. (1985) Purification and kinetic studies on narrow specificity synaptosomal membrane pyroglutamate aminopeptidase from guinea pig brain. *Eur. J. Biochem.* **150,** 47–52.
6. Doolittle, R. F. and Armentrout, R. W. (1968) Pyrrolidonyl peptidase. An enzyme for selective removal of pyrrolidonecarboxylic acid residues from polypeptides. *Biochemistry* **7,** 516–521.
7. Pfleiderer, G., Celliers, P. G., Stanulovic, M., Wachsmuth, E. D., Determann, H., and Braunitzer, G. (1964) Eizenschaftern und analytische anwendung der aminopeptidase aus nierenpartikceln. *Biochem. Z.* **340,** 552–564.
8. Personal communication. Boehringer Mannheim.
9. Podell, D. N. and Abraham, G. N. (1978) A technique for the removal of pyroglutamic acid from the amino terminus of proteins using calf liver pyroglutamate amino peptidase. *Biochem. Biophys. Res. Commun.* **81,** 176–185.

89

The Dansyl Method for Identifying N-Terminal Amino Acids

John M. Walker

1. Introduction

The reagent 1-dimethylaminonaphthalene-5-sulfonyl chloride (dansyl chloride, DNS-Cl) reacts with the free amino groups of peptides and proteins as shown in Fig. 1. Total acid hydrolysis of the substituted peptide or protein yields a mixture of free amino acids plus the dansyl derivative of the N-terminal amino acid, the bond between the dansyl group and the N-terminal amino acid being resistant to acid hydrolysis. The dansyl amino acid is fluorescent under UV light and is identified by thin-layer chromatography on polyamide sheets. This is an extremely sensitive method for identifying amino acids and in particular has found considerable use in peptide sequence determination when used in conjunction with the Edman degradation (*see* Chapter 90). The dansyl technique was originally introduced by Gray and Hartley *(1)*, and was developed essentially for use with peptides. However, the method can also be applied to proteins (*see* Note 1).

2. Materials

1. Dansyl chloride solution (2.5 mg/mL in acetone). Store at 4°C in the dark. This sample is stable for many months. The solution should be prepared from concentrated dansyl chloride solutions (in acetone) that are commercially available. Dansyl chloride available as a solid invariably contains some hydrolyzed material (dansyl hydroxide).
2. Sodium bicarbonate solution (0.2*M*, aqueous). Store at 4°C. Stable indefinitely, but check periodically for signs of microbial growth.
3. 5*N* HCl (aqueous).
4. Test tubes (50 × 6 mm) referred to as "dansyl tubes."
5. Polyamide thin layer plates (7.5 × 7.5 cm). These plates are coated on both sides, and referred to as "dansyl plates." Each plate should be numbered with a pencil in the top corner of the plate. The origin for loading should be marked with a pencil 1 cm in from each edge in the lower left-hand corner of the numbered side of the plate. The origin for loading on the reverse side of the plate should be immediately behind the loading position for the front of the plate, i.e., 1 cm in from each edge in the lower right-hand corner.
6. Three chromatography solvents are used in this method.
   Solvent 1: Formic acid:water, 1.5:100 (v/v);
   Solvent 2: Toluene:acetic acid, 9:1 (v/v);
   Solvent 3: Ethyl acetate:methanol:actic acid, 2:1:1 (v/v/v).

From: The Protein Protocols Handbook
Edited by: J. M. Walker Humana Press Inc., Totowa, NJ

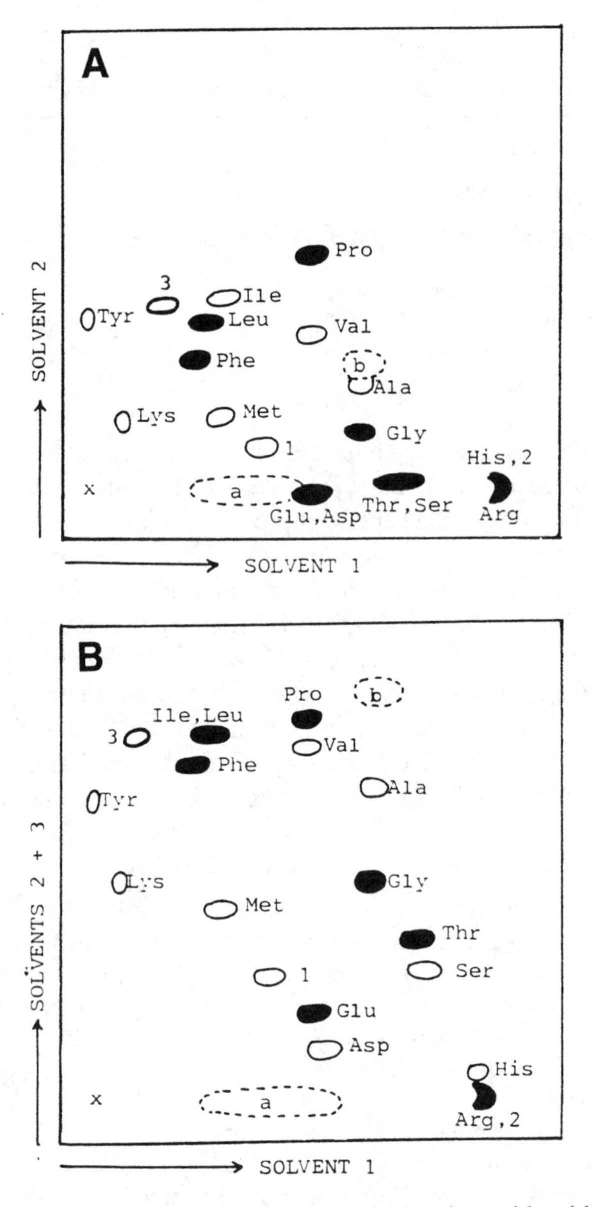

Fig. 1. Reaction sequence for the labeling of N-terminal amino acids with dansyl chloride.

7. An acetone solution containing the following standard dansyl amino acids. Pro, Leu, Phe, Thr, Glu, Arg (each approx 50 µg/mL).
8. A UV source, either long wave (265 µm) or short wave (254 µm).

3. Method

1. Dissolve the sample to be analyzed in an appropriate volume of water, transfer to a dansyl tube, and dry *in vacuo* to leave a film of peptide (1–5 nmol) in the bottom of the tube.
2. Dissolve the dried peptide in sodium bicarbonate (0.2*M*, 10 µL) and then add dansyl chloride solution (10 µL) and mix (*see* Note 2).
3. Seal the tube with parafilm and incubate at 37°C for 1 h, or at room temperature for 3 h.

4. Dry the sample *in vacuo*. Because of the small volume of liquid present, this will only take about 5 min.
5. Add 6*N* HCl (50 µL) to the sample, seal the tube in an oxygen flame, and place at 105°C overnight (18 h).
6. When the tube has cooled, open the top of the tube using a glass knife, and dry the sample *in vacuo*. If phosphorus pentoxide is present in the desiccator as a drying agent, and the desiccator is placed in a water bath at 50–60°C, drying should take about 30 min.
7. Dissolve the dried sample in 50% pyridine (10 µL) and, using a microsyringe, load 1-µL aliquots at the origin on each side of a poly-amide plate. This is best done in a stream of warm air. Do not allow the diameter of the spot to exceed 3–4 mm (*see* Note 3).
8. On the *reverse side only*, also load 0.5 µL of the standard mixture at the origin.
9. When the loaded samples are completely dry, the plate is placed in the first chromatography solvent and allowed to develop until the solvent front is about 1 cm from the top of the plate. This takes about 10 min, but can vary depending on room temperature.
10. Dry both sides of the plate by placing it in a stream of warm air. This can take 5–10 min since one is evaporating an aqueous solvent.
11. If the plate is now viewed under UV light, a blue fluorescent streak will be seen spreading up the plate from the origin, and also some green fluorescent spots may be seen within this streak. However, no interpretations can be made at this stage (*see* Note 4).
12. The dansyl plate is now developed in the second solvent, at right angles to the direction of development in the first solvent. The plate is therefore placed in the chromatography solvent so that the blue "streak" runs along the bottom edge of the plate.
13. The plate is now developed in the second solvent until the solvent front is about 1 cm from the top of the plate. This takes 10–15 min.
14. The plate is then dried in a stream of warm air. This will only take 2–3 min since the solvent is essentially organic. However, since toluene is involved, drying **must** be done in a fume cupboard.
15. The side of the plate containing the sample only should now be viewed under UV light. Three major fluorescent areas should be identified. Dansyl hydroxide (produced by hydrolysis of dansyl chloride) is seen as a blue fluorescent area at the bottom of the plate. Dansyl amide (produced by side reactions of dansyl chloride) has a blue–green fluorescence and is about one-third of the way up the plate. These two spots will be seen on all dansyl plates and seen as useful internal markers. Occasionally other marker spots are seen and these are described in the Notes section below. The third spot, which normally fluoresces green, will correspond to the dansyl derivative of the N-terminal amino acid of the peptide or protein. However, if the peptide is not pure, further dansyl derivatives will of course be seen. The separation of dansyl derivatives after solvent 2 is shown in Fig. 2. Solvent 2 essentially causes separation of the dansyl derivatives of hydrophobic and some neutral amino acids, whereas derivatives of charged and other neutral amino acids remain at the lower end of the chromatogram.
16. A reasonable identification of any faster-moving dansyl derivatives can be made after solvent 2 by comparing their positions, relative to the internal marker spots, with the diagram shown in Fig. 2. Unambiguous identification is made by turning the plate over and comparing the position of the derivative on this side with the standard samples that were also loaded on this side.

    Note that both sides of the plate are totally independent chromatograms. There is no suggestion that fluorescent spots can be seen through the plate from one side to the other.
17. Having recorded one's observations after the second solvent, the plate is now run in solvent 3 in the same direction as solvent 2. The plate is run until the solvent is 1 cm from the top, and this again takes 10–15 min.

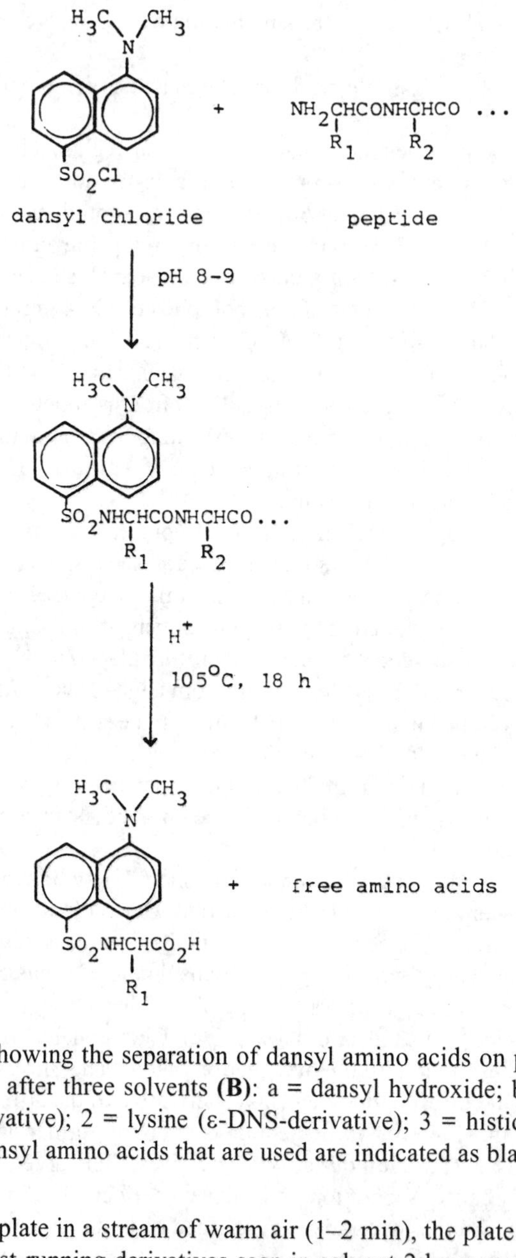

H_3C CH_3 N

SO_2Cl

dansyl chloride + $NH_2CHCONHCHCO$ ... (R_1, R_2)

peptide

pH 8-9

H_3C CH_3 N

$SO_2NHCHCONHCHCO$... (R_1, R_2)

H^+

105°C, 18 h

H_3C CH_3 N

$SO_2NHCHCO_2H$ (R_1) + free amino acids

Fig. 2. Diagrams showing the separation of dansyl amino acids on polyamide plates after two solvents **(A)**, and after three solvents **(B)**: a = dansyl hydroxide; b = dansyl amide; 1 = tyrosine (*o*-DNS-derivative); 2 = lysine (ε-DNS-derivative); 3 = histidine (*bis*-DNS-derivative). The standard dansyl amino acids that are used are indicated as black spots.

18. After drying the plate in a stream of warm air (1–2 min), the plate is again viewed under UV light. The fast-running derivatives seen in solvent 2 have now run to the top of the plate and are generally indistinguishable (hence the need to record one's observations after solvent 2). However, the slow-moving derivatives in solvent 2 have now been separated by solvent 3 and can be identified if present. The separation obtained after solvent 3 is also shown in Fig. 2. The sum of the observations made after solvents 2 and 3 should identify the number and relative intensities of N-terminal amino acids present in the original sample (*see* Notes 5–12).

4. Notes

1. The dansylation method described here was developed for use with peptides. However, this method can also be applied quite successfully to proteins, although some difficulties

arise. These are caused mainly by insolubility problems, which can limit the amount of reaction between the dansyl chloride and protein thus resulting in a lower yield of dansyl derivative, and the presence of large amounts of *o*-DNS-Tyr and ε-DNS-Lys on the chromatogram that can mask DNS-Asp and DNS-Glu. A modification of the basic procedure described here for the dansylation of proteins is described in ref. *3*.

2. It is important that the initial coupling reaction between dansyl chloride and the peptide occurs in the pH range 9.5–10.5. This pH provides a compromise between the unwanted effect of the aqueous hydrolysis of dansyl chloride and the necessity for the N-terminal amino group to be unprotonated for reaction with dansyl chloride. The condition used, 50% acetone in bicarbonate buffer, provides the necessary environment. The presence of buffer or salts in the peptide (or protein) sample should therefore be avoided to prevent altering the pH to a value outside the required range.
3. Because of the unpleasant and irritant nature of pyridine vapor, loading of samples onto the dansyl plates should preferably be carried out in a fume cupboard.
4. The viewing of dansyl plates under UV light should always be done wearing protective glasses or goggles. Failure to do so will result in a most painful and potentially damaging conjunctivitis.
5. Most dansyl derivatives are recovered in high (>90%) yield. However, some destruction of proline, serine, and threonine residues occurs during acid hydrolysis, resulting in yields of approx 25, 65, and 70%, respectively. When viewing these derivatives, therefore, their apparent intensities should be visually "scaled-up" accordingly.
6. The sensitivity of the dansyl method is such that as little as 1–5 ng of a dansylated amino acid can be visualized on a chromatogram.
7. The side chains of both tyrosine and lysine residues also react with dansyl chloride. When these residues are present in a peptide (or protein), the chromatogram will show the additional spots, *o*-DNS-Tyr and ε-DNS-Lys, which can be regarded as additional internal marker spots. The positions of these residues are shown in Fig. 2. These spots should not be confused with *bis*-DNS-Lys and *bis*-DNS-Tyr, which are produced when either lysine or tyrosine is the N-terminal amino acid.
8. At the overnight hydrolysis step, dansyl derivatives of asparagine or glutamine are hydrolyzed to the corresponding aspartic or glutamic acid derivatives. Residues identified as DNS-Asp or DNS-Glu are therefore generally referred to as Asx or Glx, since the original nature of this residue (acid or amide) is not known. This is of little consequence if one is looking for a single N-terminal residue to confirm the purity of a peptide or protein. It does, however, cause difficulties in the dansyl-Edman method for peptide sequencing (*see* Chapter 90) where the residue has to be identified unambiguously.
9. When the first two residues in the peptide or protein are hydrophobic residues, a complication can occur. The peptide bond between these two residues is particularly (although not totally) resistant to acid hydrolysis. Under normal conditions, therefore, some dansyl derivative of the first amino acid is produced, together with some dansyl derivative of the N-terminal dipeptide. Such dipeptide derivatives generally run on chromatograms in the region of phenylalanine and valine. However, their behavior in solvents 2 and 3, and their positions relative to the marker derivatives should prevent misidentification as phenylalanine or valine. Such dipeptide spots are also produced when the first residue is hydrophobic and the second residue is proline, and these dipeptide derivatives run in the region of proline. However, since some of the N-terminal derivative is always produced, there is no problem in identifying the N-terminal residue when this situation arises. A comprehensive description of the chromatographic behavior of dansyl-dipeptide derivatives has been produced *(2)*.

10. Three residues are difficult to identify in the three solvent system described in the methods section; DNS-Arg and DNS-His because they are masked by the ε-DNS-Lys spot, and DNS-Cys because it is masked by DNS-hydroxide. If these residues are suspected, a fourth solvent is used. For arginine and histidine, the solvent is 0.05*M* trisodium phosphate:ethanol (3:1 [v/v]). For cysteine the solvent is 1*M* ammonia:ethanol (1:1 [v/v]) Both solvents are run in the same direction as solvents 2 and 3, and the residues are identified by comparison with relevant standards loaded on the reverse side of the plate.
11. When working with small peptides it is often of use also to carry out the procedure known as "double dansylation." Having identified the N-terminal residue, the remaining material in the dansyl tube is dried down and the dansylation process (steps 2–4) repeated. The sample is then redissolved in 50% pyridine (10 μL) and a 1-μL aliquot is examined chromatographically. The chromtogram will now reveal the dansyl derivative of each amino acid present in the peptide. Therefore for relatively small peptides (<10 residues) a quantitative estimation of the amino acid composition of the peptide can be obtained. This method is not suitable for larger peptides or proteins since most residues will be present more than once in this case, and it is not possible to quantitatively differentiate spots of differing intensity.
12. Although the side chain DNS-derivative will be formed during dansylation if histidine is present in the peptide or protein sequence, this derivative is unstable to acid and is not seen during N-terminal analysis. Consequently, N-terminal histidine yields only the α-DNS derivative and not the *bis*-DNS compound as might be expected. The *bis*-DNS derivative is observed, however, if the mixture of free amino acids formed by acid hydrolysis of a histidine-containing peptide is dansylated and subsequently analyzed chromatographically (i.e., during "double dansylation").

References

1. Gray, W. R. and Hartley, B. S. (1963) A fluorescent end group reagent for peptides and proteins. *Biochem. J.* **89,** 59P.
2. Sutton, M. R. and Bradshaw, R. A. (1978) Identification of dansyl dipeptides. *Anal. Biochem.* **88,** 344–346.
3. Gray, W. R. (1967) in *Methods in Enzymology,* vol. XI (Hirs, C. H. W., ed.), Academic, New York, p. 149.

90

The Dansyl-Edman Method for Peptide Sequencing

John M. Walker

1. Introduction

The Edman degradation is a series of chemical reactions that sequentially removes N-terminal amino acids from a peptide or protein. The overall reaction sequence is shown in Fig. 1. In the first step (the coupling reaction) phenylisothocyanate (PITC) reacts with the N-terminal amino group of the peptide or protein. The sample is then dried and treated with an anhydrous acid (e.g., trifluoracetic acid), which results in cleavage of the peptide bond between the first and second amino acid. The N-terminal amino acids is therefore released as a derivative (the thiazolinone). The thiazolinone is extracted into an organic solvent, dried down, and then converted to the more stable phenylthiohydantoin (PTH) derivative (the conversion step). The PTH amino acid is then identified, normally by reverse-phase HPLC. This is known as the direct Edman degradation and is, for example, the method used in an automated sequencing machine. The dansyl-Edman method for peptide sequencing described here is based on the Edman degradation, but with the following modifications. Following the cleavage step the thiazolinone is extracted, but rather than being converted to the PTH derivative it is discarded. Instead, a small fraction (5%) of the remaining peptide is taken and the newly liberated N-terminal amino acid determined in this sample by the dansyl method (*see* Chapter 89). Although the dansyl-Edman method results in successively less peptide being present at each cycle of the Edman degradation, this loss of material is compensated for by the considerable sensitivity of the dansyl method for identifying N-terminal amino acids. The dansyl-Edman method described here was originally introduced by Hartley *(1)*.

2. Materials

1. Ground glass stoppered test tubes (approx 65 × 10 mm, e.g., Quickfit MF 24/0). All reactions are carried out in this "sequencing" tube.
2. 50% Pyridine (aqueous, made with AR pyridine). Store under nitrogen at 4°C in the dark. Some discoloration will occur with time, but this will not affect results.
3. Phenylisothiocyanate (5% [v/v]) in pyridine (AR). Store under nitrogen at 4°C in the dark. Some discoloration will occur with time, but this will not affect results. The phenylisothiocyanate should be of high purity and is best purchased as "sequenator grade." Make up fresh about once a month.

From: The Protein Protocols Handbook
Edited by: J. M. Walker Humana Press Inc., Totowa, NJ

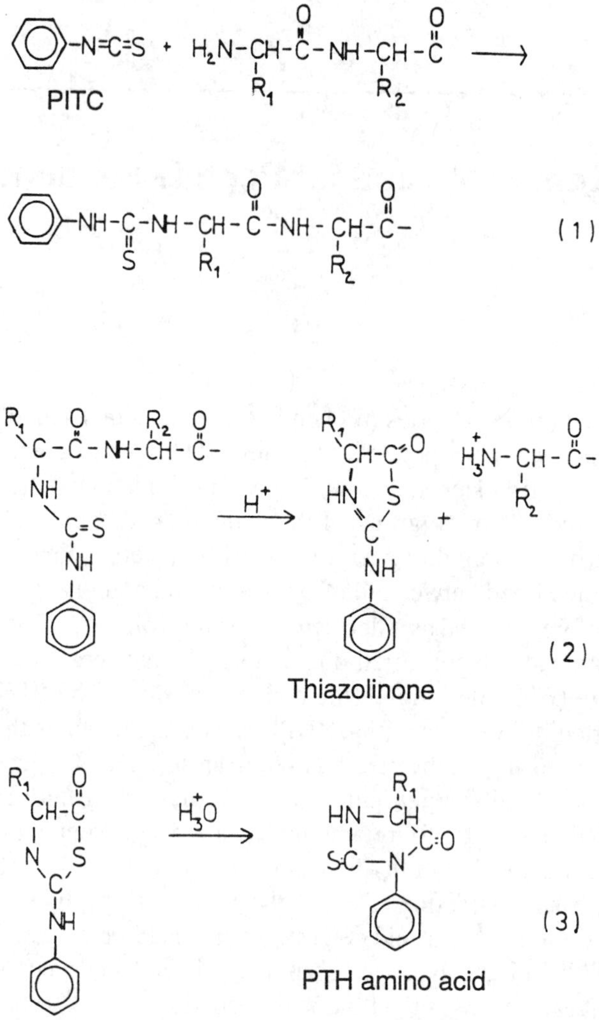

Fig. 1. The Edman degradation reactions and conversion step: (1) the coupling reaction; (2) the cleavage reaction; (3) the conversion step.

4. Water-saturated *n*-butyl acetate. Store at room temperature.
5. Anhydrous trifluoracetic acid (TFA). Store at room temperature under nitrogen.

3. Method

1. Dissolve the peptide to be sequenced in an appropriate volume of water, transfer to a sequencing tube, and dry *in vacuo* to leave a film of peptide in the bottom of the tube (*see* Notes 1 and 2).
2. Dissolve the peptide in 50% pyridine (200 μL) and remove an aliquot (5 μL) for N-terminal analysis by the dansyl method (*see* Chapter 89 and Note 3).
3. Add 5% phenylisothiocyanate (100 μL) to the sequencing tube, mix gently, flush with nitrogen and incubate the stoppered tube at 50°C for 45 min.
4. Following this incubation, unstopper the tube and place it *in vacuo* for 30–40 min. The desiccator should contain a beaker of phosphorus pentoxide to act a drying agent, and if

possible the desiccator should be placed in a water bath at 50–60°C. When dry, a white "crust" will be seen in the bottom of the tube. This completes the coupling reaction (*see* Notes 4–6).

5. Add TFA (200 µL) to the test-tube, flush with nitrogen, and incubate the stoppered tube at 50°C for 15 min.
6. Following incubation, place the test tube *in vacuo* for 5 min. TFA is a very volatile acid and evaporates rapidly. This completes the cleavage reaction.
7. Dissolve the contents of the tube in water (200 µL). Do not worry if the material in the tube does not all appear to dissolve. Many of the side-products produced in the previous reactions will not in fact be soluble.
8. Add *n*-butyl acetate (1.5 mL) to the tube, mix vigorously for 10 s, and then centrifuge in a bench centrifuge for 3 min.
9. Taking care not to disturb the lower aqueous layer, carefully remove the upper organic layer and discard (*see* Note 7).
10. Repeat this butyl acetate extraction procedure once more and then place the test-tube containing the aqueous layer *in vacuo* (with the desiccator standing in a 60°C water bath if possible) until dry (approx 60 min).
11. Redissolve the dried material in the test tube in 50% pyridine (200 µL) and remove an aliquot (5 µL) to determine the newly liberated N-terminal amino acid (the second one in the peptide sequence) by the dansyl method (*see* Chapter 89).
12. A further cycle of the Edman degradation can now be carried out by returning to step 3. Proceed in this manner until the peptide has been completely sequenced (*see* Notes 8 and 9).
13. The identification of N-terminal amino acids by the dansyl method is essentially as described in Chapter 89. However, certain observations peculiar to the dansyl-Edman method are described in Notes 10–15.

4. Notes

1. Manual sequencing is normally carried out on peptides between 2 and 30 residues in length and requires 1–5 nmol of peptide.
2. Since the manipulative procedures are relatively simple it is quite normal to carry out the sequencing procedure on 8 or 12 peptides at one time.
3. As sequencing proceeds, it will generally be necessary to increase the amount of aliquot taken for dansylation at the beginning of each cycle, since the amount of peptide being sequenced is reduced at each cycle by this method. The amount to be taken should be determined by examination of the intensity of spots being seen on the polyamide plates.
4. It is most important that the sample is completely dry following the coupling reaction. Any traces of water present at the cleavage reaction step will introduce hydrolytic conditions that will cause internal cleavages in the peptide and a corresponding increase in the background of N-terminal amino acids.
5. Very occasionally it will prove difficult to completely dry the peptide following the coupling reaction and the peptide appears oily. If this happens, add ethanol (100 µL) to the sample, mix, and place under vacuum. This should result in a dry sample.
6. Vacuum pumps used for this work should be protected by cold traps. Considerable quantities of volatile organic compounds and acids will be drawn into the pump if suitable precautions are not taken.
7. When removing butyl acetate at the organic extraction step, take great care not to remove any of the aqueous layer as this will considerably reduce the amount of peptide available for sequencing. Leave a small layer of butyl acetate above the aqueous phase. This will quickly evaporate at the drying step.

8. A repetitive yield of 90–95% is generally obtained for the dansyl-Edman degradation. Such repetitive yields usually allow the determination of sequences up to 15 residues in length, but in favorable circumstances somewhat longer sequences can be determined.
9. A single cycle takes approx 2.5 h to complete. When this method is being used routinely, it is quite easy to carry out three or four cycles on eight or more peptides in a normal day's work. During the incubation and drying steps the dansyl samples from the previous days sequencing can be identified.
10. Tryptophan cannot be identified by the dansyl method as it is destroyed at the acid hydrolysis step. However, where there is tryptophan present in the sequence, an intense purple color is seen at the cleavage (TFA) step involving the tryptophan residue that unambiguously identifies the tryptophan residue.
11. If there is a lysine residue present in the peptide, a strong ε-DNS-lysine spot will be seen when the dansyl derivative of the N-terminal amino acid is studied. However, when later residues are investigated the ε-DNS-lysine will be dramatically reduced in intensity or absent. This is because the amino groups on the lysine side chains are progressively blocked by reaction with phenylisothiocyanate at each coupling step of the Edman degradation.
12. The reactions of the lysine side chains with phenylisothiocyanate causes some confusion when identifying lysine residues. With a lysine residue as the N-terminal residue of the peptide it will be identified as *bis*-DNS-lysine. However, lysine residues further down the chain will be identified as the α-DNS-ε–phenylthiocarbamyl derivative because of the side chain reaction with phenylisothiocyanate. This derivative runs in the same position as DNS-phenylalanine in the second solvent, but moves to between DNS-leucine and DNS-isoleucine in the third solvent. Care must therefore be taken not to misidentify a lysine residue as a phenylalanine residue.
13. When glutamine is exposed as the new N-terminal amino acid during the Edman degradation, this residue will sometimes cyclize to form the pyroglutamyl derivative. This does not have a free amino group, and therefore effectively blocks the Edman degradation. If this happens, a weak DNS-glutamic acid residue is usually seen at this step, and then no other residues are detected on further cycles. There is little one can do to overcome this problem once it has occurred, although the enzyme that cleaves off pyroglutamyl derivatives (pyroglutamate aminopeptidase) is commercially available (Boehringer, Lewes, UK).
14. The amino acid sequence of the peptide is easily determined by identifying the new N-terminal amino acid produced after each cycle of the Edman degradation. However, because the Edman degradation does not result in 100% cleavage at each step, a background of N-terminal amino acids builds up as the number of cycles increases. Also, as sequencing proceeds, some fluorescent spots reflecting an accumulation of side products can be seen toward the top of the plates. For longer runs (10–20 cycles) this can cause some difficulty in identifying the newly liberated N-terminal amino acid. This problem is best overcome by placing the dansyl plates from consecutive cycles adjacent to one another and viewing them at the same time. By comparison with the previous plate, the increase of the new residue at each cycle, over and above the background spots, should be apparent.
15. The main disadvantage of the dansyl-Edman method compared to the direct Edman method is the fact that the dansyl method cannot differentiate acid and amide residues. Sequences determined by the dansyl-Edman method therefore usually include residues identified as Asx and Glx. This is most unsatisfactory since it means the residue has not been unambiguously identified, but often the acid or amide nature of an Asx or Glx residue can be deduced from the electrophoretic mobility of the peptide *(2)*.
16. Having identified any given residue it can prove particularly useful to carry out the procedure referred to as "double dansylation" on this sample (*see* Chapter 89). This double-

dansylated sample will identify the amino acids remaining beyond this residue. Double dansylation at each step should reveal a progressive decrease in the residues remaining in the peptide, and give an excellent indication of the amount of residues remaining to be sequenced at any given cycle.

References

1. Hartley, B. S. (1970) Strategy and tactics in protein chemistry. *Biochem. J.* **119,** 805–822.
2. Offord, R. E. (1966) Electrophoretic mobilities of peptides on paper and their use in the determination of amide groups. *Nature* **211,** 591–593.

91

Matrix-Assisted Laser Desorption Ionization Mass Spectrometry as a Complement to Internal Protein Sequencing

Kenneth R. Williams, Suzy M. Samandar, Kathryn L. Stone, Melissa Saylor, and John Rush

1. Introduction

Matrix-assisted laser desorption ionization mass spectrometry (MALDI-MS) is a relatively new mass-analysis method suitable for the analysis of biological molecules *(1–3)*. Compared to other mass-analysis methods, MALDI-MS is based on simpler instrumentation that is more easily operated and so is more readily accessible to laboratories without prior experience in mass spectrometry. In addition, the method is sensitive and capable of high throughput. Its use in biochemical laboratories is already widespread, despite its recent introduction and the continued development of MALDI-MS instrumentation and applications.

The primary objective of most internal protein sequencing is to provide sufficient internal sequence rapidly and inexpensively either to identify the protein or to design oligonucleotide probes for selecting appropriate clones from cDNA libraries. Most eukaryotic cytoplasmic proteins have blocked amino-termini *(4,5)*, and so are not suitable for direct amino-terminal sequencing and must be sequenced internally. One approach for obtaining peptides for internal sequencing from proteins present in low quantities is to digest the proteins in Coomassie blue-stained, SDS-PAGE gel slices *(6,7)*. The peptides produced by digestion are then eluted from the gel slices, fractionated by narrow-bore reverse-phase HPLC, and suitable fractions are then sequenced.

One difficulty with this approach to internal sequencing is evaluating the suitability of fractions for sequencing. In the past, HPLC peak symmetry has been the primary and often sole criterion for suitability, because pure peptides are expected to elute as symmetrical peaks. However, in a recent study, only 67% of symmetrical HPLC peaks provided useful sequence results *(8)*. The remaining 33% of symmetrical peaks contained peptide mixtures, protease autolysis peptides, or chemical artifacts, such as Coomassie blue, chemical modification, and detergent byproducts. To a large extent, selecting HPLC peptide fractions for sequencing is a general problem that increases

From: The Protein Protocols Handbook
Edited by: *J. M. Walker Humana Press Inc., Totowa, NJ*

the effort and expense of internal sequencing, and that is not specific to the in-gel approach used in this work, which is described in Chapter 71.

In this chapter, we describe how MALDI-MS can be used to screen HPLC fractions, prepared as described above, for suitability for internal protein sequencing. Our experience provides useful guidelines for evaluating peptide purity, and sets realistic expectations for sensitivity and mass accuracy for routine MALDI-MS performed with a commercially available instrument.

2. Materials

2.1. Preparation of Synthetic and Protease Cleavage Peptides

1. Synthetic peptides were made on an Applied Biosystems (Foster City, CA) Model 431 Peptide Synthesizer in the W. M. Keck/HHMI Biopolymer Facility at Yale University or on an Advanced ChemTech (Louisville, KY) Multiple Peptide Synthesizer in the HHMI Biopolymer Facility at the University of California at San Francisco (for which the authors gratefully acknowledge Chris Turck). All synthetic peptides were purified by reverse-phase HPLC, were quantitated by hydrolysis/amino acid analysis and, based on MALDI-MS, had the predicted mass.
2. The remaining peptides used in this study were from in-gel digests of 64 proteins submitted to the W. M. Keck/HHMI Biopolymer Facility (Yale University) for internal sequencing. Prior to digestion, the amount of protein in each gel slice was determined by subjecting 10–15% of the gel slice to acid hydrolysis and ion-exchange amino acid analysis on a Beckman Model 6300 analyzer. Despite the limitations of this quantitative method *(9)*, we find it provides a more accurate estimate of the amount of protein present and of the probability of success of the impending digestion than can be determined by relative Coomassie blue staining intensity. If the gel band contained at least 25 pmol of protein, it was then digested with either trypsin (Boehringer Mannheim, Indianapolis, IN), modified trypsin (Promega, Madison, WI), or lysyl endopeptidase (Wako Chemicals, Inc., Osaka, Japan) using a modified version (described in detail in Williams and Stone *[7]*) of the procedure initially described by Rosenfeld et al. *(6)*. Subsequently, this procedure has been further improved as described in Chapter 71. In this revised procedure, the Coomassie blue-stained gel bands (*see* Chapter 11 for methods used for SDS-PAGE) are digested in the gel by adding a final enzyme ratio of about 0.033 μg/mm^3 gel volume in 0.1*M* Tris-HCl pH 8.0, 0.1% Tween 20, and are then *S*-methylated prior to reverse-phase HPLC. A negative (i.e., an equal size section of gel that should not contain any protein) and a positive (i.e., an equivalent amount of transferrin that had also been subjected to SDS-PAGE) control digest is carried out in parallel with each sample.
3. Reverse-phase HPLC was carried out on a Hewlett Packard Model 1090 instrument equipped with a Vydac 2.1 × 250 mm column that was eluted at 150 μL/min as described in Stone and Williams *(10)*. Subsequent improvements in the HP1090 HPLC system used for these studies permit the flow rate to be decreased to 75 μL/min (*see* Chapter 72). Peptides were detected by their absorbance at 210 nm and were collected by peak detection *(8)* in 1.5-mL capless Eppendorf tubes (Sarstedt, Newton, NC). All fractions were tightly capped within an hour of collection and were stored at 10°C. Fractions selected for sequencing were loaded directly onto an Applied Biosystems Model 470A or 477 sequencer operated according to the manufacturer's recommended protocol. If the entire fraction was loaded, the empty tube was rinsed with 30 μL neat trifluoroacetic acid, which was then applied to the polybrene-coated filter containing the sample, along with 25 pmol of an internal sequencing standard *(11)*.

2.2. MALDI-MS

1. MALDI-MS analyses were carried out on Fisons Instruments VG TofSpec mass spectrometers that were operated at accelerating voltages of 20 or 25 kV for detection of positive ions in linear mode. These instruments were equipped with nitrogen lasers (337 nm) and 0.65-m linear flight tubes and they accept 15-sample targets. The location of the laser beam within each 1.5-mm sample position is controlled by varying the *X* and *Y* coordinates of the beam. Laser energy is adjusted through coarse controls that range from 1 to 4 and fine controls that range from 0 to 200 (with the higher settings providing less attenuation).
2. The α-cyano-4-hydroxycinnamic acid (α-CHCA) matrix solution (about 10–20 mg/mL in 40% CH_3CN/0.1% trifluoroacetic acid [TFA]) was vortexed and allowed to stand for a few min before using the relatively clear supernatant for target preparation. Matrix solutions were stored for no more than 2 d at –20°C.
3. The calibrants for external calibration were gramicidin S (m/z = 1142.5) and insulin (m/z = 5734.5), and were stored at –20°C as 10 pmol/μL stocks in either 50% CH_3CN/0.1% TFA, or 0.1% TFA.

3. Methods

3.1. Sample Preparation for MALDI-MS

To find at least two peptides routinely from each protein digest that would be suitable for sequencing, about six symmetrical, late-eluting absorbance peaks—that were not present in the positive or negative controls—from each HPLC collection run were analyzed by MALDI-MS. Before analysis, the total volume of the fraction was recorded (as determined by pulling the sample up in a Gilson Pipetman) and 2 × 1.5 μL aliquots of the fraction were mixed with a 1.0-μL aliquot of α-cyano-4-hydroxycinnamic acid (αCHCA) matrix solution directly on a methanol-cleaned target by pipeting. The samples were then allowed to dry at room temperature. To avoid crosscontamination, targets were used only once.

3.2. MALDI-MS

1. To obtain spectra, the laser beam is moved across each sample using a fine energy setting of 200 and a coarse energy setting of 2. When a signal is observed, the fine energy is gradually lowered to optimize resolution. About 30 shots are averaged for each spectrum, and 3–6 spectra are acquired for each sample. A mixture of 10 pmol each of gramicidin S and insulin is present on each target and is used to create an external calibration spectrum. After calibration, a representative spectrum is chosen for each sample, the centroid masses are obtained, and the predicted average masses compared to the observed mass of the singly protonated species (*see* Note 1 for data relating to the impact of MALDI-MS on internal sequencing).
2. Limit of detection studies were done on duplicate targets prepared on different days by spotting serial 10-fold dilutions (in 50% CH_3CN, 0.05% TFA) of peptide stock solutions. For this study, the limit of detection is the lowest amount of peptide that provided a calculated signal-to-noise ratio of 5 for the ionized peptide. If these duplicate determinations varied by more than fivefold, an additional determination was made and the two lowest limits of detection were averaged. Because the mean variation in the limits of detection was about eightfold, differences in limits of detection of less than an order of magnitude are not considered significant (*see* Note 2 for a discussion of the impact of differential extents of ionization and other problems on the use of MALDI-MS, and

Table 1
Summary of Results from the MALDI-MS and Sequence Analysis of 102 Peptides Isolated from 64 Proteins Submitted to the W. M. Keck Facility at Yale University

Description	Range	Median	Mean
Total amount of peptide[a] (pmol)	0.4–600	11.4	35.9
Total volume of peptide (μL)	23–405	93.8	111
%Peptide applied to MALDI-MS target	0.7–13	3.2	3.4
Amount of peptide applied to MALDI-MS target (fmol)	20–6210	292	711
Observed mass	702–2906	1569	1609
%Mass error with external calibration (±SD)	0–1.1	—	0.25 (±0.25)

[a] Based on initial sequencing yield. Because initial yields are often in the 50–75% range, the actual total amount of peptide and the amount applied to the target may be 1.3- to 2-fold above the amounts given.

Table 2
Impact of MALDI-MS Screening on Internal Sequencing

Criteria for assessing purity	*n*	% Peptides that Sequenced[a]
Symmetrical HPLC peak	105	67.1[b]
Symmetrical HPLC peak with MALDI-MS major:minor peak ratio ≥ 2 and < 10	22	63.6
Symmetrical HPLC peak with MALDI-MS major:minor peak ratio ≥ 10	75	89.3

[a] Defined as being able to assign positively at least four residues in the peptide. The mean number of positive assignments was 10.6 residues.
[b] Data Stone et al. *(8)*.

see Note 3 for a general discussion regarding the usefulness of MALDI-MS as an adjunct to internal sequencing).

4. Notes

1. Table 1 summarizes results from the MALDI-MS and sequence analysis of 102 peptides isolated from 64 proteins submitted to the W. M. Keck Facility for internal sequencing. For each protein, approximately six of the late-eluting, symmetrical absorbance peaks were analyzed by MALDI-MS and, then one or two were sequenced. Peptides were selected for sequencing based on their apparent purity as judged by MALDI-MS, with preference being given to peptides with higher mass. The volume of each fraction used for MALDI-MS was always 3 μL, about 3% of the 111-μL average volume of the peptide fractions (Table 1). Based on initial sequencing yield, the median amount of peptide in these fractions was about 11 pmol, so the median amount of peptide subjected to MALDI-MS was <300 fmol (Table 1). Because no effort was made to determine the minimum amounts of these tryptic peptides needed for MALDI-MS analysis, 300 fmol appears to be a reasonable amount to use for routine screening. Of the 102 peptides included in this study, 63 were sequenced to completion or identified by data base searching. For these 63 peptides, the average mass accuracy was ±0.25% based on external calibration, corresponding to an error of ±4 mass units on an average mass of 1600 (Table 1).

   Table 2 summarizes the positive impact that MALDI-MS screening has on selecting peptides for internal sequencing. In a previous study *(8)*, we found that only 67% of

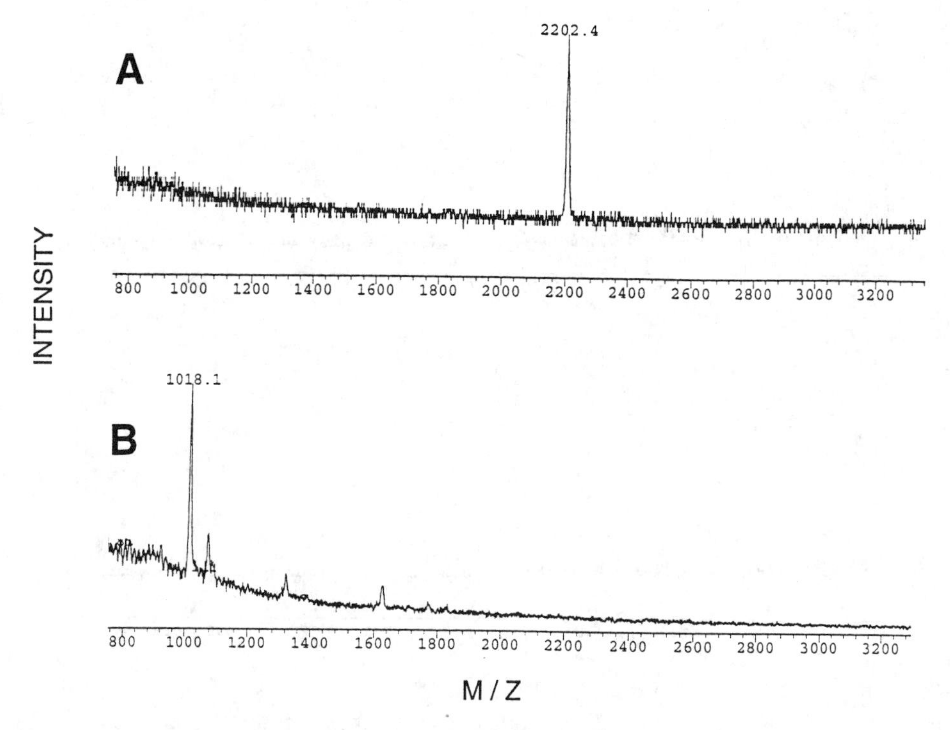

Fig. 1. Examples of MALDI-MS spectra that do (Panel A) and do not (Panel B) conform to the suggested major/minor peak ratio criterion of at least 10 for selecting peptide fractions for sequencing. The spectrum in Panel A was obtained from a tryptic peptide isolated from 53 pmol of a 44-kDa protein that was digested as described in Section 2. Consistent with the MALDI-MS peak ratio, sequencing demonstrated this peptide was indeed pure. Based on the initial sequencing yield of 600 fmol, this spectrum was obtained from about 78-fmol peptide. The spectrum in Panel B was from a tryptic peptide isolated from an in-gel digest of a 150-kDa protein (115 pmol). Because the ratio of the 1018.1 to the next major species is about 4, this fraction would not be considered suitable for amino acid sequencing. Indeed, sequencing demonstrated this fraction contained a mixture of at least three peptides, all present in the 2–5 pmol range.

symmetrical HPLC peaks derived from enzymatic digests provided usable sequence data. The remaining 33% contained peptide mixtures or were derived from protease autolysis products (that were not present in an enzyme/no substrate control), from Coomassie blue, or from other reagents that absorb at 210 nm. In this study, a similar success rate (64%) was found for 22 symmetrical HPLC peaks with MALDI-MS major to minor peak height ratios of ≥2 and <10 (Table 2). However, when the MALDI-MS criterion was raised to a major/minor peak ratio of 10 or more, 90% of peptides selected for sequencing provided useful data (Table 2).

Figure 1 shows examples of MALDI-MS spectra that do (Panel A) and do not (Panel B) meet this suggested criterion for selecting peaks for sequencing. Figure 2 (Panel A) shows the MALDI-MS spectrum for a well-separated, symmetrical HPLC absorbance peak that without the benefit of MALDI-MS would have been chosen for sequencing. Because MALDI-MS analysis indicated that the peak contained several peptides, this peak was rechromatographed on an Aquapore C-8 column *(10)*. Five major peaks were present after

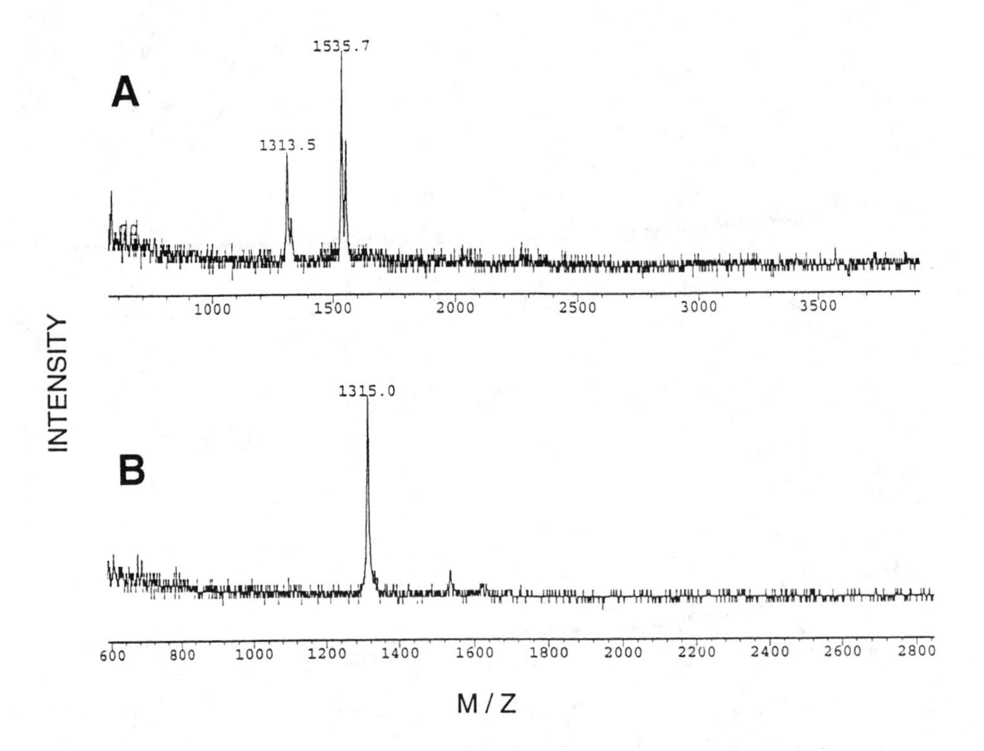

Fig. 2. MALDI-MS spectra of an HPLC peak isolated from a cyanogen bromide/tryptic digest of a 105-kDa protein (500 pmol) before (Panel A) and after (Panel B) repurification on an Aquapore C-8 column *(10)*. Although the absorbance peak from the Vydac C-18 column appeared to be sharp and symmetrical, MALDI-MS indicated the presence of multiple species (Panel A). Subsequent Aquapore C-8 chromatography resolved this fraction into at least five major peaks, and MALDI-MS analysis of the middle peak (Panel B) indicated this fraction was suitable for sequencing. This fraction gave an 11-residue sequence with a calculated mass of 1312.5, which is within 0.19% of the observed mass of 1315.0.

rechromatography (data not shown), and one of these peaks provided the MALDI-MS spectrum shown in Panel B of Fig. 2. Amino acid sequencing indicated this peak contained a nearly homogeneous 11-residue peptide.

In addition to providing a valuable criterion of peptide purity, MALDI-MS also identifies absorbance peaks resulting from reagents, such as Coomassie blue, which has a protonated mass of about 827, and from protease autolysis products. Figure 3 shows the MALDI-MS spectrum for a symmetrical peak that was not present in an enzyme control and that was assigned to residues 193–206 in trypsin based on its mass (i.e., observed mass = 1437.2, predicted mass = 1434.7, error = 0.2%). Sequencing confirmed this identification and indicated the MALDI-MS spectrum was obtained on 87 fmol of this peptide. The absence of this peak in the enzyme control suggests more extensive autolysis may occur in the absence of a "substrate" protein. Our recent results suggest that use of modified trypsin (such as that available from Promega) significantly reduces trypsin autolysis.

2. To evaluate features that influence peptide ionization, we analyzed a series of synthetic peptides by MALDI-MS. The peptides differed in sequence, length, and net charge at pH 2.0. Serial 10-fold dilutions of each peptide were analyzed to determine each one's limit

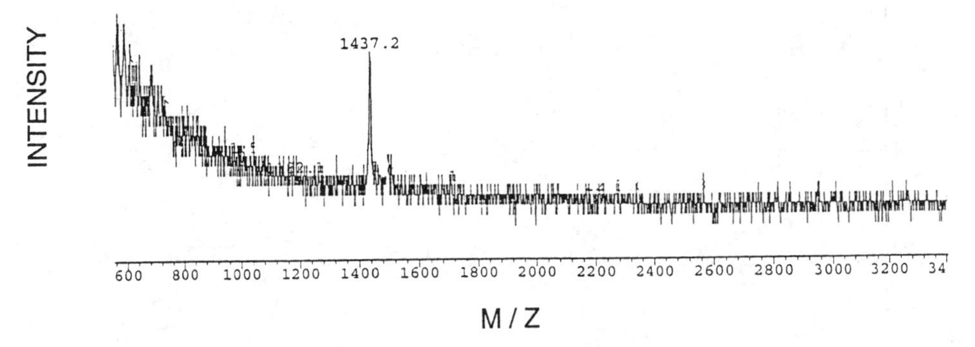

Fig. 3. MALDI-MS identification of a trypsin autolysis product from the digest of a 6800-Dalton protein (148 pmol). After reduction and carboxymethylation *(10)*, this sample was digested in solution with 26.7 pmol of trypsin (Boehringer Mamnheim) for 24 h at room temperature in 40 µL 2*M* urea, 0.1*M* NH_4HCO_3. The resulting peptides were separated by reverse-phase HPLC on a Vydac 2.1-mm (id) column, and 3 µL (7.9%) of this fraction were analyzed by MALDI-MS. Based on the observed mass of 1437.2, this peptide was identified as a trypsin autolysis fragment corresponding to residues 193–206 (predicted mass = 1434.7, which corresponds to a 0.17% error for the observed mass). This identification was confirmed by sequence analysis. Based on the sequencing yield for Ile-4 (1.1 pmol) the spectrum was obtained from 87-fmol peptide.

of detection. The data in Table 3 demonstrate there is a >5000-fold range in the limit of detection of different peptides in the 8–32 residue range. One striking observation from the data in Table 3 is that the range in MALDI-MS response narrows for those peptides that have a net positive charge at pH 2.0 of +2 or more. With the exception of peptide 13, which has an unusual stretch of Ala residues, all peptides in this table with a net charge of +2 or more have limits of detection below 150 fmol. The importance of net positive charge as a determinant of MALDI-MS response is again demonstrated in Table 4. Increasing the net charge of peptides 6(-R) and 15 from +1 to +2 and of peptide 16 from 0 to +2 lowered the limit of detection of each peptide by more than 10-fold. The data in Table 3 suggest that increasing the net positive charge above +2 has little effect on the limit of detection, except for peptide 13 (Table 4), where further increasing its net positive charge does, in fact, lower its limit of detection. By relying on proteases, such as trypsin or lysyl endopeptidase, which cleave after basic residues, nearly all peptides will have net positive charges at pH 2.0 of +2 or more, thus reducing the variability of MALDI-MS response so evident in Table 3. Under these conditions, most peptides seem to have limits of detection that span only about a 10-fold range (i.e., from 10 to 100 fmol). Based on the six peptides in Table 3 that have net positive charges of +4, hydrophobicity (as measured by the retention time of the peptide on a reverse-phase support) does not appear to have a major impact on the limit of detection.

MALDI-MS spectra with several peaks of similar intensity usually indicate peptide mixtures that are not suitable for sequencing, but there are some exceptions to this guideline. Hence, even though there are two major species in both the MALDI-MS spectra shown in Fig. 4 (*see* p. 550), both of these peptides actually provided only a single sequence. The peptide whose MALDI-MS spectrum is shown in Panel A (Fig. 4) begins with the sequence Gln-Tyr-Lys, so the –16.5 mass peak probably results from partial cyclization of the NH_2-terminal Gln. Similarly, since the peptide whose MALDI-MS spectrum is shown in Panel B of Fig. 4 was shown to contain one Met residue, the +15 peak

Table 3
Relative MALDI-MS Response of Several Synthetic Peptides

Peptide #	Sequence	# Res.	Mass	NH_2-term.	COO-term.	Basic res., #	Net charge, pH 2.0	Limit of Detection,[a] fmol		HPLC RT, min
								Yale	Harvard	
1	VHFFKNIVTPRT	12	1459.7	Free	Free	3	+4	6	21	63
2	KRLPATSMAIPLVGKY	16	1746.2	Free	Free	3	+4	9	62	69
3	L(OH-P)GL(OH-P)GP(OH-P)GPTGAKGLR	17	1633.9	Free	Free	2	+3	18	21	53
4	KVEQLSPEEEEKRRIRRERNKMAAA	25	3055.5	Free	Free	8	+9	20[b]	—	51
5	ASQKRPSQRHGSKYLA	16	1857.1	Blocked	Free	5	+5	27	21	42
Gram-S	VOLFPVOLFP	10	1142.5	Cyclic	Cyclic	2	+2	35	—	107
Peptide Y	PAEDLARYYSALRHTINITRQRY	24	2983.4	Free	Blocked	5	+6	39	—	82
6	GNFGGGRGGGFGG	13	1097.1	Free	Free	1	+2	43	62	55
7	VSWGRAPCEHDIMYKFTL	18	2153.5	Free	Blocked	3	+4	47	62	75
8	SRPVAGGPGAPPAARPPASPSPQRQAGPPQAT	32	3030.4	Free	Free	3	+4	56	—	44
9	ASQKRPSQRHG	11	1294.4	Blocked	Free	4	+4	81	62	27
10	APPAGITRSKSK	12	1213.4	Free	Free	3	+4	112	557	33
Secretin	HSDGTFTSELSRLREGARLQRLLQGLV	27	3040.4	Free	Blocked	5	+6	138[c]	—	82
11	MASSTPSPATSSNAGK	16	1536.7	Blocked	Free	1	+1	723	5000	38
12	TSTEPQYQPGENL	13	1464.5	Free	Free	0	+1	>1000	186	42
13	ARAAAAAA	8	672.8	Free	Free	1	+2	3400	—	ND
14	SNNLGPVLPP	10	1008.2	Free	Free	0	+1	7800	5000	57
15	GGGGSTSSNTIS	12	1024.0	Free	Blocked	0	+1	>50,000	5000	27
16	CSEDNADSGQ	10	1067.0	Blocked	Blocked	0	0	>50,000	>50,000	32

[a] Defined as the amount of peptide that gives a signal/noise ratio of about 5 (*see* Sections 2. and 3.).
[b] This peptide gives increasing amounts of a +16 peak as the concentration is lowered, presumably owing to Met oxidation.
[c] This peptide gives a –17.5 peak that is about 50% of the intensity of the expected peak, probably owing to cyclization of Asp-Gly and loss of NH_3.

Table 4
Effect of Differing Peptide Charge on MALDI-MS Limit of Detection

Peptide #	Sequence	# Res.	NH_2-terminus	COO-terminus	Basic res., #	Net charge pH 2.0	Limit of detection,[a] fmol		HPLC ret. time, min
							Yale	Harvard	
6(–R)	GNFGGGGGGFGG	12	Free	Free	0	+1	516		47
6	GNFGGGRGGGFGG	13	Free	Free	1	+2	43	62	55
13	ARAAAAAA	8	Free	Free	0	+2	3400		ND
13(+E)	ARAAAAAAE	9	Free	Free	1	+2	3700		ND
13(+K)	ARAAAAAAK	9	Free	Free	2	+3	380		ND
13(+L)	ARAAAAAAL	9	Free	Free	1	+2	2800		42
15	GGGGSTSSNTIS	12	Free	Blocked	0	+1	>50,000	5000	27
15(+R)	GGGGSTSSNTISR	13	Free	Blocked	1	+2	314		26
16	CSEDNADSGQ	10	Blocked	Blocked	0	0	>50,000	>50,000	32
16(+K)	CSEDNADSGQK	11	Free	Free	1	+2	2100		ND

[a] Defined as the amount of peptide that gives a signal/noise ratio of about 5.

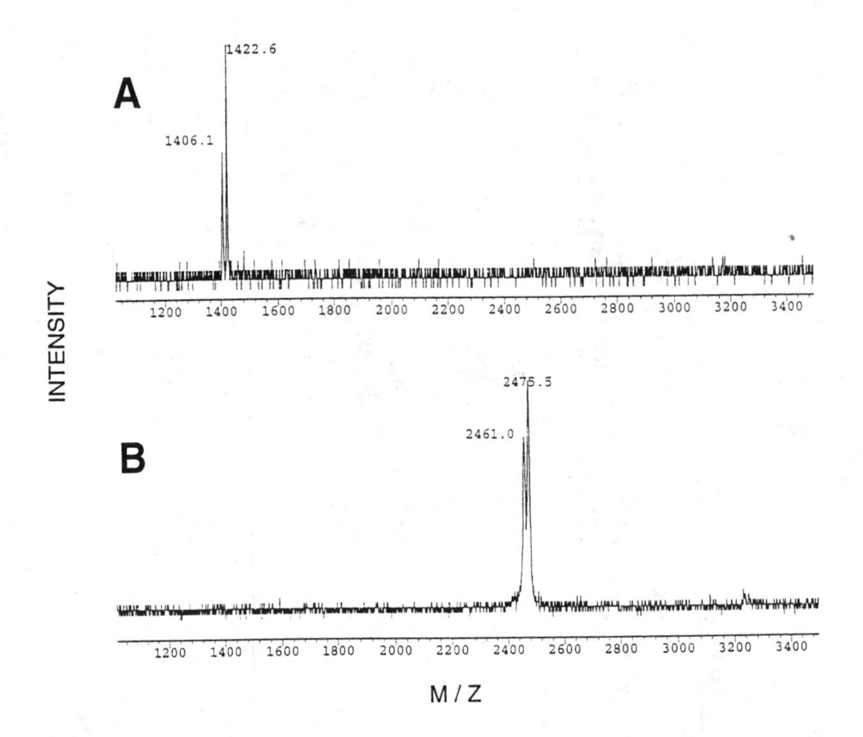

Fig. 4. Multiple MALDI-MS peaks resulting from apparent chemical modification. The MALDI-MS spectrum in Panel A was from a peptide isolated from a tryptic digest of a 20-kDa protein (250 pmol). Sequencing gave an 11-residue sequence that began with Gln-Tyr-Lys-Val … and that had a calculated mass (1421.6) that was within 0.07% of the observed mass (1422.6). The presence of the NH_2-terminal Gln and the absence of any detectable secondary amino acid sequence suggest the –16.5 mass peak present in this sample (observed mass = 1406.1) arises from a fraction of this peptide having undergone glutamine cyclization. Panel B shows the MALDI-MS spectrum for a tryptic peptide isolated from an in-gel digest of a 67-kDa protein (181 pmol). Sequencing identified residues 2–21, including a single Met at position 20, and indicated that residue 1 was Asp, Ser, Thr, Gly, or Ala. Based on the observed mass of 2461.0, residue 1 was tentatively assigned to Asp, which gives a predicted mass of 2466.7 and an error in the observed mass of 0.23%. The 2461.0 peak presumably arises from the singly protonated peptide, and the 2475.5 peak probably results from oxidation of the single Met residue in this peptide (predicted mass = 2482.7, which gives a 0.29% error for the observed mass).

probably results from partial oxidation of Met. Finally, based on sequencing, the two +15–17 peaks that are present in Panel A of Fig. 2 probably arise from the expected equilibrium between homoserine and homoserine lactone (i.e., these peptides were from a combined CnBr/tryptic digest, and the expected mass difference between homoserine and its lactone is 18).

Figure 5 illustrates another potential problem with MALDI-MS and that is sampling variation of the laser beam on the target surface. Although both spectra in this figure were obtained from the same target, the position used in Panel A provided a nearly 5:1 ratio of the major:minor species, whereas that in Panel B gave a nearly equal ratio. As a result, we routinely acquire spectra from at least three to six different positions on each target and then print out representative scans. Although Fig. 5 represents one of the more severe

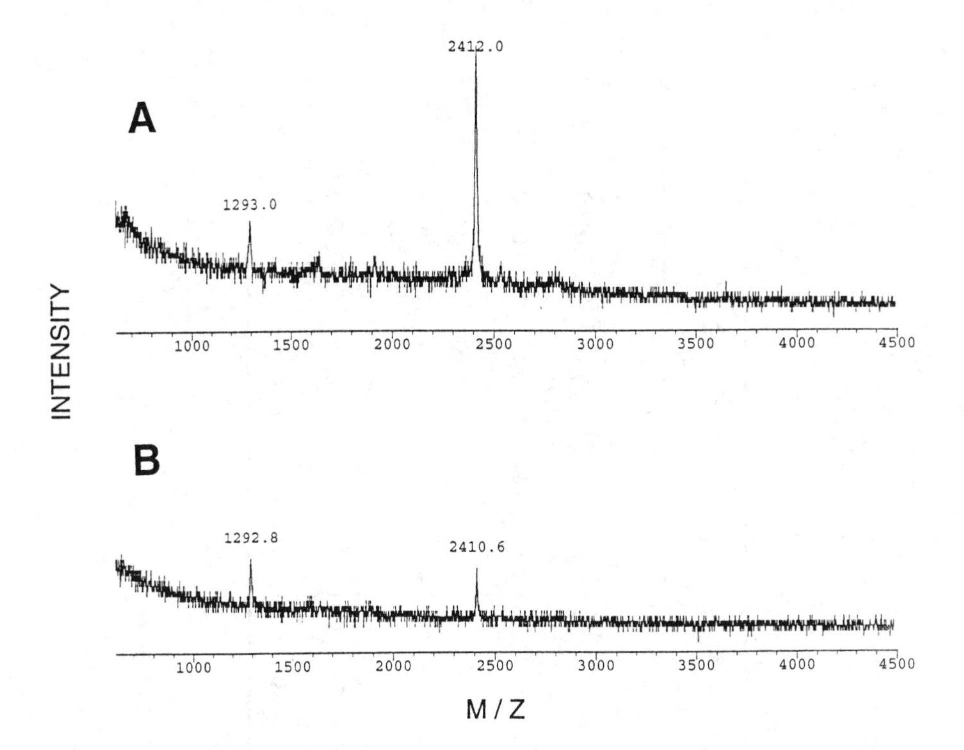

Fig. 5. Variability of MALDI-MS spectra as a function of laser position. The above spectra were obtained from two different laser positions on the same target that had been spotted with 3 μL of a peptide fraction obtained from the tryptic digest of 145 pmol of a 39-kDa protein. Note that this sample was mixed on the target (*see* Section 3.1. and Note 1), and depending on the laser position, the ratio of the 2411 to the 1293 species varies from about 5:1 in Panel A to about 1:1 in Panel B.

examples of this phenomenon, it appears to be present to some degree on nearly all targets that contain multiple species, and this factor undoubtedly contributes to variability in MALDI-MS spectra. Hence, the Yale limits of detection reported in Table 3 are the mean of two or more determinations that on average varied nearly sevenfold. Similarly, the average limits of detections varied by nearly fourfold for the Yale vs Harvard data sets, which were acquired on identical instruments. In an effort to reduce the apparent target inhomogenity shown in Fig. 5, the Keck Facility at Yale now mixes the sample and matrix solution prior to deposition on the target. Typically, samples for MALDI-MS are now prepared by mixing a 2-μL sample with a 1-μL matrix (α-CHCA at 4 mg/mL in 0.05% TFA, 50% CH_3CN) on a Sarstedt 1.5-mL tube cap, and 2 μL of the resulting suspension are loaded onto the target. Our impression is this procedure decreases the effect of laser position (within a given target) on the resulting spectra.

Finally, a major concern with interpreting MALDI-MS data is the mass accuracy that can be reasonably expected under "normal operating conditions" where, for instance, limiting amounts of material may result in far less than optimal resolution and signal/noise ratios. Because the average mass accuracy that we routinely obtain was only about ±0.25% (Table 1), studies were carried out to identify some of the critical parameters that affect routine laser desorption mass accuracy. As shown by the first two entries in Table 5, increasing the average amount of sample applied to the target from ~150 to 1300 fmol

Table 5
Variables That Affect Routine Mass Accuracy in MALDI-MS

Variables		Mode	Calib.	n	% Error	
					Mean	Range
"Unknown" samples (mass range: 702–2906)						
20–311 fmol (mean = 147 fmol)		Linear	Ext	43	0.31	0.06–1.13
312–6210 fmol (mean = 1300 fmol)		Linear	Ext	42	0.23	0.01–0.80
Single known sample (mass = 2153.5)						
HHMI/Harvard Facility	500 pmol	Linear	Ext	24	0.17	0.00–0.39
Keck Facility at Yale	500 pmol	Linear	Ext	24	0.15	0.02–0.46
Keck Facility at Yale	500 pmol	Linear	Int	24	0.03	0.00–0.092
Keck Facility at Yale	10 pmol	Linear	Int	15	0.01	0.00–0.036
Four known peptides (mass range: 1097–1857/100 fmol)						
Keck Facility at Yale		Linear	Ext	4	0.18	0.07–0.44
Company "A"		Linear	Ext	4	0.19	0.02–0.44
Company "B"		Linear	Ext	4	0.23	0.02–0.59
Company "C"		Reflectron	Ext	5	0.03	0.00–0.08

only marginally decreased the average error from about 0.3 to 0.2%, respectively. Because the average error was not significantly altered by analyzing synthetic peptides, as opposed to peptides derived from enzymatic digests, the average errors reported in Table 1 are apparently not elevated owing to the presence of unrecognized posttranslational modifications. Furthermore, a similar ~0.2% error was also found when 500 pmol of a synthetic peptide standard were analyzed at both Harvard and Yale and when 100 fmol amounts of four different synthetic peptides were analyzed with external calibration either at Yale, or at three different companies that manufacture time-of-flight laser desorption mass spectrometers (Table 5). A significant increase in mass accuracy was, however, readily achieved either by the use of internal calibration (*see* the Yale data in Table 5 where the average mass accuracy was increased to ±0.01–0.03%) or by the use of a reflectron (with external calibration; *see* the data from Company C in Table 5 where the mass accuracy was 0.03%). Of these two approaches for increasing mass accuracy, we prefer the use of a reflectron, because internal calibration standards can differentially suppress the MALDI-MS response of sample components (ref. *12* and data not shown).

3. Based on our experience, MALDI-MS has several characteristics that make it effective, not only as a mass-analysis method, but also as a complement to other methods performed in facility laboratories, such as internal protein sequencing (*see* refs. 1–3 and 12–15 for recent reviews and applications). These characteristics include high throughput, high sensitivity, and ease of use. In our laboratories, one person can reasonably analyze 24 or more samples/d, including sample target preparation and data interpretation. The data in Table 3 suggest that a sample target load of <100 fmol of peptide is within the limit of detection for most tryptic peptides. This amount represents a negligible portion of each sample analyzed by internal protein sequencing using the most sensitive methods currently available and an amount that is generally insufficient for other mass-analysis methods. Sample preparation and instrument operation are simple enough that the method had critical impact in our laboratories within a few days of the instrument's installation, even though none of the authors had prior training or experience in mass spectrometry.

   The high sensitivity of MALDI-MS makes it especially suitable for screening peptide fractions prior to sequencing. The Keck Facility performs at least 125 protein digests/yr, typically starting with 25–250 pmol of protein submitted to the facility as SDS-polyacrylamide gel slices stained with Coomassie blue. The average recovery of a tryptic peptide from the digests used in this study was about 11% based on initial sequencing yield (data not shown). Since our initial sequencing yields are often about 50%, the actual peptide recovery is probably above 20%. Assuming that 50 pmol protein were digested, this would correspond to 10 pmol peptide in a volume of about 100 μL acetonitrile/0.05% TFA. If 3-μL fractions are used for MALDI-MS analysis, the amount of peptide consumed for this step is only about 300 fmol, which is well above the expected limit of detection of most tryptic peptides and, yet, is still an extremely small quantity given the usefulness of mass analysis prior to sequencing, both for screening and for mass measurement. Actually, only a very small portion of the peptide subjected to mass analysis is consumed. Since the laser covers <4% of the sample surface area on the target and the sample can usually be analyzed repeatedly without loss of signal, most of the spectra shown here were generated by ionizing <1 fmol peptide.

   The usefulness of screening peptide fractions by MALDI-MS prior to sequencing has been shown in this work. MALDI-MS provides a valuable and necessary criterion of peptide purity and length, thus optimizing protein sequencing equipment usage. It rapidly identifies protease autolysis peptides and other digestion artifacts. Comparing an observed mass to the one calculated on the basis of sequence data increases confidence in sequence assignments, sometimes allows tentative calls to be made with more certainty, and helps identify posttranslational and chemical modifications that often cannot be identified on the basis of sequencing alone. In some cases, proteins can be identified on the basis of mass analysis alone, without sequencing, as reported by Henzel et al. *(12)* and Mann et al. *(16)*. Our evaluation of MALDI-MS mass accuracy suggests, however, that protein identification by peptide mass searching should be done with internal calibration—by overlaying samples with standards of known mass—or by reflectron analysis. The mass accuracy we observe with external calibration is 0.25%, or 4 Dalton for an average 1600-Dalton tryptic peptide (Table 1). In contrast, Henzel et al. reported an average mass accuracy of 0.09% (based an 50 peptides listed in Table 2 of ref. *12*) using internal calibration, in developing an application for identifying proteins on the basis of peptide mass analysis. As noted above, of the two approaches for increasing mass accuracy, we prefer reflectron analysis.

Protein identification by mass analysis is, however, a very worthwhile objective that can positively impact on the usage of protein sequencing instrumentation. For example, 62 of the last 88 (71%) proteins submitted to the Keck Facility for internal sequencing were identified by searching protein sequence data bases with a single tryptic peptide sequence. Most of these identified proteins proved to be contaminants, such as serum albumin, fibrinogen, tubulin, collagen, immunoglobulins, transferrins, keratin, vinculin, and ubiquitin, rather than the protein of interest. With an awareness of the limitations of MALDI-MS accuracy, these proteins could have been readily identified by mass analysis, without the expense and time needed for sequence analysis.

References

1. Hillenkamp, F., Karas, M., Beavis, R., and Chait, B. (1991) Matrix-assisted laser desorption/ionization mass spectrometry of biopolymers. *Anal. Chem.* **63,** 1193–1202.
2. Cotter, R. J. (1992) Time-of-flight mass spectrometry for the structural analysis of biological molecules. *Anal. Chem.* **64,** 1027–1039.
3. Chait, B. T. and Kent, S. H. (1992) Weighing naked proteins: practical, high-accuracy mass measurements of peptides and proteins. *Science* **257,** 1885–1894.
4. Brown, J. L. and Roberts, W. K. (1976) Evidence that approximately eighty percent of the soluble proteins from Ehrlich Asites cells are amino terminally acetylated. *J. Biol. Chem.* **251,** 1009–1014.
5. Driessen, H. P., De Jong, W. W., Tesser, G. I., and Bloemendal, H. (1985) The mechanism of N-termimal acetylation of proteins, in *Critical Reviews in Biochemistry*, vol. 18 (Fasman, G. D., ed.), CRC, Boca Raton, FL, pp. 281–325.
6. Rosenfeld, J., Capdeville, J., Guillemot, J., and Ferrara, P. (1992) In-gel digestions of proteins for internal sequence analysis after 1 or 2 dimensional gel electrophoresis. *Anal. Biochem.* **203,** 173–179.
7. Williams, K. R. and Stone, K. S. (1995) In-gel digestion of SDS PAGE-separated proteins: observations from internal sequencing of 25 proteins, in *Techniques in Protein Chemistry VI* (Crabb, J., ed.), Academic, San Diego, pp. 143-152.
8. Stone, K. L., McNulty, D. E., LoPresti, M. L., Crawford, J. M., DeAngelis, R., and Williams, K. R. (1992) Elution and internal amino acid sequencing of PVDF-blotted proteins, in *Techniques in Protein Chemistry III* (Angeletti, R., ed.), Academic, New York, pp. 22–34.
9. Williams, K. R., Kobayashi, R., Lane, W., and Tempst, P. (1993) Internal amino acid sequencing: observations from four different laboratories. *ABRF News* **4,** 7–12.
10. Stone, K. L. and Williams, K. R. (1993) Enzymatic digestion of proteins and HPLC peptide isolation, in *A Practical Guide to Protein and Peptide Purification for Microsequencing* (Matsudaira, P. T., ed.), Academic, New York, pp. 43–69.

11. Elliott, J. I., Stone, K. L., and Williams, K. R. (1993) Synthesis and use of an internal amino acid sequencing standard peptide. *Anal. Biochem.* **211,** 94–101.
12. Henzel, W. J., Billeci, T. M., Stults, J. T., Wong, S. C., Grimley, C., and Watanabe, C. (1993) Identifying proteins from two-dimensional gels by molecular mass searching of peptide fragments in protein sequence databases. *Proc. Natl. Acad. Sci. USA* **90,** 5011–5015.
13. Williams, K. R. and Carr, S. A. (1995) Protein analysis by integrated sample preparation, chemistry, and mass spectrometry, in *Encyclopedia of Molecular Biology and Biotechnology* (Meyers, R. A., ed.), VCH Publishers, New York, pp. 731–737.
14. Bornsen, K. O., Schar, M., Gassman, E., and Steiner, V. (1991) Analytical applications of matrix-assisted laser desorption ionization mass spectrometry. *Biol. Mass. Spectrom.* **20,** 471–478.
15. Germanos, S., Casteels, P., Elicone, C., Powell, M., and Tempst, P. (1994) Combined Edman-chemical and laser-desorption mass spectrometric approaches to micro peptide sequencing: optimization and applications, in *Techniques in Protein Chemistry V* (Crabb, J. W., ed.), Academic, New York, pp. 143–150.
16. Mann, M., Horup, P., and Roepstorff, P. (1993) Use of mass spectrometric molecular weight information to identify proteins in sequence databases. *Biol. Mass. Spectrom.* **22,** 338–345.

92

A Manual C-Terminal Sequencing Procedure for Peptides

The Thiocyanate Degradation Method

Franca Casagranda and John F. K. Wilshire

1. Introduction

The possibility of sequencing peptides or proteins from their *N*-terminal ends became a reality over 25 years ago with the discovery by Edman *(1,2)* of a sequential degradation method. In the intervening period, the Edman method has been automated, and now peptide and protein sequencing instruments, in particular the gas-phase sequencer, are available, which are:

1. Sensitive (very small quantities of polypeptide [pmol] can be sequenced);
2. Capable of giving *N*-terminal sequencing information rapidly; and
3. Easy to use.

When used in combination with on-line high-performance liquid chromatography (HPLC), these instruments are capable of yielding unequivocal sequence information for up to 35–70 residues routinely for a wide variety of peptides and proteins before the buildup of background signals eventually makes residue assignments difficult, if not impossible. Such signals can arise from:

1. Sample preparation (contaminants in the sample reduce the initial and repetitive yields);
2. Previous residues; and
3. Instrumental noise.

In practice, the complete sequence often has to be derived by establishing the sequences of smaller fragments (i.e., peptides) obtained from the peptide/protein by the action of either enzymes or chemical reagents *(3)*.

N-terminal sequencing by itself is often not sufficient to establish the complete sequence of a lengthy peptide because uncertainty can arise concerning the identification of the last few amino acids in the chain. Furthermore, if the *N*-terminus is blocked, e.g., if the amino group is acetylated or formylated *(4)*, as occasionally occurs, then the sequence is clearly not determinable by the Edman degradation. Other blocking reactions, which are frequently encountered, are: (1) the cyclization of *N*-terminal glutamine residues to form pyroglutamyl residues, and (2) the migration of the *O*-acyl groups of serine or threonine residues to the *N*-terminus. Moreover, C-terminal sequencing information would be of great value in the area of molecular biology, particularly for:

From: The Protein Protocols Handbook
Edited by: J. M. Walker Humana Press Inc., Totowa, NJ

1. Detecting posttranslational modifications at the carboxy terminus of the expressed gene products obtained from known DNA sequences;
2. Confirming the correct placement of initiation codons and reading frames; and
3. Providing a basis for the design of oligonucleotide probes capable of screening cDNA libraries.

It is clear, therefore, that a viable C-terminal sequencing method would become a very important tool for the characterization of proteins. Although mass spectrometry (*see* Note 1) and the specific hydrolytic capability of certain enzymes (*see* Note 2 and Chapter 93) have been used, although to a limited extent, for C-terminal sequencing, the present account, which is an updated and revised version of our recent article *(5)* on the same subject, will be devoted to a description of the traditional (but recently much improved *[6–8]*) thiocyanate degradation procedure (*see* Fig. 1). This chemical procedure is not only capable of giving useful sequence determination for peptides containing all the common hydrophobic amino acids, as well as tyr and trp (its limitations are discussed in Note 3 [Section 4.2.]), but is also experimentally reasonably straightforward. Recently, variations of the thiocyanate degradation procedure based on new reagents and new solid supports have been reported *(9–17)*. As a result of these developments, C-terminal sequencing procedures (to be described elsewhere in this volume) have been developed, which have overcome some of the limitations of the procedure to be described. Significantly, the sequencing of asp and glu residues is now possible, although that of ser, cySH, thr, and pro, in particular, continue to present difficulties. Indeed, current technology has progressed to the point where automated sequencers based on the new chemistries *(9–17)* are now being marketed. Notwithstanding the difficulties that remain to be resolved, the long search (ushered in by the work of Schlack and Kumpf nearly 70 years ago *[18]*) for a viable method for the C-terminal sequencing of peptides and proteins appears to be nearly over.

1.1. Thiocyanate Degradation Procedure

The traditional thiocyanate degradation procedure (the reactions involved are shown in Fig. 1) is a chemical method based on reactions discovered by Schlack and Kumpf *(18)* (for a discussion of alternative, but less successful chemical methods, *see* Ward *[19]*). This procedure (*see* Fig. 1), which has recently been critically reviewed by Inglis *(20)*, involves the conversion of the C-terminal amino acid of a peptide into the corresponding peptidyl thiohydantoin by reaction of the peptide with acetic anhydride/acetic acid (the activation reaction), followed by treatment with either a thiocyanate salt or a solution of thiocyanic acid in acetone (the coupling reaction). Subsequent treatment with base (the cleavage reaction) releases the C-terminal amino acid (as its thiohydantoin derivative) together with a shortened (by one residue) peptide possessing a free carboxyl group available for repetition of the procedure. This method has proven successful for the sequencing of peptides containing hydrophobic residues, and also tyr and trp residues, but not of peptides containing sensitive side chains (*see* Note 3). Degradations can be performed either with a solution of the peptide or (preferably) with the peptide attached to a solid support.

Degradations performed with the peptide in solution can provide a rapid procedure for determining the C-terminal residue, but the method is limited to those peptides that are soluble in the coupling solution. Furthermore, because removal of the cleavage and

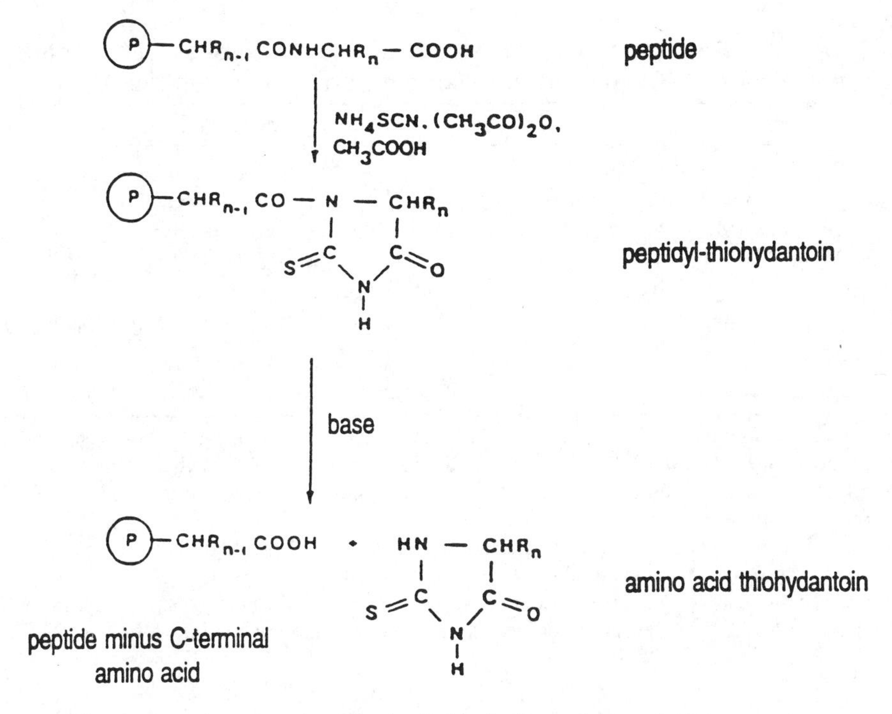

Fig. 1. Reactions involved in the Schlack-Kumpf degradation procedure.

other reagents from the peptide (gel filtration *[21]* or filtration followed by freeze-drying *[22]* has been used) can be time-consuming and incomplete, only a single cycle of reactions is usually obtained. In order to obtain more extensive sequencing information, therefore, procedures have been developed for performing degradation cycles on peptides that have been immobilized on a solid support by covalent attachment through their amino groups *(23)*. The use of this technology, together with (1) the identification of the thiohydantoin derivatives by means of HPLC *(6,24,25)* (its introduction *[24]* in 1979 revolutionized C-terminal sequencing technology) and (2) the use *(6)* of antioxidants (e.g., dithioerythritol) for the protection of the somewhat unstable thiohydantoin derivatives (*see* Note 4), has simplified the procedure and extended the number of C-terminal residues that can be sequenced (some of the sequences that have been determined are shown in Table 1).

2. Materials

High-purity reagents are necessary for optimal results when sequencing. The materials used and, where necessary, recommended purification procedures are given in this section.

1. 1 g Controlled-pore glass beads (CPG 10Å pore size, 200–400 mesh [ElectroNucleonics]) are washed by suspending them in 10% (v/v) HF for 2 min. The beads are then rinsed thoroughly with water and dried overnight in an oven at 100°C.
2. γ-Aminopropyltriethoxysilane in acetone (5% w/v).
3. Triethylamine (TEA).

Table 1
Some C-Terminal Sequences Determined Using the Traditional Thiocyanate Degradation Reaction (Cleaved Thiohydantoin Derivatives Identified)

Researchers	Sequence determined
Yamashita (1971) *(26)*	–Ala-Ser-Val[a,c]
Rangarajan and Darbre (1976) *(22)*	–His-Phe-Asp-Ala-Ser-Val[b]
Meuth et al. (1982) *(25)*	–Leu-Gly-Tyr-Gln-Gly[d]
Hawke et al. (1987) *(9)*	–Tyr-Gln-Gly[d]
Miller and Shively (1989) *(27)*	–Leu-Ala[d]
Inglis et al. (1989) *(6)*	–Leu-Ala-Ile-Tyr-Val-Met-Ala-Phe-Val[d]
Bailey and Shively (1990) *(28)*	–Glu-Leu[d]
Casagranda et al. *(29)*[e]	–Val-Leu-Phe[d]

[a]Degradation performed in solution. All other sequences were obtained using a solid-phase support.
[b]Amino acid thiohydantoins identified by gas-liquid chromatography as their trimethylsilylated derivatives, and by thin layer chromatography. Serine thiohydantoin unknown; formation inferred by gas liquid chromatography.
[c]Serine thiohydantoin unknown; formation inferred by thin-layer chromatography.
[d]Amino acid thiohydantoins identified by HPLC.
[e]C-terminal fragments of two hexapeptides containing thr and ser residues, respectively, in the next position (*see* Section 4.2.); sequencing stops at these residues and does not continue further.

4. Succinic anhydride.
5. Dimethylformamide (DMF), redistilled.
6. Trifluoroacetic acid (TFA), anhydrous.
7. 0.5*M* sodium hydrogen phosphate (Na_2HPO_4), pH 9.0 containing 0.2*M* sodium chloride.
8. 5.8*M* hydrochloric acid.
9. Trisodium citrate buffer: 19.6 g trisodium citrate ($Na_3C_6H_5O_7 \cdot 2H_2O$), 10 mL concentrated nitric acid, 40 mL methanol, and 1 g phenol are mixed cautiously together; the resultant mixture is then made up to 1 L, and the solution adjusted to pH 2.0 with nitric acid.
10. Dowex 50 × 8 (H^+ form) resin: Soak commercial Dowex 50 × 8 (H^+ form) resin (Bio-Rad, Hercules, CA) in 2*M* NaOH for 2 h, wash with water, and finally soak it in 0.1*M* aqueous EDTA solution overnight. Then wash the resin with water, soak it in HCl (2*M*) for several hours, wash with water again, and finally rinse the resin with acetone several times (to remove any water).
11. Ammonium thiocyanate.
12. Acetic anhydride.
13. Acetic acid.
14. Dithioerythritol (DTE).
15. Activation solution: acetic anhydride, acetic acid (4:1).
16. Coupling solution: thiocyanic acid in acetone solution (~1.8*M*) (*see* Section 3.1.6.).
17. Cleavage solution: 0.5*M* KOH in 33% MeOH/water containing DTE (~1 mg/10 mL).
18. HPLC buffers: buffer A: 0.25 g/L ammonium acetate, pH 4.76; buffer B: acetonitrile.
19. HPLC column: Waters Pico-Tag ODS, 300 × 2.1 mm, 5 µm pore size.

3. Methods

3.1. Manual C-Terminal Sequencing (see Notes 4–7)

The following section describes an experimental procedure (*see* Notes 4 and 5) with which the authors are familiar *(6,7)*. The peptide under investigation is covalently attached to a solid support (glass beads). With this procedure, which is experimentally simple and capable of being performed without difficulty by the individual researcher, useful sequence information for peptides (*see* Note 6) containing all the common hydrophobic amino acids, as well as Tyr and Trp, can be obtained. It has recently been automated (*see* Note 7).

3.1.1. Preparation of Carbonyldiimidazole-Activated Glass Beads

Several steps are involved in the preparation of the activated *N,N*-carbonyldiimidazole (CDI) glass beads, which are used as the solid support *(25)*. We have modified the literature method, however, in the way in which the glass beads are washed. The preparation of the activated glass beads is carried out as follows.

3.1.1.1. Preparation of Aminoalkylsilyl Beads

1. Suspend the glass beads (0.5 g) in a solution (20 mL) of γ-aminopropyltriethoxysilane in acetone (5% w/v); degas the mixture and shake for 20 h at 45–50°C.
2. Filter the aminoalkylsilyl beads, which will have become yellow, wash with acetone (4 × ~2 mL), dry in a vacuum, and store in a desiccator at –20°C.

3.1.1.2. Preparation of Succinyl Beads

1. Suspend aminoalkylsilyl glass beads (0.5 g) in a solution of acetone (12.5 mL) containing triethylamine (0.75 mL) and succinic anhydride (0.5 g). Degas the mixture, and shake for 3 h at room temperature.
2. Filter the resultant succinyl glass beads, wash with water and finally with acetone, dry in a vacuum, and store in a desiccator at –20°C.

3.1.1.3. Activation of Succinyl Beads with CDI

1. Suspend succinyl glass beads (0.5 g) in DMF (3 mL). Add CDI (2.4 g), degas the mixture, and shake overnight at room temperature.
2. Filter the beads, and thoroughly wash them with DMF and finally with dichloromethane. Dry the CDI-activated glass beads under vacuum, and store in a desiccator at –20°C.

3.1.2. Attachment of Peptide to CDI-Activated Glass Beads (see Notes 8 and 9)

1. Dissolve the peptide (~1 µmol) in anhydrous trifluoroacetic acid (TFA) (1 mL), and allow the resultant solution to stand at room temperature for 1 h under an atmosphere of nitrogen.
2. Remove TFA on a rotary evaporator using a nitrogen flush, and wash the dried residue with dichloromethane (2 × 3 mL).
3. Dissolve the resultant residue in DMF (1 mL), and add TEA (30 µL).
4. Add CDI-activated glass beads (100 mg) to the mixture, degas the resultant suspension, and cover the suspension with a nitrogen atmosphere.
5. Gently shake the suspension at room temperature overnight in order to ensure maximum covalent binding of the peptide to the beads.
6. Filter the beads, and wash successively with water, Na_2HPO_4 solution (0.5*M,* pH 9.0) containing NaCl (0.2*M*), water again, and finally with acetone. Dry the peptide-bound glass beads in a vacuum, and store in a desiccator at –20°C before use.

3.1.3. Amino Acid Analysis of Peptide-Bound Glass Beads

1. To determine the amount of peptide bound to the glass beads, treat the beads (1–2 mg) with 5.8*M* HCl (200 µL) at 108°C for 24 h under vacuum in a Waters Pico Tag Work Station.
2. Dry the beads under vacuum, add trisodium citrate buffer (0.2*M,* pH 2.0, 200 µL), briefly stir the mixture, remove the supernatant liquor, and analyze the liquor on a Waters amino acid analyzer.

3.1.4. Preparation of Thiocyanic Acid (HSCN) in Acetone Solution

A modified procedure of the literature Method II *(30)* was adopted as follows:

1. Suspend Dowex H^+ resin (*see* Section 2.) in acetone; degas the suspension under vacuum, and pour into a glass column (2 × 30 cm). After the resin particles have settled, wash the column through with acetone.
2. Adjust the flowrate to 1 drop/8 s, and add a solution of ammonium thiocyanate (6 g) in acetone (20 mL). When the eluate begins to produce a deep red color with dilute (ca. 1%) aqueous ferric chloride solution, collect fractions. When stored in a freezer at –20°C, our HSCN/acetone solutions were stable for long periods (upward of 1 yr or longer).

3.1.5. Determination of HSCN/Acetone Concentration

A modification of the Volhard method (cf *25*) is used as follows:

1. To an Erlenmeyer flask, add water (20 mL), concentrated nitric acid (2 mL), saturated ferric ammonium sulfate solution in water (10 drops), a known volume (100 μL) of an HSCN/cetone solution, and 0.1*M* silver nitrate solution (3 mL).
2. Cover the resultant mixture with toluene (3 mL), allow to react for a few minutes, and then titrate with 0.1*M* ammonium thiocyanate standard solution; the end point is reached when the solution turns a pale orange color. The concentrations of HSCN solutions in acetone vary from 0.2 to 2.5*M*. In general, 1.5–2.0*M* solutions are used for the coupling reactions.

3.1.6. C-Terminal Amino Acid Sequencing (Solid-Phase Method)

1. Place glass beads (~2–10 mg) coupled with peptide or protein (~15–30 nmol) in a screw-cap Eppendorf vial (1.5 mL). Add a solution of acetic anhydride and acetic acid (4:1, 120 μL), vortex the mixture, and heat at 80°C (water bath) for 5 min.
2. Add HSCN/acetone solution (~1.8*M,* 30 μL), stir the resultant mixture, and heat for an additional 30 min at 80°C.
3. Remove the vial from the water bath, carefully remove the supernatant liquid by pipet, and discard it.
4. Wash the glass beads with DMF (2 × 200 μL), and then with 70% acetonitrile/water (3 × 200 μL).
5. Cover the beads with a solution (50 μL) of 0.5*M* KOH in 33% MeOH/water containing DTE (0.01%); stir the suspension at room temperature for 3 min.
6. Remove the supernatant alkaline liquor by pipet, neutralize with 5% acetic acid in 20% acetonitrile (50 μL), and analyze (HPLC) the resultant solution for thiohydantoin content.
7. Immediately wash the glass beads successively (200-μL aliquots) with 5% acetic acid in 70% acetonitrile (twice), 70% acetonitrile (twice), acetonitrile (twice), and finally rinse twice with activating solution (acetic anhydride/acetic acid; 4:1) before commencing the next cycle.

3.2. Thiohydantoins: HPLC Identification and Analysis

3.2.1. Preparation of Standard Amino Acid Thiohydantoins (Crystalline)

1. Add acetylamino acid (15 mmol) and ammonium thiocyanate (1.5 g) to a stirred mixture of acetic anhydride (9 mL) and acetic acid (1 mL). Stir the suspension at 80–85°C (water bath) for 60 min, by which time the reagents will have dissolved.
2. Evaporate the solution to dryness with hydrochloric acid (1*M,* 40 mL) in an evaporating dish.
3. Dissolve the residue in water (40 mL), and again evaporate the solution to dryness. Wash the resultant solid from the evaporating dish with ice water, filter, wash with more ice water, and dry at 50°C. If desired, the thiohydantoin derivative can be crystallized either from water or from aqueous ethanol.

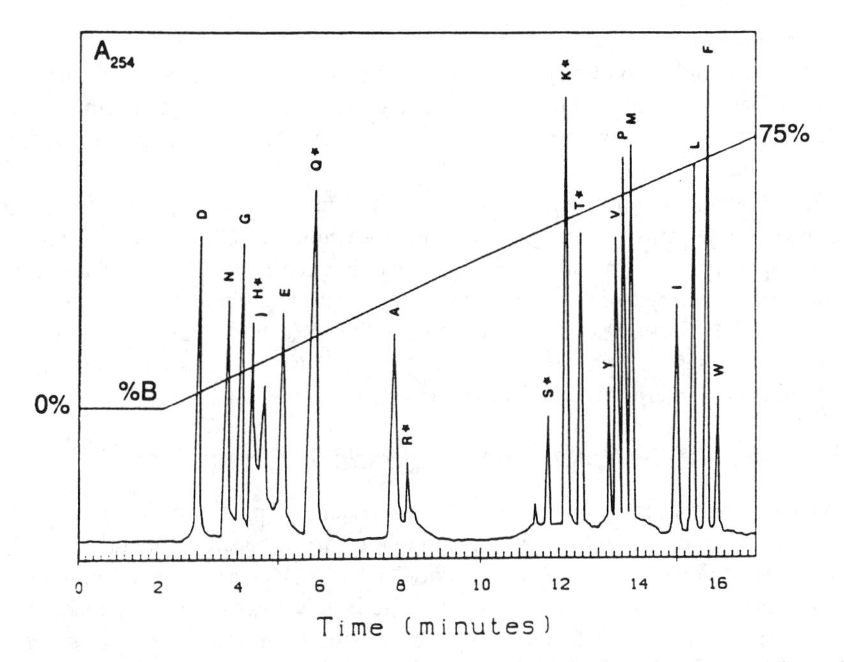

Fig. 2. Gradient elution profile of common amino acid thiohydantoins separated on a Waters Pico Tag ODS column (flowrate = 1 mL/min, column temperature = 40°C) and detected at 254 nm. The elution gradient used is shown in the figure. *Indicates amino acid "thiohydantoins" prepared *in situ*.

3.2.2. Preparation of Standard Amino Acid Thiohydantoins (In Situ)

1. Add acetylamino acid (100 nmol) to a mixture of acetic anhydride/acetic acid (4:1, 120 µL) in a screw-cap Eppendorf vial, and heat the resultant mixture for 5 min at 80°C.
2. Add HSCN/acetone solution (~1.8*M,* 30 µL), and heat the mixture at 80°C for 30 min; finally evaporate the mixture to dryness under vacuum.
3. Treat the residue at room temperature with cleavage reagent (0.5*M* KOH in 33% methanol/water), stir the mixture, immediately remove the resultant solution, and analyze for the corresponding thiohydantoin by HPLC.

3.2.3. Analysis of Standard Amino Acid Thiohydantoins

1. All the crystalline 2-thiohydantoins are resolved by gradient elution using a Waters Pico-Tag ODS column *(6–8)*. Several amino acid 2-thiohydantoins have not yet been obtained in a crystalline or a pure form; consequently, the HPLC peaks obtained by *in situ* reactions with the respective amino acids (*see* Fig. 2) may not necessarily be the corresponding thiohydantoins.
2. The following buffers are used: buffer A: ammonium acetate (0.25 g/L, pH 4.76); buffer B: acetonitrile. Both buffers contain DTE (~10 mg/L). A column temperature of 40°C and a buffer flowrate of 1.0 mL/min are used. A typical buffer gradient is shown in Fig. 2. All thiohydantoins are eluted within 16 min (*see* Fig. 2).

4. Notes

4.1. Other C-Terminal Sequencing Methods

1. Mass spectrometry has proven useful for the sequencing of certain peptides, because peptides tend to fragment in the ion source of the mass spectrometer at the peptide bond *(31–35)*. Disadvantages in the use of mass spectrometry are the need for:

a. Large amounts of peptide;
b. Extremely high resolution (and expensive) spectrometers in order to identify residues of similar or identical molecular weight, e.g., Lys and Gin, and Leu and Ile; and
c. Sophisticated operator expertise for the interpretation of the complex mass spectra produced.

2. The use of some carboxypeptidases has aroused interest because they are capable of removing amino acids one at a time from the C-terminal end of a peptide *(19,36–38)*. In practice, however, interpretation of the results obtained is not always clear-cut because the rates of cleavage are dependent on the nature of the side chain of the amino acid being removed and, to a lesser extent, on the nature of the adjacent residues. Furthermore, many carboxypeptidases undergo autodigestion; "foreign" amino acids are thereby produced, and problems in interpretation of the results arise *(19)* (*see* Chapter 93).

4.2. Limitations of the Thiocyanate Degradation Procedure

3. At present, this procedure is limited to the sequencing of peptides containing the hydrophobic amino acids, and trp and tyr. This limitation arises because only the thiohydantoin derivatives of Gly, Val, Ala, Leu, Ile, Phe, Tyr, Lys, Met, His, Trp, Asn, and Gln are readily obtainable (and unequivocally characterized as such) by the application of the original thiocyanate degradation procedure (or minor variations thereof) to the parent amino acids *(6–8,21,25,26,39)*. Much confusion still exists as to whether or not the thiohydantoins of Asp, Glu, Ser, Thr, and Arg can similarly be prepared, and that of pro has until recently been obtained only with difficulty *(40,41)*. Recently, it has been reported *(42)* that many of the previously unknown thiohydantoins can be prepared by the judicious use of appropriate protecting groups. Full details of this work, however, have yet to appear. It is relevant to note, however, that the thiohydantoins of His, Arg, and Lys (as their hydrochlorides), and of pro, were prepared more than 40 yr ago by indirect routes *(43)* not involving the Schlack-Kumpf reaction. More recently, a detailed investigation *(29)* of the thiocyanate degradation with these "difficult" amino acids has been carried out. The thiohydantoin derivatives of asp and glu could not be prepared (cf *44*), although both derivatives are obtainable by indirect methods *(45,46)*. Curiously, Asp thiohydantoin and Glu thiohydantoin have been detected during the solid-phase sequencing of certain asp *(6,7)* and glu *(8)* peptides; the reasons for this unexpected result are unknown. Although arg thiohydantoin has not yet been prepared in a pure state, an apparently pure (HPLC analysis) *(29)* product has been obtained (by an *in situ* reaction) that is expected to be a useful indicator for the presence of arg in a peptide. The Schlack-Kumpf reactions with thr and ser occur more slowly than with the hydrophobic amino acids, and give, surprisingly, thiohydantoin derivatives of β-methylcysteine (the structure of the derivative obtained was ascertained by X-ray crystallography *[47]*) and cysteine, respectively, which, however, do not survive the cleavage reaction (decomposition to give the corresponding olefinic thiohydantoins occurs to a significant extent). This slowness to react explains why sequencing of the hexapeptides, Tyr-Gly-Thr-Val-Leu-Phe and Tyr-Gly-Ser-Val-Leu-Phe, which contain the readily sequenceable residues of Tyr, Gly, Val, Leu, and Phe, stops at the threonine and serine residues, respectively *(29)* (*see also* Table 1). The reaction with Pro appears to take place readily, and a reasonably pure (melting point; cf *43*) sample of proline thiohydantoin was obtained (isolated, however, in poor yield [6%]), and fully characterized by UV and, ^{1}H and ^{13}C nmr spectroscopy. Since our work was completed, the preparation of proline thiohydantoin in good (approx 40%) yield by a modified Schlack-Kumpf reaction has been reported *(17)*, and a promising procedure for the sequencing of C-terminal proline residues developed.

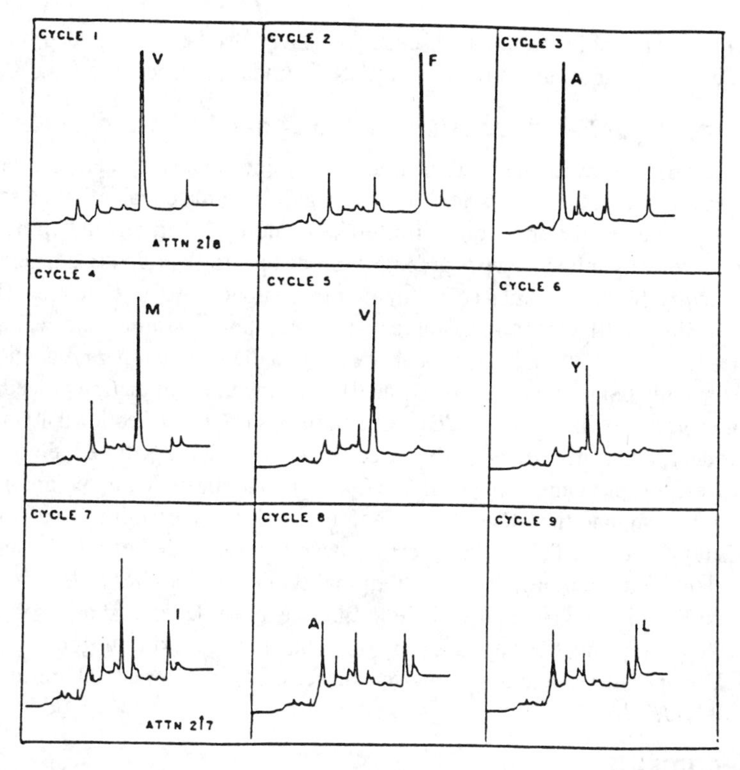

Fig. 3. C-terminal sequence (nine residues) of a synthetic decapeptide ⒼYLAIYVMAFV.

4.3. Manual Sequencing

4. The addition of a thiol antioxidant, e.g., DTE, retards decomposition of the amino acid thiohydantoin released from the C-terminus by the cleavage reagent. All washing solvents (except for the final acetonitrile rinse) and also the cleavage reagent therefore contained DTE (0.01%).
5. With this procedure, 3 d are required for the preparation of the insoluble glass beads support. Each cycle time requires 48 min, and therefore, the procedure is comparable in speed with that used in current automated *N*-terminal sequencing methodology. The method does not require any drying steps and is simple to carry out. Manual solid-phase sequencing from the C-terminal end has been used to sequence up to nine residues from a hydrophobic decapeptide peptide (Tyr-Leu-Ala-Ile-Tyr-Val-Met-Ala-Phe-Val) (*see* Fig. 3).
6. The C-terminal sequencing of proteins by this degradation method has been less well studied; nevertheless, the C-terminal amino acids of lysozyme and ribonuclease (Leu and Val, respectively) were readily identified *(8)*.
7. Automation of the thiocyanate degradation procedure would enable:
   a. The acquisition of sequence data to be accelerated;
   b. Repetitive yields and the sequencing capability to be improved;
   c. Much smaller amounts of peptide to be sequenced; and
   d. The C-terminal sequencing procedure to be simplified.

   Recently, modified *N*-terminal sequencers have been successfully used for automated C-terminal sequencing *(8,48)* by this procedure. With these sequencers, suitable programs have been developed for the sequencing of peptides covalently attached to activated glass beads and to PVDF membranes (the sequencing protocol is similar to that described above for the manual method). Automated sequencers have also been developed for use with the newly described variations of the thiocyanate degradation procedure *(12,13,16)*.

4.4. Sequencing of Peptides Attached to Other Solid Supports

8. It would clearly be advantageous if the traditional thiocyanate degradation procedure could be carried out without the necessity for attaching the peptide covalently to a solid support. However, in our experience, only limited sequencing information can be obtained by application of this thiocyanate degradation procedure to peptides (e.g., met-enkephalin) noncovalently bound to other solid (membrane) supports, e.g., Sequelon (DITC disks, MilliGen/Biosearch). Only two or three of the hydrophobic amino acid residues could be sequenced, an observation that is probably owing to the fact that there is a tendency for the electrostatically bound peptide to be washed off during the degradation procedure. Furthermore, it is probable that the basic cleavage conditions used (*see* Section 3.1.6.) also remove the peptide. If less concentrated base is used or a shorter cleavage time is employed in order to overcome this problem, then not all the C-terminal amino acid thiohydantoin is removed in the first cycle, and therefore, thiohydantoin overlap occurs in the next cycle. Interestingly, Zitex (a porous Teflon) has recently been recommended as a solid support for the noncovalent binding of proteins for C-terminal sequencing studies *(13)*.
9. A carboxylic acid-modified polyethylene film has been developed as a solid support for the covalent attachment of peptides *(13–16)*. This material exhibits good stability to base and high temperatures, and has been used successfully in automated C-terminal sequencing studies *(13–16)*.

Acknowledgments

We wish to thank A. Kirkpatrick and Auspep Pty. Ltd. (Australia) for making available to us the model peptides used in our studies.

References

1. Edman, P. (1950) Preparation of phenylthiohydantoins from some natural amino acids. *Acta Chem. Scand.* **4,** 277–282.
2. Edman, P. and Begg, G. (1967) A protein sequenator. *Eur. J. Biochem.* **1,** 80,81.
3. Aitken, A., Geisow, M. J., Findlay, J. B. C., Holmes, C., and Yarwood, A. (1989) Peptide preparation and characterisation, in *Protein Sequencing: A Practical Approach* (Findlay, J. B. C. and Geisow, M. J., eds.), Oxford University, Oxford, UK, pp. 43–68.
4. Chauhan, J. and Darbre, A. (1981) Determination of acetyl and formyl groups by gas-liquid chromatography. *J. Chromatogr.* **211,** 347–359.
5. Casagranda, F. and Wilshire, J. F. K. (1994) C-terminal sequencing of peptides: The thiocyanate degradation method, in *Methods in Molecular Biology,* vol. 32: *Basic Protein and Peptide Protocols* (Walker, J. M., ed.), Humana, Totowa, NJ, pp. 335–349.
6. Inglis, A. S., Wilshire, J. F. K., Casagranda, F., and Laslett, R. L. (1989) C-Terminal sequencing: A new look at the Schlack-Kumpf thiocyanate degradation procedure, in *Methods of Protein Sequence Analysis* (Wittmann-Liebold, B., ed.), Springer-Verlag, Berlin, pp. 137–144.
7. Casagranda, F. (1989) M. App. Sc. Thesis. Swinburne Institute of Technology, Melbourne, Australia.
8. Inglis, A. S., Casagranda, F., and Wittmann-Liebold, B. (1990) Progress in the development of a C-terminal sequencing method for proteins and peptides, in *Methods in Protein and Nucleic Acid Research* (Tschesche, H., ed.), de Gruyter, Berlin, pp. 187–211.
9. Hawke, D., Lahm, H. W., Shively, J. E., and Todd, C. W. (1987) Microsequence analysis of peptides and proteins: trimethylsilyl isothiocyanate as a reagent for COOH-terminal sequence analysis. *Anal. Biochem.* **166,** 298–307.

10. Boyd, V. L., Bossini, M., Zon, G., Noble, R. L., and Mattaliano, R. L. (1992) Sequencing of peptides and proteins from the carboxy terminus. *Anal. Biochem.* **206,** 344–352.
11. Bozzini, M. L., Zhao, J., Yuan, P. M., and Boyd, V. L. (1994) Applications using the alkylation method for carboxy-terminal protein sequencing. *Proc. 7th Protein Soc. Meeting,* San Diego, CA, pp. 13–18.
12. Bozzini, M., Ciolek, D., DeBarbieri, B., Hollfelder, K., Boyd, V. L., Yuan, P. M., Michel, H., and Pan, Y. C. E. (1994) Using an automated C-terminal sequencer in the characterization of recombinant proteins. *Proc. 7th Protein Soc. Meeting,* San Diego, CA, pp. 19–22.
13. Bailey, J. M., Shenoy, N. R., Ronk, M., and Shively, J. E. (1992) Automated carboxy-terminal sequence analysis of peptides. *Protein Sci.* **1,** 68–80.
14. Shenoy, N. R., Bailey, J. M., and Shively, J. E. (1992) Carboxylic acid-modified polyethylene: A novel support for the covalent immobilization of polypeptides for C-terminal sequencing. *Protein Sci.* **1,** 58–67.
15. Bailey, J. M., Nikfarjam, F., Shenoy, N. R., and Shively, J. E. (1992) Automated carboxy-terminal sequence analysis of peptides and proteins using diphenyl phosphoroisothiocyanatidate. *Protein Sci.* **1,** 1622–1633.
16. Shenoy, N. R., Shively, J. E., and Bailey, J. M. (1993) Studies in C-terminal sequencing; New reagents for the synthesis of peptidylthiohydantoins. *J. Protein Chem.*, **12,** 195–205.
17. Bailey, J. M., Tu, O., Issai, G., Ha, A., and Shively, J. E. (1995) Automated carboxy-terminal sequence analysis of polypeptides containing C-terminal proline. *Anal. Biochem.* **224,** 588–596.
18. Schlack, P. and Kumpf, W. (1926) On a new method for determination of the constitution of peptides. *Hoppe-Seyler's Z. Physiol. Chem.* **154,** 125–170.
19. Ward, C. W. (1986) Carboxyl terminal sequence analysis, in *Practical Protein Chemistry: A Handbook* (Darbre, A., ed.), John Wiley, New York, pp. 492–525.
20. Inglis, A. S. (1991) Chemical procedures for C-terminal sequencing of peptides and proteins. *Anal. Biochem.* **195,** 183–196.
21. Stark, G. R. (1968) Sequential degradation of peptides from their carboxyl termini with ammonium thiocyanate and acetic anhydride. *Biochemistry* **7,** 1796–1807.
22. Rangarajan, M. and Darbre, A. (1976) Studies on sequencing of peptides from the carboxyl terminus by using the thiocyanate method. *Biochem. J.* **157,** 307–316.
23. Williams, M. J. and Kassell, B. (1975) A solid-phase method for sequencing from the carboxyl terminus. *FEBS Lett.* **54,** 353–357.
24. Schlesinger, D. H., Weiss, J., and Audhya, T. K. (1979) Isocratic resolution of amino acid thiohydantoins by high-performance liquid chromatography. *Anal. Biochem.* **95,** 494–496.
25. Meuth, J. L., Harris, D. E., Dwulet, F. E., Crowl-Powers, M. L., and Gurd, F. R. N. (1982) Stepwise sequence determination from the carboxyl terminus of peptides. *Biochemistry* **21,** 3750–3757 (supplementary material).
26. Yamashita, S. (1971) Sequential degradation of polypeptides from the carboxyl-ends. I. Specific cleavage of the carboxyl-end peptide bonds. *Biochem. Biophys. Acta.* **229,** 301–309.
27. Miller, C. G. and Shively, J. E. (1989) Carboxy-terminal sequence analysis of proteins and peptides by chemical methods, in *Methods of Protein Sequence Analysis* (Wittmann-Liebold, B., ed.), Springer-Verlag, Berlin, pp. 144–151.
28. Bailey, J. M. and Shively, J. E. (1990) Carboxy-terminal sequencing: formation and hydrolysis of C-terminal peptidylthiohydantoins. *Biochemistry* **29,** 3145–3156.
29. Casagranda, F., Duggan, B. M., Kirkpatrick, A., Laslett, R. L., and Wilshire, J. F. K. (1996) Studies in thiohydontoin chemistry II: C-terminal sequencing of peptides. *Aust. J. Chem.* (in press).
30. Dwulet, F. E. and Gurd, F. R. N. (1979) Use of thiocyanic acid to form 2-thiohydantoins at the C-terminus of proteins. *Int. J. Pept. Protein Res.* **13,** 122–129.

31. Biemann, K. and Scobie, H. A. (1987) Characterization by tandem mass spectrometry of structural modifications in proteins. *Science* **237,** 992–998.
32. Johnson, R. S. and Biemann, K. (1987) The primary structure of thioredoxin from *chromatium vinosum* determined by high-performance tandem mass spectrometry. *Biochemistry* **26,** 1209–1214.
33. Biemann, K. (1989) Sequencing by mass spectrometry, in *Protein Sequencing: A Practical Approach* (Findlay, J. B. C. and Geisow, M. J., eds.), IRL, Oxford University, Oxford, UK, pp. 99–118.
34. Tsugita, A., Takamoto, K., Kamo, M., and Iwadate, H. (1992) C-terminal sequencing of proteins. A novel partial hydrolysis and analysis by mass spectrometry. *Eur. J. Biochem.* **206,** 691–696.
35. Rosnack, K. J. and Stroh, J. G. (1992) C-terminal sequencing of peptides using electrospray ionization mass spectrometry. *Rapid Commun. Mass Spectrom.* **6,** 637–640.
36. Klemm, P. (1984) Carboxy-terminal sequence determination of proteins and peptides with carboxypeptidase Y, in *Methods in Molecular Biology* (Walker, J. M., ed.), Humana, Clifton, NJ., pp. 255–259.
37. Hayashi, R. (1988) Enzymatic methods of protein/peptide sequencing from carboxy-terminal end, in *Protein/Peptide Sequencing Analysis: Current Methodologies* (Bhown, A. S., ed.), CRC, Baton Rouge, LA, pp. 145–159.
38. Dal-Degan, F., Ribadeau-Dumas, B., and Breddam, K. (1992) Purification and characterisation of two serine carboxypeptidases from Aspergillus niger and their use in C-terminal sequencing of proteins and peptide synthesis. *Appl. Environ. Microbiol.* **58,** 2144–2152.
39. Cromwell, L. D. and Stark, G. R. (1969) Determination of the carboxyl termini of proteins with ammonium thiocyanate and acetic anhydride, with direct identification of the thiohydantoins. *Biochemistry* **8,** 4735–4740.
40. Inglis, A. S., Duncan, M. W., Adams, P., and Tseng, A. (1992) Formation of proline thiohydantoin with ammonium thiocyanate: progress towards a viable C-terminal amino-acid-sequencing procedure. *J. Biochem. Biophys. Methods* **25,** 163–171.
41. Kubo, H., Nakajima, T., and Tamura, Z. (1971) Formation of thiohydantoin derivative of proline from C-terminal of peptides. *Chem. Pharmaceut. Bull.* **19,** 210,211.
42. Boyd, V. L., Hawke, D. H., and Geizer, T. G. (1990) Synthesis of the 2-thiohydantoins of amino acids using Woodward's reagent. *Tetrahedron Lett.* **31,** 3849–3852.
43. Elmore, D. T., Ogle, J. R., and Toseland, P. A. (1956) Degradative studies on peptides and proteins. part III. Synthesis of some 2-thiohydantoins as reference compounds. *J. Chem. Soc.* 192–196.
44. Swan, J. M. (1952) Thiohydantoins derived from aspartic and glutamic acids. *Aust. J. Sci. Res.* (Series A), **5,** 721–725.
45. Johnson, T. B. and Guest, H. H. (1912) The action of potassium thiocyanate on asparagine. *Am. Chem. J.* **48,** 103–111.
46. Johnson, T. B. and Guest, H. H. (1912) The action of potassium thiocyanate on pyrrolidone-carboxylic acid. 2-Thiohydantoin-4-propionic acid. *Am. Chem. J.* **47,** 242–251.
47. Mackay, M. F., Duggan, B. M., Laslett, R. L., and Wilshire, J. F. K. (1991) Structure of a substituted thiohydantoin. *Acta Cryst. (C)* **48,** 334–336.
48. Wittmann-Liebold, B., Matschull, L., Pilling, U., Bradaczek, H. A., and Graffunder, H. (1991) Modular Berlin microsequencer for the sequential degradation of proteins and peptides from the amino-and carboxyl-terminal end, in *Methods of Protein Sequence Analysis* (Jornvall, H., Hoog, I. O., and Gustavsson, A. M., eds.), Birkhauer Verlag, Basel, Switzerland, pp. 9–21.

93

C-Terminal Sequence Analysis with Carboxypeptidase Y

John M. Walker and Julia S. Winder

1. Introduction

To date there is no chemical method that provides the ability to determine extensive lengths of amino acid sequence sequentially at the C-terminus of a protein or a peptide, although the thiocyanate method described in Chapter 92 is showing great promise following recent developments (extensive sequence data from the N-terminus can of course be achieved using the Edman degradation). However, limited C-terminal sequence data can be obtained fairly quickly and easily using a class of enzymes called carboxypeptidases. Such limited information can be useful, for example, in determining positions of proteolytic cleavage in proteins, or for confirming sequence data toward the end of a peptide or protein sequence.

Carboxypeptidases are proteolytic enzymes that remove L-amino acids, one residue at a time, from the carboxyl-terminus of polypeptide chains, i.e., they are exoproteases. A number of such enzymes have been isolated from plant and animal sources, each differing in their chemical and physical properties and the rate at which they release particular amino acids. To determine the C-terminal amino acid sequence, the protein or peptide being analyzed is digested with carboxypeptidase and aliquots removed at timed intervals, and analyzed for the presence of free amino acids. The amount of each amino acid released is plotted against time, and the C-terminal sequence deduced from the relative rate of release of each amino acid. Four carboxypeptidases have been used extensively to provide peptide and protein sequence data. These are: carboxypeptidase A (EC 3.4.17.1) from bovine pancreas *(1)*, carboxypeptidase B (EC 3.4.17.2) from porcine pancreas *(1)*, carboxypeptidase C (EC 3.4.12.1) from orange leaves *(2)*, and carboxypeptidase Y (EC 3.4.16.1) from yeast *(3)*. Historically, carboxypeptidases A and B were the first to be discovered and used for sequence determination. Carboxypeptidase A releases most C-terminal amino acids, but will not cleave at arginine, proline, hydroxyproline. or lysine residues *(1)*. The specificity of carboxypeptidase B is far more restricted, cleaving only C-terminal arginine and lysine residues. Carboxypeptidases A and B, therefore, tended to be used together, but even so, exopeptidase activity was effectively blocked when a proline residue was reached. The isolation of carboxypeptidase C provided an enzyme that combined the specificity of carboxypeptidases A and B, but also cleaved at Pro residues *(2)*, i.e., it cleaves all C-terminal amino acids. Carbox-

From: The Protein Protocols Handbook
Edited by: J. M. Walker Humana Press Inc., Totowa, NJ

ypeptidase Y (CPY, isolated from baker's yeast) has the same broad specificity as carboxypeptidase C, but because of its strong action on protein substrates and its ability to work in the presence of urea and detergents, it is nowadays the enzyme of choice for C-terminal sequence analysis and will be described in this chapter.

Carboxypeptidase Y cleaves all L-amino acids one residue at a time from the C-terminal of polypeptide chains. However, the rate of release of individual amino acids varies. Catalysis is maximum when the penultimate and/or terminal residues have aromatic or aliphatic side chains *(4)*. When glycine or aspartic acid is in the terminal position, or lysine and arginine in the penultimate position, the release of the amino acid is slow *(3,5)*. The cleavage of tripeptides is difficult, and dipeptides are completely resistant to hydrolysis. C-terminal proline is a good substrate, but a proline residue on the carboxyterminal side of glycine is not likely to be released *(4)*. The optimum pH for hydrolysis of acidic amino acids is pH 5.5, whereas that for neutral and basic amino acids is pH 7.0 *(4,6)*.

2. Materials

1. Digestion buffer: 0.1*M* pyridine acetate, pH 5.6, 0.1 m*M* norleucine, 1% SDS (*see* Note 1).
2. Carboxypeptidase Y: The lyophilized enzyme is stable for 6–12 mo if stored at 4°C. A suspension of the enzyme in saturated ammonium sulfate can be stored at –20°C indefinitely. When dissolved in water and dialyzed against water to give an approx 1% solution of the enzyme, this solution can be aliquoted and stored at –20°C for at least 2 yr. Diluted solutions (<0.1 mg/mL) lose activity fairly quickly and should therefore be prepared just before use (*see* Note 2).

3. Method

1. Dissolve 20 nmol of the protein to be studied in 200 µL of digestion buffer (*see* Note 3).
2. Heat the solution at 60°C for 20 min to denature the protein (*see* Note 3).
3. After cooling, remove a 25-µL aliquot as the zero time sample. Add 2 nmol of carboxypeptidase Y in 5–10 µL of 0.1*M* pyridine acetate buffer, pH 5.6, thoroughly mix, incubate at room temperature, and remove 25-µL aliquots at T = 1, 2, 5, 10, 20, 30, and 60 min (*see* Notes 3 and 4).
4. Add 5 µL of glacial acetic acid to each sample to stop the reaction. Samples are then frozen, lyophilized (pyridine acetate is volatile), and then subjected to amino acid analysis (*see* Note 5). A graph is plotted showing the amount of each amino acid released with time and the C-terminal sequence deduced from the relative rate of release of each amino acid (*see*, for example, ref. *7*).

4. Notes

1. The norleucine is used as an internal standard for amino acid analysis to allow for compensation of any handling losses or sample errors. The concentration of norleucine should be adjusted to suit the sensitivity of your analyzer. Norleucine is an appropriate internal standard where ion-exchange chromatography with postcolumn derivatization is being used to separate the amino acids. However, for other systems, alternative internal standards are sometimes more appropriate. For example, when using Fmoc chemistry, either hydroxylysine, β-2-thienylalanine, or 2-amino-3-guanidinopropionic acid can be used. For precolumn derivatization with *o*-phthaldialdehyde, β-alanine, ethanolamine, or norvaline is used.
2. Commercial preparations of the enzyme may contain free amino acids owing to autolysis, which should be removed before use. Repeated freeze–thawing of solutions of the enzyme

or prolonged storage at room temperature can also lead to autolysis and the liberation of free amino acids. For ammonium sulfate suspensions, centrifuge and wash the pellet in saturated ammonium sulfate before dissolving in buffer. Alternatively, the dissolved enzyme can be dialyzed against the pyridine acetate buffer used for digestion.

3. If analyzing peptides, SDS can be omitted from the incubation buffer, since the C-terminal should be readily accessible. Also the heating step (step 2) can be omitted. In addition, the time intervals for sampling can be reduced, since the rate of appearance of free amino acids will be faster than that for proteins. Use T = 0, ½, 1, 2, 5, 10, and 20 min.
4. The sampling times indicated here should be appropriate for most proteins. However, should the C-terminal sequence be such that a number of slowly released amino acids are present, then the experiment may have to be repeated using longer incubation times or a higher enzyme-to-substrate ratio.
5. The small amount of SDS in each sample applied to the amino acid analysis does not interfere with the elution profile or affect the integrity of the machine.

References

1. Neurath, H. (1960) Carboxypeptidases A and B, in *The Enzymes*, vol. 4 (Boyer, P. D., Lardy, H., and Myrbäck, K., eds.), Academic, New York. pp. 11–36.
2. Sprossler, B., Heilmann, H. D., Grampp, E., and Uhlig, H. (1971) Carboxypeptidase C from orange leaves. *Hoppe-Seyler's Z. Physiol. Chem.* **352,** 1524–1530.
3. Hayashi, R., Moore, S., and Stein, W. H. (1973) Carboxypeptidase from yeast. *J. Biol. Chem.* **248,** 2296–2302.
4. Hayashi, R., Bai, Y., and Hata, T. (1975) Kinetic studies of carboxypeptidase Y. *J. Biochem.* **77,** 69–79.
5. Breddam, K. and Ottesen, M. (1987) Determination of C-terminal sequences by digestion with serine carboxypeptidases. The influence of enzyme specificity. *Carlsberg Res. Commun.* **52,** 55–63.
6. Bai, Y., Hayashi, R., and Hata, T. (1975) Kinetic studies of carboxypeptidase Y. *J. Biochem.* **78,** 617–626.
7. Klemm, P. (1984) Carboxy-terminal sequence determination of proteins and peptides with carboxypeptidase Y, in *Methods in Molecular Biology*, vol. 1: *Proteins* (Walker, J. M., ed.), Humana, Clifton, NJ, pp. 255–259.

94

Rapid Epitope Mapping by Carboxypeptidase Digestion and Immunoblotting

Philip S. Low and Jie Yuan

1. Introduction

Conventional methods for locating the epitope of an antibody on an antigen all require amino acid sequencing at some stage of the protocol. The protein footprinting approach, for example, employs arbitrary proteolysis or chemical modification to locate the sequence that is protected by the bound antibody *(1,2)*; however, the protected region still must be sequenced to identify its position on the immunogenic protein. Limited protease digestion of an antigen followed by Western blotting with the desired antibody has also been exploited for identifying crossreactive peptides *(3,4)*, but again the stained peptide must be sequenced to pinpoint its location in the original antigen. Localization of epitopes by comparison of an antibody's ability to recognize a series of highly homologous proteins has been used to relate differences in specificity with known amino acid substitutions *(5,6)*. However, such families of homologous proteins are rare, and manual generation of homologous families by site-directed mutagenesis can be very time-consuming. Finally, competitive inhibition of antibody binding by synthetic or natural peptides derived from the antigen can often disclose the desired epitope *(7,8)*; nonetheless, the sequence of the competitive peptide must still be determined to localize it within the antigen's primary structure. In a few unexpected situations, epitopes have been assigned to an enzyme's active site without sequence information when the antibody was found to block catalytic activity *(6)*. However, results of this sort may not always be reliable, since an antibody can also inhibit enzyme activity by noncompetitive mechanisms.

As an alternative to current epitope mapping protocols, we have developed a method that can reveal the position of an epitope in an antigen's primary structure without the requirement for sequence information *(9)*. Briefly, the antigen is digested with a mixture of carboxypeptidases into a continuous series of fragments containing the same *N*-terminus, but exhibiting different degrees of truncation at the C-terminus. After SDS polyacrylamide gel electrophoretic separation and immunoblotting of the mixture, the epitope is identified by observing the smallest mol-wt *N*-terminal fragment that is still recognized by the antibody (Fig. 1). Because the epitope's distance from the *N*-terminus can be directly determined from the immunoblot with the use of appropriate mol-wt standards, no sequence information is needed. Thus, as diagrammed in Fig. 1,

From: The Protein Protocols Handbook
Edited by: J. M. Walker Humana Press Inc., Totowa, NJ

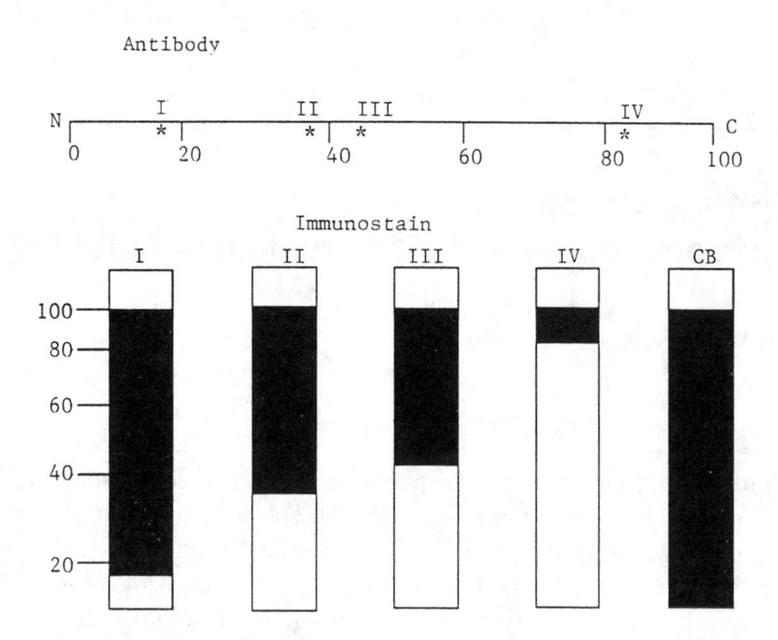

Fig. 1. Diagrammatic representation of the staining pattern generated by four hypothetical antibodies with distinct epitopes on a common 100-kDa antigen. The locations of the hypothetical epitopes are indicated in the linearized peptide diagram at the top. These epitopes are found 18 (antibody I), 38 (antibody II), 43 (antibody III), and 82 kDa (antibody IV) from the *N*-terminus of the intact antigen. Digestion of the intact antigen into a continuous series of C-terminally truncated peptides having the same *N*-terminus yields the immunoblot staining pattern shown below for each antibody. CB reveals the expected Coomassie Blue staining pattern of the carboxypeptidase digestion mixture.

the antibody (I), which recognizes an epitope ~18 kDa from the *N*-terminus, stains all *N*-terminal fragments of the antigen from the intact 100-kDa polypeptide to the 18-kDa fragment terminating with the epitope.

Further C-terminal trimming by carboxypeptidase, however, deletes the epitope and concurrently abrogates the staining. An analogous pattern is seen with antibodies recognizing epitopes 38 (antibody II), 43 (antibody III), and 82 kDa (antibody IV) from the *N*-terminus. As mentioned, the molecular weight at which the continuum of peptide fragments is no longer immunostained identifies the position in the polypeptide sequence where the critical antigenic sequence is located.

2. Materials

1. Yeast carboxypeptidase Y and porcine pancreas carboxypeptidase B (from Calbiochem, La Jolla, CA) should be used because they retain their activity in 6*M* urea *(10)*.
2. α_2-Macroglobulin immobilized on agarose beads purchased from Boehringer Mannheim (Indianapolis, IN).
3. Nitrocellulose paper from Schleicher and Schuell (Keene, NH).
4. Prestained low-mol-wt markers from Gibco-BRL (Gaithersburg, MD).

5. Electrophoresis reagents for sodium dodecyl sulfate polyacrylamide gel electrophoresis (SDS-PAGE) are available from a variety of commercial sources, and any supplier should suffice as long as the reagents are pure.
6. Low-mol-wt cutoff dialysis membranes from Spectrum (Los Angeles, CA).
7. Antigen: The protein antigen need not be pure, since it will ultimately be visualized by immunoblotting. However, all cleavage products of the antigen must be removed, since they will lead to staining of fragments shorter than those containing the true *N*-terminus. For best resolution on immunoblots, the molecular weight of the antigen should be <100,000, so that epitopes near the C-terminus can be narrowly defined.
8. Antibodies: The antibodies to be mapped can be either monoclonal or polyclonal. In our initial experiments, we used four previously mapped polyclonal antipeptide antibodies and two MAb *(4)*. Obviously, the antibodies must be capable of immunoblotting their antigens; otherwise, the mapping protocol would be impossible.
9. Buffers:
   a. Digestion buffer: 50 m*M* citrate, 10% acetonitrile, pH 6.0.
   b. Dialysis buffer: digestion buffer plus 1 m*M* dithiothreitol.
   c. Unfolding buffer: 6*M* urea, 10 m*M* dithiothreitol, 20 m*M* methylamine.
   d. Blotting buffer: 20 m*M* Tris base, 500 m*M* NaCl, pH 7.5.

3. Methods

3.1. Carboxypeptidase Selection and Treatment

Even a small amount of contaminating endoproteinase in the digestion mixture will invalidate the result. To prevent such contamination, both the antigen and the carboxypeptidase mixture must be cleared of any traces of endoproteinase activity (*see* Note 1).

1. Dissolve antigen at ~5 mg/mL in digestion buffer.
2. Dissolve 1000 U/mL of carboxypeptidase Y and 100 U/mL of carboxypeptidase B in digestion buffer.
3. Incubate the antigen and carboxypeptidase solutions separately for 30 min at room temperature with 0.1 vol of agarose beads containing immobilized α_2-macroglobulin, a potent inhibitor of virtually all endoproteinases *(11)*.
4. After incubation, remove and discard the beads.

3.2. Digestion of Antigen

1. Dialyze the treated protein antigen against unfolding buffer to promote its unfolding. In many cases, simple dilution with unfolding buffer is sufficient to induce denaturation, whereas in the case of highly stable antigens, some heating may also be required.
2. Add an aliquot of the treated carboxypeptidase mixture (10 U of Y plus 1 U of B) to 1 mL of the treated antigen solution dissolved in unfolding buffer at 1 mg/mL. We selected a carboxypeptidase Y-to-B ratio of 10:1, because it provided the best digestion result for proteins we examined. For other protein antigens, however, other combinations of carboxypeptidases may yield a more continuous distribution of peptide fragments (*see* Note 2). Below is a list of the specificities of the major carboxypeptidases that are commercially available (Table 1). The denatured protein antigen is initially allowed to digest for 1 h at room temperature in the unfolding buffer.
3. Transfer the digestion mixture to dialysis tubing suspended in dialysis buffer.
4. Treat the mixture in the tubing again with an additional 10 U of carboxypeptidase Y plus 1 U of carboxypeptidase B, and continue the digestion in the dialysis bag for 5 more hours (*see* Note 3).

Table 1
Specificity of Carboxypeptidases

Carboxypeptidase	Specificity
Y	Any amino acids, slow for Asp and Gly
B	Arg and Lys only
A	Aromatic and large aliphatic side chains only
P	Any amino acids, slow for Gly and Ser

5. If necessary, the same protease supplementation should be repeated at 12 and 24 h after the initial digestion to ensure the presence of shorter fragments in the digestion mixture.
6. In order to monitor the progress of the digestion, collect 20 μL of the digestion mixture at 1-, 3-, 6-, 12-, and 24-h intervals. Separate the fragments on a 15% Laemmli polyacrylamide gel and stain with Coomassie blue *(12)*. The digestion reaction is then terminated at the point where a continuum of antigen fragments extending from the intact protein to the smallest peptide resolvable on the gel is generated (*see* Notes 4 and 5).

3.3. Epitope mapping by Immunoblotting

1. Terminate the digestion reaction at the desired time by boiling for 5 min.
2. Separate the proteolytic fragments on a 15% polyacrylamide gel by SDS-PAGE *(12)*, and transfer the separated peptides to nitrocellulose *(13)* for 3 h at 240 mA. Then block the resulting blot for 15 min with 4% bovine serum albumin in blotting buffer (*see* Note 6).
3. Incubate the blot 3 h with ascites fluid containing MAb diluted ~1:500 in blotting buffer or with rabbit antipeptide antibody diluted ~1:100 in the same buffer.
4. Wash the nitrocellulose sheets with blotting buffer containing 0.05% Tween-20, and incubate 2 h with goat antimouse or goat antirabbit IgG-horseradish peroxidase conjugate diluted 1:500 in blotting buffer.
5. Stain the blots with 4-chloronaphthol *(14)* to visualize immunoreactive peptide bands.
6. Prestained mol-wt markers in an adjacent lane on the blot should be used to determine the molecular weight where the staining pattern terminates. From this evaluation, an estimate of the position of the antibody's epitope in the primary structure of the antigen is obtained (*see* Note 7). When applied to the red cell protein band 3, the epitopes of 6 previously mapped antibodies were correctly determined *(9)* (*see* Notes 8 and 9).

4. Notes

1. All commercial carboxypeptidases that we tested were found to be contaminated with small amounts of endoproteinases that cleaved our antigen internally. Although the most common contaminations were from trypsin and chymotrypsin, not all endoproteinase activity could be eliminated by a cocktail of common proteinase inhibitors. Consequently, proteolysis with such mildly contaminated carboxypeptidase preparations generated fragments that were predictably stained by antibodies at abnormally low-mol-wts. In contrast, when the carboxypeptidase mixture was pretreated with α_2-macroglobulin (a protein that inhibits all endoproteinases without inactivating exoproteinases), the staining pattern invariably terminated at the expected molecular weight of the known epitope. Thus, endoproteinase inhibitor cocktails do not adequately substitute for α_2-macroglobulin in inhibiting contaminating endoproteinases in commercial carboxypeptidase preparations.
2. Carboxypeptidases from different commercial vendors commonly have different specific activities and levels of endoproteinase contamination. It is therefore desirable to obtain the highest quality carboxypeptidase available and to purchase future lots of the same carbox-

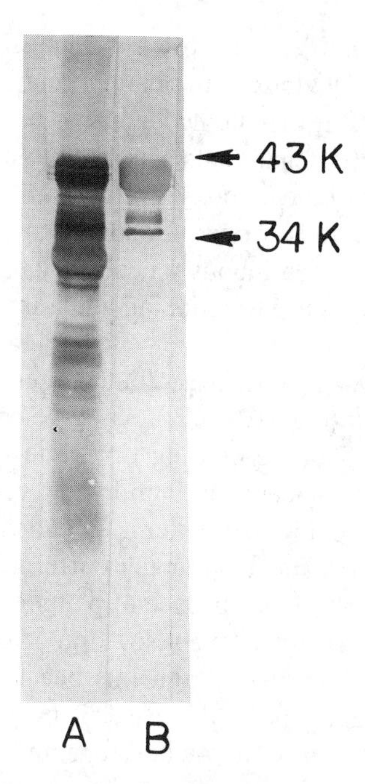

Fig. 2. Effect of denaturation of the cytoplasmic domain of band 3 (cdb3) on carboxypeptidase digestion and epitope mapping. Purified cdb3 (1.0 mg/mL) was digested with α_2-macroglobulin-treated carboxypeptidases for 1 h either in digestion buffer alone (lane B) or digestion buffer supplemented with 6*M* urea, 10 m*M* dithiothreitol, and 20 m*M* methylamine to promote protein unfolding (lane A). Following this initial digestion period, the proteolysis of both samples was continued for 23 h in unmodified digestion buffer with carboxypeptidase supplements. Fragments of cdb3 were separated on SDS-PAGE and immunoblotted with a polyclonal anti-cdb3 antipeptide antibody. (Reproduced from ref. *9* with permission from Academic Press.)

ypeptidase type from the same vendor. Unfortunately, this commercial variability requires that the type and amount of carboxypeptidase used be optimized empirically by each user.

3. The activity of carboxypeptidases tends to decrease as digestion proceeds because of the following two reasons: (a) carboxypeptidases are competitively inhibited by free amino acids, i.e., their digestion products, and (b) the carboxypeptidases can slowly digest themselves. Therefore, it is important to dialyze away the free amino acids and to supplement the digestion mixture periodically with fresh carboxypeptidases. The activity of carboxypeptidase is also enhanced by 10% acetonitrile in buffer.
4. Sometimes we found it difficult to generate a balanced continuum of antigen fragments from the intact polypeptide to the smallest *N*-terminal peptide using a single digestion mixture. In these cases, we determined the simplest solution was to conduct three separate antigen digestion reactions for different lengths of time, and then mix the contents of all three. As long as no internal digestion occurs, any number of digestion mixtures can be combined to obtain the desired continuum of peptide fragments.
5. Carboxypeptidases tend to cleave unfolded stretches of a polypeptide rapidly and halt at highly folded domains of the protein. For example, when the cytoplasmic domain of band 3 protein was digested for 24 h with the aforementioned carboxypeptidase mixture in digestion buffer lacking denaturants, the fragmentation pattern shown in Fig. 2, lane B was obtained. In contrast, when the antigen was first unfolded in 6*M* urea, 10 m*M* dithio-

threitol, and 20 m*M* methylamine, the more continuous/complete digestion pattern of Fig. 2, lane A was observed. Obviously, to obtain a relatively continuous distribution of *N*-terminal antigen fragments, it may be necessary to promote unfolding of the antigen actively. For band 3 protein, the denaturants could be gradually removed by dialysis without compromising the ability of the carboxypeptidase mixture to continue digestion after the initial 1-h proteolysis period. However, for some protein antigens, a low level of urea (1*M*) in the dialysis buffer was continuously necessary to achieve efficient digestion. In all cases, the uninterrupted presence of reducing agent is required to avoid halting digestion at disulfide bonds.

6. Depending on the molecular weight of the protein antigen, a constant acrylamide concentration (e.g., 15%) or gradient (e.g., 10–20%) may give better resolution in the low-mol-wt region of the gel. Electrophoretic blotting times must then be adjusted accordingly, since short peptides and a low-percent acrylamide both result in decreased transfer times to nitrocellulose. If the electrophoretic transfer is extended too long, short peptides can be forced through the nitrocellulose into the transfer buffer, leading to an artifactual termination of antibody staining owing to the absence of peptide on the paper.
7. For obvious reasons, we observed that the resolution of our immunoblots diminished as the molecular weights of the antigenic fragments became very small, i.e., for epitopes near the *N*-terminus of the antigen. In this case, it may be desirable to map the same epitope using aminopeptidase digestion. As before, commercial aminopeptidases must be treated with α_2-macroglobulin to eliminate contaminating endoproteinases.
8. Advantages of the protocol:
   a. No sequence information is required.
   b. The antigen need not be pure.
   c. The method is rapid, sensitive, versatile, and inexpensive.
9. Potential disadvantage of the protocol: when antibodies recognizing discontinuous (i.e., conformational) epitopes are analyzed, the immunostaining pattern on the blot will likely terminate when the most C-terminal member of the epitope is digested away.

References

1. Atassi, M. Z. (1975) Antigenic structure of myoglobin: the complete immunochemical anatomy of a protein and conclusions relating to antigenic structures of proteins. *Immunochemistry* **12,** 423–438.
2. Sheshberadaran, H. and Payne, L. G. (1989) Protein footprinting method for studying antigen–antibody interactions and epitope mapping. *Methods in Enzymol.* **178,** 746–765.
3. Pratt, L. H., Cordonnier, M., and Lagarias, J. C. (1988) Mapping of antigenic domain on phytochrome from etiolated *Avena sativa L.* by immunoblot analysis of proteolytically derived peptides. *Arch. Biochem. Biophys.* **267,** 723–735.
4. Willardson, B. M., Thevenin, B. J., Harrison, M. J., Kuster, W. M., Benson, M. D., and Low, P. S. (1989) Localization of the ankyrin-binding site on erythrocyte membrane protein, band 3. *J. Biol. Chem.* **264,** 15,893–15,899.
5. Smith-Gill, S. J., Lavoie, T. B., and Mainhart, C. R. (1984) Antigenic regions defined by monoclonal antibodies correspond to structural domains of avian lysozymes. *J. Immunol.* **133,** 384–393.
6. Benjamin, D. C. (1984) The antigenic structure of proteins: A reappraisal. *Annu. Rev. Immunol.* **2,** 67–101.
7. Arnon, R., Marcon, E., Sela, M., and Anfinsen, C. B. (1971) Antibodies reactive with native lysozyme elicited by a completely synthetic antigen. *Proc. Natl. Acad. Sci. USA* **68,** 1450–1455.

8. Maelicke, A., Plümer-Wilk, R., and Conti-Tronconi, B. (1989) Epitope mapping employing antibodies raised against short synthetic peptides. A study of the nicotinic acid acetylcholine receptor. *Biochemistry* **28,** 1396–1405.
9. Yuan, J. and Low, P. S. (1992) Epitope mapping by a method that requires no amino acid sequence information. *Anal. Biochem.* **205,** 179–182.
10. Hayashi, R., Moore, S., and Stein, W. H. (1973) Carboxypeptidase from yeast:Large scale preparation and the application to COOH-terminal analysis of peptides and proteins. *J. Biol. Chem.* **248,** 2296–2302.
11. James, K. (1980) α_2-Macroglobulin and its possible importance in immune systems. *TIBS* **5,** 43–47.
12. Laemmli, U. K. (1970) Cleavage of structural proteins during the assembly of the head of bacteriophage T4. *Nature* **227,** 680–685.
13. Burnette, W. N. (1981) Western blotting: electrophoretic transfer of proteins from sodium dodecyl sulfate polyacrylamide gels to unmodified nitrocellulose and radiographic detection with antibody and radioiodinated protein A. *Anal. Biochem.* **112,** 195–203.
14. Hawkes, R. (1982) Identification of concanavalin A-binding proteins after sodium dodecyl sulfate-gel electrophoresis and protein blotting. *Anal. Biochem.* **123,** 143–146.

95

Epitope Mapping of a Protein Using the Geysen (PEPSCAN) Procedure

J. Mark Carter

1. Introduction

1.1. General

Immunity to many diseases is dependent on the ability of the host's antibodies (Ab) to recognize foreign antigens (Ag), such as surface proteins or toxins, and bind them tightly and specifically. This binding is an important aspect of the immune response, and it is often required for subsequent immune processes that ultimately result in re-establishment of a disease-free state.

One of the problems encountered in vaccine development is that of delineating the Ab response to a protein Ag. Whereas the overall response to an Ag may involve various molecular species of Ab, each Ab molecule can bind specifically to one unique part of the Ag referred to as that Ab's epitope. Often only a subset of these epitopes is involved in blocking a protein's function, clearing of infectious organisms, or other steps in an effective immune response.

The PEPSCAN procedure, developed by Mario Geysen and marketed by Chiron Mimotopes (Victoria, Australia), is a variation of solid-phase peptide synthesis. It comprises the simultaneous synthesis and simultaneous immunochemical assay of hundreds of peptides covalently linked to plastic pins. This technology represents a major advance in the epitope mapping of protein antigens because of its ability to create the large numbers of overlapping peptides necessary for complete epitope mapping, and because of the reusability of those peptides in many ELISAs with different Ab *(1)*.

The plastic pins for attachment of peptides are commercially available from Chiron. They are now prepared for distribution according to a modification of the method originally published *(1)*. Basically, the polyethylene matrix of the rods is grafted with a proprietary acrylic-like polymer. Then the free carboxylic acid moiety on the polymer is modified with F_{moc}-glycine and capped via acetylation to block unreacted hydroxyl sites. Next the F_{moc} group is removed, and the pins are acylated with F_{moc}-β-alanine and capped via acetylation again.

1.2. History and Development of Pin Hardware

One of the limitations plaguing early application of the PEPSCAN technique was poor reproducibility in the substitution level of these derivatized pins. In tests of pins

From: The Protein Protocols Handbook
Edited by: J. M. Walker Humana Press Inc., Totowa, NJ

from the same lot, substitution levels range from 6 to 26 nmol NH_2/pin, with a mean of 12 nmol and an SD of 3 nmol. The issue here is not that this variability might prevent accumulation of worthwhile data. Rather, because of this limitation, all results from PEPSCAN must usually be considered qualitative. For confirmation, such results may be double-checked via synthesis and immunoassay of peptides via classical solid-phase methods.

Recent advances in pin design incorporate increased surface area as well as improved level and stability of pin derivatization. As the new pins become more widely available, the peptide pin methodology itself is expected to become more widely accepted and more generally utilized.

The pins for peptide synthesis are arranged in 8 × 12 arrays on 9-mm centers, like commercially available microtiter plates. This geometry allows for familiarity and simplification in a subsequent enzyme-linked immunosorbent assay (ELISA) for the detection of Ab reactivity. Many laboratory technicians are already quite familiar with standard ELISA assays, and only minor modifications to this procedure are necessary to perform a PEPSCAN ELISA. Furthermore, automated microtiter plate readers are widely available for rapid determination of absorbance data in assays performed with these 96-well plates.

1.3. Computer Automated Amino Acid Indexer

Other than a variable substitution level in the pins, the most significant problem in PEPSCAN is the logistics of the simultaneous synthesis of several hundred peptides. Clearly, computer support is required, but even a computer-generated synthesis schedule possibly leaves a large margin of error. The person performing the synthesis must manually transpose amino acid locations from the hard copy list to the microtiter wells. This procedure takes about 4 h to fill 10 microtiter plates, and it commonly results in an approx 3% error rate.

In order to address this problem, scientists at the Walter Reed Army Institute of Research (WRAIR) developed a computer-driven device that locates and identifies each of the different wells, and indicates their respective amino acid derivative requirements via illumination with LEDs. Using this computer-driven amino acid indexer, the time in filling 10 microtiter plates, for simultaneous synthesis of 960 peptides, is reduced sixfold to 40 min, and error becomes undetectable. The device is thoroughly described in VanAlbert et al. *(2)*, and it is commercially distributed by CRACO (Vienna, VA).

Chiron now distributes a similar device with the pins. It is smaller and cheaper, operating with only one microtiter plate (96 peptides) at a time. It may be more suitable for PEPSCAN users with a low synthesis throughput.

1.4. Linear vs Conformational Epitopes

PEPSCAN is particularly effective in the detection of linear (continuous) epitopes. Unfortunately, however, most Ab are probably directed against discontinuous epitopes *(3–5)*. This fact becomes especially important when PEPSCAN is used to study the specificity of monoclonal antibodies (MAb). In many cases, the results of such experiments are weak and equivocal. Nonetheless, Geysen has suggested that binding of Ab to discontinuous epitopes (such as are reported for most MAb) may be detected on peptide pins, at least in some instances. Such binding is thought to involve two or three discontinuous regions of the protein sequence, which fold into a discrete conformation on the solvent-accessible surface of the native structure of the Ag. Theoretically, the Ab should also bind, although much more weakly, to each of these subregions when presented separately. In fact, many have observed data suggesting this conclusion, but the binding so detected is often not statistically significant above background (nonspecific) binding.

Excellent results are generally obtainable using immune serum as a source of Ab. It is probably true that serum raised against a native protein Ag will contain only a limited subset of Ab reactive to linear epitopes presented by the peptide pins. However, there is usually such a large variety of reactivities represented by a polyclonal serum that a fair number of linear epitopes can be readily demonstrated by PEPSCAN ELISA. Antizera raised against peptide Ags and peptide conjugates tend to contain a greater proportion of Ab that is reactive to linear epitopes because of their more limited conformational freedom. Consequently, this type of immune serum generally gives the highest level of detected binding on PEPSCAN.

2. Materials

2.1. Synthesis

1. Prederivatized polyethylene pins and polyethylene microtiter plates.
2. *N,N,*-dimethylformamide (DMF).
3. Methanol.
4. PIP solution: 20% piperidine in DMF. (*See* Note 2).
5. Amino acid solutions: 60 m*M* amino acid derivatives, 65 m*M* 1-hydroxybenzotriazole (HOBt) in DMF. Optional: bromophenol blue (50 μ*M*). (*See* Note 6).
6. Dichloromethane (DCM). **Note: This solvent is a suspected carcinogen.**
7. Acetic acid wash solution: 0.5% acetic acid, 50% methanol in water.
8. Optional acetylation cocktail: 3% acetic anhydride and 0.5% diisopropylethylamine (DIEA) in DMF. Prepare fresh immediately before use.
9. Deblocking cocktail: 2.5% anisole and 2.5% 1,2-dithioethane in trifluoroacetic acid. Prepare fresh within 1 h of use. Note: This reagent is extremely corrosive, and it smells absolutely terrible. Wear appropriate protective devices, and use it in a fume hood.
10. Deionized water.
11. Silica gel desiccant.
12. Plastic baths and sealable bags.

2.2. Disruption

1. Sonicator. (*See* Note 12.)
2. Disruption buffer: 1% reagent-grade sodium dodecyl sulfate, 0.1% 2-mercaptoethanol and 0.1*M* sodium phosphate, pH 7.2 at 55–65°C.
3. Explosion-proof heating bath, filled with boiling ACS-grade methanol (about 60°C).
4. Silica gel.
5. Sealable bags, tongs.

2.3. ELISA Analysis

1. Peptide pins.
2. Phosphate-buffered saline (PBS): 150 m*M* NaCl, 25 m*M* phosphate, pH 7.4. Prepare in 1-L batches, filter-sterilize, and store at 4°C. PBS keeps for about 2 or 3 wk. For indefinite storage add 2 g (per liter) sodium azide.

3. PBS/Tween-20 (PBST): PBS (as above) with 0.1% Tween-20. Prepare in 1-L batches (*see* Notes 13 and 18).
4. Blocking solution (*see* Note 14). Use a commercial ELISA blocker solution or one of the following two solutions. Prepare in 1-L batches, filter-sterilize, and store at 4°C. Solution keeps for about 2 or 3 wk. For extended storage (up to 8 wk), add 2 g/L sodium azide.
   a. 1% Bovine serum albumin (BSA) and 1% chicken ovalbumin (OVA) dissolved in PBST.
   b. 2% Casein in PBST: Boil 20 g casein in 100 mL 1*N* NaOH until completely dissolved. Adjust pH to 7.4 by addition of HCl. Add PBST to make final volume 1 L.
5. Test Ab solution. For serum or ascites fluid, the concentration used should be the same as that which gives a good strong signal on a standard ELISA. If a standard ELISA titer is not available, then use a dilution of 1/500. For a purified Ab, use 1–10 μg/mL.
6. Second Ab solution (*see* Notes 19 and 20). The working concentration of the second Ab is usually specified by the manufacturer. Make the dilution in PBST.
7. Substrate solution for alkaline phosphatase. Use commercial preparation, or prepare the following buffer: 0.1*M* diethanolamine, pH 9.8, with 0.01% $MgCl_2$ and 0.02% NaN_3.

   This buffer may be stored at 4°C for several months. However, it should be allowed to warm to room temperature before use. Immediately before use, dissolve *p*-nitrophenylphosphate (the substrate) to a final concentration of 1 mg/mL.
8. Plastic boxes with tight-fitting lids (e.g., Tupperware).

3. Methods

3.1. Synthesis

Historically, Cambridge Research Biologicals (Cheshire, UK) distributed a recipe for synthesis of peptides on pins via F_{moc} chemistry. This original method has been improved. Still, the basic procedure is reminiscent of solid-phase peptide synthesis on polystyrene resin, i.e., the prederivatized polyethylene pins are deprotected, washed, neutralized, washed, and amino-acylated repeatedly until peptides of the desired length are completed. These peptides are then *N*-acetylated (optional), side chain-deblocked, and washed once more. Finally, the peptide pins are subjected to ultrasonic disruption before ELISAs are performed. All reactions are performed at room temperature in a fume hood.

1. Deprotection (removal of the N-F_{moc} group): Perform deprotection batch-wise in polyethylene boxes with lids. Scientific supply distributors sell boxes in various sizes that work well, depending on the scale of your synthesis. Pour baths to a depth of about 2.5 cm with PIP solution, and insert blocks with pins downward. Leave them for 1 h.
2. Wash: Deprotection is followed by washes in DMF (1 wash, 2 min), and then methanol (4 washes, 2 min each). Then the blocks are allowed to air-dry completely in a fume hood for 30 min to 1 h.
3. Coupling (amino acylation): After pre-equilibration in DMF bath for 5 min, pins are amino-acylated individually with 100 μL/well in the polyethylene microtiter plates. The plates bearing the peptide pins are carefully oriented and lowered so that the pins are inserted into their respective wells. In order to reduce evaporation and contamination, the reaction is allowed to proceed for 4 h (or overnight) inside a sealed zip-lock bag.
4. Wash: Following the amino-acylation reaction, pins are again washed with methanol (1 wash, 5 min), and then air-dried again for 2–5 min. If they are to be used again immediately, the pins should be soaked in DMF for 5 min before continuing. If they will be

stored overnight or longer, they should be washed in DMF for 5 min, followed by methanol (2 washes, 2 min each), and then air-dried thoroughly for at least 30 min. Store them clean, and dry.

5. Elongation: Deprotection, washing, amino acylation, and washing are repeated until peptides of the desired length are produced. After the last amino acid is coupled, final deprotection, washing, and air-drying are performed as above in step 2.
6. N-acetylation: (optional) α-amino groups on the peptides are acetylated in polyethylene microtiter plates with 100 μL/well solution of the acetylation cocktail for 90 min.
7. Wash: Pins are then washed with methanol (1 wash, 15 min) and air-dried again for at least 15 min.
8. Deblocking/cleavage: Blocking groups are removed from the peptide amino acid side chains by incubation of the pins for 3–4 h in 2.5-cm deep baths of deblocking cocktail.
9. Wash: The pins are then washed in methanol (1 wash, 10 min), acetic acid wash solution (1 wash, 60 min), methanol (2 washes, 2 min each), allowed to air-dry over silica gel overnight in zip-lock bags or a desiccator. The finished blocks of peptide pins are conveniently stored at –20°C over silica gel in zip-lock bags.
10. Disruption: Before they will perform properly in ELISA assays, the peptide pins must be disrupted, as in Section 3.2.

3.2. Disruption

After peptide synthesis is complete, ELISAs are typically unsuccessful without prior ultrasonic "disruption." In order to make the peptides on the pins accessible to Ab binding, high-power ultrasonic treatment at elevated temperature is absolutely necessary.

1. The sonicator is filled with the disruption buffer and allowed to heat to 55–65°C. The polyethylene blocks bearing peptide pins are floated in the buffer, with the pins pointing downward. The sonicator is then operated for 10 min.
2. Pins are removed from the sonicator with tongs and rinsed thoroughly in 60°C water. Chiron recommends 2 washes for 30 s each, followed by shaking on a hot water bath for another 30 min. Peptide chemists at the WRAIR put them under hot running tap water for about 1 min, until the foam (from SDS in the disruption buffer) subsides.
3. At this point, the pins may be used immediately for ELISA. If they are to be stored for more than a few minutes, they should be boiled in methanol for 2 min, air-dried for at least 15 min in the fume hood, and finally stored in zip-lock bags at –20°C over silica gel.

3.3. ELISA Analysis

A typical ELISA has five main steps: First the Ag is allowed to bind to the microtiter plate wells overnight in a dilute solution with PBS. Next the excess Ag solution may be removed, or even a brief rinse performed, before a "blocking" solution is added. After an hour or two, the blocking solution is removed, and the test or "first" Ab is added. The first Ab is usually allowed 1 or 2 h to bind the Ag on the plates. After a series of thorough washes, an enzyme-conjugated "second" Ab is added and allowed to bind to the first Ab. After an hour or two, another thorough wash is made, and then a chromogenic substrate solution for the enzyme is added and allowed to develop for a few minutes to 2 h. Finally, the results are read on an automated microtiter plate reader, which generally stores the values for absorbance for each well in a computer file.

The peptide pin ELISA is performed very much like a typical ELISA. Persons experienced in the latter generally have little trouble with the technical aspects of pin

ELISAs. There is, indeed, but one major difference between the two. In a typical ELISA, the first step comprises the adsorption of the Ag onto the bottom of a microtiter plate. This molecule acts as a solid-phase "capture" Ag for the subsequent binding of Abs. Contrarily, with peptide pins the peptide Ag remains covalently linked to the solid-phase support pin at all times.

This means that a peptide pin ELISA has only four steps. Briefly, the pins are "blocked" with a suitable buffer, they are subjected to binding of a first Ab, they are probed with an enzyme-labeled second Ab, and they are developed with a substrate. Each of these steps, as well as washing between them, is detailed below.

1. Blocking: Into each well of a microtiter plate, pipet 200 µL of blocking solution. Insert the pins, and incubate for 1 h at room temperature.
2. Test Ab: Pipet 175 µL test Ab solution into each well of a microtiter plate. Insert the pins and leave to incubate overnight at 4°C, rocking gently on a platform (*see* Notes 15 and 17).
3. Wash: Pour PBST into a clean plastic box so that the level of liquid comes at least halfway up the pins when the blocks of pins are inserted with their pins downward. Put the box with wash buffer and pins on a rotating platform for 10 min. Discard the used wash buffer down the sink. Repeat for a total of three washes.
4. Second Ab: Pipet 150 µL second Ab solution into each well of a microtiter plate, insert pins, and allow 1–2 h for binding.
5. Second wash: After the second Ab, make another thorough wash to remove excess enzyme conjugate reagent. Again, three washes of 10 min each are sufficient.
6. Substrate: Pipet 125 µL substrate solution into each well of a good-quality ELISA plate. Before inserting them into the plate, carefully orient the pins so that the numbered edges of the plates correspond with the numbered edges of the block containing the peptide pins. This will prevent confusion when the plates are being read after development.
7. Development: Allow development to proceed until the positive reactions are well colored, usually 30–60 min (*see* Note 21). Stop the development by removing the pins. Do not allow development to proceed until the negative peptides give a strong color reaction (*see* Note 22). After development is complete, remove the pins and rinse them immediately in water.
8. Plate reader: Read the plates on an automated microtiter plate reader within an hour.
9. Disruption: Disrupt as soon as possible. If this cannot be done within a couple of hours, store the pins over night in a methanol bath. Do not let any of the ELISA reagents dry onto the pins.

3.4. DATA Interpretation

3.4.1. Epitope Analysis

For each plate, individually, subtract the mean of the lowest 10–25% of absorbance readings. This is background. This is facilitated by means of a spreadsheet program. In lieu of any officially established criterion for differentiating between positive and negative reactions, positive responses are identified through the judgment of the experimenter. In general, the highest responses will be scored as positive reactions, whereas most sequences will be unreactive. Peptides with intermediate reactivity are often borne out as positive or negative after a repeat of the ELISA experiment.

There are several different combinations of Ab and peptides commonly used in the peptide pin system. Each combination may be expected to give different results, although they will all generally allow the same conclusions to be drawn.

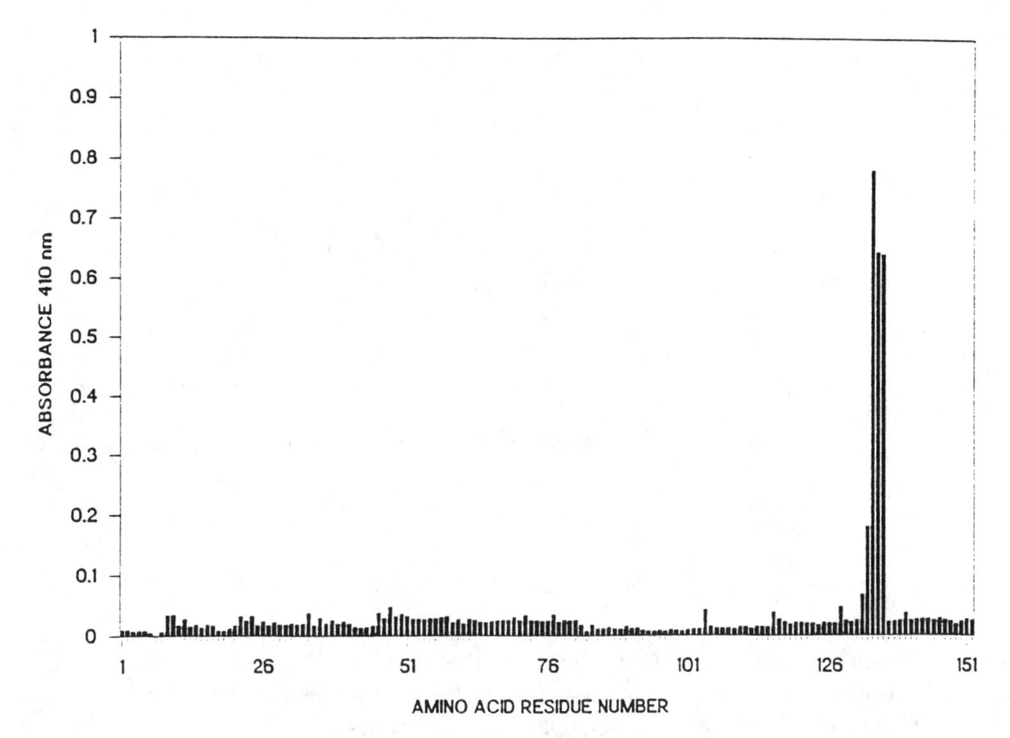

Fig. 1. Typical peptide pin ELISA. ELISA was performed according to standard methods presented in the text. Immune rabbit serum was used to probe pins bearing overlapping 8-mer peptides comprising a bacterial protein sequence. The *X*-axis of the figure represents the position in the protein sequence, and the *Y*-axis indicates absorbance (i.e., ELISA reactivity). Obviously, one epitope reacts much more strongly than the remainder of the protein. This phenomenon is referred to as immunodominance.

3.4.2. Polyclonal Antibody Epitopes

One of the most common applications of this system is epitope mapping of a full protein sequence of overlapping octamers, where the immunogen was the intact native protein (or even an entire organism). ELISA reactivity of such a polyclonal immune serum typically gives several peaks, each corresponding to an epitope. Frequently, there is one relatively strong immunodominant epitope that stands out among the others (*see* Fig. 1).

3.4.3. Epitope Overlaps

Each of the peaks of epitope recognition will typically span several pins, and therefore, several overlapping peptides. The minimal region of recognition is the sequence contained in all the recognized peptides of a given epitope (*see* Fig. 2). This vital information is only accessible through synthesis of many overlapping peptides.

3.4.4. Antipeptide Antibodies

You may wish to map the fine specificity of a serum raised against a synthetic peptide immunogen. These experiments generate the strongest *PEPSCAN* ELISA signals. However, the results are often complicated by strong reactivity to two (or more!) closely neighboring epitopes. This gives a broad peak, so that it is difficult to tell where one

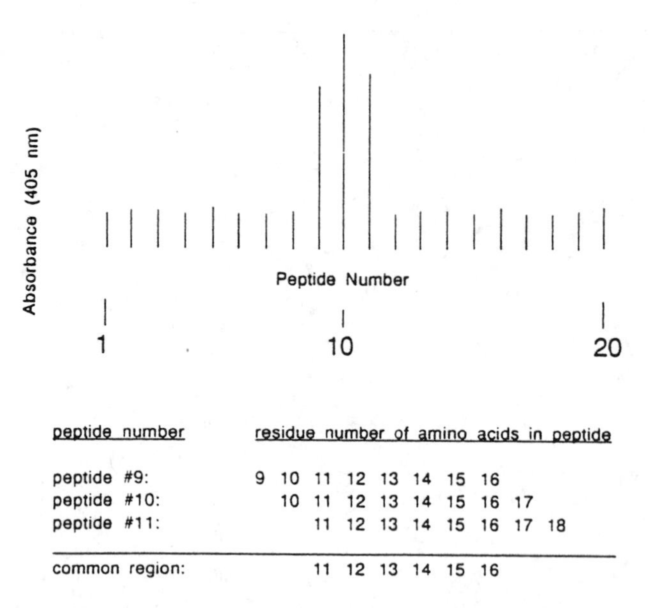

peptide number	residue number of amino acids in peptide
peptide #9:	9 10 11 12 13 14 15 16
peptide #10:	10 11 12 13 14 15 16 17
peptide #11:	11 12 13 14 15 16 17 18
common region:	11 12 13 14 15 16

Fig. 2. Schematic of overlapping peptides representing a linear epitope. Peptides 1–20 are overlapping 8-mer peptides. The figure shows their reactivity with an antibody preparation. Peptides 9, 10, and 11 are reactive. The amino acid sequence in common to these three peptides is 11–16. This sequence represents the epitope of the test antibody.

epitope ends and another one begins. Of course, this is more of a problem with sera raised against larger peptides (30+ amino acids) as immunogens.

3.4.5. MAb Epitopes

Most MAb are raised with an intact protein as immunogen. The limited reactivity of these MAb to peptide pins emphasizes the paradigm of underrepresentation of linear epitopes among the general population of Abs. Only about one out of every eight of the MAb tested reacts strongly to any of the peptides on pins. Frequently, MAb give somewhat ambiguous results with two or three peaks detected. This may seem like an artifact, since a MAb should only have one target sequence, but Geysen has suggested that these multiple regions of recognition suggest the location of the noncontiguous regions that would fold together in the native protein to give the conformational epitope for these MAb. Others have seen gross crossreactivity of MAb to several peptides with related sequences on pins *(6)*. For MAb raised against peptide immunogens, only linear epitopes are seen.

3.4.6. Human Serum Epitopes

Immunologists who regularly work with human sera are generally familiar with its idiosyncrasies. Chief among these is a remarkably high background reaction in ELISAs. This is presumably owing to the broad sensitization of humans and owing to diversity of exposure experiences, as well as a large amount of low-specificity Ab in the naive state. Monkey sera exhibit these same problems, although to a lesser degree.

A brief treatment at 56°C will kill most disease organisms and viruses, as well as neutralizing complement and many other serum proteases. However, heat treatment often increases nonspecific "stickiness" of the serum. An ELISA blocking solution based on 2% casein (dissolved in 100 m*M* NaOH) in PBS, pH 7.4, is very effective in reducing this background reactivity. Two other useful blocking cocktails you may need to try include 0.5% boiled casein, 0.5% BSA, 1% Tween-20, 2% newborn calf serum in PBS; and 1*M* glucose, 10% glycerol, 10% newborn calf serum in PBS *(7)*.

3.4.7. Differential Responses

Different species of immune animals will often react to different sets of epitomes in a given immunogen. Different individuals within a species often react differently, especially if they are "out-bred" (not genetically homogenous). Even with inbred strains, differences will arise because of heterogeneity in the animals' immune history and in injection technique.

3.5. Other Applications

3.5.1. Mimotopes

One application now widely touted by Geysen is the construction of mimotopes. These are artificial epitopes comprising peptides containing natural and nonnatural amino acids in nonnative sequences *(8,9)*. Mimotopes can attain conformations in assays that have the same binding characteristics of naturally occurring conformational epitopes. It seems probable that mimotopes may also be able to elicit Ab with affinity for naturally occurring conformational and even nonprotein (e.g., carbohydrate) epitopes *(10)*. For this reason, they are promising candidates for future vaccines.

3.5.2. Cleavable Pins

Another approach utilizes the chemical spacer built onto the peptides. This is the nonpeptide moiety that attaches the peptides to the plastic support pins. Incorporation of an acid-labile amino acid sequence (Asp-Pro) at this position in the peptide facilitates acidolytic cleavage from the pin after synthesis is completed. This results in generation of a large number of soluble peptides although in limited quantities. This technique has proven useful in studies demonstrating T-lymphocyte epitope specificity through mitogenesis assays *(11,12)*. More recently, pins bearing cleavable chemical links have been made commercially available from Chiron *(12,13)*. Depending on the linker and respective cleavage chemistry, the new pins can be used to generate peptides with C-terminal free carboxylic acids, carboxamides, or diketopiperazines. Chiron currently operates its successful custom peptide synthesis facility using special large high-loading pins and cleavable linkers.

Yet another fairly simple variation is the use of proteins other than Ab to probe the peptide pin arrays. This approach is promising for structure–function studies on biological receptor molecules. Another example takes advantage of the reversibility of binding of Abs to peptides in the typical ELISA application of the pins. By elating the bound Abs from the individual peptides, it is possible to affinity-purify small quantities of Ab. The amounts of antibody protein isolated from each pin by this technique are vanishingly small, but sufficient to be detected by means of binding to Western blots.

4. Notes

1. Before deprotection, pins may be first pre-equilibrated in DMF baths for 5 min. This step may reduce nonspecific attachment of the piperidine molecule to the polyethylene pin

matrix, facilitating its removal at later wash steps. Other chemists have increased the piperidine concentration in order to shorten the time for deprotection *(14)*.

2. The original piperidine solution may be reused in every deprotection cycle for the entire synthesis. This is possible because the reaction of piperidine with the F_{moc} groups on the peptide pins is stoichiometric: each mole of F_{moc} removed requires only 1 mol of piperidine. This means very little piperidine is actually consumed in one use of the reagent solution. With this in mind, and partly because of the difficulty and expense of obtaining and storing large quantities of piperidine (which is a controlled substance), the peptide chemistry group at WRAIR investigated the possibility of reusing the 20% piperidine solution. They found that, after 13 daily uses for deprotection of the peptide pins, the solution was as effective as it was when freshly prepared. This was in spite of the observation that, after 4 d, it began to develop a white crystalline material. This material may be a piperidine formate salt resulting from hydrolysis of the DMF solvent (by atmospheric water in tile presence of the piperidine base as a catalyst). When the piperidine/DMF reagent is stored over molecular sieves to keep it anhydrous, the crystalline material does not form, and the reagent maintains its clarity as well as its efficacy.
3. When washing, it is important to rinse both sides of the pin blocks in order to remove contamination resulting from splashes and condensation of solvents and reagent, which otherwise accumulate on their undersides. Do this at least once, with the first wash of each set.
4. From the residual odor remaining on the pins, it is apparent that the piperidine may not be completely removed by the organic solvent washes alone, as described in Section 3.1., step 2. Because piperidine is a basic, its presence is easily confirmed by testing the pH of an aqueous solution of the final wash solvent. Worried that residual piperidine would affect the peptide syntheses, the group at WRA1R now typically adds a 5-min wash in 1% acetic acid (freshly prepared) in DMF. This is intended to neutralize the piperidine base and reduce its affinity for the polyethylene. After washing with this modified protocol, a 50% solution of the final methanol wash in water demonstrated a neutral pH, indicating essentially complete removal of piperidine. Although this step leaves the nascent peptides with a partially protonated α-amino group, the acetic acid is apparently easily removed by the inherently basic DMF. Note that this acid wash step is not appropriate for chemistry modification incorporating *in situ* activation of the amino acid derivatives. Also, it has not been tested with all of the new cleavable linker chemistries currently offered by Chiron.
5. If you wish to elevate the concentration of the amino acid derivatives (and activator, if used, *see* Note 6) to 100 m*M*, and the concentration of HOBt to 120 m*M*, you may reduce the time for acylation to 2 h. This allows several amino acids to be added to the growing peptides in a 24-h period.
6. Other popular chemistries for peptide bond formation are compatible with the pins, including carbodiimide, HBTU, and BOP. For these techniques, prepare the activated amino acid derivative solutions as follows. For carbodiimide chemistry, use 60 m*M* amino acid, 60 m*M* diisopropylcarbodiimide, and 70 m*M* HOBt in DMF. For HBTU, use 60 m*M* amino acid, 60 m*M* HBTU, and 70 m*M* HOBt. For BOP, use 60 m*M* amino acid, 60 m*M* BOP, and 70 m*M* HOBt. You may also add bromophenol blue at 50 µL, as an indicator of acylation reaction end point.
7. Some have substituted dimethylacetamide for the DMF solvent in the coupling step, claiming that this improves solvation efficiency of the pin matrix *(14)*. You may also prefer to use NMP, which is nonflammable and less toxic than DMF.
8. In order to expedite and improve the accuracy of the placement of amino acid esters in the appropriate microtiter plate wells, an automated indexer is used in the laboratory at WRAIR *(2)*. Driven by menu software on a PC-type computer, this device indicates the

appropriate wells for each of the amino acids for the synthesis *(15)*. Chiron distributes a similar device that is supplied with Windows or MacIntosh software.

9. Most successful users of PEPSCAN typically make peptides 6–12 amino acids in length. Shorter molecules may not have a measurable affinity for the test Ab, whereas longer molecules will probably contain little of the full-length peptide because of the limited efficiency of the nonsequence-optimized coupling chemistry.
10. To keep the silica gel from intimate contact with the pins, pouches may be made from paper towels, filled with a generous handful of indicator-grade silica, and then stapled shut. These silica pouches may be regenerated when necessary by baking overnight at 120°C.
11. It is critical to avoid microbial contamination of the peptide pins. Amino acid analysis indicates that the peptides are rapidly destroyed by microbial action. Indeed, pins left overnight in PBS at room temperature are thereby completely ruined. It is also probably best to prevent any of the ELISA solutions from drying onto the pins.
12. The group at WRAIR uses a large ultrasonic cleaner instrument manufactured by Blackstone (Jamestown, NY) and rated for 500 W at 25 kHz. Chiron recommends a power rating of 7 kW/m^2 at 25 kHz. This sonicator has an electrical heater and a thermostat that is operated at 65°C. My personal experience includes several attempts using less powerful sonicators. They were all quite ineffective, resulting in high background signals in the ELISA and residual primary antibody protein on the pins (detected by ELISA without the usual addition of primary antibody, or via amino acid analysis). Similarly, poor results were obtained when the bath temperature was allowed to drop below 60°C. Temperatures higher than 70°C are probably not good for the peptides.
13. The sonicator used at WRAIR has a volume of 20 L, so it can fit eight blocks of peptide pins at once, floating in a single layer on the top. Although fresh 2-mercaptoethanol should be added every day, you may reuse the disruption buffer 10 or 12 times, until it begins to darken. Before discarding old disruption buffer (cooled to room temperature), you may deodorize it by addition of hydrogen peroxide.
14. To keep the silica gel from intimate contact with the pins during storage, make pouches from paper towels, filled with a generous handful of indicator-grade silica, and then stapled shut. These silica pouches may be regenerated when necessary by baking overnight in an oven at 120°C.
15. In all buffers and reagent solutions used for the pin ELISA, 0.1% Tween-20 is typically added. Tween-20 is a very mild nonionic detergent. It serves as a wetting agent, thereby improving reproducibility and helping to reduce nonspecific binding. Because Tween-20 is surface-active, all pipeting should be performed carefully so as to minimize aeration, since foaming will affect reproducibility.
16. Two percent casein gives lower background for some Ab, such as human serum. Either of the blocking buffers described will keep for 2 or 3 wk if sterility is maintained. If desired, 0.2% NaN_3 may be added. This will increase the practical storage time for the reagent to several weeks, but 1 L is typically consumed in a few days of ELISA work. Blocking is generally performed for 1 h at room temperature, but if the solution contains 0.2% sodium azide, it may be left overnight in the refrigerator.
17. As an alternative to a rocking platform, you may use a rotating (orbital) platform for incubations.
18. Remember that proper reactions for control peptides, if they are used, will probably require a different first Ab solution. They may also require a different second Ab solution.
19. For the overnight incubation with test Ab, put the filled plates into a sealable plastic box lined with a moistened paper towel to maintain humidity and minimize evaporation. If this step is allowed longer than about 12 h, evaporation and condensation may nonetheless begin to affect reproducibility, especially for the pins closest to the edge of the plate. After

the overnight incubation, the first Ab solution is usually discarded. However, you may pool and reuse this reagent at least four times without any discernible loss in signal-to-noise ratio. In that case, add a single wash step between the blocking and first Ab to minimize dilution and contamination of the valuable test Ab solution. Also store the Ab solution at 4°C.

20. For convenience in preparation of PBST, consider purchasing 10X PBS in liter bottles, and then add Tween-20 and sodium azide. You may store this 10X stock solution in a carboy at room temperature for up to 2 wk. Then from the stock, prepare 1X PBST for each day's use by diluting 1/10 with deionized water.
21. For second Ab (enzyme conjugates), you may use commercially available reagents from various sources or prepare your own conjugates. Although some prefer to use conjugates with horseradish peroxidase (HRP), alkaline phosphatase (AP) conjugates give maximum reproducibility with good sensitivity. This is probably because of irreversible reactions between hydrogen peroxide in the substrate buffer of the HRP enzyme and the peptides. You may perform incubations for second Ab binding on the laboratory bench top, or in the refrigerator, with or without a rocking platform. This incubation may also be performed overnight.
22. In some cases, it may be necessary to prepare your own enzyme-conjugated second Ab. To make an AP-conjugated goat Ab to recognize Aotus monkey Ab in ELISAs, the group at WRAIR used the following generally applicable protocol: Isolate several milligrams of nonimmune Ab from the monkey serum by protein A affinity. Use most of this protein as an immunogen to raise Ab in a goat. Isolate several milligrams of goat immune Ab by protein A affinity. Couple a few milligrams of the Aotus Ab immunogen to Sepharose. Use the immobilized Aotus Ab to affinity-purify the goat Ab vs Aortas Ab. Conjugate the purified Ab to commercial AP via glutaraldehyde. Dilute in PBS with 0.2% NaN_3. Test the second Ab conjugate at various concentrations to determine the appropriate working concentration for the reagent. Store frozen, avoiding refreezing. For details, refer to Lyon and Haynes *(16)*.
23. To aid in visualization of color development, place the plate containing the substrate solution on a piece of white paper. If you are using more than one plate for the ELISA (which is likely), number them on their outer edges. During development, avoid thermal gradients. These may be caused, for example, by drafts or sunlight. It helps to cover the pins (e.g., with an overturned empty cardboard carton) to isolate them from environmental effects.
24. After development, removal of the pins stops the color generation catalyzed by the enzyme conjugate on the pins, but the substrate is thermolabile, and the chromophore is photolabile, so avoid unnecessary delays by setting up the reader while development is still taking place. Avoid touching the bottom of the ELISA plate before it is read. Do not discard the plates until you are certain that you have two legible copies (either "soft" or "hard" copies, according to your preference) of your data. If the signals are weak, you may return the pins to the plates for further development and read them again later.

References

1. Geysen, H. M., Meloen, R. H., and Barteling, S. J. (1984) Use of peptide synthesis to probe viral antigens for epitomes to a resolution of a single amino acid. *Proc. Natl. Acad. Sci. USA* **81,** 3998–4002.
2. VanAlbert, S., Lee, J., Lyon, J. A., and Carter, J. M. (1991) Amino acid indexer for synthesis of Geysen peptides. US Patent 5,243,540.
3. Barlow, D. J., Edwards, M. S., and Thornton, J. M. (1986) Continuous and discontinuous protein antigen determinants. *Nature* **322,** 747,748.

4. Amit, A. C., Mariuzza, R. A., Phillips, S. E. V., and Poljak, R. J. (1986) Three dimensional structure of an antigen-antibody complex at 2.8-Åw resolution. *Science* **233,** 747–753.
5. Sheriff, S., Silverton, E. W., Padlan, E. A., and Cohen, G. H. (1987) Three-dimensional structure of an antigen–antibody complex. *Proc. Natl. Acad. Sci. USA* **84,** 8075–8079.
6. Burkot, T. R., Da, Z. W., Geysen, H. M., Wirtz, R. A., and Saul, A. (1991) Fine specificities of monoclonal antibodies against the *Plasmodium falciparum* circumsporozoite protein: recognition of repetitive and nonrepetitive domains. *Parasite Immunol.* **13,** 161–170.
7. Birk, H. W. and Koepsell, H. (1987) Reaction of monoclonal antibodies with plasma membrane proteins after binding on nitrocellulose: renaturation of antigenic sites and reduction of nonspecific anybody binding. *Anal. Biochem.* **164,** 12–22.
8. Geysen, H. M., Rodda, S. J., and Mason, T. J. (1986) A priori delineation of a peptide which mimics a discontinuous antigenic determinant. *Mol. Immunol.* **23,** 709–715.
9. Geysen, H. M., Rodda, S. J., and Mason, T. J. (1986) Synthetic peptides as antigens. *Ciba Foundation Symp.* **119,** 130–149.
10. Geysen, H. M., MacFarlan, R., Rodda, S. J., Tribbick, G., Mason, T. J., and Schoofs, P. G. (1987) Peptides which mimic carbohydrate antigens, in *Towards Better Carbokydrate Vaccines* (Bell, R. and Torrigiani, G., eds.), Wiley, Chinchester, UK, pp. 103–118.
11. Van der Zee, R., van Eden, W., Meloen, R. H., Noordzij, A., and van Embden, J. D. A. (1989) Epitope mapping and characterization of a T-cell epitope by the simultaneous synthesis of multiple peptides. *Eur. J. Immunol.* **19,** 43–47.
12. Bray, A., Maeji, N. J., and Geysen, H. M. (1990) The simultaneous multiple production of solution phase peptides; assessment of the Geysen method of simultaneous peptide synthesis. *Tetrahedron Lett.* **31,** 5811–5814.
13. Maeji, N. J., Bray, A. H., and Geysen, H. M. (1990) Multi-pin peptide synthesis strategy for T-cell determinant analysis. *J. Immunol. Methods* **134,** 23–33.
14. Arendt, A. and Hargrave, P. A. (1993) Optimization of peptide synthesis on polyethylene rods. *Peptide Res.* **6,** 346–352.
15. Carter, J. M., VanAlbert, S., Lee, J., Lyon, J. A., and Deal, C. D. (1992) An aid to peptide pin syntheses. *Biotechnology* **10,** 509–513.
16. Lyon, J. A. and Haynes, J. D. (1986) *Plasmodium falciparum* antigens synthesized by schizonts and stabilized at the merozoite surface when schizonts rupture in the presence of protease inhibitors. *J. Immunol.* **136,** 2245–2251.

96

Epitope Mapping of Protein Antigens by Competition ELISA

Glenn E. Morris

1. Introduction

At their most elaborate, epitope mapping techniques can provide detailed information on the amino acid residues in a protein antigen which are in direct contact with the antibody-binding site. For immunoassay purposes, however, it is often more important to know whether one antibody interferes with the binding of a second antibody against the same antigen or whether both antibodies can bind without mutual competition. Antibodies in the latter category must recognize different epitopes and are useful for two-site assays in which one antibody is used to capture the antigen from a complex mixture (e.g., plasma or tissue extracts), while a second is used to detect the captured antigen. Antibodies that do compete with each other probably recognize the same epitope or a very near neighbor *(1)*, but the possibility that they bind to widely separated epitopes and compete by altering the conformation of the antigen must always be considered. Epitope mapping is usually performed with monoclonal antibodies (MAb), since polyclonal antisera are usually directed against a number of different epitopes and results could be difficult to interpret.

The basic principle of epitope mapping by competition ELISA is simple: (1) Antigen is attached in some way to a plastic microtiter plate, and (2) the effect of increasing concentrations of one MAb on the binding of a second (labeled) MAb is studied.

The method can be applied to antibodies that recognize assembled (conformational, discontinuous, native) epitopes and to those that recognize continuous (sequential, denatured) epitopes, but it is an obvious prerequisite that all antibodies under study should bind to the antigen in the particular ELISA format used. Some proteins are partially denatured when coated directly onto plastic, and they lose many assembled epitopes. In such cases, it may be possible to coat plates with a "capture" antibody that will bind the protein to the plate indirectly in its native form. This approach may also be useful if insufficient purified antigen is available to coat plates directly. If a monoclonal "capture" antibody is used, it should obviously recognize a different epitope from those being tested, but polyclonal "capture" antibodies, if available, can work rather well, provided it is possible to do internal controls to ensure that the monoclonals are competing only with each other for antigen and are not displacing antigen from the "capture" antibodies.

From: The Protein Protocols Handbook
Edited by: J. M. Walker Humana Press Inc., Totowa, NJ

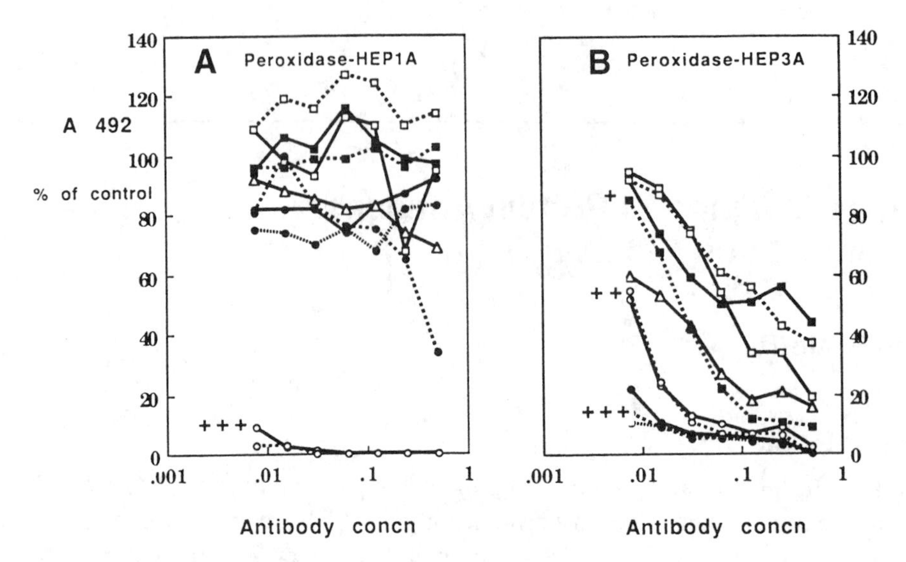

Fig. 1. Examples of competition between MAb against recombinant hepatitis B surface antigen. Limiting concentrations of peroxidase-conjugated ascites antibodies **(A)** HEP1A 1:1500 or **(B)** HEP3A 1:6000 were mixed with decreasing amounts of 10 different MAb culture supernatants for competition ELISA on plates coated with recombinant S protein. Competition was rated as +++ (>50% competition at 1:128), ++ (>50% at 1:32), or + (>50% at 1:1 dilution). Antibody concentrations are expressed as mL of culture supernatant/mL final volume. Competing antibodies are HEP1A (open circles, continuous), HEP1B (open circles, broken), HEP 2 (triangles), HEP3A (filled circles, continuous), HEP3B (filled circles, broken), HEP3C (filled circles, heavy broken), HEP4A (open squares, continuous), HEP4B (open squares, broken), HEP5A (filled squares, continuous), and HEP5B (filled squares, broken). (Reproduced with permission from ref. *4*.)

It is also essential to obtain at least a rough idea of the relative avidities of the antibodies being tested in the particular ELISA format used for epitope mapping. Provided antibody concentrations are known (or known to be similar), it should be sufficient to perform a serial dilution test in the assay format and determine the dilution required to reduce antibody binding to 50% of the plateau, or saturation, level. Low-avidity antibodies require little dilution and may fail to reach saturation, even at high concentrations. Data obtained with them should be treated with caution, since a low-avidity antibody may fail to compete with a high-avidity one, even at the same epitope. An important control in this respect is to repeat the ELISA after switching around the labeled and unlabeled antibodies (*see* Figs. 1 and 2).

Finally, it is usually necessary to have reasonably pure Ig preparations from each MAb for labeling purposes. A method using peroxidase-labeled antibodies will be described, but other labels can be used. Unless the antibodies can be purified easily from culture supernatants, this necessitates the use of ascites fluids from which a suitable Ig is easily obtained by a simple 50% saturated ammonium sulfate precipitation.

The general principle of competition among MAb can be applied flexibly to suit particular circumstances. Kordossi and Tzartos *(1)* compared the competition ELISA format to be described here with an assay in which one MAb was used to displace

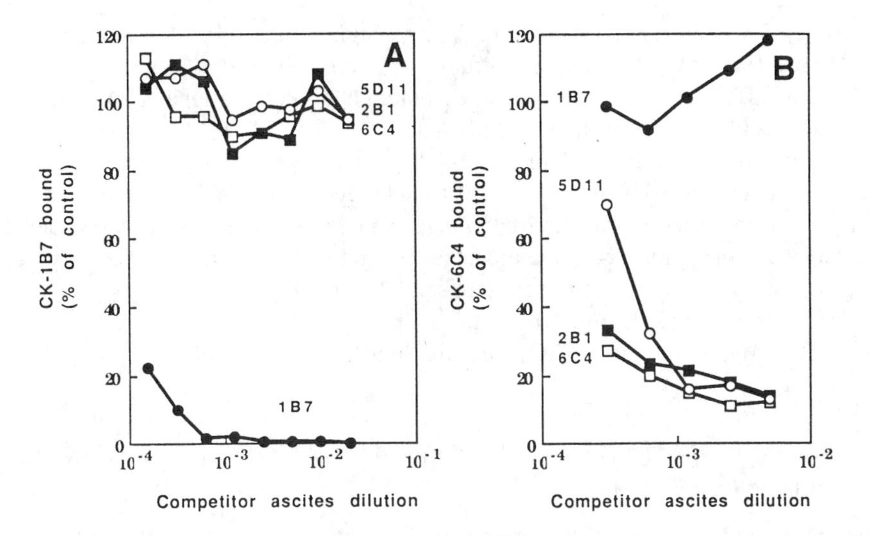

Fig. 2. An MAb against human muscle creatine kinase, CK-1B7, recognizes a different epitope from another three antibodies. **(A)** Alkaline phosphatase-conjugated CK-1B7 at a predetermined limiting dilution of 1:400 was premixed with serial dilutions of ascites fluids containing CK-1B7 itself, CK-5D11, CK–2B1, and CK-6C4 (approximate conc. range, 0.4–50 µg/mL). The mixtures were then transferred to the wells of a sandwich ELISA plate with human MM-CK as bridge. **(B)** A similar experiment using peroxidase-labeled CK-6C4 at a predetermined limiting dilution of 1:25 is shown. (Reproduced with permission from ref. *2.*)

labeled antigen from another MAb bound to Sepharose beads. We included our own MAb against CK as competitors in the standard format of the commercial Tandem-E kit (Hybritech) for cardiac-specific MB-CK dimers in order to show that three MAb in Fig. 2 (6C4, 2B1, and 5D11) competed with Hybritech anti-(M-CK subunit) MAb, whereas 1B7 recognized a quite different epitope *(2)*. One may even abandon the principle of competition with good effect. Thus, Tzartos et al. *(3)* fixed one MAb to their antigen by glutaraldehyde crosslinking to block the epitope irreversibly and were then able to test other antibodies in a subsequent step. By testing the ability of the crosslinked complex to remove a second MAb from the supernatant, they were able to avoid the need to label their antibodies. With a little ingenuity, a mapping method in which one antibody excludes or displaces others can be designed to suit most circumstances.

2. Materials

1. Microtiter plates: Any high-binding capacity plate suitable for ELISA can be used (e.g., Immulon, Dynatech Labs [Chantilly, VA]), but flexible PVC plates (e.g., M25 plates from Dynatech Labs) are easy to wash manually between steps and can still be used on automatic plate readers, though not designed for optical use.
2. Eight-channel pipeter (e.g., Labsystems or Finnpipet): Automatic plate washers and readers may be useful for frequent use, but are not essential.
3. PBS: 0.9% NaCl, 25 m*M* sodium phosphate buffer, pH 7.2.
4. Incubation buffer (IB): PBS + 0.05% Triton X-100. It is convenient to prepare this using a 5% w/v stock of Triton X-100 in PBS.

5. IB + 4% BSA (bovine albumin fraction V; e.g., Sigma [St. Louis, MO] A7906); IB + 1% BSA; IB + 1% BSA + 1% fetal calf serum + 1% horse serum.
6. Substrate buffer: Mix 25.7 mL of 0.2*M* disodium hydrogen phosphate, 24.3 mL of 0.1*M* citric acid, and 50 mL of water (final pH = 5.0).
7. *o*-phenylene diamine dihydrochloride (OPD); (Sigma) stock solutions (80X: 32 mg/mL in water) should be stored in aliquots at –20°C and used only while they remain pale yellow. This should be treated as a possible carcinogen and, in particular, the dusty powder should be weighed with appropriate containment and operator protection. Prepared substrate tablets (Sigma) may be a safer alternative in some laboratories.
8. Hydrogen peroxide (30% v/v) "100 vol."
9. 3*N* H_2SO_4: Add 8 mL conc. H_2SO_4 carefully to 92 mL of water.

3. Methods

The following method can be completed in 3–4 h, if labeled MAb of known titer are already available (*see* Note 1).

1. Coating (stage 1): For direct attachment, 50 µL of antigen at 1–10 µg/mL in PBS is placed in each well of a microtiter plate and left at room temperature for 1 h (*see* Note 2). This coating solution can often be recovered and reused several times. For indirect attachment through a "capture" antibody, purified Ig (5 µg/mL in PBS) is coated in the same way (*see* Note 3). One whole row should be left uncoated as a control for nonspecific binding in later steps.
2. Rinse once with IB. Efficient rinsing between steps is very important. Fill each well manually, invert the plate over the sink, and then flick it firmly several times face down onto a sheet of paper tissue (this is easier and less noisy with flexible PVC plates than with hard polystyrene plates). An automatic plate washer can be used, but if high backgrounds are obtained, its efficiency should be compared with the manual method.
3. Blocking (stage 2): Remaining binding sites on the plastic are blocked by filling the wells with 4% BSA in IB. It is important to fill the wells right to the top; erratic results and high backgrounds are often the result of failure to do this. Ten minutes at room temperature are usually enough. Rinse once with IB.
4. If a "capture" antibody was used, an extra step is required (otherwise go to step 5). Fifty microliters of antigen at 1–10 µg/mL in IB + 1% BSA are placed in each well for 1 h at room temperature. Then discard and rinse three times with IB.
5. Premixing competing antibodies: While the antigen is being coated or captured, labeled and unlabeled MAb may be premixed in a separate microtiter plate so that they can be added simultaneously to antigen on the main plate. A predetermined limiting dilution of the peroxidase-labeled MAb conjugate (*see* Note 4) should be used in the presence of serial dilutions of the unlabeled competitor. A good initial choice of dilutions would be 1 in 2 down to 1 in 128 for MAb culture supernatants and 1 in 100 down to 1 in 6400 for MAb acites fluid, with at least one well per row without any competitor. At least 6–8 wells on the whole plate should be without competitor to obtain an accurate average 100% value for competitor curves (Fig. 2). Dilutions are made in IB + 1% BSA + 1% fetal calf serum + 1% horse serum (*see* Note 5).
6. Competitive binding to antigen (stage 3): The premixed MAb are transferred to the ELISA plate using an eight-channel multipipeter and left for 1 h at room temperature. Then discard and rinse five times with IB. A fresh piece of paper tissue should be used after the final rinse and the flat surface of the plate wiped dry.
7. Color development (stage 4): Mix 15.8 mL of substrate buffer with 0.2 mL of 80X OPD substrate and 6.7 µL of 30% H_2O_2 immediately before use. Always test a small aliquot

with a few microliters of peroxidase conjugate; it should turn brown immediately. Place 150 μL in each well. Allow 5–30 min for the color to develop before stopping the reaction with 50 μL of 3*N* H_2SO_4. The absorbance is read at 492 nm on a plate reader or by diluting in water in a cuvet for a conventional spectrophotometer. If using a plate reader, it is a good idea to put substrate alone in an empty row (on a separate plate, if necessary) as a blank; there should be little or no visible color (*see* Note 4). The control row that was left uncoated (step 1) should show a similar lack of color; if it is significantly yellower than the blank row after 20–30 min, the efficiency of the blocking and rinsing steps should be re-examined.

8. Typical results and interpretations are illustrated in Figs. 1 and 2. Fig. 1A shows a typical straightforward result obtained when 10 unlabeled MAb competed with one peroxidase-labeled MAb for purified hepatitis B surface antigen bound directly to an ELISA plate *(4)*. Two antibodies recognize the same epitope (+++; one of them is also the labeled MAb), whereas the other eight show no significant competition. Sometimes the results are less straightforward. When a different MAb, HEP3A, was labeled with peroxidase, two other MAb competed, as well as HEP3A itself (+++; Fig. 1B), but all MAb competed to some degree (++, +). Such results cannot be interpreted simply, and an intimate knowledge of antigen, antibodies, and their interactions may be required. Figure 2 shows that good straightforward results can be obtained when antigen is "captured" onto an ELISA plate using Ig from a polyclonal antiserum. The MAb in this example will not bind at all when the antigen (creatine kinase [CK]) is bound directly to the ELISA plate, because partial denaturation of CK results in loss of the assembled epitopes that they recognize. The results show that the MAb CK-1B7 recognizes one epitope (Fig. 2A), whereas a different epitope is recognized by the other three MAb (Fig. 2B) *(2)*. Note the internal control: if any of the antibodies were displacing the antigen from the "capture" antibody, they would show competition in both Fig. 2A and 2B.

4. Notes

1. For conjugation with peroxidase, very rapid and simple one- or two-step glutaraldehyde methods may be satisfactory, though the periodate method is said to be superior; these are described in refs. *5* and *6*. Commercial kits are also available (e.g., Actizyme, Zymed Labs, San Francisco, CA).
2. We use PBS, but the nature of the coating buffer is usually not critical, in our experience, provided there are no traces of detergent or other proteins. If a denatured antigen is required, it is possible to coat plates in 8*M* urea. Peptides as short as eight amino acids can also be coated successfully.
3. For maximum sensitivity, the Ig should be as pure as possible. For MAb ascites fluid, a standard 50% ammonium sulfate Ig fraction is satisfactory, and this method may also work well for very high-titer polyclonal antisera. For many antisera, however, further purification by affinity chromatography on protein A Sepharose or Sepharose-bound antigen may be necessary *(2)*. The latter method is recommended where possible because it selects rugged, high-affinity antibodies (if they exist in the antiserum), which are excellent for antigen capture.
4. To determine a limiting dilution for the peroxidase-conjugated MAb experimentally, this whole ELISA procedure must first be carried out in the absence of competitor MAb and using serial dilutions of the labeled antibody over a row of eight wells (from 1 in 10 down to 1 in 1280 should be a suitable range for most conjugates). Any nonsaturating concentration could be used, but 50–70% of the plateau level is a good choice.
5. There is no logical reason for having the sera present in this particular assay, but we routinely add them at later stages, and their presence may help to keep backgrounds low.

When labeled second antibodies are used in other types of ELISA (e.g., peroxidase-labeled rabbit anti-[mouse Ig]), sera prevent high backgrounds owing to second antibody crossreaction with proteins from MAb culture supernatant (horse or calf) or antibody coats (usually rabbit or goat).

References

1. Kordossi, A. A. and Tzartos, S. J. (1987) Conformation of cytoplasmic segments of acetylcholine receptor α- and β-subunits probed by monoclonal antibodies: sensitivity of the antibody competition approach. *EMBO. J.* **6,** 1605–1610.
2. Nguyen thi Man, Cartwright, A. J., Andrews, K. M., and Morris, G. E. (1989) Treatment of human muscle creatine kinase with glutaraldehyde preferentially increases the immunogenicity of the native conformation and permits production of high-affinity monoclonal antibodies which recognize two distinct surface epitopes. *J. Immunol. Methods* **125,** 251–259.
3. Tzartos, S. J., Rand, D. E., Einarson, B. R., and Lindstrom, J. M. (1981) Mapping of surface structures of electrophorus acetylcholine receptor using monoclonal antibodies. *J. Biol. Chem.* **256,** 8635–8645.
4. Le Thiet Thanh, Nguyen thi Man, Buu Mat, Phan Ngoc Tran, Nguyen thi Vinh Ha, and Morris, G. E. (1991) Structural relationships between hepatitis B surface antigen in human plasma and dimers of recombinant vaccine: a monoclonal antibody study. *Virus Res.* **21,** 141–154.
5. O'Sullivan, M. J. and Marks, V. (1981) Methods for preparation of enzyme-antibody conjugates for use in enzyme immunoassay. *Methods Enzymol.* **73,** 147–166.
6. Catty, D. and Raykundalia, C. (1989) ELISA and related enzyme immunoassays, in *Antibodies: A Practical Approach,* vol. II (Catty, D., ed.), IRL, Oxford, UK, pp. 97–154.

Part VI

Glycoproteins

97

Identification of Glycoproteins on Nitrocellulose Membranes Using Lectin Blotting

Patricia Gravel and Olivier Golaz

1. Introduction

Glycoproteins result from the covalent association of carbohydrate moieties (glycans) with proteins. The enzymatic glycosylation of proteins is a common and complex form of posttranslational modification. The precise roles played by the carbohydrate moieties of glycoproteins are beginning to be understood *(1–3)*. It has been established that glycans perform important biological roles including: stabilization of the protein structure, protection from degradation, control of protein solubility, control of protein transport in cells, and control of protein half-life in blood. They also mediate the interactions with other macromolecules, and the recognition and association with viruses, enzymes, and lectins *(4–6)*.

Carbohydrate moieties are known to play a part in several pathological processes. Alterations in protein glycosylation have been observed, for example, with the membrane glycoproteins of cancer cells, with the plasma glycoproteins of alcoholic patients and patients with liver disease, with inflammation, and with infection. These changes provide the basis for more sensitive and more discriminative clinical tests *(2,7–9)*.

Lectins are carbohydrate binding proteins, and can be used to discriminate and analyze the glycan structures of glycoproteins. The lectin-blotting technique detects glycoproteins separated by electrophoresis (sodium dodecyl sulfate-polyacrylamide gel electrophoresis [SDS-PAGE] or two-dimensional polyacrylamide gel electrophoresis [2D PAGE]) and transferred to membranes. This chapter describes the protocol for the detection of glycoproteins on nitrocellulose membrane using biotinylated lectins and avidin conjugated with horseradish peroxidase (HRP) or with alkaline phosphatase (AP) (*see* Note 1). These complexes are subsequently revealed either by a chemiluminescent or a colorimetric reaction. To illustrate this method, human plasma glycoprotein patterns, obtained after the separation of proteins by 2D PAGE and different lectin incubations, are presented in Figs. 1–4 (*see* Note 2).

1.1. Classification of Glycoproteins

A useful classification of glycoproteins is based on the type of glycosidic linkage involved in the attachment of the carbohydrate to the peptide backbone. Three types of glycan–protein linkages have been described:

From: The Protein Protocols Handbook
Edited by: J. M. Walker Humana Press Inc., Totowa, NJ

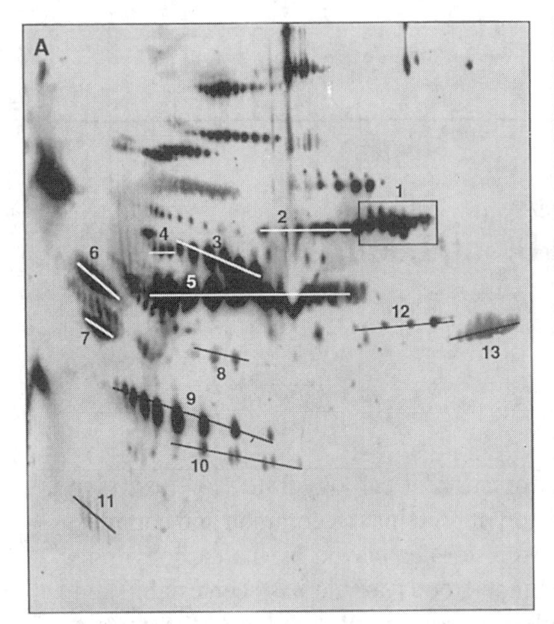

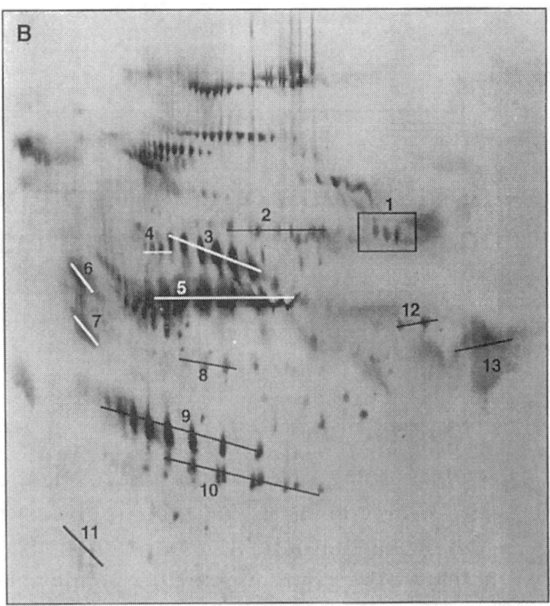

Fig. 1. Glycoprotein blot pattern of 2D PAGE separation of plasma proteins (120 μg) probed with WGA (specific for *N*-acetylglucosamine and neuraminic acid) and **(A)** revealed with chemiluminesence (15 s film exposure) and **(B)** with NBT/BCIP (20 min for the development of the color reaction). Since most of the glycoproteins in plasma contain one or more *N*-linked glycans with at least two *N*-acetylglucosamine residues, the use of WGA allows a general staining of *N*-linked glycoproteins. (1) Transferrin, (2) IgM μ-chain, (3) hemopexin, (4) α1-β-glycoprotein, (5) IgA α-chain, (6) α1-antichymotrypsin, (7) α2-HS-glycoprotein, (8) fibrinogen γ-chain, (9) haptoglobin β-chain, (10) haptoglobin cleaved β-chain, (11) apolipoprotein D, (12) fibrinogen β-chain, (13) IgG γ-chain.

1. *N*-glycosidic linkage of *N*-acetylglucosamine to the amide group of asparagine: GlcNAc (β-1,N) Asn.
2. *O*-glycosidic linkage of *N*-acetylgalactosamine to either the hydroxyl group of serine or threonine: GalNAc (α-1,3) or Ser GalNAc (α-1,3) Thr.
3. *O*-glycosidic linkage of galactose to 5-hydroxy-lysine: Gal (β-1,5) OH-Lys.

Most plasma glycoproteins bear exclusively asparagine-linked oligosaccharides *(3,10)*, and these *N*-glycosidic linkages are by far the most diverse. They can be subdivided into three groups according to structure and common oligosaccharide sequences (Table 1). They all have in common the inner-core structure presented in Fig. 5. The presence of this common core structure reflects the fact that all these asparagine-linked oligosaccharides originate from the same precursor. *O*-glycans are found frequently in mucins, but rarely in plasma glycoproteins. Table 2 illustrates the three different groups of *O*-glycosyl protein glycans.

1.2. Lectins as a Tool for Glycoprotein Detection

Lectins are a class of carbohydrate binding proteins, commonly detected by their ability to precipitate glycoconjugates or to agglutinate cells (some lectins react selectively with erythrocytes of different blood types). Lectins are present in plants, animals, and microorganisms *(11)*.

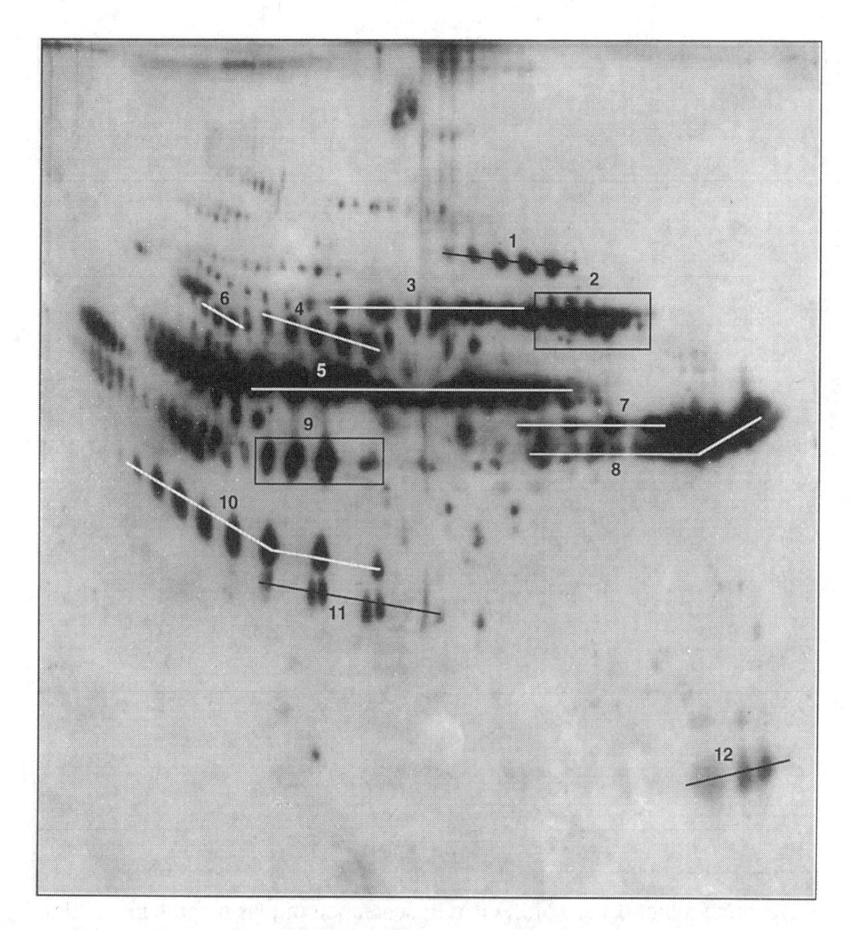

Fig. 2. Glycoprotein blot pattern of 2D PAGE separation of plasma proteins (120 µg) probed with RCA (specific for galactose) and revealed using chemiluminescence reaction (5 s film exposure). (1) Complement 3 α-chain, (2) transferrin, (3) IgM µ-chain, (4) hemopexin, (5) IgA α-chain, (6) α1-β-glycoprotein, (7) fibrinogen β-chain, (8) IgG γ-chain, (9) fibrinogen γ-chain, (10) haptoglobin β-chain, (11) haptoglobin cleaved β-chain, (12) Ig light chain.

Each lectin binds specifically and noncovalently to a certain sugar sequence in oligosaccharides and glycoconjugates. Lectins are traditionally classified into specificity groups (mannose, galactose, *N*-acetylglucosamine, *N*-acetylgalactosamine, fucose, neuraminic acid) according to the monosaccharide that is the most effective inhibitor of the agglutination of erythrocytes or precipitation of polysaccharides or glycoproteins by the lectin (fixation-site saturation method). Another method to determine the carbohydrate specificity of a given lectin consists of the determination of the association constant by equilibrium analysis.

In most cases, lectins bind more strongly to oligosaccharides (di-, tri-, and tetrasaccharides) than to monosaccharides *(12,13)*. Therefore, the concept of "lectin monosaccharide specificity" should advantageously be replaced by that of "lectin oligosaccharide specificity." Table 3 summarizes the carbohydrate specificity of lectins commonly used in biochemical/biological research, both in terms of the best monosaccharide inhibitor of the precipitation of polysaccharides or glycoproteins and in terms

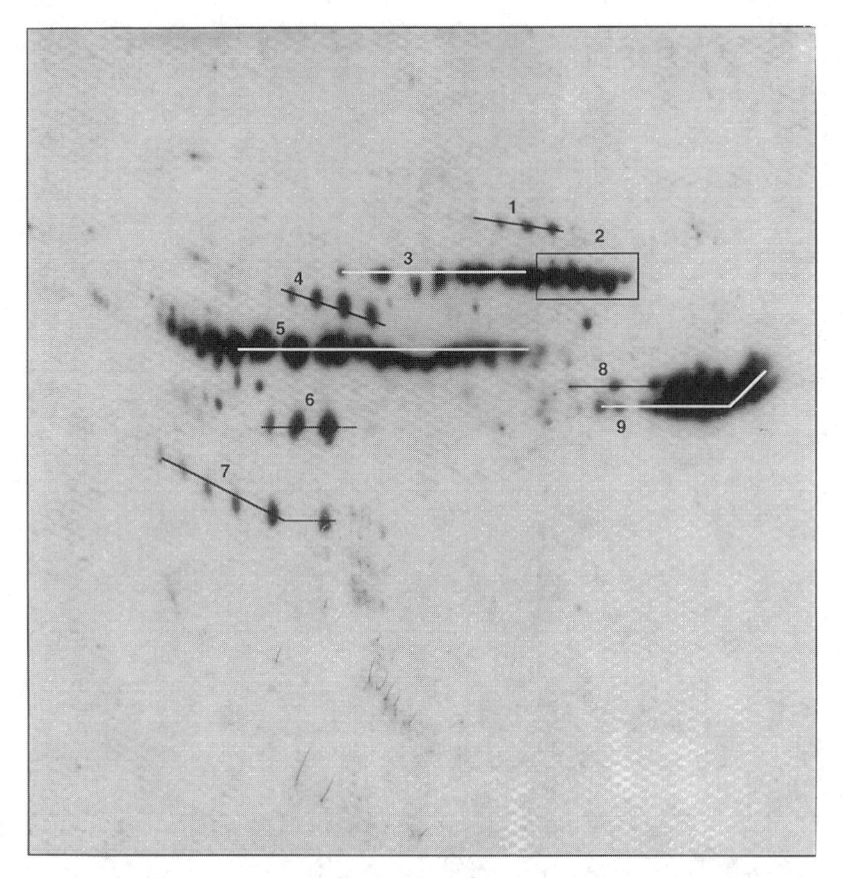

Fig. 3. Glycoprotein blot pattern of 2D PAGE separation of plasma proteins (120 μg) probed with AAA (specific for fucose) and revealed using chemiluminescence reaction (5 s film exposure). (1) Complement 3 α-chain, (2) transferrin, (3) IgM μ-chain, (4) hemopexin, (5) IgA α-chain, (6) fibrinogen γ-chain, (7) haptoglobin β-chain, (8) fibrinogen β-chain, (9) IgG γ-chain.

of the structure of oligosaccharides recognized by immobilized lectins. The interactions of lectins with glycans are complex and not fully understood. Many lectins recognize terminal nonreducing saccharides, whereas others also recognize internal sugar sequences. Moreover, lectins within each group may differ markedly in their affinity for the monosaccharides or their derivatives. They do not have an absolute specificity and therefore can bind with different affinities to a number of similar carbohydrate groups. Because lectin binding can also be affected by structural changes unrelated to the primary binding site, the results obtained with lectin-based methods must be interpreted with caution *(2)*.

Despite these limitations, lectin probes do provide some information as to the nature and composition of oligosaccharide substituents on glycoproteins. Their use together with blotting technique provides a convenient method of screening complex protein samples for abnormalities in the glycosylation of the component proteins. Lectin blotting requires low amounts of proteins, is easy to perform, and therefore is particularly indicated to be used in analyzing biological samples.

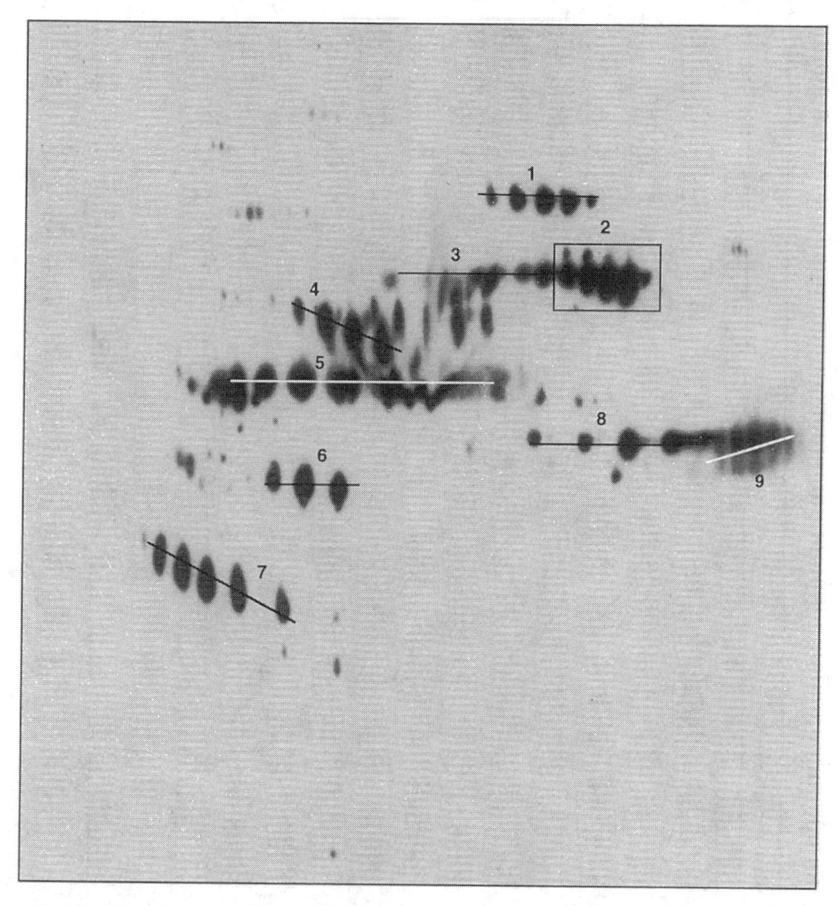

Fig. 4. Glycoprotein blot pattern of 2D PAGE separation of plasma proteins (120 μg) probed with SNA (specific for neuraminic acid linked α-2,6 to galactose) and revealed using chemiluminescence reaction (30 s film exposure). (1) Complement 3 α-chain, (2) transferrin, (3) IgM μ-chain, (4) hemopexin, (5) IgA α-chain, (6) fibrinogen γ-chain, (7) haptoglobin β-chain, (8) fibrinogen β-chain, (9) IgG γ-chain.

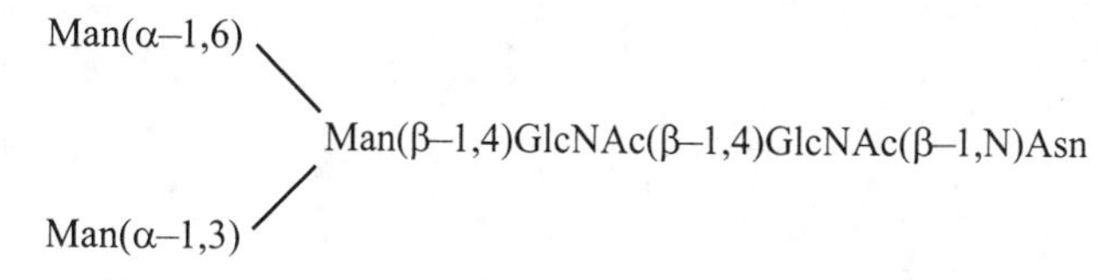

Fig. 5. Oligosaccharide inner core common to all *N*-glycosylproteins.

When the lectin-blotting method described hereafter is combined with the high resolution and reproducibility of 2D PAGE and with the sensitivity of enhanced chemiluminescence, it is possible to identify rapidly the glycoproteins of interest by comparison with a reference 2D PAGE protein map, and to obtain reliable and reproducible results *(27–29)*.

Table 1
Glycans *N*-Glycosidically Conjugated to Proteins

Type of *N*-glycosidic linkage	Examples of structures
High mannose type	Man(α–1,2)Man(α–1,3) \\ Man(α–1,6) \\ Man(α–1,2)Man(α–1,6) / Man(β–1,4)GlcNAc(β–1,4)GlcNAc(β–1,N)Asn Man(α–1,2)Man(α–1,2) Man(α–1,3) /
Complex type	NeuAc(α–2,6)Gal(β–1,4)GlcNAc(β–1,2)Man(α–1,6) \\ Man(β–1,4)GlcNAc(β–1,4)GlcNAc(β–1,N)Asn NeuAc(α–2,6)Gal(β–1,4)GlcNAc(β–1,2)Man(α–1,3) /
Hybrid type	Gal(β–1,4)GlcNAc(β–1,4) \\ Man(α–1,3) \\ GlcNAc(β–1,2) / GlcNAc(β–1,4) — Man(β–1,4)GlcNAc(β–1,4)GlcNAc(β–1,N)Asn Man(α–1,3) \\ Man(α–1,6) / Man(α–1,6) /

Man, mannose; GlcNAc, *N*-acetylglucosamine; NeuAc, neuraminic acid; Gal, galactose.

NeuAc located at the terminal position of oligosaccharide side chains is *N*-acetylated (NeuAcNAc) and often referred to as sialic acid. Sialic acid is in fact the generic term given to a family of acetylated derivatives of neuraminic acid.

Table 2
Glycans *O*-Glycosidically Conjugated to Proteins

Type of *O*-glycosidic linkage	Examples of structures
Mucin type	NeuAc(α–2,6)GalNAc(α–1,3)Ser (or) Thr
	Gal(β–1,3)GalNAc(α–1,3)Ser (or) Thr
	The presence of such *O*-glycosidically linked moieties has been established for a number of plasma proteins (IgA, IgD, hemopexin, plasminogen, apoE)
Proteoglycan type	[GalNAc(β–1,4)GlcUA(β–1,3)]$_n$GalNAc(β–1,4)GlcUA (β–1,3)Gal(β–1,3)Gal(β–1,4)Xyl(β–1,3)Ser
	They are generally linear polymers made up from repeating disaccharide units
Collagen type	Glc(β–1,2)Gal(β–1,5)OH-Lys

NeuAc, neuraminic acid; GalNAc, *N*-acetylgalactosamine; Gal, galactose; GlcUA, Glucuronic acid; Glc, glucose; Xyl, xylose.

Table 3
Specificity of Lectins for Glycan Linked to Asparagine (*N*-Linked Oligosaccharide Chains)

Lectin	Monosaccharide specificity	Oligosaccharide specificity[a]
Wheat germ (WGA)[b]	GlcNAc NeuAc	NeuAc(α–2,6)Gal(β–1,4)GlcNAc(β–1,2)Man(α–1,6) ╲ GlcNAc(β–1,4) — Man(β–1,4)GlcNAc(β–1,4)GlcNAc(β–1,N)Asn NeuAc(α–2,6)Gal(β–1,4)GlcNAc(β–1,2)Man(α–1,3) ╱
Ricinus communis (RCA1, RCA120)[c]	Gal	Gal(β–1,4)GlcNAc(β–1,6) ╲ Man(α–1,6) ╲ Gal(β–1,4)GlcNAc(β–1,2) ╱ Man(β–1,4)GlcNAc(β–1,4)GlcNAc(β–1,N)Asn Gal(β–1,4)GlcNAc(β–1,4) ╲ Man(α–1,3) ╱ Gal(β–1,4)GlcNAc(β–1,2) ╱
Concanavalin A (Con A)[d]	Man	GlcNAc(β–1,2)Man(α–1,6) ╲ Man(β–1,4)GlcNAc(β–1,4)GlcNAc(β–1,N)Asn GlcNAc(β–1,2)Man(α–1,3) ╱
Lens culinaris (LCA)[e]	Man	GlcNAc(β–1,2)Man(α–1,6) ╲ Fuc(α–1,6) | Man(β–1,4)GlcNAc(β–1,4)GlcNAc(β–1,N)Asn GlcNAc(β–1,2)Man(α–1,3) ╱
Galanthus nivalis (GNA)[f]	Terminal Man	Man (α-1,2) (α-1,3) or (α-1,6) Man

Datura stramonium (DSA)[g]	GlcNAc	[Gal(β–1,4)GlcNAc(β–1,3)]$_n$Gal(β–1,4)GlcNAc(β–1,2) \ Man(α–1,6) [Gal(β–1,4)GlcNAc(β–1,3)]$_n$Gal(β–1,4)GlcNAc(β–1,6) / Man(α–1,6) \ Man(β–1,4)GlcNAc(β–1,4)GlcNAc(β–1,N)Asn [Gal(β–1,4)GlcNAc(β–1,3)]$_n$Gal(β–1,4)GlcNAc(β–1,2) Man(α–1,3) /
Peanut agglutinin (PNA)	Gal	Gal(β–1,3)GalNAc refs. *19* and *20*
Phaseolus vulgaris (E-PHA) erythroagglutinating hemagglutinin[h]	GlcNAc	Gal(β–1,4)GlcNAc(β–1,2)Man(α–1,6) \ GlcNAc(β–1,4) — Man(β–1,4)GlcNAc(β–1,4)GlcNAc(β–1,N)Asn Gal(β–1,4)GlcNAc(β–1,2)Man(α–1,3) /
Phaseolus vulgaris (L-PHA) leukoagglutinating phyto-hemagglutinin[i]	GlcNAc	Gal(β–1,4)GlcNAc(β–1,2) \ Man(α–1,6) Gal(β–1,4)GlcNAc(β–1,6) / Man(α–1,6) \ Man(β–1,4)GlcNAc(β–1,4)GlcNAc(β–1,N)Asn Man(α–1,3) / Gal(β–1,4)GlcNAc(β–1,4) \ Man(α–1,3) Gal(β–1,4)GlcNAc(β–1,2) /
Ulex europeus 1 (UEA1)	Fuc	Fuc(α–1,6) | GlcNAc(β–1,N)Asn ref. *3*
Lotus tetragonolobus (LTA)	Fuc	Fuc(α–1,6) | GlcNAc(β–1,N)Asn; Fuc(α–1,3) | Gal(β–1,4)GlcNAc; Fuc(α–1,2) | Gal(β–1,4)GlcNAc refs. *3* and *22*

(continued)

Table 3 (continued)

Lectin	Monosaccharide specificity	Oligosaccharide specificity[a]		
Aleuria aurantia (AAA)[j]	Fuc	Fuc(α–1,6) | GlcNAc(β–1,N)Asn	Fuc(α–1,2) | Gal(β–1,4)GlcNAc	Fuc(α–1,3) | Gal(β–1,4)GlcNAc
Sambucus nigra (SNA)[k]	NeuAc	NeuAc(α–2,6)Gal		
Maackia amurensis (MAA)[l]	NeuAc	NeuAc(α–2,3)Gal		
Limulus polyphemus (LPA)	NeuAc	Terminal NeuAc ref. *26*		

Man, mannose; GlcNAc, *N*-acetylglucosamine; NeuAc, neuraminic acid; Gal, galactose; GalNAc, *N*-acetylgalactosamine; Fuc, fucose.

[a] Examples of carbohydrate structures recognized by the lectins commonly used in biochemical/biological research are presented in this table.

[b] The GlcNAc (β-1,4) Man (β-1,4) GlcNAc (β-1,4) GlcNAc (β-1,N) Asn structure is important for tight binding of glycopeptides to a WGA agarose column *(14)*.

Neuraminic acids (NeuAc) are implicated as important factors in WGA interactions, but the inhibitory effect of NeuAc is weaker than that of GlcNAc. The presence of clustering sialyl residues may be necessary for the strong interaction of sialoglycoconjugates with WGA *(15)*.

[c] RCA is specific for terminal β-galactosyl residues *(16)*. It binds primarily to the terminal Gal (β-1,4) GlcNAc sugar sequence, and much more weakly to the Gal (β-1,3) GalNAc sugar sequence *(17)*. Since galactose is the subterminal sugar in fully formed *N*-linked oligosaccharides, RCA provides a means of identifying the asialo forms of glycoproteins.

[d] Con A interacts with glycoconjugates that have at least two nonsubstituted or 2-*O*-substituted α-mannosyl residues. It detects the core portion of *N*-linked oligosaccharide chains. The most potent hapten is illustrated here *(14)*.

[e] LCA is specific of mannose (α-1,3) or (α-1,6). Fucose (α-1,6) linked to GlcNAc is required for the reaction *(3)*.

[f] GNA is particularly useful for identifying high mannose or hybrid type oligosaccharide structure because it does not react (unlike ConA) with biantennary complex type chains *(18,19)*.

[g] Datura lectin is specific for Gal-GlcNAc termini of complex oligosaccharides *(19)*.

[h] A bisecting GlcNAc residue that links (β-1,4) to the β-linked mannose residue in the core is an essential and specific determinant for high-affinity binding to E-PHA-agarose column. Without this bisecting GlcNAc residue, a complex-type glycopeptide is not retained by E-PHA-agarose *(21)*.

[i] The Gal (β-1,4) GlcNAc Man sugar sequence is essential for the binding of L-PHA (and E-PHA), since the interaction is completely abolished by the removal of the β-galactosyl residue in the peripheral portion *(21)*.

[j] The AAA lectin is particularly useful for identifying fucose bound (α-1,6) to GlcNAc. Fucose bound (α-1,6) to GlcNAc in the core region is necessary for strong binding of AAA. Fucose (α-1,3) linked to GlcNAc and fucose (α-1,2) linked to Gal (β-1,4) GlcNAc in the outer chain react only very weakly with AAA. However, fucose (α-1,3) or (α-1,2), in addition to a fucose-linked (α-1,6) to the proximal core GlcNAc enhances the binding *(14,23)*.

[k] SNA binds specifically to glycoconjugates containing the (α-2,6)-linked NeuAc, whereas isomeric structures containing terminal NeuAc in (α-2,3)-linkage are very weakly bound *(24)*.

[l] Immobilized MAA interacts with high affinity with complex-type tri- and tetra-antennary Asn-linked oligosaccharides containing outer NeuAc linked (α-2,3) to penultimate Gal residues. Glycopeptide containing NeuAc linked only (α-2,6) to Gal does not interact with MAA *(25)*.

2. Materials

1. The transfer buffer for protein blotting by the semidry method (*see* Chapter 38) or by the tank method (*see* Chapter 37) is the Towbin buffer diluted 1:2 with distilled water: 12.5 m*M* Tris, 96 m*M* glycine, and 10% (v/v) methanol *(30)*.
2. PVDF membranes (0.2 mm, 200 × 200 mm) are supplied by Bio-Rad (Hercules, CA), the filter papers (chromatography paper grade 3 mm CHr) by Whatman (Maidstone, UK), and the nitrocellulose membranes (0.45 mm) by Schleicher & Shuell (Keene, NH).
3. Blocking solution: 0.5% (w/v) Tween-20 in phosphate-buffered saline (PBS) *(31)*. All further incubations and washing steps are carried out in the same blocking solution.
4. PBS: 137 m*M* NaCl, 27 m*M* KCl, 10 m*M* Na_2HPO_4, 1.8 m*M* KH_2PO_4, pH 7.4. A 10-fold concentrated PBS solution can be prepared, sterilized by autoclaving, and stored at room temperature for many weeks (PBS 10X: 80 g NaCl, 2 g KCl, 14.4 g Na_2HPO_4, 2.4 g KH_2PO_4. Add H_2O to 1 L).
5. Biotinylated lectins are obtained from Boehringer-Mannheim (Mannheim, Germany). Sigma (St. Louis, MO) supply HRP-labeled Extravidin (a modified form of affinity-purified avidin) and AP-labeled Extravidin.
6. Two different detection methods are described: the chemiluminescent detection of HRP activity using the luminol reagent (ECL Kit, Amersham International, Buckinghamshire, UK) and the conventional colorimetric reaction of AP revealed by nitroblue tetrazolium/bromochloro-indolyl phosphate (NBT/BCIP).
   a. Stock solutions of NBT and BCIP (Fluka, Buchs, Switzerland) are prepared by solubilizing 50 mg NBT in 1 mL of 70% (v/v) dimethylformamide and 50 mg BCIP disodium salt in 1 mL dimethylformamide. These two solutions are stable when stored in closed containers at room temperature.
   b. AP buffer: 100 m*M* NaCl, 5 m*M* $MgCl_2$, and 100 m*M* Tris-HCl, pH 9.5, prepared just before the colorimetric detection.
7. For the chemiluminescent detection of HRP activity, X-ray films (X-OMAT S, 18 × 24 cm) and a cassette (X-OMATIC cassette, regular screens) are available from Kodak (Rochester, NY).
8. For the development of X-ray film, an automatic developer machine is used (Kodak RP X-OMAT Processor, Model M6B). Manual development can also be done using the developer for autoradiography films (reference number P-7042, Sigma) and the fixer (reference number P-7167, Sigma).

3. Method

The use of gloves is strongly recommended to prevent blot contamination.

1. After electrophoresis and protein-blotting procedures, the membrane is first washed with distilled water (3 × 5 min), and then treated for 1 h at room temperature and under gentle agitation with 100 mL blocking solution (PBS containing 0.5% [w/v] Tween-20) (*see* Note 3).
2. The blot is then incubated for 2 h in biotinylated lectin, at a concentration of 1 µg/mL in the blocking solution, under agitation, in a glass dish and at room temperature. Twenty-five milliliters of lectin solution are used for the incubation of a membrane of 16.5 × 21 cm (*see* Note 4).
3. A washing step is then performed for 1 h with six changes of 200 mL of PBS-Tween-20.
4. Extravidin-HRP or Extravidin-AP diluted 1:2000 in the blocking solution is added for 1 h at room temperature under agitation. The membrane is then washed for 1 h with six changes of PBS-Tween-20.
5. The colorimetric reaction with AP is carried out under gentle agitation by incubating the blot in the following solution: 156 µL BCIP stock solution, 312 µL NBT stock solution in

50 mL AP buffer. The colorimetric reaction is normally completed within 10–20 min. The blotted proteins are colored in blue.

6. The chemiluminescence detection of peroxidase activity is performed according to the manufacturer's instructions (Amersham). The enhanced chemiluminescent assay involves reacting peroxidase with a mixture of luminol, peroxide, and an enhancer, such as phenol (*see also* Chapter 50). Five milliliters of detection solution 1 are mixed with 5 mL of detection solution 2 (supplied with the ECL Kit). The washed blot is placed in a glass plate, and the 10 mL of chemiluminescent reagents are added directly to the blot with a 10-mL pipet, in order to cover all the surface carrying the proteins. The blot is incubated for 1 min at room temperature without agitation.

   The excess chemiluminescent solution is drained off by holding the blot vertically. The blot is then wrapped in plastic sheet (Saran Wrap), without introducing air bubbles and exposed (protein side up) to X-ray film in a dark room, using red safelights. The exposure time of the film depends on the amount of target proteins on the blot (*see* Note 5).

   The development of the X-ray films can be done with an automatic developer or manually with the following protocol:

   a. The developer and fixer solutions are prepared according to the manufacturer's instructions (dilution 1:4 with distilled water).
   b. In a dark room, the film is attached to the film hanger and immersed in the developer solution until the bands (or spots) appear (the maximum incubation time is 4 min). The film should not be agitated during development.
   c. The hanger is immersed into water, and the film is rinsed for 30 s to 1 min.
   d. The hanger is then placed in the fixer solution for 4–6 min. Intermittent agitation should be used throughout the fixing procedure.
   e. The film is washed in clean running water for 5–30 min.
   f. Finally, the film is dried at room temperature. All previous incubations are also undertaken at room temperature.

4. Notes

1. Instead of nitrocellulose, PVDF membrane can be used for lectin blotting.
2. Figure 1A shows the plasma glycoprotein signals detected with wheat germ agglutinin (WGA) and generated on a film after chemiluminescence detection. Figure 1B shows an identical blot stained with NBT/BCIP. As already reported *(29),* the same pattern of glycoproteins or glycoprotein subunits is revealed by both methods, but the chemiluminescent detection system shows higher sensitivity (about 10-fold) than NBT/BCIP staining. The spots in the former case are more intense, and the detection with enhanced chemiluminescence is more reliable and easier to control than the colorimetric reaction. Albumin that does not contain any carbohydrate moiety represents a negative protein control in all blots. Lectin blotting of plasma proteins with *Ricinus communis* agglutinin (RCA), with *Aleuria aurantia* agglutinin (AAA), and with *Sambucus nigra* agglutinin (SNA) is presented in Figs. 3, 4, and 5, respectively.
3. Generally, bovine serum albumin (BSA) or nonfat dry milk is used to block membranes. For lectin blotting, we have tested BSA, which has produced a very high background, probably owing to glycoprotein contamination in the commercial preparations of this protein. When low-fat dried milk was used, no glycoprotein signals were obtained on the blot after the chemiluminescent reaction. This could be attributed to the presence of biotin that competed with biotinylated lectin and avoided its binding to the sugar moieties of glycoproteins *(32).* The PBS-Tween-20 blocking solution gave a very low background both for colorimetric and chemiluminescence detections.

4. The blot should be immersed in a sufficient volume of solution to allow a good exchange of fluid over its entire surface. The use of plastic bags to incubate protein blots is not recommended because it often leads to uneven background and may cause areas without signal.
5. Most of the proteins in plasma are glycosylated. Therefore, a very short exposure time of the blot to X-ray film is needed. In general, when 120 µg of plasma proteins are loaded onto 2D PAGE, the blot is exposed for a period of 3–30 s. The difference in the exposure time depends on the lectin used. For example, WGA lectin detects most of the *N*-linked glycoproteins, and an exposure of only 3 s is sufficient *(29)*. In this case, a longer exposure time leads to a very high background with no additional spot signal.

   For an unknown sample, the blot can be exposed for 1 and 5 min as a first attempt. The enhanced chemiluminescence using luminol (ECL Kit, Amersham) leads to a "flash" of light owing to the addition of enhancers. The light emission on membrane peaks at 5–10 min after the addition of substrate and lasts for 2–3 h *(33)*. Therefore, longer exposure times (10 min to 3 h) may allow the detection of weak signals.

Acknowledgments

P.G. acknowledges a grant from the Swiss Foundation for Research on Alcohol. The authors thank Marianne Gex-Fabry for revision of the manuscript.

References

1. Varki, A. (1993) Biological roles of oligosaccharides: all of the theories are correct. *Glycobiology* **3,** 97–130.
2. Turner, G. A. (1992) *N*-glycosylation of serum proteins in disease and its investigation using lectins. *Clin. Chim. Acta* **208,** 149–171.
3. Montreuil, J., Bouquelet, S., Debray, H., Fournet, B., Spik, G., and Strecker, G. (1986) Glycoproteins in *Carbohydrate Analysis: A Practical Approach* (Chaplin, M. F. and Kennedy, J. F., eds.), Academic, Oxford, UK, pp. 143–204.
4. Baenziger, J. U. (1984) The oligosaccharides of plasma glycoproteins: synthesis, structure, and function in *The Plasma Proteins,* vol. 4 (Putnam, F. W., ed.), Academic, New York, pp. 272–315.
5. Rademacher, T. W., Parekh, R. B., and Dwek, R. A. (1988) Glycobiology. *Annu. Rev. Biochem.* **57,** 785–838.
6. Berger, E. G., Buddecke, E., Kamerling, J. P., Kobata, A., Paulson, J. C., and Vliegenthart, J. F. G. (1982) Structure, biosynthesis and functions of glycoprotein glycans. *Experimenta* **38,** 1129–1158.
7. Lundy, F. T. and Wisdom, G. B. (1992) The determination of asialoglycoforms of serum glycoproteins by lectin blotting with ricinus communis agglutinin. *Clin. Chim. Acta* **205,** 187–195.
8. Thompson, S. and Turner, G. A. (1987) Elevated levels of abnormally-fucosylated haptoglobins in cancer sera. *Br. J. Cancer* **56,** 605–610.
9. Stibler, H. and Borg, S. (1981) Evidence of a reduced sialic acid content in serum transferrin in male alcoholics. *Alcohol. Clin. Exp. Res.* **5,** 545–549.
10. Clamp, J. R. (1984) The oligosaccharides of plasma protein in *The Plasma Proteins,* vol. 2 (Putnam, F. W., ed.), Academic, NewYork, pp. 163–211.
11. Lis, H. and Sharon, N. (1986) Lectins as molecules and as tools. *Annu. Rev. Biochem.* **55,** 35–67.
12. Goldstein, I. J. and Hayes, C. E. (1978) The lectins: carbohydrate binding proteins of plants and animals. *Adv. Carbohydr. Chem. Biochem.* **35,** 127–340.

13. Goldstein, I. J. and Poretz, R. D. (1986) Isolation, physicochemical characterization, and carbohydrate-binding specificity of lectins, in *The Lectins: Properties, Functions, and Applications in Biology and Medicine* (Liener, I. E., Sharon, N., and Goldstein, I. J., eds.), Academic, Orlando, FL, pp. 35–247.
14. Osawa, T. and Tsuji, T. (1987) Fractionation and structural assessment of oligosaccharides and glycopeptides by use of immobilized lectins. *Annu. Rev. Biochem.* **56,** 21–42.
15. Bhavanandan, V. P. and Katlic, A. W. (1979) The interaction of wheat germ agglutinin with sialoglycoproteins. The role of sialic acid. *J. Biol. Chem.* **254,** 4000–4008.
16. Debray, H., Decout, D., Strecker, G., Spik, G., and Montreuil, J. (1981) Specificity of twelve lectins towards oligosaccharides and glycopeptides related to *N*-glycosylproteins. *Eur. J. Biochem.* **117,** 41–55.
17. Kaifu, R. and Osawa, T. (1976) Synthesis of *O*-β-D-galactopyranosyl-(1-4)-*O*-(2-acetamido-2-deoxy-β-D-glucopyranosyl)-(1-2)-*n*-mannose and its interaction with various lectins. *Carbohydr. Res.* **52,** 179–185.
18. Animashaun, T. and Hughes, R. C. (1989) Bowringia milbraedii agglutinin. Specificity of binding to early processing intermediates of asparagine-linked oligosaccharide and use as a marker of endoplasmic reticulum glycoproteins. *J. Biol. Chem.* **264,** 4657–4663.
19. Haselbeck, A., Schickaneder, E., Von der Eltz, H., and Hosel, W. (1990) Structural characterization of glycoprotein carbohydrate chains by using digoxigenin-labeled lectins on blots. *Anal. Biochem.* **191,** 25–30.
20. Sueyoshi, S., Tsuji, T., and Osawa, T. (1988) Carbohydrate binding specificities of five lectins that bind to *O*-glycosyl-linked carbohydrate chains. Quantitative analysis by frontal-affinity chromatography. *Carbohydr. Res.* **178,** 213–224.
21. Cummings, R. D. and Kornfeld, S. (1982) Characterization of the structural determinants required for the high affinity interaction of asparagine-linked oligosaccharides with immobilized phaseolus vulgaris leukoagglutinating and erythroagglutinating lectins. *J. Biol. Chem.* **257,** 11,230–11,234.
22. Pereira, M. E. A. and Kabat, E. A. (1974) Blood group specificity of the lectin from lotus tetragonolobus. *Annu. NY Acad. Sci.* **334,** 301–305.
23. Debray, H. and Montreuil, J. (1989) Aleuria aurantia agglutinin. A new isolation procedure and further study of its specificity towards various glycopeptides and oligosaccharides. *Carbohydr. Res.* **185,** 15–26.
24. Shibuya, N., Goldstein, I. J., Broekaert, W. F., Nsimba-Lubaki, M., Peeters, B., and Peumans, W. J. (1987) The elderberry (sambucus nigra l.) bark lectin recognizes the Neu5Ac (α2-6) Gal/GalNAc sequence. *J. Biol. Chem.* **262,** 1596–1601.
25. Wang, W. C. and Cummings, R. D. (1988) The immobilized leukoagglutinin from the seeds of Maackia amurensis binds with high affinity to complex-type Asn-linked oligosaccharides containing terminal sialic acid-linked α-2,3 to penultimate galactose residues. *J. Biol. Chem.* **263,** 4576–4585.
26. Cohen, E., Roberts, S. C., Nordling, S., and Uhlenbruck, G. (1972) Specificity of limulus polyphemus agglutinins for erythrocyte receptor sites common to M and N antigenic determinants. *Vox Sang.* **23,** 300–307.
27. Appel, R. D., Sanchez, J. C., Bairoch, A., Golaz, O., Miu, M., Vargas, J. R., and Hochstrasser, D. F. (1993) Swiss-2D PAGE: a database of two-dimensional gel electrophoresis images. *Electrophoresis* **14,** 1232–1238.
28. Jadach, J. and Turner, G. A. (1993) An ultrasensitive technique for the analysis of glycoproteins using lecting blotting enhanced chemiluminescence. *Anal. Biochem.* **212,** 293–295.
29. Gravel, P., Golaz, O., Walzer, C., Hochstrasser, D. F., Turler, H., and Balant, L. P. (1994) Analysis of glycoproteins separated by two-dimensional gel electrophoresis using lectin blotting revealed by chemiluminescence. *Anal. Biochem.* **221,** 66–71.

30. Sanchez, J. C., Ravier, F., Pasquali, C., Frutiger, S., Bjellqvist, B., Hochstrasser, D. F., and Hughes, G. J. (1992) Improving the detection of proteins after transfer to polyvinylidene difluoride membranes. *Electrophoresis* **13,** 715–717.
31. Becker, B., Salzburg, M., and Melkonian, M. (1993) Blot analysis of glycoconjugates using digoxigenin-labeled lectins: an optimized procedure. *Biotechniques* **15,** 232–235.
32. Garfin, D. E. and Bers, G. (1982) Basic aspects of protein blotting, in *Protein Blotting* (Baldo, B. A. and Tovey, E. R., eds.), Karger, Basel, Germany, pp. 5–42.
33. Durrant, I. (1990) Light-based detection of biomolecules. *Nature* **346,** 297,298.

98

A Lectin-Binding Assay for the Rapid Characterization of the Glycosylation of Purified Glycoproteins

Mohammad T. Goodarzi and Graham A. Turner

1. Introduction

Most proteins have carbohydrate chains (glycosylation) attached covalently to various sites on their polypeptide backbone. These posttranslational modifications, which are carried out by cytoplasmic enzymes, confer subtle changes on the structure and behavior of a molecule, and their composition is very sensitive to many environmental influences *(1–3)*. There is increasing interest in determining the glycosylation of a molecule because of the importance of glycosylation in affecting its reactivity *(1,4,5)*. This is particularly true in the production of therapeutic glycoproteins by recombinant methods, where glycosylation can be determined by the type of host cell used or the production process employed *(2)*. Glycosylation is also important in disease situations where changes in carbohydrate structure could be involved in the pathological processes *(3)*. Unfortunately, the glycosylation of proteins is very complex; there are variations in the sites of glycosylation, the types of amino acid—carbohydrate bond, the composition of the chains, and the particular carbohydrate sequences and linkages in each chain *(3,4)*. In addition, within any population of molecules, there is considerable heterogeneity in the carbohydrate structures (glycoforms) that are synthesized at any one time *(1)*. This is typified by some molecules showing increased branching, reduced chain length, and further addition of single carbohydrate moieties to the internal chain.

To unravel completely the complexities of the glycosylation of a molecule is a substantial task, requiring considerable effort and resources. However, in many situations, this is unnecessary, because only a limited amount of information, on a single or group of structural features, is needed. Lectins can be useful for this purpose. These are carbohydrate-binding proteins with particular specificity *(6)*. Although their specificity is not absolute, there is usually one carbohydrate or group of carbohydrates to which the lectin binds with a higher affinity than the rest of the group. The major carbohydrate specificity of a number of commonly used lectins is shown in Table 1.

Lectins have previously been used for investigating glycoprotein glycosylation by incorporating them into existing technologies, such as affinity chromatography, blotting, and electrophoresis. Although these modifications give workable methods, these frequently use large amounts of lectin, which is expensive, require a lot of technical

From: The Protein Protocols Handbook
Edited by: J. M. Walker Humana Press Inc., Totowa, NJ

Table 1
Specificity of Different Lectins

Lectin	Abbreviation	Specificity
Concanavalin A	ConA	Mannose α1–3 or Mannose α1–6
Datura stramonium agglutinin	DSA	Galactose β1–4 *N*-acetylglucosamine
Lens culinaris agglutinin	LCA	Mannose α1–3 or Mannose α1–6 (Fucose α1–6 *N*-acetylglucos-amine is also required)
Lotus tetragonolobus agglutinin	LTA	Fucose α1–2 Galactose β1–4 (Fucose α1–3) *N*-acetylglucosamine
Maackia amurensis agglutinin	MAA	*N*-acetylneuraminic acid α2–3 Galactose
Sambucus nigra agglutinin	SNA	*N*-acetylneuraminic acid α2–6 Galactose

skill, cannot handle large numbers of specimens, and only give semiquantitative results. For a more detailed discussion of these methods, the reader should refer to ref. *3* and Chapter 97.

Another approach is to use lectins in the familiar sandwich ELISA technology in multiwell plates. Procedures of this type have been described for the measurement of particular carbohydrate structures in α-fetoprotein *(7)*, haptoglobin (HP) *(8)*, immunoglobulins *(9)*, mucins *(10)*, plasminogen activator *(11)*, and transferrin (TF) *(12)*. In this method, the lectin is used either to capture the molecule of interest or to identify it. An antibody is used as the other partner in the sandwich, and the degree of binding is measured by the presence of an enzyme label on the identifier. Both configurations suffer from the disadvantage that the lectin may bind to carbohydrate determinants on the immunoglobulin used as the antibody. Furthermore, using the lectin as the capture molecule, the immobilized lectin may bind to glycans of other glycoproteins in the sample, and these will compete with the molecule of interest for the available binding sites. On the other hand, when immobilized antibody is used as the capture molecule, care must be taken to ensure that the antibody is not binding to the same determinant as that reacting with the lectin.

The lectin-binding assay (LBA) described herein *(13)* overcomes many of the disadvantages with the previous lectin immunoassays. It was developed from a previously reported procedure *(14)*. A purified glycoprotein is absorbed onto the plastic surface of a well in a microtiter plate. After removing unbound protein by washing, uncoated sites on the plate are blocked using a nonionic detergent. A lectin labeled with digoxigenin (DIG) or biotin is added and allowed to interact with the carbohydrate on the absorbed glycoprotein. Unbound lectin is removed by further washing, and the amount of bound lectin is measured by adding an anti-DIG antibody or streptavidin conjugated to an enzyme. Streptavidin has a very high affinity for biotin. Following further washing, the bound enzyme is used to develop a color reaction by the addition of the appropriate substrates. The principle of the procedure is summarized schematically in Fig. 1.

Using this method, it is possible to screen multiple specimens rapidly, with high sensitivity and excellent precision. In addition, very small amounts of lectin are used, background absorbances are low, and the procedure does not require a high degree of technical skill other than some experience with micropipets and ELISA. Because such

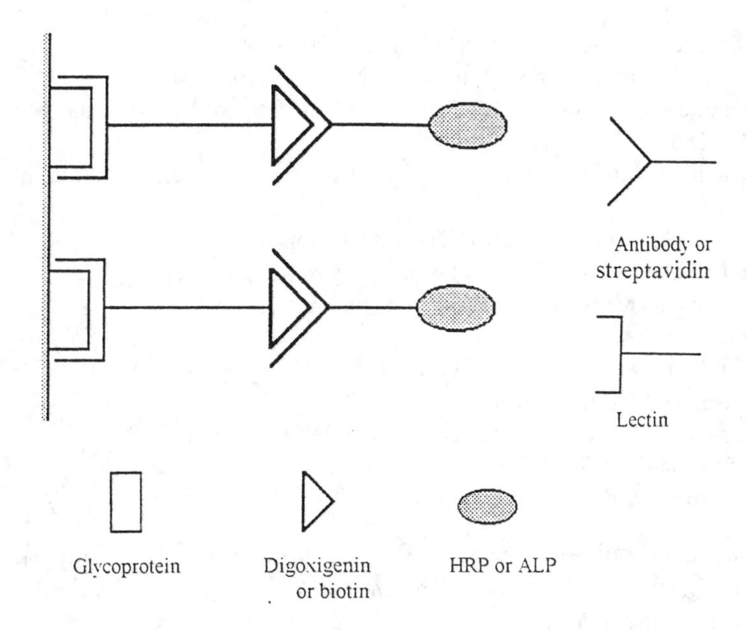

Fig. 1. Schematic diagram of the interactions in the LBA.

small amounts of glycoprotein are needed, a glycoprotein can be rapidly purified by a batch affinity chromatography method *(15,16).* The LBA has successfully been applied to the investigation of the glycosylation of purified α-1-proteinase inhibitor (API) *(16)* and HP *(17)* using Concanavalin A (ConA), *Maackia amurensis* agglutinin (MAA), and *Sambucus nigra* agglutinin (SNA). More recently, other studies of API and HP have been carried out using *Lens culinaris* agglutinin (LCA) and *Lotus tetragonolobus* (LTA) (unpublished observations). If a panel of lectins is used, an overall picture of the carbohydrate structure of a glycoprotein can be built up very quickly and cheaply. Furthermore, subtle differences in the glycosylation of the same glycoprotein in different situations can be identified, e.g., different diseases *(16,17).*

2. Materials

2.1. Reagents

1. Tris-buffered saline (TBS): 25 m*M* Tris-HCl, pH 7.5, containing 100 m*M* NaCl.
2. Control glycoprotein (*see* Notes 1 and 2).
3. Sample glycoprotein (*see* Notes 2 and 3).
4. Tween-TBS (TTBS): 25 m*M* Tris-HCl, pH 7.5, containing 100 m*M* NaCl and 0.1% (v/v) Tween-20.
5. Tris-cations (TC): 1 m*M* Tris-HCl, pH 7.5, containing 1 m*M* $CaCl_2$, 1 m*M* $MgCl_2$, 1 m*M* $MnCl_2$, and 0.1%(v/v) Tween-20.
6. DIG-labeled lectins (Boehringer Mannheim, Mannheim, Germany): Dissolve lectin-DIG in TC (*see* Note 3).
7. Biotin-labeled lectins (Sigma, St. Louis, MO): Dissolve LCA-biotin and LTA-biotin in TC (*see* Note 3).
8. Anti-DIG antibody conjugated with horseradish peroxidase (Fab fragment) (DIG/HRP) (Boehringer Mannheim): 50 mU/mL in T-TBS (*see* Note 4).

9. Streptavidin-alkaline phosphatase (S-ALP) (Sigma): 1 ng/μL (1.3 mU/μL) in T-TBS.
10. Citrate buffer: 34.8 m*M* citric acid, 67.4 m*M* Na_2HPO_4, pH 5.0.
11. Diethanolamine buffer: 100 m*M* diethanolamine, pH 9.8, containing 1 m*M* $MgCl_2$. Store in a dark bottle at 4°C and prepare monthly.
12. *O*-phenylenediamine (OPD): Prepare a 37 m*M* solution in water, and store in 1-mL aliquots at –20°C.
13. *Para*-nitrophenylphosphate (PNPP) (BDH, Atherstone, UK).
14. Hydrogen peroxide: Prepare 3% (v/v) solution by diluting concentrated (30%) H_2O_2 1:10 with deionized water. Prepare monthly, and store at 4°C.
15. Color reagent:
    a. HRP: 9 mL citrate buffer + 1 mL OPD + 50 μL H_2O_2 3% (v/v). Make up fresh 10 min before use. Add H_2O_2 just prior to use.
    b. ALP: 20 mL diethanolamine buffer + 20 mg PNPP. Prepare fresh.
16. 1.25*M* Sulfuric acid (H_2SO_4): 1.25*M*.
17. 1.0*M* Sodium hydroxide (NaOH): 1*M*.

All other reagents should be of analytical grade or better (BDH or Sigma), and prepared as required. All solutions should be prepared with double-distilled deionized water (Millipore, Bedford, MA).

2.2. Equipment

1. Multiwell plastic plate: 96-well plate (Immunolon 4, Dynatech, Chantilly, VA).
2. Plate reader: (Titerteck [McLean, VA] Multiskan MCC/340).
3. Multichannel micropipet: (Finnpipette, Labsystem, Finland).

3. Methods

1. For each lectin, add 100 μL sample glycoprotein in triplicate to a multiwell plate, and for each batch of samples, add 100 μL of a control glycoprotein or TBS in triplicate.
2. Incubate the plate for 2 h at 37°C. Pour off liquid by turning upside down, and tap on a pad of tissue paper to dry completely.
3. Add 240 μL T-TBS to each well, pour off liquid, and dry again. Repeat this operation three times.
4. Add 240 μL T-TBS to each well, and incubate for 1 h at 37°C and then overnight at 4°C.
5. Wash each well three times with 240 μL TC, removing liquid as described in step 2.
6. Add 200 μL DIG-lectin or biotin-lectin to each well, and incubate for 1 h at 37°C.
7. Wash each well four times with 240 μL T-TBS, removing liquid as described in step 2.
8. Add 150 μL DIG/HRP or S-ALP to each well, and incubate the DIG/HRP for 1 h at room temperature and the S-ALP for 1 h at 37°C.
9. Wash each well four times with T-TBS, removing liquid as described in step 2.
10. Add 100 μL color reagent to each well as appropriate.
11. Leave plate for 10–30 min (*see* Note 5) at room temperature.
12. Stop reaction:
    a. For HRP, add 100 μL 1.25*M* H_2SO_4 to each well.
    b. For ALP, add 100 μL 1*M* NaOH to each well.
13. Read the absorbance for each well with plate reader:
    a. For HRP, 492 nm.
    b. For ALP, 405 nm.
14. Calculate the mean of triplicate measurements, and subtract the absorbance value of the blank from the values of all samples and controls.

Table 2
Optimum Protein Coating Concentration for Different Lectins and Different Serum Glycoproteins

	Optimum protein concentration, µg/mL			
Lectin	API	HP	TF	AGP
ConA	0.15	0.05	0.15	2.50
SNA	0.15	0.05	0.07	0.50
MAA	1.00	0.50	3.00	0.50
LCA	1.50	0.50	NR	ND
LTA	1.50	1.00	NR	ND
DSA	0.05	0.01	0.35	0.06

NR, no reactivity; ND, not determined; AGP, α-1-acid glycoprotein.

4. Notes

1. A pool of purified glycoprotein that reacts with the lectin is recommended as a positive control in the assay, and also for preliminary experiments to develop assay conditions. Various serum glycoproteins can be obtained commercially in a purified form for this purpose, but with other glycoproteins, this may not be possible, and suitable material may have to be prepared in the laboratory.
2. Sample and control glycoproteins can be conveniently purified for assay by extracting with an antibody coupled to Sepharose. We have described a batch method that is very rapid, can handle many specimens, produces very pure protein, and works for a number of different glycoproteins (*15,16* and unpublished observations). Although the yield from this procedure can be low (5–20%), this does not matter because only low amounts of glycoprotein are needed in the assay. A similar method could be equally well adapted for the extraction of other types of glycoprotein.
3. For each combination of lectin and protein, preliminary experiments are carried out to determine the optimum concentration of protein required to coat the well. This is done by coating the well with different protein concentrations (0.01–10 µg/mL), and probing with a lectin solution of 1 µg/mL *(14)*. The protein concentration is chosen that gives an absorbance value of approx 1. If this value cannot be established, the lectin concentration is increased (e.g., 1.5 µg/mL) or decreased (e.g., 0.5 µg/mL) as required, and the protein is reassayed. Table 2 shows the optimum glycoprotein concentrations for various lectins we have previously studied. The concentrations of Con A, LCA, LTA, MAA, and SNA used in these experiments were 1, 0.5, 1, 1.5, and 1 µg/mL, respectively.
4. The activity of anti-DIG HRP varies from batch to batch, and should be routinely checked. This reagent was usually used at a dilution of 1/3000–1/6000.
5. The time required to develop the color reaction depends on the reactivity of lectin with the particular protein being investigated. For some lectins (e.g., Con A and SNA), it is between 10 and 30 min, whereas for other lectins (e.g., LTA, MAA), it is as long as 120 min.
6. To minimize background, TBS and T-TBS must be filtered through a 0.2-µm membrane (Tuffryn membrane). The absorbances (without glycoprotein, but with all other reagents) for Con A, SNA, and MAA were (mean ± SD, no. observations) 0.28 ± 0.02 (9); 0.16 ± 0.03 (9); and 0.14 ± 0.02 (9), respectively.
7. The method gives very good reproducibility. For example, in the case of API, the interassay precision using ConA, SNA, and MAA was 2.2, 4.5, and 6.4%, respectively, and the intra-assay precision for this glycoprotein with the same lectins was 1.5, 3, and 5%, respectively.

Table 3
Analysis of Purified API and Hp Using ConA, MAA, and SNA

		Lectin,[b] absorbance 492 nm		
Glycoprotein	Source	ConA	SNA	MAA
Hp	Healthy	1.40 ± 0.35	1.70 ± 0.22	0.18 ± 0.07
	Cancer[a]	0.86 ± 0.15	1.69 ± 0.29	0.45 ± 0.10
API	Healthy	1.29 ± 0.28	1.31 ± 0.19	0.47 ± 0.08
	Cancer[a]	1.87 ± 0.14	1.95 ± 0.09	0.23 ± 0.07

[a]Cancer specimen from patients with stage III/IV ovarian carcinoma.
[b]Mean ± SD calculated from the values of 8 healthy individuals or 12 cancer patients.

8. The specificity of each lectin can be checked by assaying the glycoprotein in the presence of a competitive sugar. For example, in the presence of 100 mmol/L α-D-methyl mannoside, a competitive inhibitor for ConA, the absorbance obtained for API is reduced from 2 to 0.09. Alternatively, glycoproteins can be used that are known to lack the carbohydrate grouping under investigation, e.g., carboxypeptidase Y-MAA and TF-LTA.
9. The method is only semiquantitative, and if high absorbance values are obtained for a glycoprotein, this suggests that it has high amounts of a particular carbohydrate grouping, and vice versa. Positive and negative controls must always be run in the assay, because standards are not used. The method is at its most useful when comparing different samples in the same assay. This is illustrated in Table 3, which shows the lectin-binding characteristics for Hp and API from healthy individuals and cancer patients. The results are interpreted according to the known properties of each lectin *(18)*. Therefore, in cancer, the branching of Hp is increased, there is more α 2-3 sialic acid, and the α 2-6 sialic acid content is unchanged. On the other hand, with API in cancer, the branching decreases, the α 2-6 sialic acid content increases, and the α 2-3 sialic acid content decreases. All these changes are consistent with the reported monosaccharide composition of these specimens *(14,15)* and the known carbohydrate structures present on the *N*-glycans *(4)*. It is important to emphasize that the method does not give an indication of the glycosylation of individual glycoforms in a population of glycoprotein molecules, but it represents the average glycosylation of these molecules.
10. In our hands, the only lectin not to react with HP, API, or TF was the fucose-specific lectin, Aleuria aurantia agglutinin.
11. Since OPD is carcinogenic, the noncarcinogenic substance, 3,3',5,5'-tetramethyl benzidine (dihydrochloride) can be used as a substrate for HRP *(19)*.

References

1. Rademacher, T. W., Parekh, R. B., and Dwek, R. A. (1988) Glycobiology. *Annu. Rev. Biochem.* **57,** 785–838.
2. Goochee, C. F., Gramer, M. J., Andersen, D. C., Bahr, J. B., and Rasmussen, J. R. (1991) The oligosaccharide of glycoproteins: bioprocess factors affecting oligosaccharide structure and their effect on glycoprotein properties. *Bio/Technology* **9,** 1347–1355.
3. Turner, G. A. (1992) *N*-glycosylation of serum proteins in disease and its investigation using lectins. *Clin. Chim. Acta* **208,** 149–171.
4. Kobata, A. (1992) Structure and function of the sugar chains of glycoproteins. *Eur. J. Biochem.* **209,** 483–501.
5. Varki, A. (1993) Biological role of oligosaccharides: all of the theories are correct. *Glycobiology* **3,** 97–130.

6. Lis, H. and Sharon, N. (1986) Lectins as molecules and as tools. *Annu. Rev. Biochem.* **55,** 33–67.
7. Kinoshita, N., Suzuki, S., Matsuda, Y., and Taniguchi, N. (1989) α-fetoprotein antibody-lectin enzyme immunoassay to characterize sugar chains for the study of liver diseases. *Clin. Chim. Acta* **179,** 143–152.
8. Thompson, S., Stappenbeck, R., and Turner, G. A. (1989) A multiwell lectin-binding assay using Lotus tetragonolobus for measuring different glycosylated forms of haptoglobin. *Clin. Chim. Acta.* **180,** 277–284.
9. Kinoshita, N., Ohno, M., Nishiura, T., Fujii, S., Nishikowa, A., Kawakami, Y., Uozumi, N., and Taniguchi, N. (1991) Glycosylation of Fab portion of myeloma immunoglobulin G and increased fucosylated biantennary chains: structural analysis by high-performance liquid chromatography and antibody-lectin enzyme immunoassay using Lens culinaris agglutinin. *Cancer Research* **51,** 5888–5892.
10. Parker, N., Makin, C. A., Ching, C. K., Eccleston, D., Taylor, O. M., Milton, J. D., and Rhodes, J. M. (1992) A new enzyme-linked lectin/mucin antibody sandwich assay (CAM 17.1/WGA) assessed in combination with CA 19-9 and peanut lectin binding assay for the diagnosis of pancreatic cancer. *Cancer* **70,** 1062–1068.
11. Hayashi, S. and Yamada, K. (1992) Quantitative assay of the carbohydrate in urokinase-type plasminogen activator by lectin-enzyme immunoassay. *Blood Coagulation Fibrinolysis* **3,** 423–428.
12. Pekelharing, J. M., Vissers, P., and Leijnse, B. (1987) Lectin-enzyme immunoassay of transferrin sialovariants using immobilized antitransferrin and enzyme-labelled galactose-binding lectin from *Ricinus communis*. *Anal. Biochem.* **165,** 320–326.
13. Goodarzi, M. T., Rafiq, M., and Turner, G. A. (1995) An improved multiwell immunoassay using digoxigenin-labeled lectins to study the glycosylation of purified glycoproteins. *Biochem. Soc. Trans.* **23,** 168S.
14. Katnik, I., Jadach, J., Krotkiewski, H., and Gerber, J. (1994) Investigating the glycosylation of normal and ovarian cancer haptoglobins using digoxigenin-labeled lectins. *Glycosylation Disease* **1,** 97–104.
15. Thompson, S., Dargan, E., and Turner, G. A. (1992) Increased fucosylation and other carbohydrate changes in haptoglobin in ovarian cancer. *Cancer Lett.* **66,** 43–48.
16. Goodarzi, M. T. and Turner, G. A. (1995) Decreased branching, increased fucosylation and changed sialylation of alpha-1-proteinase inhibitor in breast and ovarian cancer. *Clin. Chim. Acta* **236,** 161–171.
17. Turner, G. A., Goodarzi, M. T., and Thompson, S. (1995) Glycosylation of alpha-l-proteinase inhibitor and haptoglobin in ovarian cancer: evidence for two different mechanisms. *Glycoconjugate J* **12,** 211–218.
18. Wu, A. M., Sugii, S., and Herp, A. (1988) A table of lectin carbohydrate specificities, in *Lectins–Biology, Biochemistry, Clinical Biochemistry,* vol. 6 (Bog-Hansen, T. C. and Freed, D. I. J., eds.), Sigma Chemical Company, St. Louis, MO, pp. 723–740.
19. Bos, E. S., van der Doelen, A. A., van Rooy, N., Schuurs, A. H. W. M. (1981). 3,3',5,5'-Tetramethylbenzidine as an Ames test negative chromogen for horse-radish peroxidase in enzyme immunoassay. *J. Immunoassay* **2(384),** 187–204.

99

Staining of Glycoproteins/Proteoglycans on SDS-Gels

Holger J. Møller and Jørgen H. Poulsen

1. Introduction

Sodium dodecyl sulfate-polyacrylamide gel electrophoresis (SDS-PAGE) is a commonly employed technique for separation of proteins according to size. Among other applications, it is used for identification and characterization of proteins on the basis of mol-wt determinations. In this chapter, staining methods will be described that permit detection of highly glycosylated proteins on SDS gels at levels of a few nanograms.

Proteins with limited glycosylation are most often stained with Coomassie brilliant blue or—if high sensitivity is needed—with a silver-stain (*1,* and *see* Chapter 35). These stains, however, are far less sensitive when used for detection of highly glycosylated proteoglycans (protein-glycosaminoglycans) or glycoproteins (protein-oligosaccharides), leading to weak staining or even failure of detection. This is presumably owing to steric interference by the carbohydrates with the binding of silver ions.

Proteoglycans are traditionally stained with cationic dyes, such as alcian blue or toluidine blue *(2),* that bind to the negatively charged glycosaminoglycan side chains, whereas more neutral glycoproteins can be detected by some variation of the Schiff base reaction involving initial oxidation of carbohydrates by periodic acid and subsequent staining with Schiff's reagent (PAS) *(3),* alcian blue *(4),* or a hydrazine derivate *(5).* A protocol for PAS staining of small SDS gels is described in ref. *6.* Unfortunately, these methods are characterized by a low sensitivity, generally requiring microgram amounts of protein for detection. At the same time, they are carbohydrate specific, which means that nonglycosylated proteins are not stained.

The two methods described here are based on silver enhancement of traditional staining methods for proteoglycan and glycoprotein *(7–12),* which result in a twofold increase in sensitivity as compared with alcian blue or PAS alone. In both methods, alcian blue is used as the primary staining agent, binding either directly to the proteoglycans (Section 3.1.) or to oxidized glycoproteins (Section 3.2.), subsequently enhanced by a neutral silver-staining protocol. The alcian blue dye does not impede silver-staining of nonglycosylated proteins. Hence both glycosylated and nonglycosylated proteins will be stained. However, it is possible to exclude the staining of nonglycosylated proteins (*see* Note 7).

The high sensitivity of the methods makes them very suitable for detection of proteoglycan/glycoprotein in dilute mixed samples, and for characterization of small

From: The Protein Protocols Handbook
Edited by: J. M. Walker Humana Press Inc., Totowa, NJ

amounts of purified materials, e.g., by determination of molecular weight before and after deglycosylation *(13,14)*. In this connection, it is important to note that highly glycosylated proteins owing to heterogenity of carbohydrate substitution usually move as diffuse bands or broader smears, so that unambiguous determination of molecular weight is not always possible.

For more specific purposes, other detection systems are usually employed. Sensitive detection of glycoproteins can be achieved with lectins or specific antibodies, most often after blotting onto a membrane *(15,16)*. Proteoglycans can similarly be identified by antibodies directed toward the core protein or glycosaminoglycan structures *(17)*.

The methods described here are optimized for the small, supported gels used in the PhastSystem (Pharmacia, Uppsala, Sweden), which can be programmed for automatic staining resulting in fast and reproducible results. If other systems (larger gels) are used, generally longer time is needed in each step (*see* Note 1).

2. Materials

2.1. Staining of Proteoglycans

All chemicals used should be of analytical grade, and water should be of high purity (e.g. Maxima from Elga [High Wycombe, UK] or Milli-Q from Millipore [Bedford, MA]).

1. Washing solution I: 25% ethanol (v/v),10% acetic acid (v/v) in water.
2. Washing solution II: 10% ethanol (v/v), 5% acetic acid (v/v) in water.
3. Staining solution: 0.125% alcian blue (w/v) (e.g., Bio-Rad [Hercules, CA] no. 161-0401) in washing solution I. Stir extensively and filtrate before use.
4. Stopping solution: 10% acetic acid (v/v),10% glycerol (v/v) in water.
5. Developer stock: 2.5% (w/v) sodium carbonate in water.
6. Sensitizing solution: 5% (v/v) glutardialdehyde in water. (Prepare fresh when required.)
7. Silvering solution: 0.4% (w/v) silver nitrate in water. (Prepare fresh when required.)
8. Developer: Add formaldehyde to the developer stock to a final concentration of 0.013% (v/v), e.g., add 35 µL of a 37% formaldehyde solution to 100 mL of developer stock, and stir for a few seconds. (Prepare fresh just before use.)

2.2. Staining of Glycoproteins

1. Prepare all solutions described in Section 2.1.
2. Fixing solution: 10% (v/v) trichloroacetic acid in water.
3. Washing solution III: 5% (v/v) acetic acid in water.
4. Oxidizing solution: 1 % (w/v) periodic acid in water.
5. Reducing solution: 0.5% (w/v) potassium metabisulfite in water.

3. Methods

3.1. Alcian Blue/Silver-Staining of Proteoglycans (see Note 1)

This method stains proteoglycans, glycosaminoglycans, nonglycosylated proteins, and some glycoproteins with a high content of negatively charged groups. The method can be varied for specific staining of proteoglycan/glycosaminoglycan (*see* Note 7).

1. Immediately after electrophoresis, transfer the gel to the development chamber/staining tray (*see* Note 2), and wash the gel three times in washing solution I for 5,10, and 15 min at 50°C (*see* Note 3).

2. Add the staining solution, and stain the gel for 15 min at 50°C.
3. Wash the gel three times in washing solution I for 1, 4, and 5 min and subsequently twice in washing solution II for 2 and 4 min at 50°C (*see* Note 4).
4. Add the sensitizing solution for 6 min at 50°C (*see* Note 5).
5. Wash the gel twice in washing solution II for 3 and 5 min, and subsequently in water twice for 2 min at 50°C.
6. Submerge the gel in the silvering solution for 6.5 min at 40°C.
7. Wash the gel twice in water for 30 s at 30°C (*see* Note 6).
8. Develop the gel with the freshly prepared developer at room temperature for an initial period of 30 s and subsequently for 4–8 min, depending on the desired sensitivity, background staining, and specificity (*see* Note 7).
9. Stop the development, and preserve the gel by adding the stopping solution for 5 min (*see* Note 8).

3.2. Periodic Acid Oxidation and Subsequent Alcian Blue/Silver Staining of Glycoproteins (see Note 1)

Some glycoproteins stain weakly with alcian blue/silver alone, probably owing to the low content of negatively charged groups for binding of alcian blue. This is overcome by initial oxidation of carbohydrates by periodic acid.

1. After electrophoresis, transfer the gel to the development chamber/Petri dish, and add the fixing solution for 10 min at 30°C (*see* Note 9).
2. Wash the gel twice for 2 min in washing solution III and submerge the gel in the oxidizing solution for 20 min at 30°C.
3. Wash the gel twice for 2 min in washing solution III and subsequently in water twice for 2 min before adding the reducing solution for 12 min at 30°C.
4. Wash the gel with water twice for 2 min and subsequently with washing solution I twice for 2 min at 50°C.
5. Continue with steps 2–9 of Section 3.1. exactly as described.

4. Notes

1. The methods are optimized for the PhastSystem, which uses small, supported gels that are stained in an automatic development chamber. If larger gels are stained in staining trays at room temperature, longer incubation times are needed. For unsupported 1-mm thick gels, good results are obtained by increasing the washing with washing solution I in steps 1 and 3 (Section 3.1.) to a total of 2.5 and 1.5 h, respectively, with four changes of the solution in each step, and furthermore doubling the incubation times in all other steps.
2. Handle gels with gloves or forceps, since fingerprints stain. The staining methods are highly sensitive, and it is essential that the equipment (development chamber/staining trays) is scrupulously clean and that high-quality water is used. Use a separate tray for the staining solution. Gels should be agitated during all staining, and washing procedures.
3. This rather extensive washing procedure is necessary to remove the SDS from the gel, which otherwise precipitates alcian blue, resulting in excessively high background. If gels are run in native PAGE, one short washing step is sufficient.
4. Alcian blue is irreversibly fixed in the gel by the subsequent silver staining and results in a greenish-black background if the stain is not washed out by dilute acetic acid. Only a weak bluish nuance in the background should remain.
5. Glutaraldehyde is injurious to health. If not carried out in a closed chamber, a fume cupboard should be used.

6. This step is important for washing out excess silver ions without losing silver bound to alcian blue/proteoglycan for autocatalytic reduction in the development step. Too little washing leads to formation of metallic silver in the background, whereas too intense washing leads to decreased sensitivity.
7. The development time can be varied according to the desired sensitivity and background staining. Even if an automatic staining chamber is used, these steps can be performed advantagously in staining trays/Petri dishes. In the first development step, a dark precipitate is created and the gel is transferred to a fresh developing solution. After 1–3 min, proteoglycans will appear, whereas nonglycosylated proteins appear after 4–8 min. The gel can be photographed sequentially on a light box during the development step for identification of proteoglycans. As a control, a gel can be stained for proteins with ordinary silver (steps 3–9, Section 3.1.), in which proteoglycans will not stain.
8. The image is stable, but over time, increased background staining will develop, especially from light exposure.
9. In this staining procedure, acetic acid/ethanol solutions are not always efficient for fixation of the proteins in the gel, whereas trichloroacetic acid works well.

Acknowledgments

This work was supported by the Danish Rheumatism Association and the Danish Medical Research Council.

References

1. Switzer, R. C., Merril, C. R., and Shifrin, S. (1979) A highly sensitive silver stain for detecting proteins and peptides in polyacrylamide gels. *Anal. Biochem.* **98,** 231–237.
2. Heinegård, D. and Sommarin, Y. (1987) Isolation and characterization of proteoglycans, in *Methods in Enzymology*, vol. 144: *Structural and Contractile Proteins* (Cunningham, L. W., ed.), Academic, New York, pp. 319–372.
3. Zacharius, R. M., Zell, T. E., Morrison, J. H., and Woodlock, J. J. (1969) Glycoprotein staining following electrophoresis on acrylamide gels. *Anal. Biochem.* **30,** 148–152.
4. Wardi, A. H. and Michos, G. A. (1972) Alcian blue staining of glycoproteins in acrylamide disc electrophoresis. *Anal. Biochem.* **49,** 607–609.
5. Eckhardt, A. E., Hayes, C. E., and Goldstein, I. J. (1976) A sensitive fluorescent method for the detection of glycoproteins in polyacrylamide gels. *Anal. Biochem.* **73,** 192–197.
6. Van-Seuningen, I. and Davril, M. (1992) A rapid periodic acid-Schiff staining procedure for the detection of glycoproteins using the PhastSystem. *Electrophoresis* **13,** 97–99.
7. Min, H. and Cowman, M. K. (1986) Combined alcian blue and silver staining of glycosaminoglycans in polyacrylamide gels: Application to electrophoretic analysis of molecular weight distribution. *Anal. Biochem.* **155,** 275–285.
8. Krueger, R. C. and Schwartz, N. B. (1987) An improved method of sequential alcian blue and ammoniacal silver-staining of chondroitin sulfate proteoglycan in polyacrylamide gels. *Anal. Biochem.* **167,** 295–300.
9. Lyon, M. and Gallagher, J. T. (1990) A general method for the detection and mapping of submicrogram quantities of glycosaminoglycan oligosaccharides on polyacrylamide gels by sequential staining with azure A and ammoniacal silver. *Anal. Biochem.* **185,** 63–70.
10. Jay, G. D., Culp, D. J., and Jahnke, M. R. (1990) Silver staining of extensively glycosylated proteins on sodium dodecyl sulfate-polyacrylamide gels: enhancement by carbohydrate-binding dyes. *Anal. Biochem.* **185,** 324–330.
11. Møller, H. J., Heinegård, D., and Poulsen, J. H. (1993) Combined alcian blue and silver staining of subnanogram quantities of proteoglycans and glycosaminoglycans in sodium dodecyl sulfate-polyacrylamide gels. *Anal. Biochem.* **209,** 169–175.

12. Møller, H. J. and Poulsen, J. H. (1995) Improved method for silver staining of glycoproteins in thin sodium dodecyl sulfate polyacrylamide gels. *Anal. Biochem.* **226,** 371–374.
13. Raulo, E., Chernousov, M. A., Carey, D. J., Nolo, R., and Rauvala, H. (1994) Isolation of a neuronal cell surface receptor of heparin binding growth-associated molecule (HB-GAM). *J. Biol. Chem.* **269,** 12,999–13,004.
14. Halfter, W. and Schurer, B. (1994) A new heparan sulfate proteoglycan in the extracellular matrix of the developing chick embryo. *Exp. Cell. Res.* **214,** 285–296.
15. Furlan, M., Perret, B. A., and Beck, E. A. (1979) Staining of glycoproteins in polyacrylamide and agarose gels with fluorescent lectins. *Anal. Biochem.* **96,** 208–214.
16. Moroi, M. and Jung, S. M. (1984) Selective staining of human platelet glycoproteins using nitrocellulose transfer of electrophoresed proteins and peroxidase-conjugated lectins. *Biochem. Biophys. Acta* **798,** 295–301.
17. Heimer, R. (1989) Proteoglycan profiles obtained by electrophoresis and triple immunoblotting. *Anal. Biochem.* **180,** 211–215.

100

Chemical Methods of Analysis of Glycoproteins

Elizabeth F. Hounsell, Michael J. Davies, and Kevin D. Smith

1. Introduction

The first analysis of glycoconjugates that often needs to be carried out is to see if they indeed contain sugar. For glycoproteins in gels or oligosaccharides in solution, this can be readily achieved by periodate oxidation at two concentrations, the first to detect sialic acids, and the second, any monosaccharide that has two free vicinal hydroxyl groups *(1)*. Periodate cleaves between the hydroxyl groups to yield reactive aldehydes, which can be detected by reduction with NaB^3H_4 or coupled to high sensitivity probes available in commercial kits, e.g., from Boeringher Mannheim (Mannheim, Germany) or Oxford Glycosystems (Abingdon, UK). In solution, a quick spot assay can be carried out for the presence of any monosaccharide or oligosaccharide having a C-2 hydroxyl group by visualization with charring by phenol/sulfuric acid reagent *(2)*. These methods are relatively specific for mono/oligosaccharides *(3)*.

2. Materials

2.1. Periodate Oxidation

1. 0.1*M* Acetate buffer, pH 5.5, containing 8 or 15 m*M* sodium periodate.
2. Ethylene glycol.
3. 0.1*M* Sodium hydroxide.
4. Reducing agent: Sodium borohydride, tritiated sodium borohydride, or sodium borodeuteride.
5. Glacial acetic acid.
6. Methanol.

2.2. Phenol Sulfuric Acid Assay

1. H_2O (HPLC-grade).
2. 4% Aqueous phenol.
3. Concentrated H_2SO_4.
4. 1 mg/mL galactose.
5. 1 mg/mL mannose.

3. Methods

3.1. Periodate Oxidation

1. Dissolve 0.1–1.0 mg glycoprotein in 2 mL of acetate buffer containing sodium periodate (15 m*M* for all monosaccharides, 8 m*M* for alditols, and 1 m*M* specifically for oxidation of sialic acids).

From: The Protein Protocols Handbook
Edited by: J. M. Walker Humana Press Inc., Totowa, NJ

2. Carry out the periodate oxidation in the dark at room temperature for 1 h for oligosaccharides, or at 4°C for 48 h for alditols, or 0°C for 1 h for sialic acids (*see* Note 1).
3. Decompose excess periodate by the addition of 25 μL of ethylene glycol, and leave the sample at 4°C overnight.
4. Add 0.1*M* sodium hydroxide (about 1.5 mL) until pH 7.0 is reached.
5. Reduce the oxidized compound with 25 mg of $NaB[^3H]_4$ at 4°C overnight (*see* Note 2).
6. Add acetic acid to pH 4.0, and concentrate the sample to dryness.
7. Remove boric acid by evaporations with 3 × 100 μL methanol (*see* Note 3).

3.2. Phenol/Sulfuric Acid Assay

1. Aliquot a solution of the unknown sample containing a range of approx 1 μg/10 μL into a microtiter plate along with a range of concentrations of a hexose standard (galactose or mannose, usually 1–10 μg).
2. Add 25 μL of 4% aquenous phenol to each well, mix thoroughly, and leave for 5 min (*see* Note 4).
3. Add 200 μL of H_2SO_4 to each well, and mix prior to reading on a microtiter plate reader at 492 nm (*see* Note 5).

4. Notes

1. It is important that the periodate oxidation is carried out in the dark to avoid unspecific oxidation. The periodate reagent has to be prepared fresh, since it is degraded when exposed to light.
2. The reactive aldehydes can also be detected by coupling to an amine-containing compound, such as digoxigenin *(1)*.
3. Addition of methanol in an acidic environment leads to the formation of volatile methyl borate.
4. Do not overfill wells during the hexose assay, since the conc. H_2SO_4 will severely damage the microtiter plate reader if spilled.
5. Exercise care when adding the conc. H_2SO_4 to the phenol/alditol mature, since it is likely to "spit," particularly in the presence of salt.

References

1. Haselbeck, A. and Hösel, W. (1993) Immunological detection of glycoproteins on blots based on labeling with digoxigenin, in *Methods in Molecular Biology, vol. 14: Glycoprotein Analysis in Biomedicine* (Hounsell, E. F., ed.), Humana, Totowa, NJ, pp. 161–173.
2. Smith, K., Harbin, A. M., Carruthers, R. A., Lawson, A. M., and Hounsell, E. F. (1990) Enzyme degradation, high performance liquid chromatography and liquid secondary ion mass spectrometry in the analysis of glycoproteins. *Biomed. Chromatogr.* **4,** 261–266.
3. Hounsell, E. F. (1993) A general strategy for glycoprotein oligosaccharide analysis, in *Methods in Molecular Biology, vol. 14: Glycoprotein Analysis in Biomedicine* (Hounsell, E. F., ed.), Humana, Totowa, NJ, pp. 1–15.

101

Monosaccharide Analysis by HPAEC

Elizabeth F. Hounsell, Michael J. Davies, and Kevin D. Smith

1. Introduction

Once the presence of monosaccharides has been established by chemical methods (*see* Chapter 100), the next stage of any glycoconjugate or polysaccharide analysis is to find out the amount of sugar and monosaccharide composition. The latter can give an idea as to the type of oligosaccharides present and, hence, indicate further strategies for analysis *(1)*. Analysis by high-pH anion-exchange chromatography (HPAEC) with pulsed electrochemical detection, that is described here, is the most sensitive and easiest technique *(2)*. If the laboratory does not have a biocompatible HPLC available that will withstand high salt concentrations, a sensitive labeling technique can be used with gel electrophoresis *(3)*, e.g., that marketed by Glyko Inc. (Navato, CA) or Oxford Glycosystems (Abingdon, UK), to include release of oligosaccharides/monosaccharides and labeling via reductive amination with a fluorescence label *(4)*.

2. Materials

1. Dionex DX500 (Dionex Camberley, Surrey UK) or other salt/biocompatible gradient HPLC system (titanium or PEEK lined), e.g., 2 Gilson 302 pumps with 10-mL titanium pump heads, 802Ti manometric module, 811B titanium dynamic mixer, Rheodyne 7125 titanium injection valve with Tefzel rotor seal (Gilson Medical Electronics, Villiers-le-Bel, France).
2. CarboPac PA1 separator (4 × 250 mm) and PA1 guard column (Dionex).
3. Pulsed amperometric detector with Au working electrode (Dionex), set up with the following parameters:
   a. Time = 0 s   E = + 0.1 V
   b. Time = 0.5 s   E = + 0.1 V
   c. Time = 0.51 s   E = + 0.6 V
   d. Time = 0.61 s   E = + 0.6 V
   e. Time = 0.62 s   E = – 0.8 V
   f. Time = 0.72 s   E = – 0.8 V
4. Reagent reservoir and postcolumn pneumatic controller (Dionex).
5. High-purity helium.
6. 12.5*M* NaOH (BDH, Poole, UK).
7. Reagent grade sodium acetate (Aldrich, Poole, UK).
8. HPLC-grade H_2O.

From: The Protein Protocols Handbook
Edited by: J. M. Walker Humana Press Inc., Totowa, NJ

9. $2M$ Trifluoroacetic acid HPLC-grade.
10. $2M$ HCl.
11. Dowex 50W × 12 H^+ of cation-exchange resin.
12. 3.5-mL screw-cap septum vials (Pierce, Chester, UK) cleaned with chromic acid (2 L H_2SO_4/350 mL H_2O/100 g Cr_2O_3) (**Use care! extremely corrosive,** *see* Note 1), and coated with Repelcote (BDH, Poole, UK).
13. Teflon-backed silicone septa for 3.5-mL vials (Aldrich).

3. Method

1. Dry down the glycoprotein (10 µg) or oligosaccharide (1 µg) in a clean screw-top vial with Teflon-backed silicone lid insert (*see* Note 2).
2. Hydrolyze in an inert N_2 atmosphere for 4 h at 100°C with $2M$ HCl.
3. Dry the hydrolyzate and re-evaporate three times with HPLC-grade H_2O.
4. Purify on a 1-mL Dowex 50W × 12 H^+ column eluted in water.
5. Dry down the monosaccharides ready for injection onto the HPLC system.
6. Prepare the following eluants:
   Eluant A = 500 mL HPLC-grade H_2O.
   Eluant B = 500 mL of 50 mM NaOH, 1.5 mM sodium acetate.
   Eluant C = 100 mL of 100 mM NaOH.
7. Degas the eluants by bubbling through helium.
8. Place the postcolumn reagent in a pressurized reagent reservoir (300 mm NaOH) and use the pneumatic controller to adjust helium pressure to give a flowrate of 1 mL/min (approx 10 psi).
9. Equilibrate the column with 98% eluant A and 2% eluant B at a flowrate of 1 mL/min.
10. Add the postcolumn reagent between column and detector cell at a flowrate of 1 mL/min via a mixing tee.
11. Inject approx 100 pmol of monosaccharide and elute isocratically as follows: eluant A = 98%; eluant B = 2%; flowrate = 1 mL/min; for 30 min.
12. Calculate monosaccharide amounts by comparison with a range of known monosaccharide standards run on the same day with deoxyglucose as an internal standard. From this, it is possible to infer the type and amount of glycosylation of the glycoprotein.
13. Regenerate the column in eluant C for 10 min at 1 mL/min (*see* Note 3).
14. Re-equilibrate the column with 98%A/2%B before the next injection.
15. At the end of the analysis, regenerate the column in eluant C, and flush pumps with H_2O (*see* Note 4).

4. Notes

1. If required, an equivalent detergent-based cleaner may be used.
2. Use polypropylene reagent vessels as far as possible for HPAEC-PAD because of the corrosive nature of the NaOH, and to minimize leaching of contaminants from the reservoirs.
3. Some drift in retention times may be observed during the monosaccharide analysis. This can be minimized by thorough regeneration of the column and use of a column jacket to maintain a stable column temperature.
4. Failure to wash out the eluants from the pumps at the end of an analysis may result in crystallization and serious damage to the pump heads.

References

1. Hounsell, E. F. (1993) A general strategy for glycoprotein oligosaccharide analysis, in *Methods in Molecular Biology, vol. 14: Glycoprotein Analysis in Biomedicine* (Hounsell, E. F., ed.), Humana, Totowa, NJ, pp. 1–15.

2. Townsend, R. R. (1995) Analysis of glycoconjugates using high-pH anion-exchange chromatography. *J. Chromatog. Library* **58,** 181–209.
3. Linhardt, R. J., Gu, D., Loganathan, D., and Carter, S. R. (1989) Analysis of glycosaminoglycan-derived oligosaccharides using reversed-phase ion-pairing and ion-exchange chromatography with suppressed conductivity detection. *Anal. Biochem.* **181,** 288–296.
4. Davies, M. J. and Hounsell, E. F. (1996) Comparison of separation modes for high performance liquid chromatography for the analysis of glycoprotein- and proteoglycan-derived oligosaccharides. *J. Chromatogr.* **720,** 227–234.

102

Monosaccharide Analysis by GC

Elizabeth F. Hounsell, Michael J. Davies, and Kevin D. Smith

1. Introduction

Although the methods given in Chapter 101 can give an approximate idea of oligosaccharide amount or composition, they would not be able to distinguish the multiple monosaccharides and substituents present in nature. For this, the high resolution of gas chromatography (GC) is required *(1,2)*. The most unambiguous results are provided by analysis of trimethylsilyl ethers (TMS) of methyl glycosides with on line mass selective detection (MS).

2. Materials

1. 0.5*M* Methanolic HCl (Supelco, Bellefont, PA).
2. Screw-top PTFE septum vials.
3. Phosphorus pentoxide.
4. Silver carbonate.
5. Acetic anhydride.
6. Trimethylsilylating (TMS) reagent (Sylon HTP kit, Supelco: pyridine hexamethyldisilazane, trimethylchlorosilane; **use care; corrosive**).
7. Toluene stored over 3-Å molecular sieve.
8. GC apparatus fitted with flame ionization or MS detector and column, e.g., for TMS ethers 25 m × 0.22 mm id BP10 (SGE, Austin, TX) 30 m × 0.2 mm id ultra-2 (Hewlett Packard, Bracknell, Berkshire, UK).

3. Method

1. Concentrate glycoproteins or oligosaccharides containing 1–50 µg carbohydrate and 10 µg internal standard (e.g., inositol or perseitol) in screw-top septum vials. Dry under vacuum in a desiccator over phosphorus pentoxide.
2. Place the sample under a gentle stream of nitrogen, and add 200 µL methanolic HCl (*see* Note 1).
3. Cap immediately, and heat at 80°C for 18 h (*see* Note 2).
4. Cool the vial, open, and add approx 50 mg silver carbonate.
5. Mix the contents, and test for neutrality (*see* Note 3).
6. Add 50 µL acetic anhydride, and stand at room temperature for 4 h in the dark (*see* Note 4).
7. Spin down the solid residue (*see* Note 5), and remove the supernatant to a clean vial.
8. Add 100 µL methanol, and repeat step 7, adding the supernatants together.

From: The Protein Protocols Handbook
Edited by: J. M. Walker Humana Press Inc., Totowa, NJ

9. Repeat step 8 and evaporate the combined supernatants under a stream of nitrogen.
10. Dry over phosphorus pentoxide before adding 20 µL TMS reagent.
11. Heat at 60°C for 5 min, evaporate remaining solvent under a stream of nitrogen, and add 20 µL dry toluene.
12. Inject onto a capillary GC column with 14 psi He head pressure and a temperature program from 130 to 230°C over 20 min and held at 230°C for 20 min.
13. Calculate the total peak area of each monosaccharide by adding individual peaks and dividing by the peak area ratio of the internal standard. Compare to standard curves for molar calculation determination.

4. Notes

1. The use of methanolic HCl for cleavage of glycosidic bonds and oligosaccharide-peptide cleavage yields methyl glycosides and carboxyl group methyl esters, which gives acid stability to the released monosaccharides, and thus, monosaccharides of different chemical lability can be measured in one run. If required as free reducing monosaccharides (e.g., for HPLC), the methyl glycoside can be removed by hydrolysis. The reagent can be obtained from commercial sources or made in laboratory by bubbling HCl gas through methanol until the desired pH is reached or by adding a molar equivalent of acetyl chloride to methanol.
2. An equilibrium of the α and β methyl glycosides of monosaccharide furanose (f) and pyranose (p) rings is achieved after 18 h so that a characteristic ratio of the four possible (fα, fβ, pα, pβ) molecules is formed to aid in unambiguous monosaccharide assignment.
3. Solid-silver carbonate has a pink hue in an acidic environment, and, therefore, neutrality can be assumed when green coloration is achieved.
4. The acidic conditions remove *N*-acetyl groups, which are replaced by acetic anhydride. This means that the original status of *N*-acetylation of hexosamines and sialic acids is not determined in the analysis procedure. If overacetylation occurs, the time can be reduced.
5. Direct re-*N*-acetylation by the addition of pyridine-acetic anhydride 1:1 in the absence of silver carbonate can be achieved, but this gives more variable results.

References

1. Hounsell, E. F. (1993) A general strategy for glycoprotein oligosaccharide analysis, in *Methods in Molecular Biology, vol. 14: Glycoprotein Analysis in Biomedicine* (Hounsell, E. F., ed.), Humana, Totowa, NJ, pp. 1–15.
2. Hounsell, E. F. (1994) Physicochemical analyses of oligosaccharide determinants of glycoproteins. *Adv. Carbohyd. Chem. Biochem.* **50,** 311–350.

103

Determination of Monosaccharide Linkage and Substitution Patterns by GC-MS Methylation Analysis

Elizabeth F. Hounsell, Michael J. Davies, and Kevin D. Smith

1. Introduction

The GC or GC-MS method discussed in Chapter 102 can distinguish substituted monosaccharides, but to characterize the position of acyl groups together with the linkages between the monosaccharides, a strategy has been developed to "capture" the substitution pattern by methylation of all free hydroxyl groups. The constituent monosaccharides are then analyzed after hydrolysis, reduction, and acetylation as partially methylated alditol acetates in a procedure known as methylation analysis *(1–3)*.

2. Materials

1. DMSO.
2. Methyl iodide.
3. Sodium hydroxide (anhydrous).
4. Chloroform.
5. Acetonitrile.
6. Pyridine.
7. Ethanol.
8. Water.
9. V-bottomed reacti-vials (Pierce [Rockford, IL]) with Teflon-backed silicone lid septa.
10. Sodium borodeuteride.
11. Ammonium hydroxide.
12. Trifluoroacetic acid.
13. Acetic acid.
14. GC-MS (e.g., Hewlett Packard 5890/5972A with ultra-2 or HP-5MS capillary column).

3. Method

1. Dry 20 nmol pure, desalted oligosaccharide into "V"-bottomed reacti-vials over P_2O_5.
2. Resuspend samples into 150 µL of a suspension of powdered NaOH/anhydrous DMSO (approx 60 mg/mL) under an inert atmosphere (*see* Note 1).
3. Sonicate for 10 min (or until the glycan is dissolved).

From: The Protein Protocols Handbook
Edited by: J. M. Walker Humana Press Inc., Totowa, NJ

4. Add 75 µL of methyl iodide (care) and sonicate for 15 min (*see* Note 2).
5. Extract the permethylated glycans with 1 mL $CHCl_3$ 3 mL H_2O, washing the aqueous phase with 3 × 1 mL $CHCl_3$. Wash the combined $CHCl_3$ washes with 3 × 5 mL H_2O (*see* Note 3).
6. Dry the $CHCl_3$ phase under N_2 after taking an aliqout for liquid secondary ion mass spectrometry (LSIMS) (*see* Note 4).
7. Hydrolyze for 1 h in 2*M* TFA at 100°C.
8. Reduce samples with 50 m*M* $NaBD_4$/50 m*M* NH_4OH (*see* Note 5) at 4°C or 4 h at room temperature.
9. Evaporate with 1X AcOH and 3 × 1:10 AcOH/MeOH.
10. Re *N*-acetylate samples with 50:50 acetic anhydride/pyridine 100°C, 90 min.
11. Analyze samples by GC-MS on low-bleed 5% capillary column (e.g., HP ultra-2 or HP5-MS or equivalent), with either on-column or splitless injection and a temperature gradient from to 50–265°C over 21 min, held for a further 10 min, and a constant gas flow of 1 mL/min into the MS. Ionization is in EI mode and a mass range of 45–400 (*see* Note 6).

4. Notes

1. The NaOH can be powdered either in a mortar and pestle or glass homogenizer and thoroughly vortexed with the DMSO prior to addition to the glycan.
2. If at the end of the methylation a yellow color is present, the reaction can be stopped by adding a crystal of sodium thiosulfate with aqueous extraction.
3. The permethylated oligosaccharides may also be purified on a Sep-Pak C_{18} column (Waters, Wafford, UK) by elution of permethylated oligosaccharides with acetonitrile or acetonitrile–water mixture.
4. The reaction with the NaOH/DMSO suspension deprotonates all the free hydroxyl groups and NH of acetamido groups forming an unstable carbanion. The addition of methyl iodide then rapidly reacts with the carbanions to form O-Me groups, and thus, permethylate the oligosaccharide. When subjected to LSIMS, these permethylated oligosaccharides will fragment about their glycosidic bonds particularly at acetamido residues. This means each oligosaccharide generates a unique fragmentation pattern allowing the determination of the oligosaccharide sequence. For example, the oligosaccharide Hex-HexNAc-HexNAc-Hex will generate the following fragments: Hex-HexNAc and Hex-HexNAc-HexNAc, where Hex denotes a hexose residue and HexNAc an *N*-acetylhexosamine.
5. The reduction can also be carried out with 50 m*M* NaOH, but this is not volatile and is harder to remove prior to further derivitization.
6. The hydrolysis step generates monosaccharides with the hydroxyl groups involved in the glycosidic linkages still retaining their protons. Reduction of these monosaccharides with $NaBD_4$ will break the ring structure to form monosaccharide alditols with the anomeric (C_1) carbon being monodeuterated. Acetylation of the free hydroxyls to *O*-acetyl groups completes the derivitization. The retention times of the PMAAs on the GC allow the assignment of the monosaccharide type (galactose, *N*-acetylgalactosamine, and so on). On-line mass spectrometric detection identifies fragment ions formed by the cleavage of C—C bonds of the monosaccharide alditols with the preference: methoxy-methoxy > methoxy-acetoxy > acetoxy-acetoxy. The resulting spectra are diagnostic for the substitution pattern, and hence, the previous position of linkage, e.g., a 2-linked hexose will produce a different set of ions to a 3-linked hexose, and a 2,3-linked hexose being different again. Selected ions from the spectra of all commonly occurring linkages can be used to analyze across the chromatogram (selected ion monitoring).

References

1. Hansson, G. C. and Karlsson, H. (1993) Gas chromatography and gas chromatography-mass spectrometry of glycoprotein oligosaccharides, in *Methods in Molecular Biology, vol. 14: Glycoprotein Analysis in Biomedicine* (Hounsell, E. F., ed.), Humana, Totowa, NJ, pp. 47–54.
2. Güther, M. L. and Ferguson, M. A. J. (1993) The microanalysis of glycosyl phosphatidylinositol glycans, in *Methods in Molecular Biology, vol. 14: Glycoprotein Analysis in Biomedicine* (Hounsell, E. F., ed.), Humana, Totowa, NJ, pp. 99–117.
3. Hounsell, E. F. (1994) Physicochemical analysis of oligosaccharide determinants of glycoproteins. *Adv. Carbohyd. Chem. Biochem.* **50,** 311–350.

104

Sialic Acid Analysis by HPAEC-PAD

Elizabeth F. Hounsell, Michael J. Davies, and Kevin D. Smith

1. Introduction

The most labile monosaccharides are the family of sialic acids, which are usually chain-terminating substituents. These are therefore usually released first by either mild acid hydrolysis or enzyme digestion, and can be analyzed with great sensitivity by high pH anion-exchange chromatography with pulsed auperometric detection (HPAEC-PAD). The remaining oligosaccharide is analyzed as discussed in Chapter 103 to identify the position of linkage of the sialic acid.

2. Materials

1. HCl (0.01, 0.1, and 0.5*M*).
2. 100 m*M* NaOH.
3. 1*M* NaOAc.
4. HPLC and PA1 columns as described in Chapter 101.

3. Method

1. Dry the glycoprotein into a clean screw-top vial with a Teflon-backed silicone lid insert (*see* Note 1).
2. Release the sialic acids by hydrolysis with 0.01 or 0.1*M* HCl for 60 min at 70°C in an inert N_2 atmosphere (*see* Note 2).
3. Dry down the hydrolysate, and wash three times with HPLC-grade H_2O.
4. Prepare 500 mL of 100 m*M* NaOH, 1.0*M* sodium acetate (eluant A).
5. Prepare 500 mL of 100 m*M* NaOH (eluant B).
6. Degas eluants by bubbling helium through them in their reservoirs (*see* Note 3).
7. Regenerate the HPLC column in 100% eluant B for 30 min at a flow rate of 1 mL/min.
8. Equilibrate the column in 95% eluant B for 30 min at a flow rate of 1 mL/min (*see* Note 4).
9. Inject approx 0.2 nmol of sialic acid onto the column, and elute using the following gradient at a flow rate of 1 mL/min:
   a. Time = 0 min; 95% eluant B.
   b. Time = 4 min; 95% eluant B.
   c. Time = 29 min; 70% eluant B.
   d. Time = 34 min; 70% eluant B.
   e. Time = 35 min; 100% eluant B.
   f. Time = 44 min; 100% eluant B.
   g. Time = 45 min; 95% eluant B; 1 mL/min.

From: The Protein Protocols Handbook
Edited by: J. M. Walker Humana Press Inc., Totowa, NJ

10. Quantitate the sialic acids by comparison with known standards run on the same day.
11. When the baseline has stabilized, the system is ready for the next injection.
12. When the analyses have been completed, regenerate the column in eluant B, and flush pumps with HPLC-grade H_2O (*see* Note 5).

4. Notes

1. If problems with contaminants are encountered, it may be necessary to wash the vials with chromic acid overnight, wash them thoroughly with distilled water, and then treat with a hydrophobic coating, such as repelcoat (*see* Chapter 101).
2. 0.01*M* HCl will release sialic acids with intact *N*- or *O*-acyl groups, but without quantitative release of the sialic acids. These can also be detected by HPAEC-PAD *(3)*. At 0.1*M* HCl, quantitive release is achieved, but with some loss of *O*- and *N*-acylation. At this concentration, some fucose residues may also be labile. Alternatively, the sialic acids can be released by neuraminidase treatment, which can be specific for α2-6 or α2-3 linkage, e.g., with α-sialidase of *Arthrobacter ureafacians* for α2-6 and α-sialidase of Newcastle disease virus for α2-3 using the manufacturer's instructions.
3. Use polypropylene reagent vessels as far as possible for HPAEC-PAD because of the corrosive nature of the NaOH, and to minimize leaching of contaminants from the reservoirs.
4. For maximum efficiency of detection, always ensure that the PAD reference electrode is accurately calibrated, the working electrode is clean, and the solvents are thoroughly degassed.
5. Failure to wash out the eluants from the pumps at the end of an analysis may result in crystallization and serious damage to the pump heads.

References

1. Davies, M. J., Smith, K. D., and Hounsell, E. F. (1994) The release of oligosaccharides from glycoproteins, in *Methods in Molecular Biology, vol. 32: Basic Protein and Peptide Protocols* (Walker, J. M., ed.), Humana, Totowa, NJ, pp. 129–141
2. Townsend, R. R. (1995) Analysis of glycoconjugates using high-pH anion-exchange chromatography. *J. Chromatog. Library* **58,** 181–209.
3. Manzi, A. E., Diaz, S., and Varki, A. (1990) HPLC of sialic acids on a pellicular resin anion exchange column with pulsed amperometry. *Anal. Biochem.* **188,** 20–32.

105

Chemical Release of *O*-Linked Oligosaccharide Chains

Elizabeth F. Hounsell, Michael J. Davies, and Kevin D. Smith

1. Introduction

O-linked oligosaccharides having the core sequences shown below can be released specifically from protein via a β-elimination reaction catalyzed by alkali. The reaction is usually carried out with concomitant reduction to prevent peeling, a reaction caused by further β-elimination around the ring of 3-substituted monosaccharides *(1)*. The reduced oligosaccharides can be specifically bound by solid sorbent extraction on phenylboronic acid (PBA) columns *(2)*.

O-linked protein glycosylation core structures linked to Ser/Thr:

```
     Galβ1-3GalNAcα
  GlcNAcβ1-3GalNAcα
  GlcNAcβ1-6
            \
             GalNAcα
            /
     Galβ1-3
  GlcNAcβ1-6
            \
             GalNAcα
            /
  GlcNAcβ1-3
  GalNAcα1-3GalNAcα
  GlcNAcβ1-6GalNAcα
  GalNAcα1-6GalNAcα
```

2. Materials

1. 1*M* $NaBH_4$ (Sigma, Poole, UK) in 50 m*M* NaOH. This is made up fresh each time from 50% (w/v) NaOH and HPLC-grade H_2O.
2. Methanol (HPLC-grade containing 1% acetic acid).
3. Acetic acid.
4. 1 mL Dowex H^+ (50 W × 12) strong cation ion-exchange column (Sigma, St. Louis, MO).
5. Bond elute phenyl boronic acid column (Jones Chromatography, Hengoed, UK; activated with MeOH).
6. 0.1*M* HCl.
7. 0.2*M* NH_4OH.
8. 0.1*M* acetic acid.
9. Methanol.

From: The Protein Protocols Handbook
Edited by: J. M. Walker Humana Press Inc., Totowa, NJ

3. Method

1. Dry the glycoprotein (100 μg to 1 mg) in a screw-topped vial and resuspend in 100 μL 50 m*M* NaOH containing 1*M* $NaBH_4$ (*see* Notes 1 and 2).
2. Incubate at 55°C for 18 h.
3. Quench the reduction by the addition of ice-cold acetic acid until no further effervescence is seen.
4. Dry the reaction mixture down, and then wash and dry three times with a 1% acetic acid, 99% methanol solution to remove methyl borate.
5. Resuspend the alditols in H_2O, and pass down a 1-mL H^+ cation exchange resin. The alditols will not be retained and will elute by washing the column with water.
6. Dry the alditols and resuspend them in 100 μL of 0.2*M* NH_4OH.
7. Activate a phenyl boronic acid (PBA) column with 2 × 1 mL MeOH.
8. Equilibrate the PBA column with 2 × 1 mL 0.1*M* HCl, 2 × 1 mL H_2O, and 2 × 1 mL 0.2*M* NH_4OH.
9. Add the sample in 100 μL 0.2*M* NH_4OH and elute with 2 × 100 μL 0.2*M* NH_4OH, 2 × 100 μL H_2O, and 6 × 100 μL 0.1*M* acetic acid. Collect these fractions and test for monosaccharide (Chapter 100) or combine and analyze according to Chapter 106.
10. Regenerate the PBA column with 0.1*M* HCl and 2 × 1 mL H_2O before storing and reactivation in 2 × 1 mL MeOH.

4. Notes

1. The $NaBH_4$/NaOH solution is made up <6 h before it is required.
2. Reduction with NaB^3H_4 allows the incorporation of a radioactive label into the alditol to enable a higher degree of sensitivity to be achieved while profiling.
3. Protein degradation can be minimized by the omission of the $NaBH_4$, although this results in the degradation of sugar chains having a 3-substituted GalNAc-Ser/Thr, i.e., most types. The addition of 6 m*M* cadmium acetate, 6 m*M* Na_2EDTA to the $NaBH_4$/NaOH solution reduces protein degradation without the loss of oligosaccharide alditol.

References

1. Hounsell, E. F. (1994) Physicochemical analyses of oligosaccharides determinants of glycoproteins. *Adv. Carbohyd. Chem. Biochem.* **30,** 311–350.
2. Stoll, M. S. and Hounsell, E. F. (1988) Selective purification of reduced oligosaccharides using a phenylboronic acid bond elut column: potential application in HPLC, mass spectrometry, reductive amination procedures and antigenic/serum analysis. *Biomed. Chromatogr.* **2,** 249–253.

106

O-Linked Oligosaccharide Profiling by HPLC

Elizabeth F. Hounsell, Michael J. Davies, and Kevin D. Smith

1. Introduction

The several different core regions of *O*-linked chains (Chapter 105) can be further extended by Gal and GlcNAc containing backbones or by addition of blood group antigen-type glycosylation and the presence of sialic acid or sulfate. We sought a universal column that can be applied with high resolution to the various oligosaccharide alditols released from glycoproteins by β-elimination (*see* Note 1) and have pioneered *(1,2)* the use of porous graphitized carbon (PGC). This is an alternative to C_{18} reversed-phase HPLC and normal-phase amino-bonded columns, which can be used together in the presence and absence of high-salt buffers *(3)*.

2. Materials

1. Gradient HPLC system: e.g., 2 × 302 pumps, 802C manometric module, 811 dynamic mixer, 116 UV detector, 201 fraction collector, 715 chromatography system control software (all Gilson Medical Electronics, Villiers-le-Bel, France).
2. IBM PS-2 personal computer (or compatible model) with Microsoft Windows.
3. Hypercarb S HPLC Column (100 × 4.6 mm) (Shandon Scientific, Runcorn, Cheshire, England) or Glycosep H (100 × 3 mm, Oxford Glycosystems, Abingdon, UK).
4. HPLC-grade H_2O.
5. HPLC-grade acetonitrile.
6. HPLC-grade trifluoroacetic acid (Pierce and Warriner, Chester, UK).

3. Methods

1. Prepare eluant A: 500 mL of 0.05% TFA.
2. Prepare eluant B: 250 mL of acetonitrile containing 0.05% TFA.
3. Degas eluants by sparging with helium.
4. Equilibrate the column in 100% eluant A and 0% eluant B for 30 min at 0.75 mL/min prior to injection samples.
5. Elute 10 nmol oligosaccharide alditol (obtained as in Chapter 105) with the following gradient at a flow rate of 0.75 mL/min and UV detection at 206 nm/0.08 AUFS.
   a. Time = 0 min A = 100%; B = 0%
   b. Time = 5 min A = 100%; B = 0%
   c. Time = 40 min A = 60%; B = 40%
   d. Time = 45 min A = 60%; B = 40%
   e. Time = 50 min A = 100%; B = 0%

From: The Protein Protocols Handbook
Edited by: J. M. Walker Humana Press Inc., Totowa, NJ

6. The resulting oligosaccharide containing fractions are then derivatized for LSIMS and GC-MS or analyzed by NMR.

4. Note

1. The GalNAc residue linked to Ser/Thr in *O*-linked chains is normally substituted at least at C-3 and therefore the alkali catalyzed β-elimination reaction will also result in "peeling" of the released oligosaccharide. This is obviated by concomitant reduction to give oligosaccharide alditols ending in GalNAcol. Endo-α-*N*-acetylgalactosaminidase digestion ot hydrazinolysis under mild conditions can be used to release intact reducing sugars, which will have longer retention times on HPLC *(4)*.

References

1. Davies, M. J., Smith, K. D., Harbin, A-.M., and Hounsell, E. F. (1992) High performance liquid chromatography of oligosaccharide alditols and glycopeptides on graphitized carbon column. *J. Chromatogr.* **609,** 125–131.
2. Davies, M. J., Smith, K. D., Carruthers, R. A., Chai, W., Lawson, A. M., and Hounsell, E. F. (1993) The use of a porous graphitized carbon (PGC) column for the HPLC of oligosaccharides, alditols and glycopeptides with subsequent mass spectrometry analysis. *J. Chromatogr.* **646,** 317–326.
3. Davies, M. J. and Hounsell, E. F. (1996) Comparison of separation modes for high performance liquid chromatography of glycoprotein- and proteoglycan-derived oligosaccharides. *J. Chromatogr.* **720,** 227–234.
4. Hounsel, E. F., ed. (1993) *Methods in Molecular Biology, vol 14: Glycoprotein Analysis in Biomedicine*, Humana, Totowa, NJ.

107

O-Linked Oligosaccharide Profiling by HPAEC-PAD

Elizabeth F. Hounsell, Michael J. Davies, and Kevin D. Smith

1. Introduction

Although it can often be an advantage to be able to chromatograph neutral and sialylated oligosaccharides/alditols in one run (*see* Chapter 106), the added resolution of HPAEC and sensitivity with PAD detection means that this is an additional desirable technique for analysis. Neutral oligosaccharide alditols are poorly retained on CarboPac PA1, but can be resolved on two consecutive columns *(1)*. The carboPac PA1 column is ideal for sialylated oligosaccharides *(2)* and can also be used for sialylated alditols.

2. Materials

1. Dionex D500 (Dionex, Camberley Surrey, UK) or other salt/biocompatible gradient HPLC system (titanium or PEEK) lined, e.g., 2 Gilson 302 pumps with 10-mL titanium pump heads, 802Ti manometric module, 811B titanium dynamic mixer, Rheodyne 7125 titanium injection valve with Tefzel rotor seal, and Gilson 712 chromatography system control software (Gilson Medical Electronics, Villiers-le-Bel, France).
2. CarboPac PA1 separator (4 × 250 mm) and PA1 Guard column (Dionex).
3. Pulsed amperometric detector with Au working electrode (Dionex), set up with the following parameters:
   a. Time = 0 s E = +0.1 V
   b. Time = 0.5 s E = +0.1 V
   c. Time = 0.51 s E = +0.6 V
   d. Time = 0.61 s E = +0.6 V
   e. Time = 0.62 s E = –0.8 V
   f. Time = 0.72 s E = –0.8 V
4. Anion micromembrane suppressor 2 (AMMS2) (Dionex).
5. Autoregen unit with anion regenerant cartridge (Dionex).
6. High-purity helium.
7. NaOH 50% w/v.
8. Reagent-grade sodium acetate (Aldrich, Poole, UK).
9. HPLC-grade H_2O.
10. 500 mL of 50 m*M* H_2SO_4 (reagent-grade).

From: The Protein Protocols Handbook
Edited by: J. M. Walker Humana Press Inc., Totowa, NJ

3. Methods

1. Prepare eluant A: 100 m*M* NaOH, 500 m*M* sodium acetate.
2. Prepare eluant B: 100 m*M* NaOH.
3. Prepare the column by elution with 50% A/50% B for 30 min at a flow rate of 1 mL/min.
4. Equilibrate column is 5% eluant A/95% eluant B at a flow of 1 mL/min.
5. Connect the AMMS2 to eluant out line and autoregen unit containing 500 mL of 50 m*M* reagent-grade H_2SO_4, and pump regenerant at a flow of 10 mL/min (*see* Note 1).
6. Inject 200 pmol of each oligosaccharide or sialylated oligosaccharide alditol (more if required for NMR or LSIMS), and elute with the following gradient at a flow of 1 mL/min:
   a. Time = 0 min A = 5%; B = 95%
   b. Time = 15 min A = 5%; B = 95%
   c. Time = 50 min A = 40%; B = 60%
   d. Time = 55 min A = 40%; B = 60%
   e. Time = 58 min A = 0%; B = 100%
7. Equilibrate the column in 5%A/95%B prior to the next injection.
8. At the end of the analyses, regenerate the column in 100%, and flush at the pumps with H_2O.
9. Desalt oligosaccharide-containing fractions by AMMS, and derivatize for LSIMS and GC-MS.

4. Notes

1. A better desalting profile may be achieved with an AMMS membrane (rather than AMMS2) if a flow rate of <1 mL/min can be used. In addition, it is important that the membranes of the suppressor remain fully hydrated and that the regenerant solution is replaced about once a week.

References

1. Campbell, B. J., Davies, M. J., Rhodes, J. M., and Hounsell, E. F. (1993) Separation of neutral oligosacharide alditols from human meconium using high-pH anion-exchange chromatography. *J. Chromatogr.* **622,** 137–146.
2. Lloyd, K. O. and Savage, A. (1991) High performance anion exchange chromatography of reduced oligosaccharides from sialomucins. *Glycoconjugate J.* **8,** 493–498.

108

Release of *N*-Linked Oligosaccharide Chains by Hydrazinolysis

Tsuguo Mizuochi and Elizabeth F. Hounsell

1. Introduction

Hydrazinolysis is the most efficient method of releasing all classes of *N*-linked oligosaccharide chains from glycoproteins. Disadvantages compared to enzymic release (*see* Chapter 109) is the use of hazardous chemicals, break up of the protein backbone, and destruction of some chains. The first of these can be obviated by use of a commercial machine for hydrazinolysis—the Glyco Prep (Oxford Glycosystems, Abingdon, UK), but this is expensive, and as long as caution is exercised, the following method is excellent *(1)*.

2. Materials

1. Anhydrous hydrazine (*see* Note 1).
2. Toluene.
3. Saturated sodium bicarbonate solution prepared at room temperature.
4. Acetic anhydride.
5. 1-Octanol.
6. Lactose.
7. 1*N* Acetic acid.
8. 1*N* NaOH.
9. Methanol.
10. NaOH (0.05*N*) freshly prepared from 1*N* NaOH just before use.
11. Sodium borotritide (NaB^3H_4, approx 22 GBq/mmol) and approx 40 m*M* in dimethylformamide (silylation grade # 20672, Pierce Chemical Co., Rockford, IL).
12. 1-Butanol:ethanol:water (4:1:1, v/v).
13. Ethyl acetate:pyridine:acetic acid:water (5:5:1:3 v/v).
14. Dowex 50W-X12 (H^+ form, 50–100 mesh).
15. Whatman 3MM chromatography paper.
16. Air-tight screw-cap tube with a Teflon disk seal.
17. Dry heat block capable of maintaining 100°C.
18. Vacuum desiccator.
19. High-vacuum oil pump.
20. Descending paper chromatography tank.

From: The Protein Protocols Handbook
Edited by: J. M. Walker Humana Press Inc., Totowa, NJ

3. Method (*see* ref. *1*)

1. Add the glycoprotein (0.1–100 mg) to an air-tight screw-cap tube with a Teflon disk seal, and dry *in vacuo* overnight in a desiccator over P_2O_5 and NaOH.
2. Add anhydrous hydrazine (0.2–1.0 mL) with a glass pipet. The pipet must be dried to avoid introducing moisture into the anhydrous hydrazine (*see* Note 1).
3. Heat at 100°C for 10 h using a dry heat block. Glycoprotein is readily dissolved at 100°C.
4. Remove hydrazine by evaporation *in vacuo* in a desiccator. To protect the vacuum oil pump from hydrazine, connect traps between the desiccator and the pump in the following order: cold trap with dry ice and methanol, concentration, H_2SO_4-trap, and NaOH-trap on the desiccator side. Remove the last trace of hydrazine by coevaporation with several drops of toluene.
5. To re-*N*-acetylate, dissolve the residue in ice-cold saturated $NaHCO_3$ solution (1 mL/mg of protein). Add 10 µL of acetic anhydride, mix, and incubate for 10 min at room temperature. Re-*N*-acetylation is continued at room temperature by further addition of 10 µL (three times) and then 20 µL (three times) of acetic anhydride at 10-min intervals. The total volume of acetic anhydride is 100 µL/1 mL of saturated $NaHCO_3$ solution. Keep the solution on ice until the addition of acetic anhydride to avoid epimerization of the reducing terminal sugar.
6. To desalt, pass the reaction mixture through a column (1 mL for 1 mL of the $NaHCO_3$ solution) of Dowex 50W-X12, and wash with five column bed volumes of distilled water. Evaporate the effluent to dryness under reduced pressure at a temperature below 30°C. Addition of a drop of 1-octanol is effective in preventing bubbling over.
7. Dissolve the residue in a small amount of distilled water, and spot on a sheet of Whatman 3MM paper. Perform paper chromatography overnight using 1-butanol:ethanol:water (4:1:1, v/v) as developing solvent (*see* Note 3).
8. Cut the area 0-4 cm from the origin, recover the oligosaccharides by elution with distilled water, and then evaporate to dryness under reduced pressure. On this chromatogram, lactose migrates <4 cm from the origin.
9. To label oligosaccharides with tritium, dissolve the oligosaccharide fraction thus obtained in 100 µL of ice-cold 0.05*N* NaOH (freshly prepared from 1*N* NaOH). Verify that the pH of the oligosaccharide solution is above 11 with pH test paper using a <1-µL aliquot. If not, adjust the pH paper as 0.05*N* NaOH. After addition of NaOH solution, keep the oligosaccharide solution on ice; otherwise, part of the reducing terminal *N*-acetylglucosamine may be converted to *N*-acetylmannosamine by epimerization (*see* Note 4).
10. Add a 20*M* excess of NaB^3H_4 solution to the oligosaccharide solution, mix, and incubate at 30°C for 4 h to reduce the oligosaccharides. Then, add an equal weight of $NaBH_4$ (20 mg/mL of 0.05*M* NaOH, freshly prepared) as the original glycoprotein, and continue the incubation for an additional 1 h to reduce the oligosaccharides completely. Stop the reaction by acidifying the mixture with 1*N* acetic acid. During the reduction, and addition of acetic acid, keep the reaction mixture in a draft chamber, since tritium gas is generated (*see* Note 4).
11. To desalt, apply the reaction mixture to a small Dowex 50W-X12 column, and wash with five column bed volumes of distilled water, and then evaporate the effluent under reduced pressure below 30°C. The volume of the column should be calculated based on the amount of NaOH and $NaBH_4$, and the capacity of the resin. Then, remove the boric acid by repeated (three to five times) evaporation with methanol under reduced pressure. Dimethyl formamide used to dissolve NaB^3H_4 is usually coevaporated during the repeated evaporation.
12. Dissolve the residue in a small amount of distilled water, spot on a sheet of Whatman 3MM paper, and perform paper chromatography overnight using ethyl acetate: pyridine:acetic

acid:water (5:5:1:3, v/v) as developing solvent. This procedure is effective in removing the radioactive components originating from NaB^3H_4, which migrate a significant distance on the chromatogram.

13. Recover radioactive oligosaccharides, which migrate slower than lactitol (about 20 cm from origin), from the paper by elution with distilled water, and evaporate to dryness under reduced pressure.
14. Finally, subject the radioactive *N*-linked oligosaccharides thus obtained to high-voltage paper electrophoresis at pH 5.4 or HPLC to separate oligosaccharides by charge.

4. Notes

1. Anhydrous hydrazine is prepared by mixing 80% hydrazine hydrate (50 g), toluene (500 g), and CaO (500 g) and allowing to stand overnight. The mixture is refluxed for 3 h using a cold condenser and an NaOH tube. The mixture is then subjected to azeotropic distillation with toluene at 93–94°C under anhydrous conditions. Anhydrous hydrazine is collected from the bottom layer and stored in an air-tight screw-cap tube with a Teflon disk seal under dry conditions at 4°C in the dark. Commercially available anhydrous hydrazine (such as that from Aldrich Chemical Co., Inc., Milwaukee, WI) can also be used. It is important to check the quality with a glycoprotein of which the oligosaccharide structure has already been established before using for analysis, because contamination by trace amounts of water in some lots could modify the reducing terminal *N*-acetylglucosamine of *N*-linked oligosaccharides.

   Caution: Anhydrous hydrazine is a strong reducing agent, highly toxic, corrosive, suspected to be carcinogenic, and flammable. Therefore, great caution should be exercised during handling.
2. When stored in a small air-tight tube with a Teflon disk seal at –18°C, this NaB^3H_4 solution is stable for at least 1 y. Dimethylformamide should be stored with molecular sieves in a small screw-cap bottle with a Teflon disk seal under dry conditions.
3. This procedure is indispensable for the next tritium-labeling step of the liberated oligosaccharides. This is because oligosaccharides larger than trisaccharides remain very close to the origin, whereas the degradation products derived from the peptide moiety, which react with NaB^3H_4, move a significant distance on the paper.
4. For quantitative liberation of intact *N*-linked oligosaccharides from glycoproteins by hydrazinolysis, great care should be taken to maintain anhydrous condition until the re-*N*-acetylation step. Introduction of moisture into glycoprotein samples or anhydrous hydrazine results in diverse modifications of reducing terminal *N*-acetylglucosamine residues, especially when unsubstituted with an Fucα 1-6 group, and causes the release of *O*-linked oligosaccharides accompanied with various degradations of the reducing end.
5. When oligosaccharides are reduced by NaB^3H_4 with high specific activity (e.g., 555 GBq/mmol), the sensitivity of detection of oligosaccharides increases about 20-fold. To label oligosaccharides with tritium at high efficiency, it is recommended to keep the concentration of NaB^3H_4 high in the incubation mixture by reducing the volume of 0.05*N* NaOH (e.g., to the same volume as the NaB^3H_4 solution). A 20*M* excess of NaB^3H_4 solution is required for complete reduction of *N*-linked oligosaccharides, whereas a 5*M* excess of NaB^3H_4 solution required for complete reduction of *N*-linked oligosaccharides derived from glycoprotein samples is roughly estimated from data on the carbohydrate content or amino acid sequence. If generation of tritium gas is to be avoided, continue the incubation for an additional 1 h with large amounts of glucose to absorb excess NaB^3H_4 before acidifying the mixture.

Reference

1. Mizuochi, T. (1993) Microscale sequencing of *N*-linked oligosaccharides of glycoproteins using hydrazinalysis, Bio-Gel P-4 and sequential exoglycosidase digestion, in *Methods in Molecular Biology, vol. 14: Glycoprotein Analysis in Biomedicine* (Hounsell, E. F., ed.) Humana, Totowa, NJ, pp. 55–68.

109

Enzymatic Release of *O*- and *N*-Linked Oligosaccharide Chains

Elizabeth F. Hounsell, Michael J. Davies, and Kevin D. Smith

1. Introduction

Enzymes involved in both the synthesis and degradation of glycoconjugates are highly specific for monosaccharide, linkage position, and anomeric configuration factors further away in the oligosaccharide sequence or protein. Not withstanding this, endo- and exoglycosidases are extremely useful tools in structural analysis. The RAAM technique has automated the use of exoglycosidase digestion (Oxford Glycosystems, Abingdon, UK). That and the use of lectins in structural analysis are described in the original book *Glycoprotein Analysis in Biomedicine.* Here we discuss the release of intact oligosaccharide chains from proteins that can be further analyzed for separate functions.

2. Materials

2.1. Desalting

1. 1 mL Spectra/Chrom desalting cartridge (Orme, Manchester, UK) or Biogel P_{-2} minicolumn.
2. HPLC-grade H_2O.

2.2. Glycosidases

1. Endoglycosidase H (EC 3.2.1.96) (e.g., *E. coli,* Boehringer Mannheim, Lewes, UK). Digestion buffer: 250 m*M* sodium citrate buffer adjusted to pH 5.5 with 1*M* HCl.
2. Test-neuraminidase (EC 3.2.1.18) (e.g., *Vibrio cholerae,* Behring Ag, Marburg, Germany). Made up as 1 U/mL enzyme in digestion buffer and stored at 4°C. Digestion buffer: 50 m*M* sodium acetate, 134 m*M* NaCl, 9 m*M* $CaCl_2$.
3. Peptide-*N*-glycosidase F (EC 3.2.2.18) (e.g., *Flavobacterium meningosepticum,* Boehringer Mannheim). Digestion buffer: 40 m*M* potassium dihydrogen orthophosphate (KH_2PO_4), 10 m*M* EDTA adjusted to pH 6.2 with 1.0*M* NaOH.
4. *O*-glycosidase (EC 3.2.1.97) (e.g., *Diplococcus pneunomiae,* Boehringer Mannheim). Digestion buffer: 40 m*M* KH_2PO_4/10 m*M* EDTA adjusted to pH 6.0 with 1.0*M* NaOH.
5. Ice-cold ethanol.
6. Toluene.

3. Methods

3.1. Desalting

1. Wash the cartridge with 5 mL of HPLC-grade H_2O.
2. Load the sample onto the cartridge in a volume between 50 and 200 μL H_2O.

From: The Protein Protocols Handbook
Edited by: J. M. Walker Humana Press Inc., Totowa, NJ

3. Elute the column with 200 µL of H_2O (including sample load).
4. Elute the glycoprotein in 350 µL of H_2O.
5. Elute the salt with an additional 1 mL of H_2O.

3.2. Glycosidase Digestions

1. Dissolve 1 nmol of glycoprotein in 100 µL of H_2O, and boil for 30 min to denature. Remove a 10% aliquot for sodium dodecyl sulfate-polyacrylamide gel electrophoresis (SDS-PAGE), as a control for detection of enzyme digestion. Dry the remainder by lyophilization (*see* Note 1).
2. Resuspend the glycoprotein in 500 µL of Endo H digestion buffer and 5 µL of toluene. Add 1 mU of Endo H/l nmol glycoprotein, and incubate at 37°C for 72 h (*see* Notes 2 and 3).
3. Precipitate the protein with a twofold excess of ice-cold ethanol and centrifuge at 15,000*g* for 20 min. Wash the pellet three more times with ice-cold ethanol. Dry the protein pellet, take up in water, and aliquot 10% (relative to original amount) for SDS-PAGE.
4. Dry the remaining pellet, and resuspend in neuraminidase/neuraminidase digestion buffer at a concentration of 2 nmol of glycoprotein/10 µL of buffer. Incubate for 18 h at 37°C, and then ethanol-precipitate and aliquot as in step 3.
5. Resuspend the remaining glycoprotein in 500 µL of PNGase F digestion buffer, 5 µL toluene, and 1 U PNGase F/10 nmol glycoprotein. Incubate at 37°C for 72 h before precipitation and aliquoting as in step 3 (*see* Note 4).
6. Digest the final pellet with *O*-glycosidase under the same conditions as for the PNGase F digestion. Precipitate the pellet from ethanol washing and dry.
7. Apply all the pellet to SDS-PAGE.
8. The supernatants containing *N*- and *O*-linked oligosaccharides can be analyzed as discussed in Chapters 103, 110, and 111.

4. Notes

1. The described procedure assumes approx 10% glycosylation of the glycoprotein. The amount of glycoprotein treated may have to be increased to obtain oligosaccharides for further analysis with less highly glycosylated glycoproteins.
2. PNGase F is stored at –20°C, and all other enzymes at 4°C. Endo H removes high-mannose oligosaccharide chains, but not complex type.
3. The toluene is added to prevent bacterial growth.
4. The PNGase F digestion can be performed directly on the boiled glycoprotein if all *N*-linked glycoprotein chains (both high mannose and complex) are required to be removed. Digests may also be performed in 0.2*M* sodium phosphate buffer, pH 8.4, but this will result in the release of sialic acid residues as monosaccharides.

References

1. Yamamoto, K., Tsuji, T., and Osawa, T. (1993) Analysis of asparagine-linked oligosaccharides by sequential lectin-affinity chromatography, in *Methods in Molecular Biology, vol. 14: Glycoprotein Analysis in Biomedicine* (Hounsell, E. F., ed.), Humana, Totowa, NJ, pp. 17–34.
2. Davies, M. J., Smith, K. D., and Hounsell, E. F. (1994) The release of oligosaccharides from glycoproteins, in *Methods in Molecular Biology, vol. 32: Basic Protein and Peptide Protocols* (Walker, J. M., ed.), Humana, Totowa, NJ, pp. 129–141.

110

N-Linked Oligosaccharide Profiling by HPLC on Porous Graphitized Carbon (PGC)

Elizabeth F. Hounsell, Michael J. Davies, and Kevin D. Smith

1. Introduction

The vast array of possible *N*-linked oligosaccharides demands high-resolution HPLC columns for their purification *(1,2)*. Reverse-phase (C_{18}) and normal-phase (NH_2) columns have been used for the separation (singly or in concert) of many *N*-linked oligosaccharides. The porous graphitized carbon (PGC) column described in Chapter 106 for *O*-linked alditol separation will give improved *N*-linked oligosaccharide resolution over C_{18} columns, and has the advantage of using salt-free buffers for preparative work *(3,4)*.

2. Materials

1. Biocompatible gradient HPLC system, e.g., 2 × 302 pumps, 802C manometric module, 811 dynamic mixer, 116 UV detector, 201 fraction collector, 715 chromatography system control software (all Gilson Medical Electronics, France).
2. IBM PS-2 personal computer (or compatible model) with Microsoft Windows.
3. Hypercarb S HPEC Column (100 × 4.6 mm) (Shandon Scientific, Runcorn, Cheshire, England) or Glyco H (OGS, Abingdon, Oxon, UK).
4. HPLC-grade H_2O.
5. HPLC-grade acetonitrile.
6. HPLC-grade trifluoroacetic acid (Pierce and Warriner, Chester, UK).

3. Methods

1. Prepare eluant A: 500 mL of 0.05% TFA.
2. Prepare eluant B: 250 mL of acetonitrile containing 0.05% TFA.
3. Degas eluants by sparging with helium.
4. Equilibrate the column in 100% eluant A, and 0% eluant B for 30 min at 0.75 mL/min prior to injection of samples.
5. Elute 10 nmol oligosaccharides with the following gradient at a flow rate of 0.75 mL/min and UV detection at 206 nm/0.08 AUFS (*see* Note 1).
   a. Time = 0 min A = 100%; B = 0%
   b. Time = 5 min A = 100%; B = 0%
   c. Time = 40 min A = 60%; B = 40%
   d. Time = 45 min A = 60%; B = 40%
   e. Time = 50 min A = 100%; B = 0%

From: The Protein Protocols Handbook
Edited by: J. M. Walker Humana Press Inc., Totowa, NJ

6. The resulting oligosaccharide containing fractions are then derivatized for LSIMS and GC-MS or analyzed by NMR.

4. Notes

1. Reverse-phase or normal-phase HPLC may also be required for the complete separation of some oligosaccharide isomers.
2. To prevent anomerization, the oligosaccharides can be reduced to their alditols (either after PNGase F digestion or hydrazinolysis). The inclusion of a ^{3}H-label on reduction will give increased sensitivity over UV. Alternatively, the oligosaccharides can be fluorescently labeled at their reducing terminus (with 2-amino-benzamide or 2-amino-pyridine) by reductive amination to give a fluorescent chromophore and increased sensitivity. Sensitivity of detection may also be increased by the postcolumn addition of 300 m*M* NaOH and pulsed amperometric detection as described for HPAEC-PAD.

References

1. Hase, S. (1993) Analysis of sugar chains by pyridylamination, in *Methods in Molecular Biology, vol. 14: Glycoprotein Analysis in Biomedicine* (Hounsell, E. F., ed.), Humana, Totowa, NJ, pp. 69–80.
2. Kakehi, K. and Honda, S. (1993) Analysis of carbohydrates in glycoproteins by high performance liquid chromatography and high performance capillary electrophoresis, in *Methods in Molecular Biology, vol. 14: Glycoprotein Analysis in Biomedicine* (Hounsell, E. F., ed.), Humana, Totowa, NJ, pp. 81–98.
3. Smith, D. D., Davies, M. J., Hounsell, E. F. (1994) Structural profing of oligosaccharides of glycoproteins, in *Methods in Molecular Biology, vol. 32: Basic Protein and Peptide Protocols* (Walker, J. M., ed.), Humana, Totowa, NJ, pp. 143–155.
4. Davies, M. J. and Hounsell, E. F. (1995) Comparison of separation modes for high performance liquid chromatography of glycoprotein- and proteoglycan-derived oligosaccharides. *J. Chromatogr.* **720,** 227–234.

111

N-Linked Oligosaccharide Profiling by HPAEC-PAD

Elizabeth F. Hounsell, Michael J. Davies, and Kevin D. Smith

1. Introduction

High-pH anion-exchange chromatography with pulsed amperometric detection (HPAEC-PAD) is a very powerful tool for the profiling of *N*-linked oligosaccharides *(1,2)*, only being limited by the concentrated sodium hydroxide and sodium acetate required to achieve the separation and sensitive detection. Oligosaccharides are separated on the basis of charge (i.e., the number of sialic acid residues) and the linkage isomers present. Neutral oligosaccharide alditols are most weakly retained, retention increasing with increasing sialylation, and NeuGc- and sulfate-bearing oligosaccharides being most strongly retained. Fucosylation results in a shorter retention. If the oligosaccharides can be effectively desalted after HPAEC-PAD, this remains the method of choice for oligosaccharide purification, and is a very powerful analytical tool.

2. Materials

1. Dionex DX300 or salt/biocompatible gradient HPLC system (titanium or PEEK lined), e.g., 2 Gilson 302 pumps with 10-mL titanium pump heads, 802Ti manometric module, 811B titanium dynamic mixer, Rheodyne 7125 titanium injection valve with Tefzel rotor seal, and Gilson 715 chromatography system control software (all Gilson Medical Electronics, Villiers-le-Bel, France).
2. IBM PS-2 personal computer (or compatible model) with Microsoft Windows.
3. CarboPac PA100 separator (4 × 250 mm) and PA100 guard column (Dionex, Camberley, UK).
4. Pulsed amperometric detector with Au working electrode (Dionex), set up with the following parameters:
   a. Time = 0 s E = +0.1 V
   b. Time = 0.5 s E = +0.1 V
   c. Time = 0.51 s E = +0.6 V
   d. Time = 0.61 s E = +0.6 V
   e. Time = 0.62 s E = –0.8 V
   f. Time = 0.72 s E = –0.8 V
5. High-purity helium.
6. NaOH 50% (w/v).
7. ACS-grade sodium acetate (Aldrich, UK).
8. HPLC-grade H_2O.

From: The Protein Protocols Handbook
Edited by: J. M. Walker Humana Press Inc., Totowa, NJ

3. Methods

1. Prepare eluant A: 100 m*M* NaOH, 500 m*M* sodium acetate.
2. Prepare eluant B: 100 m*M* NaOH.
3. Prepare the column by elution with 50% A/50% B for 30 min at a flow rate of 1 mL/min.
4. Equilibrate column in 5% eluant A/95% eluant B at a flow of 1 mL/min.
5. Inject 200 pmol of each oligosaccharide or sialylated oligosaccharide alditol (more if required for NMR or LSIMS) and elute with the following gradient at a flow of 1 mL/min:
   a. Time = 0 min A = 5%; B = 95%
   b. Time = 15 min A = 5%; B = 95%
   c. Time = 50 min A = 40%; B = 60%
   d. Time = 55 min A = 40%; B = 60%
   e. Time = 58 min A = 0%; B = 100%
6. Equilibrate the column in 5%A/95%B prior to the next injection.
7. At the end of the analyzes, regenerate the column in 100%B, and flush the pumps with H_2O.
8. Desalt oligosaccharide-containing fractions and derivatize for LSIMS, GC-MS, or NMR (*see* Note 1).

4. Notes

1. Desalting can either be achieved by means of a ion-suppression system (e.g., Dionex AMMS or SRS system, *see* Chapter 107) or by off-line desalting on Biogel P2 mini-columns or H^+ cation-exchange resins. Fractions containing multiply sialylated oligosaccharides will probably contain too much salt to be totally desalted by an ion-suppressor, and column methods will be required. This can lead to losses of minor oligosaccharides.
2. All previous notes for HPAEC-PAD methods in this volume also apply.

References

1. Townsend, R. R. (1995) Analysis of glycoconjugates using high pH anion-exchange chromatography. *J. Chromatogr. Library* **58,** 181–209.
2. Smith, K. D., Davies, M. J., and Hounsell, E. F. (1994) Structural profiling of oligosaccharides of glycoproteins, in *Methods in Molecular Biology, vol. 32: Basic Protein and Peptide Protocols* (Walker, J. M., ed.), Humana, Totowa, NJ, pp. 143–155.

Part VII

Immunochemical Techniques

112

The Chloramine T Method for Radiolabeling Protein

Graham S. Bailey

Many different substances can be labeled by radioiodination. Such labeled molecules are of major importance in a variety of investigations, e.g., studies of intermediary metabolism, determinations of agonist and antagonist binding to receptors, quantitative measurements of physiologically active molecules in tissues and biological fluids, and so forth. In most of those studies, it is necessary to measure very low concentrations of the particular substance, and that in turn, implies that it is essential to produce a radioactively labeled tracer molecule of high specific radioactivity. Such tracers, particularly in the case of proteins, can often be conveniently produced by radioiodination.

Two γ-emitting radioisotopes of iodine are widely available, ^{125}I and ^{131}I. As γ-emitters, they can be counted directly in a well-type crystal scintillation counter (commonly referred to as a γ counter) without the need for sample preparation in direct contrast to β-emitting radionuclides, such as ^{3}H and ^{14}C. Furthermore, the count rate produced by 1 g atom of ^{125}I is approx 75 and 35,000 times greater than that produced by 1 g atom of ^{3}H and ^{14}C, respectively. In theory, the use of ^{131}I would result in a further sevenfold increase in specific radioactivity. However, the isotopic abundance of commercially available ^{131}I rarely exceeds 20% owing to contaminants of ^{127}I, and its half-life is only 8 d. In contrast, the isotopic abundance of ^{125}I on receipt in the laboratory is normally at least 90% and its half-life is 60 d. Also, the counting efficiency of a typical well-type crystal scintillation counter for ^{125}I is approximately twice that for ^{131}I. Thus, in most circumstances, ^{125}I is the radionuclide of choice for radioiodination.

Several different methods of radioiodination of proteins have been developed *(1,2)*. They differ, among other respects, in the nature of the oxidizing agent for converting $^{125}I^-$ into the reactive species $^{125}I_2$ or $^{125}I^+$. In the main, those reactive species substitute into tyrosine residues of the protein, but substitution into other residues, such as histidine, cysteine, and tryptophan, can occur in certain circumstances. It is important that the reaction conditions employed should lead on average to the incorporation of one radioactive iodine atom/molecule of protein. Greater incorporation can adversely affect the biological activity and antigenicity of the labeled protein.

The chloramine T method, developed by Hunter and Greenwood *(3)*, is probably the most widely used of all techniques of protein radioiodination, and is used extensively for the labeling of antibodies. It is a very simple method in which the radioactive iodide

From: The Protein Protocols Handbook
Edited by: J. M. Walker Humana Press Inc., Totowa, NJ

is oxidized by chloramine T in aqueous solution. The oxidation is stopped after a brief period of time by addition of excess reductant. Unfortunately, some proteins are denatured under the relatively strong oxidizing conditions, so other methods of radioiodination that employ more gentle conditions have been devised, e.g., the lactoperoxidase method (*see* Chapter 113).

2. Materials

1. Na ^{125}I: 37 MBq (1 mCi) concentration 3.7 GBq/mL (100 mCi/mL).
2. Buffer A: 0.5*M* sodium phosphate buffer, pH 7.4 (*see* Note 1).
3. Buffer B: 0.05*M* sodium phosphate buffer, pH 7.4.
4. Buffer C: 0.01*M* sodium phosphate buffer containing 1*M* sodium chloride, 0.1% bovine serum albumin, and 1% potassium iodide, final pH 7.4.
5. Chloramine T solution: A 2 mg/mL solution in buffer B is made just prior to use (*see* Note 2).
6. Reductant: A 1 mg/mL solution of sodium metabisulfite in buffer C is made just prior to use.
7. Protein to be iodinated: A 0.5–2.5 mg/mL solution is made in buffer B.

3. Method

1. Into a small plastic test tube (1 × 5.5 cm) are added successively the protein to be iodinated (10 µL), radioactive iodide (5 µL), buffer A (50 µL), and chloramine T solution (25 µL) (*see* Notes 3 and 4).
2. After mixing by gentle shaking, the solution is allowed to stand for 30 s to allow radioiodination to take place (*see* Note 5).
3. Sodium metabisulfite solution (500 µL) is added to stop the radioiodination, and the resultant solution is mixed. It is then ready for purification as described in Chapter 116.

4. Notes

1. The pH optimum for the iodination of tyrosine residues of a protein by this method is about pH 7.4. Lower yields of iodinated protein are obtained at pH values below about 6.5 and above about 8.5. Indeed, above pH 8.5 the iodination of histidine residues appears to be favored.
2. If the protein is seriously damaged by the use of 50 µg of chloramine T, it may be worthwhile repeating the radioiodination using much less oxidant (10 µg or less). Obviously, the minimum amount of chloramine T that can be used will depend, among other factors, on the nature and amount of protein to be iodinated.
3. The total volume of the reaction should be as low as practically possible to achieve a rapid and efficient incorporation of the radioactive iodine into the protein.
4. It is normal to carry out the method at room temperature. However, if the protein is especially labile, it may be beneficial to run the procedure at a lower temperature and for a longer period of time.
5. Because of the small volumes of reactants that are employed, it is essential to ensure adequate mixing at the outset of the reaction. Inadequate mixing is one of the most common reasons for a poor yield of radioiodinated protein by this procedure.
6. It is possible to carry out this type of reaction using an insoluble derivative of the sodium salt of *N*-chloro-benzene sulfonamide as the oxidant. The insoluble oxidant is available commercially (Iodo-Beads, Pierce, Rockford, IL). It offers a number of advantages over the employment of soluble chloramine T. It produces a lower risk of oxidative damage to the protein, and the reaction is stopped simply by removing the beads from the reaction mixture, thus avoiding any damage caused by the reductant.

References

1. Bolton, A. E. (1985) *Radioiodination Techniques,* 2nd ed. Amersham International, Amersham, Bucks, UK.
2. Bailey, G. S. (1990) In vitro labeling of proteins, in *Radioisotopes in Biology* (Slater, R. J., ed.), IRL, Oxford, UK, pp. 191–205.
3. Hunter, W. M. and Greenwood, F. C. (1962) Preparation of iodine-131 labeled human growth hormone of high specific activity, *Nature* **194,** 495,496.

113

The Lactoperoxidase Method for Radiolabeling Protein

Graham S. Bailey

1. Introduction

This method, introduced by Marchalonis *(1),* employs lactoperoxidase in the presence of a trace of hydrogen peroxide to oxidize the radioactive iodide $^{125}I^-$ to produce the reactive species $^{125}I_2$ or $^{125}I^+$. These reactive species substitute mainly into tyrosine residues of the protein, although substitution into other amino acid residues can occur under certain conditions. The oxidation can be stopped by simple dilution. Although the technique should result in less chance of denaturation of susceptible proteins than the chloramine T method, it is more technically demanding and is subject to a more marked variation in optimum reaction conditions.

2. Materials

1. $Na^{125}I$: 37 MBq (1 mCi) concentration 3.7 GBq/mL (100 mCi/mL).
2. Lactoperoxidase: available from various commercial sources. A stock solution of 10 mg/mL in 0.1*M* sodium acetate buffer, pH 5.6, can be made and stored at –20°C in small aliquots. A working solution of 20 µg/mL is made by dilution in buffer just prior to use.
3. Buffer A: 0.1*M* sodium acetate buffer, pH 5.6 (*see* Note 1).
4. Buffer B: 0.05*M* sodium phosphate buffer containing 0.1% sodium azide, final pH 7.4.
5. Buffer C: 0.05*M* sodium phosphate buffer containing 1*M* sodium chloride 0.1% bovine serum albumin and 1% potassium iodide, final pH 7.4.
6. Hydrogen peroxide: A solution of 10 µg/mL is made by dilution just prior to use.
7. Protein to be iodinated: A 0.5–2.5 mg/mL solution is made in buffer A.

It is essential that none of the solutions except buffer B contain sodium azide as antibacterial agent, since it inhibits lactoperoxidase.

3. Method

1. Into a small plastic test tube (1 × 5.5 cm) are added in turn the protein to be iodinated (5 µL), radioactive iodide (5 µL), lactoperoxidase solution (5 µL), and buffer A (45 µL).
2. The reaction is started by the addition of the hydrogen peroxide solution (10 µL) with mixing (*see* Note 2).
3. The reaction is stopped after 20 min (*see* Note 3) by the addition of buffer B (0.5 mL) with mixing.
4. After 5 min, buffer C (0.5 ml) is added with mixing. The solution is then ready for purification as described in Chapter 116 (*see* Notes 4 and 5).

From: The Protein Protocols Handbook
Edited by: J. M. Walker Humana Press Inc., Totowa, NJ

4. Notes

1. The exact nature of buffer A will depend on the properties of the protein to be radioiodinated. Proteins differ markedly in their pH optima for radioiodination by this method *(2)*. Obviously the pH to be used will also depend on the stability of the protein, and the optimum pH can be established by trial and error.
2. Other reaction conditions, such as amount of lactoperoxidase, amount and frequency of addition of hydrogen peroxide, and so forth, also markedly affect the yield and quality of the radioiodinated protein. Optimum conditions can be found by trial and error.
3. The longer the time of the incubation, the greater the risk of potential damage to the protein by the radioactive iodide. Thus, it is best to keep the time of exposure of the protein to the radioactive iodide as short as possible, but commensurate with a good yield of radioactive product.
4. Some of the lactoperoxidase itself may become radioiodinated, which may result in difficulties in purification if the enzyme is of a similar size to the protein being labeled. Thus, it is best to keep the ratio of the amount of protein being labeled to the amount of lactoperoxidase used as high as possible.

References

1. Marchalonis, J. J. (1969) An enzymic method for trace iodination of immunoglobulins and other proteins. *Biochem. J.* **113,** 299–305.
2. Morrison, M. and Bayse, G. S. (1970) Catalysis of iodination by lactoperoxidase. *Biochemistry* **9,** 2995–3000.

114

The Bolton and Hunter Method for Radiolabeling Protein

Graham S. Bailey

1. Introduction

This is an indirect method in which an acylating reagent (*N*-succinimidyl-3[4-hydroxyphenyl]propionate, the Bolton and Hunter reagent), commercially available in a radioiodinated form, is covalently coupled to the protein to be labeled *(1)*. The [^{125}I] Bolton and Hunter reagent reacts mostly with the side-chain amino groups of lysine residues to produce the labeled protein. It is the method of choice for radiolabeling proteins that lack tyrosine and histidine residues, or where reaction at those residues affects biological activity. It is particularly suitable for proteins that are sensitive to the oxidative procedures employed in other methods (*see* Chapters 112, 113, and 115).

2. Materials

1. [^{125}I] Bolton and Hunter reagent: 37 MBq (1 mCi), concentration 185 MBq/mL (5 mCi/mL) (*see* Note 1).
2. Buffer A: 0.1*M* sodium borate buffer, pH 8.5.
3. Buffer B: 0.2*M* glycine in 0.1*M* sodium borate buffer, pH 8.5.
4. Buffer C: 0.05*M* sodium phosphate buffer containing 0.25% gelatin.
5. Protein to be iodinated: A 0.5–2.5 mg/mL solution is made in buffer A.

It is essential that none of the solutions contain sodium azide or substances with free thiol or amino groups (apart from the protein to be labeled), since the Bolton and Hunter reagent will react with those compounds.

3. Method

1. The [^{125}I] Bolton and Hunter reagent (0.2 mL) is added to a small glass test tube (1 × 5.5 cm) and is evaporated to dryness under a stream of dry nitrogen.
2. All reactants are cooled in iced water (*see* Note 2).
3. The protein to be iodinated (10 µL) is added, and the tube is gently shaken for 15 min (*see* Note 2).
4. Buffer B (0.5 mL) is added (*see* Note 3). The solution is mixed and allowed to stand for 5 min.
5. Buffer C (0.5 mL) is added with mixing (*see* Note 3). The resultant solution is then ready for purification.

From: The Protein Protocols Handbook
Edited by: J. M. Walker Humana Press Inc., Totowa, NJ

4. Notes

1. [^{125}I] Bolton and Hunter reagent is available from Amersham International (Little Chalfort, UK) and Dupont NEN (Stevenage, UK). The Amersham product is supplied in anhydrous benzene containing 0.2% dimethylformamide. Aliquots can be easily withdrawn from the vial. However, the Dupont NEN is supplied in dry benzene alone, and dry dimethylformamide (about 0.5% of the sample volume) must be added to the vial with gentle shaking to facilitate the removal of aliquots.
2. [^{125}I] Bolton and Hunter reagent is readily hydrolyzed in aqueous solution. Under the described conditions, its half-life is about 10 min.
3. Buffer B stops the reaction by providing an excess of amino groups (0.2*M* glycine) for conjugation with the [^{125}I] Bolton and Hunter reagent. Thus, the carrier protein (0.25% gelatin) in buffer C does not become labeled.
4. This method of radioiodination has been used extensively, and various modifications of the described procedure, including time and temperature of the reaction, have been reported *(2)*. For example, it is possible first to acylate the protein with the Bolton and Hunter reagent, and then carry out radioactive labeling of the conjugate using the chloramine T method. However, in general, this procedure does not seem to offer any advantage over the method described here. Also, the time and temperature of the reaction can be altered to achieve optimal labeling.

References

1. Bolton, A. E. and Hunter, W. M. (1973) The labeling of protiens to high specific radioactivities by conjugation to a ^{125}I-containing acylating agent. *Biochem. J.* **133,** 529–538.
2. Langone, J. J. (1980) Radioiodination by use of the Bolton-Hunter and related reagents, in *Methods in Enzymology,* vol. 70 (Van Vunakis, H. and Langone, J. J., eds.), Academic, New York, pp. 221–247.

115

The Iodogen Method for Radiolabeling Protein

Graham S. Bailey

1. Introduction

In this method, a water-insoluble oxidant [1,3,4,6-tetrachloro-3α,6α-diphenyl glycoluril, Iodogen] *(1)* is employed to limit any damage to the protein by oxidation during radiolabeling. The oxidant, which is commercially available from Pierce (Rockford, IL), is dissolved in an organic solvent and is coated on the walls of the glass reaction tube. Radioiodination is then initiated by the addition of protein and Na ^{125}I in aqueous solution and, after a set time, is terminated by removal of the reaction mixture.

2. Materials

1. Na ^{125}I: 37 MBq (1 MCi) concentration 3.7 GBq/mL (100 mCi/mL).
2. Buffer A: 0.5*M* sodium phosphate buffer, pH 7.4 (*see* Note 1).
3. Buffer B: 0.01*M* sodium phosphate buffer containing 1*M* sodium chloride, final pH 7.4.
4. Buffer C: 0.01*M* sodium phosphate buffer containing 0.1% bovine serum albumin and 1% potassium iodide, final pH 7.4.
5. Iodogen: A 1 mg/10 mL solution in dichloromethane is made just prior to use.
6. Protein to be iodinated: A 0.5–1.0 mg/mL solution is made in buffer A.

3. Method

1. The Iodogen solution (20 µL) is added to a small glass test tube (1 × 5.5 cm) and is evaporated to dryness using a stream of nitrogen.
2. Buffer A (10 µL), protein to be labeled (10 µL) and radioactive iodide (5 µL) are added to the coated tube.
3. The tube is gently agitated at 1-min intervals over a 10-min period. (*see* Note 2).
4. Buffer B (250 µL) is added, mixed, and the resultant mixture is transferred to another tube.
5. The mixture is allowed to stand for 10 min (*see* Note 3), and then buffer C (250 µL) is added with mixing. The solution is then ready for purification as described in Chapter 116.

4. Notes

1. This method has been found to be highly efficient for the iodination of a range of proteins over the pH range 6–8.5 *(2)*.
2. If the protein is particularly labile, the reaction may be run at 4°C, possibly increasing the reaction time to 15 min.

From: The Protein Protocols Handbook
Edited by: J. M. Walker Humana Press Inc., Totowa, NJ

3. It is very important to allow the mixture to stand for 10–15 min before the addition of the carrier protein (buffer C) to allow active iodine species to decay and thus, avoid radioiodination of the carrier.

References

1. Fraker, P. J. and Speck, J. C. (1978) Protein and cell membrane iodinations with a sparingly soluble chloramide 1,3,4,6-tetrachloro-3α,6α-diphenylglycoluril. *Biochem. Biophys. Res. Comm.* **280,** 849–857.
2. Paus, E., Bormer, O., and Nustad, K. (1982) Radioiodination of proteins with the Iodogen method, in *R1A and Related Procedures in Medicine,* International Atomic Energy Agency, Vienna, pp. 161–171.

116

Purification and Assessment of Quality of Radioiodinated Protein

Graham S. Bailey

1. Introduction

At the end of a radioiodination procedure, the reaction mixture will contain the labeled protein, unlabeled protein, radioiodide, mineral salts, enzyme (in the case of the lactoperoxidase method), and possibly some protein that has been damaged during the oxidation. For most uses of radioiodinated proteins, it is essential to have the labeled species as pure as possible. In theory, any of the many methods of purifying proteins can be employed *(1)*. However, the purification of the radioiodinated protein should be achieved as rapidly as possible. For that purpose, the most widely used of all separation techniques is gel filtration.

One of the most important parameters used to assess the quality of a purified labeled protein is its specific radioactivity, which is the amount of radioactivity incorporated/unit mass of protein. It can be calculated in terms of the total radioactivity employed, the amount of the iodination mixture transferred to the gel-filtration column, and the amount of radioactivity present in the labeled protein, in the damaged components, and in the residual radioiodine. However, in practice, the calculation does not usually take into account damaged and undamaged protein. The specific activity is thus calculated from the yield of the radioiodination procedure, the amount of radioiodide, and the amount of protein used, assuming that there are no significant losses of those two reactants. The yield of the reaction is simply the percentage incorporation of the radionuclide into the protein.

It is obviously important that the radioiodinated protein should as far as possible have the same properties as the unlabeled species. Thus, the behavior of both molecules can be checked on electrophoresis or ion-exchange chromatography. The ability of the two species to bind to specific antibodies can be assessed by radioimmunoassay.

This chapter will describe a protocol for the purification of a radiolabeled protein and an example of a calculation of specific radioactivity.

2. Materials

1. Sephadex G75 resin.
2. Buffer A: 0.05*M* sodium phosphate containing 0.1% bovine serum albumin and 0.15*M* sodium chloride, final pH 7.4.

From: The Protein Protocols Handbook
Edited by: J. M. Walker Humana Press Inc., Totowa, NJ

3. Specific antiseum to the protein.
4. Buffer B: 0.1*M* sodium phosphate buffer containing 0.15*M* sodium chloride and 0.01% sodium azide, final pH 7.4.
5. γ-globulin solution: 1.4% bovine γ-globulins (Sigma [St. Louis, MO] G7516) in buffer B.
6. Polyethylene glycol/potassium iodide solution: 20% polyethylene glycol 6000 and 6.25% potassium iodide in buffer B.

3. Method

1. An aliquot (10 μL) of the mixture is retained for counting and the rest is applied to a column (1 × 20 cm) of Sephadex G75 resin (*see* Notes 1 and 2).
2. Elution is carried out at a flow rate of 20 mL/h, and fractions (0.6 mL) are automatically collected.
3. Aliquots (10 μL) are counted for radioactivity.
4. An elution profile of radioactivity against fraction number (for a typical profile, *see* Fig. 1) is plotted.
5. Immunoreactive protein is identified by reaction with specific antiserum to that protein in the following manner (steps 6–13) (*see* Note 3).
6. Aliquots (10 μL) of fractions making up the different peaks are diluted so that each gives 10,000 counts/min/100 μL.
7. Each diluted aliquot (100 μL) is incubated with the specific antiserum (100 μL) at 4°C for 4 h.
8. Protein bound to antibody and excess antibody are precipitated by the addition to each sample at 4°C of γ-globulin solution (200 μL) and polyethylene glycol/potassium iodide solution (1 mL) (*see* Note 4).
9. Each tube is vortexed and is allowed to stand at 4°C for 15 min.
10. Each tube is centrifuged at 4°C at 5000*g* for 30 min.
11. The supernatants are carefully removed by aspiration at a water pump, and the precipitates at the bottom of the tubes are counted for radioactivity.
12. Estimates of the yield of the radioiodination and specific radioactivity of the iodinated protein may then be made (*see* Note 5).
13. When the fractions containing radioiodinated protein have been identified, they are split into small aliquots that can be rapidly frozen or freeze-dried for storage (*see* Note 6).

4. Notes

1. Gel filtration by high-performance liquid chromatography (HPLC) provides a more rapid and efficient purification of iodinated protein than the conventional liquid chromatography described here, but it does entail the use of expensive columns and apparatus with the attendant danger of their contamination with radioactivity *(2)*.
2. A wide range of gel-filtration resins are available. In choosing a resin, the relative molecular masses (M_r) of the protein and other reactants and products must be borne in mind. Sephadex G-25 resin will separate the labeled protein from the low-mol-wt reagents, such as oxidants and reductants. However, if the labeled protein is contaminated with damaged protein (e.g., aggregated species), then a gel-filtration resin of higher porosity, such as Sephadex G-100, may produce a more efficient separation of the undamaged, mono-iodinated protein.
3. The occurrence of immunoreactive protein in more than one peak indicates the presence of polymeric or degradated forms of the iodinated protein. Ideally, iodinated protein should be present in a single, sharp, symmetrical peak. If this is not the case, then it is probably best to repeat the radioiodination under milder conditions or use a different method of iodination.

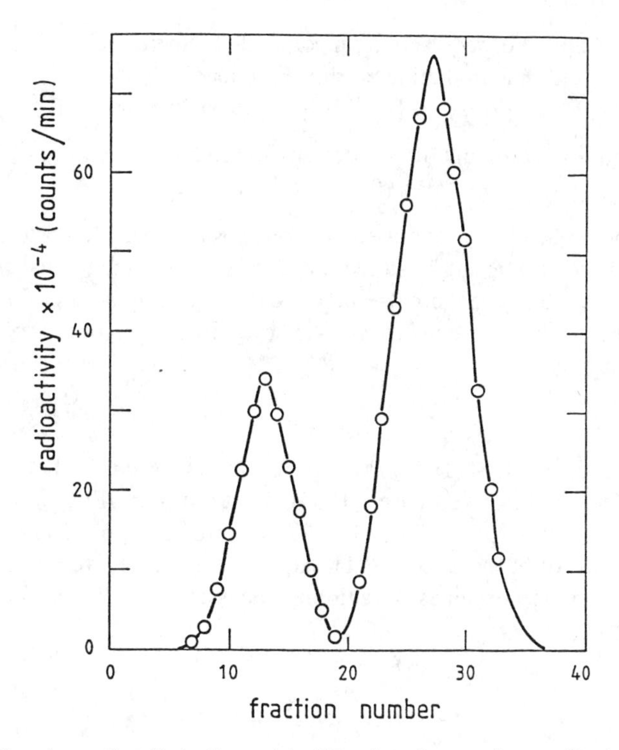

Fig. 1. Gel filtration of radioiodinated kallikrein of rat submandibular gland. The pure enzyme (10 μg) was iodinated with ^{125}I (18.5 MBq) by the chloramine T method. It was then purified on a column (1 × 20 cm) of Sephadex G-75 resin at a flow rate of 20 mL/h and collecting fractions of 0.6 mL. Aliquots (10 μL) of each fraction were measured for radioactivity. By radioimmunoassay, immunoreactive protein was found only in the first peak, and more than 90% of that radioactivity was bound by the antiserum to kallikrein from rat submandibular gland.

4. Polyethylene glycol produces precipitation of antibody and antibody-bound protein with little precipitation of unbound protein. Potassium iodide decreases the precipitation by polyethylene glycol of unbound protein *(3)*.
5. A typical example of the calculation of yield of iodination and specific radioactivity is as follows:

Counts of 10 μL incubation mixture prior to gel filtration = 1,567,925 counts/10 s (1)

Some of the radioiodinated protein is likely to bind to the reaction vessel and other surfaces, so it is best to calculate the radioactivity associated with the labeled protein in terms of the difference between total radioactivity applied to the column and the radioactivity associated with the unreacted iodide.

Counts in 10-μL aliquots of iodide peak = 388,845 counts/10 s (2)

Counts in 10-μL aliquots of protein peak =
1,567,925–388,845 counts/10 s = 1,179,080 counts/10 s (3)

Yield of radioiodination reaction = % incorporation of ^{125}I into protein
(1,179,080 × 100)/1,567,925 = 75.2% (4)

$$\text{Amount of radioactivity incorporated into protein} = \\ \% \text{ incorporation} \times \text{original radioactivity} = \\ 75.2\% \times 18.5 \text{ MBq} = 13.9 \text{ MBq} \quad (5)$$

$$\text{Specific radioactivity} = (\text{amount of radioactivity in protein/amount of protein used}) = \\ (13.9 \text{ MBq}/10\ \mu\text{g}) = (1.4 \text{ MBq}/\mu\text{g}) \quad (6)$$

6. Each aliquot should be thawed and used only once. Radioiodinated proteins differ markedly in their stability. Some can be stored for several wk (though it must be borne in mind that the half-life of ^{125}I is about 60 d), whereas others can only be kept for several days. If necessary, the labeled protein can be repurified by gel-filtration or ion-exchange chromatography prior to use.

References

1. Harris, E. L. V. and Angal, S. (1989) *Protein Purification Methods,* IRL, Oxford, UK.
2. Welling, G. W. and Welling-Webster, S. (1989) Size-exclusion HPLC of proteins, in *HPLC of Macromolecules* (Oliver, R. W. A., ed.), IRL, Oxford, UK, pp. 77–89.
3. Desbuquois, B. and Aurbach, G. D. (1971) Use of polyethyleneglycol to separate free and antibody-bound peptide hormones in radioimmunoassays. *J. Clin. Endocrin. Metab.* **33,** 732–738.

117

Conjugation of Peptides to Carrier Proteins via Glutaraldehyde

J. Mark Carter

1. Introduction

1.1. Basic Immunology and Immunochemistry of Peptide Conjugation

There are three common purposes for conjugation of peptides. The most common is induction of humoral immunity *(1)*. This is the production of antibodies capable of binding to the peptide immunogen. The antibodies are elaborated by plasma cells, which are terminally differentiated B-lymphocytes. However, in order for immunity to be successfully induced in a secondary anamnestic response, the immunogen must also react with T-lymphocytes. Many peptides contain B-cell epitopes, but not T-cell epitopes. (Such molecules are called haptens.) Coupling these molecules to a large carrier protein containing T-cell epitopes allows the induction of a B-cell response to the entire immunogen, including the peptide. New synthetic peptides thus offer promise as vaccines *(2)*.

The next most important reason for conjugation of peptides is to create an effective "capture antigen." Capture antigens are molecules used in enzyme-linked immunosorbent assays (ELISA) for binding of antibodies. However, these compounds must also be capable of binding to the plastic microtiter plate in a nonspecific fashion. Whereas larger peptides (>20 residues) usually bind well to the ELISA plates, often small- to medium-sized peptides do so poorly. The carrier proteins used in capture antigens are large enough to exhibit good nonspecific binding, and the peptides they carry on their surfaces are available for specific binding to Ab molecules.

Enzymes may also be conjugated to peptides, for the generation of non-ELISA homogenous immunodiagnostics. This system, which resembles enzyme-multiplied immunoassay technology (EMIT) with small organic haptens *(3)*, relies on the inhibition of the enzyme by Ab binding to the conjugated peptide antigen. This constitutes a third function of peptide conjugates.

Chapters 117–119 do not purport to be a comprehensive index of conjugation techniques. Rather they contain descriptions of three versatile and popular reactions capable of performing most conjugations. For a more exhaustive list of recipes, *see* ref. *(4)*.

From: The Protein Protocols Handbook
Edited by: J. M. Walker Humana Press Inc., Totowa, NJ

1.2. Antigen Orientation/Conformation for Native-Like Presentation

In each of the above functions of peptide conjugation it is necessary that the peptide attain a native-like orientation and conformation. Indeed, in most cases in which peptides are used as antigens, they must mimic certain regions of large intact protein molecules. The goal of experiments with these molecules is usually production of antibodies reactive to the native protein *(5)*. Although a free peptide in solution has many more conformational possibilities than the corresponding region of the native protein, it is assumed that the conjugate is able to assume the immunologically important structures normally exhibited by the native protein. However, this is not always true, and that same sequence, isolated from the rest of the parent protein molecule, may assume a different conformation when free in solution. For this reason, it is sometimes very important to conjugate the peptide in a manner that will encourage appropriate native-like conformations for antibody induction and binding.

1.3. Coupling at Termini

Conjugating a peptide through its terminal amino acids is often the best choice. One reason is that this avoids steric problems in epitope presentation and antibody binding of the folded peptide structure. Indeed, studies indicate that the free (nonconjugated) end of a peptide molecule thus coupled through one of its termini is much more antigenic than the bound end (site of conjugation) *(6)*.

The second reason is that some purification of the peptide product from the impurities in a crude synthesis mixture can be achieved by selection of particular N-terminal amino acids and conjugation protocols. If the N-terminal amino acid contains a moiety for conjugation not expressed in the remainder of the molecule, then only those molecules bearing this amino acid will be conjugated. To effect purification, this conjugation strategy is employed after a capping (acetylation) synthesis protocol. Incomplete couplings during peptide synthesis are terminated via acetylation, so that only full-length molecules bear the N-terminal amino acid. If the N-terminal amino acid is the one used for conjugation, then only full-length peptides will be coupled to the carrier protein. Dialysis or other size–dependent separation methods remove the truncated peptides from the macromolecular conjugates before the conjugates are used for immunochemistry *(7)*.

1.4. Circular Peptides

Another approach, especially successful in stabilization of native-like antigenic conformation in synthetic peptide bend and loop structures, involves the use of circular peptides. Circularization is generally achieved most easily via formation of a disulfide bond between the N- and C-termini of the peptide. This may be accomplished by synthesis of the peptide with an extra (nonphysiological) cysteine at each end of the sequence. Following cleavage, gentle air oxidation of a dilute solution of the peptide will generally produce the desired intramolecular disulfide bond.

On the other hand, if exactly one cysteine residue is naturally contained in the peptide sequence, then it is probably involved in a disulfide bond in the native protein structure. In the peptide molecule, this cysteine should therefore be utilized in conjugation to a carrier protein in order to mimic involvement in an internal disulfide bond.

Other fairly simple possibilities for circularization and stabilization of structure in a molecule include formation of a peptide or thioether bond between the N- and C-termini. Alternatively, the hydrogen bonds normally stabilizing a loop or bend configuration in the native protein may be replaced by covalent bonds, such as hydrazone-ethane *(8)*.

Finally, noncovalent forces must not be overlooked. One of the earliest generally successful immunization protocols for peptides (Freund's adjuvant) incorporated a lipid emulsion, and strangely, one of the latest great breakthroughs hailed in vaccine development (the liposome) also incorporates lipid emulsion. These methods are probably successful at encouraging the peptide to attain a native-like conformation via their water–lipid interface *(9)*. Because the native protein comprises an amphipathic environment, with a hydrophobic interior and a hydrophilic exterior, a peptide in a lipid emulsion probably experiences many of the same forces as in its native protein. In emulsion, the peptides probably align and insert themselves along the water–lipid interface much the same way as they do in the native protein structure.

1.5. Carrier Proteins

For immunological reasons, the choice for carrier proteins usually ends at bovine serum albumin (BSA), ovalbumin (OVA), or keyhole limpet hemocyanin (KLH). BSA or OVA are usually chosen when low price is important. As albumins they have good solubility characteristics and are well-described molecules. However BSA crossreacts immunologically with other serum albumin proteins, so it is not generally used in immunogens. On the other hand, it has high nonspecific binding, which makes it ideally suitable for use in immunoassay capture antigens. OVA does not crossreact as strongly with serum albumins as BSA, and it has been successfully used in immunogens as well as capture antigen.

KLH is the preferred choice for good immunogenicity. This is probably because of its relatively distant phylogenetic removal from the vertebrates generally used for antibody production. Other invertebrate proteins (e.g., *Limulus polyphemus* hemocyanin) are similarly immunogenic in experimental animals, and are often considerably less expensive. Unfortunately, such data are not widely published.

KLH functions for the mollusk as an oxygen-transport metalloprotein, existing as a family of soluble oligomers in the animal's hemolymph. It is usually purchased as a green–gray lyophilized powder. Dried KLH is virtually insoluble in water and does much better in PBS. For maximum solubility, dissolve it in 4*M* guanidine, followed by overnight dialysis vs PBS. As with any protein solution, do not mix vigorously (e.g., on a high-speed vortex mixer) in an attempt to increase dissolution. This will cause denaturation of the protein at the air interface, observed as foaming, which may lead to a substantial loss of soluble material.

1.6. Choice of Coupling Chemistry and Reagents

Literally dozens of different peptide and protein-coupling agents are commercially available, but most of these bear one or two of only a few active groups. The common moieties include: *N*-hydroxysuccinimide, maleimide, 2–pyridyldisulfide, haloacetate, and imidoesters. (Photoreactive agents, which may also be used for peptide conjugation, usually bear arylazides.) Investigators interested in studies of varying linker lengths, varying hydrophobicitics, or varying chemistries will find many such reagents available from Pierce (Rockford, IL). On the other hand, investigators wishing simply to produce an immunochemical reagent without optimization may prefer to follow one or two of the procedures that follow.

In this chapter and the following two chapters, I have detailed three methods of conjugation: glutaraldehyde, carbodiimide (EDAC), and *m*-maleimidobenzoyl-*N*-hydroxysuccinimide ester (MBS). These methods are popular, inexpensive, and effective. Occasionally, for immunological reasons that are not well understood, any given coupling scheme may be successful at producing a conjugate, yet be unsuccessful at generating an Ab *(10)*. Even more often, the conjugate may produce Ab reactive against the peptide, but not the native configuration of the protein from which the peptide sequence was taken. In these cases, alternate conjugation schemes may prove effective. If the sequence is short or not predicted to be highly antigenic, it may be a good idea to expend some extra energy and possibly save some time by performing the conjugation with two or even all three of these techniques and trying out each of them.

In some cases, your choice of method may be limited by the reactive moieties available on your peptide antigen. There are two considerations here. First, you must have groups available for reaction in the respective conjugation chemistry scheme. For example, you must have free primary amino groups available for effective use of glutaraldehyde. Equally important, and often unconsidered, you must avoid destroying those moieties critical for antigenic determinance. To use the example above, if the lysine you utilize for conjugation is important for antigenicity, then your glutaraldehyde conjugate will probably not behave correctly immunologically.

1.7. Peptide Solubility

Many synthetic peptides are poorly soluble in mild aqueous solutions. This characteristic can often be anticipated for longer peptides containing few charged and hydrophilic amino acids. This does not cause problems in synthesis and characterization of the compounds, because polar organic solvents are effective in maintaining solubility of the peptides and other reactants. However, carrier proteins are usually incompatible with organic solvents. For example, depending on the individual characteristics of the protein preparation used, KLH and BSA may not be soluble in >10% DMF or >70% DMSO. To maintain the solubility of the carrier proteins, therefore, most conjugations are performed at near-neutral pH in mild aqueous buffers.

For many difficult peptides, a little bit of DMF or DMSO may be enough to coax them into solution without interfering with the intended reaction. For example, conjugation with glutaraldehyde is unaffected by the presence of DMF, DMSO, or alcohols. On the other hand, DMSO will oxidize the side chain of methionine, so it should be avoided with peptides bearing this amino acid.

Another potential variable to improve the aqueous solubility of hydrophobic peptides is pH. This is especially true of peptides bearing no basic amino acids (arginine or lysine). Such weakly acidic hydrophobic peptides will often go into solution if the pH is raised to 8.0. This pH is fine for most conjugation chemistries. Consult the theoretical titration curve of the peptide to determine the charge state of the molecule at various pHs. Solubility is generally minimal at the isoelectric point.

Use of detergents to solubilize peptides is not recommended. They are often impossible to remove after the completion of conjugation, and they may be undesirable in the final (immunogen or assay antigen) preparation.

Some immunogenic sequences refuse to dissolve in aqueous buffers at any reasonable pH or temperature. These peptides may be resynthesized with the incorporation of two or three lysine residues at either end of the molecule. The increased number of positive charges on the molecule in neutral aqueous solution in turn increases the solubility of the sequence. Also, the new ε-amino groups comprise conjugation sites that are far from the antigenic center of the molecule, and do not seem to perturb antigenicity.

1.8. Optimization of Conjugation

1.8.1. Substitution Determination

Several studies suggest that maximal antibody response to a peptide-carrier complex will occur if there are about 20 peptide molecules/carrier protein molecule. However, choice of coupling chemistry may not facilitate such a level of substitution. Consider, for example, a peptide to be coupled through free thiol groups on a protein, where the protein contains only 4 disulfide bonds/molecule. Clearly no coupling can occur until the carrier protein is first reduced. Even afterward, however, only eight free sulfhydryl groups will be present, and some of them may be sterically inconvenient for the conjugation reaction. In such cases, the carrier may be derivatized to create an increased number of reactive moieties if optimum immunogenicity is desired. Note that for most peptide–carrier protein combinations, a 20:1 ratio is achieved with nearly equal masses of peptide and carrier.

After the conjugation reaction, most investigators choose to use the immunoconjugate without determining the precise level of substitution of the peptide on the carrier. This simplistic approach is appropriate for experiments requiring only the generation of Ab to the peptide, without comparisons of different immunogen preparations. However, when peptide vaccines are being considered, investigators should bear in mind the importance of reproducibility for approval of any compounds for use in humans. In these cases, substitution is important and must be documented.

There are a number of means of determining the number of peptide molecules coupled to a carrier protein. Regardless of which method is employed, it is critical that the conjugate preparation be free of unconjugated peptide. This may be achieved by extensive dialysis or gel filtration.

One of the most common techniques is quantitative amino acid analysis (AAA) of the conjugate. Comparing the amino acid profiles of the unconjugated carrier with that of the peptide-conjugated carrier allows quantitative differences to be established. For this method, the amino acid composition of the carrier protein, peptide, and conjugate must be known. This is generally determined via AAA. Calculations are greatly facilitated if some amino acids present in the carrier are not also present in the peptide.

A possibility for simplification of quantitation of substitution via AAA is introduction of nonnatural amino acids into the peptide as an internal standard. β-alanine, norleucine, ε-aminocaproic acid, and α-isobutyric acid *(11)* have commonly been utilized for this approach. Obviously, the nonnatural amino acids are not present in the carrier. They may thus be readily quantitated in the conjugate hydrolysate without interference from the carrier.

Some conjugation methods introduce a novel amino acid via unique reaction chemistry *(12)*, but AAA may not discriminate between these reaction products and the products of intramolecular reaction within the activated carrier molecule, which do not actually incorporate the peptide.

Another possibility for quantitation utilizes the release of a chromophore by the coupling reaction. This method is usually used for those chemistries driven by the release of 2-pyridylthione. One likely drawback of this type of quantitation is its sensitivity to interference by other chromophores in the assay buffer. Also, the chromophore may be released by intramolecular reaction.

A similar approach involves titration of reactive groups on the carrier, both before and after the reaction. This method is particularly useful in titrating free thiols via Ellman's reagent. Unfortunately, it also suffers from susceptibility to interference from the consumption of thiols by intramolecular reaction.

The fastest and simplest method of quantitation involves the incorporation of radiolabeled peptide into the conjugates. The peptide in the conjugate is quantitated via trace label calculation from the specific activity of the labeled peptide preparation. Unfortunately, the simplest radiolabeling methods use ^{125}I iodination of tyrosine moieties, and this modification produces a neo-antigen. The neo-antigen is not the same as the native tyrosine-containing antigen, and it may often prove immunodominant, obscuring the immunogenicity of the unmodified peptide in the preparation.

Another problem stems from the persistence of the radionuclide. "Hot" conjugates produce "hot" experimental animals. The "hot" animals, in turn, produce "hot" sera. The radioactive sera, as well as relatively large amounts of animal wastes, require special handling and disposal.

1.8.2. Storage

After the conjugate is purified (e.g., via dialysis) and the substitution level is determined (if desired), a method of storage must be considered. This will depend on the character of the conjugate preparation and the storage length requirement.

Many conjugates will precipitate during the coupling reaction. Although some will redissolve on dialysis, at least some compound often remains insoluble. Experience shows that when suspended along with the soluble conjugate, this precipitate causes no problems with animal health or peptide immunogenicity. However, it is conceivable that the aggregated form of the compounds may exhibit steric interference in regard to some epitopes. Also, it is unlikely that reproducible results would be obtained from such an inherently nonhomogenous mixture. This may prevent such vaccines from being approved for use in humans.

If the conjugate is mostly soluble, it may be sterilized and clarified (before substitution determination) via filtration on a 0.4-μ syringe filter. This will enhance its stability with respect to microbial degradation. For brief storage of filter-sterilized preparations, conjugates may be kept at 4°C. In a typical buffer, under aseptic conditions, they will last for at least 2 wk.

For longer storage of either soluble or insoluble preparations, freezing is appropriate; –20°C is sufficient for periods of a few months, whereas –70°C or lower is probably better for longer storage. Often, frozen conjugates exhibit increased precipitate on thawing, especially repeated freeze–thaw cycles.

For extremely long storage (over 1 yr), lyophilization is recommended. Unfortunately, on redissolving these compounds, formation of some amount of precipitate is virtually inevitable. This is probably a result of irreversible denaturation of the carrier protein.

1.8.3. Comparative Stability of Different Coupling Types

At least one study has been performed to examine the stability of different linking groups in conjugates, under different conditions of preparation and storage *(13)*. Although intuitive, the results suggest that increasing the number of bonds between the peptide and the carrier increases the temporal stability of the conjugate. Also, certain types of bonds are inherently more stable than others. For preparation of immunogens that must be stored for long periods of time, these considerations may be important. However, for most experiments, involving a limited number of animals studied over a period of a few months, the long-term chemical stability of the conjugate is probably not an important issue.

1.9. Glutaraldehyde Conjugation

Glutaraldehyde is an example of homobifunctional crosslinker in that it generally couples through amino groups, including lysine ε-amino and free α-amino groups. Cysteine sulfhydryls are also reactive, and tyrosine and histidine participate to a lesser extent. Because of this reaction selectivity, the glutaraldehyde method is especially appropriate for peptides containing no side-chain-reactive groups or only one lysine at either the N- or C-terminus.

This technique is excellent because of its simplicity, speed, and effectiveness. However, the reaction mechanisms and subsequent chemical linker structures remain poorly characterized. If a high degree of reproducibility is necessary, in either product performance or substitution level, then other methods should be considered.

The crosslinking reaction probably proceeds via a glutaraldehyde-lysine adduct Schiff base. However, rearrangement occurs rapidly, leading to a nonhydrolyzable bond. UV spectra suggest that a quaternary pyridinium structure is formed involving 4 glutaraldehyde molecules/lysine amino *(14)*. Otherwise, a Michael addition or aldimine condensation may occur *(15)*. The conjugation reaction may be also performed in the presence of sodium borohydride. This will reduce the Schiff base intermediate to an amine bond before rearrangement occurs.

2. Materials

1. Glutaraldehyde, 20 m*M* in water, freshly prepared. Specially purified grades are available, and they will give better reproducibility. However, technical-grade glutaraldehyde is at least as reactive as purified material.
2. Peptide: 6 mg, bearing one or two moieties available for reaction (N-terminal amino group, lysine ε-amino group, cysteine thiol group) Check for adequate solubility in buffer.
3. Carrier protein, e.g., KLH or ovalbumin, 6 mg.
4. Suitable buffer: PBS is usually quite effective. Alternatively, you may use phosphate buffer, pH 7.0-8.0, without saline, or use borate, pH 7.0–8.0. Avoid Tris and other amine-containing species. You may add a little DMSO or DMF if necessary to improve solubility of the peptide, but be careful not to precipitate the carrier protein. If pH is raised above 8.0, the glutaraldehyde itself may precipitate.
5. Optional: sodium borohydride or glycine.

3. Methods

3.1. Glutaraldehyde Conjugation

This method is based on that described by Baron and Baltimore *(16)*.

1. Weigh out 6 mg carrier and 6 mg peptide. Dissolve them together in 2–3 mL buffer. Over a 1–2 min period, add 1 mL glutaraldehyde solution, dropwise, mixing gently with a magnetic stir bar. Continue mixing, and allow to react for 1 h at room temperature. The formation of some yellowish or milky precipitate is possible. However this does not seem to reduce immunogenicity.
2. If you wish to perform the conjugation in the presence of sodium borohydride, add 10 mg/mL before adding the glutaraldehyde. This does not improve immunogenicity, although it does improve solubility of the end product, and it probably also improves the reproducibility of the conjugation reaction.
3. Another common modification is addition of glycine to 10 m*M* as a scavenger for unreacted glutaraldehyde after the conjugation is complete. Again, addition of glycine does not improve immunogenicity of the conjugate, although it does improve solubility. This is probably because the glycine prevents the relatively slow reaction that may continue after the 1 h incubation period, eventually crosslinking the conjugate molecules into large insoluble complexes.

3.2. Purification

For most peptide–carrier combinations, the glutaraldehyde will be completely consumed by the reaction above. However, it is prudent to remove any unreacted glutaraldehyde and peptide from the conjugate in order to prevent undesired reactions when the reagent is used with other biological materials. This is especially important for experimental vaccines, in which the biological material may comprise live animals.

1. For vaccines, separation is usually best performed via dialysis. In this way, any suspended solid material, which is often extremely immunogenic, is retained. A single overnight dialysis vs 4 L PBS (or other buffer) at 4°C is usually sufficient.
2. If the conjugate must be available as a soluble material, you may filter it after dialysis or perform the separation via gel filtration. For gel filtration, use at least 40-mL bed volume column packed with Sephadex G-25 (equilibrated in PBS or other buffer) for 3–4 mL of conjugate solution, so that the conjugate will elute in the void volume (*see* Section 1.8.1.).

4. Note

1. No matter your purpose, before you actually use the conjugate, it is prudent to characterize it, determining the number of peptide molecules bound per carrier molecule.

References

1. Walter, G. (1986) Production and use of antibodies against synthetic peptides. *J. Immunol. Methods* **88,** 149–161.
2. Patarroyo, M. E., Amador, R., Clavijo, P., Moreno, A., Guzman, F., Romero, P., Tascon, R., Franco, A., Murillo, L. A., Ponton, G., and Trujillo, G. (1988) A synthetic vaccine protects humans against challenge with asexual blood stages of *Plasmodium falciparum* malaria. *Nature* **332,** 158–161.
3. Rubenstein, K. E., Schneider, R. S., and Ullman, E. F. (1971) Homogenous enzyme immunoassay, a new immunochemical technique. *Biophys. Biochem Res. Commun.* **47,** 846–851.

4. VanRegenmortel, M. H. V., Briand, J. P., Muller, S., and Plaue, S. (1988) *Laboratory Techniques in Biochemistry and Molecular Biology*, vol. 19 (Burdon, R. H. and Van Knippenberg, P. H., eds.), Elsevier, Amsterdam.
5. Lerner, R. A., Green N., Alexander, H., Liu, F.-T., Sutcliffe, J. G., and Shinnick, T. M. (1981) Chemically synthesized peptides predicted from the nucleotide sequence of the hepatitis B virus genome elicit antibodies reactive with the native envelope protein of dane particles. *Proc. Natl. Acad. Sci. USA* **78,** 3403–3407.
6. Dryberg, T. and Oldstone, M. B. A. (1986) Peptides as antigens. *J. Exp. Med.* **164,** 1344–1349.
7. Ponsati, B., Giraldt, E., and Andreu, D. (1989) A synthetic strategy for simultaneous purification–congujation of antigenic peptides. *Anal. Biochem.* **181,** 389–395.
8. Satterthwait, A. C., Arrhenius, T., Hagopian, R. A., Zavala, F., Nussenzweig, V., and Lerner, R. A. (1988) Conformational restriction of peptidyl immunogens with covalent replacements for the hydrogen bond. *Vaccine* **6,** 99–103.
9. Alving, C. R., Richards, R. L., Moss, J., Alving, L. I., Clements, J. D., Shiba, T., Kotani, S., Wirtz, R. A., and Hockmeyer, W. T. (1986) Effectiveness of liposomes as potential carriers of vaccines: applications to cholera toxin and human malaria sporozoite antigen. *Vaccine* **4,** 166–172.
10. Schaaper, W. M. M., Lankohof, H., Pujik, W. C., and Meleon, R. H. (1989) Manipulation of antipeptide immune response by varying the coupling of the peptide with the carrier protein. *Mol. Immunol.* **26,** 81–85.
11. Tsao, J., Lin, X, Lackland, H., Tous, G., Wu, Y., and Stein, S. (1991) Internally standardized amino acid analysis for determining peptide/carrier protein coupling ratio. *Anal. Biochem.* **197,** 137–142.
12. Kolodny, N. and Robey, F. A. (1990) Conjugation of synthetic peptides to proteins: quantitation from *S*–carboxymethylcysteine released upon acid hydrolysis. *Anal. Biochem.* **187,** 136–140.
13. Peeters, J. M., Hazendonk, T. G., Beuvery, E. C., and Tesser, G. I. (1989) Comparison of four bifunctional reagents for coupling peptides to proteins and the effect of the three moieties on the immunogenicity of the conjugates. *J. Immunol. Methods* 120, 133–143.
14. Reichlin, M. (1980) Use of glutaraldehyde as a coupling agent for proteins and peptides. *Methods Enzymol.* **70,** 159–165.
15. Kirkeby, S., Jakobsen, P., and Moe, D. (1987) Glutaraldehyde—"pure and impure." A spectroscopic investigation of two commercial glutaraldehyde solutions and their reaction products with amino acids. *Anal. Lett.* **20(2),** 303–315.
16. Baron, M. H. and Baltimore, D. (1982) Antibodies against the chemically synthesized genome-Linked protein of poliovirus react with native virus-specific proteins. *Cell* **28,** 395–404.

118

Conjugation of Peptide to Carrier Proteins via *m*–Maleimidobenzoyl–*N*–Hydroxysuccinimide Ester (MBS)

J. Mark Carter

1. Introduction

m-Maleimidobenzoyl-*N*-hydroxysuccinimide ester (MBS) is a heterobifunctional agent that links a thiol group to an amino group at neutral pH. For peptide conjugation, the peptide usually provides the thiol group in the form of a cysteine residue, whereas the carrier provides amino groups in the form of lysine residues. The reaction proceeds in two steps. First the carrier is activated by reaction of its amino group with the succinimide moiety. Then the thiol group on the peptide reacts with the maleimide moiety of the activated carrier.

Many synthetic peptides do not contain cysteine or any other thiol donor. When such compounds are desired to be conjugated via MBS or other thiol-type chemistry, an extra cysteine may be incorporated during peptide synthesis, at either the N- or C-terminal. Whether or not a spacer is also included between the cysteine and the peptide, this generally does not adversely affect the immunogenicity of the sequence. Indeed, addition of the cysteine at the N-terminal affords a simple scheme for purification via conjugation.

As described in the introduction to Chapter 117, if the synthesis is performed with capping after each amino acid coupling, only full-length molecules will bear the N-terminal cysteine thiol. Thus, only full-length peptides will be conjugated to the carrier. Then the truncated synthesis side products may be easily removed from the conjugate by size-dependent methods, such as gel filtration or dialysis, as an adjunct to reverse-phase HPLC.

1.1. Peptide Thiolation

Quite often it is necessary to introduce a thiol group into a peptide after synthesis in order to effect conjugation. Whenever possible, introduce the extra cysteine during peptide assembly, as discussed above. Otherwise use 2-iminothiolane, also known as Traut's Reagent. This reagent will introduce a free thiol onto any free amino group, usually the peptide N–terminus or a lysine side chain. The spacer moiety introduced between the modified side chain and the new thiol is three carbons (6.8-Å) long.

From: The Protein Protocols Handbook
Edited by: J. M. Walker Humana Press Inc., Totowa, NJ

To perform the thiolation, first prepare a buffer, pH 8.0, with no free amines (e.g., avoid Tris). Degas the buffer and 1 m*M* EDTA to help prevent oxidation of the new thiol. Dissolve 10 mg peptide, add 1 mg 2-iminothiolane, and allow to react for 30–60 min at room temperature. Remove unreacted 2-iminothiolane via desalting on Sephadex G-10 equilibrated in degassed PBS with 1 m*M* EDTA, and immediately use the thiolated peptide for conjugation, or desalt via reverse-phase HPLC and lyophilize the thiolated peptide.

Another possibility for thiol introduction is *N*-hydroxysuccinimidyl 3-(2-pyridyldithio)propionate *(1)*. SPDP introduces an activated thiol group, the 2-pyridyldithio moiety, which may be easily deprotected. In this case, the linker is four carbons (8.1-Å) long, with retention of the positive charge.

To use SPDP in thiolation, add 25 µL of a stock solution of SPDP (20 m*M* in DMSO) to 10 mg peptide in pH 8.0 buffer with no amines. Allow 30–60 min to react, and then add 12 mg DTT to deprotect the thiol. Desalt after another 30–60 min.

SPDP may also be used as a heterobifunctional (amino-to-thiol) coupling reagent *(2,3)*.

1.2. Indications

In MBS conjugation, side reactions are generally not observed. The chemistry and immunology for this reagent are well characterized, providing a stable intermolecular chemical bond. Furthermore, because the reaction is carried out in two steps, intramolecular crosslinking is minimized. Although there exist a number of possible side reactions involving other amino acids, such as histidine, the desired reaction is quite fast, and problems owing to side reactivity are rare.

Note that the linker in MBS, a benzyl moiety, is not suitable for use in humans because of probable toxicity. However, other reagents, similar to MBS, but bearing less toxic linkers, are readily available. In the following protocol, they may substitute for MBS on an equimolar basis. These include succinimidyl 4-(*N*-maleimidomethyl) cyclohexane-1-carboxylate and *N*-(γ-maleimidobutyryloxy)succinimide ester.

Somewhat recently, water–soluble sulfonated analogs of all these reagents have been made commercially available. These allow the possibility of elimination of organic solvent from the protocol in cases where small amounts might interfere, for example, by precipitating the carrier protein. However, the sulfonated derivatives are also soluble in organic solvents, so they may substitute directly for the nonsulfonated derivatives in this protocol, on an equimolar basis.

A detailed discussion of peptide conjugation can be found in the introduction to Chapter 117.

2. Materials

1. MBS: 10 mg/mL in DMF or DMSO. This solution may be sealed and stored at –20°C for several months.
2. Reduced peptide: 5 mg, bearing one free thiol group.
3. Carrier protein, e.g., keyhole limpet hemocyanin (KLH) or ovalbumin, 5 mg.
4. Coupling solvent: PBS with 1 m*M* EDTA, degassed. Sodium borate, 20 m*M*, pH 6.5–7.5, may be used, but you must avoid buffers containing free amino groups, such as Tris. You may add a little DMF to improve solubility of the peptide, but be careful not to precipitate the carrier protein. Note: in order to maintain maleimide's selectivity for thiol above amine, it is very important to limit pH below 8.0.

5. Gel-filtration column: At least 10 mL bed volume of Sephadex G-25 equilibrated in coupling solvent 2.

3. Methods

The following recipe is an adaptation of Liu et al. *(4)* and Lerner et al. *(5)*:

3.1. Peptide Reduction

1. MBS conjugation requires the presence of a free (reduced) thiol. The peptide should thus be reduced immediately before use. To 1 mL PBS or DMF add 10 mg peptide to dissolve. Then add 10 mg DTT or 10 µL 2-mercaptoethanol. Allow 30 min for reduction at room temperature.
2. Desalt by gel filtration on a 10-mL bed volume of Sephadex G-10 equilibrated in degassed PBS/EDTA. For best results, use the reduced peptide solution immediately. Otherwise store overnight at 4°C under nitrogen.

3.2. Crosslinking

1. Prepare a solution of 5 mg KLH in 0.5 mL coupling solvent. As noted in Chapter 117, this may require overnight dialysis. Add 0.1 mL of the MBS solution. Allow to react for 30 min at room temperature, stirring gently.
2. Separate the MBS-activated KLH from free MBS via gel filtration on the G-25 column. The MBS-KLH should elute in the void volume, generally appearing slightly cloudy. Separation may be confirmed by measuring A_{280} of the fractions. The first peak will comprise MBS-activated KLH, whereas the second peak will be unreacted MBS. Pool the MBS-KLH fractions.
3. Dissolve 5 mg peptide in 1 mL coupling solvent. Combine the peptide solution with the MBS-KLH solution. Allow to react with gentle stirring for 3 h at room temperature.
4. Because MBS (and the other heterobifunctional agents listed above) have low acute toxicity, the conjugate may be used for experimental vaccines without further purification to remove free peptide. However, for other types of assays, it may be necessary to remove unreacted peptide, which might otherwise interfere with the desired reaction of the conjugate. This separation may be effected by dialysis or gel filtration.
5. For dialysis, a single overnight dialysis vs 4 L PBS (or other buffer) at 4°C is usually suf-ficient. For gel filtration, use at least 40-mL bed volume column packed with Sephadex G-25 (equilibrated in PBS or other buffer), so that the conjugate will elute in the void volume.

4. Note

1. No matter your purpose, before you actually use the conjugate, it is prudent to characterize it, determining the number of peptide molecules bound per carrier molecule (*see* Chapter 117, Section 1.8.1.).

References

1. Carlsson, J., Drevin, H., and Axen, R. (1978) Protein thiolation and reversible protein-protein conjugation. *Biochem. J.* **173,** 723–737.
2. Gordon, R. D., Fieles, W. E., Schotland, D. L., Hogue-Angeletti, R., and Barchi, R. L. (1987) Topographical localization of the C–terminal region of the voltage-dependent sodium channel from electrophorus electricus using antibodies raised against a synthetic peptide. *Proc. Natl. Acad. Sci. USA* **84,** 308–312.

3. Peeters, J. M., Hazendonk, T. G., Beuvery, E. C., and Tesser, G. I. (1989) Comparison of four bifunctional reagents for coupling peptides to proteins and the effect of the three moieties on the immunogenicity of the conjugates. *J. Immunol. Methods* **120,** 133–143.
4. Liu, F.-T., Zinnecker, M., Hamaoka, T., and Katz, D. H. (1979) New procedures for preparation and isolation of conjugates of proteins and a synthetic copolymer of D-amino acids and immunochemical characterization of such conjugates. *Biochemistry* **18(4),** 690–697.
5. Lerner, R. A., Green N., Alexander, H., Liu, F.-T., Sutcliffe, J. G., and Shinnick, T. M. (1981) Chemically synthesized peptides predicted from the nucleotide sequence of the hepatitis B virus genome elicit antibodies reactive with the native envelope protein of dane particles. *Proc. Natl. Acad. Sci. USA* **78,** 3403–3407.

119

Conjugation of Peptides to Carrier Protein via Carbodiimide

J. Mark Carter

1. Introduction

One of the most commonly used carbodiimide reagents is the water-soluble reagent, 1-ethyl-3-(dimethylaminopropyl) carbodiimide, also known as EDAC. Like other carbodiimides, EDAC couples an amino group to a carboxyl group (although side reactions involving cysteine sulfhydryl and tyrosine aryl hydroxyl are also reported). For peptide conjugation, usually the carboxyl group comprises the C-terminal of the peptide, and the amino group is an ε-amino group on a lysine residue contained in the carrier protein. Thus, peptides with more than one carboxyl group (i.e., peptides containing aspartate or glutamate) are not recommended for EDAC conjugation. Furthermore, the peptide preparations should be quite free of residual acetic acid and trifluoroacetic acid from synthesis, cleavage, and work-up.

Although it is generally not as effective, a two-vessel reaction is also possible, in which the carrier donates the carboxyl group, and the peptide provides the reactive amino group. Note that peptides with more than one amino group (i.e., those peptides bearing lysine residues) may react at more than one site, possibly affecting antigenicity. Also, the reaction should be performed on ice to help minimize rearrangement of the *O*-acyl iso urea intermediate to an unreactive *N*-acyl urea *(1)*.

1.1. Indications

The carbodiimide method is versatile, and creates a well-characterized and stable amide bond. It introduces no linker, and so minimizes generation of neo-antigens. However, the method is fairly susceptible to formation of intramolecular cross-links, in both the peptide and carrier protein. This may reduce the efficiency of the conjugation reaction and the solubility of the product, although it does not seem to reduce immunogenicity of the conjugate when used as an experimental vaccine.

A detailed discussion of peptide conjugation can be found in the introduction to Chapter 117.

2. Materials

1. EDAC, 25 mg: The carbodiimide reagent must be used fresh or stored desiccated and frozen.
2. Peptide: 5 mg, bearing one (or possibly two) free carboxyl moieties available for reaction. Check to confirm adequate solubility in coupling solvent.

From: The Protein Protocols Handbook
Edited by: J. M. Walker Humana Press Inc., Totowa, NJ

3. Carrier protein, e.g., keyhole limpet hemocyanin (KLH) or ovalbumin 5 mg.
4. Coupling solvent: PBS. Alternatively, use phosphate buffer, pH 7.0–8.0, without saline, or borate, pH 7.0–8.0. Avoid Tris and other amine-containing species, and similarly avoid using buffers containing carboxylates (such as acetate). An excellent buffering solvent for the two-step reaction is 20 m*M* TES, pH 6.5. Also, you may add a little DMF to improve solubility of the peptide, but be careful not to precipitate the carrier protein.
5. Optional: glacial acetic acid, dialysis tubing.

3. Method

This procedure is adapted from Bauminger and Wilchek (1).

1. For the one-step reaction, dissolve 5 mg peptide in 1 mL of solvent. Very carefully adjust the pH to 5 by the addition of 0.1*N* HCl. While stirring gently with a magnetic stir bar, add 5 mg EDAC.
2. Continue stirring, allowing the reaction to proceed for 30–60 min on ice.
3. Dissolve 5 mg of carrier protein in 1 mL of solvent, and add it to the peptide solution. Adjust the pH to 7.0–8.0 by addition of 0.1*N* NaOH.
4. Continue stirring, and allow the reaction to proceed for 1 hour on ice, and then for another hour at room temperature.
5. For the two-vessel protocol, dissolve 5 mg of carrier protein in 1 mL of solvent. Adjust the pH to 5 by addition of 0.1*N* HCl. Add 5 mg EDAC, stirring gently for 30 min on ice with a magnetic stir bar.
6. Add a drop or two of acetic acid to quench the reaction, and then dialyze vs coupling solvent overnight.
7. Dissolve 5 mg peptide in 1 mL solvent, and then combine the peptide solution with the solution of activated carrier. Allow to react for 1 h on ice, and then for another hour at room temperature.
8. Residual unreacted EDAC undergoes hydrolysis fairly rapidly, forming a urea with presumed low acute toxicity. Thus, the conjugate may immediately be used for experimental vaccines without purification to remove free peptide. However, for assays, it may be necessary to remove unreacted peptide from the conjugate mixture. As with other conjugation techniques, such a separation may be effected by simple dialysis or rapid gel filtration.
9. For dialysis, a single overnight dialysis vs 4 L PBS (or other buffer) at 4°C is usually sufficient. For gel filtration, use at least 40-mL bed volume column packed with Sephadex G-25 (equilibrated in PBS or other buffer), so that the conjugate will elute in the void volume.

4. Note

1. No matter your purpose, before you actually use the conjugate, it is prudent to characterize it, determining the number of peptide molecules bound per carrier molecule (*see* Chapter 117, Section 1.8.1).

Reference

1. Bauminger, S. and Wilcheck, M. (1980) The use of carbodiimides in the preparation of immunizing conjugates. *Methods Enzymol.* **70,** 151–159.

120

Raising of Polyclonal Antisera

Graham S. Bailey

1. Introduction

Suitable antisera are essential for use in all immunochemical procedures. Three important properties of an antiserum are avidity, specificity, and titer. The avidity of an antiserum is a measure of the strength of the interactions of its antibodies with an antigen. The specificity of an antiserum is a measure of the ability of its antibodies to distinguish the immunogen from related antigens. The titer of an antiserum is the final (optimal) dilution at which it is employed in the procedure; it depends on the concentrations of the antibodies present and on their affinities for the antigen. The values of those parameters required for a particular antiserum very much depend on the usage to which the antiserum will be put. For example, for use in radioimmunoassay, it is best to have a monospecific antiserum of high avidity, whereas for use in immunoaffinity chromatography, the monospecific antiserum should not possess too high an avidity. Otherwise, it may prove impossible to elute the desired antigen without extensive denaturation.

A substance that, when injected into a suitable animal, gives rise to an immune response is called an immunogen. The immunogenicity of a substance is dependent on many factors, such as on its size, shape, chemical composition, and structural difference from any related molecular species indigenous to the injected animal. Normally, cellular (particulate) materials are very immunogenic and induce a rapid immune response. However, the resultant antisera do not usually possess a high degree of specificity and do not store well *(1)*.

Soluble immunogens differ widely in their ability to produce an immune response. In general, polypeptides and proteins of molecular weight above about 5000 and certain large polysaccharides can be effective immunogens. Smaller molecules, such as peptides, oligosaccharides, and steroids, can often be rendered immunogenic by chemical coupling to a protein that by itself will produce an immune response (and *see* Chapters 117–119). For most situations, it is best to use the most highly purified sample available for injecting into the animal. Furthermore, it is usual to inject a mixture of the potential immunogen and an adjuvant that will stimulate antibody production.

Classically, the immune response is described as occurring in two phases. Initial administration of the immunogen induces the primary phase (response) during which only small amounts of antibody molecules are produced as the antibody-producing

From: The Protein Protocols Handbook
Edited by: J. M. Walker Humana Press Inc., Totowa, NJ

system is primed. Further administration of the immunogen results in the secondary phase (response) during which large amounts of antibody molecules are produced by the large number of specifically programmed lymphocytes. In practice, the time scale and nature of immunization process do not lead to a clear recognition of two distinct phases.

All of the factors that influence antibody production have not yet been elucidated. Thus, the raising of a polyclonal antiserum is, to some extent, a hit or miss affair. Individual animals can respond quite differently to the same process of immunization, and thus, it is best to use a number of animals. The species of animal chosen for immunization will depend on particular circumstances, but, in general, rabbits are often used.

Many different methods of producing polyclonal antisera have been described, varying in amount of immunogen required, route of injection, and frequency of injection *(2)*. Monoclonal antibodies (MAb) can be used in place of polyclonal antisera in immunological assays. However, MAb are expensive to produce, and their use in assays employing labeled ligand is reported to offer little, if any, advantage over polyclonal antisera *(3)*. This chapter will describe a method of antiserum production *(4)* that has been successfully utilized in the author's laboratory using small doses (microgram quantities) of soluble protein as immunogen *(5)*.

2. Materials

2.1. Raising the Antiserum

1. Six rabbits each of 2 kg body wt. Various types can be used, e.g., New Zealand whites, Dutch, and so forth.
2. Solution (40–400 µg/mL) of the purified immunogen in a buffer of a pH at which the immunogen is stable.
3. Complete and incomplete Freund's adjuvant, available from various commercial sources.
4. Heat lamp.

3. Method

The following procedure is carried out for each rabbit in turn (*see* Note 1).

1. Thoroughly mix 1 vol of the immunogen solution with 3 vol of complete Freund's adjuvant with the aid of a glass pestle and mortar. Initially, the mixture is very viscous, but after about 5 min, the viscosity becomes less. The mixture can then be transferred to a glass syringe. The syringe is emptied and refilled with the mixture a number of times, resulting in the formation of a stable emulsion that can be injected intradermally into the prepared rabbit. The emulsion should be used within 1 h of preparation.
2. The rabbit is prepared by cutting away the long hair along the center of its back. The short hair is removed by shaving. The emulsion of immunogen and complete Freund's adjuvant is injected via a 1-mL glass syringe plus 21-gage needle into two rows of five sites equidistantly spaced along the rabbit's back, each row being about 2 cm from the backbone, such that each site receives 0.1 mL of emulsion containing 1–10 µg immunogen, i.e., a total dose of 10–100 µg immunogen/rabbit (*see* Note 2).
3. After 8–10 wk, a test bleeding is carried out on the rabbit. The fur on the back of one ear is removed by shaving. The eyes of the rabbit are protected while the shaven ear is heated for <1 min with a heat lamp to expand the vein. The expanded vein just below the surface of the back of the ear is nicked with a scalpel blade. Blood is collected into a glass vessel (up to 20 mL can be collected in 10 min) removing the clot from the puncture wound by

rubbing the ear with cotton wool from time to time. The blood is allowed to clot standing at room temperature for a few hours. The serum is separated from the clot by centrifugation and can be stored at 4°C in the presence of 0.1% sodium azide or 0.01% thermerosal as antibacterial agent until tested (*see* Notes 3 and 4).

4. If on testing (*see* step 10) the serum shows the characteristics required for its particular usage, then further bleeding can be carried out. Up to three bleedings can be made on successive days. After that, it is best to allow the rabbit to rest for about a month before further bleeding.
5. If the original antiserum is unsatisfactory or if the quality of the bleedings taken over a period of several weeks or months starts to become unsatisfactory, then the rabbit can be boostered, i.e., receive a second injection of immunogen.
6. For the booster injection, half the original dose of immunogen is administered in incomplete Freund's adjuvant. The emulsion is prepared as detailed before, but is injected subcutaneously into the rabbit (for example, into the fold of skin of the neck).
7. After 10 d, a test bleeding is obtained from the rabbit, and the serum so produced is analyzed for the presence of antibodies using Ouchterlony double immunodiffusion (*see* Chapter 135).
8. Further bleedings (*see* step 4) can be carried out over a period of several months if the boostered antiserum is satisfactory.
9. If after boostering, the antiserum is not of the desired titer or avidity (*see* step 10), it is best to disregard that rabbit. Hopefully, one or more of the other rabbits in the group will have produced good antisera either directly or after boostering. However, in some cases, particularly with weak immunogens, it may be necessary to repeat the process of immunization with a new group of rabbits or other animals.
10. Each sample of antiserum can be tested for its ability to form an immune precipitate with the immunogen by carrying out Ouchterlony double immunodiffusion (*see* Chapter 135) (*see* Note 5). The titer and a measure of the avidity of the antiserum can be obtained by radioimmunoassay *(6)*. The specificity of the antibodies can be determined by running the antiserum against the immunogen and related antigens in Ouchterlony double immunodiffusion.

4. Notes

1. Animals other than rabbits can be used for immunization, e.g., mice, rats, guinea pigs, sheep, goat, and so on. The rabbit is often a good initial choice, but if results are unsatisfactory, other species can be tried. Obviously, only small volumes of the antisera can be generated in the smaller species, whereas large volumes can be obtained from the larger species. The latter though do require more immunogen and are more expensive to maintain.
2. The dose of immunogen employed can be of crucial importance in many procedures for antibody production. A state of tolerance can be induced in the animal with little or no production of antibody if too much or too little immunogen is repeatedly given over a relatively short period of time. The method described in this chapter should not suffer from that effect, since there is a gap of at least 10-wk between the initial and booster injections. In general, the lower the dose of antigen, the greater is the avidity of the antiserum.
3. A clean sample of antiserum should be straw-colored. Pink coloration is owing to partial hemolysis, but should not affect the properties of the antiserum.
4. Many antisera can be satisfactorily stored at 4°C in the presence of an antibacterial agent for many months. After some time, the antiserum solution may become turbid and even contain a precipitate (mostly of denatured lipoprotein). Even so, there should be no significant reduction in the quality of the antiserum. If necessary, the solution can be clari-

fied by membrane filtration. For prolonged storage, the antiserum can be kept at –20°C in small quantities so as to avoid repeated thawing and freezing.

5. Since the antisera produced by conventional methods (polyclonal antisera) consist of mixtures of different antibody molecules, it is to be expected that the properties of the antisera collected during the prolonged period of immunization may change. Thus, each bleeding should be tested for specificity, titer, and avidity.

References

1. Hurn, B. A. L. and Chantler, S. M. (1980) in *Production of Reagent Antibodies in Methods in Enzymology,* vol. 70 (Van Vunakis, H. and Langone, J. J., eds.), Academic, New York, pp. 104–141.
2. Dresser, D. W. (1986) Immunization of experimental animals, in *Handbook of Experimental Immunology,* vol. 1 (Weir, D. M., Herzenberg, L. A., Blackwell, C., and Herzenberg, L. A., eds.), 4th ed. Blackwell, Oxford, UK, pp. 8.1–8.21.
3. Chard, T. (1987) Requirements for binding assays–antibodies and other binders, in *Laboratory Techniques in Biochemistry and Molecular Biology,* vol. 6, part 2, 3rd rev. ed. Elsevier, Amsterdam, pp. 88–110.
4. Vaitukaitis, J. L. (1981) Production of antisera with small doses of immunogen: multiple intradermal injections, in *Methods in Enzymology,* vol. 73 (Langone, J. J. and Van Vunakis, H., eds.), Academic, New York, pp. 46–57.
5. Al-Hamidi, A. A. A. and Bailey, G. S. (1991) Purification of prokallikrein from bovine pancreas. *Biochim. Biophys. Acta* **1075,** 88–92.
6. Bailey, G. S. (1994) Radioimmunoassay of peptides and proteins, in *Methods in Molecular Biology*, vol. 32 (Walker, J. M., ed.), Humana, Totowa, NJ, pp. 449–459.

121

Elution of SDS-PAGE Separated Proteins from Immobilon Membranes for Use as Antigens

Donald F. Summers and Boguslaw Szewczyk

1. Introduction

The great analytical power of sodium dodecyl sulfate-polyacrylamide gel electrophoresis (SDS-PAGE) makes it one of the most effective tools of protein chemistry and molecular biology (*see* Chapter 11). In the past, there have been many attempts to convert the technique from analytical to preparative scale because, by SDS-PAGE, one can resolve more than one hundred protein species in 5–6 h. The number of papers that describe preparative elution from polyacrylamide gels is immense (e.g., *see* refs. *1–5*). In spite of the numerous variations in the procedure of elution, none of the available methods is entirely satisfactory. Some of the methods are very laborious, and others lead to loss of resolution or poor recovery.

In general, the elution of proteins above 100 kDa from polyacrylamide gels always presents considerable problems. Another of the serious limitations of elution from gels is owing to the elastic nature of preparative polyacrylamide gels. The precise excision of a protein band from a complex mixture is difficult and the slice may contain portions of other protein bands located close to the band of interest. To overcome some of the limitations of elution from gels, Parekh et al. *(6)* and Anderson *(7)* attempted to elute proteins from nitrocellulose replicas of SDS-PAGE gels. Binding of proteins to nitrocellulose is, however, so strong that the dissociating reagents (acetonitrile, pyridine) partly or completely dissolve the membrane. When such preparations are used for immunization, they may cause adverse effects in animals. We have found that when a polyacrylamide gel replica is made on Immobilon membrane and not on nitrocellulose, then the conditions for elution are much milder. Often, there is no need for concentration of the sample or for the removal of elution agents prior to immunization. Furthermore, elution from Immobilon is nearly independent of protein mol wt and recoveries of 70–90% are routinely obtained. We have also shown that, following elution using the technique described herein plus the use of *E. coli* thioredoxin to catalyze protein renaturation, one can recover significant enzymatic activity for some large complex enzymes such as *E. coli* RNA polymerase *(8)* and influenza A virus RNA polymerase *(9)*. Proteins are first separated by SDS-PAGE, and then are electroblotted to Immobilon membranes and stained with amido black or Ponceau S. The protein bands of interest are excised and are then eluted from the membrane with detergent-containing buffers at pH 9.5.

From: The Protein Protocols Handbook
Edited by: J. M. Walker Humana Press Inc., Totowa, NJ

2. Materials

1. Protein stains:
   a. 0.01% Amido black in water.
   b. 0.5% Ponceau S in 1% acetic acid.
2. Elution buffers:
   a. 1% Triton X-100 in 50 m*M* Tris-HCl, pH 9.5.
   b. 1% Triton X-100, 2% SDS in 50 m*M* Tris-HCl, pH 9.5.
3. Immobilon (polyvinylidene fluoride) membrane from Millipore Corp. (Bedford, MA); Whatman 3MM (Maidstone, UK) filter paper; Scotch Brite pads.
4. SDS-PAGE apparatus.
5. Transfer apparatus (e.g., Trans Blot Cell from Bio-Rad Laboratories, Richmond, CA).
6. Glass vessels with flat bottom (e.g., Pyrex baking dishes)
7. Rocker platform.
8. Microfuge.
9. Small dissecting scissors.

3. Method

1. Apply a mixture of proteins containing immunogen to be purified (*see* Notes 1 and 2) to an SDS-PAGE gel and run the gel.
2. Electroblot the gel onto a PVDF membrane using one of the methods in Chapters 37 and 38. Note that it is not necessary to use methanol in the transfer buffer as it does not improve the binding of proteins to this membrane (*see* Note 3).
3. After transfer, stain the membrane with amido black solution for 20–30 min or with Ponceau S for 5 min (*see* Notes 4 and 5).
4. Destain the membrane with distilled water.
5. Excise the band(s) of interest with small dissecting scissors and place it in an Eppendorf tube (*see* Note 6).
6. Add 0.2–0.5 mL of elution buffer/cm^2 of Immobilon strip. Two buffers that we used are:
   a. 1% Triton X-100 in 50 m*M* Tris-HCl, pH 9.5.
   b. 2% SDS/1% Triton in 50 m*M* Tris-HCl, pH 9.5.

   The first buffer is less effective (50–75% of total protein eluted) than the second one, but the eluted protein can be injected into animals without the necessity of Triton X-100 removal. On the other hand, the 2% SDS/1% Triton X-100 mixture leads to the complete elution of bound protein from Immobilon, but SDS has to be removed before injections (*see* Notes 7–9).
7. Mix well by vortexing the Immobilon in eluant for 10 min. Spin down (5 min) the Immobilon. Use the supernatant directly for injections (elution with Triton X-100 only) or after protein precipitation with acetone (if SDS and Triton were included in the elution buffer). Protein precipitation is carried out in a dry ice bath. Add 4 vol of cold acetone to 1 vol of protein solution. After 2 h at –20°C, pellet the protein, solubilize, and inject into animals by standard procedures (*10* and *see* Chapter 120) (*see* Note 10).

4. Notes

1. The method has been used to obtain a variety of sera against bacterial, viral, and eukaryotic proteins. The amount of immunogen needed to stimulate high levels of antibodies varies for different proteins, but generally, 50–500 μg of protein is sufficient to induce the formation of high levels of specific antibodies.
2. The Immobilon matrix should not be overloaded with protein to prevent its deep penetration into the membrane. The protein band excised from a single electrophoretic lane (about 1 cm in length) should not contain more than 10–20 μg of protein.

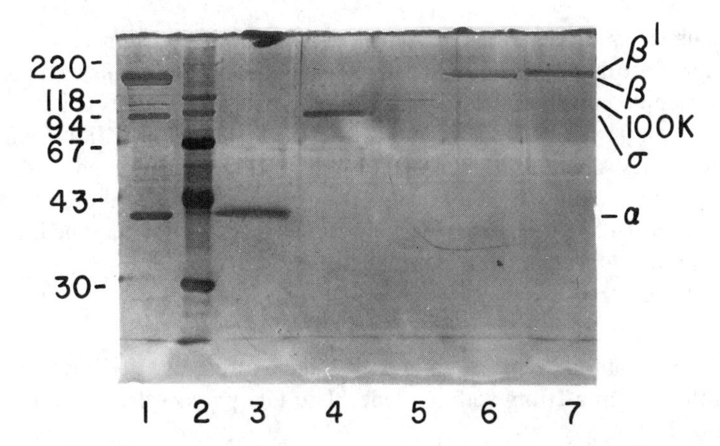

Fig. 1. SDS-PAGE pattern of *Escherichia coli* RNA polymerase subunits eluted from Immobilon membrane. *E. coli* RNA polymerase (1 µg) was resolved by SDS-PAGE and transferred to an Immobilon membrane. The proteins on the membrane were stained with amido black in water; the polymerase subunits were excised and eluted from the membrane with Triton X-100 at pH 9.0. After precipitation with acetone, the subunits were subjected to analytical SDS-PAGE and silver-stained to ascertain their purity. Lane 1, original preparation of *E. coli* RNA polymerase. Lane 2, mixture of Sigma high-mol-wt protein calibration standards; their molecular weights × 10^{-3} are given at the left-hand side of the figure. Lanes 3–7, individual subunits of the polymerase eluted from the Immobilon membrane; their designations are given at the right-hand side of the figure (100 K protein is not a constituent of the polymerase complex but, it is present in commercial preparations of the enzyme).

3. Transfer of proteins from the gel to Immobilon should not be done at elevated temperatures (above 30°C), as the force of protein binding to Immobilon apparently increases with temperature. Therefore, it is advisable to make transfers in a cold room at 4°C or use precooled transfer buffer.
4. Depending on the supplier and batch of amido black, the sensitivity of protein detection with this reagent may vary. If the sensitivity of staining is not satisfactory, it is advisable to dilute the amido black solution 5–10 times with water rather than to increase its concentration.
5. Staining with Ponceau is done in 1% acetic acid. This may lead to partial denaturation of proteins bound to Immobilon. In this case, 2% SDS/1% Triton X-100 in 50 m*M* Tris-HCl, pH 9.5 should be used as the elution buffer.
6. The method described here allows for very precise excision of protein bands from the Immobilon matrix. As an example, the elution of *E. coli* RNA polymerase subunits is shown in Fig. 1. Such precise excision of protein bands is much more difficult when they are cut out of polyacrylamide gels.
7. The elution from Immobilon is strictly pH-dependent. At pH 7.0, there is practically no elution; the maximum efficiency of elution is reached at pH 8.5–9.5 *(8)*.
8. The efficiency of elution is only slightly dependent on the mol wt of the protein. In our hands, a protein of 200 kDa was eluted with only 10% lower efficiency than a protein of 70 kDa when elution was performed in the buffer with Triton X-100 as the eluting agent.
9. Proteins that are insoluble in standard aqueous solutions and are solubilized with SDS before subjecting to electrophoresis may sometimes require special treatment. For example, *E. coli* β-galactosidase (mol wt about 120 K) can be readily eluted from Immobilon

membranes under mild conditions. However, β-galactosidase fusion proteins with short segments of some viral proteins are insoluble in aqueous salt solutions, and must be solubilized with sample buffer for SDS-PAGE. In this case, the proteins can be eluted from the Immobilon only by 2% SDS/1% Triton X-100 in 50 mM Tris-HCl, pH 9.5.

10. Probably the easiest method to obtain emulsions of protein in Freund's adjuvant is by subjecting the mixture placed in an Eppendorf microfuge tube to a short ultrasonic treatment (3–4 pulses 10–20 s each time) in an ultrasonic disintegrator equipped with a microprobe (end diameter of the probe around 1/8 in.).

References

1. Tuszynski, N. Y., Damsky, C. H., Fuhrer, J. P., and Warren, L. (1977) Recovery of concentrated protein samples from sodium dodecyl sulfate-polyacrylamide gels. *Anal. Biochem.* **83,** 119–129.
2. Nguyen, N. Y., DiFonzo, J., and Chrambach, A. (1980) Protein recovery from gel slices by steady-state stacking: an apparatus for the simultaneous extraction and concentration of ten samples. *Anal. Biochem.* **106,** 78–91.
3. Hager, D. A. and Burgess, R. R. (1980) Elution of proteins from sodium dodecyl sulfate-polyacrylamide gels, removal of sodium dodecyl sulfate, and renaturation of enzymatic activity: results with sigma subunit of *Escherichia coli* RNA polymerase, wheat germ DNA topoisomerase, and other enzymes. *Anal. Biochem.* **109,** 76–86.
4. Stralfors, P. and Belfrage, P. (1983) Electrophoretic elution of proteins from polyacrylamide gel slices. *Anal. Biochem.* **128,** 7–10.
5. Hunkapiller, M. W., Lujan, E., Ostrander, F., and Hood, L. E. (1983) Isolation of microgram quantities of proteins from polyacrylamide gels for amino acid sequence analysis. *Methods Enzymol.* **91,** 227–236.
6. Parekh, B. S., Mehta, H. B., West, M. D., and Montelaro, R. C. (1985) Preparative elution of proteins from nitrocellulose membranes after separation by sodium dodecyl sulfate-polyacrylamide gel electrophoresis. *Anal. Biochem.* **148,** 87–92.
7. Anderson, P. J. (1985) The recovery of nitrocellulose-bound protein. *Anal. Biochem.* **148,** 105–110.
8. Szewczyk, B. and Summers, D. F. (1988) Preparative elution of proteins blotted to immobilon membranes. *Anal. Biochem.* **168,** 48–53.
9. Szewczyk, B., Laver, W. G., and Summers, D. F. (1988) Purification, thioredoxin renaturation, and reconstituted activity of the three subunits of the influenza A virus RNA polymerase. *Proc. Natl. Acad. Sci. USA* **85,** 7907–7911.
10. Manson, M. (ed.) (1992) *Methods in Molecular Biology, vol. 10: Immunochemical Protocols.* Humana, Totowa, NJ.

122

Production of Highly Specific Polyclonal Antibodies Using a Combination of 2D Electrophoresis and Nitrocellulose-Bound Antigen

Monique Diano and André Le Bivic

1. Introduction

Highly specific antibodies directed against minor proteins, present in small amounts in biological fluids, or against nonsoluble cytoplasmic or membraneous proteins, are often difficult to obtain. The main reasons for this are the small amounts of protein available after the various classical purification processes and the low purity of the proteins.

In general, a crude or partially purified extract is electrophoresed on an SDS polyacrylamide (SDS-PAGE) gel; then the protein band is lightly stained and cut out. In the simplest method, the acrylamide gel band is reduced to a pulp, mixed with Freund's adjuvant, and injected. Unfortunately, this technique is not always successful. Its failure can probably be attributed to factors such as the difficulty of disaggregating the acrylamide, the difficulty with which the protein diffuses from the gel, the presence of SDS in large quantities resulting in extensive tissue and cell damage, and finally, the toxicity of the acrylamide.

An alternative technique is to extract and concentrate the proteins from the gel by electroelution (*see* Chapter 32), but this can lead to loss of material and low amounts of purified protein (*see* Chapter 32).

Another technique is to transfer the separated protein from an SDS-PAGE gel to nitrocellulose. The protein-bearing nitrocellulose can be solubilized with dimethyl sulfoxide (DMSO), mixed with Freund's adjuvant, and injected into a rabbit. However, although rabbits readily tolerate DMSO, mice do not, thus making this method unsuitable for raising monoclonal antibodies.

The monoclonal approach has been considered as the best technique for raising highly specific antibodies, starting from a crude or partially purified immunogen. However, experiments have regularly demonstrated that the use of highly heterogenous material for immunization never results in the isolation of clones producing antibodies directed against all the components of the mixture. Moreover, the restricted specificity of a monoclonal antibody that usually binds to a single epitope of the antigenic molecule is not always an advantage. For example, if the epitope is altered or modified (i.e., by fixative, Lowicryl embedding, or detergent), the binding of the monoclonal antibody might be compromised, or even abolished.

From: The Protein Protocols Handbook
Edited by: J. M. Walker Humana Press Inc., Totowa, NJ

Because conventional polyclonal antisera are complex mixtures of a considerable number of clonal products, they are capable of binding to multiple antigenic determinants. Thus, the binding of polyclonal antisera is usually not altered by slight denaturation, structural changes, or microheterogeneity, making them suitable for a wide range of applications. However, to be effective, a polyclonal antiserum must be of the highest specificity and free of irrelevant antibodies directed against contaminating proteins, copurified with the protein of interest and/or the proteins of the bacterial cell wall present in the Freund's adjuvant. In some cases, the background generated by such irrelevant antibodies severely limits the use of polyclonal antibodies.

A simple technique for raising highly specific polyclonal antisera against minor or insoluble proteins would be of considerable value.

Here, we describe a method for producing polyclonal antibodies, which avoids both prolonged purification of antigenic proteins (with possible proteolytic degradation) and the addition of Freund's adjuvant and DMSO. Two-dimensional gel electrophoresis leads to the purification of the chosen protein in one single, short step. The resolution of this technique results in a very pure antigen, and consequently, in a very high specificity of the antibody obtained. It is a simple, rapid, and reproducible technique. 2D electrophoresis with ampholines for the IEF is still considered by most as time consuming and technically demanding. The introduction, however, of precast horizontal Ampholines or Immobilines IEF gels has greatly reduced this objection. Moreover, the development of 2D gel databases and the possibility to link it to DNA databases (Swiss-2D PAGE database *[1]*) further increases the value of using 2D electrophoresis to purify proteins of interest.

A polyclonal antibody, which by nature cannot be monospecific, can, if its titer is very high, behave like a monospecific antibody in comparison with the low titers of irrelevant antibodies in the same serum. Thus, this method is faster and performs better than other polyclonal antibody techniques while retaining all the advantages of polyclonal antibodies.

2. Materials

1. For 2D gels, materials are those described by O'Farrell *(2,3)* and Laemmli *(4)* (*see* Chapters 20–22). It should be noted that for IEF, acrylamide and *bis*-acrylamide must be of the highest level of purity, and urea must be ultrapure (enzyme grade).
2. Ampholines with an appropriate pH range, e.g., 5–8 or 3–9.
3. Transfer membranes: 0.45-μm BA 85 nitrocellulose membrane filters (from Schleicher and Schüll GmBH, Kassel, Germany); 0.22-μm membranes can be used for low-mol-wt antigenic proteins.
4. Transfer buffer: 20% Methanol, 150 m*M* glycine, and 20 m*M* Tris base, pH 8.3.
5. Phosphate buffered saline (PBS), sterilized by passage through a 0.22-μm filter.
6. Ponceau Red: 0.2% in 3% trichloroacetic acid.
7. Small scissors.
8. Sterile blood-collecting tubes, with 0.1*M* sodium citrate, pH 6, at a final concentration of 3.2%.
9. Ultrasonication apparatus, with 100 W minimum output. We used a 100-W ultrasonic disintegrator with a titanium exponential microprobe with a tip diameter of 3 mm (1/8 in.). The nominal frequency of the system is 20 kc/s, and the amplitude used is 12 μ.

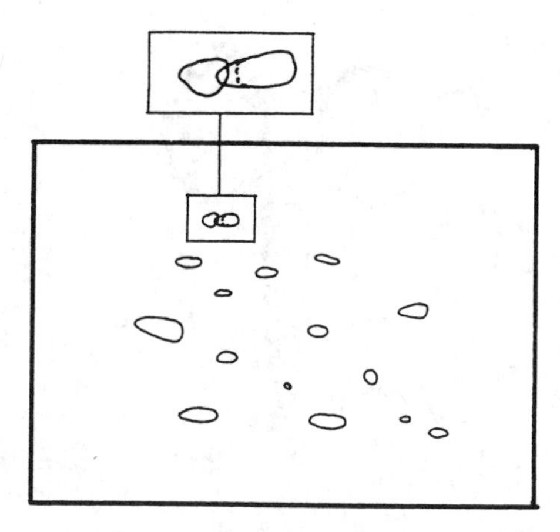

Fig. 1. Excision of the spot containing the antigen. Cut inside the circumference, for instance, along the dotted line for the right spot.

3. Method

This is an immunization method in which nitrocellulose-bound protein purified by 2D electrophoresis is employed and in which *neither DMSO nor Freund's adjuvant* are used, in contrast to the method described by Knudsen *(5)*. It is equally applicable for soluble and membrane proteins.

3.1. Purification of Antigen

Briefly, subcellular fractionation of the tissue is carried out to obtain the fraction containing the protein of interest. This is then subjected to separation in the first dimension by IEF with Ampholines according to O'Farrell's technique or by covalently bound Immobilines gradients. Immobilines allow a better resolution by generating narrow pH gradients (<1 pH unit), a larger loading capacity, and a greater tolerance to salt and buffer concentrations. At this point, it is important to obtain complete solubilization of the protein (*see* Note 1).

Separation in the second dimension is achieved by using an SDS polyacrylamide gradient gel (*see* Chapters 20–22 and refs. *2* and *4*; *see also* Note 2).

The proteins are then transferred from the gel to nitrocellulose (*see* Chapters 37–39 and ref. *6*). It is important to work with gloves when handling nitrocellulose to avoid contamination with skin keratins.

3.2. Preparation of Antigen for Immunization

1. Immerse the nitrocellulose sheet in Ponceau red solution for 1–2 min, until deep staining is obtained, then destain the sheet slightly in running distilled water for easier detection of the spots. *Never let the nitrocellulose dry out.*
2. Carefully excise the spot corresponding to the antigenic protein. Excise inside the circumference of the spot to avoid contamination by contiguous proteins (*see* Fig. 1).
3. Immerse the nitrocellulose spot in PBS in an Eppendorf tube (1-mL size). The PBS bath should be repeated several times until the nitrocellulose is thoroughly destained. The last bath should have a volume of about 0.5 mL, adequate for the next step.
4. Cut the nitrocellulose into very small pieces with scissors. Then rinse the scissors into the tube with PBS to avoid any loss (*see* Fig. 2).

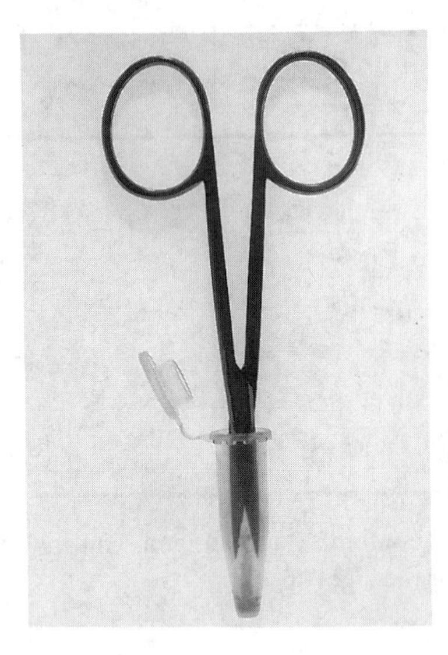

Fig. 2. Maceration of nitrocellulose.

5. Macerate the nitrocellulose suspension by sonication. The volume of PBS must be proportional to the surface of nitrocellulose to be sonicated. For example, 70–80 µL of PBS is adequate for about 0.4 cm^2 of nitrocellulose (*see* Notes 3 and 4).
6. After sonication, add about 1 mL of PBS to the nitrocellulose powder to dilute the mixture, and aliquot it in 500, 350, and 250-µL fractions and freeze these fractions at –80°C until use. Under these storage conditions, the aliquots may be used for immunization for up to one year or, may be longer. Never store the nitrocellulose without buffer. Never use sodium azide because of its toxicity.

3.3. Immunization

1. Shave the backs of the rabbits. Routinely inject two rabbits with the same antigen.
2. Thaw the 500-µL fraction for the first immunization and add 1.5–2 mL of PBS to reduce the concentration of nitrocellulose powder.
3. Inoculate the antigen, according to Vaitukaitis *(7),* into 20 or more sites (Vaitukaitis injects at up to 40 sites). Inject subcutaneously, holding the skin between the thumb and forefinger. Inject between 50 and 100 µL—a light swelling appears at the site of injection. As the needle is withdrawn, compress the skin gently. An 18-g hypodermic needle is routinely used, though a finer needle (e.g., 20- or 22-g) may also be used (*see* Note 5). Care should be taken over the last injection; generally, a little powder remains in the head of the needle. Thus, after the last injection, draw up 1 mL of PBS to rinse the needle, resuspend the remaining powder in the syringe, and position the syringe vertically to inject.
4. Three or four weeks after the first immunization, the first booster inoculation is given in the same way. The amount of protein injected is generally less, corresponding to two-thirds of that of the first immunization.
5. Ten days after the second immunization, bleed the rabbit (*see* Note 6). A few milliliters of blood suffice, i.e., enough to check the immune response against a crude preparation of the injected antigenic protein. The antigen is revealed on a Western blot with the specific serum diluted at 1:500 and a horseradish peroxidase-conjugated second antibody. We used 3,3'-Diaminobenzidine tetrahydrochloride (DAB) for color development of peroxidase

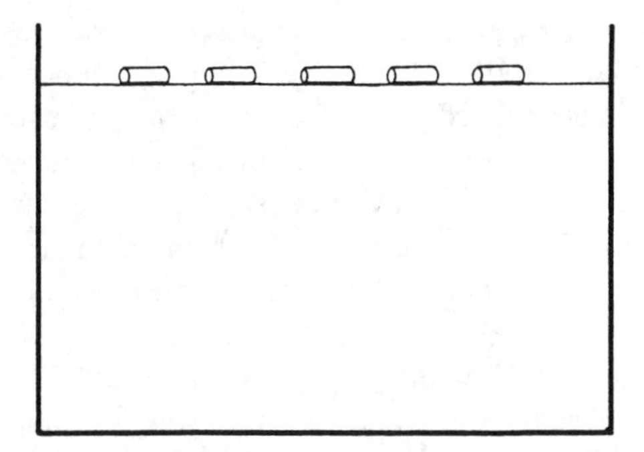

Fig. 3. Second dimension with several IEF gels. Several IEF gels are cut, 0.5 cm above and 0.5 cm below the isoelectric point of the protein of interest. They are placed side by side at the top of the second dimension slab gel. Thus, only one SDS gel is needed to collect several spots of interest.

activity. (*See* Chapter 48) If the protein is highly antigenic, the beginning of the immunological response is detectable.

6. Two weeks after the second immunization, administer a third immunization in the same way as the first two, even if a positive response has been detected. If there was a positive response after the second immunization, one-half of the amount of protein used for the original immunization is sufficient.
7. Bleed the rabbits 10 days after the third immunization and every week thereafter. At each bleeding, check the serum as after the second immunization; but the serum should be diluted at 1:4000 or 1:6000. Bleeding can be continued for as long as the antibody titer remains high (*see* Note 7).

   Another booster should be given when the antibody titer begins to decrease if it is necessary to obtain a very large volume of antiserum (*see* Note 7).
8. After bleeding, keep the blood at room temperature until it clots. Then collect the serum and centrifuge for 10 min at 3000*g* to eliminate microclots and lipids. Store aliquots at –22°C.

4. Notes

1. To ensure solubilization, the following techniques are useful:
   a. The concentration of urea in the mixture should be 9–9.5*M,* i.e., close to saturation.
   b. The protein mixture should be frozen and thawed at least six times. Ampholines should be added only after the last thawing because freezing renders them inoperative.
   c. If the antigenic protein is very basic and outside the pH range of the ampholines, it is always possible to carry out NEPHGE (nonequilibrium pH gradient electrophoresis) for the first dimension *(8)*.
   d. A significant improvement using horizontal IEF is the possibility to load the sample at any place on the gel allowing to carry out in a single experiment both NEPHGE and IEF.
2. If the antigenic protein is present in small amounts in the homogenate, it is possible to save time by cutting out the part of the IEF gel where the protein is located and depositing several pieces of the first-dimension gel side by side on the second-dimension gel slab (*see* Fig. 3).
3. Careful attention should be paid to temperature during preparation of the antigen; always work between 2 and 4°C. Be particularly careful during sonication; wait 2–3 min between consecutive sonications.

4. This is a crucial point in the procedure. If too much PBS is added, the pieces of nitrocellulose will swirl around the probe and disintegration does not occur. In this case, the nitrocellulose pieces should be allowed to settle to the bottom of the tube before sonication and the excess buffer drawn off with a syringe or other suitable instrument (70–80 µL of PBS is sufficient for about 0.4 cm^2 of nitrocellulose). For these quantities, one or two 10-s cycles suffice to get powdered nitrocellulose. We mention the volume as a reference since the surface of nitrocellulose-bound antigen may vary. In every case the volume of PBS must be adjusted.
5. What is an appropriate amount of antigenic protein to inject? There is no absolute answer to this question. It depends both on the molecular weight of the protein, and also on its antigenicity. It is well known that if the amount of antigen is very low (0.5–1 µg for the classic method with Freund's adjuvant), there is no antibody production; if the amount of antigen is very high (e.g., several hundred micrograms of highly antigenic protein), antibody production might also be blocked.

   It would appear that in our method, a lower amount of antigen is required for immunization; the nitrocellulose acts as if it progressively releases the bound protein, and thus, the entire amount of protein is used progressively by the cellular machinery.

   Our experiments show that a range of 10–40 µg for the first immunization generally gives good results, although, in some cases, 5 µg of material is sufficient. The nitrocellulose powder has the additional effect of triggering an inflammatory process at the sites of injection, thus enhancing the immune response, as does Freund's adjuvant by means of the emulsion of the antigenic protein with the tubercular bacillus; macrophages abound in the inflamed areas.
6. It is perhaps worth noting that careful attention should be paid to the condition of the rabbit at time of bleeding. We bleed the rabbits at the lateral ear artery. When the rabbit is calm, 80–100 mL of blood may be taken. The best time for bleeding is early in the morning, and after the rabbit has drunk. Under these conditions, the serum is very clear. It is essential to operate in a quiet atmosphere. If the rabbit is nervous or under stress, the arteries are constricted so strongly that only a few drops of blood can be obtained. Note that to avoid clotting, the needle is impregnated with a sterile sodium citrate solution by drawing the solution into the syringe three times.
7. When the effective concentration required corresponds to a dilution of 1:2000, the titer is decreasing. Serum has a high titer if one can employ a dilution over 1:2000 and if there is a strong specific signal without any background.
8. We have also immunized mice with nitrocellulose-bound protein by intraperitoneal injection of the powder. This is convenient when time and material are very limited, since very little protein is needed to induce a response (3–5 times less than for a rabbit) and since the time-lag for the response is shorter (the second immunization was two weeks after the first, and the third immunization, 10 days after the second). Mice have a high tolerance for the nitrocellulose powder. Unfortunately, the small amount of serum available is a limiting factor. This technique for immunizing mice can, of course, be used for the preparation of monoclonal antibodies. For people who need to raise antibodies against Drosophila, melanogaster proteins rats are better hosts than rabbits (mice too with the above restrictions). Nitrocellulose is injected on the top of the shoulder of the rats. The foot pad is also adequate as a site of priming injection for small quantities. Blood is collected from the tail.
9. Utilization of serum. The proper dilutions are determined. We routinely use 1:4000 for blots, 1:300–1:200 for immunofluorescence, and 1:50 for immunogold staining. Serum continues to recognize epitopes on tissue proteins after Lowicryl embedding. Labeling is highly specific, and gives a sharp image of *in situ* protein localization. *There is no need to*

purify the serum. IgG purified from serum by whatever means usually gives poorer results than those obtained with diluted serum. Purification procedures often give rise to aggregates and denaturation, always increase the background, and result in loss of specific staining.

10. Bacterial antigenic protein. When antibodies are used in screening cDNA libraries in which the host is *E. coli,* the antibodies produced against bacterial components of Freund's adjuvant may also recognize some *E. coli* components. An advantage of our technique is that it avoids the risk of producing antibodies against such extraneous components.
11. Is nitrocellulose antigenic? Some workers have been unable to achieve good results by immunization with nitrocellulose-bound protein. They reproducibly obtain antisera directed against nitrocellulose. We found that in every case, this resulted from injecting powdered nitrocellulose in Freund's adjuvant; using adjuvant actually increases the production of low affinity IgM that binds nonspecifically to nitrocellulose. We have never observed this effect in our experiments when the technique described here was followed strictly.
12. The purification step by 2D electrophoresis implies the use of denaturing conditions (SDS), and thus, is not appropriate for obtaining antibodies directed against native structures. For that purpose, the protein should be transferred onto nitrocellulose after purification by classical nondenaturing methods and gel electrophoresis under nondenaturing conditions.

    However, it should be pointed out that, following the method of Dunn *(9),* it is possible partially to renature proteins with modifications of the composition of the transfer buffer.
13. Second dimension electrophoresis can be carried out with a first electrophoresis under native conditions, followed by a second electrophoresis under denaturing conditions, i.e., with SDS.

    Because the resolution provided by a gradient is better, it should always be used in the 2D electrophoresis. Agarose may also be used as an electrophoresis support.
14. If only a limited amount of protein is available, and/or if the antigen is weakly immunogenic, another procedure may be used. The first immunization is given as a single injection, of up to 0.8 mL, into the popliteal lymphatic ganglion *(10),* using a 22-gage needle, i.e., the finest that can be used to inject nitrocellulose powder. In this case, the antigen is delivered immediately into the immune system. If necessary, both ganglions can receive an injection. The small size of the ganglions limits the injected volume. The eventual excess of antigen solution is then injected into the back of the rabbit, as described above. The boosters are given in the classic manner, i.e., in several subcutaneous injections into the rabbit's back.

    If the amount of protein available is even more limited, a guinea pig may be immunized (first immunization in the lymphatic ganglion, and boosters, as usual).
15. The advantage of getting a high titer for the antibody of interest is that the amount of irrelevant antibodies is, by comparison, very low and, consequently, does not generate any background. Another advantage of using a crude serum with a high antibody titer, is that this serum may be used without further purification to screen a genomic library *(11).*
16. The time required for transfer and the voltage used is dependent on the molecular weight and the nature of the protein to be transferred. During transfer, the electrophoresis tank may be cooled with tap water.

References

1. Pennington, S. (1994) 2-D protein gel electrophoresis: an old method with future potential. *Trends in Cell Biol.* **4,** 439–441.
2. O'Farrel, P. H. (1975) High resolution two-dimensional electrophoresis of proteins. *J. Biol. Chem.* **250,** 4007–4021.

3. O'Farrel, P. Z., Goodman, H. M., and O'Farrel, P. H. (1977) High resolution two-dimensional electrophoresis of basic as well as acidic proteins. *Cell* **12,** 1133–1142.
4. Laemmli, U. K. (1970) Cleavage of structural proteins during the assembly of the head of bacteriophage T4. *Nature (London)* **227,** 680–685.
5. Knudsen, K. A. (1985) Proteins transferred to nitrocellulose for use as immunogens. *Anal. Biochem.* **147,** 285–288.
6. Burnette, W. N. (1981) Electrophoretic transfer of proteins from sodium dodecylsulfate polyacrylamide gels to unmodified nitrocellulose and radiographic detection with antibody and radioiodinated protein A. *Anal. Biochem.* **112,** 195–203.
7. Vaitukaitis, J., Robbins, J. B., Nieschlag, E., and Ross, G. T. (1971) A method for producing specific antisera with small doses of immunogen. *J. Clin. Endocrinol.* **33,** 980–988.
8. Dunn, M. J. and Patel, K. (1988) 2-D PAGE using flat bed IEF in the first dimension, in *Methods in Molecular Biology, vol 3: New Protein Techniques* (Walker, J. M., ed.), Humana, Clifton, NJ. pp. 217–232.
9. Dunn, S. D. (1986) Effects of the modification of transfer buffer composition and the renaturation of proteins in gels on the recognition of proteins on Western blots by monoclonal antibodies. *Anal. Biochem.* **157,** 144–153
10. Sigel, M. B., Sinha, Y. N., and Vanderlaan, W. P. (1983) Production of antibodies by inoculation into lymph nodes. *Methods Enzymol.* **93,** 3–12.
11. Preziosi, L., Michel, G. P. F., and Baratti, J. (1990) Characterisation of sucrose hydrolising enzymes of *Zymomonas mobilis. Arch. Microbiol.* **153,** 181–186.

123

Production and Characterization of Antibodies Against Peptides

J. Mark Carter

1. Introduction

Chapters 117–119 describe methods for conjugating peptides to carrier molecules. This chapter describes immunization procedures for raising antibodies against such conjugates and describes an ELISA method for analyzing the antibodies thus produced.

1.1. Immunization Protocols

Experimental immunizations with peptide conjugates may be performed according to a large number of protocols. These vary according to the choice of experimental animals, adjuvants, immunization schedule, and bleeding schedule. Each of these issues is discussed in Section 4., although the choices are generally not critical, and variations, personal preferences, and opinions abound. One generally effective scheme is presented. For a variety of alternate immunization schedules, *see* ref. *(1)*.

Note that you will almost definitely be required to work within the constraints of your local bureaucracy (e.g., a laboratory animal care and use committee). This may limit your choices. However, it may also offer some helpful guidance, such as the availability of proven experimental protocols or hands–on training in animal handling.

1.2. Enzyme–Linked Immunosorbent Assay (ELISA)

When it is necessary to determine the presence of specific antibody in a solution, ELISA is one of the simplest methods to use. ELISA can easily and reliably detect small quantities (of the order of 1 μg/mL) of antibody in hybridoma culture supernatant solution, ascites fluid, or immune serum. In hybridoma screening, often only the presence of the antibody is to determined. On the other hand, in serum or ascites, a rough quantitation of antibody concentration and avidity can be made by determining the titer of the antibody solution. Although other definitions are sometimes used, titer is traditionally defined as the reciprocal of the dilution of antibody preparation that gives a half-maximal response in the assay.

ELISA is based on a series of molecular binding reactions, usually performed on an optically clear, flat-bottom, 96-well microtiter plate. First, the antigen is coated onto the plate. After blocking remaining sites for nonspecific binding, the experimental antibody is allowed to bind to the immobilized antigen. This antibody protein is then

From: The Protein Protocols Handbook
Edited by: J. M. Walker Humana Press Inc., Totowa, NJ

probed by binding of an enzyme–conjugated second antibody. Finally, the enzyme-conjugated antibody is detected by means of a chromogenic substrate for the enzyme.

2. Materials

2.1. Immunization and Bleeding

1. Two rabbits: preferably young (about 1 kg) females (*see* Note 1).
2. Peptide immunogen: 10 mg peptide or 20 mg conjugate as a solution or suspension in 10 mL PBS. Store this solution frozen (*see* Chapters 117–119).
3. Freund's complete adjuvant (FCA), 2 mL (*see* Note 2).
4. Freund's incomplete adjuvant. 8 mL (*see* Note 2).
5. Sterile syringes and a union to join their fittings.
6. Sterile hypodermic needles. 18- or 19-gage, 1–2 in. length.

2.2. ELISA

1. Antigen solution: 1 µg/mL (not counting carrier protein) in a mild aqueous buffer, such as PBS, borate, or carbonate.
2. Blocking solution: PBST containing 2% bovine serum albumin (*see* Note 7).
3. PBS-Tween (PBST): Phosphate-buffered saline containing 0.1% Tween-20.
4. Second (enzyme labelled) antibody: This protocol describes the use of an alkaline phosphatase-linked second antibody. This antibody must be directed against the IgG of the species in which the first antibody was raised (*see* Note 9).
5. Substrate solution: For alkaline phosphatase, 1 mg/mL *p*-nitrophenyl phosphate in 0.1*M* diethanolamine, pH 9.8, 0.01% $MgCl_2$, 0.02% NaN_3. The buffer can be stored at 4°C for several months, but substrate solution should be made fresh as required. This substrate produces a yellow color, the absorbance of which should be read at about 405 nm.
6. Stopping solution: 0.01*M* EDTA (*see* Note 10).

3. Methods

3.1. Immunization and Bleeding

1. Collect at least 5 mL normal (preimmune) serum from each animal before it is immunized. Store it at –70°C. The preimmune serum is an important reagent for subsequent experiments. A portion of it will be used as a control in every experiment using the immune serum.
2. Put 2 mL Freund's into one syringe, and 2 mL peptide solution into the other one. Couple the two syringes together tightly. Force the mixture back and forth through the small orifices between the syringes. This generally causes complete emulsion within a few min.
3. Inject each animal with 2 mL of emulsion, as follows. Inject 0.5 mL into each of two subcutaneous sites on the animal's lower back or rump area. Inject another 0.5 mL emulsion into each of two intramuscular sites in the middle of the posterior of the animal's thigh (*see* Note 3).
4. One month later, make a second group of injections. The immunogen should be prepared as before, but use Freund's incomplete adjuvant for this and any subsequent group of injections. If abscesses are noted (especially at the subcutaneous sites), make the new injection at least 1 cm away.
5. Another month later, and every month subsequently, make test bleeds of about 5 mL from each animal (*see* Note 4). Draw blood from the marginal ear vein. This causes a minimum of discomfort for the animal. For your convenience, you may make another round of injections (in incomplete adjuvant) on the same day as the test bleeds.

6. By the third or fourth injection, the animals should begin to produce antibody. This should be confirmed via ELISA results. As soon as any animal begins to produce high-titer antibody, it need not be injected anymore. However, it should be bled 20–25 mL every 2 wk. When antibody titer drops off, the animal should be retired.

3.2. Isolation of Serum and Antibody

1. The blood should be allowed to clot normally, for 10–20 min at room temperature. Leave a thin wooden stick in the blood to provide an adherent surface for clot formation.
2. To maximize recovery of serum, wring the clot by scraping it around the sides of the collection vessel, allowing contraction of the thrombin fiber complex. After another 10–20 min remove the stick, with or without the clot attached.
3. Pellet the remaining clot on a centrifuge, and decant the straw–colored serum. Serum may be stored frozen for long periods of time: at least 6 mo at –20, and several years at –70°C.
4. Methods for purifying antibodies from serum are described in Chapters 125–133.

3.3. ELISA

Times, volumes, and concentrations of reagents may require some adjustment in order to give good, reproducible results. However, the general scheme outlined below is likely to give a reliable yes-or-no result on the first attempt.

1. Coating with antigen: Place 100 µL of antigen solution into each well of the microtiter plate, and allow to incubate overnight in a humidified chamber in the refrigerator. Put an irrelevant peptide into some of the wells to act as a negative control. Alternatively, put blocking solution in the negative control wells (*see* Note 5).
2. Wash once or twice with 200 µL PBST/well (*see* Note 6).
3. Blocking: 200 µL/well of a 2% solution of blocking protein are allowed to bind for 1 h at 4°C (*see* Note 7).
4. Wash: Perform three or four washes with 200 µL PBST/well, as before.
5. Test antibody: Pipet 50 µL antibody solution into each of the microtiter plate wells, with the duplicates side by side. Tap the side of the plate gently to mix and spread the solution, and allow to incubate for 1 h at 4°C (*see* Note 8).
6. Wash: Wash three times as before, with 200 µL PBST.
7. Second (enzyme-labeled) antibody: Pipet 50 µL of the second antibody solution into each well. Tap the side of the plate gently to mix and spread the solution, and allow to incubate for 1 h at 4°C (*see* Note 9).
8. Wash: again, perform three washes with 200 µL PBST.
9. Substrate solution: Pipet 100 µL substrate solution into each well. Tap the side of the plate gently to mix and spread the solution.
10. Development: Allow to develop for 30–60 min at room temperature. If necessary, terminate color development by addition of 100 µL of stop solution to each well. The stop solution is added to the substrate solution already in the well. It is not usually necessary to tap the plate to mix these solutions (*see* Note 10).
11. Collect data: Read the plates at 405 nm on an automated ELISA plate reader.

4. Notes

1. A great variety of experimental animals have been exploited in the production of antibody. Many studies require only the production of immune sera, containing polyclonal antibody. In these cases, almost any animal on a valid protocol may be used. Females are the most popular because their generally more docile disposition makes for easier han-

dling. Always use at least 2 animals/immunogen. This will ensure against the failure of one of the animals owing to mishap or biological variability.

Popular species include mouse, rabbit, goat, and horse. Useful immunochemical reagents (such as second antibody) are generally widely available for these species and may also be available for others. Larger animals require more expensive overhead, but they also produce a larger volume of antiserum. Yield from one animal will thus range from a few microliters in a test bleed of a mouse to several liters in the exsanguination of a horse. In addition to these common experimental animals, special animals from practically any of various disease models may also be used.

2. The most popular adjuvant is FCA. This is a preparation of paraffin oil containing a suspension of killed mycobacteria. Freund's incomplete adjuvant is simply the oil without the mycobacteria. FCA is extremely effective for priming, but booster injections must be given in incomplete adjuvant in order to minimize formation of cysts and chronic inflammation. Even then significant irritation is common. For this reason, Freund's adjuvants are not allowed for use in humans. Emulsions in either FCA or incomplete adjuvant are not physically stable and should be prepared immediately before use. If necessary, they may be kept at 4°C overnight, but sterility is difficult to maintain, and contaminated preparations will result in health problems for the experimental animals.
3. In general, injections may be made in any of a number of sites: subcutaneously (sc), intramuscularly (im), intraperitoneally (ip), or intravenously (iv). sc or im sites are usually preferred for early immunizations, but boosters are frequently given ip or iv.

   sc injections are probably most effective when made into the footpad for small experimental animals, such as rodents and rabbits. However, this is quite painful for the animals and should be avoided, whenever possible, for that reason. sc injections at multiple sites in the back or rump region are also very effective for smaller animals, up to rabbit size, whereas larger animals do well with sc injections around the neck. im injections should be administered into the large thigh musculature, being careful to avoid damage to the major blood vessels there.

   ip injections must be made carefully into the abdominal cavity without piercing any vital organs. Nonetheless, because of their speed and simplicity, ip boosters are popular for use with rodents and sometimes also rabbits. Although iv boosters are difficult to administer in smaller animals, they may be given into the tail veins of rodents, the marginal ear veins of rabbits, or the jugular veins of larger animals.
4. Some investigators insist that it is best to bleed animals 2 wk after each immunization in order to allow an immune response to each injected dose of antigen. This belief is based on very early studies of the immune response to soluble antigen, and it probably applies primarily to antigen preparations presented in solution. With oily adjuvants, such as FCA, the antigen is presented in an emulsified depot that produces slow release. This allows successful bleeding at almost any convenient date, as long as it is at least 2 wk after the second injection (first boost).

   Blood may be drawn from the tail veins of rodents, or the jugular veins of larger animals, using a heparinized needle. For rabbits, the marginal ear veins are usually used, either with a needle or by means of a small open incision. Rabbits and smaller animals are often bled via cardiac puncture, using a needle and syringe. Note that cardiac puncture, even when performed properly, is often fatal.

   Test the sera for antibody using an ELISA (Section 3.3.). Animals producing antibody generally need not be further immunized, whereas those unproductive of specific antibody may be further boosted. After 6 mo, unproductive animals may be retired; they will probably never produce antibody. It is unfortunate that many animals will invariably fall into this latter group. However, if fewer than half the animals make antibody, the immunogen should probably be changed.

Once an animal is producing high levels of antibody (titer > 10,000), it should be bled regularly. Twenty milliliters of blood may be drawn from a rabbit every 2 wk without consequences to the animal's health. After about 10 mo, even productive animals will begin to show a significant decline in antibody production, and they may be "bled out" (exsanguinated) or re-boosted.

5. Many peptides will bind directly to microtiter plates without modification. Others, particularly small molecules <20 residues, require conjugation to an irrelevant carrier protein in order for them to stick. If a conjugate is used as an ELISA "capture" antigen, it should be made with a carrier protein other than the one chosen for the immunogen. This is very important to prevent false-positive reactions resulting from crossreactivity of the antiserum with the ELISA antigen carrier. Note that it may be necessary to test the experimental antibody for binding to the parent protein as well as the peptide immunogen from which its sequence was derived. The procedure for this is exactly the same, substituting the native protein antigen for the peptide in the first step, coating the plate.

   For antigen binding and subsequent incubations more elevated temperatures (e.g., room temperature or 37°C) may be used successfully if microbial growth and proteolysis are suitably inhibited. This will allow the binding reaction to reach equilibrium in 15–30 min. On the other hand, slightly stronger binding is sometimes observed with incubation at 4°C.

   To prevent uneven evaporation rates (edge effects), use a humidified chamber for incubations longer than 1 h. To make a humidified chamber, use any sort of sealable container, such as a plastic box with a tight-fitting lid or a desiccator. Put one or two water-soaked paper towels in the bottom of the chamber before placing the ELISA plates carefully inside and sealing. Change the paper towels at least weekly to avoid mold growth.

   Many scientists allow the antigen to dry onto the ELISA plate wells. This option allows extended storage of the plates before use. Dried plates, stored desiccated and frozen, may be used effectively for at least 1 yr after preparation.

6. After each incubation, washes are performed to remove excess reagent. For the wash step, pipet 200 µL wash solution into the microtiter plate well, and then either flick it out into the laboratory sink or aspirate it, being careful not to disturb the bottom of the wells. Automated and semiautomated apparatus ("microtiter plate washer" or "ELISA plate washer") are commercially available for simplification and acceleration of the wash steps. When properly used, such equipment improves reproducibility of washes. However, these machines must be well maintained to preserve accuracy in liquid deliveries.

   Some investigators perform washes and other ELISA incubations on a rocker or rotating platform. This is usually unnecessary.

   Note that in many buffers and reagent solutions used for ELISA, 0.1% Tween 20 is added to the PBS. Tween 20 is a mild nonionic detergent that reduces nonspecific binding. Handling of Tween 20 solutions should be performed carefully so as to minimize aeration, since foaming will affect reproducibility.

7. After attaching the peptide antigen to the plastic microtiter plate wells, you must block remaining nonspecific binding sites by treatment with a solution of irrelevant (inexpensive) protein. Bovine serum albumin (BSA) is commonly used, unless it was utilized as a carrier protein in the immunogen conjugate. In that case, egg albumin or boiled casein may be substituted with equivalent results. The blocking solution is made in PBST. It may be prepared in 1-L batches, filter-sterilized, and stored at 4°C for up to 2 wk.

8. Make dilutions of the test antibody in PBST. If you are testing a purified antibody, use 5 µg/mL. If you are testing a serum or ascites fluid (typically 100–1000 µg/mL specific

antibody, and around 10 mg/mL irrelevant antibody), your ELISA should include serial dilutions of the antibody to determine titer.

At the very least, test the following 10-fold dilutions: $1/10^2$, $1/10^3$, $1/10^4$, and $1/10^5$. You may prefer to make half-log dilutions: $1/10^2$, $1/10^{2.5}$, $1/10^3$, $1/10^{3.5}$,$1/10^4$, and so forth, for greater resolution of the titer. On subsequent tests you may even wish to make twofold serial dilutions or extend the range of dilutions tested. In any case, make enough of each antibody dilution to run the assay in duplicate or triplicate.

9. The second antibody is conjugated to an enzyme. With sera from commonly chosen experimental animals, you will probably use a commercially available second antibody preparation. These are usually antibody conjugates of either alkaline phosphatase (AP) or horseradish peroxidase (HRP). For the commercial reagents, the working concentration of the second antibody is specified by the manufacturer. Note that HRP is inhibited by sodium azide.

   In some cases, it may be necessary to prepare your own enzyme conjugated second antibody. This is fairly easy to do, but it takes a long time. Methods for preparing enzyme-antibody conjugates are described in Chapters 41–43.

10. In order to allow all the test wells of the ELISA to incubate for the same time, it is a good idea to use a special multiple pipettor for addition of the substrate solution and stop solution. These tools pipet 8 or 12 wells at a time. In addition to synchronizing additions of reagent, these tools can also cut down on time and work involved in other liquid handling steps of the ELISA.

    To aid in visualization of color development, place the plates on a piece of white paper. Allow development to proceed until the positive reactions are fairly well colored. This usually takes 30–60 min. During development, avoid thermal gradients, which may be caused, for example, by drafts or sunlight. Many scientists choose to cover the plates to ensure their isolation from environmental effects.

    Do not allow development to proceed until the negative samples give a strong color reaction. Once the ELISA has developed some color, read all wells on an automated ELISA plate reader.

    If you are processing several plates at once, then it may be necessary for you to add a stop solution. For the nitrophenyl phosphate substrate, use 0.01*M* EDTA as a stop solution. In this case, chelation of the magnesium cofactor results in complete inhibition of the enzyme without compromising optical quality of the chromogen. For the HRP substrate, use 1% sodium dodecyl sulfate. Here denaturation of the enzyme results in its inhibition.

Reference

1. Weir, D. M. (ed.) (1986) *Handbook of Experimental Immunology*, vol. 1, Blackwell, Oxford.

124

Production of Antibodies Using Proteins in Gel Bands

Sally Ann Amero, Tharappel C. James, and Sarah C. R. Elgin

1. Introduction

A number of methods for preparing proteins as antigens have been described *(1)*. These include solubilization of protein samples in buffered solutions (ref. *2* and *see* Chapter 120), solubilization of nitrocellulose filters to which proteins have been adsorbed (ref. *3* and *see* Chapter 122), and emulsification of protein bands in polyacrylamide gels for direct injections *(4–8)*. The latter technique can be used to immunize mice or rabbits for production of antisera or to immunize mice for production of monoclonal antibodies *(9–11)*. This approach is particularly advantageous when protein purification by other means is not practical, as in the case of proteins insoluble without detergent. A further advantage of this method is an enhancement of the immune response, since polyacrylamide helps to retain the antigen in the animal and so acts as an adjuvant *(7)*. The use of the protein directly in the gel band (without elution) is also helpful when only small amounts of protein are available. For instance, in this laboratory, we routinely immunize mice with 5–10 μg total protein using this method; we have not determined the lower limit of total protein that can be used to immunize rabbits. Since polyacrylamide is also highly immunogenic, however, it is necessary in some cases to affinity-purify the desired antibodies from the resulting antiserum or to produce hybridomas that can be screened selectively for the production of specific antibodies, to obtain the desired reagent.

2. Materials

1. Gel electrophoresis apparatus; acid-urea polyacrylamide gel or SDS-polyacrylamide gel.
2. Staining solution: 0.1% Coomassie brilliant blue-R (Sigma, St. Louis, MO, B-0630) in 50% (v/v) methanol/10% (v/v) acetic acid.
3. Destaining solution: 5%-(v/v) methanol/7% (v/v) acetic acid.
4. 2% (v/v) glutaraldehyde (Sigma G-6257).
5. Transilluminator.
6. Sharp razor blades.
7. Conical plastic centrifuge tubes and ethanol.
8. Lyophilizer and dry ice.
9. Plastic, disposable syringes (3- and 1-mL).
10. 18-gage needles.

From: The Protein Protocols Handbook
Edited by: J. M. Walker Humana Press Inc., Totowa, NJ

11. Spatula and weighing paper.
12. Freund's complete and Freund's incomplete adjuvants (Gibco Laboratories, Grand Island, NY).
13. Phosphate-buffered saline solution (PBS): 50 m*M* sodium phosphate, pH 7.25/150 m*M* sodium chloride.
14. Microemulsifying needle, 18-g (Becton Dickinson, Rutherford, NJ).
15. Female Balb-c mice, 7–8 wk old, or New Zealand white rabbits.

3. Method

1. Following electrophoresis (*see* Note 1), the gel is stained by gentle agitation in several volumes of staining solution for 30 min. The gel is partially destained by gentle agitation in several changes of destaining solution for 30–45 min. Proteins in the gel are then cross-linked by immersing the gel with gentle shaking in 2% glutaraldehyde for 45–60 min *(12)*. This step minimizes loss of proteins during subsequent destaining steps and enhances the immunological response by polymerizing the proteins. The gel is then completely destained, usually overnight (*see* Note 2).
2. The gel is viewed on a transilluminator, and the bands of interest are cut out with a razor blade. The gel pieces are pushed to the bottom of a conical plastic centrifuge tube with a spatula and pulverized. The samples in the tubes are frozen in dry ice and lyophilized.
3. To prepare the dried polyacrylamide pieces for injection, a small portion of the dried material is lifted out of the tube with a spatula and placed on a small square of weighing paper. In dry climates it is useful to first wipe the outside of the tube with ethanol to reduce static electricity. The material is then gently tapped into the top of a 3-mL syringe to which is attached the microemulsifying needle (Fig. 1A). Keeping the syringe horizontal, 200 µL of PBS solution is carefully introduced to the barrel of the syringe, and the plunger is inserted. Next, 200 µL of Freund's adjuvant is drawn into a 1-mL tuberculin syringe and transferred into the needle end of a second 3-mL syringe (Fig. 1B). This syringe is then attached to the free end of the microemulsifying needle. The two plungers are pushed alternatively to mix the components of the two syringes (Fig. 1C). These will form an emulsion within 15 min; it is generally extremely difficult to mix the material any further.
4. This mixture is injected intraperitoneally or subcutaneously into a female Balb-c mouse, or subcutaneously into the back of the neck of a rabbit (*see* refs. *13* and *14*). Since the emulsion is very viscous, it is best to use 18–g needles and to anesthesize the animals. For mice, subsequent injections are administered after 2 wk and after 3 more wk. If monoclonal antibodies are desired, the animals are sacrificed 3–4 d later, and the spleen cells are fused with myeloma cells (*13* and *see* Chapter 140). The immunization schedule for rabbits calls for subsequent injections after 1 mo or when serum titers start to diminish. Antiserum is obtained from either tail bleeds or eye bleeds from the immunized mice, or from ear bleeds from immunized rabbits. The antibodies are assayed by any of the standard techniques (*see* Chapters 141 and 142).

4. Notes

1. We have produced antisera to protein bands in acetic acid–urea gels (*see* Chapter 14), Triton–acetic acid–urea gels *(15,16)* (*see* Chapter 15), or SDS–polyacrylamide gels (*see* Chapter 11). In our experience, antibodies produced to proteins in one denaturing gel system will crossreact to those same proteins fractionated in another denaturing gel system and will usually crossreact with the native protein. We have consistently obtained antibodies from animals immunized by these procedures.
2. It is extremely important that all glutaraldehyde be removed from the gel during the destaining washes, since any residual glutaraldehyde will be toxic to the animal. Residual glutaraldehyde can easily be detected by smell. It is equally important to remove all acetic

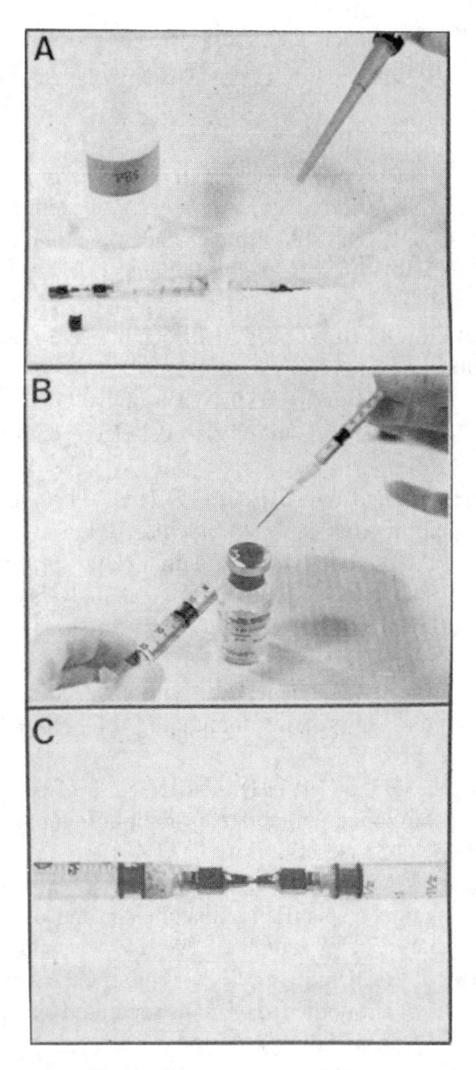

Fig. 1. Preparation of emulsion for immunizations. To prepare proteins in gel bands for injections, an emulsion of Freund's adjuvant and dried polyacrylamide pieces is prepared. **(A)** Dried polyacrylamide is resuspended in 200 μL of PBS solution in the barrel of a 3-mL syringe to which is attached a microemulsifying needle. **(B)** Freund's adjuvant is transferred into the barrel of a second 3-mL syringe. **(C)** An emulsion is formed by mixing the contents of the two syringes through the microemulsifying needle.

acid during lyophilization. Monoacrylamide is also toxic, whereas polyacrylamide is not. We do observe, however, that approx 50 mm^2 of polyacrylamide per injection is the maximum that a mouse can tolerate.

3. Freund's complete adjuvant is used for the initial immunization; Freund's incomplete adjuvant is used for all subsequent injections. The mycobacteria in complete adjuvant enhance the immune response by provoking a local inflammation. Additional doses of mycobacteria may be toxic.
4. High-titer antibodies have been produced from proteins in polyacrylamide gel by injecting the gel/protein mixture into the lumen of a perforated plastic golf ball implanted subcutane-

ously in rabbits *(17)*. This approach places less stress on the animal, as complete adjuvants need not be used, and bleeding is eliminated. The technique has also been used in rats.

References

1. Chase, M. W. (1967) Production of antiserum, in *Methods in Immunology and Immunochemistry,* vol. I (Williams, C. A. and Chase, M. W., eds.), Academic, New York, pp. 197–200.
2. Maurer, P. H. and Callahan, H. J. (1980) Proteins and polypeptides as antigens, in *Methods in Enzymology*, vol. 70 (Van Vunakis, H. and Langne, J. J., eds.), Academic, New York, pp. 49–70.
3. Knudson, K. A. (1985) Proteins transferred to nitrocellulose for use as immunogens. *Anal. Biochem.* **147,** 285–288.
4. Tjian, R., Stinchcomb, D., and Losick, R. (1974) Antibody directed against *Bacillus subtilis* σ factor purified by sodium dodecyl sulfate slab gel electrophoresis. *J. Biol. Chem.* 2**50,** 8824–8828.
5. Elgin, S. C. R., Silver, L. M., and Wu, C. E. C. (1977) The *in situ* distribution of drosophila non-histone chromosomal proteins, in *The Molecular Biology of the Mammalian Genetic Apparatus*, vol. 1 (Ts'o, P.O.P., ed.), North Holland, New York, pp. 127–141.
6. Silver, L. M. and Elgin, S. C. R (1978) Immunological analysis of protein distributions in *Drosophila* polytene chromosomes, in *The Cell Nucleus* (Busch, H., ed.), Academic, New York, pp. 215–262.
7. Bugler, B., Caizergucs-Ferrer, M., Bouche, G., Bourbon, H., and Alamric, F. (1982) Detection and localization of a class of proteins immunologically related to a 100-kDa nucleolar protein. *Eur. J. Biochem.* **128,** 475–480.
8. Wood, D. M. and Dunbar, B. S. (1981) Direct detection of two-cross-reactive antigens between porcine and rabbit zonae pellucidae by radioimmunoassay and immunoelectrophoresis. *J. Exp. Zool.* **217,** 423–433.
9. Howard, C. D., Abmayr, S. M., Shinefeld, L. A., Sato, V. L., and Elgin, S. C. I. (1981) Monoclonal antibodies against a specific nonhistone chromosomal protein of *Drosophila* associated with active genes. *J. Cell Biol.* **88,** 219–225.
10. Tracy, R. P., Katzmann, J. A., Kimlinger, T. K., Hurst, G. A., and Young, D. S. (1983) Development of monoclonal antibodies to proteins separated by two dimensional gel electrophoresis. *J. Immunol. Meth.* **65,** 97–107.
11. James, T. C. and Elgin, S. C. R. (1986) Identification of a nonhistone chromosomal protein associated with heterochromatin in *Drosophila melanogaster* and its gene. *Mol. Cell Biol.* **6,** 3862–3872.
12. Reichli, M. (1980) Use of glutaraldehyde as a coupling agent for proteins and peptides, in *Methods in Enzymology,* vol. 70 (Van Vunakis, H. and Langone, J. J., eds.), Academic, New York, pp. 159–165.
13. Campbell, A. M. (1984) *Monoclonal Antibody Technology*. Elsevier, New York.
14. Silver, L. M. and Elgin, S. C. R. (1977) Distribution patterns of three subfractions of *Drosophila* nonhistone chromosomal proteins: Possible correlations with gene activity. *Cell* **11,** 971–983.
15. Alfagame, C. R., Zweidler, A., Mahowald, A., and Cohen, L. H. (1974) Histones of *Drosophila* embryos. *J. Biol. Chem.* **249,** 3729–3736.
16. Cohen, L. H., Newrock, K. M., and Zweidler, A. (1975) Stage-specific switches in histone synthesis during embryogenesis of the sea urchin. *Science* **190,** 994–997.
17. Ried, J. L., Everad, J. D., Diani, J., Loescher, W. H., and Walker-Simmons, M. K. (1992) Production of polyclonal antibodies in rabbits is simplified using perforated plastic golf balls. *BioTechniques* **12,** 661–666.

125

Purification of IgG by Precipitation with Sodium Sulfate or Ammonium Sulfate

Mark Page and Robin Thorpe

1. Introduction

Addition of appropriate amounts of salts, such as ammonium or sodium sulfate, causes precipitation of IgG *(1)* from all mammals, and can be used for serum, plasma, ascites fluid, and hybridoma culture supernatant. Although such IgG is usually contaminated with other proteins, the ease of these precipitation procedures coupled with the high yield of IgG has led to their wide use in producing enriched IgG preparations. They are suitable for many immunochemical procedures, e.g., production of immunoaffinity columns, and as a starting point for further purification. It is not suitable however for conjugating with radiolabels, enzymes, or biotin since the contaminating proteins will also be conjugated, thereby reducing the efficiency of the labeling and the quality of the reagent. The precipitated IgG is usually very stable, and such preparations are ideally suited for long-term storage or distribution and exchange between laboratories.

Ammonium sulfate precipitation is the most widely used and adaptable procedure, yielding a 40% pure preparation; sodium sulfate can give a purer preparation for some species, e.g., human and monkey.

2. Materials

2.1. Ammonium Sulfate Precipitation

1. Saturated ammonium sulfate solution: Add excess $(NH_4)_2SO_4$ to distilled water (about 950 g to 1 L), and stir overnight at room temperature. Chill at 4°C, and store at this temperature. This solution (in contact with solid salt) is stored at 4°C.
2. PBS: 0.14*M* NaCl, 2.7 m*M* KCl, 1.5 m*M* KH_2PO_4, 8.1 m*M* Na_2HPO_4. Store at 4°C.

2.2. Sodium Sulfate Precipitation

This requires solid sodium sulfate.

3. Methods

3.1. Ammonium Sulfate Precipitation

1. Prepare saturated ammonium sulfate at least 24 h before the solution is required for fractionation. Store at 4°C.

From: The Protein Protocols Handbook
Edited by: J. M. Walker Humana Press Inc., Totowa, NJ

2. Centrifuge serum or plasma for 20–30 min at 10,000g_{av} at 4°C. Discard the pellet (*see* Note 1).
3. Cool the serum or plasma to 4°C, and stir slowly. Add saturated ammonium sulfate solution dropwise to produce 35–45% final saturation (*see* Note 2). Alternatively, add solid ammonium sulfate to give the desired saturation (2.7 g of ammonium sulfate/10 mL of fluid = 45% saturation). Stir at 4°C for 1–4 h or overnight.
4. Centrifuge at 2000–4000g_{av} for 15–20 min at 4°C (alternatively for small volumes of 1–5 mL, microfuge for 1–2 min). Discard the supernatant, and drain the pellet (carefully invert the tube over a paper tissue).
5. Dissolve the precipitate in 10–20% of the original volume in PBS or other buffer by careful mixing with a spatula or drawing repeatedly into a wide-gage Pasteur pipet. When fully dispersed, add more buffer to give 25–50% of the original volume and dialyze against the required buffer (e.g., PBS) at 4°C overnight with two to three buffer changes. Alternatively, the precipitate can be stored at 4 or –20°C if not required immediately.

3.2. Sodium Sulfate Precipitation (see Note 3)

1. Centrifuge the serum or plasma at 10,000g_{av} for 20–30 min. Discard the pellet, warm the serum to 25°C, and stir.
2. Add solid Na_2SO_4 to produce an 18% w/v solution (i.e., add 1.8 g/10 mL), and stir at 25°C for 30 min to 1 h.
3. Centrifuge at 2000–4000g_{av} for 30 min at 25°C.
4. Discard the supernatant, and drain the pellet. Redissolve in the appropriate buffer as described for ammonium sulfate precipitation (Section 3.1., step 5).

4. Notes

1. If lipid contamination is excessive in ascites fluids, thereby compromising the salt precipitation, add silicone dioxide powder (15 mg/ mL) and centrifuge for 20 min at 2000g_{av} *(2)* before adding the ammonium or sodium salt.
2. The use of 35% ammonium sulfate will produce a pure IgG preparation, but will not precipitate all the IgG present in serum or plasma. Increasing saturation to 45% causes precipitation of nearly all IgG, but this will be contaminated with other proteins, including some albumin. Purification using $(NH_4)_2SO_4$ can be improved by repeating the precipitation, but this may cause some denaturation. Precipitation with 45% $(NH_4)_2SO_4$ is an ideal starting point for further purification steps, e.g., ion-exchange or affinity chromatography and FPLC purification (*see* Chapters 127 and 128).
3. Sodium sulfate may be used for precipitation of IgG instead of ammonium sulfate. The advantage of the sodium salt is that a purer preparation of IgG can be obtained, but this must be determined experimentally. The disadvantages are that yield may be reduced depending on the IgG characteristics of the starting material, IgG concentration, and composition. Fractionation must be carried out at a precise temperature (usually 25°C), since the solubility of Na_2SO_4 is very temperature dependent. Sodium sulfate is usually employed only for the purification of rabbit or human IgG.

References

1. Heide, K. and Schwick, H. G. (1978) Salt fractionation of immunoglobulins, in *Handbook of Experimental Immunology,* 3rd ed. (Weir, D. M., ed.), chap. 7. Blackwell Scientific, Oxford.
2. Neoh, S. H., Gordon, C., Potter, A., and Zola, H. (1986) The purification of mouse MAb from ascitic fluid. *J. Immunol. Methods* **91,** 231.

126

Purification of IgG Using Caprylic Acid

Mark Page and Robin Thorpe

1. Introduction

Caprylic (octanoic) acid can be used to purify mammalian IgG from serum, plasma, ascites fluid, and hybridoma culture supernatant by precipitation of non-IgG protein *(1)* (*see* Note 1). Other methods have been described where caprylic acid has been used to precipitate immunoglobulin depending on the concentration used. The concentration of caprylic acid required to purify IgG varies according to species (*see* Section 3., step 2). For MAb, it is usually necessary to determine experimentally the quantity required to produce the desired purity/yield. Generally, the product is of low to intermediate purity but this will depend on the starting material. Caprylic acid purified IgG preparations can be used for most immunochemical procedures, such as coating plates for antigen capture assays and preparation of immunoaffinity columns, but would not be suitable for conjugation with radioisotopes, enzymes, and biotin where contaminating proteins will reduce the specific activity.

2. Materials

1. 0.6*M* sodium acetate buffer, pH 4.6. Adjust pH with 0.6*M* acetic acid.
2. Caprylic acid (free acid).

3. Methods

1. Centrifuge the serum at 10,000g_{av} for 20–30 min. Discard the pellet and add twice the volume of 0.06*M* sodium acetate buffer, pH 4.6.
2. Add caprylic acid dropwise while stirring at room temperature. For each 25 mL of serum, use the following amounts of caprylic acid: human and horse, 1.52 mL; goat, 2.0 mL; rabbit, 2.05 mL; cow, 1.7 mL. Stir for 30 min at room temperature.
3. Centrifuge at 4000g_{av} for 20–30 min. Retain the supernatant and discard the pellet. Dialyze against the required buffer (e.g., PBS) at 4°C overnight with two or three buffer changes.

4. Note

1. The method can be used before ammonium sulfate precipitation (*see* Chapter 125) to yield a product of higher purity.

Reference

1. Steinbuch, M. and Audran, R. (1969) The isolation of IgG from mammalian sera with the aid of caprylic acid. *Arch. Biochem. Biophys.* **134,** 279–284.

From: The Protein Protocols Handbook
Edited by: J. M. Walker Humana Press Inc., Totowa, NJ

127

Purification of IgG Using DEAE-Sepharose Chromatography

Mark Page and Robin Thorpe

1. Introduction

IgG may be purified from serum by a simple one-step ion-exchange chromatography procedure. The method is widely used and works on the principle that IgG has a higher or more basic isoelectric point than most serum proteins. Therefore, if the pH is kept below the isoelectric point of most antibodies, the immunoglobulins do not bind to an anion exchanger and are separated from the majority of serum proteins bound to the column matrix. The high capacity of anion-exchange columns allows for large-scale purification of IgG from serum. The anion-exchange reactive group, diethylaminoethyl (DEAE) covalently linked to Sepharose (e.g., DEAE Sepharose CL-6B, Pharmacia, Uppsala, Sweden) is useful for this purpose. It is provided preswollen and ready for packing into a column, and is robust and has high binding capacity. Furthermore, it is relatively stable to changes in ionic strength and pH. Other matrices (e.g., DEAE cellulose) are provided as solids, and will therefore require preparation and equilibration *(1)*.

This procedure does not work well for murine IgG or preparations containing mouse or rat MAb, since these do not generally have the high pI values that IgGs of other species have. Other possible problems are that some immunoglobulins are unstable at low-ionic strength, e.g., mouse IgG_3, and precipitation may occur during the ion-exchange procedure. The product is of high purity (>90%) and can be used for most immunochemical procedures including conjugation with radioisotopes, enzymes, and so on, where pure IgG is required.

2. Materials

1. DEAE Sepharose CL-6B.
2. 0.07*M* sodium phosphate buffer, pH 6.3.
3. 1*M* NaCl.
4. Sodium azide.
5. Chromatography column (*see* Note 1).

3. Methods

1. Dialyze the serum (preferably ammonium sulfate fractionated; *see* Chapter 125) against 0.07*M* sodium phosphate buffer, pH 6.3, exhaustively (at least two changes over a 24-h period) at a ratio of at least 1 vol of sample to 100 vol of buffer.

From: The Protein Protocols Handbook
Edited by: J. M. Walker Humana Press Inc., Totowa, NJ

2. Apply the sample to the column, and wash the ion exchanger with 2 column volumes of sodium phosphate buffer. Collect the wash, which will contain IgG, and monitor the absorbance of the eluate at 280 nm (A_{280}). Stop collecting fractions when the A_{280} falls to baseline.
3. Regenerate the column by passing through 2–3 column volumes of phosphate buffer containing 1*M* NaCl.
4. Wash thoroughly in phosphate buffer (2–3 column volumes), and store in buffer containing 0.1% NaN_3.
5. Pool the fractions from step 2 and measure the A_{280} (*see* Note 2).

4. Notes

1. The column size will vary according to the user's requirements or the amount of antibody required. Matrix binding capacities are given by manufacturers and should be used as a guide.
2. The extinction coefficient ($E^{1\%}_{280}$) of human IgG is 13.6 (i.e., a 1 mg/mL solution has an A_{280} of 1.36). This can be used as an appropriate value for IgGs from other sources.

Reference

1. Johnstone, A. and Thorpe, R. (1996) *Immunochemistry in Practice,* 3rd ed. Blackwell Scientific, Oxford, UK.

128

Purification of IgG Using Ion-Exchange HPLC/FPLC

Mark Page and Robin Thorpe

1. Introduction

Conventional ion-exchange chromatography separates molecules by adsorbing proteins onto the ion-exchange resins that are then selectively eluted by slowly increasing the ionic strength (this disrupts ionic interactions between the protein and column matrix competitively) or by altering the pH (the reactive groups on the proteins lose their charge). Anion-exchange groups (such as diethylaminoethyl; DEAE) covalently linked to a support matrix (such as Sepharose) can be used to purify IgG in which the pH of the mobile-phase buffer is raised above the pI of IgG, thus allowing most of the antibodies to bind to the DEAE matrix. Compare this method with that described in Chapter 127 in which the IgG passes through the column. The procedure can be carried out using a laboratory-prepared column that is washed and eluted under gravity *(1)*; however, high-performance liquid chromatography (HPLC) and fast-performance liquid chromatography (FPLC) provide improved reproducibility (because of the sophisticated pumps and accurate timers), speed (because of the small, high-capacity columns), and increased resolution (because of the fine resins and control systems). FPLC is a variant of HPLC that has proven useful in the purification of murine MAb, although the technique is applicable to IgG preparations from all species.

2. Materials

2.1. Anion-Exchange HPLC

1. Anion-exchanger (e.g., Anagel TSK DEAE, Anachem, Luton, UK).
2. Buffer A: 0.05*M* Tris-HCl buffer, pH 8.5.
3. Buffer B: Buffer A containing 0.5*M* Na_2SO_4.
4. 0.3*M* NaOH.
5. Sodium azide.

2.2. Anion-Exchange FPLC

1. Anion-exchanger (e.g., Mono-Q HR 5/5 or HR 10/10, Pharmacia, Uppsala, Sweden).
2. Buffer A: 0.02*M* triethanolamine, pH 7.7.
3. Buffer B: Buffer A containing 1*M* NaCl.
4. 2*M* NaOH.
5. Sodium azide.

From: The Protein Protocols Handbook
Edited by: J. M. Walker Humana Press Inc., Totowa, NJ

3. Methods

3.1. Anion-Exchange HPLC

1. Prepare IgG fraction from serum/ascitic fluid by ammonium sulfate preciptiation (45% saturation; *see* Chapter 125). Dialyze the sample (1–10 mL) against 0.05*M* Tris-HCl, pH 8.5. Dilute the sample at least 1:1 in 0.05*M* Tris-HCl, and filter (*see* Note 1) before use (0.2 µm).
2. Assemble the HPLC system according to the manufacturer's instructions.
3. Wash the column with 0.05*M* Tris-HCl, pH 8.5 (buffer A), at an optimal flow rate (1.0 mL/min) until absorbance at 280 nm is stable.
4. Run a blank salt gradient 0–100% buffer B (buffer A + 0.5*M* Na_2SO_4) over 20 min to complete preparation of the matrix. Finally, re-equilibrate the column with buffer A before sample application, so that the A_{280} is at baseline.
5. Add the sample manually using a 1–15 mL loop, and pass buffer for at least 10 min before applying the salt gradient. The sample volume will depend on the column size; refer to the manufacturer's specifications for loading capacities of the column. Small columns of approx 5 × 1 cm are for analytical purposes and larger columns are usually required for preparative work.
6. Apply salt gradient from 0–100% (buffer B) over 45 min, collecting protein peaks. The length of the run will depend on column size. A larger column will take longer to apply the salt gradient.
7. Purge the column with 100% 0.5*M* Na_2SO_4 for 10 min.
8. Wash the column with 0.3*M* NaOH for 10 min.
9. Wash the column with sample buffer until absorbance is stable, and store in distilled water containing 0.02% sodium azide.

3.2. Anion-Exchange FPLC

1. Prepare serum by ammonium sulfate precipitation (45% saturation; *see* Chapter 125). Redissolve the precipitate in 0.02*M* triethanolamine buffer, pH 7.7, and dialyze overnight against this buffer at 4°C. Filter the sample (*see* Note 1) before use (0.2 µm).
2. Assemble the FPLC system according to the manufacturer's instructions for use with the Mono-Q ion-exchange column.
3. Equilibrate the column with 0.02*M* triethanolamine buffer, pH 7.7 (buffer A). Run a blank gradient form 0 to 100% buffer B (buffer A + 1*M* NaCl). Use a flow rate of 4–6 mL/min for this and subsequent steps.
4. Load the sample depending on column size. Refer to the manufacturer's instructions for loading capacities of the columns.
5. Equilibrate the column with buffer A for at least 10 min.
6. Set the sensitivity in the UV monitor control unit, and zero the baseline or chart recorder.
7. Apply a salt gradient from 0 to 28% buffer B for about 30 min (*see* Note 2). Follow with 100% 1*M* NaCl for 15 min to purge the column of remaining proteins.
8. Wash the Mono-Q ion-exchange with at least 3 column volumes each of 2*M* NaOH followed by 2*M* NaCl.
9. Store the Mono-Q ion-exchange column in distilled water containing 0.02% NaN_3.

4. Notes

1. It is essential that the sample and all buffers be filtered using 0.2-µm filters. For HPLC, all buffers must be degassed by vacuum pressure.

2. For FPLC, IgG elutes between 10 and 25% buffer B, usually around 15%. When IgG elutes at 25% (dependent on pI), then it will tend to coelute with albumin, which elutes at around 27%. When this occurs, alternative purification methods should be employed. For HPLC, IgG elutes between 15 and 25% buffer B.

Reference

1. Johnstone, A. and Thorpe, R. (1996) *Immunochemistry in Practice*, 3rd ed., Blackwell Scientific, Oxford, UK.

129

Purification of IgG by Precipitation with Polyethylene Glycol (PEG)

Mark Page and Robin Thorpe

1. Introduction

PEG precipitation works well for IgM, but is less efficient for IgG; salt precipitation methods are usually recommended for IgG. PEG precipitation may be preferred in multistep purifications that use ion-exchange columns, because the ionic strength of the Ig is not altered. Furthermore, it is a very mild procedure that usually results in little denaturation of antibody. This procedure is applicable to both polyclonal antisera and most MAb containing fluids.

2. Materials

1. PEG solution: 20% (w/v) PEG 6000 in PBS.
2. PBS: 0.14*M* NaCl, 2.7 m*M* KCl, 1.5 m*M* KH_2PO_4, 8.1 m*M* Na_2HPO_4.

3. Methods

1. Cool a 20% w/v PEG 6000 solution to 4°C.
2. Prepare serum/ascitic fluid, and so forth, for fractionation by centrifugation at $10,000g_{av}$ for 20–30 min at 4°C. Discard the pellet. Cool to 4°C.
3. Slowly stir the antibody containing fluid, and add an equal volume of 20% PEG dropwise (*see* Note 1). Continue stirring for 20–30 min.
4. Centrifuge at $2000–4000g_{av}$ for 30 min at 4°C. Discard the supernatant, and drain the pellet. Resuspend in PBS or other buffer as described for ammonium sulfate precipitation (*see* Chapter 125).

4. Note

1. Although the procedure works well for most antibodies, it may produce a fairly heavy contamination with non-IgG proteins. If this is the case, reduce the concentration of PEG in 2% steps until the desired purification is achieved. Therefore, carry out a pilot-scale experiment before fractionating all of the sample.

From: The Protein Protocols Handbook
Edited by: J. M. Walker Humana Press Inc., Totowa, NJ

130

Purification of IgG Using Protein A or Protein G

Mark Page and Robin Thorpe

1. Introduction

Some strains of *Staphylococcus aureus* synthesize protein A, a group-specific ligand that binds to the Fc region of IgG from many species *(1,2)*. Protein A does not bind all subclasses of IgG, e.g., human IgG_3, mouse IgG_3, sheep IgG_1, and some subclasses bind only weakly, e.g., mouse IgG_1. For some species, IgG does not bind to protein A at all, e.g., rat, chicken, goat, and some MAbs show abnormal affinity for the protein. These properties make the use of protein A for IgG purification limited in certain cases, although it can be used to an advantage in separating IgG subclasses from mouse serum *(3)*. Protein G (derived from groups C and G *Streptococci*) also binds to IgG Fc with some differences in species specificity from protein A. Protein G binds to IgG of most species, including rat and goat, and recognizes most subclasses (including human IgG_3 and mouse IgG_1), but has a lower binding capacity. Protein G also has a high affinity for albumin, although recombinant DNA forms now exist in which the albumin-binding site has been spliced out, and are therefore very useful for affinity chromatography.

Another bacterial IgG-binding protein (protein L) has been identified *(4)*. Derived from *Peptostreptococcus magnus,* it binds to some κ (but not λ) chains. Furthermore, protein L binds to only some light-chain subtypes, although immunoglobulins from many species are recognized. Its full potential is yet to be established.

Finally, hybrid molecules produced by recombinant DNA procedures, comprising the appropriate regions of IgG-binding proteins (e.g., protein L/G, protein L/A) also have considerable scope in immunochemical techniques. These proteins are therefore very useful in the purification of IgG by affinity chromatography. Columns are commercially available (MabTrap G II, Pharmacia, Uppsala, Sweden) or can be prepared in the laboratory. The product of this method is of high purity and is useful for most immunochemical procedures including affinity chromatography and conjugation with radioisotopes, enzymes, biotin, and so forth.

2. Materials

1. PBS: 0.14*M* NaCl, 2.7 m*M* KCl, 1.5 m*M* KH_2PO_4, 8.1 m*M* Na_2HPO_4.
2. Sodium azide.
3. Dissociating buffer: 0.1*M* glycine-HCl, pH 3.5. Adjust the pH with 2*M* HCl.
4. 1*M* Tris.

From: The Protein Protocols Handbook
Edited by: J. M. Walker Humana Press Inc., Totowa, NJ

5. Binding buffer: Optimal binding performance occurs using a buffer system between pH 7.5 and 8.0. Suggested buffers include 0.1*M* Tris-HCl, 0.15*M* NaCl, pH 7.5; 0.05*M* sodium borate, 0.15*M* NaCl, pH 8.0; and 0.1*M* sodium phosphate, 0.15*M* NaCl, pH 7.5.
6. IgG preparation: serum, ascitic fluid, or hybridoma culture supernatant.
7. Protein A, G column.

3. Methods

Refer to Chapter 132 for CNBr activation of Sepharose and coupling of protein A, G.

1. Wash the column with an appropriate binding buffer.
2. Pre-elute the column with dissociating buffer, 0.1*M* glycine-HCl, pH 3.5.
3. Equilibrate the column with binding buffer.
4. Prepare IgG sample: If the preparation is serum, plasma, or ascitic fluid, dilute it at least 1:1 in binding buffer and filter through 0.45-μm filter. Salt-fractionated preparations (*see* Chapter 125) do not require dilution, but the protein concentration should be adjusted to approx 1–5 mg/mL. Hybridoma culture supernatants do not require dilution.
5. Apply sample to column at no more than 10 mg IgG/2-mL column.
6. Wash the column with binding buffer until the absorbance at 280 nm is <0.02.
7. Dissociate the IgG-ligand interaction by eluting with dissociating buffer. Monitor the absorbance at 280 nm, and collect the protein peak. Neutralize immediately with alkali (e.g., 1*M* Tris, unbuffered).
8. Wash the column with binding buffer until the pH returns to that of the binding buffer. Store the column in buffer containing at least 0.15*M* NaCl and 0.1% sodium azide.
9. Dialyze the IgG preparation against a suitable buffer (e.g., PBS) to remove glycine/Tris.

References

1. Lindmark, R., Thorén-Tolling, K., and Sjöquist, J. (1983) Binding of immunoglobulins to protein A and immunoglobulin levels in mammalian sera. *J. Immunol. Methods* **62,** 1–13.
2. Hermanson, G. T., Mallia, A. K., and Smith, P. K. (1992) *Immobilized Affinity Ligand Techniques.* Academic, San Diego, CA.
3. Ey, P. L., Prowse, S. J., and Jenkin, C. R. (1978) Isolation of pure IgG_1, IgG_{2a}, and IgG_{2b} immunoglobulins from mouse serum using protein A-sepharose. *Immunochemistry* **15,** 429–436.
4. Kerr, M. A., Loomes, L. M., and Thorpe, S. J. (1994) Purification and fragmentation of immunoglobulins, in *Immunochemistry Labfax* (Kerr, M. A. and Thorpe, R., eds.), Bios Scientific, Oxford, UK, pp. 83–114.

131

Purification of IgG Using Gel-Filtration Chromatography

Mark Page and Robin Thorpe

1. Introduction

In gel filtration, a protein mixture (the mobile phase) is applied to a column of small beads with pores of carefully controlled size (the stationary phase). The movement of the solute is dependent on the flow of the mobile phase, and the Brownian motion of the solute molecules causes their diffusion into and out of the chromatographic bed. Large proteins, above the "exclusion limit" of the gel, cannot enter the pore and, hence, are eluted in the "void volume" of the column (*see* Note 1). Small proteins enter the pores and are therefore eluted in the "total volume" of the column, and intermediate-size proteins are eluted between the void and total volumes. Proteins are therefore eluted in order of decreasing molecular size. Column matrices are available in a number of fractionation ranges, allowing the user to select the appropriate column for their particular application.

Gel filtration is not especially effective for the purification of IgG, which tends to elute in a broad peak and is usually contaminated with albumin (mainly derived from dimeric albumin, M_r 135,000). The technique is more useful for the purification of IgM or may be used as an adjunct to ion-exchange chromatography. Some IgGs (monoclonal), however, possess pIs that make them unsuitable for fractionation using ion-exchange chromatography. In such cases, it may be desirable to use gel filtration as a method of fractionation *(1)*.

2. Materials

1. Gel-filtration buffer (PBS): 0.14*M* NaCl, 2.7 m*M* KCl, 1.5 m*M* KH_2PO_4, 8.1 m*M* Na_2HPO_4. Store at 4°C.
2. Gel-filtration column, e.g., Bio-gel P 200 (Bio-Rad, Hemel Hempstead, UK), Sephacryl (Pharmacia, Uppsala, Sweden) (*see* Notes 2 and 3).
3. IgG preparation: serum, ascitic fluid, hybridoma culture supernatant.

3. Methods

3.1. Preparation and Equilibration of Gel-Filtration Column (see Notes 3 and 4)

1. Gently stir the filtration medium (enough to fill the column plus 10%) into two times the column volume of buffer.

From: The Protein Protocols Handbook
Edited by: J. M. Walker Humana Press Inc., Totowa, NJ

2. Degas the slurry using a Buchner side-arm flask under vacuum for 1–2 h with periodic swirling. Do not use a magnetic stirrer, since this may damage the beads.
3. Resuspend the slurry in approx 5 times its volume of buffer, and leave to stand until most of the beads have settled. Remove the fines by aspirating the supernatant down to about 1.5 times the settled slurry volume.
4. Carefully pack a clean column by first filling with a third column volume of buffer. Swirl the slurry to resuspend it evenly, and pour it down a glass rod onto the inside wall to fill the column. Allow to settle under gravity for 0.5–1 h and to let air bubbles escape.
5. Adjust the height of the outlet end of the column so that the vertical distance between it and the top of the column is less than the maximum operating pressure for the gel (*see* Note 5). Unclamp the bottom of the column and allow the gel to pack under this pressure.
6. Top up the column periodically by syphoning off excess supernatant, stirring the top of the gel (if it has settled completely), and filling the column up to the top with resuspended slurry.
7. Once the column is packed (gel bed just runs dry), connect the top of the column to a buffer reservoir, remove any air bubbles in the tube, and allow 1 column volume of buffer to run through the column.

3.2. Sample Application and Elution (see Note 6)

1. Disconnect the top of the column from the buffer reservoir, and allow the gel to just run dry.
2. Apply the IgG sample (*see* Note 2) carefully by running it down the inside wall of the column so that the gel bed is not disturbed.
3. When the sample has entered the bed, gently overlay the gel with buffer and reconnect the column to the buffer reservoir.
4. Collect fractions (4–6 mL) if using a 100-cm column, and monitor the absorbance at 280 nm (*see* Notes 7 and 8).
5. IgG-containing fractions may be identified and their purity assessed by electrophoresis in polyacrylamide gels (*see* Chapter 134).

4. Notes

1. The void volume of the column is the volume of liquid between the beads of the gel matrix and usually amounts to about one-third of the total column volume.
2. The choice of column size will depend on the sample size. As a general rule, the sample size should not exceed 5% of the total column volume. Ten to thirty milligrams of protein/cm^2 cross-sectional area is a suitable loading (equivalent to 10–30 mg of protein/100 mL of gel for a 100-cm column). When selecting a column for gel filtration, the column should be of controlled diameter glass with as low a dead space as possible at the outflow. Column tubing should be about 1 mm in diameter, which helps reduce dead space volume. A useful length for a gel-filtration column is about 100 cm, and the choice of cross-sectional area is governed by sample size, both in terms of volume and amount of protein. As a rough rule of thumb, the sample volume should not be >5% of the total column volume (1–2% gives better resolution). More protein will increase yield, but decrease resolution and, hence, purity, whereas less protein loaded improves resolution, but reduces the yield. Commercial columns suitable for gel filtration are available from a number of manufacturers. Ensure that columns are clean before use by washing with a weak detergent solution and rinsing thoroughly with water.
3. There are a number of criteria to consider when setting up a gel-filtration system, e.g., choice of gel, column, and buffer. The gel of choice may be composed of beads of carbo-

hydrate or polyacrylamide, and is available in a wide variety of pore sizes and, hence, fractionation ranges. Useful gels for the separation of IgG include Ultrogel AcA 44 (mixtures of dextran and polyacrylamide) and Bio-Gel P 200 (polyacrylamide). For IgM fractionation, Sephacel S-300 (crosslinked dextran) is useful.

Gel filtration can be performed using a wide variety of buffers generally of physiological specification, i.e., pH 7.2, 0.1*M*.

4. Gel-filtration columns may also be packed by using an extension reservoir, attached directly to the top of the column. In this case, the total volume of gel and buffer can be poured and allowed to settle without continual topping up. When using an extension reservoir, leave the column to pack until the gel bed just runs dry, and then remove excess gel.
5. The operating pressure of the gel is the vertical distance between the top of the buffer in the reservoir and the outlet end of the tube; this should never exceed the manufacturer's recommended maximum for the gel.
6. When loading commercial columns with flow adaptors touching the gel bed surface, the most satisfactory way to apply the sample is by transferring the inlet tube from the buffer reservoir to the sample. The sample enters the tube and gel under operating pressure, and then the tube must be returned to the buffer reservoir. Ensure that no air bubbles enter the tube. In the case of homemade columns, it is also possible to load the sample directly onto the gel bed by preparing it in 5% w/v sucrose. Elution of samples is either by pressure from the reservoir (operating pressure) or by a peristaltic pump between the reservoir and the top of the column. In general, the flow rate should be the volume contained in 2–4 cm height of column/hour. In order to prevent columns without pumps from running dry, the inlet tube should be arranged so that part of it is below the outlet point of the column. Eluted fractions should be collected by measured volume rather than time (this prevents fluctuations in flow rate from altering fraction size). In general, for a 100-cm column, a column volume of eluent should be collected in about 100 fractions, and the void volume is eluted at about fraction 30–35.
7. The protein yield for gel-filtration chromatography should be >80% and is often as high as 95%. Yield will improve after the first use of a gel, because new gel adsorbs protein nonspecifically in a saturable fashion. When very small amounts of protein are being fractionated, the column should be saturated with an irrelevant protein, e.g., albumin, prior to use.
8. Columns should be stored at room temperature or 4°C, usually in the presence of a bacteriostatic agent, e.g., 0.02% sodium azide.

References

1. Guse, A. H., Milton, A. D., Schulze-Koops, H., Müller, B., Roth, E., Simmer, B., Wächter, H., Weiss, E., and Emmrich, F. (1994) Purification and analytical characterization of an anti-CD4 monoclonal antibody for human therapy. *J. Chromatogr.* **661,** 13–23.

132

Purification of IgG Using Affinity Chromatography on Antigen-Ligand Columns

Mark Page and Robin Thorpe

1. Introduction

Affinity chromatography is a particularly powerful procedure, which can be used to purify IgG, subpopulations of IgG, or the antigen binding fraction of IgG present in serum/ascitic fluid/hybridoma culture supernatant. This technique requires the production of a solid matrix to which a ligand having either affinity for the relevant IgG or vice versa has been bound *(1)*. Examples of ligands useful in this context are:

1. The antigen recognized by the IgG (for isolation of the antigen-specific fraction of the serum/ascitic fluid, and so forth).
2. IgG prepared from an anti-immunoglobulin serum, e.g., rabbit antihuman IgG serum or murine antihuman IgG MAb for the purification of human IgG (*see* Note 1).
3. IgG-binding proteins derived from bacteria, e.g., protein A (from *Staphylococcus aureus* Cowan 1 strain) or proteins G or C (from *Streptococcus* and *see* Chapter 130).

The methods for production of such immobilized ligands and for carrying out affinity-purification of IgG are essentially similar, regardless of which ligand is used. Sepharose 4B is probably the most widely used matrix for affinity chromatography, but other materials are available. Activation of Sepharose 4B is usually carried out by reaction with cyanogen bromide (CNBr); this can be carried out in the laboratory before coupling, or ready-activated lyophilized Sepharose can be purchased. The commercial product is obviously more convenient than "homemade" activated Sepharose, but it is more expensive and may be less active.

2. Materials

1. Sepharose 4B.
2. Sodium carbonate buffer: 0.5*M* Na_2CO_3, pH 10.5. Adjust pH with 0.1*M* NaOH.
3. Cyanogen bromide. (**Warning: CNBr is toxic and should be handled in a fume hood.)**
4. Sodium hydroxide: 1*M*; 4*M*.
5. Sodium citrate buffer: 0.1*M* trisodium citrate, pH 6.5. Adjust pH with 0.1*M* and citric acid.
6. Ligand solution (2–10 mg/mL in 0.1*M* sodium citrate buffer, pH 6.5).
7. Ethanolamine buffer: 2*M* ethanolamine.
8. PBS: 0.14*M* NaCl, 2.7 m*M* KCl, 1.5 m*M* KH_2PO_4, 8.1 m*M* Na_2HPO_4.

From: The Protein Protocols Handbook
Edited by: *J. M. Walker Humana Press Inc., Totowa, NJ*

9. IgG preparation—serum, ascitic fluid, hybridoma culture supernatant.
10. PBS containing 0.1% sodium azide.
11. Disassociating buffer: 0.1*M* glycine, pH 2.5. Adjust pH with 1*M* HCl.
12. 1*M* Tris-HCl buffer, pH 8.8. Adjust pH with 1*M* HCl.

3. Method

3.1. Activation of Sepharose with CNBr and Preparation of Immobilized Ligand

Activation of Sepharose with CNBr requires the availability of a fume hood and careful control of the pH of the reaction—failure to do this may lead to the production of dangerous quantities of HCN as well as compromising the quality of the activated Sepharose. CNBr is toxic and volatile. All equipment that has been in contact with CNBr and residual reagents should be soaked in 1*M* NaOH overnight in a fume hood and washed before discarding/returning to the equipment pool. Manufacturers of ready activated Sepharose provide instructions for coupling (*see* Note 2).

1. Wash 10 mL (settled volume) of Sepharose 4B with 1 L of water by vacuum filtration. Resuspend in 18 mL of water (do not allow the Sepharose to dry out).
2. Add 2 mL of 0.5*M* sodium carbonate buffer, pH 10.5, and stir slowly. Place in a fume hood and immerse the glass pH electrode in the solution.
3. **Carefully** weigh 1.5 g of CNBr into an air-tight container (Note: weigh in a fume hood; wear gloves)—remember to decontaminate equipment that has contacted CNBr in 1*M* NaOH overnight.
4. Add the CNBr to the stirred Sepharose. Maintain the pH between 10.5 and 11.0 by dropwise addition of 4*M* NaOH until the pH stabilizes and all the CNBr has dissolved. If the pH rises above 11.5, activation will be inefficient, and the Sepharose should be discarded.
5. Filter the slurry using a sintered glass or Buchner funnel, and wash the Sepharose with 2 L of cold 0.1*M* sodium citrate buffer, pH 6.5—do not allow the Sepharose to dry out. Carefully discard the filtrate (use care: this contains CNBr).
5. Quickly add the filtered washed Sepharose to the ligand solution (2–10 mg/mL in 0.1*M* sodium citrate, pH 6.5), and gently mix on a rotator ("windmill') at 4°C overnight (*see* Note 3).
7. Add 1 mL of 2*M* ethanolamine solution, and mix at 4°C for a further 1 h—this blocks unreacted active groups.
8. Pack the Sepharose into a suitable chromatography column (e.g., a syringe barrel fitted with a sintered disk) and wash with 50 mL of PBS. Store at 4°C in PBS containing 0.1% sodium azide.

3.2. Sample Application and Elution

1. Wash the affinity column with PBS. "Pre-elute" with dissociating buffer, e.g., 0.1*M* glycine-HCl, pH 2.5. Wash with PBS; check that the pH of the eluate is the same as the pH of the PBS (*see* Note 4).
2. Apply the sample (filtered through a 0.45-µm membrane) to the column. As a general rule, add an eqiuvalent amount (mole:mole) of IgG in the sample to that of the ligand coupled to the column. Close the column exit, and incubate at room temperature for 15–30 min (*see* Note 5).
3. Wash non-IgG material from the column with PBS; monitor the A_{280} as an indicator of protein content.

4. When the A_{280} reaches a low value (approx 0.02), disrupt the ligand–IgG interaction by eluting with dissociating buffer. Monitor the A_{280}, and collect the protein peak into tubes containing 1*M* Tris-HCl, pH 8.8 (120 µL/1 mL fraction) to neutralize the acidic dissociating buffer.
5. Wash the column with PBS until the eluate is at pH 7.4. Store the column in PBS containing 0.1% azide. Dialyze the IgG preparation against a suitable buffer (e.g., PBS) to remove glycine/Tris.

4. Notes

1. The use of subclass-specific antibodies or MAb allows the immunoaffinity isolation of individual subclasses of IgG.
2. Coupling at pH 6.5 is less efficient than at higher pH, but is less likely to compromise the binding ability of immobilized ligands (especially antibodies).
3. Check the efficiency of coupling by measuring the A_{280} of the ligand before and after coupling. Usually at least 95% of the ligand is bound to the matrix.
4. Elution of bound substances is usually achieved by using a reagent that disrupts noncovalent bonds. These vary from "mild" procedures, such as the use of high salt or high or low pH, to more drastic agents, such as 8*M* urea, 1% SDS or 5*M* guanidine hydrochloride. Chaotropic agents, such as 3*M* thiocyanate or pyrophosphate, may also be used. Usually an eluting agent is selected that is efficient, but does not appreciably denature the purified molecule; this is often a compromise between the two ideals. In view of this, highly avid polyclonal antisera obtained from hyperimmune animals are often not the best reagents for immunoaffinity purification, as it may be impossible to elute the IgG in a useful form. The 0.1*M* glycine-HCl buffer, pH 2.5, will elute most IgG, but may denature some MAb. "Pre-elution" of the column with dissociating reagent just before affinity chromatography ensures that the isolated immunoglobulin is minimally contaminated with ligand.
5. The column will only bind to its capacity and therefore some IgG may not be bound; however, this can be saved and passed through the column again. The main problem is with back pressure or even blockage, but this can be reduced by diluting the sample by at least 50% with a suitable buffer (e.g., PBS). Incubation of the IgG containing sample with the ligand matrix is not always necessary, but this will allow maximal binding to occur. Alternatively, slowly recirculate the sample through the column, typically at <0.5 mL/min.

References

1. Hermanson, G. T., Mallia, A. K., and Smith, P. K. (1992) *Immobilized Affinity Ligand Techniques.* Academic, San Diego, CA.

133

Purification of IgG Using Thiophilic Chromatography

Mark Page and Robin Thorpe

1. Introduction

Immunoglobulins recognise sulphone groups in close proximity to a thioether group *(1)*, and, therefore, thiophilic adsorbents provide an additional chromatographic method for the purification of immunoglobulins that can be carried out under mild conditions preserving biological activity. A thiophilic gel is prepared by reducing divinylsulfone (coupled to Sepharose 4B) with β-mercaptoethanol. The product is of intermediate purity and would be useful for further processing, e.g., purification by ion-exchange, size exclusion, and/or affinity chromatography.

2. Materials

1. Sepharose 4B.
2. 0.5*M* sodium carbonate.
3. Divinylsulfone.
4. Coupling buffer: 0.1*M* sodium carbonate buffer, pH 9.0.
5. β-mercaptoethanol.
6. Binding buffer: 0.1*M* Tris-HCl, pH 7.6, containing 0.5*M* K_2SO_4.
7. IgG preparation: serum, ascitic fluid, or hybridoma culture supernatant.
8. 0.1*M* ammonium bicarbonate.

3. Methods

Caution: Divinylsulfone is highly toxic and the column preparation procedures should be carried out in a well-ventilated fume cabinet.

1. Wash 100 mL Sepharose 4B (settled volume) with 1 L of water by vacuum filtration.
2. Resuspend in 100 mL of 0.5*M* sodium carbonate, and stir slowly.
3. Add 10 mL of divinylsulfone dropwise over a period of 15 min with constant stirring. After addition is complete, slowly stir the gel suspension for 1 h at room temperature.
4. Wash the activated gel thoroughly with water until the filtrate is no longer acidic (*see* Note 1).
5. Wash activated gel with 200 mL of coupling buffer using vacuum filtration and resuspend in 75 mL of coupling buffer.
6. In a well-ventilated fume cabinet, add 10 mL of β-mercaptoethanol to the gel suspension with constant stirring, and continue for 24 h at room temperature (*see* Note 2).

From: The Protein Protocols Handbook
Edited by: J. M. Walker Humana Press Inc., Totowa, NJ

7. Filter and wash the gel thoroughly. The gel may be stored at 4°C in 0.02% sodium azide.
8. Pack 4 mL of the gel in a polypropylene column (10 × 1 cm) and equilibrate with 25 mL of binding buffer.
9. Perform chromatography at 4°C. Mix 1 mL of IgG containing sample with 2 mL of binding buffer, and load onto the column.
10. After the sample has entered the gel, wash non-IgG from the column with 20 mL of binding buffer; monitor the A_{280} as an indicator of protein content in the wash until the absorbance returns to background levels.
11. Elute the bound IgG with 0.1*M* ammonium bicarbonate, and collect into 2-mL fractions. Monitor the protein content by absorbance at 280 nm, and pool the IgG containing fractions (i.e., those with protein absorbance peaks). Dialyze against an appropriate buffer (e.g., PBS) with several changes, and analyze by gel electrophoresis under reducing conditions (*see* Chapter 134).

4. Notes

1. The activated gel can be stored by washing thoroughly in acetone and kept as a suspension in acetone at 4°C.
2. Immobilized ligands prepared by the divinylsulfone method are unstable above pH 8.0.

References

1. Porath, J., Maisano, F., and Belew, M. (1985) Thiophilic adsorption—a new method for protein fractionation. *FEBS Lett.* **185,** 306–310.

134

Analysis of IgG Fractions by Electrophoresis

Mark Page and Robin Thorpe

1. Introduction (*see* Note 1)

After using a purification procedure it is necessary to obtain some index of purity obtained. One of the simplest methods for assessing purity of an IgG fraction is by sodium dodecyl sulfate-polyacrylamide gel electrophoresis (SDS-PAGE). Although "full-size" slab gels can be used with discontinuous buffer systems and stacking gels, the use of a "minigel" procedure, using a Tris-bicine buffer system *(1),* rather than the classical Tris-glycine system, is quicker and easier, and gives improved resolution of immunoglobulin light-chains (these are usually smeared with the Tris-glycine system). Gel heights can be restricted to <10 cm and are perfectly adequate for assessing purity and monitoring column fractions.

2. Materials

1. Gel solution: 2.0 mL of 1*M* Tris, 1*M* bicine; 4.0 mL 50% w/v acrylamide containing 2.5% w/v bis-acrylamide; 0.4 mL 1.5% w/v Ammonium persulfate; 0.2 mL 10% w/v SDS.
   Make up to 20 mL with distilled water.
2. Gel running buffer: 2.8 mL 1*M* Tris, 1*M* bicine; 1.4 mL 10% w/v SDS.
   Make up to 140 mL with distilled water.
3. Sample buffer: 1.0 g Sucrose, 0.2 mL 1*M* Tris, 1*M* bicine, 1.0 mL 10% SDS, 0.25 mL 2-mercaptoethanol.
   Make up to 3 mL with distilled water, and add 0.001% w/v bromophenol blue. Store at –20°C.
4. Coomassie blue R stain: Add Coomassie brilliant blue R (0.025 g) to methanol (50 mL), and stir for 10 min. Add distilled water (45 mL) and glacial acetic acid (5 mL). Use within 1 mo.
5. Destain solution: Glacial acetic acid (7.5 mL) and methanol (5 mL). Make up to 100 mL with distilled water.
5. *N,N,N',N'*-Tetramethylethylenediamine (TEMED).
7. Molecular-weight markers (mol-wt range 200,000–14,000).
8. Purified IgG or column fraction samples.

3. Methods

1. Prepare sample buffer.
2. Adjust antibody preparation to 1 mg/mL in 0.1*M* Tris, 0.1*M* bicine (*see* Note 2).

From: The Protein Protocols Handbook
Edited by: J. M. Walker Humana Press Inc., Totowa, NJ

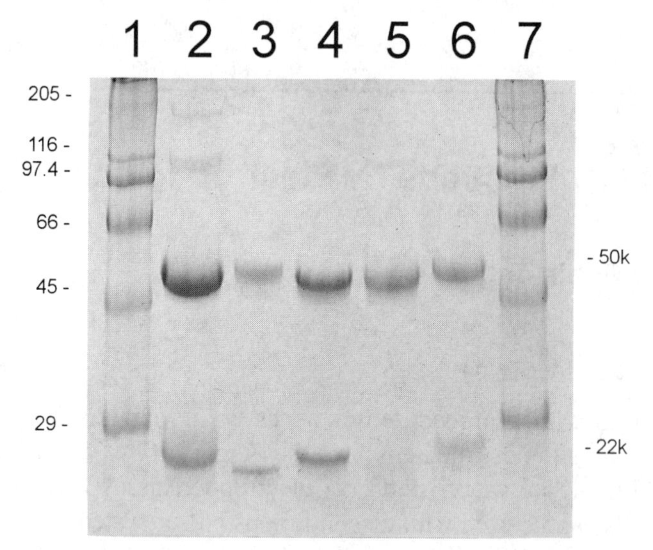

Fig. 1. SDS-PAGE minigel depicting purified IgG preparations derived from human serum (lane 2), mouse ascitic fluid (lanes 3 and 4), rabbit serum (lane 5), and sheep serum (lane 6). Samples are electrophoresed under reducing conditions and stained with Coomassie blue. All IgGs consist predominantly of two bands comprising heavy (50,000 M_r) and light (22,000 M_r) chains with no major contaminating proteins. Molecular-weight markers are shown in lanes 1 and 7 and their molecular weights (in thousands) given on the left.

3. Mix the sample in the ratio 2:1 with sample buffer.
4. Heat at 100°C for 2–4 min.
5. Prepare gel solution and running buffer as described in Section 2.
6. Assemble gel mold according to manufacturer's instructions.
7. Add 30 µL TEMED to 10 mL of gel solution, pour this solution between the plates to fill the gap completely, and insert the comb in the top of the mold (there is no stacking gel with this system). Leave for 10 min for gel to polymerize.
8. Remove comb and clamp gel plates into the electrophoresis apparatus. Fill the anode and cathode reservoirs with running buffer.
9. Load the sample(s) (30–50 µL/track), and run an IgG reference standard and/or mol-wt markers in parallel.
10. Electrophorese at 150V for 1.5 h.
11. Remove gel from plates carefully, and stain with Coomassie blue R stain for 2 h (gently rocking) or overnight (stationary) (*see* Note 3).
12. Pour off the stain, and rinse briefly in tap water.
13. Add excess destain to the gel. A piece of sponge added during destaining absorbs excess stain. Leave until destaining is complete (usually overnight with gentle agitation).

4. Notes

1. The "minigel" is easily and quickly prepared consisting of a resolving (separating) gel only and takes approx 1.5 h to run once set up. The IgG sample is prepared for electrophoresis under reduced conditions, and is run in parallel with either a reference IgG preparation or with standard mol-wt markers. The heavy chains have a characteristic relative molecular weight of approx 50,000 and the light-chains a molecular weight of 22,000 (Fig. 1).

2. The extinction coefficient ($E^{1\%}_{280}$) of human IgG is 13.6 (i.e., a 1 mg/mL solution will have an A_{280} of 1.36).
3. Mark the gel uniquely prior to staining so that its orientation is known. By convention, a small triangle of the bottom left- or right-hand corner of the gel is sliced off.

Acknowledgment

We would like to thank Chris Ling for expert technical assistance.

References

1. Johnstone, A. and Thorpe, R. (1996) *Immunochemistry in Practice,* 3rd ed., Blackwell Scientific, Oxford, UK.

135

Ouchterlony Double Immunodiffusion

Graham S. Bailey

1. Introduction

Immunodiffusion in gels encompasses a variety of techniques that are useful for the analysis of antigens and antibodies *(1)*. The fundamental immunochemical principles behind their use are exactly the same as those that apply to antigen–antibody interactions in the liquid state. Thus, an antigen will react with its specific antibody to form a complex, the composition of which will depend on the nature, concentrations, and proportions of the initial reactants. As increasing amounts of a multivalent antigen are allowed to react with a fixed amount of antibody, precipitation occurs, in part because of extensive crosslinking between the reactant molecules. Initially, the antibody is in excess, and all the added antigen is present in the form of soluble antigen–antibody adducts. Addition of more antigen leads to the formation of an immune precipitate (precipitin) where all of the antigen and antibody molecules are in an extensive lattice of antigen–antibody complex. Further addition of antigen produces an excess of antigen and leads to a reduction in the amount of the precipitate as soluble antigen–antibody adducts are formed again. The analysis of such interactions occurring in gels is of much higher sensitivity and resolution than that for the liquid state, thus explaining the extensive use of immunochemical gel techniques.

The methods employing immunodiffusion in gels are often classified as simple (single) diffusion or double diffusion. In single diffusion (*see* Chapter 136), one of the reactants (often the antigen) is allowed to diffuse from solution into a gel containing the corresponding reactant, whereas in double diffusion, both antigen and antibody diffuse into the gel. A variety of information, both qualitative and quantitative, can be obtained from the numerous techniques *(2)*. This chapter describes Ouchterlony double immunodiffusion. It is used extensively to check antisera for the presence and specificity of antibodies for a particular antigen (*see* Notes 1 and 2).

2. Materials

1. Agarose.
2. 0.07*M* barbitone buffer, pH 8.6, containing 0.01% thimerosal as antibacterial agent. The buffer is prepared by dissolving sodium barbitone (14.5 g), disodium hydrogen phosphate decahydrate (7.16 g), boric acid (6.2 g), and thimerosal (sodium ethyl mercurithiosalicylate) (0.1 g) in distilled water and making the final volume up to 1 L. Barbitone is a controlled substance, and its use is strictly regulated (*see* Note 3).

From: The Protein Protocols Handbook
Edited by: J. M. Walker Humana Press Inc., Totowa, NJ

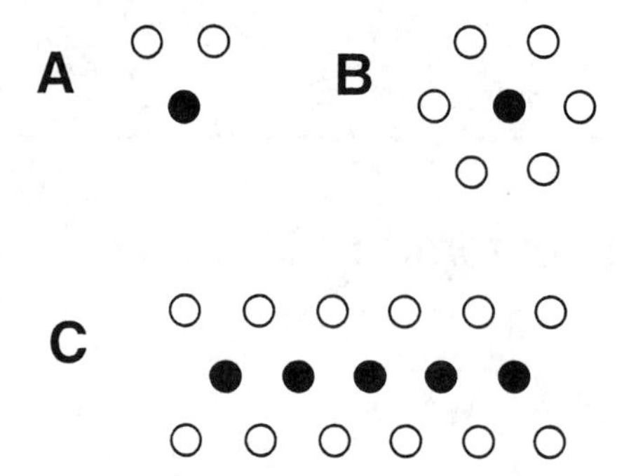

Fig. 1. Patterns of wells often used in double immunodiffusion: ○, well contains the antigen; ●, well contains the antiserum.

3. Solution of antigen and antiserum (*see* Section 3.).
4. Plastic or glass Petri dishes or rectangular plates.
5. Gel punch and template. Suitable gel punches of various sizes (2-, 2.5-, and 4-mm diameter) and templates of various designs can be obtained from commercial sources (Pharmacia Biotech, Uppsala, Sweden).
6. Flat level surface.
7. Humidity chamber at constant temperature.
8. 0.1*M* sodium chloride in distilled water.
9. Staining solution. The solution is prepared by mixing ethanol (90 mL), glacial acetic acid (20 mL), and distilled water (90 mL), and then Coomassie brilliant blue R-250 (1 g) is added with stirring for 6 h. The solution is filtered before use and can be reused several times.
10. Destaining solution. The composition is the same as that of the staining solution, but without the dye.

3. Method

1. Agarose (1 g) is dissolved in the barbitone buffer (100 mL) by heating to 90°C on a water bath with constant stirring for about 15 min (*see* Note 4).
2. The agarose solution is poured to a depth of 1–2 mm into the Petri dishes or onto the rectangular plates that had previously been set on a horizontal level surface. The gels are allowed to form on cooling and when set (5–10 min) can be stored at 4°C in a moist atmosphere for at least 1 wk (*see* Note 5).
3. A template of the desired pattern, according to the number of samples to be analyzed, is positioned on top of the gel. Commonly used patterns are shown in Fig. 1. The gel punch, which is connected to a water vacuum pump, is inserted into the gel in turn through each hole of the template so that the wells are cleanly formed as the resultant agarose plugs are sucked out. If a gel punch or a vacuum pump is unavailable, wells can be made using a glass Pasteur pipet and teat (*see* Note 6).
4. The wells are filled with the solutions of antigen and antisera until the meniscus just disappears. The concentrations of antigen solutions and the dilution of the antiserum to be used have to be established largely by trial and error by running pilot experiments with solutions of different dilutions. However, as a rough guide, for the analysis of antigen, the following concentrations of antigen solutions can be run against undiluted antiserum: for a pure antigen, 1 mg/mL; for a partially pure antigen, 50 mg/mL; for a very impure antigen, 500 mg/mL, using 5 μL samples of antigen solutions and antiserum.

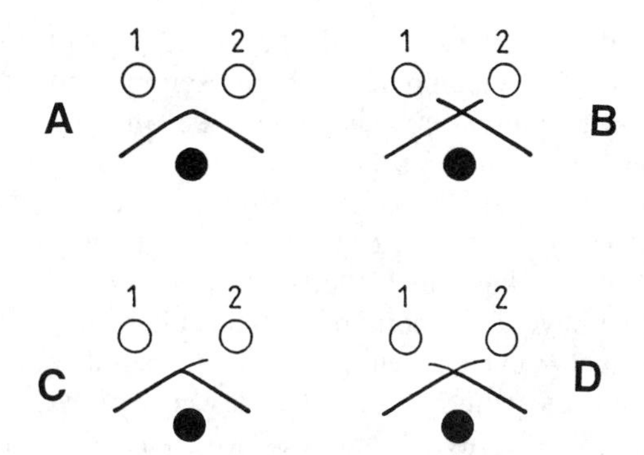

Fig. 2. Patterns of precipitin lines: (A) pattern of coalescence; (B) pattern of absence of coalescence; (C) pattern of partial coalescence of one antigen; (D) pattern of partial coalescence of two antigens. Well 1 contains antigen 1; well 2 contains antigen 2; well ● contains antiserum; —precipitin line.

5. The gel plate is then left in a moist atmosphere, e.g., in a humidity chamber at a constant temperature of 20°C for 24 h. A sealed plastic sandwich box containing wet tissue paper at one end can act as a suitable humidity chamber (*see* Notes 7 and 8).
6. The precipitin lines can then be recorded either directly by photography with dark-field illumination or by drawing the naked eye observation on suitable oblique illumination of the plate against a dark background or after staining. Prior to staining, excess moisture is first removed from the gel plate by application of a 1-kg weight (e.g., a liter beaker full of water) to a wad of filter paper (the filter paper in contact with the gel having been wetted with distilled water to prevent it from sticking to the gel surface) placed on top of the gel for 15 min. Remaining soluble protein is then removed by washing the gel (3 × 15 min) in 0.1*M* sodium chloride solution followed by further pressing. The gel is then dried using cold air from a hair dryer and is then placed in the staining solution for 5 min. The plate is finally washed with distilled water and placed in the destaining solution for about 10 min.

4. Notes

1. Ouchterlony double diffusion is frequently used for comparing different antigen preparations. If the antigen solution contains several different antigens that can react with the antibodies of the antiserum, then multiple lines of precipitation will be produced. The relative position of each line is determined by the local concentration of each antigen and antibody in the gel. In turn, those concentrations depend not only on the initial concentrations of the reactants in the wells, but on the rates at which they diffuse through the gel and, hence, are also dependent on molecular size.

   If different antigen preparations, each containing a single antigenic species capable of reacting with the antiserum used, are allowed to diffuse from separate wells, then the degree of similarity of the antigens can be assessed by observation of the geometrical pattern produced. For example, Fig. 2 shows the four basic precipitin patterns that can be produced in a balanced system by two related antigens interacting with an antiserum that contains antibodies that can recognize both antigens.

   Pattern (a) is called the "pattern of identity" or "pattern of coalescence," and indicates that the antibodies in the antiserum employed cannot distinguish the two antigens, i.e., the two antigens are immunologically identical as far as that antiserum is concerned.

Pattern (b) is called the "pattern of nonidentity" or "pattern of absence of coalescence," and indicates that none of the antibodies in the antiserum employed react with antigenic determinants that may be present in both antigens, i.e., the two antigens are unrelated as far as that antiserum is concerned.

Pattern (c) is called a "pattern of partial identity" or "pattern of partial coalescence of one antigen," and indicates that more of the antibodies in the antiserum employed react with one antigen (that diffusing from the left-hand well in Fig. 2c) than the other antigen. The "spur," extending beyond the point of partial coalescence, is thought to result from the determinants present in one antigen but lacking in the other antigen.

Pattern (d) is called "the pattern of partial coalescence of two antigens" and is another type of "pattern of partial identity." It indicates that some of the antibodies in the antiserum employed react with both antigens, whereas other antibodies only react with one or other of the antigens. Thus, the two antigens have at least one antigenic determinant in common, but also differ in other antigenic determinants.

2. Ouchterlony double immunodiffusion is also frequently used to check the specificity of a polyclonal antiserum. If the antiserum contains antibody molecules that can react with different antigens, then multiple lines of precipitation will be seen on double immunodiffusion. If, however, the antiserum is monospecific for a particular antigen, only a single line of precipitation will be seen on immunodiffusion whenever that antiserum is tested against samples containing the crossreacting antigen. No line of precipitation will be formed between the monospecific antiserum and a sample that does not contain the crossreacting antigen.
3. Buffers of different compositions and pHs from the described buffer can be used if more convenient, e.g., 0.05*M* phosphate buffer, pH 7.2, containing 0.1*M* sodium chloride.
4. The agarose gel solution, once made, can be separated into test tubes and stored at 4°C for several weeks. Approximately 12 mL of 1% agarose gel solution is required for an area of 75 mL. The contents of each test tube can be simply melted as required.
5. If glass Petri dishes or rectangular plates are used for this technique, it normally is necessary for them to be precoated with a 0.5% agarose gel prior to formation of the main gel to prevent the main gel from becoming detached during the washing and staining procedures. A sufficient amount of the 0.5% agarose solution is pipeted onto the level surface of a clean, dry plate or Petri dish to cover the surface completely. The gel solution is allowed to set at room temperature.
6. The sensitivity of the Ouchterlony technique is largely determined by the relative concentrations of the antigens and antibodies, and by the separation of the wells. In the latter case, the closer the wells, the greater the sensitivity. It should be possible to detect protein solutions of 10 μg/mL concentration.
7. Diffusion of antigens and antibodies is more rapid at higher temperatures. Thus, double immunodiffusion is often run for 48 h at 4°C, 24 h at room temperature, or for as little as 3 h at 37°C. It is important to maintain a constant temperature. Otherwise, artifacts may be produced.
8. Other artifacts can be produced by refilling of the wells or by denaturation of the antigen during diffusion. If the latter is a possibility, it is best to run the immunodiffusion at 4°C.

References

1. Oudin, J. (1980) Immunochemical analysis by antigen–antibody precipitation in gels, in *Methods in Enzymology,* vol. 70 (Van Vunakis, H. and Langone, J. J., eds.), Academic, New York, pp. 166–198.
2. Ouchterlony, O. and Nilsson, L. A. (1986) Immunodiffusion and immunoelectrophoresis, in *Handbook of Experimental Immunology*, vol. 1 (Weir, D. M., Herzerberg, L. A., Blackwell, C., and Herzerberg, L. A., eds.), 4th ed., Blackwell, Oxford, UK, pp. 32.1–32.50.

136

Single Radial Immunodiffusion

Graham S. Bailey

1. Introduction

Single radial immunodiffusion is used extensively for the quantitative estimation of antigens *(1)*. In this method, the antigen–antibody precipitation is made more sensitive than in double immunodiffusion (*see* Chapter 135) by the incorporation of the antiserum in the agar solution before the gel is made *(2)*. Thus, the antiserum is uniformly distributed throughout the agar gel. Antigen is then allowed to diffuse from wells cut into the agar gel. This is an example of single (simple) immunodiffusion. Initially, as the antigen diffuses out of the well, its concentration is relatively high and it forms relatively soluble antigen–antibody adducts. However, as it diffuses further and further from the well, its concentration decreases. When its concentration becomes equivalent to that of the antibody in the gel, a disk of antigen–antibody precipitate (precipitin) is formed. The greater the initial concentration of antigen in the well, the greater the diameter of the precipitin disk. Thus, by running a range of known antigen concentrations on the gel and by measuring the diameters of their precipitin disks, a calibration graph can be constructed. The antigen concentrations of unknown samples run on the same gel can then be found by simple interpolation having measured the diameters of the respective precipitin disks.

2. Materials

1. Agarose.
2. 0.07*M* barbitone buffer, pH 8.6, containing 0.01% thimerosal as antibacterial agent. The buffer is prepared by dissolving sodium barbitone (14.5 g), disodium hydrogen phosphate decahydrate (7.16 g), boric acid (6.2 g), and thimerosal (sodium ethyl mercurithiosalicylate) (0.1 g) in distilled water, and making the final volume up to 1 L. Barbitone is a controlled substance, and its use is strictly regulated.
3. Solutions of antigen and antiserum.
4. Plastic or glass rectangular plates (84 × 94 mm).
5. Gel punch and template: Suitable gel punches of various sizes (2-, 2.5-, and 4-mm diameter) and templates of various designs can be obtained from commercial sources (Pharmacia Biotech, Uppsala, Sweden).
6. Flat level surface.
7. Humidity chamber at constant temperature: This can simply be a sealed plastic sandwich box with a wet tissue paper at one end.
8. 0.1*M* sodium chloride in distilled water.

From: The Protein Protocols Handbook
Edited by: J. M. Walker Humana Press Inc., Totowa, NJ

9. Staining solution: The solution is prepared by mixing ethanol (90 mL), glacial acetic acid (20 mL), and distilled water (90 mL), and then Coomassie brilliant blue R-250 (1 g) is added with stirring for 6 h. The solution is filtered before use and can be reused several times.
10. Destaining solution: The composition is the same as that of the staining solution, but without the dye.

3. Method

1. Agarose (2 g) is dissolved in the barbitone buffer (100 mL) by heating to 90°C on a water bath with constant stirring for about 30 min.
2. The solution is placed in a water bath and allowed to cool to 55°C. An appropriate volume is mixed with an equal volume of a suitable dilution of the monospecific antiserum also at 55°C. The optimal dilution of the antiserum has to be found from pilot experiments using different dilutions of the antiserum. The dilution chosen depends in part on the range of antigen concentrations that is required to be measured. It is suggested that a 1:100 dilution be initially employed.
3. The agarose solution containing the antiserum (12 mL) is poured onto the rectangular plate (84 × 94 mm) that had previously been set on a horizontal level surface. The gel is allowed to form on cooling. Gel plates are normally used within 24 h, but can be stored for longer periods at 4°C in a humidity chamber.
4. Using a gel punch and a template, a row of a desired number of wells of 2-mm diameter is cut into the gel.
5. A constant volume of standard solutions of antigen of different concentrations (e.g., 2.5, 7.5, and 22.5 µg/mL) is added to at least three of the wells. The same volume of solutions containing the antigen at unknown concentrations is added to the other wells. Addition should be carried out using a microsyringe. The wells must not be completely filled in order to avoid overflow onto the gel surface.
6. The gel plate is then left in a humidity chamber for at least 24 h. As the antigen diffuses into the gel containing the antiserum, a disk of immune precipitate is formed. The final, maximum diameter of that disk is directly proportional to the initial concentration of the antigen in the well (and inversely proportional to the antibody concentration in the gel). The time required to achieve maximum precipitation depends on the velocity of diffusion of the antigen. That, in turn, is dependent on temperature and molecular size of the antigen. In general, diffusion at room temperature must be allowed to continue for a few days, taking measurements of the diameters of the disks every 24 h until no further increase takes place.
7. The diameter of each disk can be measured directly with the aid of a magnifying glass on suitable oblique illumination of the plate positioned over a dark background. For increased resolution of the disks, the plate can be placed in 1% tannic acid for 3 min prior to viewing.
8. At the end of the diffusion process, the plate can be stained. Soluble protein is removed from the gel by washing for 2 d with several changes of 0.1*M* sodium chloride solution. The plate is then pressed, dried, stained, and destained as detailed for Ouchterlony double immunodiffusion (*see* Chapter 135).
9. A standard graph is constructed by plotting the diameters of the disks against the logarithm of the antigen solutions of known concentration. The concentrations of the antigen in the test samples can then be determined by simple interpolation (*see* Fig. 1).

4. Notes

1. If samples in neighboring wells are too concentrated, their precipitin disks will overlap and, in that case, the samples have to be run again at higher dilutions.

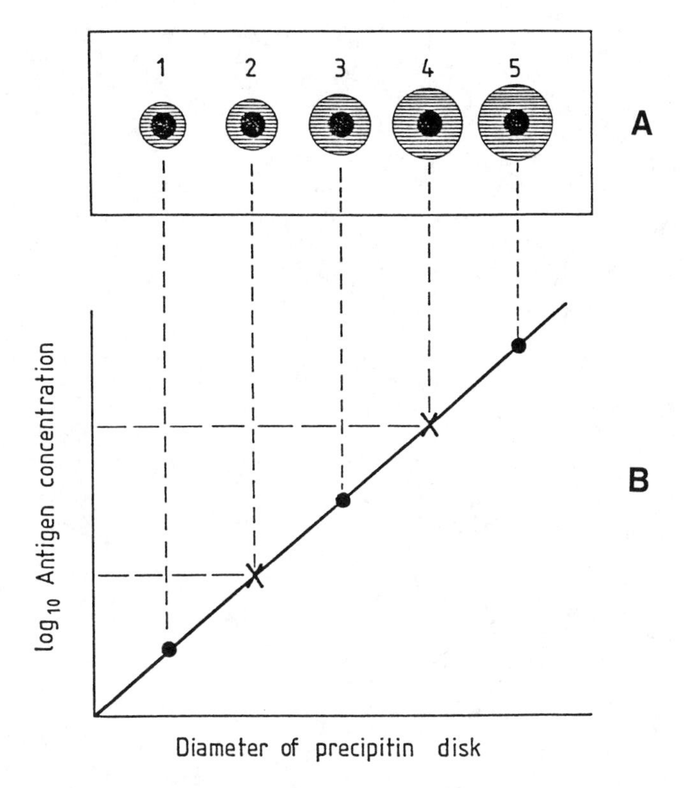

Fig. 1. Single radial diffusion. **(A)** Precipitin disks formed at termination of diffusion of antigen into gel containing monospecific antiserum. Wells 1, 3, and 5 contained standard antigen solutions of increasing concentration. Wells 2 and 4 contained samples of the antigen at unknown concentrations. Shaded area, immunoprecipitate. **(B)** Semilog plot of diameter of the precipitin disks of standard antigen solutions (●) against concentration. Measurement of the diameters of the precipitin disks of the unknown solutions (X) allows an estimation of the antigen concentration to be made by simple interpolation.

2. If the antiserum employed is not monospecific for the required antigen in an impure sample, a number of precipitin disks will be produced by the different antigens interacting with their corresponding antibodies. The problem then is to identify the disk produced by the antigen under study. One solution is to rerun each sample in duplicate with one well containing the sample plus a fixed amount of standard antigen. The disk caused by the antigen under study can then be identified by its larger diameter around the duplicate well.
3. For maximum sensitivity, the antiserum used in the gel should be diluted with nonimmune serum, i.e., serum from a nonimmunized animal. Furthermore, the dimensions of the wells and volumes of samples and standards applied can be increased. It should be possible to measure protein solutions of 1 μg/mL concentration.

References

1. Vaerman, J. P. (1981) Single radial immunodiffusion, in *Methods in Enzymology,* vol. 73 (Langone, J. J. and Van Vunakis, H., eds.), Academic, New York, pp. 291–305.
2. Roitt, I. (1994) *Essential Immunology,* 8th ed., Blackwell Scientific, Oxford, UK, pp. 114,115.

137

Rocket Immunoelectrophoresis

John M. Walker

1. Introduction

Rocket electrophoresis (also referred to as electroimmunoassay or electroimmunodiffusion) is a simple, quick, and reproducible method for determining the concentration of a specific protein in a protein mixture. The method, originally introduced by Laurell *(1)*, involves a comparison of the sample of unknown concentration with a series of dilutions of a known concentration of the protein, and requires a monospecific antiserum against the protein under investigation. The samples to be compared are loaded side-by-side in small circular wells along the edge of an agarose gel that contains the monospecific antibody. These samples (antigen) are then electrophoresed into the agarose gel, where interaction between antigen and antibody takes place. As the protein antigen starts to leave the well and enter the gel, antigen molecules will start to interact, and bind with, antibody molecules. However, at this early stage, there is considerable antigen excess over antibody and no precipitation occurs. However, as the antigens sample electrophoreses further through the gel, more antibody molecules are encountered that interact with the antigen, until eventually there is sufficient antibody–antigen crosslinking such that "equivalence" is reached and the antigen–antibody complex precipitates. One might therefore expect to see some sort of precipitation line appear in the gel. In practice, a "rocket" shape is seen (Fig. 1). The majority of the antibody–antigen precipitate is indeed at the head of this rocket, but the fine precipitation lines up the side of the rockets are formed by a small amount of antigen diffusing sideways as the antigen passes through the gel. This small amount of antigen very quickly meets sufficient antibody to reach equivalence and precipitate. The greater the amount of antigen loaded in a well, the further the antigen will have to travel through the gel before it can interact with sufficient antibody to form a precipitate. Therefore, if a series of wells are loaded with increasing antigen concentration, then a series of rockets of increasing height should be produced, with the area under the curve (rocket) being proportional to the amount of antigen in the well. However, since the rockets are nearly perfect isosceles triangles, the height of the rocket is also proportional to the area under the curve (and hence the antigen concentration), and it is this easier-to-measure parameter that is normally recorded. Rocket electrophoresis is therefore carried out by loading a series of samples of different antigen concentrations, with one sample being the

From: The Protein Protocols Handbook
Edited by: J. M. Walker Humana Press Inc., Totowa, NJ

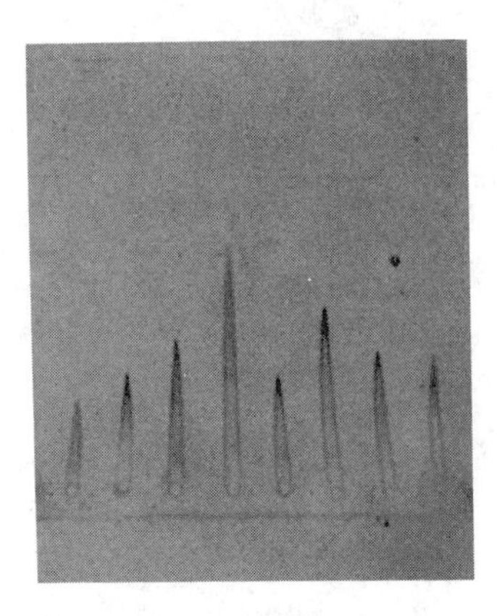

Fig. 1. A rocket immunoelectrophoresis gel (5 × 5 cm) run at 20 mA for 2 h, and stained for protein with Coomassie brilliant blue. The gel contained anti-bovine serum albumin (50 μL, anti-BSA) and the sample loadings (2 μL/well) were (left to right): 1. 40 ng BSA; 2. 67 ng BSA; 3. 100 ng BSA; 4. 200 ng BSA; 5. a 1:1200 dilution of bovine serum; 6. 130 ng BSA; 7. 100 ng BSA.

unknown. The unknown sample can be a highly complex mixture of proteins (e.g., serum sample, tissue extract, urine sample, cerebrospinal fluid, and so forth), but only one rocket will be produced from this sample owing to the interaction of the antibody in the gel with the antigen of unknown concentration. A calibration graph of protein concentrations vs peak height is then constructed and, knowing the peak height of the unknown sample, the concentration can be read off from the graph. Concentrations of proteins as little as 1 μg/mL can be measured in this manner (requiring as little as 20 ng of protein to be loaded in a well). An excellent detailed review of this subject has been published *(2)*. References *(3–8)* give some idea of the range of applications of this technique. This chapter will describe the construction of a simple calibration curve using bovine serum albumin (BSA) as antigen and anti-BSA. This is an ideal trial system for the first-time user and will give the user confidence to use the technique with his or her own, often more precious, antigen and antibody.

2. Materials

1. Electrophoresis buffer: 0.03*M* barbitone buffer, pH 8.4: Take 5.15 g of sodium barbitone, 0.92 g barbitone, and make up to 1 L with distilled water (*see* Notes 1 and 2).
2. 1% Agarose in electrophoresis buffer: Fully dissolve the agarose by boiling for 2–3 min. This can be most conveniently done in a microwave oven. **Care!** When this boils, it can become very frothy and flow out the top of the flask. Ensure you have a small volume in a large flask, e.g., 50 mL in 250-mL flask. When the agarose has dissolved, place the flask in a 55°C water bath (*see* Note 3). Use agarose that has low electroendosmosis.
3. 5 × 5 cm glass plates, 1–2 mm thick (*see* Note 4).
4. Anti-BSA: commercially available, e.g., Sigma (St. Louis, MO), Vector.
5. The following dilutions, made in electrophoresis buffer, of a 40 mg/mL solution of BSA: 1:2000; 1:1000; 1:600; 1:400; 1:200; 1:150; 1:75.
6. A microsyringe for loading samples in the wells.
7. A leveling plate.
8. Gel well-puncher: These are commercially available, although the author successfully uses a Pasteur pipet.

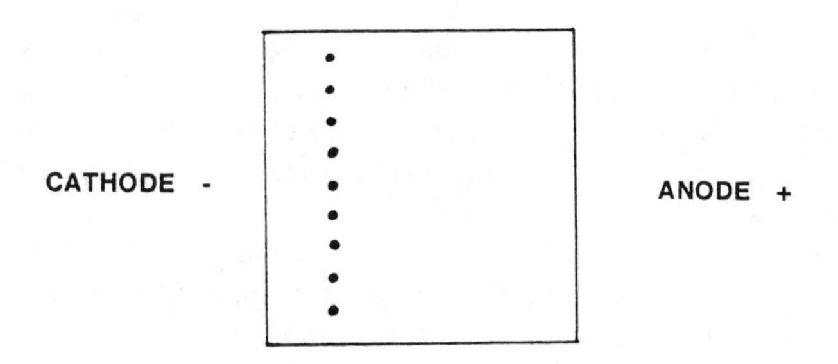

Fig. 2. Preparation of electrophoresis plates for rocket immunoelectrophoresis.

9. Electrode wicks: These can be six thicknesses of filter paper, cut exactly to the width of the glass plate and soaked in electrophoresis buffer (*see* Note 5).
10. Flat-bed electrophoresis apparatus, preferably with a cooling plate.
11. Gel bond (optional).
12. Protein stain: 0.1% Coomassie brilliant blue in 50% methanol/10% glacial acetic acid. (Dissolve the stain in the methanol/water component first and **then** add the acetic acid.)

3. Method

1. Melt the 1% agarose as described in Section 2., step 2, and then place the flask in a 55°C water bath. Stand a 10-mL pipet in this solution, and allow time for the agarose to cool to 55°C. At the same time, the pipet will be warming.
2. Place a test tube in the water bath, allow it to warm up for a couple of minutes, and then add 5.5 mL of 1% agarose to the tube. This transfer should be made as quickly as possible to avoid the agarose setting in the pipet. Using a pipet with the end cut off to give a larger orifice allows a more rapid transfer of the viscous agarose solution. Some setting of agarose in the pipet is inevitable, but is of no consequence as long as you transfer approx 5.5 mL (±0.2 mL) of agarose solution to the tube. Briefly mix the contents of the tube and return to the water bath.
3. Thoroughly clean a 5 × 5 cm glass plate with methylated spirit (*see* Note 6). When dry, put the glass plate on a leveling plate or level surface.
4. Add antiserum (50 µL, but *see* Notes 7 and 8) to the diluted agarose solution, and **briefly** mix to ensure even dispersion of the antiserum.
5. Pour the contents of the tube onto the glass plate. Keep the neck of the tube close to the center of the plate, and pour the contents of the tube onto the middle of the plate. The liquid will spread to the edges of the glass plate, but surface tension will prevent the liquid from running off the edge of the plate. Alternatively, tape can be used to form an edging to the plate. The final gel will be approximately 2 mm thick (*see* Notes 9–11).
6. Allow the gel to set for 5 min (when set the gel will appear slightly opalescent) and then make holes (1 mm diameter) at 0.5 cm spacings, 1 cm from one edge of the plate. This is most easily done by placing the gel plate over a predrawn (dark ink) template when well positions can easily be seen through the gel (*see* Fig. 2). The wells can be made using a Pasteur pipet or a piece of metal tubing attached to a weak vacuum source (e.g., a water pump).
7. Place the gel plate on the cooling plate of an electrophoresis tank, and place the electrode wicks (six layers of Whatman No. 1 paper, prewetted in electrode buffer) over the edges of the gel so that they overlap the gel by 2–3 mm. Ensure that there is firm contact between the wicks and the gel. Take care not to overlap the sample wells, and ensure that these wells are nearest the cathode (*see* Notes 12–14).

8. Each well is now filled with 2 μL of sample. It is important to ensure that all samples are loaded in the same volume. The loading of wells should be carried out as quickly as possible to minimize diffusion from the wells. Some workers prefer to load wells with a current (1–2 mA) passing through the gel, which should overcome any diffusion problems **(but take great care!)**.
9. Once the samples are loaded, electrophoresis is commenced. For fast runs, a current of 20 mA (~200 V, 10 V/cm across the plate) is passed through the gel for about 3.0 h. Alternatively, gels can be run at 2–3 mA overnight. It is important that water cooling be used to dissipate heat generated during electrophoresis (*see* Notes 15–16).
10. After this time, remove the plate from the electrophoresis apparatus and observe the protein precipitation peaks using a dark-background illuminator. Plot a graph of peak height against protein concentration.
11. A permanent record can be made by drying the gel and staining for protein. It is not possible to stain the gel for protein directly, since all the background serum proteins in the gel will stain giving a totally blue gel (there is no reason why the stain should preferentially pick out the precipitate only). One way to remove background protein is to copiously wash the gel in saline. However, a much quicker way is to press the gel dry while at the same time absorbing the background protein out of the gel using filter paper.

    Place the gel to be dried either on a glass plate or the hydrophilic side of a sheet of gel bond. Cover the gel with six sheets of thick filter paper (3MM) that are larger in size than the gel to be dried. Then place a heavy weight on top of this and leave for a minimum of 1 h. Carefully remove the filter papers (the last one may be stuck to the gel–wet it with distilled water before removing it) to reveal a squashed, opaque gel. Complete the drying process by blowing warm air (hairdryer) on the gel for about 5 min. The gel should now appear as a glassy film. Immerse the gel in Coomassie stain for 15 min, and then rinse under tap water (destaining is **not** required). You now have a permanent record.

4. Notes

1. Barbitone buffer was the original buffer used for this method and has been traditionally used ever since. However, because they are barbiturates, some workers may have problems gaining access to these reagents. The author has found that a suitable alternative buffer is 0.1*M* Tris-borate, pH 7.4.
2. The pH of the barbitone buffer is carefully chosen at pH 8.4. The pH is close to the isoelectric point of the IgG molecule (each different IgG molecule will have a slightly different pI). The IgG molecule has a very slight positive to zero charge and will basically be immobile or possibly will move extremely slowly toward the cathode. By using the pH value, this ensure that the advancing antigen molecules do meet the IgG, rather than the IgG molecule also moving toward the anode, when antigen and antibody would not meet.
3. Unused agarose solution can be stored at 4°C and then reused at a later date. Each time the agarose is reused, check for signs of microbial contamination.
4. Where possible, the gel plates used should be as thin as possible to maximize the effect of the cooling plate. This is particularly important in the case of the shorter run time (*see* Section 3., step 9) when considerable heat is evolved, which can cause distortion of rockets.
5. This chapter has described the use of six thicknesses of filter paper as the electrode wicks, but any highly absorbent, inert material can be used (e.g., lint, cotton-wool, and so on). What is important is that these wicks absorb a large amount of buffer and therefore have a low resistance. If, for example, one thin sheet of filter paper was used, this would absorb little buffer and therefore would be a region of high resistance. This has two disadvantages. Remembering Ohm's law, where: V = I R (V = potential difference [volts], I =

current, R = resistance), to obtain a given current, if the resistance is high, then a very high voltage will be needed. This may not be possible with your power pack, but in any event, working with high voltage is discouraged on safety grounds. Second, with a high resistance, excessive heat will be generated: Power (W) = I^2R . Excessive heat will cause distortion of your rockets and can also cause evaporation from the gel resulting in the gel ultimately drying out.

6. The size of gel described here (5 × 5 cm) is suitable for routine use. However, for highly accurate determinations, larger plates should be used (7.5 × 7.5 cm) with larger wells (5–10 μL volumes) so that sample loadings can be made with greater accuracy, and the correspondingly larger peaks measured more accurately. The volume of agarose and antisera used should of course be increased accordingly.
7. The amount of antiserum to be used in the gel depends of course on the antibody titer, and is best determined by trial and error. If rockets are too small, use less antiserum, and vice versa. The amount quoted in Section 3. (50 μL) is the volume of commercially available anti-BSA used in our laboratory when measuring BSA levels (*see* Fig. 1). This level of antiserum (1%) in the gel is a good starting point for most workers.
8. In general, MAb cannot be used in immunoprecipitation techniques, since they cannot achieve the crosslinked matrix of antigen-antibody reactions required at equivalence. This can only be achieved by a monoclonal antibody if the antigen has two or more copies of the epitope recognized by the antibody, e.g., when the antigen comprises two identical subunits.
9. It is important to note that 55°C is not far above the setting temperature of agarose, so when working with the solution out of the water bath, one should work very quickly to avoid the gel setting prematurely. It may seem tempting to use a higher temperature, but this will only cause denaturation (and hence precipitation on the gel) of the serum proteins from the antiserum.
10. If your antibody is in short supply, it is possible to prepare a gel where the region covered by the electrode wicks is free of antibody. First, pour a gel, **without** antibody, as described in Section 3., steps 1–5. When the gel has set, cut out the central portion of the gel, just leaving a 0.5-mm length of gel at both the anode and cathode ends. Now prepare 4.4 mL of agarose containing 40 μL of antibody, and pour this on the central region of the gel.
11. A few air bubbles may rise to the surface of the gel, but these should not cause any problems. However, some workers prefer to "flame" the surface of the gel briefly with a yellow Bunsen flame to burst and dispel these bubbles. Also some workers prefer to flame the glass plate to warm it up prior to pouring the gel. This ensures that the gel sets more slowly than it would if poured onto a cold plate, and allows a little more time for manipulation.
12. It is important that the electrode wicks are cut to exactly the same width as the glass plate, and that they make firm contact all along the gel. If the wicks are too wide, too short, or only making contact with part of the gel, then there will be an uneven force field across the gel, resulting in distorted rockets.
13. Because of the thickness of the electrophoresis wicks used, they are quite heavy when wet and have a tendency to "slip off" the gel during the course of the run. This can be prevented by placing thick glass blocks on top of the wicks where they join the gel, or by placing a heavy glass sheet across both wicks. This, by also covering the gel, has the added advantage of reducing evaporation from the gel.
14. When setting up the electrophoresis wicks, it is important that they be kept well away from the electrodes. Since the buffer used is of low ionic strength, if electrode products are allowed to diffuse into the gel, they can ruin the gel run. Most commercially available apparatus is therefore designed such that the platinum electrodes are masked from the wicks and also separated by a baffle. To minimize these effects, the use of large tank buffer volumes (~800 mL/reservoir) is also encouraged.

15. This technique has been described assuming your protein antigen runs to the anode at pH 8.4 (i.e., it has a pI <8.4). However, highly basic proteins, for example, that have a pI >8.4, will run in the opposite direction. In this case, samples must be loaded at the anode end.
16. In general, we find that we get better results with the lower current.
17. The times given here for electrophoresis are appropriate for measuring BSA. However, BSA is nearly the most highly charged serum protein at pH 8.6 and therefore moves rapidly to the anode. Your own protein of interest may have a very weak positive charge at pH 8.4 (depending on how close its pI is to 8.4), and therefore, in the times given here may not have moved sufficiently far through the gel to reach equivalence. Therefore, if you do not obtain rockets, try running the gel for longer. It is not possible to overrun this method. Once the rockets are formed, they are unaffected by further electrophoresis. For molecules that do have a slow electrophoretic mobility, and indeed for small protein antigen, side diffusion can be marked, giving rise to broad rockets. There is nothing that can be done about this.
18. This method is ideal for measuring antigens down to concentrations of about 1–10 μg/mL, and at this level, the method is to be recommended over setting up an ELISA that can take many weeks to perfect. (If large numbers of samples are to be analyzed, however, ELISA is probably more appropriate.) Below this level, the sensitivity of an ELISA or radioimmunoassay will be required.
19. If a given sample produces more than one rocket precipitate, this could be because:
    a. The antiserum is not monospecific (i.e., it contains antibodies against more than one protein.
    b. There is electrophoretic homogeneity of the protein being studied.
    c. Other proteins present share common antigenic determinants.

References

1. Laurell, C. B. (1966) Quantitative estimation of proteins by electrophoresis in agarose gel containing antibodies. *Anal. Biochem.* **15,** 45–52.
2. Laurell, C. B. and McKay, E. J. (1981) Electroimmunoassay, in *Methods in Enzymology* (Langone, J. J. and Van Vunakis, H., eds.), Academic, London, pp. 339–369.
3. Tessier, F., Quentin, C., Capdepuy, M., and Guinet, R. (1993) Antigenic relationships among bacteroids species studied by rocket immunoelectrophoresis. *Int. J. Systematic Bacteriol.* **43, No. 2,** 191–195.
4. Soohoo, C. K. and Hollocher, T. C. (1990) Loss of nitrous oxide reductase in pseudomonas aeruginosa cultured under N_2 as determined by rocket immunoelectrophoresis. *Appl. Environ. Microbiol.* **56, No. 11,** 3591–3592.
5. Espersen, G. T., Lageland, B., and Grunnet, N. (1990) Comparative study of assays detecting complement activation—split product C3D (Rocket immunoelectrophoresis) and C3D neodeterminants (ELISA). *Scand. J. Clin. Lab. Invest.* **50, No. 4,** 389–393.
6. Lassiter, M. O., Kindle, J. C., Hobbs, L. C., and Gregory, R. L. (1989) Estimation of immunoglobulin protease activity by quantitative rocket immunoelectrophoresis. *J. Immunol. Methods* **123, No. 1,** 63–69.
7. Wahl, R., Oliver, J. D., Hauck, P. R., and Roig, J. (1989) Rocket immunoelectrophoresis: a useful screening method for house dust extracts. *Ann. Allergy* **63, No. 2,** 137–141.
8. Jacobson, T., Poulsen, O. M., and Hau, J. (1989) Enzyme activity electrophoresis and rocket immunoelectrophoresis for the qualitative and quantitative analysis of *Geotrichum candidum* ligase activity. *Electrophoresis* **10, No. 1,** 49–52.

138

Two-Dimensional Immunoelectrophoresis

John M. Walker

1. Introduction

Two-dimensional (2D) immunoelectrophoresis, also known as crossed immunoelectrophoresis, is a particularly useful technique for the quantitation of one or more proteins in a mixture of proteins and the analysis of the composition of protein mixtures. The method consists of two sequential electrophoretic steps:

1. The first dimension, during which the protein mixture to be analyzed is separated by electrophoresis in an agarose gel.
2. The second dimension, during which the separated proteins are electrophoresed at right angles into a freshly applied layer of agarose containing a predetermined amount of antibody.

During the second electrophoresis step, the proteins migrate into the gel until they are completely precipitated by the antiserum, each protein forming a separate precipitate peak. The area under any peak is directly proportional to the concentration of that protein in the sample and inversely proportional to the concentration of specific antibody in the gel.

This technique was originally introduced as a method for quantitatively measuring changes in serum protein patterns in human disease and, although not often used nowadays, it proved particularly successful in this respect, with as many as 40 different serum proteins being quantitated in a single experiment *(1–4)*. However, this method can also be applied to a range of problems in the field of molecular biology. It is important to remember that the choice of protein(s) to be quantitated is controlled by the antiserum used in the second dimension. For example, isoenzymes will normally separate during electrophoresis in the first dimension. Using the appropriate antiserum in the second dimension (which will recognize all isoforms of the enzyme) will result in overlapping or adjacent peaks corresponding to each isozyme, the area under each peak being proportional to the concentration of each isozyme. Similarly, the relative proportions of different proteins at different stages of tissue development can be quantitated with this technique by including in the gel appropriate antisera against the proteins that you wish to compare. References *(5–9)* give an indication of some of the wider applications of this techniques. This chapter describes the production of a 2D serum profile. This is an ideal trial system for the first-time user and will give the user confidence to use the technique.

From: The Protein Protocols Handbook
Edited by: J. M. Walker Humana Press Inc., Totowa, NJ

2. Materials

1. Electrophoresis buffer: 0.03*M* barbitone buffer, pH 8.4. Take 5.15 g of sodium barbitone, 0.92 g barbitone, and make up to 1 L with distilled water (*see* Notes 1 and 2).
2. 1% Agarose in electrophoresis buffer. Fully dissolve the agarose by boiling for 2–3 min. This can be most conveniently done in a microwave oven. **Care!** When this boils, it can become very frothy and flow out the top of the flask. Ensure you have a small volume in a large flask, e.g., 50 mL in a 250-mL flask. When the agarose has dissolved, place the flask in a 55°C water bath (*see* Note 3). Use agarose that has low electroendosmosis.
3. Glass plates, 5 × 5 cm, 1.0–2.0 thick (*see* Note 4).
4. Bovine serum.
5. Reference serum: bovine serum containing one drop of concentrated Bromophenol blue solution.
6. Anti-bovine serum.
7. Flat-bed electrophoresis apparatus with cooling plate.
8. A microsyringe for loading samples in the wells.
9. A leveling plate.
10. Gel well-puncher: These are commercially available, although the author successfully uses a Pasteur pipet attached to a weak vacuum source.
11. Electrode wicks: These can be six thicknesses of filter paper, cut exactly to the width of the glass plate and soaked in electrophoresis buffer (*see* Note 5).
12. Protein stain: 0.1% Coomassie brilliant blue in 50% methanol and 10% acetic acid. Dissolve the stain in the methanol component first, and then add the appropriate volumes of acetic acid and water.

3. Method

1. Dissolve the 1% agarose as described in Section 2., item 2, and then place the flask in a 55°C water bath. Stand a 10-mL pipet in this solution, and allow time for the agarose to cool to 55°C. At the same time, the pipet will be warming.
2. Place a test tube in the 55°C water bath, allow it to warm up for a few minutes and then add 5.5 mL of 1% agarose solution to the test tube. This transfer should be made as quickly as possible to avoid the agarose setting in the pipet. Briefly Whirlymix the contents and return to the water bath.
3. Thoroughly clean a 5 × 5 cm glass plate with methylated spirit. When dry, put the glass plate on a leveling plate or level surface.
4. Remove the test tube from the water bath and pour the contents of the tube onto the glass plate. Keep the neck of the tube close to the plate, and pour the contents of the tube onto the middle of the plate. Surface tension will keep the liquid on the surface of the plate. Alternatively, tape can be used to form an edging to the plate.
5. Allow the gel to set for 5 min. While the gel is setting, pour 0.03*M* barbitone buffer into each reservoir of the electrophoresis tank and completely wet the wicks. The wicks are prepared from six layers of Whatman No. 1 filter paper (but *see* Note 5). They should be cut to exactly the width of the gel to be run, and should be well wetted with barbitone buffer.
6. Make two holes (1 mm diameter) in the gel 0.8 cm in from one edge of the plate and approx 1.5 cm in from the side of the plate. This is most easily done by placing the gel plate over a predrawn (dark ink) template (such as the one shown in Fig. 1), when well positions can easily be seen through the gel. The wells can be made using a Pasteur pipet or a piece of metal tubing attached to a weak vacuum source (e.g., a water pump).

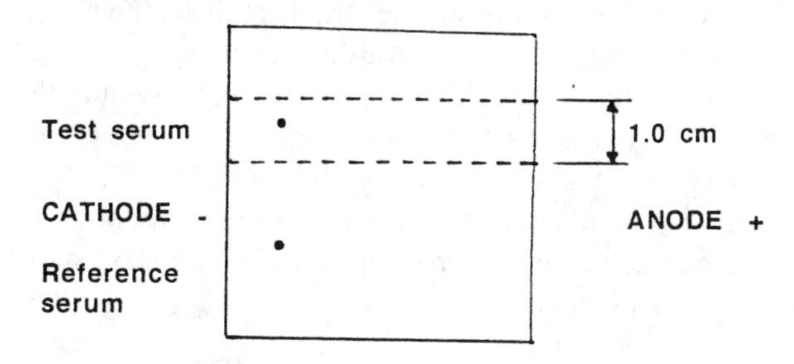

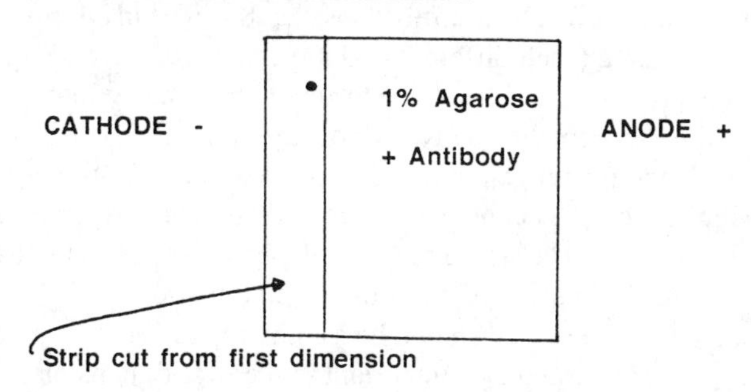

Fig. 1. Diagram showing sample loading and gel-slicing positions.

7. Place 0.5 µL of serum (bovine) in the test well and 0.5 µL of reference serum in the other well. "Top up" the wells with electrode buffer (~1 µL).
8. Place the gel plate on the cooling plate of an electrophoresis tank, and place the electrode wicks over the edges of the gel. Take care not to overlap the sample wells, and ensure that these wells are nearest the cathode (*see* Notes 6–8).
9. Immediately pass a current of 20 mA (~250 V) until the blue marker dye in the reference sample (which is bound to the albumin) reaches or just enters the anode wick. This will take approx 40 min. (This is about 10 V/cm across the plate.)
10. Toward the end of this run, repeat step 2 using a new test tube, but this time use only 4.4 mL of agarose. (You will be covering a smaller surface of glass plate with this second-dimension solution.)
11. When the plate has run, return the plate to the leveling table, and cut the gel as shown in Fig. 1. Transfer the gel slice (containing the test sample, **not** the reference sample!) to a second clean glass plate (*see* Note 9).
12. To the tube in the water bath, now add 150 µL of rabbit antibovine serum, and briefly Whirlymix to ensure even dispersion of antibody. Return briefly to the water bath to allow any bubbles to settle out (*see* Notes 10 and 11).
13. Pour the agarose and antibody mixture onto the second plate adjacent to the strip of gel, and allow the gel to set for 5 min.

14. Place the plate in the electrophoresis apparatus as before and run at 3.0 mA/plate overnight (1 V/cm), making sure that the sample strip is at the cathode end (i.e., protein separated in the first dimension now moves toward the anode).
15. At the end of the overnight run, precipitation peaks will be seen in the gel. These are not always easy to visualize and are best observed using oblique illumination of the gel over a dark background (e.g., using a dark background illuminator).
16. A more clear result can be obtained by staining these precipitation peaks with a protein stain. Follow the procedure described in Chapter 137, Section 3., step 11. A typical result (stained for protein) is shown in Fig. 2.

4. Notes

1. Barbitone buffer was the original buffer used for this method and has been traditionally used ever since. However, because they are barbiturates, some workers may have problems gaining access to these reagents. The author has found that a suitable alternative buffer is 0.1*M* Tris-borate, pH 7.4.
2. The pH of the barbitone buffer is carefully chosen at 8.4. The pH is close to the isoelectric point of the IgG molecule (each different IgG molecule will have a slightly different pI). The IgG molecule has a very slight positive to zero charge, and will basically be immobile or possibly will move extremely slowly towards the cathode. By using the pH value, this ensures that the advancing antigen molecules do meet the IgG, rather than the IgG molecule also moving toward the anode, where antigen and antibody would not meet.
3. Unused agarose solution can be stored at 4°C and then reused at a later date. Each time the agarose is reused, check for signs of microbial contamination.
4. When possible, the gel plates used should be as thin as possible to maximize the effect of the cooling plate. This is particularly important in the first-dimension run when considerable heat is evolved.
5. The author has described the use of 6 thicknesses of filter paper as the electrode wicks, but any highly absorbent, inert material can be used (e.g., lint, cotton-wool, and so on). What is important is that these wicks absorb a large amount of buffer and therefore have a low resistance. If, for example, one thin sheet of filter paper was used, this would absorb little buffer and therefore would be a region of high resistance. This has two disadvantages. Remembering Ohm's law, where: $V = I R$ (V = potential difference [volts], I = current, R = resistance) to obtain a given current, if the resistance is high, then a very high voltage will be needed. This may not be possible with your power pack, but in any event, working with high voltage is discouraged on safety grounds. Second, with a high resistance, excessive heat will be generated: Power (W) = I^2R .
6. Because of the thickness of the electrophoresis wicks used, they are quite heavy when wet and have a tendency to "slip off" the gel during the course of the run. This can be prevented by placing thick glass blocks on top of the wicks, where they join the gel, or a heavy glass sheet across both wicks. This, by also covering the gel, has the added advantage of reducing evaporation from the gel.
7. A uniform field strength over the entire gel is critical in rock immunoelectrophoresis. For this reason, the wick should be exactly the width of the gel.
8. When setting up the electrophoresis wicks, it is important that they be kept well away from the electrodes. Since the buffer used is of low ionic strength, if electrode products are allowed to diffuse into the gel, they can ruin the gel run. Most commercially available apparatus is therefore designed such that the platinum electrodes are masked from the wicks and also separated by a baffle. To minimize these effects, the use of large tank buffer volumes (800 mL/reservoir) is also encouraged.

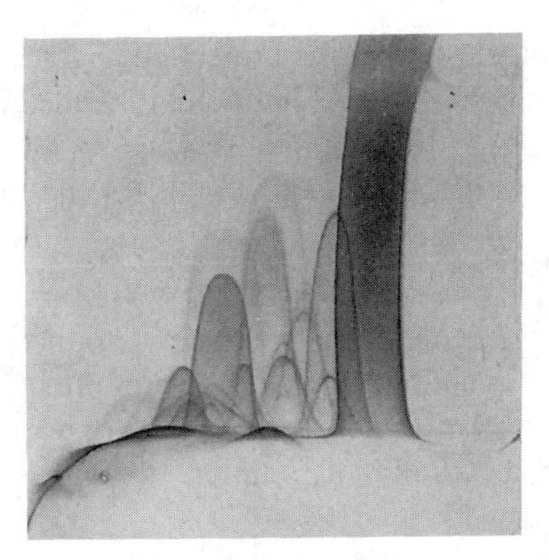

Fig. 2. 2D Immunoelectrophoresis pattern for bovine serum, produced using the method described in this chapter.

9. The working temperature for the agarose gel (55°C) is just above the setting temperature of agarose, and great care must be taken to ensure that agarose solutions are not allowed to drop below this temperature prior to pouring the gel. This temperature is chosen in order to minimize the chance of denaturing the antiserum, which would be considerably less stable at higher temperatures, but of course the agarose must stay liquid. In any case, once the antiserum has been added to the agarose, the gel should be poured immediately to prevent the antibody from being exposed unnecessarily to high temperatures.
10. When cutting the gel strip from the first dimension, make sure that sufficient gel remains adjacent to the sample track to allow for the placing on a wick on this gel in the second dimension, i.e., cut a gel slice with the sample to one side of this slice (*see* Fig. 1). It is important that the wicks do not cover the sample when running the second dimension.
11. The amount of antiserum to be used in the gel depends of course on the antibody titer and the amount of protein run in the first dimension, and is best determined by trial and error. If peaks go off the top of the gel, then either the antiserum concentration must be increased in the second dimension or the amount of protein run in the first dimension reduced. The reverse applies if the peaks produced are too small for accurate measurement. The amount of antiserum quoted (150 μL) is the volume of commercially available antibovine serum used in the author's laboratory when producing 2-D immunoelectrophoresis gels from bovine serum. This is an ideal test system for someone setting up the technique for the first time.
12. The time quoted for electrophoresis is suitable for most samples. For proteins with low electrophoresis mobility, however, the time given for the electrophoresis runs may be too short. In the protocol described, the bromophenol blue spot that is visualized in the first dimension is not "free" bromophenol blue, but the dye bound to albumin. Since albumin is one of the fastest-running proteins in the first dimension (only a few minor prealbumin proteins run faster), when the blue dye reaches the anode, the majority of the serum proteins are now spread along the full length of the gel. When studying other protein mixtures, however, the appropriate time for the first dimension run should be determined in initial trial experiments. Free bromophenol blue can be used as a marker dye to indicate the status of an electrophoresis run at any given time. Similarly, in the second dimension, the time of electrophoresis given may not be long enough to allow for complete development of precipitation peaks for proteins with low electrophoresis mobility. Do note that it is not possible to "overrun" the second-dimension electrophoresis. Once the precipitation peaks are formed, they are quite stable to further electrophoresis.

13. If no results are obtained, check that the proteins being analyzed run to the anode under the conditions used. This assumption is implicit in the method described here.
14. Some workers include 3% w/v polyethylene glycol 6000 in the gel, which is said to enhance precipitation of antigen–antibody complexes and thus increase sensitivity.

References

1. Laurell, C.-B. (1965) Antigen–antibody crossed electrophoresis. *Anal. Biochem.* **10,** 358.
2. Peeters, H. (ed.) (1967) A quantitative immunoelectrophoresis method, in *Protides of the Biological Fluids*. Elsevier, Amsterdam.
3. Clarke, H. and Freemann, T. (1968) Quantitative immunoelectrophoresis of human serum proteins. *Clin. Sci.* **35,** 403–413.
4. Weeke, B. (1970) The serum proteins identified by means of the Laurell cross electrophoresis. *Scand. J. Clin. Lab. Invest.* **25,** 269–275.
5. Nielsen, J. L., Poulsen, O. M., and Abildtrup, A. (1994) Studies on serum protein complexes with nickel using crossed immunoelectrophoresis . *Electrophoresis* **15, No. 5,** 666–671.
6. Hatanaka, K. (1994) A case-report of deficiency in an inhibitor of calcium dependent association of protein-S with C4B-binding protein suggested by a modified cross-immunoelectrophoresis . *Thrombosis Res.* **74, No. 6,** 643–654.
7. Misset, M. T. and Fonternelle, C. (1993) Antigenic relations among species of the genus ulex L determined by crossed-immunoelectrophoresis. *Evolutionary Trends in Plants* **7, No. 1,** 23–28.
8. Karpatkin, S., Shulman, S., and Howard, I. (1992) Crossed immunoelectrophoresis of human platelet membranes. *Methods Enzymol.* **215,** 440–455.
9. Goetz, D. W., Whisman, B. A., and Goetz, A. D. (1992) Crossed immunoelectrophoresis identification of nut cross-reactivities. *J. Allergy Clin. Immunol.* **91, No. 1, Pt. 2,** 360.

Part VIII

Monoclonal Antibodies

139

Immunogen Preparation and Immunization Procedures for Rats and Mice

Mark Page and Robin Thorpe

1. Introduction

A high-titer antibody response usually requires use of an adjuvant for the first (priming) immunization. For most purposes, the immunogen is prepared by emulsification in a mineral oil containing heat-killed mycobacterium (Freund's complete adjuvant—FCA). The emulsion ensures that the antigen is released slowly into the animal's circulation, and the bacteria stimulate the animal's T-helper cell arm of the immune system. Further booster (secondary) immunizations are almost always necessary for production of high antibody levels, and these are given either in phosphate-buffered saline (PBS) or as an oil emulsion (bacteria are not normally included in the boosting injections; a suitable oil adjuvant is Freund's incomplete adjuvant—FIA). Immunization with substances with molecular weights <3000 (such as peptides) are not normally immunogenic and will require conjugation to a carrier protein (*see* ref. *1* and Chapters 117–119), such as purified protein derivative (PPD) or keyhole limpet hemocyanin (KLH).

2. Materials

1. Freund's complete adjuvant (FCA).
2. Freund's incomplete adjuvant (FIA).
3. Phosphate-buffered saline (PBS).
4. Immunogen preparation.
5. 2-mL glass Luer lock syringes (two).
6. Syringe coupler, Luer lock with female inlet and outlet ports (Sigma, St. Louis, MO).
7. Rat/mouse.

3. Methods

3.1. Immunogen Preparation (see Notes 1–5)

1. Dilute immunogen in physiological buffer (e.g., PBS without sodium azide) so that the preparation will contain 10–100 μg of protein in approx 300–600 μL/animal (Note: the final quantity will be 5–50 μg, since the preparation will be diluted with an equal volume of adjuvant). Mix the immunogen solution with an equal volume of FCA and draw up into a glass syringe or prepare directly in the syringe barrel.

From: The Protein Protocols Handbook
Edited by: J. M. Walker Humana Press Inc., Totowa, NJ

2. Remove excess air from the syringe barrel, and connect to a second glass syringe with its plunger fully depressed via a double-hub connector.
3. Ensure the connections are tight and not leaking, and transfer the oil and immunogen solutions from one syringe to the other. Continue this action until the mixture is fully emulsified when it should appear as a creamy, white thick liquid.
4. Transfer emulsion into one of the glass syringes, remove from double-hub connector and empty syringe, and fit a small-diameter needle (the size of which will depend on the animal to be immunized and route of immunization).

3.2. Immunization Procedure

1. Prime mice or rats by immunizing with immunogen subcutaneously on the flanks and neck (0.1 mL/site, 3–5 sites). Do his by raising the skin between thumb and forefinger, and inserting the needle into the raised area at a shallow angle. A short narrow-diameter needle is preferred (0.4 × 27 mm) to avoid injection into the deeper body layers/cavities. The result should be a discrete lump under the skin.
2. Boost intraperitoneally after 14–28 d using the immunogen prepared in PBS via a short narrow-diameter needle. Administer the immunogen at one site in no more than 0.5 mL using the same total dose as that used for priming (usually 5–50 μg).
3. Three days after the booster immunization, withdraw blood from the tail vein with a needle (0.4 mm) and syringe, and use this as a positive control in screening assays during hybridoma production and as a check on the success of the immunization. Sacrifice mouse, and remove spleen aseptically.

4. Notes

1. Shake the complete adjuvant before use to disperse the Mycobacterium particles fully.
2. The emulsion is very difficult to recover completely from the walls of vessels, and so forth; therefore, it is inevitable that some will be lost during preparation. To minimize this, prepare the emulsion in glass syringes, one of which can be used for the immunization. If possible, prepare slightly more than required to compensate for losses, which normally amount to around 10%.
3. If Luer connectors are not available, the emulsion can be prepared by vigorous shaking or mixing using a whirlimixer. This is less efficient at producing an emulsion and requires larger volumes of immunogen.
4. When using connected syringes, keep the hub connector as short as possible to avoid emulsion loss.
5. Use glass syringes to prepare the emulsion since the rubber seals of plastic syringes are not compatible with the mineral oil of the adjuvant.

Reference

1. Thorpe, R. (1994) Producing antibodies, in *Immunochemistry Labfax* (Kerr, M. A. and Thorpe, R., eds.), Bios Scientific, Oxford, UK, pp. 63–81.

140

Hybridoma Production

Mark Page and Robin Thorpe

1. Introduction

Köhler and Milstein *(1)* introduced technology for the production of MAb in vitro by the construction of hybridomas. These hybridomas are formed by the fusion of neoplastic B-cells (normally a B-cell line derived from a tumor) with spleen cells from an immune animal. Cells can be induced to fuse by mixing them together at high density in the presence of polyethylene glycol (some viruses also induce cell fusion). The efficiency of fusion is usually fairly low, but hybridomas can be selected for if the parent neoplastic cell line is conditioned to die in selective medium. The tumor cells are killed by the selective medium, normal nonfused spleen cells die after a period in culture, but hybridomas inherit the ability to survive in the selective medium from the normal parent cell. Usually, medium containing hypoxanthine, aminopterin, and thymidine (HAT) is used. There are two pathways available to the cell for synthesis of nucleic acid: (1) *de novo* synthesis and (2) synthesis by salvaging nucleotides produced by breakdown of nucleic acid. Aminopterin (and thus HAT medium) inhibits *de novo* synthesis of nucleic acid, but this is not lethal for normal cells, since the salvage pathway can still function (the hypoxanthine and thymidine present in HAT medium ensures that there is no deficiency of nucleotides). However, the enzyme hypoxanthine-guanine phosphoribosyl transferase (HGPRTase) is essential for the operation of the salvage pathway, so if the tumor cell line is deficient in this enzyme, it will be unable to synthesize nucleic acid and die. HGPRTase-deficient cell lines are produced by selection in medium containing 8-azaguanine. Cell possessing HGPRTase incorporate the 8-azaguanine into their DNA and die, whereas HGPRTase-deficient cells survive.

If individual hybridomas are isolated by cloning, it is possible to produce large quantities of the secreted MAb (all the immunoglobulin secreted by a clone has identical antigen-binding specificities, allotype, heavy chain subclass, and so forth).

Hybridoma technology requires facilities for, and knowledge of, tissue culture, freezing and storing viable cells, and methods of screening for antibody secretion. It is advisable to learn the cell culture, fusion, and cloning techniques from an experienced worker rather than trying to set them up in isolation. The exact methodology for hybridoma production varies considerably between laboratories; one such protocol is given below.

From: The Protein Protocols Handbook
Edited by: J. M. Walker Humana Press Inc., Totowa, NJ

2. Materials

1. Media: RPMI-1640 or DMEM. Add penicillin (60 mg/L) and streptomycin (50 mg/L) to medium.
2. Selective media supplements: These reagents contain hypoxanthine (H), aminopterin (A), and thymidine (T), and are utilized for the selective growth of hybridomas. HAT supplement and HT supplement are both supplied in 50X concentrated stock solutions.
3. Polyethylene glycol 1500 (PEG 1500) in 7.5 m*M* HEPES (PEG 50% w/v).
4. Fetal bovine serum (FBS) (*see* Note 1).
5. NSO myeloma cells: The NSO cell line is a mouse myeloma line that does not synthesize or secrete immunoglobulin. Grow in 3–5% FBS-RPMI at a density of $2–5 \times 10^5$ cell/mL. Approximately $3–5 \times 10^7$ cells are required for a mouse spleen fusion (*see* Note 2).
6. Dimethylsulfoxide (DMSO).

3. Method

Cells must be grown and processed using aseptic conditions.

3.1. Fusion Protocol

1. Warm stock solutions of medium, FBS, PEG, and 50X HAT by placing stock bottles in a water bath set at 37°C.
2. Estimate cell numbers of myeloma cells (NSO) using a hemocytometer. The cells should be in log growth at $2–4 \times 10^5$/mL.
3. Remove the spleen from an immunized mouse (*see* Chapter 139) under sterile conditions. If rats are used, then the Y3 myeloma cell line should be used as fusion partner.
4. Transfer the spleen to a glass homogenizer, and add 15 mL of warmed RPMI. Push the pestle down the homogenizer, twist four times, and then pull the pestle up. Repeat three times. Alternatively, transfer the spleen to a Petri dish containing 15 mL RPMI. Hold the spleen at one end with forceps, and pierce the other end of the spleen with a bent needle attached to a 5-mL syringe. Tease out the cells by stroking the spleen with the bent needle, and then pass the cell suspension three to four times through the needle and syringe.
5. Transfer the cells to a 50-mL tube, add warm RPMI, and centrifuge for 5 min at 300*g*. At the same time, pellet the NSO cells by centrifugation.
6. Decant the supernatant. Assume the mouse spleen contains about 10^8 cells (a rat spleen contains 2×10^8 cells).
7. Mix the NSO and spleen cells together at a ratio of 1 NSO cell:2 spleen cells in 50 mL RPMI.
8. Centrifuge and decant the supernatant. Remove the last drop of supernatant with a pipet.
9. Add 1 mL 50% of PEG over 1 min (30 s for rat). Stir gently with the pipet while adding the PEG.
10. Leave for 1 min.
11. a. Add 1 mL RPMI over 1 min while stirring gently.
    b. Add 2 mL RPMI over 1 min while stirring gently.
    c. Add 5 mL RPMI over 1 min while stirring gently.
    d. Add 10 mL RPMI over 1 min while stirring gently.
    Top up the tube with RPMI and spin at 300*g* for 5 min.
12. Resuspend in 200 mL HAT medium (4 mL 50X HAT + 196 mL 15% FBS/RPMI).
13. Distribute into 8 × 24-well culture plates, 1 mL/well.
14. Feed with HAT medium every 6–8 d—remove half of the medium from each well, and replace with an equal amount of fresh HAT medium. Check for hybridoma growth and tumor cell death using an inverted microscope.

15. After 10–14 d, feed the wells containing hybridomas with warm HT medium (2 mL 50X HT + 98 mL 15% FBS/RPMI).
16. Screen the wells containing hybridomas for antibody secretion (*see* Chapters 141 and 142) when the medium becomes yellow-orange (acid) and hybridomas are nearly confluent.
17. Expand positive wells into further plates or bottles (after a further 5–8 d, the cells will grow in medium containing 10% v/v FBS). Clone or cryopreserve in liquid nitrogen.
18. To cryopreserve, resuspend about 10^7 hybridomas in 0.5–1.0 mL of medium containing 20% v/v FBS and 10% v/v DMSO in cryotubes. The freezing mixture should be cooled to approx 4°C during use by storing the solution in a refrigerator and placing the cryotubes in ice water. Slowly freeze the cells in nitrogen vapour for at least 2 h, followed by storage in liquid nitrogen (*see* Note 3). Hybridomas are thawed rapidly in a 37°C water bath, washed in ~50 mL medium to remove DMSO, spun down, and resuspended in fresh medium/FBS.

3.2. Cloning (see Note 4)

1. Weigh out 0.5 g agar into a 100-mL autoclave bottle.
2. Add 10 mL distilled water, and autoclave for 15 min.
3. Transfer to a water bath set at 50–55°C, and mix with 90 mL warmed medium (15% FBS/RPMI at 50°C).
4. Add 12–14 mL of the agar solution to each Petri dish, and allow to set for 20 min at room temperature (make up three dishes for each line to be cloned).
5. Suspend hybridomas in a well or flask, and transfer 2–4 × 10^3 cells in 0.2 mL of medium to one well of a fresh 24-well plate. Add 0.8 mL medium to this and make two more serial 1:5 dilutions (i.e., 0.2 + 0.8 mL medium). Discard 0.2 mL from the final well.
6. Allow the agar solution to cool to about 45–47°C, and add 1 mL of the agar to each well. Mix and transfer the contents of each well into separate agar containing Petri dishes (avoiding introducing bubbles). Incubate at 37°C in a CO_2 incubator.
7. Check daily for cell growth with an inverted microscope, and discard plates containing overgrown or no cells. When clones grow into colonies of 20–100 cells, pick them individually from the plate using a Pasteur pipet, and transfer into 1 mL of medium in separate wells in a 24-well plate.
8. Feed every 2–4 d, and when cells are 75–100% confluent, test for antibody secretion (*see* Chapters 141 and 142) and expand positive wells. Hybridomas may be weaned onto 10% FBS/RPMI at this stage.
9. Repeat the cloning procedure.

4. Notes

1. Test several batches of FBS before ordering a large quantity. Check for toxicity and sterility, and if possible, compare with a previous batch of FBS that is known to support hybridoma growth. Alternatives to FBS are available, such as CLEX, Dextran Products, Canada.
2. To reduce the possibility of contamination of the NS0 line with mycoplasma, bacteria, fungi, yeast, or virus, always:
   a. Use a separate bottle of medium for NS0.
   b. Do not manipulate other cell lines in the hood at the same time as NS0.
3. Do not leave cells in DMSO at room temperature since they will die.
4. If hybridomas are in HAT medium, carry out cloning in HT medium.

Reference

1. Köhler, G. and Milstein, C. (1975) Continuous cultures of fused cells secreting antibody of predefined specificity. *Nature* **256,** 495.

141

Screening Hybridoma Culture Supernatants Using Solid-Phase Radiobinding Assay

Mark Page and Robin Thorpe

1. Introduction

A large number of hybridoma culture supernatants (up to 200) need to be screened for antibodies at one time. The assay must be reliable so that it can accurately identify positive lines, and it must be relatively quick so that the positive lines, which are 75–100% confluent, can be fed and expanded as soon as possible after the assay results are known. Solid-phase binding assays are appropriate for this purpose and are commonly used for detection of antibodies directed against soluble antigens *(1)*. The method involves immobilizing the antigen of choice onto a solid phase by electrostatic interaction between the protein and plastic support. Hybridoma supernatants are added to the solid phase (usually a 96-well format) in which positive antibodies bind to the antigen. Detection of the bound antibodies is then achieved by addition of an antimouse immunoglobulin labeled with radioactivity (usually ^{125}I) and the radioactivity counted in a γ counter.

2. Materials

1. Antigen: 0.5–5 µg/mL in phosphate-buffered saline (PBS) with 0.02% sodium azide.
2. PBS: 0.14*M* NaCl, 2.7 m*M* KCl, 1.50 m*M* KH_2PO_4, 8.1 m*M* Na_2HPO_4.
3. Blocking buffer: PBS containing either 5% v/v pig serum, 5% w/v dried milk powder, or 3% w/v hemoglobin (*see* Note 1).
4. Hybridoma culture supernatants.
5. Negative control: irrelevant supernatant or culture medium.
6. Positive control: serum from immunized mouse from which spleen for hybridoma production was derived.
7. Antimouse/rat immunoglobulin, ^{125}I-labeled (100 µCi/5 µg protein).

3. Method

1. Pipet 50 µL of antigen solution into each well of a 96-well microtiter plate and incubate at 4°C overnight. Such plates can be stored at 4°C for several weeks. Seal plates with cling film to prevent the plates from drying out (*see* Note 2).
2. Remove antigen solution from wells by either a Pasteur pipet (cover the tip with a small piece of plastic tubing to prevent scratching the antigen-coated wells) or by shaking the contents from the plate in one quick movement (*see* Note 3).

From: The Protein Protocols Handbook
Edited by: J. M. Walker Humana Press Inc., Totowa, NJ

3. Wash the plate by filling the wells with PBS and rapidly discarding the contents. Repeat this twice more, tap the plate dry, then fill the wells with blocking buffer, and incubate at room temperature for 30 min to 1 h. Wash three more times with PBS.
4. Pipet 100 μL of neat hybridoma supernatant into the wells, and incubate for approx 3 h at room temperature. Include the negative and positive controls.
5. Wash with blocking buffer three times.
6. Dilute the ^{125}I-labeled antimouse immunoglobulin in blocking buffer to give 1×10^6 cpm/mL. Add 100% μL to each well, and incubate for 1 h at room temperature.
7. Wash with blocking buffer.
8. Cut out individual wells using scissors or a hot nichrome wire plate cutter, and determine the radioactivity bound using a γ counter. A positive result would have counts that are four to five times greater than the background. Normally, the background counts would be around 100–200 cpm with a postive value of at least 1000 cpm (*see* Note 4).

4. Notes

1. Filter the milk and hemoglobin solutions coarsely through an absorbent cloth (e.g., Kimnet, Kimberley Clark) to remove lumps of undissolved powder.
2. Antigens (diluted in distilled water) can be dried onto the plates by incubation in a warm room (37°C) without sealing. This method is usually preferred if the antigen is a peptide improving binding to the plastic; however, the background signal may be increased.
3. The antigen solution may be reused several times to coat additional plates, since only a small proportion of the protein adheres to the plastic.
4. Counting radioactivity in a 96-well format can also be performed by using plates designed for use in scintillation counters. The assay is performed as described, except that in the final step, a scintillation fluid (designed to scintillate with ^{125}I, such as Microscint 20, Packard Instrument Co.) is added to each well, and the scintillation counted in a purpose-built machine (e.g., Topcount, Packard Instrument Co.).

Reference

1. Johnstone, A. and Thorpe, R. (1996) *Immunochemistry in Practice,* 3rd ed. Blackwell Scientific, Oxford, UK.

142

Screening Hybridoma Culture Supernatants Using ELISA

Mark Page and Robin Thorpe

1. Introduction

Enzyme-linked immunosorbent assay (ELISA) is a widely used method for the detection of antibody and is appropriate for use for screening hybridoma supernatants *(1)*. As with radiobinding assays, it is a solid-phase binding assay that is quick, reliable, and accurate. The method is often preferred to radioactive assays, since the handling and disposal of radioisotopes is avoided and the enzyme conjugates are more stable than radioiodinated proteins. However, most of the substrates for the enzyme reactions are carcinogenic or toxic and, hence, require handling with care.

Enzymes are selected that show simple kinetics, and can be assayed by a simple procedure (normally spectrophotometric). Cheapness, availability, and stability of substrate are also important considerations. For these reasons, the most commonly used enzymes are alkaline phosphatase, β-D-galactosidase, and horseradish peroxidase.

2. Materials

1. PBS-Tween: Phosphate-buffered saline (PBS) (0.14*M* NaCl, 2.7 m*M* KCl, 1.5 m*M* KH_2PO_4, 8.1 m*M* Na_2HPO_4) containing 0.05% Tween-20.
2. Antigen: 1–5 μg/mL in PBS.
3. Blocking buffers: PBS-Tween containing either 5% v/v pig serum or 5% w/v dried milk powder (*see* Note 1).
4. Hybridoma culture supernatants.
5. Alkaline phosphatase conjugated antimouse immunoglobulin.
6. Carbonate buffer: 0.05*M* sodium carbonate, pH 9.6.
7. *p*-Nitrophenyl phosphate, disodium hexahydrate: 1 mg/mL in carbonate buffer containing 0.5 m*M* magnesium chloride.
8. 1*M* sodium hydroxide.

3. Method

1. Pipet 50 μL of antigen solution into each well of a 96-well plate, and incubate overnight at 4°C. Such plates can be stored for 4°C for several weeks. Seal plates with cling film to prevent the plates from drying out (*see* Note 2).
2. Remove antigen solution from wells by either a Pasteur pipet (cover the tip with a small piece of plastic tubing to prevent scratching the antigen-coated wells) or by shaking the contents from the plate in one quick movement (*see* Note 3).

From: The Protein Protocols Handbook
Edited by: J. M. Walker Humana Press Inc., Totowa, NJ

3. Wash the plate by filling the wells with PBS and rapidly discarding the contents. Repeat this twice more, then fill the wells with blocking buffer, and incubate at room temperature for 30 min to 1 h. Wash three more times with PBS.
4. Pipet 100 µL of neat hybridoma supernatant into the wells, and incubate for approx 3 h at room temperature. Include the negative and positive controls (*see* Note 4).
5. Wash with blocking buffer three times.
6. Prepare 1:1000 dilution of alkaline phosphatase-conjugated antimouse immunoglobulin in PBS-Tween, and add 200 µL of this to each well. Cover, and incubate at room temperature for 2 h.
7. Shake-off conjugate into sink and wash three times with PBS.
8. Add 200 µL *p*-nitrophenyl phosphate solution to each well, and incubate at room temperature for 20–30 min.
9. Add 50 µL sodium hydroxide to each well, mix, and read the absorbance of each well at 405 nm (*see* Note 5). Typical background readings for absorbance should be less than 0.2 with positive readings usually three or more standard deviations above the average background value.

4. Notes

1. Filter the milk solution coarsely through an absorbent cloth (e.g., Kimnet, Kimberley Clark) to remove lumps of undissolved powder.
2. Antigens (diluted in distilled water) can be dried onto the plates by incubation in a warm room (37°C) without sealing. This method is usually preferred if the antigen is a peptide improving binding to the plastic; however, the background signal may be increased.
3. The antigen solution may be reused several times to coat additional plates, since only a small proportion of the protein adheres to the plastic.
4. All assays should be carried out in duplicate or triplicate.
5. Purpose-built plate readers provide a very rapid and convenient means of determining the absorbance.

Reference

1. Johnstone, A. and Thorpe, R. (1996) *Immunochemistry in Practice,* 3rd ed. Blackwell Science, Oxford, UK.

143

Growth and Purification of Murine Monoclonal Antibodies

Mark Page and Robin Thorpe

1. Introduction

Once a hybridoma line has been selected and cloned, it can be expanded and seed stocks cryopreserved for future use. Relatively large amounts of purified MAb may also be required. There are a variety of procedures for this that ensure the establishment of a stable cell line secreting high levels of specific immunoglobulin. High concentrations of antibody can be generated by growing the line in the peritoneal cavity of mice/rats of the same strain as the tumor cell line donor and spleen cell donor. Antibody is secreted into the ascitic fluid formed within the cavity at a concentration up to 10 mg/mL. However, the ascites will contain immunoglobulins derived from the recipient animal that can be removed by affinity chromatography if desired. Several in vitro culture methods using hollow fibres or dialysis tubing *(1)* have been developed and are commercially available; this avoids the use of recipient mice/rats and contamination by host immunoglobulins, although contamination with culture medium-derived proteins may be a problem.

2. Materials

1. RPMI-1640/DMEM medium.
2. Fetal bovine serum (FBS).
3. Phosphate-buffered saline (PBS): 0.14*M* NaCl, 2.7 m*M* KCl, 1.50 m*M* KH_2PO_4, 8.1 m*M* Na_2HPO_4.
4. HT medium (1 mL 50X HT supplement in 50 mL 10% FBS/RPMI).
5. HT supplement (Gibco, Paisley, UK).

3. Method

3.1. Growth of Hybridomas

1. After cloning, grow hybridomas in 1 mL HT medium in separate wells of a 24-well plate. Check growth every day and feed with 0.5–1 mL HT medium if supernatant turns yellow (*see* Note 1).
2. When cells are 75–100% confluent (*see* Note 2), expand into two further wells with fresh HT medium. When these are confluent, transfer into a 25-cm^3 flask and feed with approx 10 mL medium (*see* Note 3). Subsequently, cell lines should be weaned off HT medium by

From: The Protein Protocols Handbook
Edited by: J. M. Walker Humana Press Inc., Totowa, NJ

increasing the dilution of HT medium stepwise with normal medium lacking the HT supplement, eventually culturing the cells in normal medium. Do this by reducing the amount of HT medium by 25–50% at each feed for expansion of the hybridomas. For example, reduce the amount of HT from 100 to 75 to 50 to 25 to 0% replacing with normal medium.
3. The FBS may be reduced to 5% or lower if hybridoma growth is strong.

3.2. Purification of MAb

MAb purification can be carried out by a number of methods as described in Chapters 125–134.

4. Notes

1. Phenol red is included in the medium as a visual indicator of pH. A yellow supernatant indicates acid conditions as a result of dissolved CO_2 (carbonic acid) derived from active cell growth (respiration). Normally, a yellow supernatant will indicate that the cells should be split (passaged) because the active cell growth will have exhausted the medium of nutrients and hence will not sustain cell viability.
2. Cells are confluent when the bottom of the flask/well is completely covered. Usually, the cells will require feeding or expanding at this stage (*see* Note 1).
3. To prevent inadvertant loss of precious cell lines, it may be prudent to freeze cloned lines before expansion into flasks.

Reference

1. Pannell, R. and Milstein, C. (1992) An oscillating bubble chamber for laboratory scale production of MAb as an alternative to ascitic tumors. *J. Immunol. Methods* **146,** 43–48.

144

A Rapid Method for Generating Large Numbers of High-Affinity Monoclonal Antibodies from a Single Mouse

Nguyen thi Man and Glenn E. Morris

1. Introduction

Since the first description by Kohler and Milstein *(1)*, many variations on this method for the production of monoclonal antibodies (MAb) have appeared (e.g., *see* refs. *2–4* and Chapter 143), and it may seem superfluous to add another. The variation we describe here, however, includes a number of refinements that enable rapid (6–10 wk) production from a single spleen of large numbers *(20–30)* of cloned, established hybridoma lines producing antibodies of high-affinity. We have applied this method to recombinant fusion proteins containing fragments of the muscular dystrophy protein dystrophin *(5,6)*, and dystrophin-related proteins *(7)*, to hepatitis B surface antigen *(8)* and to the enzyme creatine kinase (CK) *(9)*, and have used the MAb thus produced for immunodiagnosis, epitope mapping, and studies of protein structure and function *(5–12)*. Epitopes shared with other proteins are common, so availability of several MAb against different epitopes on a protein can be important in ensuring the desired specificity in immunolocalization and Western blotting studies *(7)*.

In the standard Kohler and Milstein method, Balb/c mice are immunized with the antigen over a period of 2–3 mo, and spleen cells are then fused with mouse myelomas using polyethylene glycol (PEG) to immortalize the B-lymphocytes secreting specific antibodies. The hybrid cells, or hybridomas, are then selected using medium containing hypoxanthine, aminopterin, and thymidine (HAT medium) *(13)*. All unfused myelomas are killed by HAT medium, and unfused spleen cells gradually disappear in culture; only hybridomas survive.

Special features of the detailed method to be described here include:

1. A culture medium for rapid hybridoma growth without feeder layers;
2. Screening early to select high-affinity antibodies;
3. Cloning without delay to encourage hybridoma growth and survival; and
4. The use of round-bottomed microwell plates to enable rapid monitoring of colony growth with the naked eye.

In typical fusion experiments, cells are distributed over eight microwell plates (768 wells) and 200–700 wells show colony growth. Of these hybridomas, up to 50% (100–

From: The Protein Protocols Handbook
Edited by: J. M. Walker Humana Press Inc., Totowa, NJ

Table 1
Examples of Colony and MAb Yields Using This Method

Antigen	Wells with growth	Wells binding to antigen strongly	Wells with desired affinity/specificity	Final no. of MAb
108-kDa rod fragment of dystrophin in fusion protein *(5)*	587	136	27	16
59-kDa rod fragment of dystrophin in fusion protein *(6)*	669	147	30	25
55-kDa C-terminal fragment of dystrophin in fusion protein *(6)*	700+	42	20	17
37-kDa C-terminal fragment of a dystrophin-related protein in fusion protein *(7)*	186	54	24	19

300) produce antibodies that show some antigen binding, though many may be too weak to be useful. After further screening for desired affinity and specificities, we normally select 25–30 as a convenient number to clone. Only occasionally, in our experience, are two or more MAb produced that are indistinguishable from each other when their epitope specificities are later examined in detail. Table 1 illustrates the results obtained from four recent fusions by this method.

In the early stages after fusion, a particularly good cell-culture system is needed to promote rapid hybridoma growth in HAT medium. Horse serum, selected for high cloning efficiency, is a much better growth promoter than most batches of fetal calf serum, and HAT medium is prepared using myeloma-conditioned medium and human endothelial culture supernatant, a hybridoma growth promoter. Under these conditions, colonies are visible with the naked eye 7–10 d after fusion, and screening should start immediately. Even unimmunized mouse spleens will produce large numbers of colonies, so the use of hyperimmunized mice, with high serum titers in the screening assay, is essential for generating a high proportion of positive colonies. The screening assays are also of critical importance, because they determine the characteristics of the MAb produced. Both these points are illustrated in Fig. 1. CK, a fairly typical "globular" enzyme, is denatured and inactivated when attached directly to ELISA plastic, but it can be captured onto ELISA plates in its native form by using Ig from a polyclonal antiserum *(9)*. When mice were immunized with untreated CK, antibody titers against denatured CK were high, whereas titers of antibody against native CK were insufficient to produce a plateau in the capture ELISA, even at high serum concentrations (Fig. 1A). Seven fusions from such mice (over 1000 colonies screened) produced only

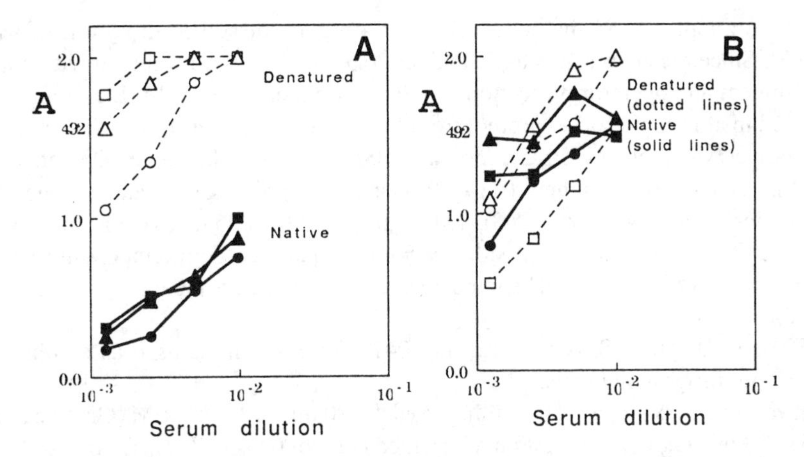

Fig. 1. Effect of aggregation of CK on the immune response in Balb/c mice. Test sera were taken from three mice immunized with untreated CK **(A)**: circles, squares, and triangles represent individual mice) or with glutaraldehyde-aggregated CK **(B)**, and serial dilutions were used as primary antibody in either direct ELISA for denatured CK (dotted lines) or sandwich ELISA for native CK (solid lines). (Reproduced with permission from ref. *9*.)

two, low-affinity MAb specific for native CK. In contrast, when we immunized with CK aggregated by glutaraldehyde treatment (Fig. 1B), titers were higher for native CK antibodies than for those against denatured CK. In a fusion performed with one of these mice, only 2% of the fusion wells were positive in the direct ELISA (denatured CK), but 13% recognized only native CK in the capture ELISA *(9)*. This illustrates the importance of having both high-titer mice and a suitable screen.

Finally, although the method is rapid and efficient, it is also labor-intensive, especially during screening and cloning. When embarking on the method, one should not be daunted by the prospect of having over 70 culture plates in the incubator simultaneously by the time that the later cloning stages of a single experiment are reached.

2. Materials

1. One or two vertical laminar flow sterile hoods (e.g., Gelaire BSB4).
2. 37°C CO_2 incubator (e.g., LEEC, Nottingham, UK; *see* Note 1).
3. Transtar 96 (Costar, Cambridge, MA) for plating fusions and removing hybridoma supernatants.
4. Round-bottomed 96-well tissue-culture microwell plates (NUNC/Gibco, Life Technologies, Gaithersburg, MD, cat. no. 1–63320) (*see* Note 2).
5. Sterile tissue-culture flasks 50 mL (NUNC/Gibco cat. no. 1–63371), 260 mL (NUNC/Gibco cat. no. 1-53732), and 24-well plates (NUNC/Gibco cat. no. 1-43982), 1.8-mL cryotubes (NUNC/Gibco cat. no. 368632), and sterile 30-mL universal bottles.
6. Freund's adjuvant, complete (Sigma [St. Louis, MO] no. F-4258) stored at 4°C.
7. Freund's adjuvant, incomplete (Sigma no. F-5506) stored at 4°C.
8. Pristane (2,6,10,14 tetramethyl pentadecane; Aldrich Chemical Co. [Milwaukee, WI], no. T2280–2) stored at room temperature.
9. 500X HT: 6.8 mg/mL hypoxanthine (Sigma no. H-9377) and 1.95 mg/mL thymidine (Sigma no. T-9250) (*see* Note 3). Filter to sterilize. Store in 5-mL aliquots in sterile plastic universals at –20°C (stable for a few years).

10. 1000X Aminopterin (Sigma no. A–2255): 4.4 mg are dissolved in 25 mL of water (*see* Note 3). Since the chemical is toxic and teratogenic, it must be weighed with appropriate containment and operator protection. Filter to sterilize. Store in 2-mL aliquots at –20°C, wrapped in aluminum foil to protect from light (stable for years).
11. 50% polyethylene glycol (PEG), mol wt 1500 (BDH-Marck, Poole, Dorset, UK, Prod. 29575), is made up by weighing 10 g of PEG in a glass universal and autoclaving. Although the solution is still hot (about 60°C), add 10 mL of DMEM/25 m*M* HEPES prewarmed in a 37°C water bath. Mix well, and store in 5-mL aliquots at room temperature in the dark; stable at least for 6 mo. The pH should be slightly alkaline (as judged by phenol red) when used for fusion.
12. DMEM/25 m*M* HEPES. (Gibco cat. no. 041–2320H). This is used as a "physiological saline," not for growing cells.
13. DMEM/20% HS is prepared by adding to each 100 mL of 1X DMEM (Gibco cat. no. 041-1885): 1.3 mL L-glutamine (200 m*M*) (Gibco cat. no. 043-5030H), 1.3 mL sodium pyruvate (100 m*M*) (Gibco cat. no. 066–01840E), 1.3 mL penicillin-streptomycin (Gibco cat. no. 043-5140H), 0.6 mL nystatin (Gibco cat. no. 043-5340H), 1.3 mL nonessential amino acids (MEM) (Gibco cat. no. 043-1140H), and 25 mL selected horse serum (HS). For routine growth and maintenance of both myelomas and established hybridomas, serum is reduced to 5%.
14. Cloning medium: DMEM/20% HS supplemented with 1X HT and 5% human endothelial culture supernatant (HECS) (Costar no. M712).
15. HAT medium: Myeloma-conditioned DMEM/20% HS with 1X HAT and 5% HECS.
16. Medium for feeding the fusion: DMEM/20% HS with 5% HECS and 5X HT.
17. Medium for freezing down cells: 92.5% HS/7.5% dimethyl sulfoxide (DMSO) (Sigma no. D-5879).
18. Trypan blue (0.1%) (Sigma no. T6146) in PBS.

3. Methods

3.1. Immunization of Balb/c Mice

1. Purify the protein antigen. With purified antigen, screening is faster and easier, and the yield of MAb is higher, but proteins of 50% purity or less on SDS-PAGE may still give good results.
2. Each of three Balb/c mice (6–8 wk old) is given an ip injection of 50–100 µg of antigen in 0.1 mL PBS emulsified with an equal volume of Freund's complete adjuvant by sucking up and down many times into a disposable sterile plastic 2-mL syringe with a sterile 21-gage 1-1/2 in. no. 2 needle (*see* Note 4).
3. Four weeks later, an ip boost of 50–100 µg antigen in Freund's incomplete adjuvant is given.
4. Ten days later, blood (ca. 0.1 mL) is taken from the tail vein, allowed to clot, and centrifuged at 10,000 rpm for 5 min. The serum is tested by the method to be used for screening hybridomas. If the titers are high in this assay (*see* Fig. 1), the mice should be left to rest for an additional month before fusion.
5. The best mouse is boosted with 50–100 µg of antigen in PBS, ip, and/or iv in the tail vein, 4, 3, and 2 d before fusion.

3.2. Selection of Horse Serum

1. Most suppliers will provide test samples of different serum batches and keep larger amounts on reserve. All sera are heat-inactivated at 56°C for 30 min before use, though

this is not known to be essential. Thaw serum in 37°C water bath and swirl to mix before incubating in a 56°C water bath for 30 min with occasional swirling.

2. Prepare cells of any hybridoma or myeloma line at 1000 cells/mL in DMEM and dilute to 40 cells/mL, 20 cells/mL, 10 cells/mL in DMEM/20% HS of different batches (cloning of established cell lines does not require feeder layers or HT and HECS). For each test serum, plate one 96-well round-bottomed microplate with four rows at four cells in 0.1 mL/well, four rows at two cells in 0.1 mL/well, and four rows at one cell in 0.1 mL/well.
3. From the fifth day onward, note the number of colonies at each cell density for each test serum, as well as the size of the clones. After 2 wk, choose the batch of serum with the highest number and the largest clones. Serum may be stored at –70°C for at least 1–2 yr without loss of activity.

3.3. Growth and Maintenance of Myelomas

NSO/1 *(2)* and SP2/O *(14)* myeloma lines grow very fast and do not synthesize immunoglobulins They should be kept in logarithmic growth in DMEM/5% HS for at least a wk before fusion by diluting them to 5×10^4/mL when the cell density reaches 4 or 5×10^5/mL. If thawing myelomas from liquid nitrogen, it is advisable to begin at least 2–3 wk before fusion (*see* Note 5). We have never found it necessary to treat lines with azaguanine to maintain their aminopterin sensitivity, which can be checked by plating in HAT medium at 100 cells/well.

3.4. Fusion with Spleen Cells

1. All dissection instruments are sterilized by dry heat (160°C for 5 h), autoclaving (120°C for 30 min) or dipping in 70% ethanol, and flaming in a Bunsen burner. Two separate fusions are done with each spleen as a precaution against accidents.
2. Two days before fusion, set up 2 × 260-mL flasks with 40 mL each DMEM 20% HS and 1×10^5 myeloma cells/mL (NSO/1 or Sp2/O).
3. On the day of fusion, place in a 37°C water bath 100 mL of DMEM/25 m*M* HEPES, 10 mL of DMEM/20% HS, and 1 bottle of PEG1500.
4. Count the myeloma cells in a hemocytometer after mixing an aliquot with an equal volume of 0.1% trypan blue in PBS; cell density should be $4–6 \times 10^5$ with viability 100% and no evidence of contamination.
5. Transfer the myeloma cells to four universals and centrifuge for 7 min at 300*g* (MSE 4 L). Remove the supernatants with a 10-mL pipet and retain them for plating the fusion later (myeloma-conditioned medium). Resuspend each pellet in 5 mL of DMEM/HEPES, and combine in two universals (2 × 10 mL).
6. The immunized mouse is killed by cervical dislocation outside the culture room and completely immersed in 70% ethanol (15–30 s). Subsequent procedures are performed in sterile hoods. The mouse is pinned onto a "sterile" surface (polystyrene covered with aluminum foil and swabbed with 70% ethanol), and a small incision in the abdominal skin is made with scissors. The peritoneum is then exposed by tearing and washed with 70% ethanol before opening it and pinning it aside. The spleen is lifted out, removing the attached pancreas with scissors, and placed in a sterile Petri dish (*see* Note 6).
7. After removing as much surrounding tissue as possible, the spleen is placed in 10 mL of DMEM/25 m*M* HEPES and held at one end with forceps while making deep longitudinal cuts with a sterile, curved scalpel to release the lymphocytes and red blood cells. This process is completed by scraping with a Pasteur pipet sealed at the end until only connective tissue is left (2–3 min).

8. The spleen cell suspension is transferred to a 25-mL universal, with pipeting up and down five to six times to complete the dispersal. Large lumps are allowed to settle for 1 min before transferring the supernatant to a second universal and centrifuging for 7 min at 300*g*. Resuspend the pellet in 5 mL of DMEM/25 m*M* HEPES, and keep at room temperature. Usually about 1×10^8 cells/spleen are obtained (*see* Note 7).
9. Add 2.5 mL spleen cell suspension to each 10 mL of myeloma cell suspension, and mix gently. Centrifuge at 300*g* for 7 min at 20°C. Remove supernatants with a pipet, and resuspend pellets in 10 mL DMEM/25 m*M* HEPES using a pipet. Centrifuge at 300*g* for 7 min. Remove the supernatant completely from one pellet (use a Pasteur pipet for the last traces), and then loosen the pellet by tapping the universal gently (*see* Note 8).
10. Remove the DMEM/25 m*M* HEPES and the 50% PEG from the water bath just before use. Take 1 mL of PEG in a pipet, and add dropwise to the cell pellet over a period of 1 min, mixing between each drop by shaking gently in the hand. Continue to shake gently for another minute. Add 10 mL DMEM/25 m*M* HEPES dropwise with gentle mixing, 1 mL during the first minute, 2 mL during the second, and 3.5 mL during the third and fourth minutes. Centrifuge at 300*g* for 7 min. (This procedure can be repeated with the second cell pellet, while the first is spinning.)
11. Remove supernatant, and resuspend pellet gently in 5 mL of DMEM/20% HS. Place both 5-mL cell suspensions in their universals in the CO_2 incubator for 1–3 h.
12. During this time, take the ca. 80 mL of myeloma-conditioned medium and add 4.5 mL of HECS, 90 µL of aminopterin, and 180 µL of hypoxanthine/thymidine solutions. Filter 2×40 mL through 22-µm filters (47 mm; Millipore [Bedford, MA]) to resterilize, and remove any remaining myeloma cells.
13. To each 40 mL, add the 5-mL fusion mixture from the CO_2 incubator and distribute in 96-well microtiter plates (4 plates/fusion; 100 µL/well) using a Transtar 96 or a plugged Pasteur pipet (3 drops/well) (*see* Note 9). Put the plates from each fusion in a separate lunch box in the CO_2 incubator and leave for 3 d.
14. On d 4, add to each well 80 µL of DMEM/20% HS supplemented with HECS and 5X HT. Replace in the CO_2 incubator as quickly as possible.
15. By d 10–14, a high proportion of the wells should have a clear, white, central colony of cells, easily visible with the naked eye. If you want to select for high-affinity MAb (and you would be well advised to do so, unless you have some very special objectives), do not delay screening. If you are using a sensitive screening method, such as ELISA, immunofluorescence, or Western blotting, you will detect high-affinity antibodies from even the very small colonies. Do not wait for the medium to turn yellow, or your cells may start to die (if you are using a less sensitive screen, such as an inhibition assay, you *may* need to wait longer). A high proportion of the 768 wells (50–100%) should have colony growth by 14 d. If it is <10%, try reducing the PEG concentration. If you regularly obtain over 70%, use 6–8 plates/fusion instead of 4. There should be no microbial contamination (*see* Note 10).

3.5. Screening

It would be quite wrong to describe one screening method in detail, since this must be chosen to suit the purpose for which MAb are required. We recommend ELISA as the simplest technique for screening 8×96 wells in 1 d, but for coating ELISA plates directly, 10–50 µg of reasonably pure antigen are required. Alternatively, plates can be coated with a capture antibody; we have used this approach to capture CK in its native conformation *(9)* and to capture hepatitis B surface antigen from human plasma *(8)*. The availability of multichannel miniblotters makes direct screening by Western blotting feasible, and with two miniblotters, 224 microwells can be screened in one long day. Immunohistochemistry on cultured cells grown in microwells is also feasible, though we have not used it ourselves. A Transtar 96 apparatus is a rapid and sterile way to remove 10–50 µL of culture supernatant from all 96 wells of a microwell plate. The culture plates have to be open inside the hood for only a few seconds. They should always be wiped with paper tissue soaked in 70% ethanol before opening. Screening the whole plate, including wells without cell growth, by ELISA avoids possible errors when trying to keep track of individual wells. If ELISA-positive wells are to be screened further (e.g., by Western blotting using a multichannel miniblotter), this first 50 µL of supernatant may be diluted into 100 µL or more of PBS; multiple sampling of fusion wells should be minimized to avoid contamination. We usually screen by ELISA on d 10, carry out further screening of ELISA-positive wells on d 11 and 12, and clone about 30 of the best wells on d 12, 13, and 14.

3.6. Cloning by Limiting Dilution

1. Using a plugged Pasteur pipet, transfer the positive clone to a 25-well plate containing 0.5 mL of cloning medium (*see* Note 11).
2. Take an aliquot of these cells, dilute with an equal volume of 0.1% trypan blue in PBS, and count cells on at least two chambers of a hemocytometer.
3. Prepare 6 mL of 160 cells/mL and perform serial dilutions to 40 cells/mL and 10 cells/mL. Plate four rows at 16 cells in 0.1 mL/well, four rows at four cells in 0.1 mL/well, and four rows at one cell in 0.1 mL/well. Three drops from a plugged Pasteur pipet are about 0.1 mL.
4. Eight to ten days after plating, clones at 1 cell/well are visible, and screening can start immediately. It is best to screen at least 16–24 wells from each cloning plate. As soon as positive clones have been identified, they should be cloned a second time, but at 4, 1, and 0.5 cells/well instead of 16, 4, and 1.
5. When these plates are screened again after another 8–10 d, at least one dilution should have <50% of the wells with growth, and *all wells with growth* should be positive in the screening. If any colonies are negative, cloning should be repeated until all wells are positive (*see* Note 12).

3.7. Preservation of Hybridoma Lines

1. Hybridoma clones should be expanded slowly from microwells, first into 0.5 mL in 25-well plates, feeding to 1 or 2 mL before transferring to a 50-mL flask in 5 mL of medium.
2. To preserve cell lines indefinitely, centrifuge 1–3 × 10^6 growing cells in a universal, remove all supernatant, suspend cell pellet in 0.3–0.5 mL of ice-cold 7.5% DMSO/HS, and transfer the suspension into a 1.8-mL cryotube.
3. Surround the cryotube with 1 in. of polystyrene, and place in a –70°C freezer. Transfer it the next day to a liquid nitrogen container.
4. It is advisable to freeze several vials of the same line and thaw one after a week to ensure viability, while maintaining the cells in culture.
5. To recover cells from liquid nitrogen, thaw the cryovial quickly in a 37°C water bath until only a tiny piece of ice is visible. Transfer the cell suspension immediately to 10 mL of ice-cold DMEM/20% HS. Centrifuge at 300*g* for 7 min, remove supernatant completely, and resuspend the cell pellet in 3–5 mL of cloning medium. Transfer the cell suspension into a 50-mL flask, reserving about 0.1 mL to count viable cells, and place flask in the CO_2 incubator. If the viable cell count is <1 × 10^4/mL, cloning by limiting dilution should be carried out immediately (*see* Note 13).

3.8. Antibody Production

1. Antibody can be generated as culture supernatant by starting a flask culture at 1×10^5/mL in DMEM/20% HS and, when the cell density reaches $4–6 \times 10^5$/mL, diluting to 1×10^5/mL. The concentration of horse serum is gradually reduced to 10% and then 5% in the process of reaching the volume required (*see* Note 14). The cells are then left to grow undisturbed until they are all dead (*see* Note 15). Culture supernatant can then be harvested by centrifugation, and stored in aliquots at –20°C until required. For routine use, keep an aliquot of 1–2 mL with 0.1% sodium azide at 4°C
2. For ascites fluid, an adult Balb/c mouse is primed with 0.2 mL of pristane ip.
3. After 7–10 d, $1–3 \times 10^6$ hybridoma cells in logarithmic growth (100% viability) are centrifuged at 300*g* for 7 min at 20°C and washed once with DMEM/25 m*M* HEPES to remove all serum. The cell pellet is resuspended in 0.2 mL of DMEM/HEPES, and injected ip into a primed mouse using a 21-gage no. 2 needle (green).
4. Ascites development becomes evident by external examination within 7–14 d and mice must be examined twice a day over this period. Mice are killed by cervical dislocation as soon as they show signs of discomfort (advice should be sought and followed). The peritoneum is exposed as in Section 3.4., step 6, and a 21-gage needle on a 5-mL syringe is inserted about 1 cm so that the tip remains visible through the peritoneum and fluid can be withdrawn without blockage. Last traces of fluid are removed after opening the cavity with scissors. The final volume obtained is usually 2–3 mL, though blood from the heart and thoracic cavity can also be collected (about 0.5 mL) and processed separately as a source of antibody.
5. Fluids are allowed to clot and then centrifuged at 3000*g* for 20 min. Store in aliquots of 200 µL at –20°C or preferably –70°C.

4. Notes

1. Fusions from a single mouse spleen aimed at producing 20–30 cloned hybridoma lines by the method described here can eventually fill a standard size bench-top incubator completely. The floor of the incubator is filled with autoclaved, double-distilled water. Do not use the heater on the inner glass door; a soaking wet door is a good check for 100% humidity. The interior of the incubator must be kept as free of contamination as possible; this may necessitate removing the contents, changing the water, and wiping the interior with 70% ethanol on several occasions during hybridoma growth and cloning. Even so, we use large plastic lunch boxes to insulate unsealed culture vessels (e.g., microwell plates) from the turbulent incubator environment (a fan is desirable for uniform temperature and humidity). Holes plugged with cotton wool at each end of the lunch box to admit CO_2 are optional; an alternative is to leave the lids ajar for 1–2 min to equilibrate before closing them. They can be wiped regularly, inside and out with 70% ethanol. *Never* use toxic chemicals (e.g., bleach) to decontaminate an incubator. *Never* handle culture plates without gloves.
2. Cells roll to the bottom giving a white colony in the exact center of each well, visible even at early stages; flat-bottomed plates have initially transparent colonies, often at the side of the well and easily overlooked. Use of round-bottomed plates makes it very easy to monitor colony growth after fusion without using a microscope.
3. The mixture does not dissolve by itself, so add 1*N* NaOH dropwise while stirring until the solution goes clear.
4. Avoid syringes with rubber plunger tips; when the emulsion thickens, they come off the plunger.

5. The myeloma cells should look uniform and healthy by this time with no sign of cell debris.
6. We normally perform all steps up to this point in a separate sterile hood. The mouse is nonsterile externally, and particular care is taken to avoid hairs; the outer skin is torn, rather than cut, for this reason. Mice should not be introduced into the hood used for cell culture.
7. Some protocols remove red blood cells by differential lysis at this point, but we have not found it necessary.
8. The final concentration of PEG during fusion is thought to be critical for the yield of colonies. If initial yields are low, try adding increasing (but small) amounts of DMEM/25 m*M* HEPES to the pellet before adding PEG.
9. Never use unplugged micropipet tips for adding to culture plates; sterile tips can be used for *removing* culture supernatant only.
10. Bacterial and yeast contaminations rarely occur and are usually the result of a major failure in sterile technique (e.g., inadequate sterilization of culture medium or glassware). Sources of any sporadic fungal contamination should be tracked down and eliminated.
11. Some protocols recommend freezing down uncloned or partly cloned cells as a precautionary measure. Following our rapid cloning protocol, however, the first round of cloning and screening is often complete before the original colony is sufficiently expanded for freezing. It is certainly advisable to keep the original culture alive, however, by adding 0.1 mL of cloning medium back to the fusion well and by feeding the 24-well culture, if necessary. Cloning should not be regarded as an ordeal suitable only for healthy cells, but rather as a means of invigorating a failing culture. We always clone twice for this reason, even if the line is evidently clonal after the first round. *As a general maxim*: if in doubt or trouble, clone immediately.
12. There are rare exceptions to the general principle that cloning should be continued until all colonies are positive in the screens. We were once performing an initial screen by ELISA and then testing ELISA-positive wells by immunofluorescence microscopy (IMF) on muscle sections. After cloning one well that was positive in both assays, we found that only half the clones were ELISA-positive and very few of these were also IMF-positive. We recloned wells that were positive in both assays again with the same result. We thought we had come across our first "unstable" hybridoma, but by chance we tested ELISA-negative wells in IMF and found they were all IMF-positive. Only then did we realize that the original fusion well had contained two different hybridomas, one ELISA-positive and one IMF-positive. The purpose of cloning is to separate such lines, but by selecting wells that were positive in both assays (and hence still had two clones), we had been systematically defeating this objective.
13. When very few cells survive after being kept in liquid nitrogen, "cloning" by limiting dilution at about 100 total cells/well may be the only way to recover the cell line.
14. Cells may also be collected by centrifugation and resuspended in 5% fetal calf serum at this stage, if a culture supernatant with low levels of nonmouse Ig is required.
15. Most MAb are stable for long periods in sterile culture medium, but there are undoubtedly some that lose activity rapidly even at 4°C. These are perhaps best avoided, but, if required, bulk culture would have to be monitored regularly and supernatants harvested when their antibody activity is still high.

Acknowledgment

We thank C. J. Chesterton (King's College, London) for sharing with us his enthusiasm for, and experience of, hybridoma technology in 1981.

Further Reading

1. Langone, J. J. and Van Vunakis, H. (eds.) (1986) Methods in enzymology, in *Immunochemical Techniques,* vol. 121, part I. Academic, New York.
2. Goding, J. W. (1986) *Monoclonal Antibodies: Principles and Practice.* Academic, New York.

References

1. Kohler, G. and Milstein, C. (1975) Continuous cultures of fused cells secreting antibody of predefined specificity. *Nature* **256,** 495–497.
2. Galfre, G. and Milstein, C. (1981) Preparation of Monoclonal antibodies: strategies and procedures. *Methods Enzymol.* **73,** 3–46.
3. Fazekas de St. Groth, S. and Scheidegger, D. (1980) Production of monoclonal antibodies: strategy and tactics. *J. Immunol. Methods* **35,** 1–21.
4. Zola, H. and Brooks, D. (1982) Techniques for the production and characterization of monoclonal hybridoma antibodies, in *Monoclonal Antibodies: Techniques and Applications* (Hurrell, J. G., ed.), CRC, FL, pp. 1–57.
5. Nguyen thi Man, Cartwright, A. J., Morris, G. E., Love, D. R., Bloomfield, J. F., and Davies, K. E. (1990) Monoclonal antibodies against defined regions of the muscular dystrophy protein, dystrophin. *FEBS Lett.* **262,** 237–240.
6. Ellis, J. M., Nguyen thi Man, Morris, G. E., Ginjaar, I. B., Moorman, A. F. M., and van Ommen, G.-J. B. (1990) Specifity of dystrophin analysis improved with monoclonal antibodies. *Lancet* **336,** 881,882
7. Nguyen thi Man, Ellis, J. M., Love, D. R., Davies, K. E., Gatter, K. C., Dickson, G., and Morris, G. E. (1991) Localization of the DMDL-gene-encoded dystrophin-related protein using a panel of 19 monoclonal antibodies. Presence at neuromuscular junctions, in the sarcolemma of dystrophic skeletal muscle, in vascular and other smooth muscles and in proliferating brain cell lines. *J. Cell Biol.* **115,** 1695–1700.
8. Le Thiet Thanh, Nguyen thi Man, Buu Mat Phan, Ngoc Tran Nguyen, thi Vinh Ha, and Morris, G. E. (1991) Structural relationships between hepatitis B surface antigen in human plasma and dimers of recombinant vaccine: a monoclonal antibody study. *Virus Res.* **21,** 141–154.
9. Nguyen thi Man, Cartwright, A. J., Andrews, K. M., and Morris, G. E. (1989) Treatment of human muscle creatine kinase with glutaraldehyde preferentially increases the immunogenicity of the native conformation and permits production of high-affinity monoclonal antibodies which recognize two distinct surface epitopes. *J. Immunol. Methods* **125,** 251–259.
10. Nguyen thi Man, Cartwright, A. J., Osborne, M., and Morris, G. E. (1991) Structural changes in the C-terminal region of human brain creatine kinase studied with monoclonal antibodies. *Biochim. Biophys. Acta* **1076,** 245–251.
11. Morris, G. E. and Cartwright, A. J. (1990) Monoclonal antibody studies suggest a catalytic site at the interface between domains in creatine kinase. *Biochim. Biophys. Acta* **1039,** 318–322.
12. Sedgwick, S. G., Nguyen thi Man, Ellis, J. M., Crowne, H., and Morris, G. E., (1991) Rapid mapping by transposon mutagenesis of epitopes on the muscular dystrophy protein, dystrophin. *Nucleic Acids Research* **19,** 5889–5894.
13. Littlefield, J. W. (1964) Selection of hybrids from matings of fibroblasts in vitro and their presumed recombinants. *Science* **145,** 709,710.
14. Shulman, M., Wilde, C. D., and Kohler, G. (1978) A better cell line for making hybridomas secreting specific antibodies. *Nature* **276,** 269,270.

Index

B

POINT LOMA NAZARENE COLLEGE
RYAN LIBRARY